G_{sec}	Secant shear modulus
G_{50}	Shear modulus of 50% of failure shear stress
G_s	Specific gravity
h_p	Pressure head
h_z	Elevation head
H	Head
H_o	Height
H_{dr}	Drainage path
i	Hydraulic gradient
i_{cs}	Critical hydraulic gradient
I_D	Material index
j_s	Seepage force per unit volume
I	Influence factor
I_z	Influence factor for vertical stress
I_s	Settlement influence factor
k	Hydraulic conductivity or coefficient of permeability
k_h	Horizontal acceleration
k_v	Vertical acceleration
k_s	Subgrade modulus
K_a	Active earth lateral pressure coefficient
K_D	Horizontal stress index
K_p	Passive lateral earth pressure coefficient
K_o	Lateral earth pressure coefficient at rest
K_r	Relative foundation–soil stiffness
K_{aE}	Seismic active earth pressure coefficient
K_{pE}	Seismic passive earth pressure coefficient
L	Length
M	Critical state friction factor
M_c	Compression friction factor
M_E	Extension friction factor
m_v	Modulus of volume compressibility
n	Porosity
N	Standard penetration number
N_c, N_q, N_γ	Bearing capacity factors
NCL	Normal consolidation line
OCR	Overconsolidation ratio
P	Load
P_L	Probability of liquefaction
P_p	Probability
P_f	Probability function
p	Mean stress
p_{atm}	Atmospheric pressure
p'_c	Preconsolidation mean effective stress
q	Deviatoric stress or shear stress
q_a	Allowable bearing capacity
q_s	Surface stress
q_c	Cone resistance
q_v	Flow rate
q_{ult}	Ultimate gross bearing capacity
q_u	Ultimate net bearing capacity
Q	Load
Q_b	End bearing or point resistance
Q_f	Skin or shaft friction
Q_{ult}	Ultimate load capacity

Foundations and Earth Retaining Structures

The Wiley Bicentennial–Knowledge for Generations

Each generation has its unique needs and aspirations. When Charles Wiley first opened his small printing shop in lower Manhattan in 1807, it was a generation of boundless potential searching for an identity. And we were there, helping to define a new American literary tradition. Over half a century later, in the midst of the Second Industrial Revolution, it was a generation focused on building the future. Once again, we were there, supplying the critical scientific, technical, and engineering knowledge that helped frame the world. Throughout the 20th Century, and into the new millennium, nations began to reach out beyond their own borders and a new international community was born. Wiley was there, expanding its operations around the world to enable a global exchange of ideas, opinions, and know-how.

For 200 years, Wiley has been an integral part of each generation's journey, enabling the flow of information and understanding necessary to meet their needs and fulfill their aspirations. Today, bold new technologies are changing the way we live and learn. Wiley will be there, providing you the must-have knowledge you need to imagine new worlds, new possibilities, and new opportunities.

Generations come and go, but you can always count on Wiley to provide you the knowledge you need, when and where you need it!

William J. Pesce
President and Chief Executive Officer

Peter Booth Wiley
Chairman of the Board

FOUNDATIONS AND EARTH RETAINING STRUCTURES

MUNI BUDHU
Professor, Department of Civil Engineering & Engineering Mechanics
University of Arizona

BICENTENNIAL 1807 WILEY 2007 BICENTENNIAL

JOHN WILEY & SONS, INC.

VICE PRESIDENT AND EXECUTIVE PUBLISHER	Don Fowley
ACQUISITIONS EDITOR	Jenny Welter
VICE PRESIDENT AND DIRECTOR OF MARKETING	Susan Elbe
EXECUTIVE MARKETING MANAGER	Chris Ruel
PRODUCTION EDITOR	Nicole Repasky
DESIGNER	Hope Miller
PRODUCTION MANAGEMENT SERVICES	Ingrao Associates
EDITORIAL ASSISTANT	Lindsay Murdock/Mark Owens
MEDIA EDITOR	Lauren Sapira
COVER PHOTO	John Geiven Kamp, University of Arizona
WILEY 200th ANNIVERSARY LOGO DESIGN	Richard J. Pacifico

This book was set in Times New Roman by Thomson Digital Limited and printed and bound by Hamilton Printing. The cover was printed by Phoenix Color.

This book is printed on acid free paper. ∞

To order books or for customer service, please call 1-800-CALL WILEY (225-5945).

Library of Congress Cataloging in Publication Data:

Budhu, M.

Foundations and earth retaining structures / Muni Budhu.

p. cm.

ISBN 978-0471-47012-0 (cloth)

1. Soil mechanics. 2. Foundations. I. Title.

TA710.B763 2007

624.1′5—dc22

2007018089

Printed in the United States of America

10 9 8 7 6 5 4 3 2 1

PREFACE

OBJECTIVES

There are three primary objectives for this textbook: first, to present basic concepts and fundamental principles that are necessary to understand the background of the methods employed in foundation design; second, to inform students on the values and limitations of popular methods of analysis in foundation engineering; and third, to provide a framework for students to carry out simple foundation design and appreciate the design process.

BACKGROUND

Foundation engineering is becoming more challenging as the transportation, housing, industrial, environmental, and commercial needs of the population increase and solid ground is becoming scarcer. Foundation engineers have to deal with these challenges through

1. Better understanding of the fundamentals
2. Utilization of modern technology such as remote sensing, geographic information systems, novel in-situ testing equipment, and numerical simulation
3. Better appreciation of the social, economic, and environmental context of foundation engineering
4. Accumulation of experience

There is a trend toward more usage of computer software in engineering practice because of advancement in computer technologies, ready availability of this software, and its applications to a wide range of problems. Computer software can increase efficiency, save resources, and allow designers to analyze increasingly complex foundation systems. Constitutive models are at the heart of computer software used for numerical analysis. These models are the domain of experts. However, a growing number of novices (young and inexperienced engineers) are using these models for daily work without understanding the fundamental bases for them or proper applications to the problem or system that is intended to be modeled. This is an undesirable situation because foundation engineering works impact lives, properties, and the environment. Improper use of these models could lead to disastrous outcomes. This textbook provides students with the basic background information to better understand foundation analysis and to assist them in interpreting the results for design.

AUDIENCE

This textbook is written primarily for an undergraduate course in foundation analysis and design. It should also appeal to graduate students and practicing engineers. The coverage of topics in this textbook does not follow any particular curriculum because different institutions have different educational missions and the student body normally comprises students with different abilities. Instructors can sequence selected topics to structure different learning pathways depending on the educational mission of the institution. Two learning pathways (tracks) are suggested for a semester course in Foundation Engineering (Geotechnical Engineering II). Track I, **basic learning pathway**, provides students with the basic knowledge of foundation engineering with only rudimentary theoretical background. Track II, **advanced learning pathway**, provides students with the fundamentals to help them appreciate the basis and the limitations of foundation analysis and to initiate learning that will allow them to later model complex foundation problems. The table below shows the suggested coverage for the two learning pathways:

Learning Pathways Chapters

Learning pathways	1	2	3	4	5	6	7	8	9	10	11
Track I	Entire chapter	2.0 – 2.6	Skip	Entire chapter	Entire chapter except 5.10	Entire chapter	7.0 – 7.3.1, 7.4, 7.6–7.8, 7.11 –7.13	Entire chapter except 8.15	9.0 – 9.6	Entire chapter except 10.10 and 10.12	Entire chapter
Track II	Entire chapter	Entire chapter	Entire chapter	Entire chapter	Entire chapter	Entire chapter	Entire chapter	Entire chapter	Entire chapter	Entire chapter	Entire chapter

APPROACH

This is a textbook and not a design manual. Consequently, it emphasizes fundamentals rather than procedures. However, practical procedures, where appropriate, are included to allow students to transit into "office" design. It is expected that when students complete this textbook, they should have a firm grounding in the principles of the practice of foundation analysis and should be able to plan, analyze, and design simple shallow and deep foundations and retaining walls. The topics are sequenced so as not to rush the students into design but to build a solid foundation in the fundamentals so that they can understand the implications of the assumptions in the design.

The importance of a particular topic is declared and the topic is presented with sufficient details using, wherever appropriate, the students' background in engineering mechanics and soil mechanics. Example problems have been solved to provide further insights of the basic concepts, to demonstrate the application of the fundamental principles in problem solving, and to foster understanding of the design process. Each solution provides a step-by-step procedure to guide the student in problem solving techniques. Students are encouraged to use software modules available at www.wiley.com/college/budhu to carry out iterative calculations. Most of these software modules were developed using Excel to give students control over the calculation process. Students can modify the Excel spreadsheets to solve similar classes of problems. The author suggests that students be encouraged to solve problems using "hand calculations" and use the spreadsheets to carry out sensitivity analyses (parametric studies).

Typical soil parameters and correlations are grouped together in the appendices. This allows students to easily find these parameters and correlations for design use.

ORGANIZATION

This textbook is organized into 11 chapters divided into two main parts with a transition. The first five chapters (part 1), are fundamental materials including soil mechanics review materials. All the soil mechanics review topics are directly connected with foundation engineering. Some of them may appear to be unwarranted, but they are needed since most students tend to dispose of their soil mechanics textbook after taking the course. Chapter 6—the transition—deals with uncertainties and design philosophy. The last five chapters (part 2) present practical applications and procedures for the design of foundations and earth retaining walls.

Chapter 1 is a review of basic soil mechanics concentrating on physical soil parameters.

Chapter 2 deals with loads, stresses, strains, stress states, and stress paths. Much of the material in this chapter is taught in soil mechanics courses. However, some additional materials, such as Cerrutti's and Mindlin's solutions for horizontal loads (e.g., wind loads) and a vertical point load in the interior of a half space (e.g., end bearing pile load) respectively and stress paths, which may not be covered in some undergraduate soil mechanics courses but are important fundamental learning materials, are included.

Chapter 3 deals with fundamental issues such as yielding and failure of soils at a level that can, hopefully, be easily assimilated by undergraduates. These are important topics because students need to understand the difference between yielding and failure, and the differences among the failure criteria that are in use to interpret soil failure (Chapter 4). The stress-strain behavior of ideal materials is presented to help students appreciate that the analyses used in foundation engineering are approximations of soil behavior. The popular limit equilibrium and limit analyses, pervasive in analysis of foundations, are presented using materials that the students would have learned in courses in statics and engineering materials. Numerical analyses are briefly discussed. The aim is to give undergraduate students, who may use numerical methods in foundation design soon after graduation, some background material without the mathematical rigor and provide some guidance on the use and misuse of these methods. Most institutions have courses dedicated to numerical analyses wherein the mathematics is expounded.

Chapter 4 is about the interpretation of settlement, strength, and deformation parameters from laboratory tests for foundation design. Much of the material in this chapter may have been completed by some undergraduate students. However, in this textbook soil is treated as a dilatant-frictional material rather than a cohesive-frictional material, so the presentation is concentrated on how the parameters required for foundation design are interpreted from the appropriate laboratory tests for soil as a dilatant-frictional material using the appropriate failure criteria. Popular empirical relationships among settlement, strength, and deformation parameters and simply obtained soil parameters are summarized. While it is possible that the topics in this chapter could have been integrated into Chapter 3, the author believes that Chapter 3 would have been too long and students might be weary of it. In addition, treating settlement, strength, and deformation parameters separately rightly establishes their importance in the context of foundation design.

Chapter 5 deals with site characterization. Here again some of the material may have been covered in some undergraduate soil mechanics courses. In-situ tests are described and data reduction and interpretation presented. Evaluation of liquefaction potential is dealt with here because empirical procedures in practical use have been developed using SPT and CPT data.

Chapter 6 is the bridge between fundamentals and application. Students are informed about uncertainties in foundation design to prepare them to make informed compromises in foundation design. The two methods of design—allowable stress and load and resistance factor—are discussed and compared.

Chapter 7 begins foundation design with consideration of shallow foundations. Mat foundations are not considered here because coupling mats with piles is becoming a popular type of foundation, especially for high-rise buildings. Mat foundations are then presented after pile foundations so that piled rafts can be discussed.

Chapter 8 deals with pile foundations. Drilled shafts are not considered separately because they are pile foundations also. Both vertical and lateral loads are included in the presentation.

Chapter 9 is about mat foundations. The key issue with mat foundations is the structural design. Because this textbook is intended primarily for undergraduates, a broad-brush approach is taken by giving the governing equations and alerting students that computer programs are available commercially to solve the equations. Some broad discussions on the assumptions made in the analyses of mat foundations by these programs, in general, are presented. Empirical procedures for mat foundations used for expansive and collapsible soils are included because some students, soon after graduating, may be practicing in areas where expansive or collapsible soils are present.

Chapter 10 deals with earth pressures and designing of rigid and flexible retaining structures, and braced excavations. This chapter is included because topics on retaining walls and braced excavations are covered in foundation engineering courses at some institutions.

Chapter 11 briefly covers mechanical stabilized earth walls and other walls such as soil nail walls. These walls are becoming increasingly popular.

ONLINE TEACHING AND LEARNING AIDS

The following resources are available for students' use at www.wiley.com/college/budhu. Click on this text, **Foundations and Earth Retaining Structures**

- **Software tools (e.g., Excel worksheets)**
- **Answers to selected problems**

ONLINE MATERIALS FOR INSTRUCTORS

The author's lecture material (PowerPoint slides), quizzes, and software tools are available at www.wiley.com/college/budhu for instructors' use. The quizzes are *.pdf files converted from the electronic version originally developed using tools in a learning management system. Online material is upgraded periodically.

ACKNOWLEDGMENTS

I am grateful to following persons, who helped with various tasks in the preparation of this textbook.

My former graduate students, Sandeep Mahajan and Melissa Cox, have assisted in reviewing the first draft of the text. Ted Smithwick, a former undergraduate student, helped me with the typing of some parts of the first draft of the MS.

I am grateful to the following faculty members at various higher educational institutions, who have reviewed the manuscript for this textbook and have offered many valuable suggestions for improving it.

Andrew Budek, Texas Tech University
Mandar Dewoolkar, University of Vermont
Robert Easton, Rochester Institute of Technology
Carsten Floess, Rensselaer Polytechnic University
William Gergis, West Virginia University
Narsilio Guillermo, Georgia Institute of Technology
Andrew Heydinger, University of Toledo
Bryson Lindsey, Ohio University
Kunchi Madhavan, Christian Brothers University
Chopra Manoj, University of Central Florida
Diaz Manuel, University of Texas at San Antonio
Robert Parsons, The University of Kansas
Hryciw Roman, University of Michigan—Ann Arbor
Radhey Sharma, Louisiana State University

Ms. Jenny Welter and the staff of John Wiley & Sons, and Suzanne Ingrao of Ingrao Associates were very helpful in all aspects of the production of this book.

CONTENTS

Answers to Selected Problems can be found on the web at www.wiley.com/college/budhu.

Foundations and Earth Retaining Structures

CHAPTER 1

REVIEW OF BASIC SOIL MECHANICS

1.0 INTRODUCTION

The purpose of this chapter is to review some basic soil mechanics concepts that you will use in foundation analysis and design. Soils are geological materials that have the vagaries of the natural world. One of the primary tasks of a geotechnical engineer is to collect, classify, and investigate the physical properties of soils to use in design and construction. Certain properties, parameters, and relationships have been established to accomplish this task. In this chapter, we will summarize the important physical properties and parameters that will be used later in foundation analysis and design.

When you complete this chapter, you should be able to:

- Understand the importance of geology for foundations.
- Calculate and use basic physical soil parameters.

1.1 DEFINITIONS OF KEY TERMS

Soils are materials that are derived from the weathering of rocks.

***Water content* (w)** is the percentage weight of water relative to the weight of solids.

***Void ratio* (e)** is the ratio of the volume of void space to the volume of solids.

***Porosity* (n)** is the ratio of the volume of voids to the total volume of soil.

***Degree of saturation* (S)** is the ratio of the volume of water to the volume of void.

***Bulk unit weight* (γ)** is the weight density (i.e., the weight of a soil per unit volume).

***Saturated unit weight* (γ_s)** is the weight of a saturated soil per unit volume.

***Dry unit weight* (γ_d)** is the weight of a dry soil per unit volume.

***Effective unit weight* (γ')** is the weight of soil solids in a submerged soil per unit volume.

***Relative density* (D_r)** is an index that quantifies the degree of packing between the loosest and densest state of coarse-grained soils.

***Effective particle size* (D_{10})** is the average particle diameter of the soil for which 10% of the particles have smaller diameters.

***Average particle diameter* (D_{50})** is the average particle diameter of the soil.

***Liquid limit* (LL)** is the water content at which a soil changes from a plastic state to a liquid state.

***Plastic limit* (PL)** is the water content at which a soil changes from a semisolid to a plastic state.

Shrinkage limit (SL) is the water content at which a soil changes from a solid to a semisolid state without further change in volume.

Hydraulic conductivity (k) is a proportionality constant to determine the flow of water through soils.

Maximum dry unit weight **($\gamma_{d(\max)}$)** is the maximum unit weight that a soil can attain using a specified means of compaction.

Optimum water content **(w_{opt})** is the water content required to allow a soil to attain its maximum dry unit weight.

Compaction is the densification of a soil by expelling air from the soil pores.

Consolidation is the time-dependent settlement (densification) of fine-grained soils resulting from the expulsion of water from the soil pores.

Porewater pressure (u) is the water pressure in the soil pores.

1.2 ENGINEERING GEOLOGY

Importance *Soils are notoriously variable. You need to understand the geological setting of your project site and the soils that exist there to provide a qualitative understanding of their possible use and potential problems. In this section, basic information on soil formation is reviewed.*

1.2.1 Soil Formation

Soils are geological materials formed from the physical and chemical weathering of rocks. Physical weathering involves reduction of size without any change in the original composition of the parent rock. The main agents responsible for this process are exfoliation, unloading, erosion, freezing, and thawing. Chemical weathering causes both reductions in size and chemical alteration of the original parent rock. The main agents responsible for chemical weathering are hydration, carbonation, and oxidation. Often, chemical and physical weathering take place in concert.

Soils that remain at the site of weathering are called *residual soils.* These soils retain many of the elements that comprise the parent rock. Alluvial soils, also called fluvial soils, are soils that were transported by rivers and streams.

1.2.2 Soil Fabric

The composition of these soils depends on the environment under which they were transported and is often different from the parent rock. The environment under which deposition occurs influences the structural framework that is formed. In particular, the electrochemical environment has the greatest influence on the kind of soil fabric that is formed during deposition.

The two common types of soil fabrics are *flocculated* and *dispersed.* A flocculated structure, formed under saltwater environment, comprises many particles that tend to orient parallel to each other. A flocculated structure, formed under freshwater environment, comprises many particles that tend to orient perpendicular to each other. A dispersed structure comprises a majority of the particles orienting parallel to each other. Any loading (tectonic or otherwise) during or after deposition permanently alters the soil fabric or structural arrangement in a way that is unique to that particular loading condition. Consequently, the history of loading and changes in the environment are imprinted in the soil fabric. The soil fabric is the brain; it retains the memory of the birth of the soil and subsequent changes that occur.

1.2.3 Soil Minerals

The response of soils (essentially fine-grained soils, see Section 1.3) to loads, known as the *mechanical behavior,* depends on the type of predominant minerals present. Minerals are crystalline materials and make up the solids constituent of a soil. The mineral particles of fine-grained soils are platy. Minerals are classified according to chemical composition and structure. Most minerals of interest to geotechnical engineers are composed of oxygen and silicon—two of the most abundant elements on earth. Silicates are a group of minerals with a structural unit called the *silica tetrahedron.*

The predominant minerals in fine-grained soils are kaolinite, illite, and montmorillonite. Kaolinite has a structure that consists of one silica sheet and one alumina sheet bonded together into a layer about 0.72 nm thick and stacked repeatedly. The layers are held together by hydrogen bonds. Tightly stacked layers result from numerous hydrogen bonds. Kaolinite is common in clays in humid tropical regions. Illite consists of repeated layers of one alumina sheet sandwiched by two silicate sheets. The layers, each of thickness 0.96 nm, are held together by potassium ions.

Montmorillonite has a similar structure to illite, but the layers are held together by weak van der Waals forces and exchangeable ions. Water can easily enter the bond and separate the layers in montmorillonite, causing swelling. Montmorillonite is often called *swelling* or *expansive clay.*

The* key points *are:

1. ***Knowledge of geology is important for the successful practice of geotechnical engineering.***
2. ***Soils are derived from the weathering of rocks.***
3. ***Physical weathering causes reduction in size of the parent rock without change in its composition.***
4. ***Chemical weathering causes reduction of size and chemical composition different from the parent rock.***
5. ***The in-situ structure of a soil depends on the deposition environment and any past mechanical loading.***
6. ***Fine-grained soils (clays) are composed of three main types of minerals: kaolinite, illite, and montmorillonite.***
7. ***The clay minerals consist of silica and alumina sheets that are combined to form layers. The bonds between the layers play a very important role in the mechanical behavior of clays. The bond between the layers in montmorillonite is very weak compared with kaolinite and illite. Water can easily enter between the layers in montmorillonite, causing swelling.***
8. ***The soil fabric (structural arrangement of the soil grains) is the brain; it retains the memory of the birth of the soil and subsequent changes that occur.***

1.3 SOIL TYPES AND SOIL DESCRIPTION

Importance *Geotechnical engineers have categorized soils into various types and common descriptive terms have evolved to aid them in getting a quick, qualitative estimate on the soils' strength and deformation properties and variability. Some of the key soil types and common descriptive terms are reviewed.*

1.3.1 General Soil Types and Their Qualities for Engineering Applications

Soils are described as coarse-grained (gravels and sands) and fine-grained (silts and clays). Coarse-grained soils feel gritty and hard. Fine-grained soils feel smooth. In general, coarse-grained soils have good load-bearing capacities and good drainage qualities, and their strength and volume change characteristics are not significantly affected by change in moisture conditions. They are practically incompressible when dense, but significant volume changes can occur when they are loose. Vibrations accentuate volume changes in loose coarse-grained soils by rearranging the soil fabric into a dense configuration.

Fine-grained soils have poor load-bearing capacities compared with coarse-grained soils. Fine-grained soils are practically impermeable. They change volume and strength with variations in moisture conditions and are frost susceptible. The engineering properties of coarse-grained soils are controlled mainly by the grain size of the particles and their structural arrangement. The engineering properties of fine-grained soils are controlled by mineralogical factors rather than grain size. Thin layers of fine-grained soils, even within thick deposits of coarse-grained soils, have been responsible for many geotechnical failures and therefore you need to pay special attention to fine-grained soils.

1.3.2 Common Soil Types

There are many different types of soils. Some soil types are listed below:

- ***Alluvial soils*** are fine sediments that have been eroded from rock, transported by water, and have settled on river and stream beds.
- ***Calcareous soil*** contains calcium carbonate and effervesces when treated with hydrochloric acid.
- ***Caliche*** consists of gravel, sand, and clay cemented together by calcium carbonate.
- ***Collovial soils*** are found at the base of mountains that have been eroded by the combination of water and gravity.
- ***Expansive soils*** are clays that undergo large volume changes (expansion) from cycles of wetting and drying.
- ***Collapsible soils*** are generally low-plasticity soils that undergo large volume changes (settle suddenly) from wetting and loading.
- ***Eolian soils*** are sand-sized particles deposited by wind.
- ***Glacial soils*** are mixed soils consisting of rock debris, sand, silt, clays, and boulders.
- ***Glacial till*** is a soil that consists mainly of coarse particles.
- ***Glacial clays*** are soils that were deposited in ancient lakes and subsequently frozen. The thawing of these lakes reveals a soil profile of neatly stratified silt and clay, sometimes called *varved clay*. The silt layer is light in color and was deposited during summer periods while the thinner, dark clay layer was deposited during winter periods.
- ***Lacustrine soils*** are mostly silts and clays deposited in glacial lake waters.
- ***Lateritic soils*** are residual soils that are cemented with iron oxides and are found in tropical regions.
- ***Loam*** is a mixture of sand, silt, and clay that may contain organic material.
- ***Loess*** is a wind-blown, uniform fined-grained soil.
- ***Marine soils*** are sand, silts, and clays deposited in salt or brackish water.
- ***Mud*** is clay and silt mixed with water into a viscous fluid condition.

1.3.3 Soil Description and Identification

Common descriptive terms and methods of identification of soils are as follows:

Color: Color is not directly related to engineering properties of soils, but is related to the soil mineralogy and texture.

Gray and bluish: unoxidized soils

White and cream: calcareous soils

Red and yellow: oxidized soil

Black and dark brown: soil containing organic matter

Moisture: Appearance due to water is described as wet, dry, and moist.

Structure:

Homogeneous: Color and texture feel the same throughout.

Nonhomogeneous: Color and texture vary.

Shape: Angular, subangular, subrounded, rounded, flaky.

Weathering: Fresh, decomposed, weathered.

Carbonate: Effervesces with acid. Add a small amount of hydrochloric acid and check if soil effervesces. If it does, it contains carbonate.

Odor: Organic soils give off a strong odor that intensifies with heat. Nonorganic soils have a subtle odor with the addition of water.

Feel: Use feel to distinguish between sand, silts, and clays.

Sand has a gritty feel.

Silt has a rough feel similar to fine sandpaper.

Clay feels smooth and greasy. It sticks to fingers and is powdery when dry.

Consistency:

Very stiff: Finger pressure barely dents soil but cracks under significant pressure.

Stiff: Finger pressure dents soil.

Firm: Soil can be molded using strong finger pressure.

Soft: Easily molded by finger.

Very soft: Soil flows between fingers when fist is closed.

Dilatancy: Place a small amount of the soil in your palm and shake horizontally. Then strike it with the other hand. If the surface is slurry and water appears, the soil probably has a large amount of silt.

Packing: Coarse-grained soils are described as:

Very loose: collapses with slight disturbance, open structure

Loose: collapses upon disturbance, open structure

Medium dense: indents when pushed firmly

Dense: barely deforms when pushed by feet or with stomping

Very dense: impossible to depress with stomping

The* key points *are:

1. ***There are many soil types that you need to distinguish.***
2. ***Common descriptive terms have evolved to get a quick, qualitative estimate on the soils' strength and deformation properties and variability.***

1.4 SOIL PHASES

Importance *Soils are complex materials. Like all complex engineering materials, we have to decompose soils into simpler forms to understand them. Geotechnical engineers have simplified soils into three constituents. The proportion of each constituent has significant influences on the strength and deformation properties. In this section, we will review fundamental relationships among the three constituents.*

It is convenient to decompose soils into three constituents; solids, liquids, and gases. The solid phase may be mineral, organic matter, or both. The liquid phase is predominantly water, and the gaseous phase is air. The spaces between the solids (soil particles) are called *voids*. The water within the voids is commonly called *porewater* and plays a very important role in the behavior of soils under load. If all the voids are filled by water, the soil is saturated. Otherwise, the soil is unsaturated. If all the voids are filled with air, the soil is said to be dry. (See Fig. 1.1.)

Several physical parameters are used to relate the various phases. The essential ones are as follows:

1. ***Specific gravity*** (G_s) is the ratio of the weight of the soil solids to the weight of water of equal volume:

$$G_s = \frac{W_s}{V_s \gamma_w} \tag{1.1}$$

where $\gamma_w = 9.8$ kN/m^3 is the unit weight of water. The specific gravity of soils does not vary significantly. For most soils, a value of 2.7 is sufficiently accurate for practical purposes.

2. ***Water content*** (w) is the ratio, often expressed as a percentage, of the weight of water to the weight of solids:

$$w = \frac{W_w}{W_s} \times 100\% \tag{1.2}$$

The water content of soils varies from nearly zero to over 200%. The usual range of water content for granular soils is 0.1 to 40% while the range for fine-grained soils is 0.1 to 75%. Some fine-grained soils, for example, soils with an abundance of sodium montmorillonite mineral, can have water contents in excess of 700%.

3. ***Void ratio*** (e) is the ratio of the volume of void space to the volume of solids. Void ratio is usually expressed as a decimal quantity:

$$e = \frac{V_v}{V_s} \tag{1.3}$$

The void ratio of a soil is not the same in all directions.

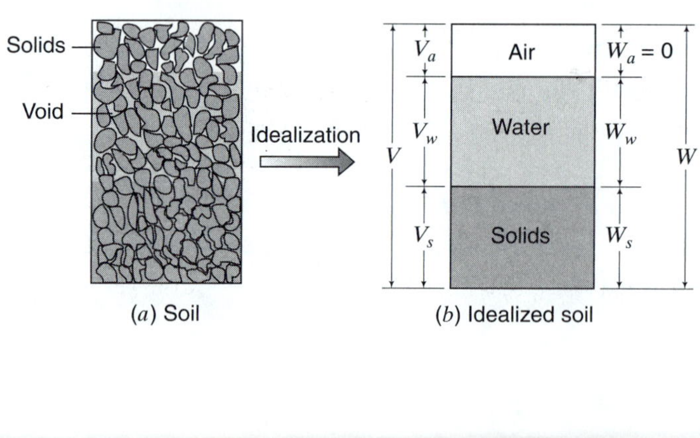

FIGURE 1.1 Soil phases.

4. ***Porosity*** (n) is the ratio of the volume of voids to the total volume. Porosity is usually expressed as a percentage:

$$n = \frac{V_v}{V} \tag{1.4}$$

Porosity and void ratio are related by the expression

$$n = \frac{e}{1+e} \tag{1.5}$$

5. ***Degree of saturation*** (S) is the ratio, often expressed as a percentage, of the volume of water to the volume of voids:

$$S = \frac{V_w}{V_v} = \frac{wG_s}{e} \quad \text{or} \quad Se = wG_s \tag{1.6}$$

If $S = 1$ or 100%, the soil is saturated. If $S = 0$, the soil is bone dry. It is practically impossible to obtain a soil with $S = 0$.

6. ***Unit weight*** is the weight of a soil per unit volume. We will use the term *bulk unit weight*, γ, to denote unit weight:

$$\gamma = \frac{W}{V} = \left(\frac{G_s + Se}{1+e}\right)\gamma_w \tag{1.7}$$

Special Cases

(a) Saturated unit weight $(S = 1)$:

$$\gamma_{sat} = \left(\frac{G_s + e}{1+e}\right)\gamma_w \tag{1.8}$$

(b) Dry unit weight $(S = 0)$:

$$\gamma_d = \frac{W_s}{V} = \left(\frac{G_s}{1+e}\right)\gamma_w = \frac{\gamma}{1+w} \tag{1.9}$$

(c) Effective or buoyant unit weight is the weight of a saturated soil, surrounded by water, per unit volume of soil:

$$\gamma' = \gamma_{sat} - \gamma_w = \left(\frac{G_s - 1}{1+e}\right)\gamma_w \tag{1.10}$$

Typical values of unit weight of soils are given in Appendix A, Table A.1.

7. ***Relative density*** (D_r) is an index that quantifies the degree of packing between the loosest and densest possible state of a clean, coarse-grained soil as determined by experiments:

$$D_r = \frac{e_{max} - e}{e_{max} - e_{min}} \tag{1.11}$$

where e_{max} = maximum void ratio (loosest condition), e_{min} = minimum void ratio (densest condition), and e = current void ratio. A description of coarse-grained soils based on relative density is given in Appendix A, Table A.2.

The key points are:
1. ***Soils are complex materials. They are conveniently decomposed into three constituents: solids, liquids, and gases.***
2. ***The proportion of each constituent influences the strength and deformation properties.***

EXAMPLE 1.1 *Calculation of Basic Soil Parameters*

A cylindrical sample of clay, 50 mm (2 in) × 100 mm (4 in) long, was placed in a container and weighed. The weight of the sample was 3.5 N (0.79 lb). The clay in its container was placed in an oven for 24 hours at 105°C. The sample weight reduced to a constant value of 2.9 N (0.65 lb). If $G_s = 2.7$, determine the following:

(a) Water content
(b) Void ratio
(c) Bulk unit weight
(d) Dry unit weight
(e) Effective unit weight
(f) Degree of saturation

Strategy Write down what is given and then use the appropriate equations to find the unknowns. You are given the weight of the natural soil, sometimes called the *wet weight*, and its dry weight. The difference between these will give the weight of water and you can find the water content by using Eq. (1.2). SI units will be used in the solution of this problem.

Solution 1.1

Step 1: Write down what is given.

$$\text{Wet weight of sample, } W = 3.5 \text{ N}$$

$$\text{Dry weight of sample, } W_s = 2.9 \text{ N}$$

$$\text{Volume of wet soil} = \frac{\pi \times 0.05^2 \times 0.1}{4} = 196.4 \times 10^{-6} \text{m}^3$$

Step 2: Determine the weight of water.

$$\text{Weight of water, } W_w = 3.5 - 2.9 = 0.6 \text{ N}$$

Step 3: Determine the water content.

$$w = \frac{W_w}{W_s} \times 100 = \frac{0.6}{2.9} \times 100 = 20.7\%$$

Step 4: Determine the bulk unit weight.

$$\gamma = \frac{W}{V} = \frac{3.5 \times 10^{-3}}{196.4 \times 10^{-6}} = 17.8 \text{ kN/m}^3$$

Step 5: Determine the dry unit weight.

$$\gamma_d = \frac{\gamma}{(1 + w)} = \frac{17.8}{1 + 0.207} = 14.7 \text{ kN/m}^3$$

Step 6: Determine the void ratio.

$$\gamma_d = \left(\frac{G_s}{1+e}\right)\gamma_w; \therefore 14.7 = \frac{2.7}{1+e} \times 9.8$$

$$e = 0.8$$

Step 7: Determine the degree of saturation.

$$S = \frac{wG_s}{e} = \frac{0.207 \times 2.7}{0.8} = 0.7 = 70\%$$

Step 8: Determine the saturated unit weight.

$$\gamma_{sat} = \left(\frac{G_s + e}{1+e}\right)\gamma_w = \left(\frac{2.7 + 0.8}{1 + 0.8}\right) \times 9.8 = 19.1\,\text{kN/m}^3$$

Step 9: Determine the effective unit weight.

$$\gamma' = \gamma_{sat} - \gamma_w = 19.1 - 9.8 = 9.3\,\text{kN/m}^3$$

■

1.5 PARTICLE SIZE OF SOILS

Importance *The particle sizes and their distribution within a soil mass can indicate the appropriate engineering use of that soil and give qualitative estimates of the soil's load-bearing capacity. In this section, the determination of the particle sizes for soils is described.*

The particle or grain sizes of soils are conveniently divided into two groups: coarse-grained soils (sands and gravels) and fine-grained soils (silts and clays). The distribution of grain sizes or average grain diameters of coarse-grained soils—gravels and sands—is obtained by screening a known weight of the soil through a stack of sieves of progressively finer mesh size. An illustrated particle size distribution plot is shown in Fig. 1.2. On the abscissa is the grain size or sieve number (each sieve number corresponds to the number of square openings per square inch of mesh area). There is no theoretical basis for this. Rather, the log scale is used for convenience because of the large range of grain sizes of soils. The ordinate is the percent finer, which is the percentage of particles passing a particular sieve.

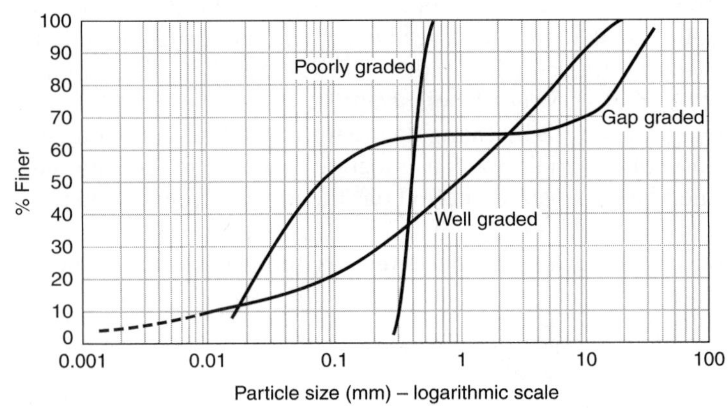

FIGURE 1.2 Particle size distribution curves.

The screening process, also called *sieve analysis*, cannot be used for fine-grained soils (silts and clays) because of their extremely small size (0.002 mm to 0.075 mm). Traditionally, a hydrometer test is conducted to determine the distribution of grain size in fine-grained soils.

The laboratory method to determine the grain-size distribution is described in the American Society for Testing and Materials (ASTM) test methods ASTM D422 and ASTM D1140.

The* key points *are:

1. ***A sieve analysis is used to determine the grain-size distribution of coarse-grained soils.***
2. ***A hydrometer test is used to determine the grain-size distribution of fine-grained soils.***
3. ***Particle-size distribution is represented on a semilogarithmic plot of % finer (ordinate, arithmetic scale) versus particle size (abscissa, logarithmic scale).***
4. ***The particle-size distribution plot is used to delineate the different soil textures (percentages of gravel, sand, silt, and clay) in a soil.***

1.6 PHYSICAL SOIL STATES AND INDEX PROPERTIES OF FINE-GRAINED SOILS

Importance *The proportion of water in a soil matrix can change the way a soil behaves. This is particularly important for fine-grained soils. In this section, we review how water changes fine-grained soils.*

The physical and mechanical behavior of fine-grained soils is linked to four distinct states: solid, semisolid, plastic, and liquid in order of increasing water content. The water contents at which these changes in states occur are called the *index properties* of soils. There are three index properties: liquid limit, plastic limit, and shrinkage limit. The water content at which the soil changes from a liquid to a plastic state is called the *liquid limit* (LL). As the soil dries below the liquid limit, it behaves in a plastic manner—it can be molded into any desired shape without rupture—until the semisolid state is reached. The water content at which the soil changes from a plastic to a semisolid is known as the *plastic limit* (PL). The range of water contents over which the soil deforms plastically is known as the *plasticity index* (PI).

$$\mathrm{PI} = \mathrm{LL} - \mathrm{PL} \tag{1.12}$$

As the soil continues to dry, it comes to a final state called the *solid state*. At this state, no further volume change occurs since nearly all the water in the soil has been removed. The water content at which the soil changes from a semisolid to a solid is called the *shrinkage limit* (SL). The shrinkage limit is useful for the determination of the swelling and shrinking capacity of soils. The liquid and plastic limits are called the *Atterberg limits*, named after their originator, Swedish soil scientist, A. Atterberg (1911).

We have changed the states of fine-grained soils by changing the water content. Since engineers are interested in the strength and deformation of materials, we can associate specific strength characteristics to each of the soil states. At one extreme, the liquid state, the soil has the lowest strength and the largest deformation. At the other extreme, the solid state, the soil has the largest strength and the lowest deformation. A qualitative measure of soil strength using the Atterberg limits is known as the *liquidity index* (LI) and is expressed as:

$$\mathrm{LI} = \frac{w - \mathrm{PL}}{\mathrm{PI}} \tag{1.13}$$

The liquidity index is the ratio of the difference in water content between the natural or in-situ water content of a soil and its plastic limit to its plasticity index. A description of soil strength based on values of LI is given in Appendix A, Table A.3. Typical values for the Atterberg limits for soils are shown in Appendix A, Table A.4.

The Atterberg limits depend on the type of predominant mineral in the soil. If montmorillonite is the predominant mineral, the liquid limit can exceed 100%, because the bond between the layers in montmorillonite (smectite) is weak and large amounts of water can easily infiltrate the spaces between the layers. Sodium smectite can have liquid limits as high as 1100%. In the case of kaolinite, the layers are held relatively tightly and water cannot easily infiltrate between the layers in comparison with montmorillonite. Therefore, you can expect the Atterberg limits for kaolinite to be, in general, much lower than either montmorillonite or illite.

Skempton (1953) showed that for soils with a particular mineralogy, the plasticity index is linearly related to the amount of the clay fraction. He coined the term *activity* (A) to describe the importance of the clay factions on the plasticity index. The equation for A is

$$A = \frac{\text{PI}}{\text{Clay fraction (\%)}} \tag{1.14}$$

You should recall that the clay fraction is the amount of particles less than 2 microns.

The laboratory tests to determine the index properties are described in ASTM D4318 and ASTM D427.

The key points are:

1. ***Fine-grained soils can exist in one of four states: solid, semisolid, plastic, and liquid.***
2. ***Water is the agent that is responsible for changing the states of soils.***
3. ***A soil gets weaker if its water content increases.***
4. ***Three limits are defined based on the water content that causes a change of state. These are the liquid limit (the water content that caused the soil to change from a liquid to a plastic state); the plastic limit (the water content that caused the soil to change from a plastic to a semisolid), and the shrinkage limit (the water content that caused the soil to change from a semisolid to a solid state). All of these limiting water contents are found from laboratory tests.***
5. ***The plasticity index defines the range of water content for which the soil behaves like a plastic material.***
6. ***The liquidity index gives a qualitative measure of strength.***

EXAMPLE 1.2 *Calculation of Plasticity Index and Liquidity Index*

Atterberg limit tests were conducted on a saturated clay of water content 36%. The liquid limit is 74% and the plastic limit 22%. Determine the plasticity index and the liquidity index, and describe the possible qualitative strength of the clay.

Strategy Use the appropriate equations and Appendix A, Table A.3.

Solution 1.2

Step 1: Calculate the plasticity index.

$$PI = LL - PL = 74 - 22 = 52\%$$

Step 2: Calculate the liquidity index.

$$LI = \frac{w - PL}{PI} = \frac{36 - 22}{52} = 0.27$$

Step 3: Estimate the strength.
Appendix A, Table A.3: The soil will behave in a plastic manner. ■

1.7 SOIL CLASSIFICATION

Importance *We have reviewed some basic soil data—particle size and index properties from quick, simple tests. The question that arises is, what do we do with this data? Engineers would like to use this data to get a first impression on the use and possible performance of the soil for a particular purpose such as a foundation for a skyscraper. This is currently achieved by classification systems. A classification system or scheme provides a method of grouping soils that would likely exhibit similar characteristics. We will review one popular classification scheme.*

Various classification systems have evolved over the years to describe soils based on their particle sizes. Each system was developed for a specific engineering purpose. In this book, we will use the Unified Soil Classification System (USCS). The USCS separates soils into two categories. One category is coarse-grained soils that are delineated if more than 50% of the soil is greater than 0.075 mm. The other category is fine-grained soils that are delineated if more than 50% of the soil is finer than 0.075 mm (no. 200 sieve). Coarse-grained soils are subdivided into gravels and sands while fine-grained soils are divided into silts and clays. Each soil type—gravel, sand, silt, and clay—is identified by grain size as shown in Appendix A, Table A.5. The USCS does not differentiate silts from clays based on grain size. Clays have particle sizes

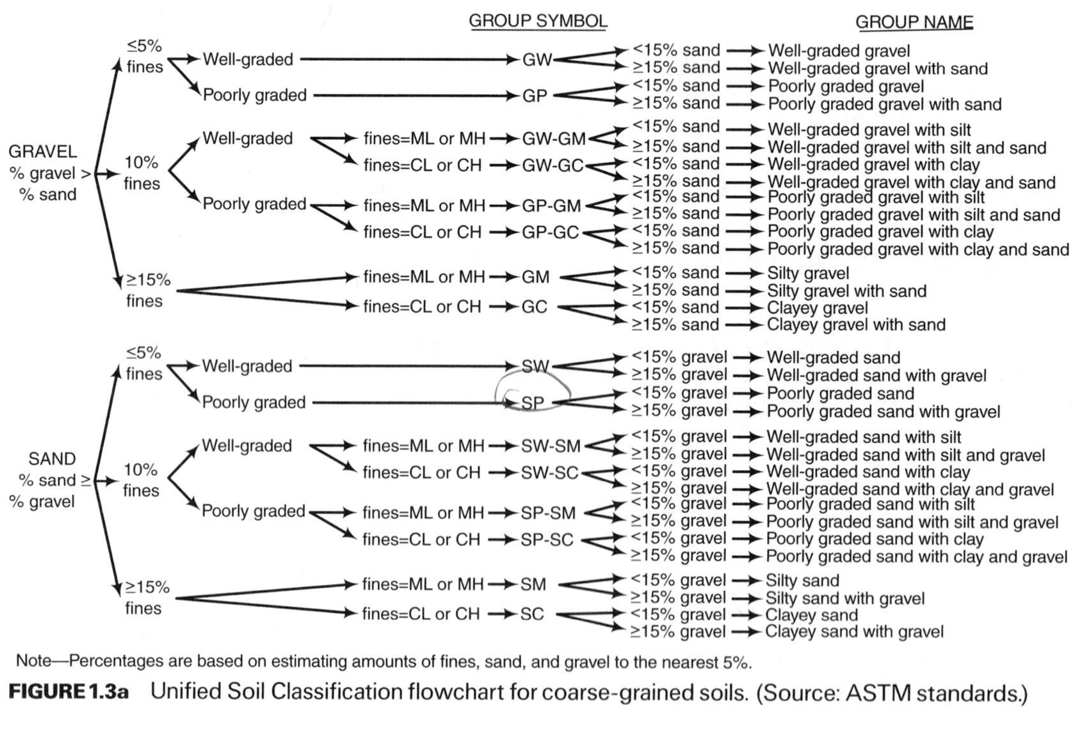

FIGURE 1.3a Unified Soil Classification flowchart for coarse-grained soils. (Source: ASTM standards.)

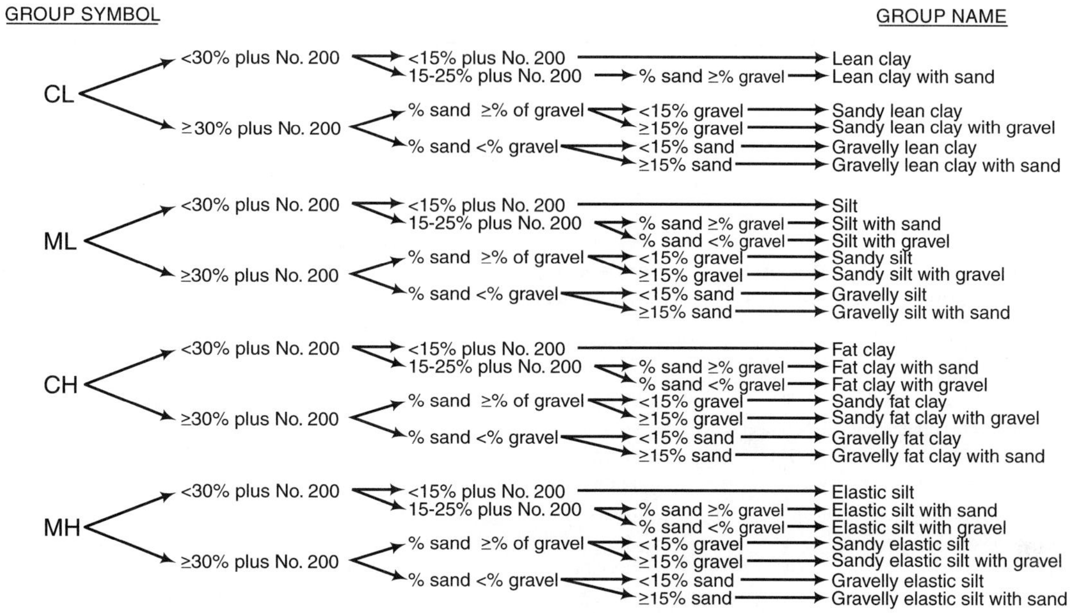

NOTE—Percentages are based on estimating amounts of fines, sand, and gravel to the nearest 5%.

FIGURE 1.3b Unified Soil Classification flowchart for fine-grained soils. (Source: ASTM standards.)

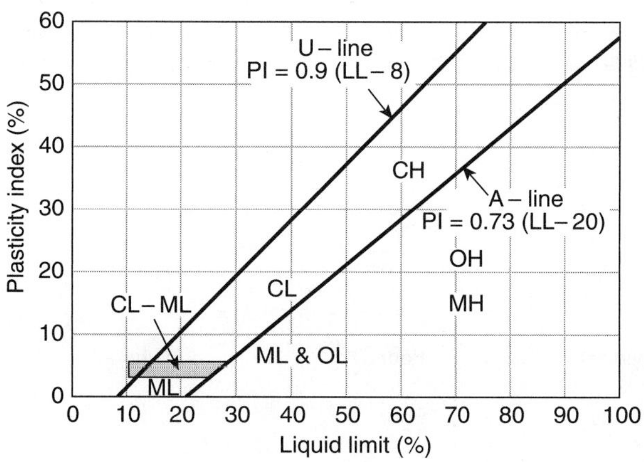

FIGURE 1.3c Plasticity chart for fine-grained soils.

less than 0.002 mm (2 microns). ASTM identifies a third category, called *highly organic soils*. We will not deal with this category of soils in this text.

The USCS uses symbols for the particle size groups. These symbols and their representations are: G—gravel, S—sand, M—silt, C—clay. These are combined with other symbols expressing gradation characteristics (W for well graded, P for poorly graded) and plasticity characteristics (H for high and L for low), and a symbol, O, indicating the presence of organic material. A typical classification of CL means a clay soil with low plasticity, while SP means a poorly graded sand. The flowcharts shown in Figs. 1.3a,b are systematic means of classifying soils according to USCS.

In addition to USCS, a plasticity chart shown in Fig. 1.3c is used exclusively for fine-grained soils. This chart was obtained by plotting experimental data from fine-grained soils tested from different parts of the world.

TABLE 1.1 Engineering Use Chart

		Important Properties			
Typical names of soil groups	Group symbols	Permeability when compacted	Shearing strength when compacted and saturated	Compressibility when compacted and saturated	Workability as a construction material
Well-graded gravels, gravel-sand mixtures, little or no fines	*GW*	Pervious	Excellent	Negligible	Excellent
Poorly graded gravels, gravel-sand mixtures, little or no fines	*GP*	Very pervious	Good	Negligible	Good
Silty gravels, poorly graded gravel-sand-silt mixtures	*GM*	Semipervious to impervious	Good	Negligible	Good
Clayey gravels, poorly graded gravel-sand-clay mixtures	*GC*	Impervious	Good to fair	Very low	Good
Well-graded sands, gravelly sands, little or no fines	*SW*	Pervious	Excellent	Negligible	Excellent
Poorly graded sands, gravelly sands, little or no fines	*SP*	Pervious	Good	Very Low	Fair
Silty sands, poorly graded sand-silt mixtures	*SM*	Semipervious to impervious	Good	Low	Fair
Clayey sands, poorly graded sand-clay mixtures	*SC*	Impervious	Good to fair	Low	Good
Inorganic silts and very fine sands, rock flour, silty or clayey fine sands with slight plasticity	*ML*	Semipervious to impervious	Fair	Medium	Fair
Inorganic clays of low to medium plasticity, gravelly clays, sandy clays, silty clays, lean clays	*CL*	Impervious	Fair	Medium	Good to fair
Organic silts and organic silt-clays of low plasticity	*OL*	Semipervious to impervious	Poor	Medium	Fair
Inorganic silts, micaceous or diatomaceous fine sandy or silty soils, elastic silts	*MH*	Semipervious to impervious	Fair to poor	High	Poor
Inorganic clays of high plasticity, fat clays	*CH*	Impervious	Poor	High	Poor
Organic clays of medium to high plasticity	*OH*	Impervious	Poor	High	Poor
Peat and other highly organic soils	*Pt*	—	—	—	—

SOURCE: Wagner, 1957.

Geotechnical engineers have prepared charts based on USCS and on experience to assist you in selecting a soil for a particular construction purpose. One such chart is shown in Table 1.1. The numerical values 1 to 9 are ratings, with number 1 the best. The chart should be used only to provide guidance and to make a preliminary assessment of the suitability of a soil for a particular use. You should not rely on such descriptions as "excellent" shear strength or "negligible" compressibility to make final design and construction decisions.

Relative Desirability for Various Uses										
							Roadways			
Rolled earth dams			Canal sections		Foundations		Fills			
Homogeneous embankment	Core	Shell	Erosion resistance	Compacted earth lining	Seepage important	Seepage not important	Frost heave not possible	Frost heave possible	Surfacing
—	—	1	1	—	—	1	1	1	3
—	—	2	2	—	—	3	3	3	—
2	4	—	4	4	1	4	4	9	5
[illegible]	1	—	3	1	2	6	5	5	1
—	—	3 if gravelly	6	—	—	2	2	2	4
—	—	4 if gravelly	7 if gravelly	—	—	5	6	4	—
4	5	—	8 if gravelly	5 Erosion critical	3	7	8	10	6
3	2	—	5	2	4	8	7	6	2
6	6	—	—	6 Erosion critical	6	9	10	11	—
5	3	—	9	3	5	10	9	7	7
[illegible]	8	—	—	7 Erosion critical	7	11	11	12	—
[illegible]	9	—	—	—	8	12	12	13	—
[illegible]	7	—	10	8 Volume change critical	9	13	13	8	—
10	10	—	—	—	10	14	14	14	—
—	—	—	—	—	—	—	—	—	—

The key points are:

1. ***A classification scheme provides a method of identifying soils in a particular group that would likely exhibit similar characteristics.***
2. ***Soil classification is used to specify a certain soil type that is best suited for a given application.***

EXAMPLE 1.3 *Interpreting a Soil Classified According to USCS*

In a geotechnical report, a soil was classified according to USCS as SP-SC. What type of soil is this? Would this be a potentially excellent material for a foundation?

Strategy Use the ASTM flowchart and Table 1.1.

Solution 1.3

From Fig. 1.3a, the soil is poorly graded sand with clay. If the percent of gravel were given, you could then add gravel to the description for gravel content greater than or equal to 15%. From Table 1.1, the rating for this soil as a foundation material is 5 (rating of 1 is best). So it is not a potentially excellent foundation material. ■

1.8 ONE-DIMENSIONAL FLOW OF WATER THROUGH SOILS

Importance *Water can cause instability, and many geotechnical structures (e.g., roads, bridges, dams, and excavations) have failed because of instability from flow of water. The key physical property that governs flow of water in soils is permeability or hydraulic conductivity. In this section, we will review one-dimensional flow through soils.*

Water can flow through the interconnected voids with a soil. The flow of water through soils is governed by *Darcy's law*. Darcy (1856) proposed that average flow velocity through soils is proportional to the gradient of the total hydraulic head. The total hydraulic head or simply total head is the sum of the pressure head, the velocity head and the elevation head. The velocity head in soils is generally small and is neglected. The flow in any direction, j, is

$$v_j = k_j \frac{dH}{dx_j} \tag{1.15}$$

where v is the average flow velocity, k is a coefficient of proportionality called the *hydraulic conductivity*, and dH is the change in total head over a distance dx. The unit of measurement for k is length/time (i.e., cm/s or ft/min). If ΔH is the change in total head over a distance l, then the hydraulic gradient is

hydraulic gradient:

$$i = \frac{\Delta H}{l} \tag{1.16}$$

and the average velocity of flow through a cross-sectional area normal to the direction of flow is

Darcy's law:

$$v = k_j \frac{\Delta H}{l} = k_j i \tag{1.17}$$

Flow through soils, however, occurs only through the interconnected voids. The velocity through the void spaces is called seepage velocity (v_s) and is obtained by dividing the average velocity by the porosity of the soil:

seepage velocity:

$$v_s = \frac{k_j}{n} i \tag{1.18}$$

accounts for water only flowing through voids

The volume rate of flow (q) or, simply, flow rate is the product of the average velocity and the cross-sectional area:

$$q = v_j A = A k_j i \tag{1.19}$$

The unit of measurement for q is m^3/sec or cm^3/sec. The conservation of flow (law of continuity) stipulates that the volume rate of inflow (q_{in}) into a soil element must equal the volume rate of outflow (q_{out}) or, simply, inflow must be equal to outflow (i.e., $q_{in} = q_{out}$). Typical ranges of values of hydraulic conductivity for vertical flow are given in Appendix A, Table A.6.

The hydraulic conductivity depends on the following:

1. ***Soil type:*** Coarse-grained soils have higher hydraulic conductivities than fine-grained soils. The water in the double layer in fine-grained soils significantly reduces the seepage pore space.
2. ***Particle size:*** Hydraulic conductivity depends on D_{50}^2 (or D_{10}^2) for coarse-grained soils.
3. ***Pore fluid properties, particularly viscosity:*** $k_1:k_2 \approx \mu_2:\mu_1$, where subscripts 1 and 2 denote two types of pore fluids in a given soil.
4. ***Void ratio:*** $k_1:k_2 \approx e_1^2:e_2^2$, where subscripts 1 and 2 denote two types of soil fabric for coarse-grained soils. This ratio is useful in comparing the hydraulic conductivities of similar soils with different void ratios.
5. ***Pore size:*** The greater the pore size, the higher the hydraulic conductivity.
6. ***Homogeneity, layering, and fissuring:*** Water tends to seep quickly through loose layers, through fissures, and along the interface of layered soils. Catastrophic failures can occur from such seepage.
7. ***Entrapped gases:*** Entrapped gases tend to reduce the hydraulic conductivity. It is often very difficult to get gas-free soils. Even soils that are under groundwater level and are assumed to be saturated may still have some entrapped gases.
8. ***Validity of Darcy's law:*** Darcy's law is valid only for laminar flow (Reynold's number less than 2000).

For a homogeneous soil, the hydraulic conductivity depends predominantly on void ratio. You should recall that the void ratio is dependent on the soil fabric or structural arrangement of the soil grains. A number of empirical relationships have been proposed linking k to void ratio and grain size for coarse-grained soils. Hazen (1930) conducted permeability tests on sands with D_{10} ranging from 0.1 mm (0.004 in) to 3 mm (0.012 in) and $\text{Cu} < 5$ (Cu is the uniformity coefficient; $\text{Cu} = D_{60}/D_{10}$) and proposed that

$$k = CD_{10}^2 \text{ cm/sec} \tag{1.20}$$

where C is a constant varying between 0.4 and 1.2 if the unit of measurement of D_{10} is mm. Typically, $C = 1.0$.

The void ratio is generally not the same in each direction, so k is not the same in each direction. The hydraulic conductivity for horizontal flow parallel to soil layers is generally greater than the hydraulic conductivity for vertical flow normal to the soil layers. For layered soils, an equivalent hydraulic conductivity given as

$$k_{\text{eq}} = \sqrt{k_{x(\text{eq})} k_{z(\text{eq})}} \tag{1.21}$$

is used, where $k_{x(\text{eq})}$ and $k_{z(\text{eq})}$ are the equivalent hydraulic conductivities for horizontal flow parallel to the soil layers and vertical flow normal to the soil layers, respectively.

Two types of tests are used to determine the hydraulic conductivity. One, the constant head test (ASTM D2434), is used for coarse-grained soils. The other, the falling head test, is used for fine-grained soils.

The* key points *are:

1. ***The flow of water through soils is governed by Darcy's law, which states that the average flow velocity is proportional to the hydraulic gradient.***
2. ***The proportionality coefficient in Darcy's law is called the hydraulic conductivity, k.***
3. ***The value of k is influenced by (among other factors) the void ratio, particle size distribution, and the wholeness of the soil mass.***
4. ***Homogeneous clays are practically impervious, while sands and gravels are pervious.***

EXAMPLE 1.4 *Estimating the Hydraulic Conductivity of a Soil*

The hydraulic conductivity of a sand with $D_{10} = 0.5$ mm (0.02 in) near a project is 2.0×10^{-3} cm/s (0.8×10^{-3} in/s). At the project site, a similar soil exists but $D_{10} = 1.5$ mm (0.06 in). You, the geotechnical engineer, need to make a quick estimate of the hydraulic conductivity. What is your estimate?

Strategy Since the D_{10} is within the range of the tests conducted by Hazen, we can use his empirical relationship to obtain a quick estimate. SI units will be used in the solution of this problem.

Solution 1.4

$$\frac{k_{\text{at site}}}{k_{\text{near project}}} = \frac{(CD_{10}^2)_{\text{at site}}}{(CD_{10}^2)_{\text{near project}}} = \frac{1.5^2}{0.5^2} = 9$$

$$k_{\text{at site}} = 2 \times 10^{-3} \times 9 = 18 \times 10^{-3} \text{cm/s}$$

■

1.9 TWO-DIMENSIONAL FLOW OF WATER THROUGH SOILS

Access an interactive computer program to draw flow nets using a finite difference solution of Laplace's equation at http://www.wiley.com/college/budhu. Click Foundations and Earth Structures, Chapter 1, 2Dflow.xls.

Importance *Many catastrophic failures in geotechnical engineering can be traced to the instability of soil masses due to groundwater flow. Lives are lost, infrastructures are damaged or destroyed, and major economic losses are incurred. In this section, we briefly review two-dimensional flow of water through soils. The information presented will be particularly useful in designing earth-retaining structures (Chapter 10).*

The two-dimensional flow of water through soils is described by Laplace's equation as

$$k_x \frac{\partial^2 H}{\partial x^2} + k_z \frac{\partial^2 H}{\partial z^2} = 0 \tag{1.22}$$

where H is the total hydraulic head, and k_x and k_z are the hydraulic conductivities in the x (lateral) and z (vertical) directions. Laplace's equation expresses the condition that the change of hydraulic gradient in one direction is balanced by changes in the other directions.

The assumptions in Laplace's equation are:

- Darcy's law is valid.
- The soil is homogeneous and saturated.
- The soil and water are incompressible.
- No volume change occurs.

If the soil were an *isotropic* material, then $k_x = k_z$ and Laplace's equation would become

$$\frac{\partial^2 H}{\partial x^2} + \frac{\partial^2 H}{\partial z^2} = 0 \tag{1.23}$$

The solution of Laplace's equation for the spatial (x, z) variation of the hydraulic head is found using a graphical technique called *flow net sketching*, or with numerical methods such as finite difference, finite element, and boundary element, or by using physical models to attempt to replicate the flow through the real structure. The solution is found by transformation of the head into two functions. One is the velocity potential (ξ), which describes the variation of total head in a soil mass as $\xi = kH$ where k is a generic hydraulic conductivity. The other is a stream function, $\Psi(x, z)$. The application of the velocity potential and potential functions leads to the following condition:

1. The velocity of flow (v) is normal to lines of constant total head (also called constant *piezometric* head or *equipotential* lines) as illustrated in Fig. 1.4. The direction of v is in the direction of decreasing total head. The head difference between two equipotential lines is called a *potential drop* or *head loss*.
2. If lines are drawn that are tangents to the velocity of flow at every point in the flow field in the x–z plane, we will get a series of lines that are normal to the equipotential lines. These tangential lines are called *streamlines* or *flow lines* (Fig. 1.4). A flow line represents the flow path that a particle of water is expected to take in steady-state flow.
3. Since flow lines are normal to equipotential lines, there can be no flow across flow lines. The rate of flow between any two flow lines is constant. The area between two flow lines is called a *flow channel* (Fig. 1.4). Therefore, the rate of flow is constant in a flow channel.

There are several computer programs available to find solutions for Laplace's equation. Geotechnical engineers often used them in conjunction with hand-drawn flow nets. The procedure to sketch flow nets for isotropic soils is as follows:

1. Draw the structure and soil mass to a suitable scale.
2. Identify impermeable and permeable boundaries. The soil–impermeable boundary interfaces are flow lines because water can flow along these interfaces. The soil–permeable boundary interfaces are equipotential lines because the total head is constant along these interfaces.
3. Sketch a series of flow lines (4 or 5) and then sketch an appropriate number of equipotential lines such that the area between a pair of flow lines and a pair of equipotential lines (cell) is approximately

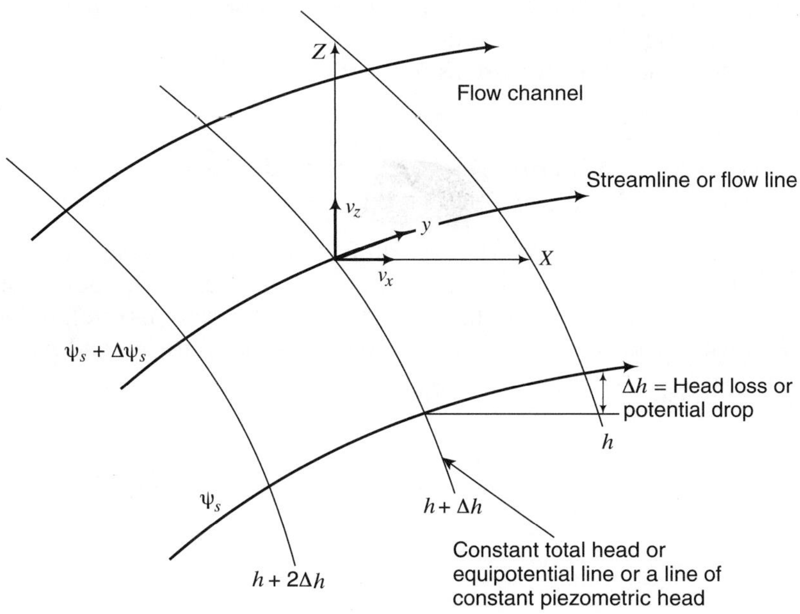

FIGURE 1.4 Illustration of two-dimensional flow terms.

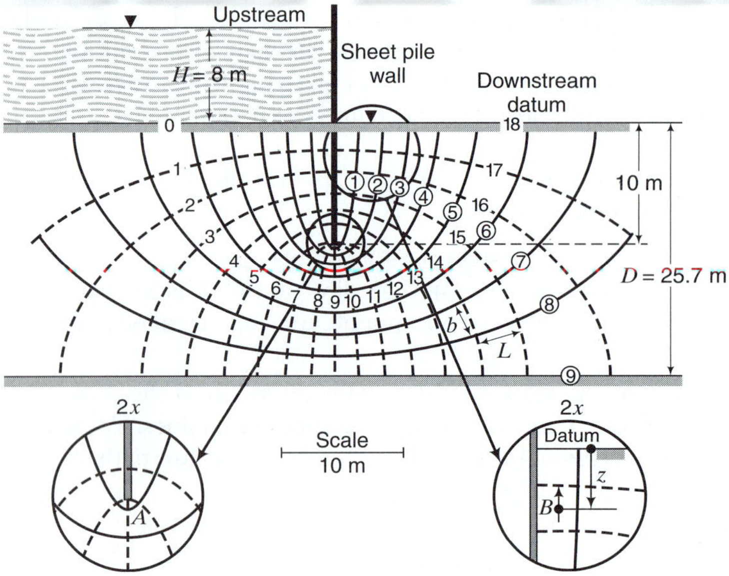

FIGURE 1.5 Flow net for a sheetpile retaining wall.

a curvilinear square. You would have to adjust the flow lines and equipotential lines to make curvilinear squares. You should check that the average width (b) and the average length (l) of a cell are approximately equal. You should also sketch the entire flow net before making adjustments.

The flow net in confined areas between parallel boundaries usually consists of flow lines and equipotential lines that are elliptical in shape and symmetrical (Fig. 1.5). Try to avoid making sharp transitions between straight and curved sections of flow and equipotential lines. Transitions should be gradual and smooth. For some problems, portions of the flow net are enlarged and are not curvilinear squares, and do not satisfy Laplace's equation. For example, the portion of the flow net below the bottom of the sheet pile retaining wall in Fig. 1.5 does not consist of curvilinear squares. For an accurate flow net, you should check these portions to ensure that repeated bisection results in a point. In the case of anisotropic soils, the same procedure as for isotropic soil is followed except that you must draw the structure and flow domain by multiplying the horizontal distances by $\sqrt{k_z/k_x}$.

The key parameters calculated from a flow net are as follows:

Flow Rate

$$q = k\sum_{i=1}^{N_f}\left(\frac{\Delta H}{N_d}\right)_i = k\Delta H\frac{N_f}{N_d} \tag{1.24}$$

where the total head loss across the flow domain is ΔH (i.e., the difference in upstream and downstream water level elevation), N_d is the number of equipotential drops (i.e., the number of equipotential lines minus one), and N_f, is the number of flow channels (i.e., the number of flow lines minus one). In Fig. 1.5, $\Delta H = 8$ m, $N_d = 18$, and $N_f = 9$. The ratio N_f/N_d is called the *shape factor*. Both N_f and N_d can be fractional. In the case of anisotropic soils, the flow is

$$q = \Delta H\frac{N_f}{N_d}\sqrt{k_x k_z} \tag{1.25}$$

Hydraulic Gradient

$$i = \frac{\Delta H}{N_d L} = \frac{\Delta h}{L}; \quad \Delta h = \frac{\Delta H}{N_d} \tag{1.26}$$

where i is the hydraulic gradient over each square. The hydraulic gradient is the head loss per cell (curvilinear square), which is the same for all cells by construction, divided by the length, L, of the cell, You should notice from Fig. 1.5 that L is not constant. Therefore, the hydraulic gradient is not constant. The maximum hydraulic gradient occurs where L is a minimum:

$$i_{\max} = \frac{\Delta h}{L_{\min}} \tag{1.27}$$

where $L_{\min}$ is the minimum length of the cells within the flow domain. Usually, $L_{\min}$ occurs at exit points or around corners (e.g., point A in Fig. 1.5) and it is at these points we can get the maximum hydraulic gradient.

Critical Hydraulic Gradient

$$i_{cr} = \frac{\gamma'}{\gamma_w} = \left(\frac{G_s - 1}{1 + e}\right)\frac{\gamma_w}{\gamma_w} = \frac{G_s - 1}{1 + e} \tag{1.28}$$

The critical hydraulic gradient is the hydraulic gradient that brings a soil mass (essentially, coarse-grained soils) to static liquefaction—the soil behaves like a viscous fluid. Because G_s is constant, the critical hydraulic gradient is solely a function of the void ratio of the soil. In designing structures that are subjected to steady-state seepage, it is essential to ensure that the critical hydraulic gradient would not develop.

Porewater Pressure Distribution The pore water pressure at any point, j, is calculated as follows:

1. Select a datum. Let us choose the downstream water level as datum (Fig. 1.5).
2. Determine the total head at j, $H_j = \Delta H - (N_d)_j \Delta h$, where $(N_d)_j$ is the number of equipotential drops at point j; $(N_d)_j$ can be fractional. For example, at B, $H_B = \Delta H - 16.5\Delta h$.
3. Subtract the elevation head at point j from the total head, H_j, to get the pressure head. For point B, (Fig. 1.5), the elevation head, h_z, is $-z$ (point B is below the datum). The pressure head is then

$$(h_p)_j = \Delta H - (N_d)_j \Delta h - h_z \tag{1.29}$$

For point B, $(h_p)_B = \Delta H - 16.5\Delta h - (-z) = \Delta H - 16.5\Delta h + z$

4. The porewater pressure is

$$u_j = (h_p)_j \gamma_w \tag{1.30}$$

The key points are:

1. ***Streamlines or flow lines represent flow paths of particles of water.***
2. ***The area between two flow lines is called a flow channel.***
3. ***The rate of flow in a flow channel is constant.***
4. ***Flow cannot occur across flow lines.***
5. ***The velocity of flow is normal to the equipotential line.***
6. ***Flow lines and equipotential lines are orthogonal (perpendicular) to each other.***
7. ***The difference in head between two equipotential lines is called the potential drop or head loss.***

EXAMPLE 1.5 *Flow Under a Dam*

The flow net for a dam is shown in Fig. E1.5a. The average hydraulic conductivity of the silty sand is 2.0×10^{-4} cm/sec. Assume that the silty sand is homogeneous and isotropic.

(a) Calculate the flow rate, q.

(b) Calculate and draw the porewater pressure distribution at the base of the dam.

(c) Determine and draw the porewater pressure distribution on the upstream and downstream faces of the sheetpile wall.

(d) Determine the resultant lateral force on the sheetpile wall due to the pore water.

(e) Determine the maximum hydraulic gradient.

(f) Will piping occur if the void ratio of the silty sand is 0.8?

(g) What is the effect of reducing the depth of penetration of the sheetpile wall?

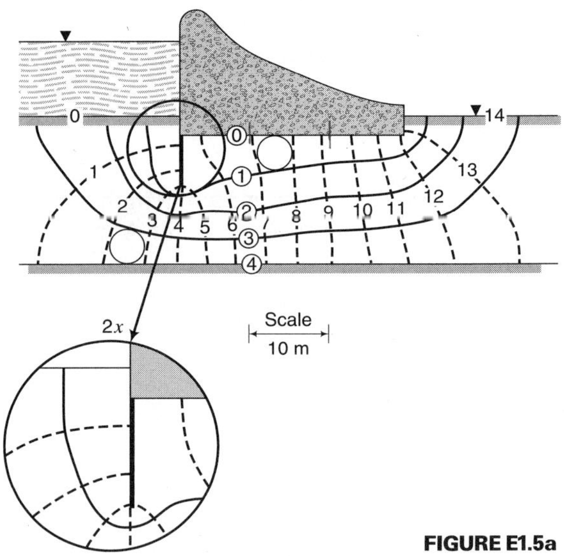

FIGURE E1.5a

Strategy Extract the flow channels and equipotential drops as discussed in this section and perform the calculation for the various quantities. SI units are used in the solution of this problem.

Solution 1.5

Step 1: Extract desired parameters.

Select the downstream end, EF, as datum:

$$\Delta H = 10 \text{ m}, \quad N_d = 14, \quad N_f = 4$$

Step 2: Calculate the flow rate.

$$q = k\Delta H \frac{N_f}{N_d} = 2 \times 10^{-4} \times (10 \times 10^2) \times \frac{4}{14} = 0.057 \text{ cm}^3\text{/sec}$$

Step 3: Determine porewater pressure under the base of the dam.

Divide the base into a convenient number of equal intervals. Let us use 10 intervals:

$$\Delta x = \frac{30.6}{10} = 3.06 \text{ m}$$

Determine the porewater pressure at each nodal point. Use a table for convenience or, better yet, use a spreadsheet:

$$\Delta h = \frac{\Delta H}{N_d} = \frac{10}{14} = 0.714 \text{ m}$$

Under Base of Dam

x (m)	0	3.06	6.12	9.18	12.24	15.3	18.36	21.42	24.48	27.54	30.6
N_d (m)	5.60	5.80	6.20	6.90	7.40	8.00	8.80	9.40	10.30	11.10	12.50
$N_d \Delta h$ (m)	4.00	4.14	4.43	4.93	5.28	5.71	6.28	6.71	7.35	7.93	8.93
h_z (m)	−2.40	−2.40	−2.40	−2.40	−2.40	−2.40	−2.40	−2.40	−2.40	−2.40	−2.40
$h_p(m) = \Delta H - N_d \Delta h - h_z$	8.40	8.26	7.97	7.47	7.12	6.69	6.12	5.69	5.05	4.47	3.48
u(kPa) $= h_p \gamma_w$	82.3	80.9	78.1	73.2	69.7	65.5	59.9	55.7	49.4	43.9	34.1

Plot porewater pressure distribution. See Fig. E1.5b.

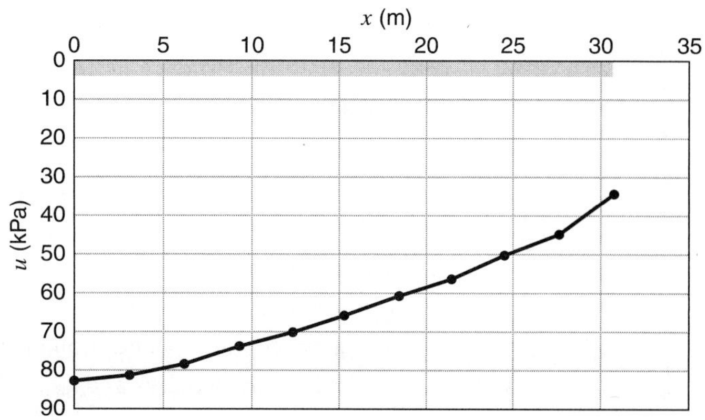

FIGURE E1.5b

Step 4: Determine the porewater pressure distribution on the sheet pile wall.

Divide the front face of the wall into 6 intervals of $\frac{7}{6} = 1.17$ m and the back face into 1 interval.

Six intervals were chosen because it is convenient for the scaling using the scale that was used to draw the flow net. The greater the intervals, the greater the accuracy. Only one interval is used for the back face of the wall because there is no equipotential that meets there. Use a spreadsheet to compute the porewater pressure distribution and the hydrostatic forces. The distributions of porewater pressure at the front and back of the wall are shown in Fig. E1.5c. Use *Simpson's rule*, $P_w = \frac{\Delta x}{3}(u_1 + u_n + 2\sum_{\substack{i=3\\ \text{odd}}}^{n} u_i + 4\sum_{\substack{i=2\\ \text{even}}}^{n} u_i)$, to calculate the hydrostatic force on the front face of the wall. The porewater pressure distribution at the back face is a trapezoid and the area is readily calculated.

Step 5: Determine the maximum hydraulic gradient.

The shortest value of L occurs at the exit. By measurement, $L_{\min} = 2$ m:

$$i_{\max} = \frac{\Delta h}{L_{\min}} = \frac{0.714}{2} = 0.36$$

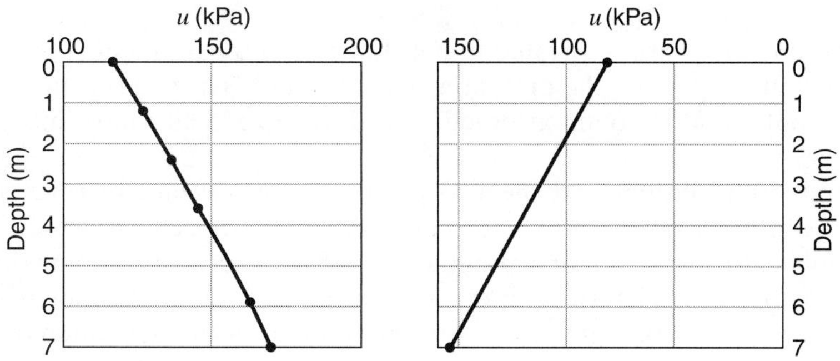

FIGURE E1.5c

Sheetpile

	Front of wall							Back of wall	
z(m)	0	1.17	2.33	3.50	4.67	5.83	7.00	7.00	0.00
N_d (m)	0.70	1.00	1.30	1.60	1.90	2.40	3.00	5.00	5.60
$N_d \Delta h$ (m)	0.50	0.71	0.93	1.14	1.36	1.71	2.14	3.57	4.00
h_z (m)	−2.40	−3.57	−4.73	−5.90	−7.07	−8.23	−9.40	−9.40	−2.40
$h_p(\text{m}) = \Delta H - N_d \Delta h - h_z$	11.90	12.85	13.81	14.76	15.71	16.52	17.26	15.83	8.40
$u\,(\text{kPa}) = h_p \gamma_w$	116.6	126.0	135.3	144.6	154.0	161.9	169.1	155.1	82.3
	Front	Back	Difference						
P_w (kN/m)	1011.7	830.9	180.8						

Step 6: Determine if piping would occur.

$$i_{cr} = \frac{G_s - 1}{1 + e} = \frac{2.7 - 1}{1 + 0.8} = 0.89$$

Since $i_{max} < i_{cr}$, piping would not occur.

The factor of safety against piping is $\frac{0.89}{0.36} = 2.5$.

Step 7: Effect of reducing the depth of penetration of sheetpile wall.

If depth is reduced, the value of Δh increases and i_{max} is likely to increase. ■

1.10 SOIL COMPACTION

Importance *Soil compaction is a method of improving the strength and settlement performance of soils and of lowering the hydraulic conductivity. It is one of the most popular methods of improving soils for foundation. In this section, the Proctor compaction test is reviewed.*

Water and mechanical energy are traditionally used to expel air from the voids and force the soil grains closer together. A laboratory test, called the *Proctor test* (ASTM D698, D1557), was developed to deliver a standard amount of mechanical energy (compactive effort) to determine the maximum dry unit weight of a soil. In the standard Proctor test, a dry soil sample is mixed with water and compacted in a cylindrical mold of volume of 9.44×10^{-4} m^3 (0.0333 ft^3) by repeated blows from the mass of a hammer, 2.5 kg (5.5 lb), falling freely from a height of 305 mm (12 in). The soil is compacted in three layers, each of which is subjected to 25 blows.

Compaction curves for some soil types are illustrated in Fig. 1.6. The two parameters of practical importance are the maximum dry unit weight and the optimum water content. The soil is generally not saturated at the optimum water content. It is common practice to draw a zero air voids line (100% saturation) on the compaction curve.

Knowledge of the optimum water content and the maximum dry unit weight of soils is very important for construction specifications of soil improvement by compaction. Specifications for earth structures (embankments, footings, etc.) usually call for a minimum of 95% of Proctor maximum dry unit weight. This level of compaction can be attained at two water contents: one before the attainment of the maximum dry unit weight or *dry of optimum*, and the other after attainment of the maximum dry unit

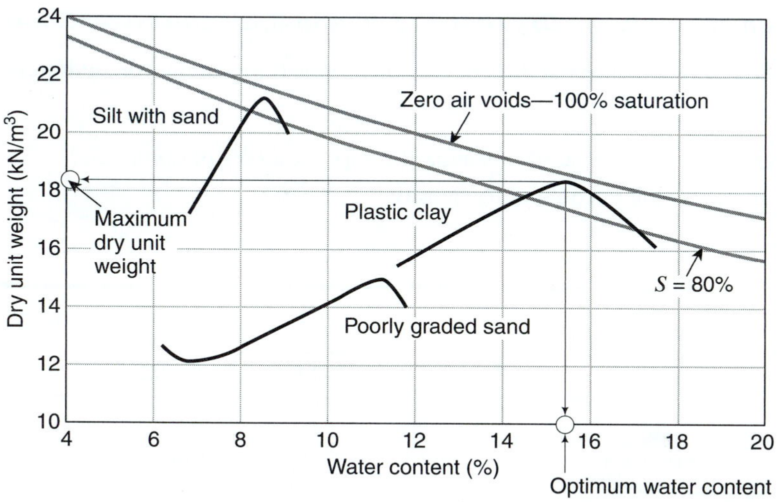

FIGURE 1.6 Dry unit weight–water content curves.

weight or *wet of optimum*. Normal practice is to compact the soil dry of optimum except for swelling and expansive soils.

A modified Proctor test was developed for compaction of airfields to support heavy aircraft loads. In the modified Proctor test, a hammer with a mass 4.54 kg (10 lb) falls freely from a height of 457 mm (18 in). The specimen is compacted in five layers of soil with 25 blows per layer in the standard Proctor mold. The higher level of compactive energy achieves a higher maximum dry unit weight at a lower optimum water content than the standard test. The degree of saturation is also lower at higher levels of compaction than the standard compaction test.

A variety of mechanical equipment is used to compact soils in the field. You may have seen various types of rollers being used in road construction. Each type of roller has special mechanical systems to effectively compact a particular soil type. For example, a sheepsfoot roller is generally used to compact fine-grained soils while a drum-type roller is generally used to compact coarse-grained soils.

A geotechnical engineer needs to check that field compaction meets specifications. Various types of equipment are available to check the amount of compaction achieved in the field. Three popular pieces of equipment are: (1) sand cone (ASTM D1556), (2) balloon (ASTM D2167), and (3) nuclear density meter (ASTM D2922).

The key points are:

1. ***Compaction is the densification of a soil by the expulsion of air and the rearrangement of particles.***
2. ***The Proctor test is used to determine the maximum dry unit weight and the optimum water content, and serves as the reference for field specifications of compaction.***
3. ***Higher compactive effort increases the maximum dry unit weight and reduces the optimum water content.***
4. ***Compaction increases strength, lowers compressibility, and reduces the permeability of soils.***
5. ***A variety of field equipment is used to check the dry unit weights achieved in the field. The popular types of field equipment are: the sand cone apparatus, the balloon apparatus, and the nuclear density meter.***

EXAMPLE 1.6 *Determination of Maximum Dry Unit Weight and Optimum Water Content*

The results of a standard compaction test are shown in the table below. Determine the maximum dry unit weight and optimum water content.

(a) What is the dry unit weight and water content at 95% standard compaction (dry of optimum)?

(b) Determine the degree of saturation at the maximum dry density.

(c) Plot the zero air voids line.

Water content (%)	6.2	8.1	9.8	11.5	12.3	13.2
Bulk unit weight—kN/m^3 (pcf)	16.9 (107.5)	18.7 (118.9)	19.5 (124)	20.5 (130.4)	20.4 (129.7)	20.1 (127.8)

Strategy Compute γ_d, plot the results of w (%) versus γ_d, and then extract the required information. SI units will be used in the solution of this problem.

Solution 1.6

Step 1: Use a table or a spreadsheet program to tabulate γ_d.

			Zero air voids	
Water content (%)	Bulk unit weight (kN/m^3)	Dry unit weight $\gamma_d = \frac{\gamma}{1+w}$ (kN/m^3)	Water content (%)	Dry unit weight $\gamma_d = \left(\frac{G_s}{1+\frac{wG_s}{S}}\right)\gamma_w$ (kN/m^3)
6.2	16.9	15.9	6	22.8
8.1	18.7	17.3	8	21.8
9.8	19.5	17.8	10	20.8
11.5	20.5	18.4	12	20.0
12.3	20.4	18.2	14	19.2
13.2	20.1	17.8		

Step 2: Plot graphs as shown in Fig. E1.6.

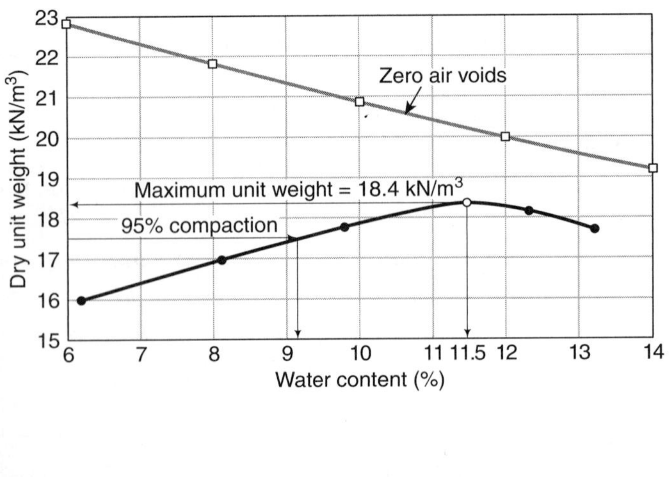

FIGURE E1.6 Compaction test results.

Step 3: Extract desired values.

$$(\gamma_d)_{max} = 18.4 \text{ kN/m}^3, \quad w_{opt} = 11.5\%$$

At 95% compaction, $\gamma_d = 18.4 \times 0.95 = 17.5 \text{ kN/m}^3$, $w = 9.2\%$ (from graph).

Step 4: Calculate the degree of saturation at maximum dry unit weight.

$$S = \frac{wG_s \frac{(\gamma_d)_{max}}{\gamma_w}}{G_s - \frac{(\gamma_d)_{max}}{\gamma_w}} = \frac{0.115 \times 2.7 \times \frac{18.4}{9.8}}{2.7 - \frac{18.4}{9.8}} = 0.71 = 71\%$$

■

1.11 GEOSTATIC STRESSES AND THE PRINCIPLE OF EFFECTIVE STRESS

Importance *The effective stress in a soil is the most important parameter in foundation analysis because it influences the strength and deformations of the soils. In this section, the calculation of total and effective stresses due to gravity (geostatic) is reviewed.*

The total stress in a soil mass not subjected to external loads is found from the unit weight of the soil and the depth of interest. Consider a soil element at a depth z below the ground surface and the groundwater level (GWL) is at ground surface (Fig. 1.7a).

The vertical total stress is:

$$\sigma = \gamma_{sat} z \tag{1.31}$$

The porewater pressure is:

$$u = \gamma_w z \tag{1.32}$$

and the vertical effective stress is:

$$\sigma' = \sigma - u = \gamma_{sat} z - \gamma_w z = (\gamma_{sat} - \gamma_w) z = \gamma' z \tag{1.33}$$

If the GWL is at a depth z_w below ground level (Fig. 1.7b), then

$$\sigma = \gamma z_w + \gamma_{sat}(z - z_w) \quad \text{and} \quad u = \gamma_w (z - z_w)$$

The vertical effective stress is

$$\sigma' = \sigma - u = \gamma z_w + \gamma_{sat}(z - z_w) - \gamma_w (z - z_w)$$
$$= \gamma z_w + (\gamma_{sat} - \gamma_w)(z - z_w) = \gamma z_w + \gamma'(z - z_w)$$

As water flows through soil it exerts a frictional drag on the soil particles. The frictional drag is called *seepage force*. It is often convenient to define seepage as the seepage force per unit volume (it has units

FIGURE 1.7 Soil element at a depth z with ground water level (a) at ground level and (b) below ground level.

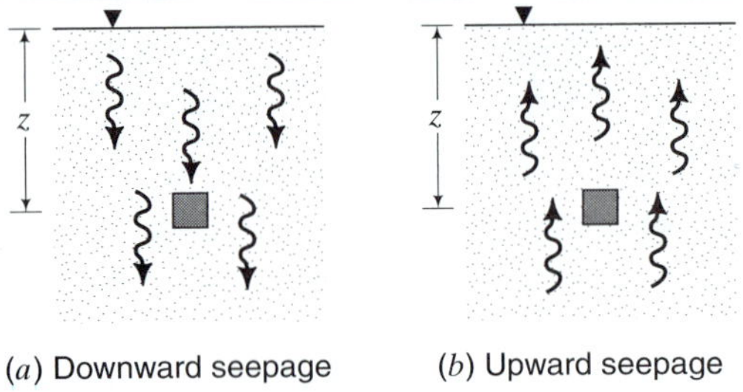

FIGURE 1.8 Seepage in soils.

similar to unit weight), which we will denote by j_s. The seepage force per unit volume is

$$j_s = i\gamma_w \tag{1.34}$$

where i is the hydraulic gradient.

If seepage occurs downward (Fig. 1.8a), then the seepage stresses are in the same direction as the gravitational effective stresses. From static equilibrium the resultant vertical effective stress is:

$$\sigma'_z = \gamma' z + iz\gamma_w = \gamma' z + j_s z \tag{1.35}$$

If seepage occurs upward, (Fig. 1.8b), then the seepage stresses are in opposite direction to the gravitational effective stresses. From static equilibrium the resultant vertical effective stress is:

$$\sigma'_z = \gamma' z - iz\gamma_w = \gamma' z - j_s z \tag{1.36}$$

Seepage forces play a very important role in destabilizing geotechnical structures.

The key points are:

1. *The effective stress represents the average stress carried by the soil solids and is the difference between the total stress and the porewater pressure.*
2. *The effective stress principle applies only to normal stresses and not to shear stresses.*
3. *Deformations of soils are due to effective, not total, stress.*
4. *Downward seepage increases the resultant effective stress; upward seepage decreases the vertical effective stress.*

EXAMPLE 1.7 *Calculation of Vertical Total and Vertical Effective Stresses*

Calculate the vertical effective stress at a depth A shown in Fig, E1.7.

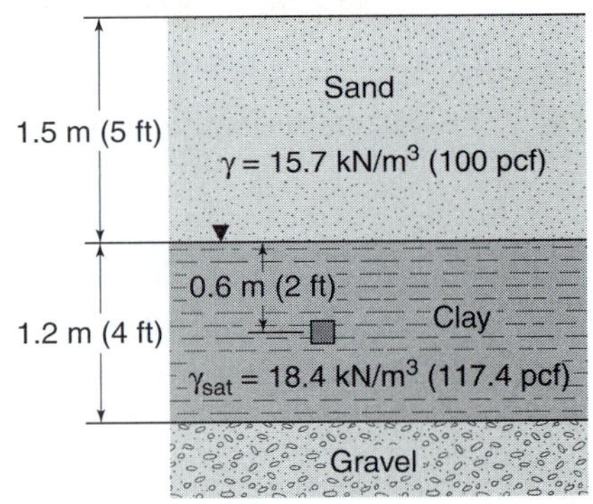

FIGURE E1.7

Strategy Use Eqs. (1.31–1.33). US units will be used in the solution of this problem.

Solution 1.7

Step 1: Calculate the vertical total stress.

$$\sigma = (\gamma z)_{\text{sand}} + (\gamma_{\text{sat}} z)_{\text{clay}} = 100 \times 5 + 117.4 \times 2 = 734.8 \text{ psf}$$

Step 2: Calculate the porewater pressure.

$$u = \gamma_w z = 62.4 \times 2 = 124.8 \text{ psf}$$

Step 3: Calculate the vertical effective stress.

$$\sigma' = \sigma - u = 734.8 - 124.8 = 610 \text{ psf}$$

■

1.12 SUMMARY

Soil is a natural material that can vary widely over short distances. A brief review of the basic physical soil parameters was presented. These parameters are for a homogeneous soil mass and may not represent the actual soil as a whole in the field. Consequently, experience and judgment are needed to use these parameters for practical applications.

EXERCISES

Assume $G_s = 2.7$, where necessary, for solving the following problems.

1.1 The mass of a sample of saturated soil is 500 grams (1.1 lb). The dry mass, after oven drying, is 450 grams (0.99 lb). Determine the (a) water content, (b) void ratio, (c) saturated unit weight, and (d) effective unit weight.

1.2 A saturated silty-clay encountered in a deep excavation is found to have a water content of 38%. Determine its porosity and bulk unit weight.

1.3 A sandy soil has a natural water content of 8% and a bulk unit weight of 16.5 kN/m^3 (105 pcf). The void ratios corresponding to the densest and loosest state of this soil are 0.51 and 0.87. Find the relative density and degree of saturation for this soil.

1.4 In a soils report, a soil is classified as CL-ML.

(a) Describe the soil.

(b) What additional information is needed to describe this soil more accurately?

(c) Would this be an excellent material for a foundation for a water tank?

(d) What soil type, using USCS, would be an excellent foundation material?

Justify all your answers.

1.5 The liquid limit of a soil is 120% and its plastic limit is 32%. The natural water content is 80%.

(a) What is likely the predominant clay mineral in this soil?

(b) Determine the plastici and liquidity indexes.

(c) Do you expect a brittle type of failure for this soil? Why?

1.6 The hydraulic conductivity from a falling head test on a soil (USCS is CH) is 2×10^{-7} cm/sec (0.4 in/sec).

(a) Is this a clay or a sand?

(b) Would this soil be a suitable drainage material?

(c) List some concerns regarding the use of this soil as a foundation for a building. [*Hint:* Table 1.1 should be of help.]

(d) If a thin horizontal layer of sandy silt were to bisect the clay layer, what effects would this have on the value of the hydraulic conductivity reported?

Justify all your answers.

1.7 A Proctor compaction test in the laboratory on a borrow pit soil gives a maximum dry unit weight of 19 kN/m^3 (120 pcf) and optimum water content of 11.5%. The water content at 95% compaction is 11% on the dry side and 12% on the wet side of the compaction curve. The bulk unit weight and water content of the soil in the borrow pit are 17 kN/m^3 (108 pcf) and 8.2% respectively. A highway fill is to be constructed using this soil. The specifications require the fill to be compacted to 95% Proctor compaction.

(a) What is the minimum unit weight of the soil that you would specify for the compaction?

(b) Describe how you would ensure that your compaction specification is met.

(c) If the soil is expansive, would you specify compaction on the dry side or the wet side of optimum water content? Justify your answer.

(d) How many truck loads of soil will be required for a 100,000 m^3 (3,500,000 ft^3) highway embankment? Each truck has a capacity of 10 m^3 (350 ft^3).

1.8 A borehole at a site for a proposed building reveals the following soil profile:

0–5 m (15 ft)	Dense sand, $\gamma = 16.5$ kN/m^3 (105 pcf), $\gamma_{sat} = 16.8$ kN/m^3 (106.8 pcf)
at 4 m (12 ft)	Groundwater level
5–10 m (15–30 ft)	Soft normally consolidated clay, $\gamma_{sat} = 17$ kN/m^3 (108 pcf)
10–15 m (30–45 ft)	Fractured (semi-impervious) rock
Below 15 m (45 ft)	Bedrock

(a) Sketch the soil profile.

(b) Plot the vertical total and vertical effective stresses, and porewater pressure distribution with depth.

(c) What is the magnitude of the vertical effective stress at the center of the soft, normally consolidated clay?

1.9 Water at a hydraulic gradient of 0.2 seeps upward through a saturated soil at depth 3 m (10 ft) below the surface. If the void ratio of the soil is 0.6, calculate the vertical effective stress with and without seepage.

1.10 The sheetpile retaining wall (Fig. P1.10) supports 6 m (20 ft) of water. A clay (almost impervious) blanket of 3 m (10 ft) wide on the downstream side was formed from deposits of a drainage system a few years after construction.

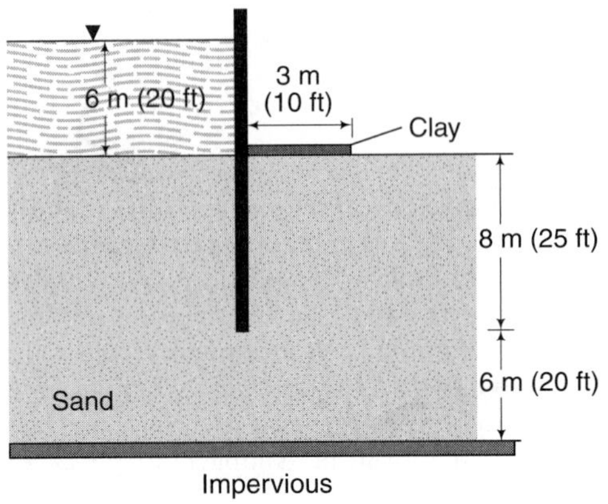

FIGURE P1.10

(a) Draw the flow net with and without the drainage blanket using the computer program at http://www.wiley.com/college/budhu. Click Foundations and Earth Structures, Chapter 1, 2Dflow.zip.

(b) Determine the flow rate with and without the drainage blanket if $k = 0.0004$ cm/sec. (2×10^{-5} in/sec)

(c) Determine the porewater pressure distributions on the upstream and downstream faces of the wall with and without the drainage blanket.

(d) Would piping occur if the drainage blanket were absent? Assume $e = 0.55$.

1.11 The soil profile at a project site consists of 20 m (60 ft) of very soft saturated clay followed by a deep deposit of mixed sand and gravel. Groundwater level is at ground surface. The water content of the clay is 60% and the specific gravity is 2.65.

(a) Calculate the vertical total stress, the porewater pressure, and the vertical effective stress of the clay in situ at a depth of 10 m (30 ft).

(b) A clay sample at 10 m (30 ft) was extracted and placed on a table. Calculate the vertical total stress, the porewater pressure, and the vertical effective stress at the instant the soil sample was placed on the table.

(c) Would you expect the soil sample to collapse and flow as soon as it is placed on the table? Justify your answer.

(d) What would you expect to happen to the soil sample if the porewater was allowed to drain from it? Would the soil sample disintegrate? Justify your answers.

CHAPTER 2

FOUNDATION LOADS, STRESSES, AND STRAINS IN SOILS

2.0 INTRODUCTION

In this chapter, we consider the loads that act on a geosystem and the stress and strains induced in the soil. A geosystem (geotechnical system) or a geotechnical structure is a single component or multiple integrated components that transmit loads from a structure to the soil. Examples of geosystems are foundations for buildings and bridges, tunnels, retaining walls, and dams.

One important task of a geotechnical engineer is to determine the stresses and strains that are imposed on a soil mass by external loads. It is customary to assume that the strains in the soils are small and this assumption allows us to apply our knowledge of mechanics of elastic bodies to soils. Small strains mean infinitesimal strains. For a realistic description of soils, elastic analysis is not satisfactory. We need soil models that can duplicate the complexity of soil behavior. However, even for complex soil models, an elastic analysis is a first step. In this chapter, we will deal with stresses and strains in soils assuming that they are elastic materials.

When you complete this chapter, you should be able to:

- Understand loads on a geosystem.
- Understand stresses and strains induced in soils.
- Understand stress and strain states and stress and strain paths.

2.1 DEFINITIONS OF KEY TERMS

Loads are forces imposed on a structure.

Dead loads are static (immovable) loads.

Live loads are movable loads resulting from occupancy.

Dynamic loads are loads that change with time.

Stress **(σ)** or intensity of loading is the load per unit area. The fundamental definition of stress is the ratio of the force ΔP acting on a plane ΔS to the area of the plane ΔS when ΔS tends to zero.

Strain **(ϵ)** or intensity of deformation is the ratio of the change in a dimension to the original dimension or the ratio of change in length to the original length.

Stress (strain) state at a point is a set of stress (strain) vectors corresponding to all planes passing through that point. Mohr's circle is used to graphically represent stress (strain) state for two-dimensional bodies.

***Mean stress* (*p*)** is the average stress on a body or the average of the orthogonal stresses in three dimensions.

***Deviatoric* (*q*)** stress is the shear or distortional stress or stress difference on a body.

Stress path is a graphical representation of the locus of stresses on a body.

Isotropic means the same material properties in all directions and also the same loading in all directions.

Anisotropic means the material properties are different in different directions and also the loadings are different in different directions.

Elastic materials are materials that return to their original configuration on unloading and obey Hooke's law.

2.2 LOADS

Importance *Geosystems are often subjected to a variety of loads. As a geotechnical engineer, you must know or anticipate the type and magnitude of the loads that your geosystem must support. Each type of load or certain combinations of them could cause your system to respond differently. Chapter 16 of the International Building Code (2006) gives the provisions for structural loads. In most practices, the geotechnical engineer is supplied with the loads from the superstructure (the structure that the foundation supports). However, there are additional loads that the foundation must support that are not included in the superstructural loads and the combinations of loads for worst-case structural response may not be the same for the soil. Also, for in situ soil characterization (Chapter 5), the loads to be supported by the foundation are often not known. You will then have to estimate the load based on the proposed structure at the site to help you plan the number and depth of boreholes. In this section, the loads on foundation systems are discussed.*

There are two broad classes of loads: static and dynamic. Fig. 2.1 illustrates typical loads on a structure.

2.2.1 Static Loads

Static loads are any single force or combination of forces that are applied slowly on a structure with only small fluctuations in magnitude, direction, and location until the ultimate value is reached. The maximum deformation that occurs under static loading is achieved when the ultimate static load acts on the geosystem.

Static loads consist of the following.

1. ***Dead loads:*** These consist of the self-weight of the structure (e.g., super or above-ground structures and soils or below-ground structures), the weight of attached building components such as walls, mechanical equipment, and electrical equipment. Dead loads are obtained from manufactures of the materials to be used.
2. ***Live loads:*** These are movable loads resulting from the intended use (occupancy) of the structure. Examples of live loads are the weights of people, stored materials, equipment, furnitures, appliances, and so on. Codes, for example, the International Building Code (IBC, 2006), list recommended live loads. For a school classroom, IBC (2006) lists a live load of either a uniformly distributed load of 1.9 kPa (40 lb/ft^2) or a concentrated load of 4.45 kN (1000 lb), whichever produces the worse effect.

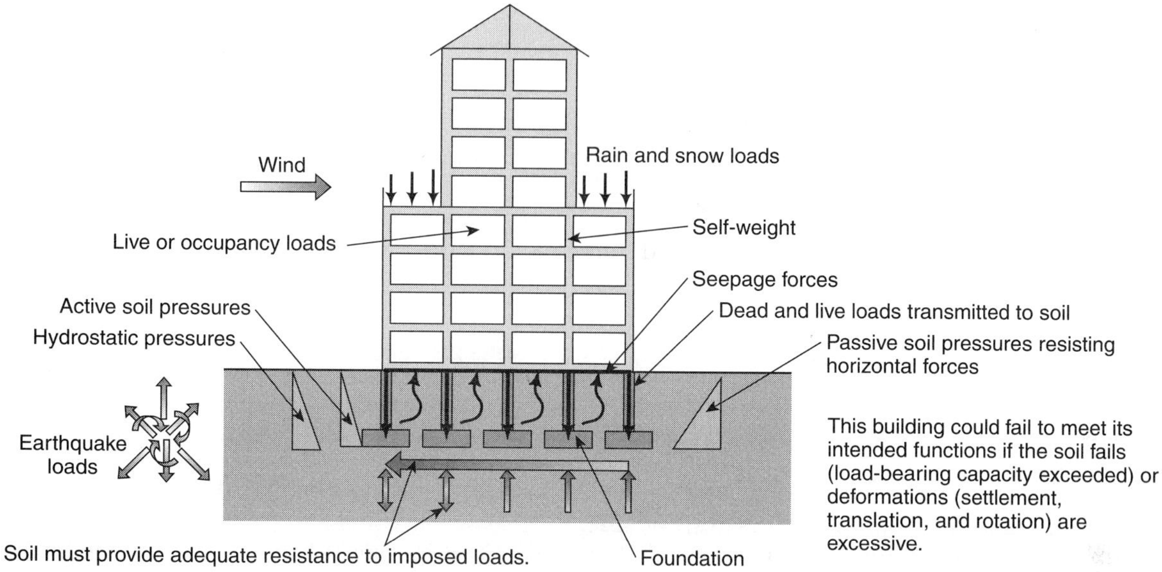

FIGURE 2.1 Loads on a typical geotechnical system.

3. ***Earth loads:*** These are loads imposed on the geosystem by the soil.
4. ***Hydrostatic loads:*** These loads are due to water pressures. Water pressures can be imposed vertically and laterally and can cause uplift of a geosystem. If there are differences in water levels along a geosystem or component, seepage forces will be present. These are internal forces within the soil mass.
5. ***Snow loads:*** These loads result from accumulation of snow on roofs. Snow loads vary with geographic location of the structure. Recommended snow loads are listed in codes.
6. ***Rain loads:*** These loads result from accumulation of water from rain on roofs and other exposed structural components. Recommended rain loads are obtained from empirical equations in codes.

2.2.2 Dynamic Loads

Dynamic loads are applied quickly on a structure and usually change magnitude, direction, and location. The maximum deformation is time dependent and may not correspond to the maximum peak load. There are four classes of dynamic loads that are of practical importance to geotechnical engineers:

1. ***Impact loads:*** These loads are applied suddenly and are of short duration. A force from a hammer is an impact load. Impact loads are usually treated as static loads by increasing or amplifying their magnitude. Common amplification factors range from 1.10 to 1.25.
2. ***Machine operation loads:*** These loads are cyclically applied during the operation of machines. Frequency (cycles per second, denoted by Hertz, Hz, or megahertz, MHz) of loading is important and depends on the operation of the machinery. The deformation depends on the frequency, duration of loading, and location of the machine.
3. ***Wind loads:*** Wind loads result from the kinetic energy of air mass on a structure, which is transferred from the structure to the foundation. The wind loads act against the sides of the

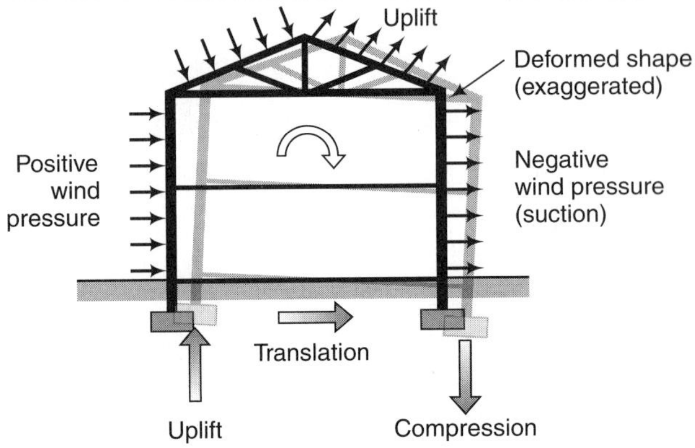

FIGURE 2.2 Effects of wind loads on a foundation of a building.

structure and consist of lateral and vertical loads that can result in sliding, uplift, overturning, and differential settlement of a structure (Fig. 2.2).

Wind loads are treated as imposed static forces considering the maximum expected wind velocity, exposure condition, building geometry (particularly height and orientation), gust, and importance of the structure. Recommended wind loads are obtained from codes. Wind loads depend on the geometry of the structure (shape, size, roof type, and orientation with respect to the wind direction), wind speed, type of terrain at and near project site, porosity of the structure, and its importance.

4. ***Earthquake loads:*** Earthquakes are natural occurrences and have been responsible for large-scale destruction and untold human tragedy. The occurrence of an earthquake cannot be predicted accurately. We know the mechanism that causes earthquakes and can broadly predict whether an earthquake is likely to occur in a certain region. When the earthquake would actually occur is speculative.

The earth consists of a series of plates that move relative to each other. Abrupt ruptures or movements along the earth's plates cause earthquakes. These ruptures produce vibrations, which propagate in all directions. The vibrations are recorded as waves. *Body waves* and *surface waves* are the two basic waves produced in a seismic event. Body waves consist of P waves (compression waves) and S waves (shear waves). P waves cause movement in the direction of the wave (Fig 2.3a). S waves cause movement perpendicular to the wave and produce distortion (Fig. 2.3b). Surface waves consist of Rayleigh and Love waves. Rayleigh waves create a rolling type of motion due to a combination of longitudinal and transverse vibrations (Fig. 2.3c). Love waves are created when the waves travel from a layer of low velocity propagation to a layer of high velocity propagation. Love waves are possible only in a nonuniform material and travel horizontally (Fig 2.3d). For geosystems, the S waves are critical because they distort the soil mass. The velocity of propagation of the S waves is called the *shear wave velocity*.

The shaking motion propagating from the soil to the structure causes the structural mass to be accelerated back and forth, up and down, and sideways. The largest force will occur where the structural mass is the greatest. Geotechnical engineers are particularly interested in the response of soils to ground motion that could potentially destroy life and the built environment (Fig. 2.4). We are mostly concerned with amplitude (particularly, peak acceleration, and peak shear wave velocity), frequency, and duration of ground motions. Earthquakes generally produce acceleration amplitudes less than 1 g, where g = 9.81 m/sec^2(32.2 ft/sec^2) is the acceleration due to gravity, at frequencies less than 5 Hz, and last for only a few seconds. An acceleration history for a site in California during the 1989 Loma Prieta earthquake is shown in Fig. 2.5. The maximum acceleration is about 0.35 g and the duration of the major part of the motion is about 3 seconds.

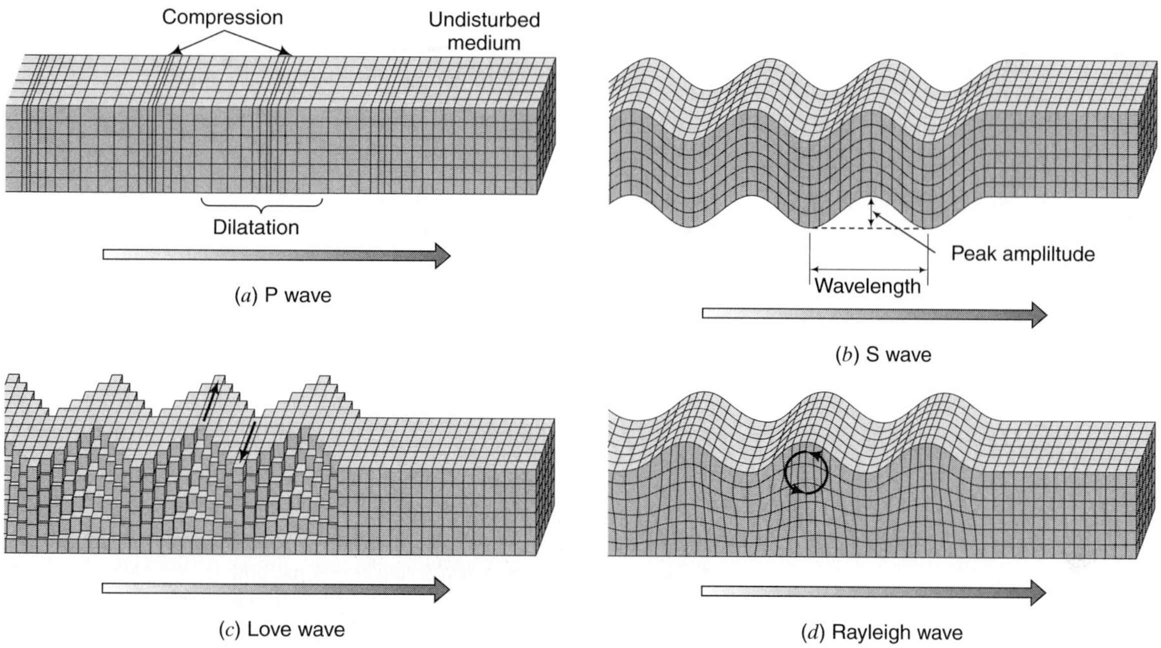

FIGURE 2.3 Body waves (P wave and S wave) and surface waves (R wave and L wave).

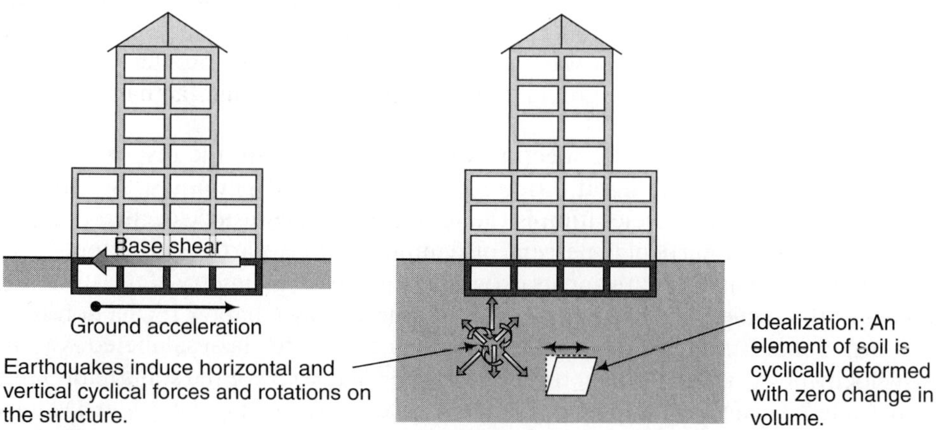

FIGURE 2.4 Forces from an earthquake and idealized displacement of an element of soil.

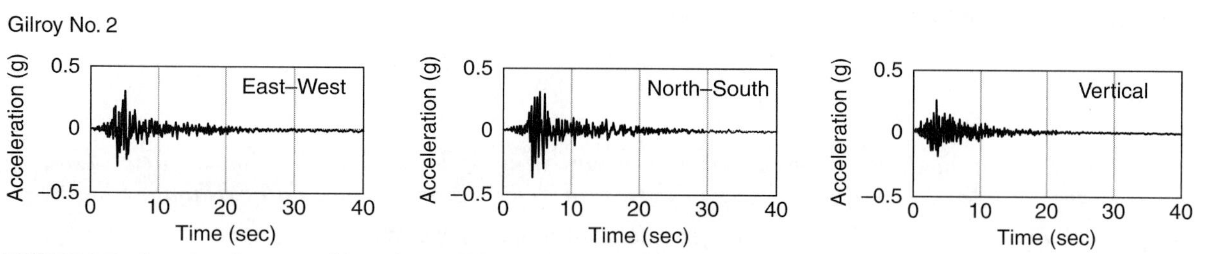

FIGURE 2.5 Acceleration record from Loma Prieta earthquake (Gilroy No. 2).

TABLE 2.1 Site Class

Site class	Soil profile name	Average properties in top 30 m: Soil shear wave velocity, V_s (m/s)	Standard penetration resistance, N (blows/ft)	Soil undrained shear strength, s_u (kPa)
A	Hard rock	$V_s > 1500$	Not applicable	Not applicable
B	Rock	$750 < V_s \leq 1500$	Not applicable	Not applicable
C	Very dense soil and soft rock	$375 < V_s \leq 750$	$N > 50$	$s_u \geq 100$
D	Stiff soil profile	$180 < V_s \leq 375$	$15 \leq N \leq 50$	$50 \leq s_u \leq 100$
E	Soft soil profile	$V_s > 180$	$N < 15$	$s_u < 50$
E	—	Any profile with more than 3 m of soil having the following characteristics: 1. Plasticity index PI $> 20\%$ 2. Moisture content with $\geq 40\%$ 3. Undrained shear strength, $s_u < 25$ kPa		
F	—	Any profile containing soils having one or more of the following characteristics: 1. Soils vulnerable to potential failure or collapse under seismic loading such as liquefiable soils, quick and highly sensitive clays, collapsible weakly cemented soils 2. Peats and/or highly organic clays ($H > 3$ m of peat and/or highly organic clay where H = thickness of soil) 3. Very high plasticity clays ($H > 8$ m with plasticity index PI $> 75\%$) 4. Very thick soft/medium stiff clays ($H > 35$ m)		

Ground motions are represented by response spectra. Response spectra are the responses of an ideal single-degree-of-freedom system oscillating at different frequencies and periods with different degrees of damping. The spectral response acceleration is the maximum acceleration reached at a site with particular soil characteristics for a particular percent of damping over a particular period.

Seismic loads depend on the type of structure and its geographic location, geometry of the structure (shape and size), importance of the structure, soil type, and the structural mass. Chapter 16, Section 1615 (Table 1615.1.1) of the International Building Code (IBC, 2006) defines six site classes, labeled A to F, based on the shear wave velocity or the standard penetration test (see Chapter 5) or the soil's undrained shear strength as shown in Table 2.1. The author has modified this table by converting the original values given in a foot-pound system to SI units and rounding the resulting values.

IBC (2006) uses as a base the spectral response for site class B (rock) at a period of 1.0 second (S_1) and at short periods (S_s) less than 1.0 second. Spectral maps are available in Chapter 16, Section 1615 of IBC (2006) for various locations. The spectral response accelerations for sites other than site B are calculated from formulas given in IBC (2006).

The key points are:

1. ***Loads are divided into different categories consisting of dead loads, live loads, earth loads, hydrostatic loads, wind loads, snow loads, rain loads, impact loads, and earthquake loads.***
2. ***Each type of load could cause the geosystem to react differently.***

2.3 STRESSES AND STRAINS

2.3.1 Normal Stresses and Normal Strains

Importance *Loads applied to a geosystem are transferred as stresses to the soil. These stresses in turn cause strains in the soil. The geotechnical engineer must determine the type and magnitude of the stresses and strains imposed on the soil. Some types of stresses may cause a soil to compress while other types of stresses may cause the soil to distort. Most often these stresses and strains are not distributed uniformly. This could result in progressive failure due to some parts of the soil mass reaching a failure state while other parts are in a prefailure state. In this section, we consider basic definitions of stresses and strains. Most of the content in this section is review material.*

Consider a cube of dimensions $x = y = z$ that is subjected to forces P_x, P_y, P_z, normal to three adjacent sides as shown in Fig. 2.6. The normal stresses are:

$$\sigma_z = \frac{P_z}{xy}, \quad \sigma_x = \frac{P_x}{yz}, \quad \sigma_y = \frac{P_y}{xz} \tag{2.1}$$

Stresses that cause compression are positive values.

Let us assume that under the forces in Fig. 2.6 the cube is compressed by Δx, Δy, and Δz in the X, Y and Z directions. The normal strains in these directions, assuming they are small (infinitesimal), are:

$$\varepsilon_z = \frac{\Delta z}{z}, \quad \varepsilon_x = \frac{\Delta x}{x}, \quad \varepsilon_y = \frac{\Delta y}{y} \tag{2.2}$$

The volumetric strain is:

$$\varepsilon_p = \varepsilon_x + \varepsilon_y + \varepsilon_z \tag{2.3}$$

2.3.2 Shear Stresses and Shear Strains

Let us consider, for simplicity, the XZ plane and apply a force F that causes the square to distort into a parallelogram as shown in Fig. 2.7. The force F is a shearing force and the shear stress is:

$$\tau = \frac{F}{xy} \tag{2.4}$$

If the maximum horizontal displacement is Δx, the shear strain or simple shear strain, γ_{zx}, is

$$\gamma_{zx} = \tan^{-1}\left(\frac{\Delta x}{z}\right) \tag{2.5}$$

For small strains, $\tan(\gamma_{zx}) = \gamma_{zx}$ and therefore,

$$\gamma_{zx} = \frac{\Delta x}{z} \tag{2.6}$$

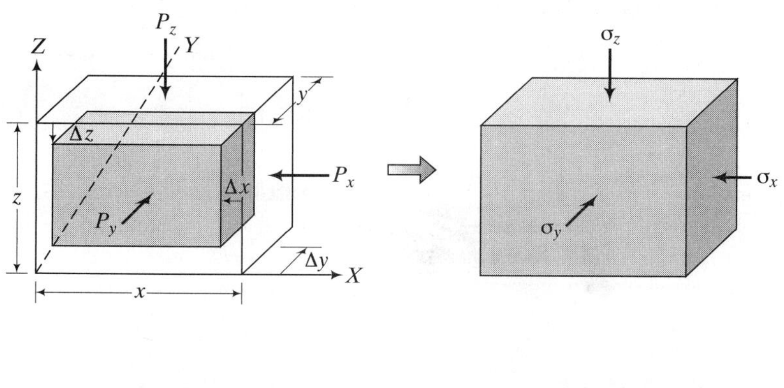

FIGURE 2.6 Stresses and displacements due to applied loads.

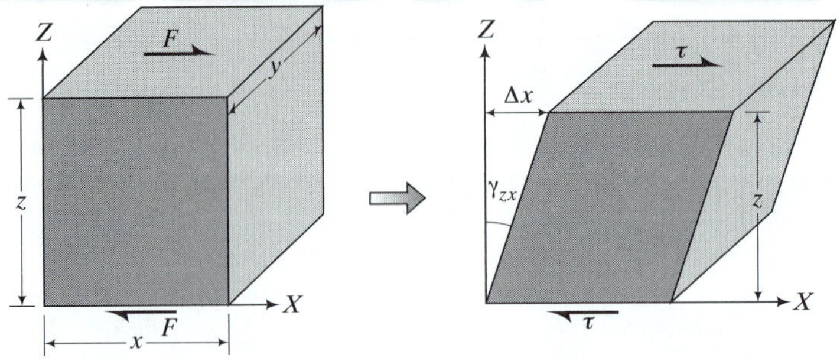

FIGURE 2.7 Shear stresses and shear strains.

Simple shear strain is a measure of the angular distortion of a body by shearing forces. If the shear stress on a plane is zero, the normal stress on that plane is called a *principal stress*. We will discuss principal stresses later. Soils cannot sustain any appreciable tensile stresses and we normally assume that the tensile strength of soils is negligible. Strains can be compressive or tensile.

The* key points *are:

1. ***Normal stress is the load divided by the area of the plane normal to the load.***
2. ***Strain is the change in a dimension divided by the original dimension.***
3. ***Volumetric strain is the sum of the normal strains in all directions.***
4. ***Shear strains distort a body.***
5. ***Compressive stresses are positive. Strains can be compressive or tensile.***
6. ***Soils cannot sustain tension.***

EXAMPLE 2.1 *Calculation of Shear Stress and Shear Strain*

A soil sample 0.1 m (0.3 ft) × 0.1 m (0.3 ft) × 0.02 m (0.06 ft) high is subjected to a shear force 1 kN (225 lb). The horizontal displacement is 0.2 mm (0.008 in). Calculate the shear stress and the shear strain. SI units is used in solving this example problem

Strategy Draw a diagram representing the problem and then use Eqs. (2.4) and (2.6). SI units will be used in solving this example problem.

Solution 2.1

Step 1: Draw a diagram of the problem. See Fig. E2.1.

Step 2: Calculate the shear stress.

$$\tau = \frac{F}{xy} = \frac{1}{0.1 \times 0.1} = 100\,\text{kPa}$$

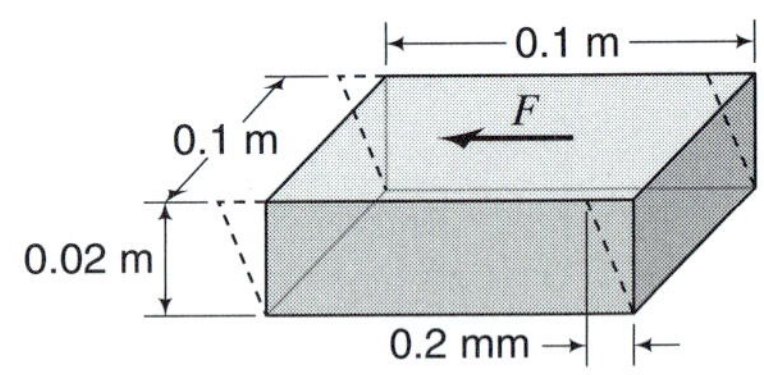

FIGURE E2.1

Step 3: Calculate the shear strain.

$$\gamma_{xy} = \frac{\Delta x}{z} = \frac{0.0002}{0.02} = 0.01$$

■

2.4 STRESSES IN SOILS FROM SURFACE AND INTERIOR LOADS

Access the interactive computer program called *STRESS* to compute stress increases beneath foundations from surface loads at http://www.wiley.com/college/budhu. Click Foundations and Earth Structures, Chapter 2, STRESS.zip.

Importance *Various types of surface loads are applied to soils. For example, an oil tank on a flexible foundation will impose a uniform, vertical stress on the surface of the soil while an unsymmetrical building subjected to static and dynamic loads may impose nonuniform stresses. We would like to know how the surface stresses are distributed within the soil and the resulting displacements to estimate whether the soil under our geosystem would fail or the resulting displacements are excessive or whether nearby structures would be negatively affected. The calculation of the distribution of stresses within the soil from surface loads requires assumptions that grossly simplify soil behavior. In this section, we consider these assumptions and the distribution of vertical stress increases due to some typical surface loads.*

The distribution of surface stresses within a soil is determined by assuming that the soil is a semi-infinite, homogeneous, linear, isotropic, elastic material. A semi-infinite mass is bounded on one side and extends infinitely in all other directions. This is also called an *elastic half space*. For soils, the horizontal surface is the bounding side. Equations and charts for several simple types of surface loads in foundation engineering based on the above assumptions are presented. Only the equations for change in vertical stresses and the vertical displacements are presented in this chapter since these are most often required in practice. The program http://www.wiley.com/college/budhu. Click Foundations and Earth Structures, Chapter 2, STRESS.zip gives the stresses in all three Cartesian directions. A comprehensive set of equations for a variety of loading situations is available in Poulos and Davis (1974).

Because we are assuming soil is an elastic material, we can use the principle of superposition to determine the stress distribution for complex loadings by decomposing the complex loading into simple loads (e.g., rectangular or circular) and then add the solution for each of these simple loads. The stress increases from surface loads are total stress increases. If the soil is dry or very permeable (e.g., clean coarse-grained soils), then you can assume that the stress increases are effective stress increases.

2.4.1 Vertical Point Load on Soil Surface—Boussinesq's Solution

Boussinesq (1885) presented a solution for the distribution of stresses for a point load applied on the soil surface, assuming that the soil is an elastic half space. An example of a point load is the vertical load transferred to the soil from an electric power line pole.

The increase of vertical stress on a soil element located at point A (Fig. 2.8a) is

$$\Delta\sigma_z = \frac{3Q}{2\pi z^2}\frac{1}{[1+(r/z)^2]^{\frac{5}{2}}} \tag{2.7}$$

Eq. 2.7 can be written as

$$\Delta\sigma_z = \frac{Q}{z^2} I \tag{2.8}$$

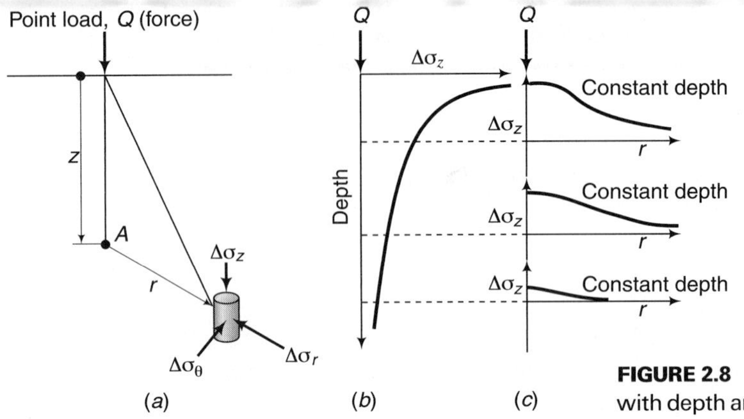

FIGURE 2.8 Point load and vertical stress distribution with depth and radial distance.

where I is an influence factor, and

$$I = \frac{3}{2\pi}\frac{1}{[1+(r/z)^2]^{\frac{5}{2}}} \tag{2.9}$$

The distributions of the increase in vertical stress (Fig. 2.8) reveal that the increase in vertical stress decreases with depth (Fig. 2.8b) and radial distance (Fig. 2.8c).

The vertical elastic settlement is

$$\Delta z = \frac{Q(1+\nu)}{2\pi E z\left[1+\left(\frac{r}{z}\right)^2\right]^{½}}\left[2(1-\nu)+\frac{1}{1+\left(\frac{r}{z}\right)^2}\right] \tag{2.10}$$

where E is Young's modulus (also called *elastic modulus*). A discussion on the elastic modulus used in foundation engineering to estimate settlement is presented in Chapter 4.

2.4.2 Strip Surface Loads

A strip load is the load transmitted by a structure when the ratio of width of the load to the length of the load approaches zero. Two types of strip loads are common in foundation engineering. One is a load that imposes a uniform stress on the soil, for example, the middle section of a long embankment (Fig. 2.9a). The other is a load that induces a triangular stress distribution over an area of width B (Fig. 2.9b). An example of a strip load with a triangular distribution of stress is the slope of the side of an embankment.

The increase in vertical stresses due to a normal surface stress, q_s (force/area), is as follows:

- Area transmitting a uniform stress (Fig. 2.9a):

$$\Delta\sigma_z = \frac{q_s}{\pi}\{\alpha + \sin\alpha\cos(\alpha+2\beta)\} \tag{2.11}$$

The vertical displacement due to a strip loading is useful only as relative displacement between two points not located at infinity.

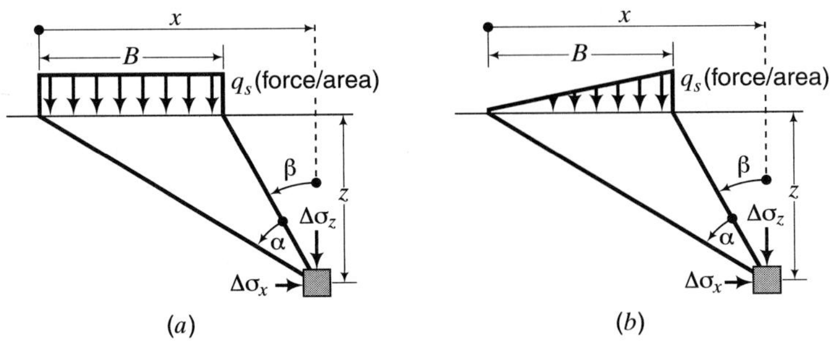

FIGURE 2.9 Strip load imposing (a) a uniform stress (b) linear varying stress.

The relative vertical elastic settlement between the center of the strip load (0, 0) and a point at the surface $(x,\ 0)$ is

$$\Delta z(x,\,0) - \Delta(0,\,0) = \frac{2q_s(1-\nu^2)}{\pi E}\left\{\left(x-\frac{B}{2}\right)\ln\left|x-\frac{B}{2}\right| - \left(x+\frac{B}{2}\right)\ln\left|x+\frac{B}{2}\right| + B\ln\left(\frac{B}{2}\right)\right\} \tag{2.12}$$

- Area transmitting a triangular stress (Fig. 2.9b):

$$\Delta\sigma_z = \frac{q_s}{\pi}\left(\frac{x}{B}\alpha - \frac{1}{2}\sin 2\beta\right) \tag{2.13}$$

The relative vertical displacement between the center of the strip load (0, 0) and a point at the surface $(x,\ 0)$ is

$$\Delta z(x,\,0) - \Delta(0,\,0) = \frac{q_s(1-\nu^2)}{\pi E\left(\frac{B}{2}\right)}\left\{\frac{B^2}{2}\ln B - \frac{x^2}{2}\ln x + \left(\frac{x^2}{2} - \frac{B^2}{2}\right)\ln|B-x| + \frac{B}{2}x\right\} \tag{2.14}$$

2.4.3 Uniformly Distributed Circular Surface Load

An example of a circular area that transmits a uniform stress to a soil mass is a rigid circular foundation of an oil or water tank. The increase of vertical and radial stresses under the center of the circular area of radius, r_o, is

z → depth of soil particle (?)

$$\Delta\sigma_z = q_s\left[1 - \left\{\frac{1}{1+(r_o/z)^2}\right\}^{3/2}\right] = q_s I_c \tag{2.15}$$

where $I_c = \left[1 - \left\{\frac{1}{1+(r_o/z)^2}\right\}^{3/2}\right]$ is an influence factor.

The vertical elastic settlement at the surface due to a circular loaded area is:

Below center of loaded area: $$\Delta z = \frac{q_s D(1-\nu^2)}{E} \tag{2.16}$$

Below edge: $$\Delta z = \frac{2}{\pi}\frac{q_s D(1-\nu^2)}{E} \tag{2.17}$$

where $D = 2r_o$ is the diameter of the loaded area.

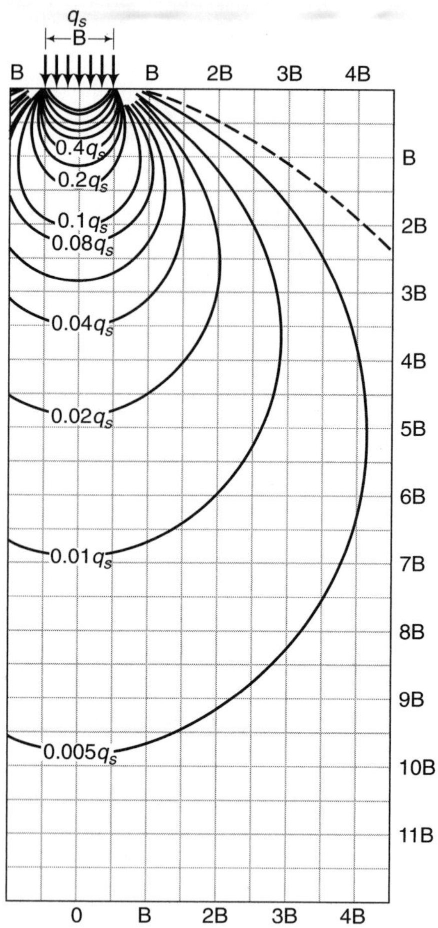

FIGURE 2.10 Contours of vertical stress increase beneath a square foundation.

2.4.4 Uniformly Distributed Rectangular Surface Load

Many structural foundations are squares or rectangles. The increase in vertical stresses below the corner of a rectangular area of width B and length L are:

$$\Delta\sigma_z = \frac{q_s}{2\pi}\left[\tan^{-1}\frac{LB}{zR_3} + \frac{LBz}{R_3}\left(\frac{1}{R_1^2} + \frac{1}{R_2^2}\right)\right] \tag{2.18}$$

where $R_1 = (L^2 + z^2)^{1/2}$, $R_2 = (B^2 + z^2)^{1/2}$, and $R_3 = (L^2 + B^2 + z^2)^{1/2}$.

Eq. 2.18 can be written as

$$\Delta\sigma_z = q_s I_z \tag{2.19}$$

where I_z denotes influence factor. The influence factor for the vertical stress increase is

$$I_z = \frac{1}{4\pi}\left[\frac{2mn\sqrt{m^2+n^2+1}}{m^2+n^2+m^2n^2+1}\left(\frac{m^2+n^2+2}{m^2+n^2+1}\right) + \tan^{-1}\left(\frac{2mn\sqrt{m^2+n^2+1}}{m^2+n^2-m^2n^2+1}\right)\right] \tag{2.20}$$

where $m = \frac{B}{z}$, $n = \frac{L}{z}$, B is the width, and L is the length of the load. The contours of vertical stress increase beneath a square foundation are shown in Fig. 2.10.

You can program your calculator or use a spreadsheet program to find I_z. You must be careful in the last term ($\tan^{-1}$) in programming. If $m^2 + n^2 + 1 < m^2n^2$, then you have to add π to the bracketed quantity in the last term. A chart for I_z produced by a spreadsheet program is shown in Fig. 2.11. You would have to calculate $m = \frac{B}{z}$, $n = \frac{L}{z}$ and read I_z from the chart; m and n are interchangeable. In general, the vertical stress increase is less than 10% of the surface stress when $z > 3B$.

The vertical elastic settlement at the ground surface under a rectangular flexible surface load is

$$\Delta z = \frac{q_s B(1-\nu^2)}{E} I_s \tag{2.21}$$

where I_s is a settlement influence factor that is a function of the L/B ratio (L is length and B is width). Setting $\xi_s = L/B$, the equations for I_s are:

At the center of a rectangle (Giroud, 1968):

$$I_s = \frac{2}{\pi}\left\{\ln\left(\xi_s + \sqrt{1+\xi_s^2}\right) + \xi_s \ln\frac{1+\sqrt{1+\xi_s^2}}{\xi_s}\right\} \tag{2.22}$$

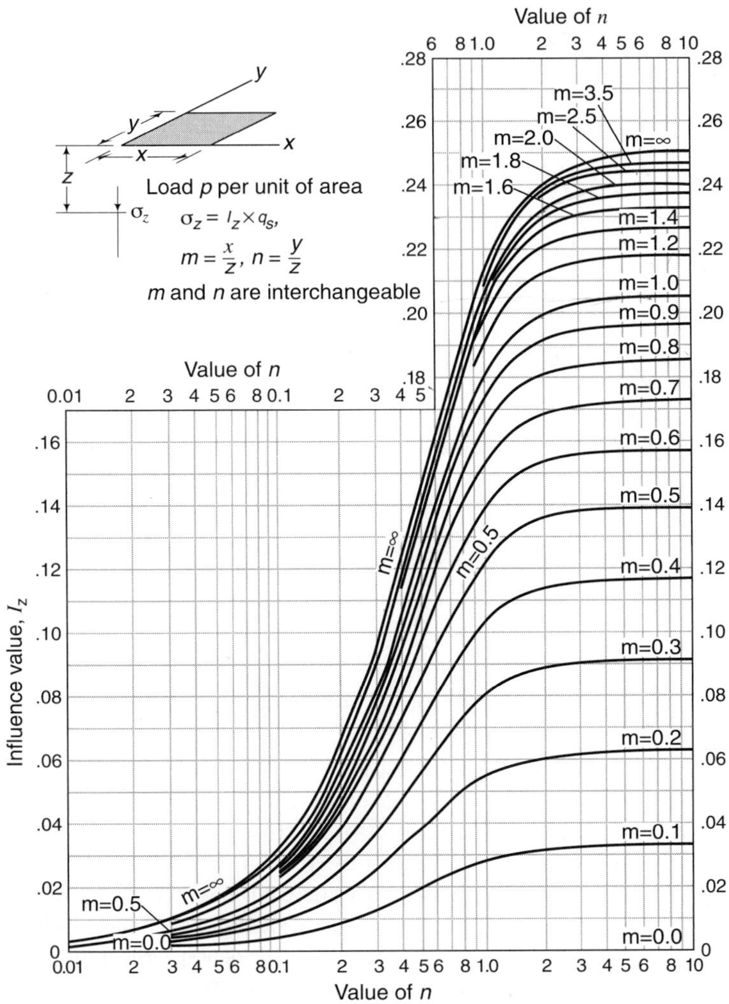

FIGURE 2.11 Influence factor for calculating the vertical stress increase under the corner of a rectangle.

At the corner of a rectangle (Giroud, 1968)

$$I_s = \frac{1}{\pi}\left\{ \ln\left(\xi_s + \sqrt{1+\xi_s^2}\right) + \xi_s \ln\frac{1+\sqrt{1+\xi_s^2}}{\xi_s} \right\} \tag{2.23}$$

The above equation can be simplified to the following for $\xi_s \geq 1$

At the center of a rectangle: $I_s \cong 0.62 \ln(\xi_s) + 1.12$ (2.24)

At the corner of a rectangle: $I_s \cong 0.31 \ln(\xi_s) + 0.56$ (2.25)

2.4.5 Embankment Loads

The vertical stress increase beneath an embankment can be found by superposition of the solution of a strip load imposing a uniform strain and a strip load imposing a triangular stress. A chart for the vertical stress increase beneath an embankment is shown in Fig. 2.12.

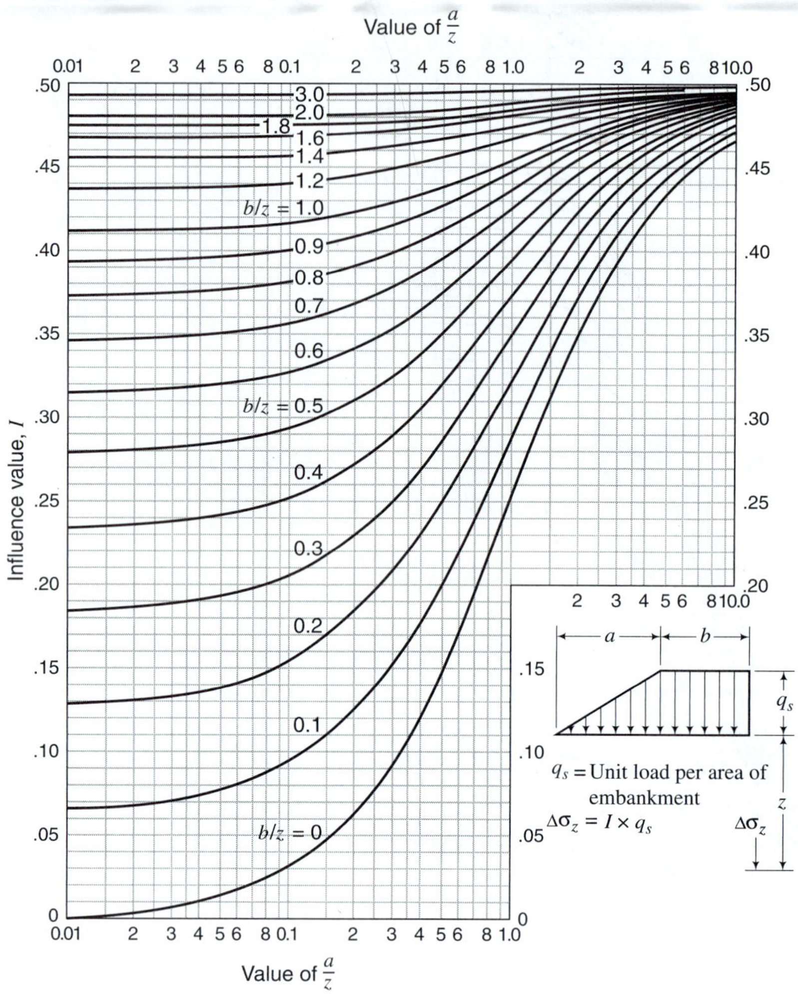

FIGURE 2.12 Influence factor for calculating vertical stress increase for an embankment.

2.4.6 Approximate Method for Rectangular Loads

In preliminary analyses of vertical stress increases under the center of rectangular loads, geotechnical engineers often use an approximate method (sometimes called the 2:1 method). The surface load on an area BL is dispersed at a depth z over an area $(B + z)(L + z)$ as illustrated in Fig. 2.13. The vertical stress increase under the center of the load is:

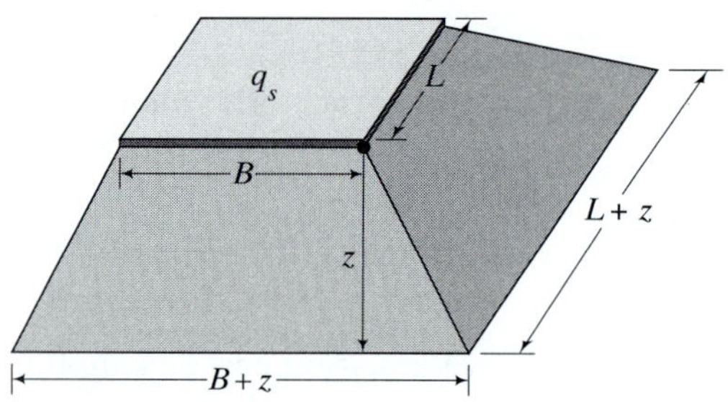

FIGURE 2.13 Dispersion of load for approximate increase in vertical stress under a rectangle.

$$\Delta\sigma_z = \frac{q_s BL}{(B + z)(L + z)} \tag{2.26}$$

The approximate method is reasonably accurate (compared with Boussinesq's elastic solution) when $z > B$.

2.4.7 Uniformly Distributed Surface Load on Irregularly Shaped Area—Newmark's Solution

Newmark (1942) developed a chart to determine the increase in vertical stress due to a uniformly loaded area of any shape. The chart consists of concentric circles divided by radial lines (Fig. 2.14). The area of each segment represents an equal proportion of the applied surface stress at a depth z below the surface. If there are 10 concentric circles (only 9 are shown because the 10th extends to infinity) and 20 radial lines, the stress on each circle is $\frac{1}{10}q_s$ and on each segment is $\frac{1}{10\times 20}q_s$. The radius to depth ratio of the first (inner) circle is found by setting $\Delta\sigma_z = 0.1q_s$ in Eq. 2.14; that is, $0.1q_s = q_s\left[1 - \left\{\frac{1}{1+(r_o/z)^2}\right\}^{3/2}\right]$ from which $\frac{r}{z} = 0.27$. For the other circles, substitute the appropriate value for $\Delta\sigma_z$ (e.g., for the second circle $\Delta\sigma_z = 0.2q_s$), and find $\frac{r}{z}$. The chart is normalized to the depth (i.e., all dimensions are scaled by a factor initially determined for the depth). Every chart should show a scale and an influence factor I_N, which for our case is $\frac{1}{10\times 20} = 0.005$.

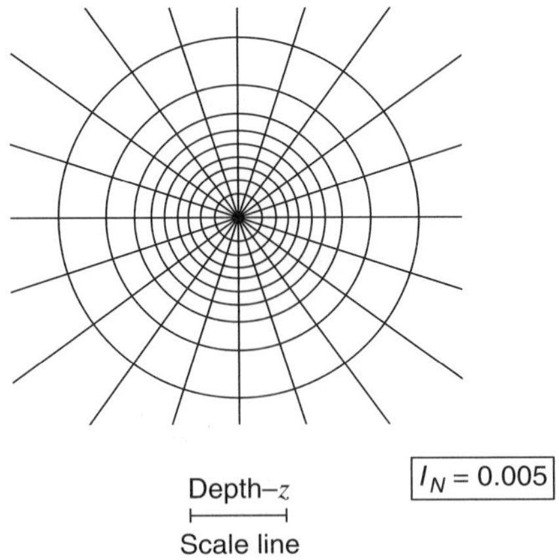

FIGURE 2.14 Newmark's chart for increase in vertical stress.

The procedure for using the Newmark chart is as follows:

1. Set the scale, shown on the chart, equal to the depth at which the increase in vertical stress is required. We will call this the *depth scale*.
2. Identify the point on the loaded area below which the stress is required. Let us say this point is A.
3. Plot the loaded area using the depth scale with point A at the center of the chart.
4. Count the number of segments (N) covered by the scaled loaded area. If certain segments are not fully covered, you can estimate what fraction is covered.
5. Calculate the increase in vertical stress as $\Delta\sigma_z = qI_N N$.

Access the computer program called *STRESS* at http://www.wiley.com/college/budhu. Click Foundations and Earth Structures, Chapter 2, STRESS.zip to plot Newmark charts.

2.4.8 Uniform Loads of Large Lateral Extent

Uniform loads of large lateral extent are assumed to be transferred to the soil as a uniformly distributed vertical stress throughout the depth. For example, if a fill of unit weight 15 kN/m^3 (95 pcf) and height 2 m (6 ft) is placed on the surface of a soil, then the vertical stress at any depth below the surface is $2 \times 15 = 30$ kPa ($6 \times 95 = 570$ psf).

2.4.9 Horizontal Point Load Parallel to Soil Surface—Cerutti's Solution

Access the spreadsheet at http://www.wiley.com/college/budhu. Click Foundations and Earth Structures, Chapter 2, Cerutti_mindlin.xls to calculate stress increases and displacements for horizontal point loads and vertical interior point loads.

The solution for the increase in vertical stress due to a horizontal surface load parallel to the surface (Fig. 2.15) was derived by Cerutti. Wind load at the base of a column is an example of a horizontal point load.

The vertical stress increase due to a horizontal load F parallel to the ground surface is:

$$\Delta\sigma_z = \frac{3Fxz^2}{2\pi R^5} \tag{2.27}$$

The vertical settlement is

$$\Delta z = \frac{F(1+\nu)}{2\pi ER}\left[\frac{xz}{R^2} + \frac{(1-2\nu)x}{R+z}\right] \tag{2.28}$$

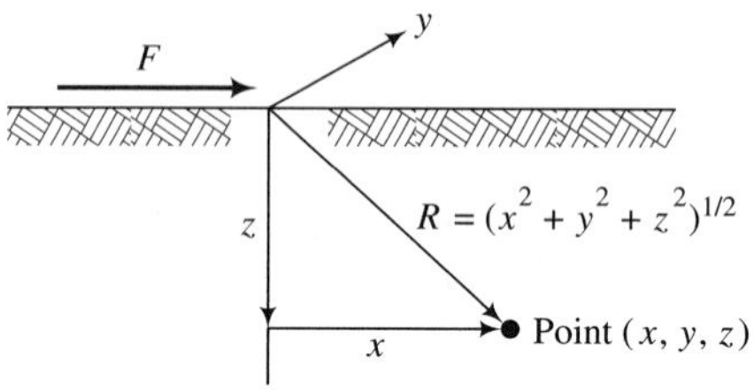

FIGURE 2.15 Horizontal surface load parallel to the surface of the soil.

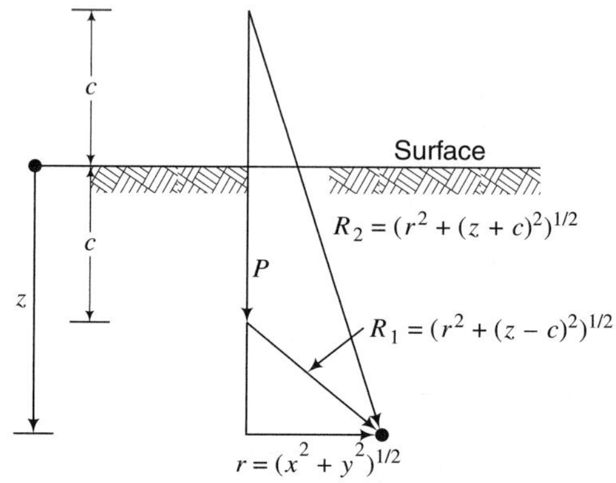

FIGURE 2.16 Vertical point load located beneath the soil surface.

2.4.10 Vertical Point Load Within Soil Mass—Mindlin's Solution

Mindlin (1936) derived solutions for the stress increases and displacements for a vertical point load P located beneath the soil surface (Fig. 2.16). A pile foundation is an example of a vertical point load within a soil mass.

The vertical stress increase is given as

$$\Delta\sigma_z = \frac{-P}{8\pi(1-\nu)}\left[\begin{array}{l} -\dfrac{(1-2\nu)(z-c)}{R_1^3} + \dfrac{(1-2\nu)(z-c)}{R_2^3} - \dfrac{3(z-c)^2}{R_1^5} \\ -\dfrac{3(3-4\nu)z(z+c)^2 - 3c(z+c)(5z-c)}{R_2^5} - \dfrac{30cz(z+c)^3}{R_2^7} \end{array}\right] \tag{2.29}$$

The vertical settlement is

$$\Delta z = \frac{P}{16\pi G(1-\nu)}\left[\begin{array}{l} \dfrac{(3-4\nu)}{R_1} + \dfrac{8(1-\nu)^2 - (3-4\nu)}{R_2} + \dfrac{(z-c)^2}{R_1^3} + \\ \dfrac{(3-4\nu)(z+c)^2 - 2cz}{R_2^3} + \dfrac{6cz(z+c)^2}{R_2^5} \end{array}\right] \tag{2.30}$$

The key points are:

1. ***The increase in stresses below a surface load is found by assuming the soil is an elastic, semi-infinite mass.***
2. ***Various equations are available for the increase of stresses from the various types of surface loading.***
3. ***The vertical stress increase at any depth depends on the shape, magnitude, and distribution of the surface load.***

4. ***A stress applied at the surface of a soil mass by a loaded area decreases with depth and lateral distance away from the center of the loaded area.***
5. ***The increase in vertical stresses is generally less than 10% of the surface stress when the depth-to-width ratio is greater than about 2.***
6. ***The increases in stresses below a static surface load are increases in total stresses except for free-draining, coarse-grained soils. In these soils, the increases can be assumed to be increases in effective stresses.***

EXAMPLE 2.2 *Calculation of Vertical Stress Increase Below a Loaded Rectangular Foundation*

A rectangular concrete slab, 3 m (10 ft) × 4.5 m (15 ft), rests on the surface of a soil mass (Fig. E2.2a,b). The load on the slab is 2025 kN (455 kips). Determine the vertical stress increase at a depth of 3 m (10 ft) (a) under the center of the slab, point A, (b) under point B, and (c) at a distance of 1.5 m (5 ft) from a corner, point C.

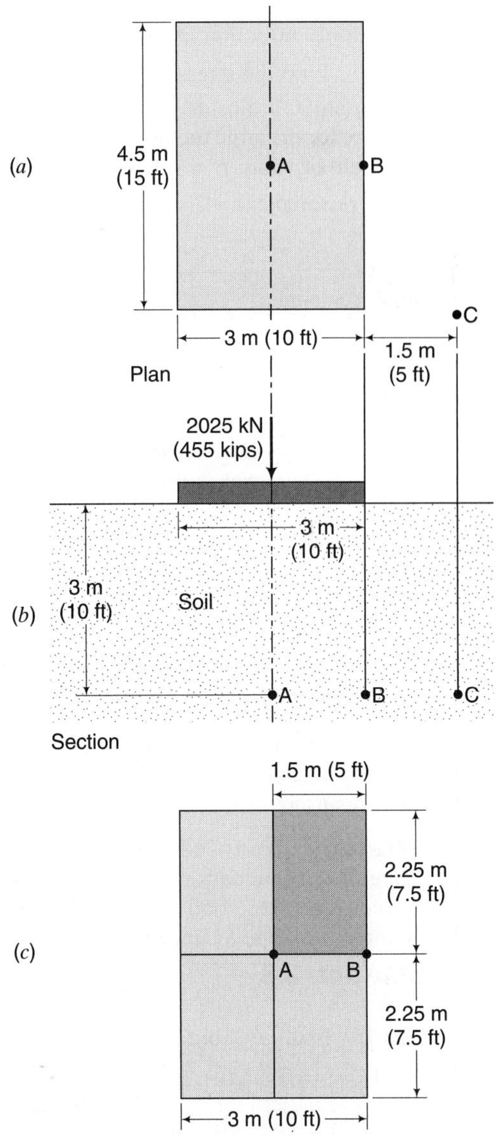

FIGURE E2.2a,b,c

Strategy The slab is rectangular and the equations for uniformly loaded rectangular area are for the corner of the area. You should divide the area so that the point of interest is a corner of a rectangle(s). You may have to extend the loaded area if the point of interest is outside it (loaded area). The extension is fictitious so you have to subtract the fictitious increase in stress for the extended area. US units will be used in solving this example problem.

Solution 2.2

Step 1: Identify loading type.

Uniformly loaded rectangle.

Step 2: Divide rectangle so that the center is a corner. In this problem, all four rectangles after the subdivision are equal, so you need only find the stress increase for one rectangle of size $B = 5$ ft, $L = 7.5$ ft and multiply the results by four.

$$m = \frac{B}{z} = \frac{5}{10} = 0.5; \; n = \frac{L}{z} = \frac{7.5}{10} = 0.75$$

From chart (Fig. 2.11), $I_z = 0.105$.

Step 3: Find the stress increase at the center of the slab (point A, Fig. E2.2c).

$$q_s = \frac{455}{10 \times 15} = 3\,\text{ksf}$$

$$\Delta\sigma_z = 4q_sI_z = 4 \times 130 \times 0.105$$

$$= 1.26\,\text{ksf} = 1260\,\text{psf}$$

The approximate method (Eq. 2.26) gives

$$\Delta\sigma_z = \frac{q_s BL}{(B+z)(L+z)} = \frac{455}{(10+10)(15+10)}$$
$$= 0.91\text{ ksf} = 910\text{ psf}$$

which is about 30% less than the elastic solution.

Step 4: Find the stress increase for point B.

Point B is at the corner of two rectangles each of width 10 ft and length 7.5 ft. You need to find the stress increase for one rectangle and multiply the result by two.

$$m = \frac{10}{10} = 1; \quad n = \frac{7.5}{10} = 0.75$$

From chart (Fig. 2.11), $I_z = 0.158$.

$$\Delta\sigma_z = 2 \times 3 \times 0.158 = 0.948\text{ ksf} = 948\text{ psf}$$

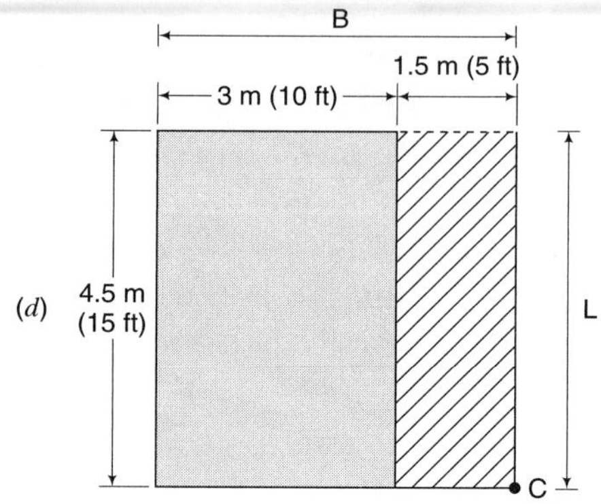

FIGURE E2.2d

You should note that the vertical stress increase at B is lower than A, as expected.

Step 5: Find the stress increase for point C.

Stress point C is outside of the rectangular slab. You have to extend the rectangle to C (Fig. E2.2d) and find the stress increase for the large rectangle of width $B = 15$ ft, length $L = 15$ ft and then subtract the stress increase for the smaller rectangle of width $B = 5$ ft and length 15 ft.

Large rectangle:

$$m = \frac{15}{10} = 1.5, \quad n = \frac{15}{10} = 1.5; \quad \text{from chart (Fig. 2.11)}, I_z = 0.22$$

Small rectangle:

$$m = \frac{5}{10} = 0.5, \quad n = \frac{15}{10} = 1.5; \quad \text{from chart (Fig. 2.11)}, I_z = 0.13$$

$$\Delta\sigma_z = 3 \times (0.22 - 0.13) = 0.27\text{ ksf} = 270\text{ psf}$$

The result for point C using the computer program STRESS is as follows. (1) Input the values $q_s = 3000$, $L = B = 15$, and $z = 10$ and click submit. (2) Replace $B = 1.5$ by $B = 5$, select subtract, then click submit. The answer is $\Delta\sigma_z = 253$ psf (final value).

The difference between the results from the program STRESS and those using Fig. 2.11 is due to interpolation error. The results from the program STRESS are more accurate but you must input the correct values and follow the procedures accessible from the help button. ■

EXAMPLE 2.3 *Calculation of Vertical Stress Increase Below a Loaded Irregularly Shaped Foundation Using Newmark's Chart*

The plan of a foundation of uniform thickness for a building is shown in Fig. E2.3a. Determine the vertical stress increase at a depth of 4 m (13 ft) below the centroid. The foundation applies a vertical stress of 200 kPa (4200 psf) on the soil surface.

Strategy You need to locate the centroid of the foundation, which you can find using the given dimensions. The shape of the foundation does not fit neatly into one of the standard shapes (e.g., rectangles or circles) discussed. The convenient method to use for this (odd) shape foundation is Newmark's chart. SI units will be used in solving this example problem. SI units will be used in the solution of this problem.

Solution 2.3

Step 1: Find centroid.

$$\bar{x} = \frac{1.0 \times 10.0 \times 5.0 + 1.5 \times 2.0 \times 1.0 + \frac{1}{2} \times 8.0 \times 1.5 \times \left(2.0 + \frac{1}{3} \times 8.0\right)}{(1.0 \times 10.0) + (1.5 \times 2.0) + \frac{1}{2} \times 8.0 \times 1.5} = \frac{81}{19} = 4.26\text{ m}$$

$$\bar{z} = \frac{1.0 \times 10.0 \times 0.5 + 1.5 \times 2.0 \times \left(1.0 + \frac{1.5}{2}\right) + \frac{1}{2} \times 8.0 \times 1.5 \times \left(1.0 + \frac{1.5}{3}\right)}{(1.0 \times 10.0) + (1.5 \times 2.0) + \frac{1}{2} \times 8.0 \times 1.5} = \frac{16.25}{19} = 0.86\,\text{m}$$

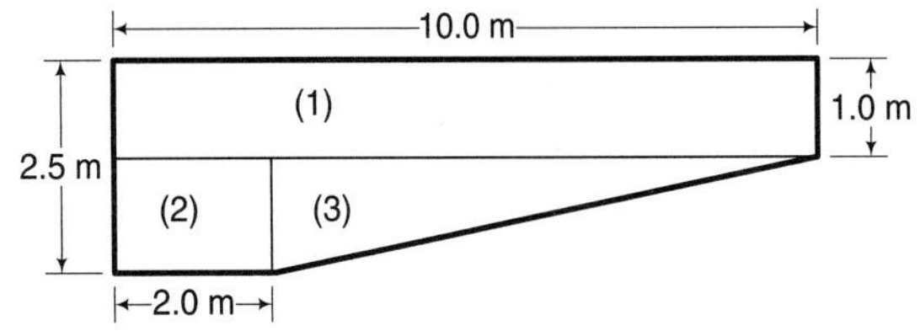

(a)

Step 2: Create a Newmark chart using the program STRESS. Scale and plot foundation on chart.

The scale on the chart is set equal to the depth. The centroid is located at the center of the chart and the foundation is scaled using the depth scale (Fig. E2.3b).

Step 3: Count the number of segments covered by the foundation.

$$N = 61$$

Step 4: Calculate vertical stress increase.

$$\Delta\sigma_z = q_s I_N N = 200 \times 0.005 \times 61 = 61\,\text{kPa}$$

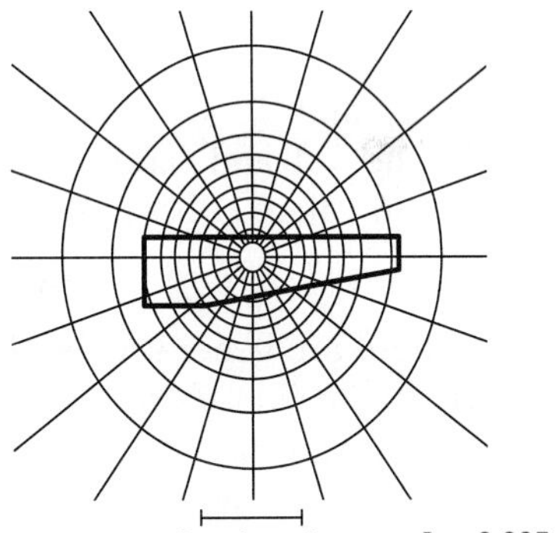

(b)

■

EXAMPLE 2.4 *Calculation of Vertical Stress Increase and Vertical Displacement Due to Horizontal Load Using Cerutti's Solution*

A wind load of 25 kN (5.6 kips) acts at the base of a column. Determine (a) the vertical stress increase at a depth of 1 m (3 ft) and lateral (x) distance of 0.5 m (1.5 ft) from the center line of the column and (b) the vertical elastic settlement. The soil parameters are $E = 50$ MPa (1044 ksf) and $v = 0.3$.

Strategy Use the Cerutti-Mindlin spreadsheet. US units will be used in solving this example problem.

Solution 2.4

Step 1: Identify input parameters.

$$x = 1.5\,\text{ft}, \quad y = 0,\ z = 3\,\text{ft}, \quad F = 5.6\,\text{kips}, \quad E = 1044\,\text{ksf}, \quad \text{and} \quad v = 0.3$$

Step 2: Calculate the stress increase and displacement.

Using the spreadsheet Cerrutti_Mindlin.xls with the input values in Step 1 gives

$$\Delta\sigma_z = 85\,\text{psf},\ \Delta z = 0.00015\,\text{ft} = 0.0018\,\text{in}$$

Check vertical stress increase using Eq. 2.27.

$$\Delta\sigma_z = \frac{3Fxz^2}{2\pi R^5}$$

$$R^5 = \left[x^2 + y^2 + z^2\right]^{\frac{5}{2}} = \left[1.5^2 + 0 + 3^2\right]^{\frac{5}{2}} = 424.5\,\text{ft}^5$$

$$\Delta\sigma_z = \frac{3 \times 5.6 \times 1.5 \times 3^2}{2\pi \times 424.5} \times 1000 = 85\,\text{psf}$$

■

2.4.11 Influence of Foundation-Soil Relative Stiffness on Stress Distribution

Importance *The relative rigidity of a foundation to the soil influences the stress distribution within the soil. The elastic solutions presented are for flexible foundations and do not account for the relative rigidity of the soil-foundation system. If the foundation is rigid, the stress increases are generally lower than those calculated from the elastic solutions presented. In this subsection, an expression to determine relative soil-foundation stiffness and suggestions on how to modify the elastic solutions for foundation flexiblity are presented.*

The stress increase in a soil from a foundation depends on the relative stiffness between the soil and the foundation. The relative foundation-soil stiffness K_r for a circular foundation is

$$K_r = \frac{E_{fnd}(1 - \nu_{soil}^2)}{E_{soil}} \left(\frac{t}{r_o}\right)^3 \quad (2.31)$$

where E_{fnd} and E_{soil} are the Young's modulus of the foundation and soil respectively, ν_{soil} is the Poisson's ratio of the soil, and t and r_o are the thickness and radius of the foundation. For noncircular foundation, use an equivalent r_o based on an equivalent area. For example, for a square foundation of width B the equivalent r_o is $\sqrt{\frac{B \times B}{\pi}} = 0.56B$.

There are generally two limiting classes of soil-foundation relative stiffness: rigid ($K_r > 10$) and flexible ($K_r < 0.1$). A rigid foundation imposes uniform displacements to the soil but the stresses are nonuniformly distributed. The vertical stress distribution beneath a rigid footing, assuming that the soil is an elastic solid, is shown in Fig. 2.17a. Experiments on footings resting on coarse-grained and fine-grained soils show vertical stress distributions as illustrated in Fig 2.17b,c. A flexible foundation (e.g., earthen embankment) is assumed to induce uniform stresses but the displacements are nonuniform. Foundation displacements in coarse-grained soils tend to follow a convex shape while those in fine-grained soils are concave (Fig. 2.18).

Foundations with $0.1 < K_r < 10$, such as very large, relatively thin concrete slabs, may not rest completely on the soil. The vertical stress increase beneath shallow rigid foundations calculated using Boussinesq's method could be 15% to 30% less for clays and 20% to 30% less for sands. Traditionally, the vertical stress increases from the elastic solutions are not adjusted because soil behavior is nonlinear and it is better to err on the conservative side. However, you may adjust the vertical stress increase for rigid foundation based on experience.

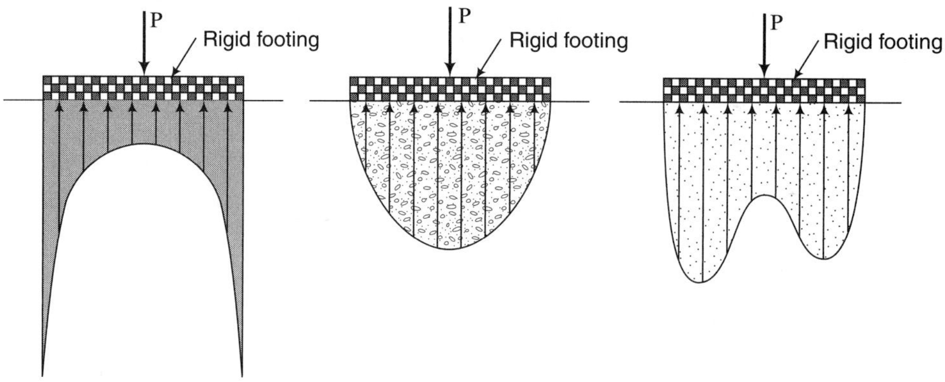

(*a*) Ideal elastic material (*b*) Course-grained soil (*c*) Fine-grained soil

FIGURE 2.17 Vertical stress increase ($\Delta\sigma_z$) distribution under a rigid footing.

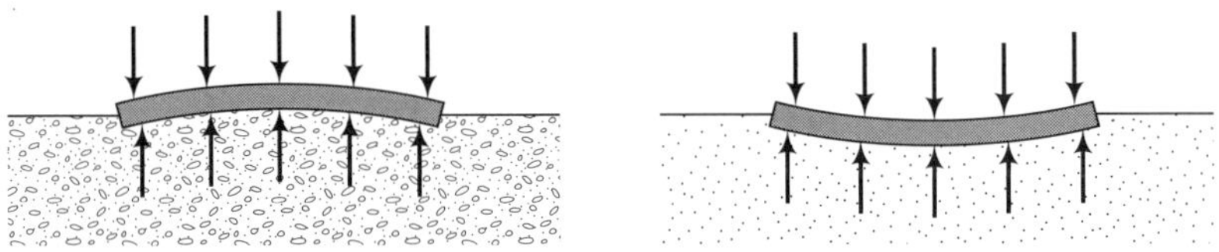

(*a*) Coarse-grained soils (*b*) Fine-grained soils

FIGURE 2.18 Displacements under flexible foundations.

The key points are:

1. ***There are generally two limiting classes of soil-foundation relative stiffness: rigid ($K_r > 10$) and flexible ($K_r < 0.1$).***
2. ***The vertical stress due to surface load is affected by the relative stiffness of the soil-foundation system. The vertical stress increase beneath shallow rigid foundations calculated using Boussinesq's method could be 15% to 30% less for clays and 20% to 30% less for sands.***

EXAMPLE 2.5 *Effects of Soil-Foundation Relative Stiffness on Vertical Stress Increase*

A circular foundation of radius 5 m (16 ft) and a thickness of 1 m (3 ft) rests on a clay soil. Estimate the vertical stress increase at a depth of 1 m (3 ft) below the center of the foundation from a surface stress of 10 kPa (210 psf). The soil and foundation parameters are $E_{soil} = 15$ MPa (310 ksf), $V_{soil} = 0.3$, $E_{fnd} = 25$ GPa (522,000 ksf).

Strategy Find K_r and $\Delta\sigma_z$ for a flexible foundation. Adjust $\Delta\sigma_z$ for rigidity. SI units will be used in solving this example problem.

Solution 2.5

Step 1: Find K_r.

$$K_r = \frac{E_{fnd}(1 - v_{soil}^2)}{E_{soil}}\left(\frac{t}{r_o}\right)^3 = \frac{25000(1 - 0.3^2)}{15}\left(\frac{1}{5}\right)^3 = 12 > 10$$

The foundation is rigid.

Step 2: Find the vertical stress increase for a flexible foundation.

Using the program STRESS with $q_s = 10\,\text{kPa}$, $r_o = 5\,\text{m}$, and $z = 3$ m, we get $\Delta\sigma_z = 8.6\,\text{kPa}$.

Step 3: Adjust stress increase for rigidity.

For clays, the reduction ranges from 15% to 30%. Assuming 20% decrease as a conservative estimate, we get $\Delta\sigma_z = 0.8 \times 8.6 = 6.7\,\text{kPa}$. ■

2.5 LATERAL EFFECTIVE STRESS COEFFICIENTS

Importance *Geosystems constructed beneath the ground surface will be subjected to lateral earth pressures, and lateral and uplift pressures from groundwater. Knowledge of these pressures is important for the design of a stable geosystem. We consider lateral stresses in this section. More information on lateral stresses is given in Chapter 10.*

If an elastic soil is confined laterally and is vertically compressed under drained conditions, the ratio of the lateral stress to the vertical stress is

$$\frac{\sigma_x}{\sigma_z} = \frac{\sigma'_x}{\sigma'_z} = \frac{\sigma_3}{\sigma_1} = \frac{\sigma'_3}{\sigma'_1} = \frac{\nu'}{1-\nu'} \tag{2.32}$$

where ν' is Poisson's ratio for effective stress condition. Eq. 2.32 indicates that for soils the changes in horizontal stresses are lower than the changes in vertical stresses for vertical loads. For example, if $\nu' = 0.3$, then $\frac{\Delta\sigma_x}{\Delta\sigma_z} = 0.4$ (i.e, the horizontal stress increase is 40% of the vertical stress increase). The ratio of the horizontal principal effective stress to the vertical principal effective stress is called the *lateral earth pressure coefficient at rest* (K_o):

$$K_o = \frac{\sigma'_3}{\sigma'_1} \tag{2.33}$$

For a soil that was never subjected to effective stresses higher than its current effective stress (normally consolidated soil) $K_o = K_o^{nc}$ is reasonably predicted by an equation suggested by Jaky (1944) as

$$K_o^{nc} \approx 1 - \sin\phi'_{cs} \tag{2.34}$$

where ϕ'_{cs} is a fundamental frictional soil constant.

The value of K_o^{nc} is constant. During unloading or reloading, the soil stresses must adjust to be in equilibrium with the applied stress. This means that stress changes take place not only vertically but also horizontally. For a given surface stress, the changes in horizontal total stresses and vertical total stresses are different but the porewater pressure changes in every direction are the same. Therefore, the current effective stresses are different in different directions. A soil in which the current effective stress is lower than the past maximum stress is called an *overconsolidated* soil. The K_o values for overconsolidated soils are not constants. We will denote K_o for overconsolidated soils as K_o^{oc}. Various equations have been suggested linking K_o^{oc} to K_o^{nc}. One equation that is popular and found to match test data reasonably well is an equation proposed by Meyerhof (1976) as

$$K_o^{oc} = K_o^{nc}(\text{OCR})^{1/2} = (1 - \sin\phi'_{cs})(\text{OCR})^{1/2} \tag{2.35}$$

where OCR is the overconsolidation ratio (see Chapter 4 for more information) defined as the ratio of the past vertical effective stress to the current vertical effective stress. For normally consolidated soil, OCR = 1, while for overconsolidated soil, OCR > 1.

When a rigid foundation is pushed laterally into a soil, it is resisted by passive earth pressures that vary linearly from the soil surface to the base of the foundation. If the rigid foundation were to move laterally away from the soil, active earth pressures would develop that vary linearly from the soil surface to the base of the foundation. For a flexible foundation, the active and passive earth pressures are nonlinearly distributed with depth.

The passive lateral earth pressure for a horizontal soil surface and a frictionless soil-foundation interface is

$$(\sigma_x)_p = K_p\sigma'_z \tag{2.36}$$

where $K_p = \frac{1+\sin\phi'}{1-\sin\phi'} = \tan^2(45 + \frac{\phi'}{2})$ is called the passive earth pressure coefficient and ϕ' is the friction angle of the soil ($\phi' = \phi'_p$ for peak state and $\phi' = \phi'_{cs}$ for critical state; see Chapter 4), and where z is the depth below the soil surface.

The passive lateral force is

$$P_p = \frac{1}{2}(\sigma'_x)_p z = \frac{1}{2}K_p\gamma' z^2 \tag{2.37}$$

Similarly, for the active lateral earth pressure

$$(\sigma'_x)_a = K_a\sigma'_z = K_a\gamma' z \tag{2.38}$$

where $K_a = \frac{1-\sin\phi'}{1+\sin\phi'} = \tan^2(45 - \frac{\phi'}{2})$ is called the *active earth pressure coefficient*. The active lateral force is

$$P_a = \frac{1}{2}(\sigma'_x)_a z = \frac{1}{2}K_a\gamma' z^2 \tag{2.39}$$

The active and passive earth pressure coefficients strictly apply only to effective principal stresses.

Groundwater applies lateral hydrostatic pressure as well as uplift pressures on foundations. The lateral hydrostatic pressure (porewater pressure) is

$$u_o = \gamma_w z_w \tag{2.40}$$

where γ_w is the unit weight of water and z_w is the depth from the groundwater surface. The lateral hydrostatic force is

$$P_w = \frac{1}{2}u_o z = \frac{1}{2}\gamma_w z_w^2 \tag{2.41}$$

The active and passive earth pressures are affected by the geometry of the soil surface and the soil-foundation interface friction. Poncelet (1840) developed equations to account for the geometry of the soil surface and the soil-foundation interface friction (see Budhu, 2006). Further discussion of lateral earth pressures as they apply to retaining walls is presented in Chapter 10.

The* key points *are:

1. ***Foundations can be subjected to lateral pressures from soil and water.***
2. ***The ratio of lateral (minor) principal effective stress to vertical (major) principal effective stress is called the lateral earth pressure coefficient.***
3. ***Three lateral earth pressure coefficients are used in practice; the at-rest coefficient, K_o, which refers to the current or existing soil state; the active coefficient, K_a, which refers to foundation moving slightly away from the soil; and the passive coefficient, K_p, which refers to foundation moving slightly toward the soil.***
4. ***The lateral earth pressure coefficients apply to effective principal stresses. Pressures due to water must be added to the lateral effective stresses to give the total lateral stresses.***

EXAMPLE 2.6 *Calculation of Lateral Stresses*

Determine the lateral effective and lateral total stresses on a normally consolidated soil for:

(a) Time: $t = t_1$, $\Delta u = 20$ kPa (400 psf)

(b) Time: $t \to \infty$, $\Delta u = 0$

The vertical total stress is 100 kPa (2000 psf), shear stress is zero, and the frictional constant $\phi'_{cs} = 30°$.

Strategy The lateral earth pressure coefficient must be applied to the vertical principal effective stress, not the vertical total principal stress. You need to calculate the vertical principal effective stress, then the lateral principal effective stress. Add the excess porewater pressure to the lateral principal effective stress to find the lateral principal total stress. Since the shear stress is zero, the vertical total stress is a principal stress. U.S. units will be used in solving this example problem.

Solution 2.6

Step 1: Calculate the vertical effective stresses.

$$\sigma'_z = \sigma_z - \Delta u$$

(a) $\sigma'_z = 2000 - 400 = 1600$ psf

(b) $\sigma'_z = 2000 - 0 = 2000$ kPa

Step 2: Calculate the lateral effective stress.

$$K_o^{nc} = 1 - \sin \phi'_{cs} = 1 - \sin 30° = 0.5$$

$$\sigma'_x = K_o^{nc} \sigma'_z$$

(a) $\sigma'_x = 0.5 \times 1600 = 800$ psf

(b) $\sigma'_x = 0.5 \times 2000 = 1000$ psf

Step 3: Calculate the total lateral stresses.

$$\sigma_x = \sigma'_x + \Delta u$$

(a) $\sigma_x = 800 + 400 = 1200$ psf

(b) $\sigma_x = 1000 + 0 = 1000$ psf ■

EXAMPLE 2.7 *Determination of the Depth of a Foundation Required to Resist a Lateral Load*

A foundation must resist a horizontal load of 124 kN (28 kips). Determine the minimum depth of embedment to resist the lateral load with a factor of safety of 2. The friction angle of the soil is 30°, its unit weight is 18 kN/m^3 (114 pcf) and the base resistance is 200 kN (45 kips). Groundwater is 10 m (33 ft) below the surface (Fig. E2.7a).

Strategy Calculate the active and passive earth forces for an assumed depth and then use equilibrium to solve for the unknown depth. SI units will be used in solving this example problem.

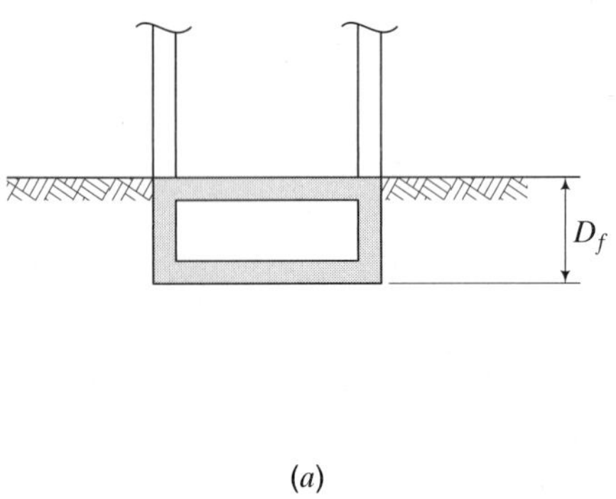

FIGURE E2.7a

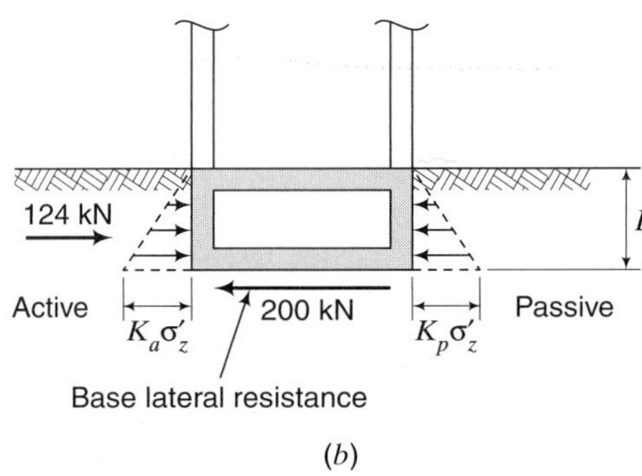

FIGURE E2.7b

Solution 2.7

Step 1: Assume a depth of embedment, D_f (see Fig. E2.7b).

Step 2: Calculate the active and passive earth pressure coefficients.

$$K_a = \tan^2\left(45 - \frac{\phi'}{2}\right) = \tan^2\left(45 - \frac{30}{2}\right) = \frac{1}{3}$$

$$K_p = \tan^2\left(45 + \frac{\phi'}{2}\right) = \tan^2\left(45 + \frac{30}{2}\right) = 3$$

Step 3: Calculate active and passive forces.

Assume that the depth D_f is above the groundwater level. All stresses are principal stresses.

$$(\sigma'_z) = \gamma D_f = 18D_f \text{ kPa}$$

$$(\sigma'_x)_a = K_a\sigma'_z = \frac{1}{3} \times 18D_f = 6D_f \text{ kPa}$$

$$(\sigma'_x)_p = K_p\sigma'_z = 3 \times 18D_f = 54D_f \text{ kPa}$$

$$\text{Active force: } P_a = \frac{1}{2}(\sigma'_x)_a D_f = \frac{1}{2} \times 6D_f \times D_f = 3D_f^2$$

$$\text{Passive force : } P_p = \frac{1}{2}(\sigma'_x)_p D_f = \frac{1}{2} \times 54D_f \times D_f = 27D_f^2$$

Step 4: Use equilibrium equation to solve for D_f.

Factor of safety, $FS = 2$.

$$\overset{+}{\rightarrow} \Sigma F_{\text{horizontally}} = 0 : 124 + 3D_f^2 = \frac{27D_f^2 + 200}{FS} = \frac{27D_f^2 + 200}{2} = 13.5D_f^2 + 100$$

Solving for D_f, we get $D_f = 1.51$ m.
Since $D_f < 10$ m, the groundwater will not affect the calculated stresses. ■

2.6 STRESS AND STRAIN STATES

You can use the interactive computer program *Mohr circle* at http://www.wiley.com/college/budhu. Click Foundations and Earth Structures, Chapter 2, mohrcircle.zip to plot Mohr's circle.

Importance *A foundation can be subjected to loads that induce stresses on various planes in a soil mass. Geotechnical engineers need to know the stress and strain states at a point due to applied loads, because failure is generally initiated at points of high stresses and/or large strains. In this section, we review Mohr's circle of stresses and strains to determine the stress and strain states for a two-dimensional system of stresses and strains.*

The stress state at a point is the set of stress vectors corresponding to all planes passing through that point. One way of calculating the stress state for two-dimensional stresses is by using *Mohr's circle*, which you would have studied in a strength of materials or engineering mechanics course.

2.6.1 Mohr's Circle for Stress States

Suppose a sample of soil is subjected to the stresses shown in Fig. 2.19a. We would like to know what the stresses are at a point within the sample due to the applied stresses. Let us draw the Mohr's circle. First, we have to choose a sign convention. We have already decided that compressive stresses are positive for soils. We will assume counterclockwise shear is positive and $\sigma_z > \sigma_x$. The two coordinates of the circle are (σ_z, τ_{zx}) and (σ_x, τ_{xz}). Recall from your strength of materials course that for equilibrium, $\tau_{xz} = -\tau_{zx}$; these are called complementary shear stresses and are orthogonal to each other. Plot these two coordinates on a graph of shear stress (ordinate) and normal stress (abscissa) as shown by A and B in Fig. 2.19b. Draw a circle with AB as the diameter. The circle crosses the normal stress axis at 1 and 2. The stresses at these points are the major principal stress, σ_1, and the minor principal stress, σ_3.

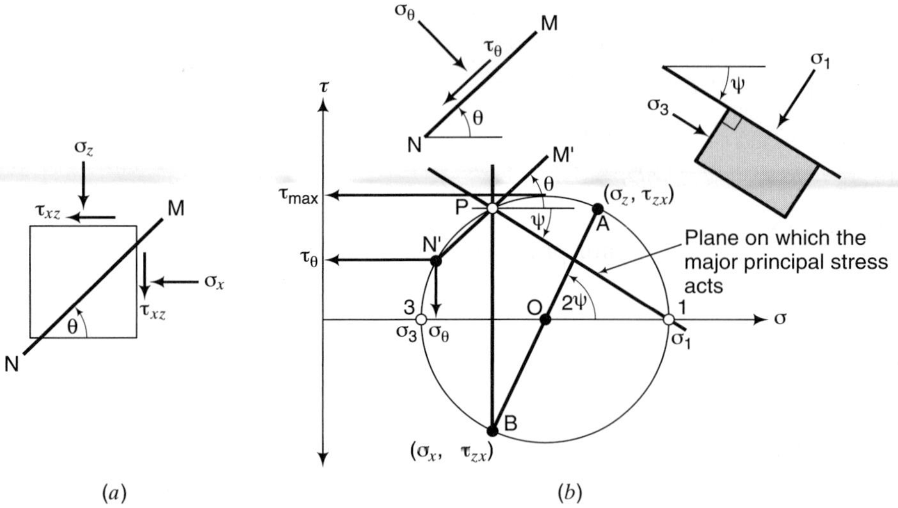

FIGURE 2.19 (a) Stresses on a two-dimensional element and (b) Mohr's circle.

The principal stresses are related to the stress components σ_z, σ_x, τ_{zx} by

$$\sigma_1 = \frac{\sigma_z + \sigma_x}{2} + \sqrt{\left(\frac{\sigma_z - \sigma_x}{2}\right)^2 + \tau_{zx}^2} \tag{2.42}$$

$$\sigma_3 = \frac{\sigma_z + \sigma_x}{2} - \sqrt{\left(\frac{\sigma_z - \sigma_x}{2}\right)^2 + \tau_{zx}^2} \tag{2.43}$$

The angle between the major principal stress plane and the horizontal plane (ψ) is:

$$\tan \psi = \frac{\tau_{zx}}{\sigma_1 - \sigma_x} \tag{2.44}$$

The stresses on a plane oriented at an angle θ to the major principal stress plane are:

$$\sigma_\theta = \frac{\sigma_1 + \sigma_3}{2} + \frac{\sigma_1 - \sigma_3}{2} \cos 2\theta \tag{2.45}$$

$$\tau_\theta = \frac{\sigma_1 - \sigma_3}{2} \sin 2\theta \tag{2.46}$$

The stresses on a plane oriented at an angle θ to the horizontal plane are:

$$\sigma_\theta = \frac{\sigma_z + \sigma_x}{2} + \frac{\sigma_z - \sigma_x}{2} \cos 2\theta + \tau_{zx} \sin 2\theta \tag{2.47}$$

$$\tau_\theta = \tau_{zx} \cos 2\theta - \frac{\sigma_z - \sigma_x}{2} \sin 2\theta \tag{2.48}$$

In the above equations, θ is positive for counterclockwise orientation.

The maximum shear stress is at the top of the circle with magnitude

$$\tau_{max} = \frac{\sigma_1 - \sigma_3}{2} \tag{2.49}$$

The stress σ_z acts on the horizontal plane and σ_x acts on the vertical plane for our case. If we draw these planes in the Mohr's circle, they intersect at a point P. Point P is called the *pole* of the stress circle. It is a special point because any line passing through the pole will intersect the Mohr's circle at a point that represents the stresses on a plane parallel to the line. Let us see how this works. Suppose we want to find the stresses on a plane inclined at an angle θ from the horizontal plane (θ is counterclockwise from the horizontal plane) as depicted by MN in Fig. 2.19a. Once we locate the pole P we can draw a line parallel to MN through P as shown by M′N′ in Fig. 2.19b. The line M′N′ intersects the circle at N′ and the coordinate of N′, (σ_θ, τ_θ), represents the normal and shear stresses on MN.

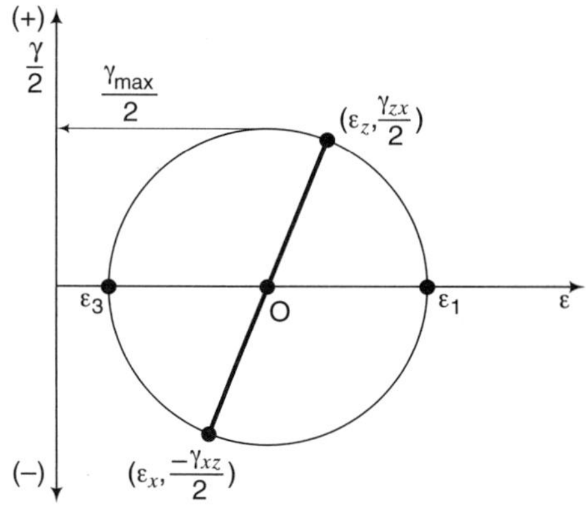

FIGURE 2.20 Mohr's circle of strain.

2.6.2 Mohr's Circle for Strain States

So far we have studied stress states. The strain state is found in a manner similar to the stress state. With reference to Fig. 2.20, the principal strains are major principal strain

$$\varepsilon_1 = \frac{\varepsilon_z + \varepsilon_x}{2} + \sqrt{\left(\frac{\varepsilon_z - \varepsilon_x}{2}\right)^2 + \left(\frac{\gamma_{zx}}{2}\right)^2} \tag{2.50}$$

and minor principal strain

$$\varepsilon_3 = \frac{\varepsilon_z + \varepsilon_x}{2} - \sqrt{\left(\frac{\varepsilon_z - \varepsilon_x}{2}\right)^2 + \left(\frac{\gamma_{zx}}{2}\right)^2} \tag{2.51}$$

where γ_{zx} is called the engineering shear strain or simple shear strain.

The maximum simple shear strain is

$$\gamma_{max} = \varepsilon_1 - \varepsilon_3 \tag{2.52}$$

In soils, strains can be compressive or tensile. There is no absolute reference strain. For stresses, we can select atmospheric pressure as the reference, but not so for strains. Usually, we deal with changes or increments of strains resulting from stress changes.

The* key points *are:

1. ***Mohr's circle is used to find the stress state or strain state from a two-dimensional set of stresses or strains on a soil.***
2. ***The pole on a Mohr's circle identifies a point through which any plane passing through it will intersect the Mohr's circle at a point that represents the stresses on that plane.***

EXAMPLE 2.8 *Determination of Stress State from Mohr's Circle of Stress*

A sample of soil, 100 mm × 100 mm, is subjected to the forces as shown in Fig. E2.8a. Determine σ_1, σ_3, and ψ, the maximum shear stress, the stresses on a plane oriented at 30° counterclockwise from the major principal plane.

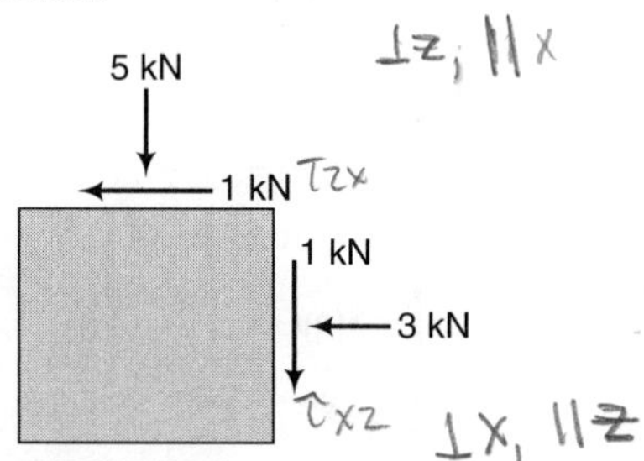

FIGURE E2.8a

Strategy There are two approaches to solve this problem. You can use Mohr's circle or the appropriate equations. Both approaches will be used here. SI units is used in solving this example problem.

Solution 2.8

Step 1: Find the area.

$$\text{Area: } A = 100 \times 100 = 10^4\,\text{mm}^2 = 10^{-2}\,\text{m}^2$$

Step 2: Calculate stresses.

$$\sigma_z = \frac{\text{Force}}{\text{Area}} = \frac{5}{10^{-2}} = 500\,\text{kPa}$$

$$\sigma_x = \frac{3}{10^{-2}} = 300\,\text{kPa}$$

$$\tau_{zx} = \frac{1}{10^{-2}} = 100\,\text{kPa}\,;\ \tau_{xz} = -\tau_{zx} = -100\,\text{kPa}$$

τ_{zx} is positive because it causes the soil element to rotate counterclockwise with reference to its center. Similarly, τ_{xz} is negative because it causes the element to rotate clockwise.

Step 3: Draw Mohr's circle and extract σ_1, σ_3, and τ_{max}.

Mohr's circle is shown in Fig. E2.8b.

$$\sigma_1 = 540\,\text{kPa}, \quad \sigma_3 = 260\,\text{kPa}, \quad \tau_{max} = 140\,\text{kPa}$$

Step 4: Draw the pole on Mohr's circle.

The pole of Mohr's circle is shown by point P in Fig. E2.8b.

Step 5: Determine ψ.

Draw a line from P to σ_1 and measure the angle between the horizontal plane and this line.

$$\psi = 22.5^\circ$$

Alternatively:

The angle AOC $= 2\psi = 45^\circ$. Therefore, $\psi = 22.5^\circ$.

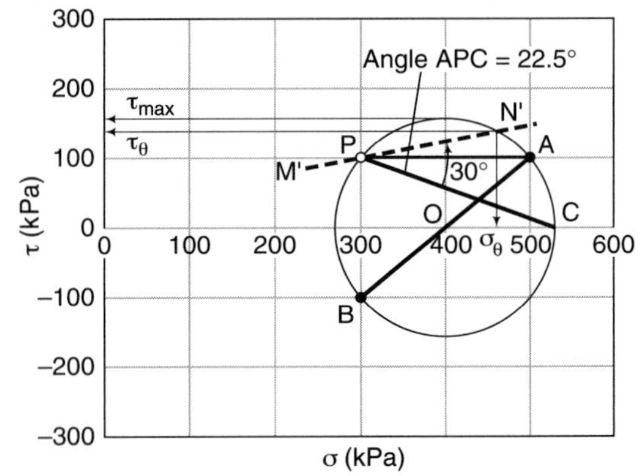

FIGURE E2.8b

Step 6: Determine stresses on a plane inclined at 30° counterclockwise from the major principal stress plane.

Draw a line, M′N′, through P with an inclination of 30° to the major principal stress plane, angle CPN′. The coordinate at point N′ is (470, 120).

Alternatively:

Step 1: Use Eqs. (2.42) to (2.44) and (2.49) to find σ_1, σ_3, ψ, and τ_{max}.

$$\sigma_1 = \frac{500 + 300}{2} + \sqrt{\left(\frac{500 - 300}{2}\right)^2 + 100^2} = 541.4\,\text{kPa}$$

$$\sigma_3 = \frac{500 + 300}{2} - \sqrt{\left(\frac{500 - 300}{2}\right)^2 + 100^2} = 258.6\,\text{kPa}$$

$$\tan\psi = \frac{\tau_{yx}}{\sigma_1 - \sigma_x} = \frac{100}{541.4 - 300} = 0.414$$

$$\therefore \Psi = 22.5^\circ$$

$$\tau_{max} = \frac{\sigma_1 - \sigma_3}{2} = \frac{541.4 - 258.6}{2} = 141.4\,\text{kPa}$$

Step 2: Use Eqs. (2.47) and (2.48) to find σ_θ and τ_θ.

$$\sigma_\theta = \frac{541.4 + 258.6}{2} + \frac{541.4 - 258.6}{2}\cos(2 \times 30^\circ) = 470.7\,\text{kPa}$$

$$\tau_\theta = \frac{541.4 - 258.6}{2}\sin(2 \times 30^\circ) = 122.5\,\text{kPa}$$

■

2.7 STRESS AND STRAIN INVARIANTS

Importance *The stresses we discussed so far act on planes referenced to the Cartesian axis system, which is an arbitrary selection. If we could define stresses and strains such that they are independent of any axis system, then we would have a set of measures that are unbiased and we could avoid misunderstanding in the interpretation of analytical, numerical, laboratory, and field test results. In this section, stress and strain invariants that are useful in foundation engineering are discussed.*

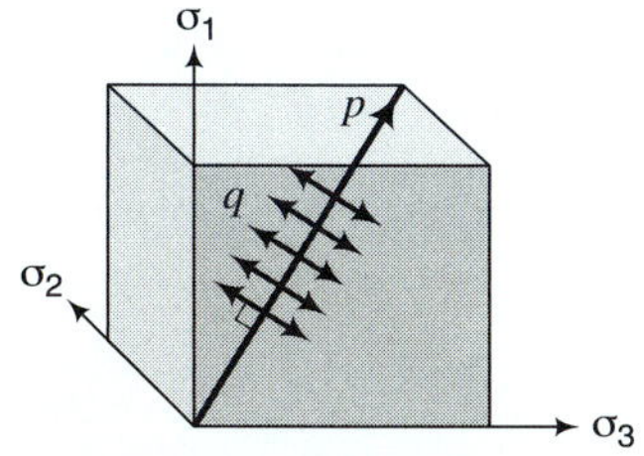

FIGURE 2.21 Mean and deviatoric stresses.

Stress and strain invariants are measures that are independent of the axis system. We will define stress invariants that provide measures of (1) mean stress and (2) deviatoric or distortional or shear stress, and strain invariants that provide measures of (1) volumetric strains and (2) deviatoric or distortional or shear strains. The mean stress p is defined as

$$p = \frac{\sigma_1 + \sigma_2 + \sigma_3}{3} = \frac{\sigma_x + \sigma_y + \sigma_z}{3} \tag{2.53}$$

On a graph with orthogonal principal stress axes, $\sigma_1, \sigma_2, \sigma_3$, the mean stress is the space diagonal, that is, a line oriented at equal angles to all three axes (Fig. 2.21). Mean stress causes volume changes.

A line normal to the mean stress as illustrated in Fig. 2.21 represents the deviatoric stress. Deviatoric stress causes distortions or shearing of a soil mass. The deviatoric strain, q, is

$$q = \frac{1}{\sqrt{2}}\left[(\sigma_1 - \sigma_2)^2 + (\sigma_2 - \sigma_3)^2 + (\sigma_3 - \sigma_1)^2\right]^{1/2} \tag{2.54}$$

Volumetric strain is the sum of all the normal strains:

$$\varepsilon_p = \varepsilon_1 + \varepsilon_2 + \varepsilon_3 = \varepsilon_x + \varepsilon_y + \varepsilon_z \tag{2.55}$$

Deviatoric strains (shear strains) are measures of distortion of the soil. The deviatoric strain, ε_q, is deformed as

$$\varepsilon_q = \frac{\sqrt{2}}{3}\left[(\varepsilon_1 - \varepsilon_2)^2 + (\varepsilon_2 - \varepsilon_3)^2 + (\varepsilon_3 - \varepsilon_1)^2\right]^{1/2} \tag{2.56}$$

For these two conditions, the stress and strain invariants are as follows:

Axisymmetric condition:

$$\sigma_2' = \sigma_3' \text{ or } \sigma_2 = \sigma_3;\ \varepsilon_2 = \varepsilon_3$$

$$p' = \frac{\sigma_1' + 2\sigma_3'}{3} \quad \text{and} \quad p = \frac{\sigma_1 + 2\sigma_3}{3} \tag{2.57}$$

$$p' = p - u$$

$$q = \sigma_1 - \sigma_3$$

$$q' = \sigma_1' - \sigma_3' = (\sigma_1 - \Delta u) - (\sigma_3 - \Delta u) = \sigma_1 - \sigma_3 \tag{2.58}$$

Therefore $q = q'$; shear is unaffected by porewater pressures.

$$\varepsilon_p = \varepsilon_1 + 2\varepsilon_3 \tag{2.59}$$

$$\varepsilon_q = \frac{2}{3}(\varepsilon_1 - \varepsilon_3) \tag{2.60}$$

An example of axisymmetric condition is a soil element under the center of an oil tank (Fig. 2.22).

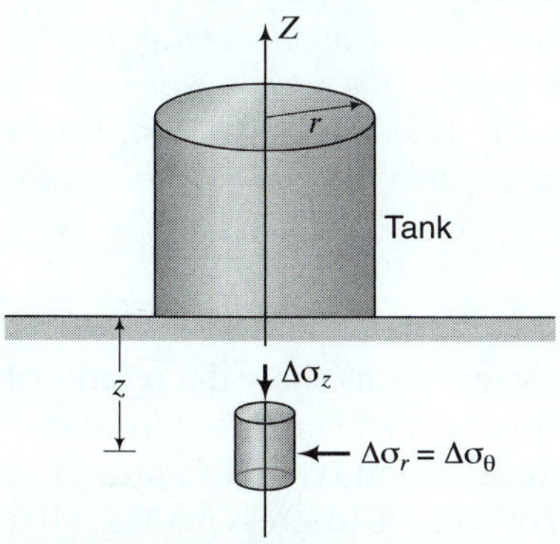

FIGURE 2.22 Axisymmetric condition: a soil element under the center of an oil tank.

Plane strain condition:

$$\varepsilon_2 = 0$$

$$p' = \frac{\sigma_1' + \sigma_2' + \sigma_3'}{3}, \quad \text{and} \quad p = \frac{\sigma_1 + \sigma_2 + \sigma_3}{3} \tag{2.61}$$

$$p' = p - u \tag{2.62}$$

$$q' = q = \frac{1}{\sqrt{2}}\left[(\sigma_1' - \sigma_2')^2 + (\sigma_2' - \sigma_3')^2 + (\sigma_3' - \sigma_1')^2\right]^{1/2} \tag{2.63}$$

$$\text{or } \frac{1}{\sqrt{2}}\left[(\sigma_1 - \sigma_3)^2 + (\sigma_2 - \sigma_3)^2 + (\sigma_3 - \sigma_1)^2\right]^{1/2}$$

$$\varepsilon_p = \varepsilon_1 + \varepsilon_3 \tag{2.64}$$

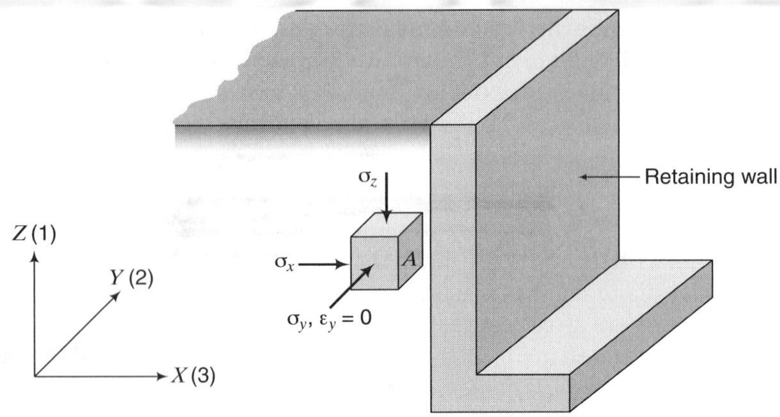

FIGURE 2.23 Plane strain condition: a soil element near a retaining wall.

$$\varepsilon_q = \frac{2}{3}\left(\varepsilon_1^2 + \varepsilon_3^2 - \varepsilon_1\varepsilon_3\right)^{1/2} \tag{2.65}$$

An example of plane strain condition is an element of soil next to a retaining wall (Fig. 2.23).

The* key points *are:

1. ***Stress and strain invariants are independent of the chosen axis system.***
2. ***Stress and strain invariants are convenient measures to determine the effects of a general state of stresses and strains on soils.***
3. ***Mean stress represents the average stress on a soil, while deviatoric stress represents the average shear or distortional stress.***

2.8 STRESS PATHS

Importance *The responses of a soil depend on the stress directions, which vary because of the variety of loads that are transmitted to it. Some of these stress directions may bring the soil to failure; others may cause excessive settlements and instability. Ideally, we would like to test soil samples following similar stress paths to those expected in the field. Consequently, it is necessary to estimate the stress paths that the applied loads will impose on the soil. In this section, two methods of illustrating stress paths are discussed.*

A stress path is a graphical representation of stresses in stress space. Stress paths show the relationship between stress parameters and provide a convenient way to allow a geotechnical engineer to study the changes in stresses in a soil caused by loading conditions. The response, stability, and failure of soils depend on the stress path. In soil testing, we should ideally subject the soil sample to stress paths similar to those expected in the field. We can plot a two-dimensional graph of σ_1 versus σ_3 or σ_2 that will gives us a relationship between these stress parameters. However, the stress invariants, being independent of the axis system, are more convenient to use.

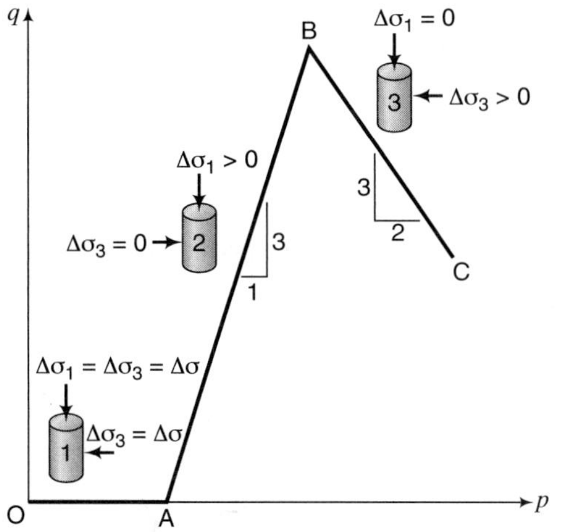

FIGURE 2.24 Stress paths.

2.8.1 Basic Concept

Let us consider a cylindrical soil sample for illustrative purposes and subject it to several loading conditions. Let us apply equal increments of axial and radial stresses ($\Delta\sigma_z = \Delta\sigma_r = \Delta\sigma$) to an initially stress-free sample as illustrated in inset figure 1 in Fig. 2.24. Since we are not applying any shearing stresses on the horizontal and vertical boundaries, the axial and radial stresses are principal stresses (i.e., $\Delta\sigma_z = \Delta\sigma_1$ and $\Delta\sigma_r = \Delta\sigma_3$).

The loading condition we are applying is called *isotropic compression*; that is the stresses in all directions are equal ($\Delta\sigma_1 = \Delta\sigma_2 = \Delta\sigma_3$). We will call this loading condition *loading 1*. It is often convenient to work with increments of stresses in determining stress paths. Consequently, we are going to use the incremental form of the stress invariants. The stress invariants for isotropic compression are:

$$\Delta p_1 = \frac{\Delta\sigma_1 + 2\Delta\sigma_3}{3} = \frac{\Delta\sigma_1 + 2\Delta\sigma_1}{3} = \Delta\sigma_1$$
$$\Delta q_1 = \Delta\sigma_1 - \Delta\sigma_3 = \Delta\sigma_1 - \Delta\sigma_1 = 0$$

The subscript 1 on p and q denotes loading 1.

Let us now prepare a graph with axes p (abscissa) and q (ordinate), as depicted in Fig. 2.24. We will call this graph the (p, q) plot. The initial stresses on the soil sample are zero (i.e., $p_o = 0$ and $q_o = 0$). The stresses at the end of loading 1 are:

$$p_1 = p_o + \Delta p_1 = 0 + \Delta\sigma_1 = \Delta\sigma_1$$
$$q_1 = q_o + \Delta q_1 = 0 + 0 = 0$$

and shown as coordinate A in Fig. 2.24. The line OA is called the *stress path for isotropic compression*. The slope of OA is:

$$\frac{\Delta q}{\Delta p} = 0$$

Let us now apply loading 2 by keeping σ_3 constant (i.e., $\Delta\sigma_3 = 0$) but continue to increase σ_1 (i.e., $\Delta\sigma_1 > 0$). (See inset figure 2 in Fig. 2.24.) The increases in stress invariants for loading 2 are:

$$\Delta p_2 = \frac{\Delta\sigma_1 + 2 \times 0}{3} = \frac{\Delta\sigma_1}{3}$$
$$\Delta q_2 = \Delta\sigma_1 - 0 = \Delta\sigma_1$$

and the stress invariants at the end of loading 2 are:

$$p_2 = p_1 + \Delta p_2 = \Delta\sigma_1 + \frac{\Delta\sigma_1}{3} = \frac{4}{3}\Delta\sigma_1$$
$$q_2 = q_2 + \Delta q_2 = 0 + \Delta\sigma_1 = \Delta\sigma_1$$

Point B in Fig. 2.24 represents (p_2, q_2) and the line AB is the stress path for loading 2. The slope of AB is:

$$\frac{\Delta q_2}{\Delta p_2} = \frac{\Delta\sigma_1}{\left(\frac{\Delta\sigma_1}{3}\right)} = 3$$

The path OAB represents the stress path followed during isotropically consolidated $(O \to A)$ triaxial compression (A → B) tests.

Let us make another change to the loading conditions. We will now keep σ_1 constant ($\Delta\sigma_1 = 0$) and then increase $\sigma_3(\Delta\sigma_3 > 0)$, as illustrated by inset figure 3 in Fig. 2.24. The increases in stress invariants are:

$$\Delta p_3 = \frac{0 + 2\Delta\sigma_3}{3} = \frac{2\Delta\sigma_3}{3}$$
$$\Delta q_3 = 0 - \Delta\sigma_3 = -\Delta\sigma_3$$

The stress invariants at the end of loading 3 are:

$$p_3 = p_2 + \Delta p_3 = \frac{4}{3}\Delta\sigma_1 + \frac{2\Delta\sigma_3}{3}$$
$$q_3 = q_2 + \Delta q_3 = \Delta\sigma_1 - \Delta\sigma_3$$

The stress path for loading 3 is shown as BC in Fig. 2.24. The slope of BC is

$$\frac{\Delta q_3}{\Delta p_3} = \frac{-\Delta\sigma_3}{\frac{2\Delta\sigma_3}{3}} = -\frac{3}{2}$$

You should note that q decreases but p increases for stress path BC. The stress path BC is the path followed in triaxial extension (TE) tests.

So far, we have not discussed whether the soil was allowed to drain. You will recall that the soil solids and the porewater must carry the applied increase in stresses. If the porewater were allowed to drain from the soil sample, the increase in stress carried by the porewater, called *excess porewater pressure* (Δu), will continuously decrease to zero and the soil solids will have then to completely support the increase in applied stresses. We will assume that during loading 1, the excess porewater was allowed to drain (called *drained condition* in geotechnical engineering). The type of loading imposed by loading 1 is called *isotropic consolidation*. Since the excess porewater pressure (Δu_1) dissipates as the porewater drains from the soil, the mean effective stress at the end of each increment of loading 1 is equal to the mean total stress:

$$\Delta p_1' = \Delta p_1 - \Delta u_1 = \Delta p_1 - 0 = \Delta p$$

The effective stress path (ESP) and the total stress path (TSP) are the same and represented by OA in Fig. 2.25. You should note that we have used dual labels p', p for the horizontal axis in Fig. 2.25. This dual labeling allows us to use one plot to represent both the effective and total stress paths.

We will assume that for loadings 2 and 3, the porewater was prevented from draining out of the soil. The implication is that the volume of our soil sample remains constant. By preventing the water within the pores from draining, a pressure is built up in the porewater. In geotechnical engineering, the term *undrained* is used to denote a loading situation in which the porewater cannot drain from the soil. Under undrained conditions, the change in mean effective stress is zero for a linear, isotropic, elastic soil. Consequently, $\Delta p' = 0$, and the effective stress path is vertical as shown by AB′ and BC′ in Fig. 2.25. For elastic-plastic soils (see Chapter 5), the ESP for loading 2 would bend toward the left as excess porewater pressures become greater when the soil starts behaving like a plastic material. The horizontal distance (Δp) separating the TSP from the ESP is the excess porewater pressure. In this book, we will represent TSP as dotted lines and ESP as solid lines, because failure of a soil is due to effective stresses, not total stresses.

For two-dimensional stresses, we can use an alternative stress path presentation based on Mohr's circle. We can define

$$t = \frac{\sigma_1 - \sigma_3}{2} = \frac{\sigma_1' - \sigma_3'}{2} \tag{2.66}$$

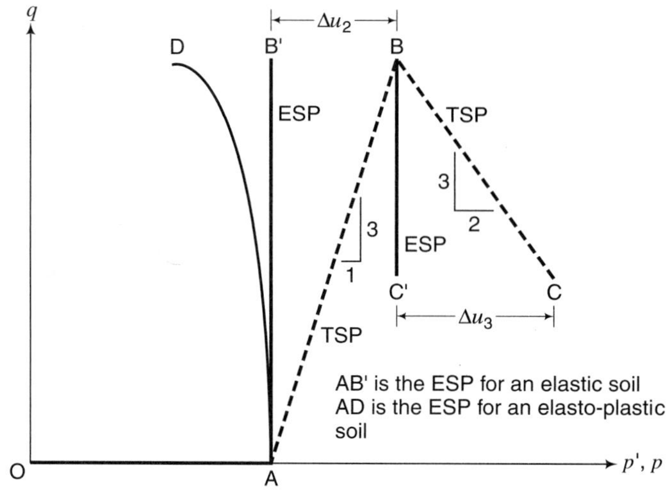

FIGURE 2.25 Total and effective stress paths.

$$s = \frac{\sigma_1 + \sigma_3}{2}; \; s' = \frac{\sigma_1' + \sigma_3'}{2} \tag{2.67}$$

where t and s are radius and center of Mohr's circle respectively. This representation of stress in the stress space (s' or s, t) neglects the effects of the intermediate principal stresses and is appropriate for plane stress condition. It is also convenient for plane strain condition, because we often do not know the value of the intermediate stress mobilized from conventional laboratory test equipment.

The predicted changes in excess porewater pressure, which depend on mean stress p, would be different for the two types of stress path presentations. For example, let us consider the triaxial compression test for a linear, isotropic, elastic soil for which the TSP is represented by AB and the ESP is represented by AB′ (Figs. 2.25 and 2.26). The predicted change in excess porewater pressure for (p, q) space (Fig. 2.25) is

$$\Delta u_2 = \Delta p = \frac{\Delta\sigma_1}{3}$$

For the (s, t) space (Fig. 2.26),

$$\Delta u_2 = \Delta s = \frac{\Delta\sigma_1}{2}$$

Thus, interpreting the excess porewater pressure from the stress path in (s, t) space would lead to a 50% greater excess porewater pressure than from the stress path in (p, q) space, because the intermediate stress is not accounted for in the (s, t) space. The slope of the TSP is also different for the two-stress-path space. For the (p, q) space, the TSP for triaxial compression (TC) is $\frac{\Delta q}{\Delta p} = 3$, while in ($s$, t) space it is $\frac{\Delta t}{\Delta s} = 1$. In the literature p or p' and q are sometimes used to denote the stress state characterized by s or s' and t.

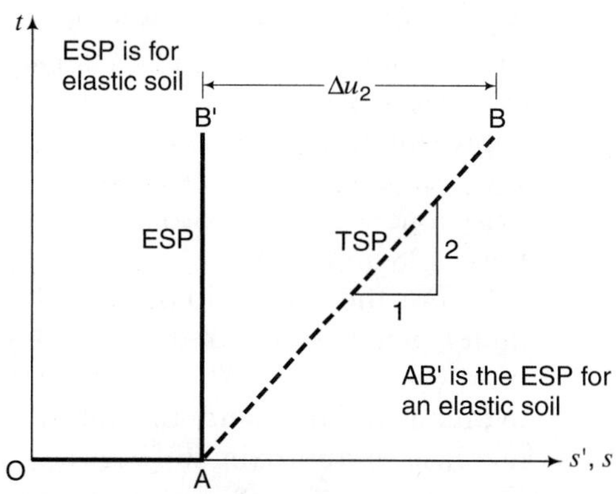

FIGURE 2.26 Total and effective stress path in (s, t) space.

Let us consider the total stress paths for (1) a soil element under the center of an oil tank, (2) a soil element under the center of an excavation to construct a foundation, and (3) a soil element adjacent to the side of the

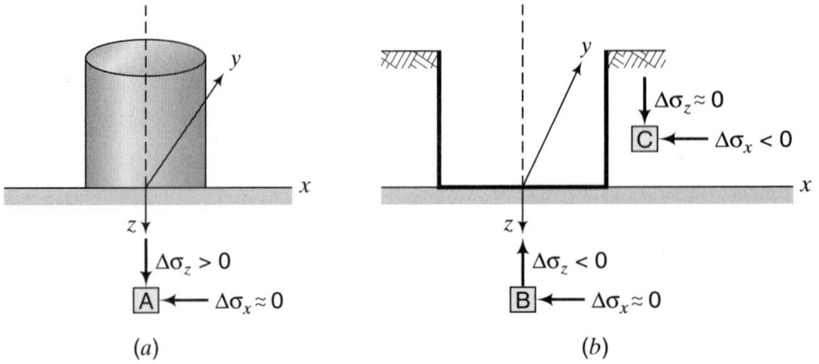

FIGURE 2.27 Stresses in soil elements (a) beneath an oil tank and (b) near an excavation.

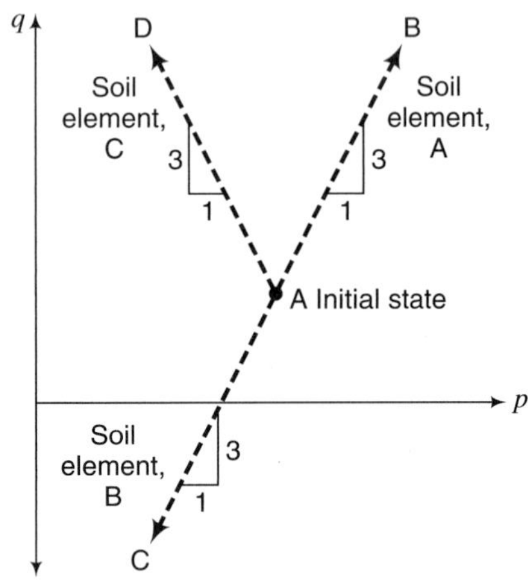

FIGURE 2.28 Total stress path in (*p*, *q*) space.

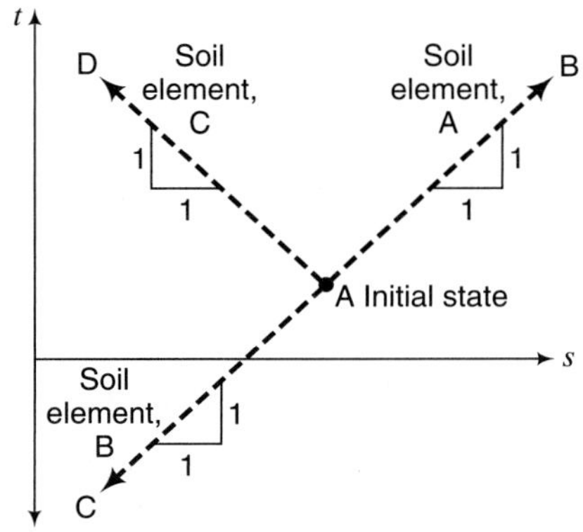

FIGURE 2.29 Total stress path in (*s*, *t*) space.

excavation as shown in Fig. 2.27. For plotting the stress paths, we will assume that changes occur in small increments. The element A under the center of the oil tank is under axisymmetric condition and will be subjected to increases in vertical and lateral stresses. The increases in lateral stresses are much lower than the vertical stresses as shown for a linear, elastic soil in Section 2.5. We can assume, as a first approximation, that the lateral stress increase is small and can be neglected. For element A, $\Delta p = \frac{\Delta\sigma_z}{3}$, $\Delta q = \Delta\sigma_z$, $\frac{\Delta q}{\Delta p} = \frac{3}{1}$ or $\Delta s = \frac{\Delta\sigma_z}{2}$, $\Delta t = \frac{\Delta\sigma_z}{2}$, $\frac{\Delta s}{\Delta t} = \frac{1}{1}$. Consequently, element A can be simulated by an axisymmetric compression test (triaxial compression). The TSP for element A is represented by AB in Figs. 2.28 and 2.29.

For the excavation, a soil element below the center of the excavation, soil element B, will be under plane strain condition and will experience reductions in vertical stresses and small changes in lateral stresses. The latter increases occur because as the soil element is invaded it becomes overconsolidated and the lateral stress coefficient increases. Assuming that the lateral stress increase is approximately zero, then for soil element B, $\Delta p = \frac{-\Delta\sigma_z}{3}$, $\Delta q = -\Delta\sigma_z$, $\frac{\Delta q}{\Delta p} = \frac{3}{1}$ or $\Delta s = \frac{-\Delta\sigma_z}{2}$, $\Delta t = \frac{-\Delta\sigma_z}{2}$, $\frac{\Delta s}{\Delta t} = \frac{1}{1}$. This stress path can be simulated by an axisymmetric (triaxial, see Chapter 4) extension test. The TSP for soil element B is represented by AC in Figs. 2.28 and 2.29.

Soil element C at the side of the excavation will experience reductions in lateral stresses and no change in vertical stresses. For soil element C, $\Delta p = \frac{-\Delta\sigma_x}{3}$, $\Delta q = \Delta\sigma_x$, $\frac{\Delta q}{\Delta p} = \frac{3}{1}$ or $\Delta s = \frac{-\Delta\sigma_x}{2}$, $\Delta t = \frac{\Delta\sigma_x}{2}$, $\frac{\Delta s}{\Delta t} = \frac{1}{1}$. The TSP for this element is shown as AD in Figs. 2.28 and 2.29. This stress path can be simulated by an axisymmetric test by keeping the vertical stress constant and reducing the lateral stress on the soil sample.

For a long excavation, elements B and C are best represented by plane strain condition. You should note how different these stress paths are for the selected soil elements. In fact, if you were to consider a soil element at the edge of the excavation, the stress path would be very complex and cannot be simulated by available laboratory tests. Each of these TSPs would cause the soil to behave differently. Generally, soil elements beneath a foundation would be subjected to different stress paths, and the effective stresses may cause some elements to reach the failure state while other elements are in a prefailure state.

The* key points *are:

1. ***A stress path is a graphical representation of stresses in stress space. For convenience, stress paths are plotted in (p and/or p', q) space or (s and/or s', t) space.***
2. ***The effective stress path for a linear, elastic soil under undrained condition is vertical (i.e., $\Delta p' = 0$).***
3. ***The mean stress difference between the total stress path and the effective stress path is the excess porewater pressure.***
4. ***The response, stability, and failure of soils depend on stress paths.***
5. ***Stress paths are important in understanding how the soil will respond to loading.***

2.8.2 Plotting Stress Paths

A summary of the procedure for plotting stress paths in (p and/or p', q) space is given below. For plotting stress paths in (s and/or s', t) space, replace p by s, p' by s', q by t and do not calculate $\Delta\sigma_2$ in the following.

1. Determine the loading condition (i.e., drained or undrained or both).
2. Calculate the initial values of p'_o, p_o, and q_o.
3. Set up a graph of p' (and p, if you are going to also plot the total stress path) as abscissa and q as ordinate. Plot the initial values of (p'_o, q_o) and (p_o, q_o).
4. Determine the increase in stresses, $\Delta\sigma_1$, $\Delta\sigma_2$ and $\Delta\sigma_3$. These stresses can be negative.
5. Calculate the increase in stress invariants, $\Delta p'$, Δp, and Δq. These stress invariants can be negative.
6. Calculate the current stress invariants as $p' = p'_o + \Delta p'$, $p = p_o + \Delta p$ and $q = q_o + \Delta q$. The current values of p' and p cannot be negative but q can be negative.
7. Plot the current stress invariants (p', q) and (p, q).
8. Connect the points identifying effective stresses and do the same for total stresses.
9. Repeat items 4 to 8 for the next loading condition.
10. The excess porewater pressure at a desired level of deviatoric stress is the mean stress difference between the total stress path and the effective stress path.

Remember that for drained loading condition, ESP = TSP, and for undrained condition, the ESP for a linear, elastic soil is vertical.

EXAMPLE 2.9 *Stress Paths Under a Loaded Circular Foundation for an Oil Tank*

An oil tank of diameter 10 m (30 ft) will impose a vertical surface stress of 100 kPa (2000 psf) on a homogeneous, saturated, normally consolidated, clay soil of unit weight 20 kN/m^3 (127 pcf) and $\phi'_{cs} = 30°$. The groundwater level is at ground surface. For an element 2 m (6 ft) below the ground,

(a) Plot the initial TSP and ESP.

(b) Plot the TSP and ESP (assuming elastic response) due to the surface stress.

(c) Estimate the excess porewater pressure due to the surface stress assuming elastic response.

(d) Compare the results for $(p,\ q)$ space and $(s,\ t)$ space.

Strategy Calculate the stresses for the initial and final state and then plot the results. The increases in stress due to the surface stresses are computed from the elastic solution for a circular foundation. The soil element is under axisymmetric condition. SI units will be used in solving this example problem.

Solution 2.9

Step 1: Find K_o^{nc}.

Since soil is normally consolidated,

$$K_o^{nc} = 1 - \sin\phi'_{cs}$$
$$K_o^{nc} = 1 - \sin 30° = 0.5$$

Step 2: Find the initial stresses. All initial stresses are principal stresses.

$$\sigma_{zo} = 2 \times 20 = 40\,\text{kPa}$$
$$u_o = 2 \times 9.8 = 19.6\,\text{kPa}$$
$$\sigma'_{zo} = 40 - 19.6 = 20.4\,\text{kPa}$$
$$\sigma'_{xo} = K_o^{nc}\sigma'_{zo} = 0.5 \times 20.4 = 10.2\,\text{kPa}$$
$$\sigma_{xo} = \sigma'_{xo} + u_o = 10.2 + 19.6 = 29.8\,\text{kPa}$$

Step 3: Calculate the initial stress invariants p, p' and q.

$$p_o = \frac{\sigma_{zo} + 2\sigma_{xo}}{3} = \frac{40 + 2 \times 29.8}{3} = 33.2\,\text{kPa}$$
$$p'_o = p_o - u_o = 33.2 - 19.6 = 13.6\,\text{kPa}$$
$$q_o = q'_o = \sigma_{zo} - \sigma_{xo} = 40 - 29.8 = 10.2\,\text{kPa}$$
$$\text{Slope}\ \frac{q_o}{p_o} = \frac{10.2}{33.2} = 0.31,\ \text{slope}\ \frac{q_o}{p'_o} = \frac{10.2}{13.6} = 0.75$$

Step 4: Calculate the initial stress invariants s, s', and t.

$$s_o = \frac{\sigma_{zo} + \sigma_{xo}}{2} = \frac{40 + 29.8}{2} = 34.9\,\text{kPa}$$
$$s'_o = s_o - u_o = 34.9 - 19.6 = 15.3\,\text{kPa}$$
$$t_o = t'_o = \frac{\sigma_{zo} - \sigma_{xo}}{2} = \frac{40 - 29.8}{2} = 5.1\,\text{kPa}$$
$$\text{Slope}\ \frac{t_o}{s_o} = \frac{5.1}{34.9} = 0.15,\ \text{slope}\ \frac{t_o}{s'_o} = \frac{5.1}{15.3} = 0.33$$

Step 5: Calculate stress increases on the element at 2 m.

Using the program STRESS with $q_s = 100$ kPa, $r_o = 5$ m, and $z = 2$ m, we get

$$\Delta\sigma_z \cong 95\,\text{kPa}, \quad \Delta\sigma_r \cong 47\,\text{kPa}$$

$$\Delta p = \frac{\Delta\sigma_z + 2\Delta\sigma_r}{3} = \frac{95 + 2(47)}{3} = 63\,\text{kPa}$$

$$\Delta q = \Delta\sigma_z - \Delta\sigma_r = 95 - 47 = 48\,\text{kPa}$$

$$\frac{\Delta q}{\Delta p} = \frac{48}{63} = 0.762$$

$$\Delta s = \frac{\Delta\sigma_z + \Delta\sigma_r}{2} = \frac{95 + 47}{2} = 71\,\text{kPa}$$

$$\Delta s = \frac{\Delta\sigma_z - \Delta\sigma_r}{2} = \frac{95 - 47}{2} = 24\,\text{kPa}$$

$$\frac{\Delta q}{\Delta p} = \frac{24}{71} = 0.338$$

Step 6: Calculate the excess porewater pressure.

Since the soil is elastic, $\Delta p' = 0$, so the ESP is vertical.

$$(p, q)\text{ space: } \Delta u = \Delta p = 63\,\text{kPa}$$

$$(s, t)\text{ space: } \Delta u = \Delta s = 71\,\text{kPa}$$

The two excess porewater pressure predictions are different.

Porewater pressure changes are the results of changes in mean stress. Consequently, $\Delta u = \Delta p$ is a better theoretical prediction than $\Delta u = \Delta s$.

See Fig. E2.9 for the plots of the stress paths.

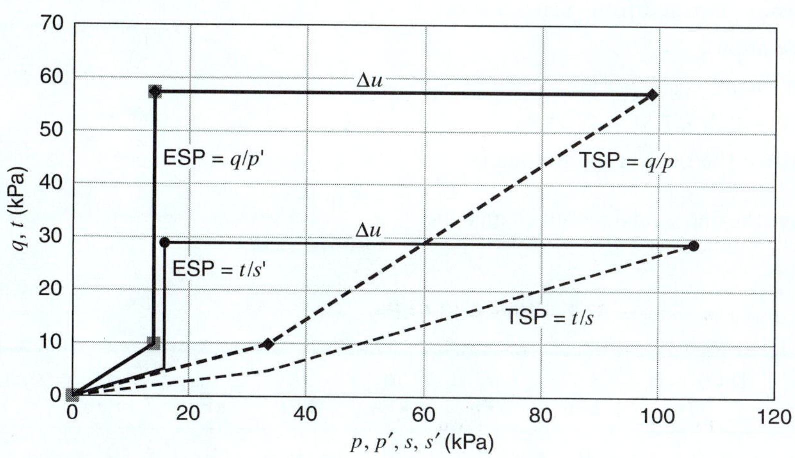

FIGURE E2.9 ■

2.9 SUMMARY

Loads are divided into different categories consisting of dead loads, live loads, earth loads, hydrostatic loads, wind loads, snow loads, rain loads, impact loads, and earthquake loads. Most structures are subjected to a combination of these loads. Elastic theory is used as a first approximation to estimate the stresses imposed on the soil by the applied loads assuming the soil is an elastic material. The induced stresses are distributed such that their magnitude decreases with depth and distance away from the point of load application.

Stress paths provide a useful means through which the history of loading of a soil can be followed. The effective stress changes for a linear, isotropic, elastic soil are zero under undrained loading, and the effective stress path is a vector parallel to the deviatoric stress axis with the q ordinate equal to the corresponding state on the total stress path. Porewater pressures can be computed from the difference in mean stress between the total stress path and the effective stress path at a desired value of deviatoric stress.

Practical Example

EXAMPLE 2.10 *Stress Distribution in a Soil Mass.*

The soil profile at a site is shown in Figure E2.10a. Plot (a) the depth–vertical stress increase relationship due to the vertical pressure from the fill, (b) the effective overburden pressure versus depth, and (c) the increase in vertical stress due to the footing load. The plots should extend up to the bottom of the soft silty clay layer. Groundwater level is at 2.5 m (8 ft) below the existing ground surface.

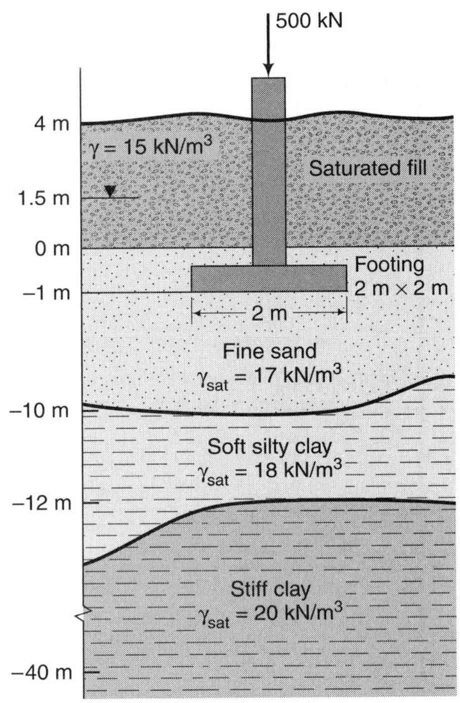

FIGURE E2.10a

Strategy It is best to set up a table to do the calculations or use a spreadsheet. The fill is normally assumed to impose a uniform stress throughout the soil layers. SI units will be used in solving this example problem.

Solution 2.10

Step 1: Set up a spreadsheet.

Table E2.10 is a table of the stresses extracted from a spreadsheet.

Step 2: Check spreadsheet calculations.

The vertical effective stress from the fill is

$$(15 \times 4) - (9.8 \times 1.5) = 45.3\,\text{kPa}$$

The effective overburden pressure at the base of the footing is $17 \times 1 - 9.8 \times 1 = 7.2\,\text{kPa}$.

Note that the height of water above the fine sand does not change the effective overburden pressure.

TABLE E2.10

	Depth m	Thickness m	γ kPa	σ_z kPa	u kPa	σ'_z kPa	σ_{fill} kPa	$\Delta\sigma_z$ kPa	$\sigma_{fill} + \Delta\sigma_z$ kPa
Ground surface	0	0	0	0	0	0	0	0	0
Fill	2.5	2.5	15	37.5	0	0	37.5	0	37.5
Fill	4	1.5	15	60.0	14.7	0	45.3	0	45.3
Fine sand (footing base)	5	1	17	31.7	24.5	7.2	45.3	125	170.3
Fine sand	7	2	17	65.7	44.1	21.6	45.3	42	87.3
Fine sand	9	2	17	99.7	63.7	36	45.3	13.5	58.8
Fine sand	11	2	17	133.7	83.3	50.4	45.3	6.4	51.7
Fine sand	13	2	17	167.7	102.9	64.8	45.3	3.6	48.9
Fine sand	14	1	17	184.7	112.7	72	45.3	2.9	48.2
Soft silty clay	15	1	18	202.7	122.5	80.2	45.3	2.3	47.6
Soft silty clay	16	1	18	220.7	132.3	88.4	45.3	1.9	47.2

The effective overburden pressure at the bottom of the soft silty clay is $[(17 \times 10) + (18 \times 2) + (9.8 \times 1.5)] - (9.8 \times 13.5) = [(17 \times 10) + (18 \times 2)] - (9.8 \times 2) = 88.4\,\text{kPa}$

The vertical stress at the base of the footing is $q_s = \frac{500}{2 \times 2} = 125\,\text{kPa}$. Use the program STRESS to find the vertical stress increase at various depths below the base. At 1 m below the footing base, m and n = 2/2 = 1. The STRESS program gives a vertical stress increase of 10.5 kPa. You need to multiply this value by 4 since the vertical stress increases are calculated for a corner of a rectangle. (See Example 2.2.)

Step 3: Plot results.

The results are shown in Fig. E2.10b.

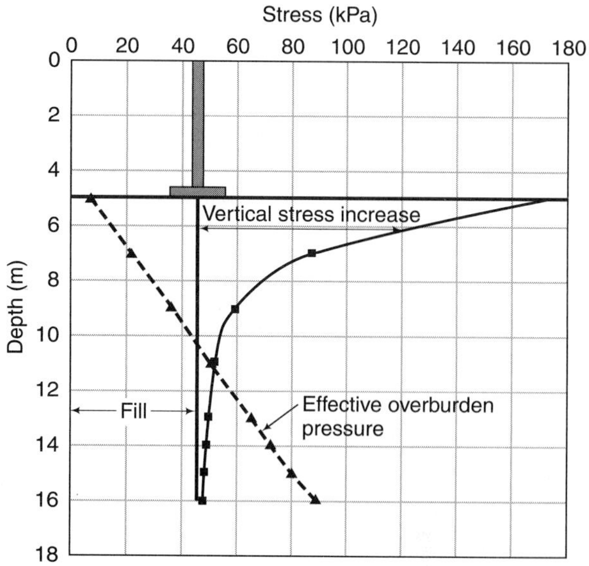

FIGURE E2.10b

EXERCISES

2.1 For a four-story office building in your area, determine the type of loads that will be applied to a foundation buried 2 m (6 ft) from the surface. According to IBC (2006) or similar, what are the values of spectral acceleration (site class B, period of 0.2 sec, and damping of 5%) and the site class for your area?

2.2 What are the differences between wind loads and seismic loads?

2.3 Discuss the assumptions made in developing the solution of the stress distribution from surface loads. How would these assumptions affect the design of a foundation?

2.4 The foundation shown in Fig. P2.4 is located on the surface of a homogeneous soil with a saturated unit weight of 18 kN/m^3 (114 pcf). The ground water is at 2 m (6 ft) below the surface. The applied vertical stress over DEFG is 150 kPa (3000 psf) and 100 kPa (2000 psf) over ABCG. Plot the distribution of (a) the current vertical effective stress and (b) the increase in vertical stress with depth up to a depth of 5 m (16 ft) under point D.

2.5 The wind pressure on a silo (Fig. P2.5) imposes a horizontal load of 400 kN (90 kips) at the base of the circular foundation of diameter 8 m (26 ft). The vertical load is 5 MN (1125 kips). Assume that the horizontal load is a point load at the center of the base of the silo.

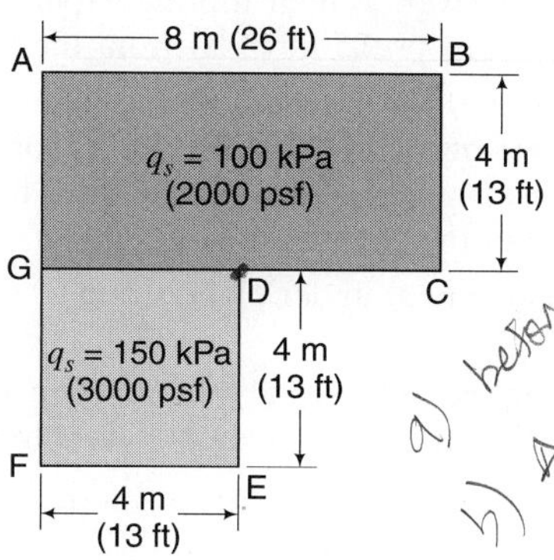

FIGURE P2.4

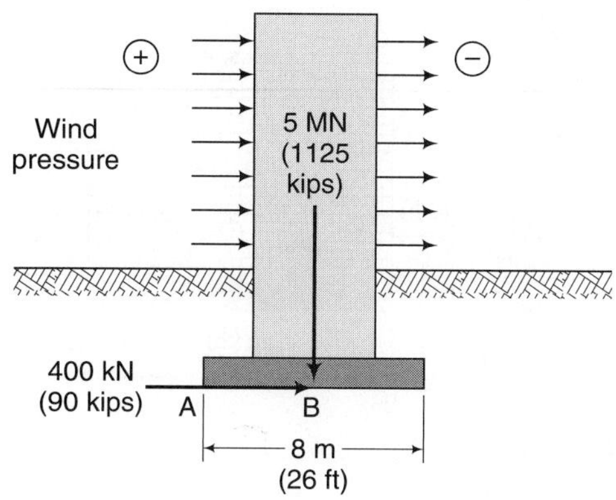

FIGURE P2.5

(a) Determine the vertical stress increases at the edge A and at the center B of the foundation at a depth of 0.5 m (2 ft) below the base.

(b) Determine the vertical elastic settlement at A and B, if $E = 30$ MPa (600 ksf) and Poisson's ratio $\nu = 0.35$.

(c) What other load condition (force, moment, etc.) is not accounted for in Fig. P2.5?

2.6 A pile (a long, slender structural element) 10 m (33 ft) long subjects the soil to a point load of 1 MN (225 kips) as shown in Fig. P2.6. Calculate the increase in the vertical and lateral stresses and the elastic settlement at a point 1 m below the centerline of the pile (Point A). Assume $E = 40$ MPa (800 ksf) and Poisson's ratio $\nu = 0.4$.

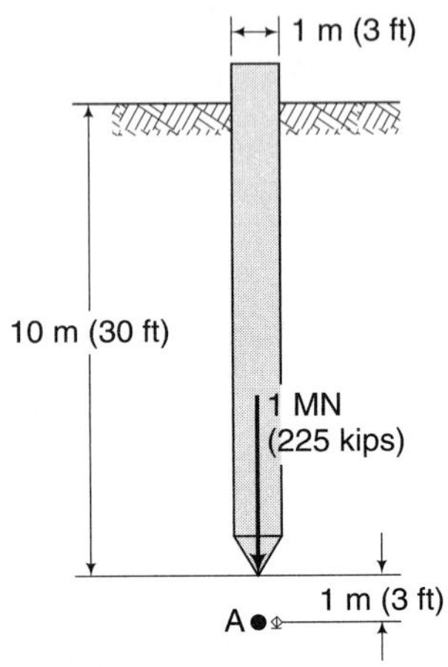

FIGURE P2.6

2.7 A square foundation 4 m (12 ft) × 4 m (12 ft) for a proposed building is to be located next to an existing foundation 10 m (30 ft) × 10 m (30 ft) (Fig. P2.7).

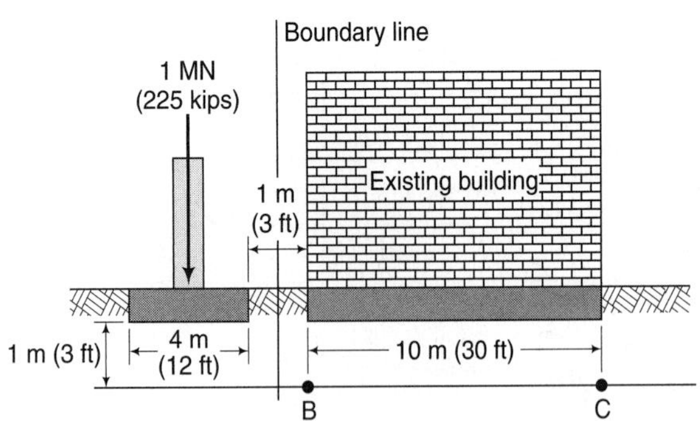

FIGURE P2.7

(a) Determine the vertical stress increases from the proposed foundation at a depth 1 m (3 ft) below the corner edges B and C of the existing foundation.

(b) Calculate the elastic settlement due to the vertical stress increases from the proposed foundation at B and C. Assume $E = 20$ MPa (400 ksf) and Poisson's ratio $\nu = 0.35$.

(c) Calculate the elastic tilt (degrees) of the existing foundation.

2.8 Plot the distribution of vertical effective stresses and the vertical total stress increase with depth up to a depth of 5 m (16 ft) under the center of the strip footing and the square footing shown in Fig. P2.8. You may use the program at http://www.wiley.com/college/budhu. Click Foundations and Earth Structures, Chapter 2, STRESS. zip or use the equations in Chapter 2 in a spreadsheet. Assume the soil is a deep deposit of saturated dense sand. Would you expect the elastic settlement of the two footings to be the same? Assume Poisson's ratio $\nu = 0.35$. Explain your answer. Which footing experiences higher stresses at a depth of 0.5 m (2 ft) from the base?

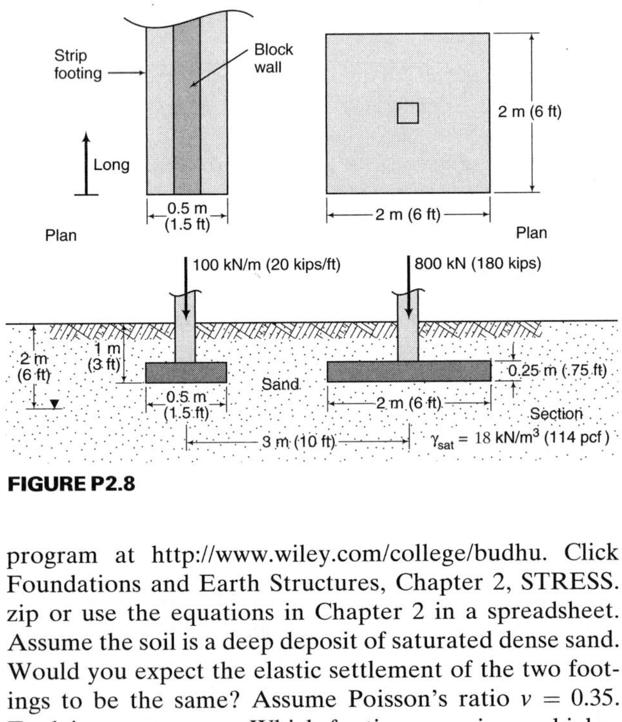

FIGURE P2.8

2.9 A circular surface foundation, 1 m (3 ft) in diameter, transmits a load of 280 kN (60 kips) to the soil as shown in Fig. P2.9. Consider a soil element at a depth 2 m (6 ft) below the centerline of the foundation.

(a) Is the soil element under plane strain or axisymmetric condition?

(b) Determine the initial total and effective stresses (vertical and lateral) on the soil element before the foundation was constructed.

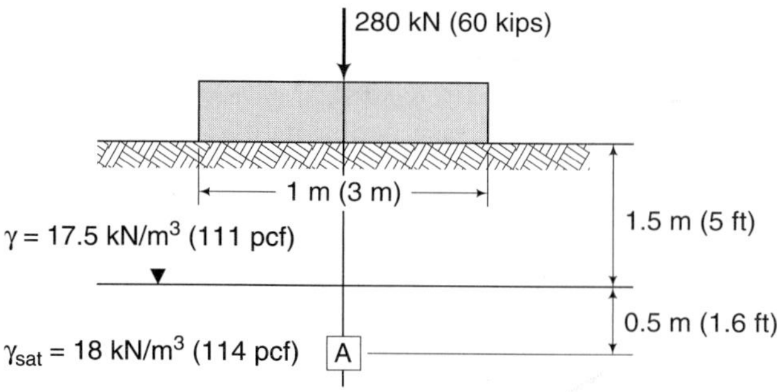

FIGURE P2.9

(c) Assume the soil is elastic, determine the stress increases (vertical, lateral, and shear) on the soil element due to the foundation load. The distribution of vertical stresses at the base of the foundation is uniform. You may use the program STRESS.

(d) Determine the principal stresses on element A after the foundation is constructed. You may use the program at http://www.wiley.com/college/budhu. Click Foundations and Earth Structures, Chapter 2, mohrcicle.zip.

(e) How much did the plane on which the major principal stress acts rotate after the foundation was constructed?

(f) Plot the total stress path for element A in (p, q) and (t, s) spaces.

(g) Plot the effective stress path for element A assuming the soil to be elastic. You should use dual labels p', p and s, s' on the x-axis.

(h) Calculate the excess porewater pressures for both (p, q) and (s, t) spaces.

2.10 An embankment 6 m (20 ft) high with a top width of 14 m (45 ft) and base width of 22 m (75 ft) is to be constructed from a soil compacted to a unit weight of 18 kN/m^3 (114 pcf). The saturated unit weight of the soil beneath the embankment is 20 kN/m^3 (125 pcf) and the groundwater level is at the soil surface. Assume the soil under the embankment is linearly elastic.

(a) Sketch a diagram of the embankment and the soil.

(b) Calculate the mean total stress increase, Δp, and the deviatoric strain increase, Δq, for an element of soil 1 m (3 ft) below the center of the embankment. Use Fig. 2.12 in calculating the vertical stress increase.

(c) Plot the total stress paths in (p, q) and (s, t) spaces for the soil element.

(d) Plot the effective stress paths on (p', q) and (s', t) spaces for the soil element.

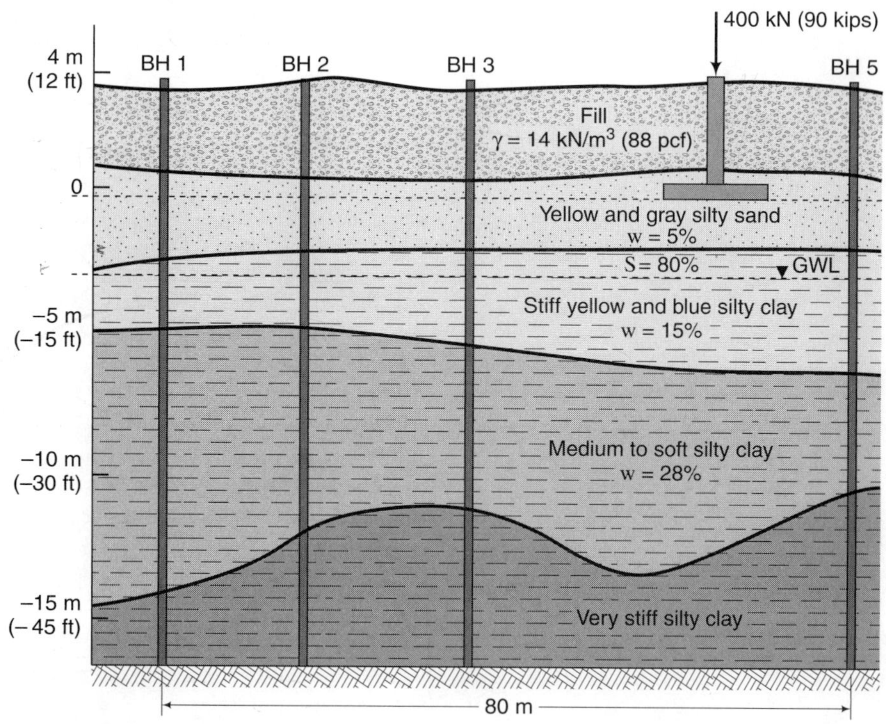

FIGURE P2.11

(e) Calculate the excess porewater pressure from the stress paths in (p, q) and (s, t) spaces.

2.11 A square footing of width 2 m (6 ft) is located in the soil profile at a site as shown in Fig. P2.11. Plot the distributions of (a) the vertical stress due to the vertical pressure from the fill, (b) the overburden effective pressure, and (c) the increase in vertical stress with depth due to the footing load up to a depth of 10 m (30 ft).

CHAPTER 3

ANALYTICAL, NUMERICAL, AND OBSERVATIONAL METHODS FOR FOUNDATION DESIGN

3.0 INTRODUCTION

The fundamental basis of all analyses in geotechnical engineering is engineering mechanics. In engineering mechanics, we decompose complex bodies into various idealized components (free bodies) to conduct the analyses. In foundation analyses, we do the same. However, soils are complex natural materials and finding the appropriate idealization to capture the soil responses when loaded by a foundation system is often difficult. Consequently, we often resort to simple idealizations and call on experience to make decisions on foundation design.

In this chapter, we will briefly discuss some popular methods of foundation analyses. This discussion is intended to provide you with the basis for some of the equations that you will later use in foundation analyses and design. Regardless of the type of analyses that you use, you should bear in mind that all the analyses involve assumptions that (grossly) simplify soil behavior. Foundation analyses are merely tools to give you indications about the geometry of a geotechnical system with a certain risk of failure. When you complete this chapter, you should be able to:

- Understand the concept of yielding and failure criteria for soils.
- Understand the basic features and assumptions of some popular methods of analyses.
- Conduct foundation analyses using limit equilibrium and limit analyses.
- Understand the use and implications of numerical analyses such as the finite difference method and the finite element method.
- Understand the importance of observational methods.

3.1 DEFINITIONS OF KEY TERMS

Collapse is a sudden decrease in the shear strength of a soil.

Elastic (ideal) materials return to their original configuration when unloaded.

Failure is a constant stress state and constant volume condition achieved by a soil under continued shearing.

Isotropic means the same response or loading in all directions.

Plastic (ideal) materials do not return to their original configuration on unloading.

Yielding occurs when the elastic limit is exceeded.

Limit equilibrium is a method of analysis through which you speculate on a failure mechanism and then use statics to determine the failure or the collapse loads.

Limit analysis is a method by which a failure criterion and the stress state in a body are used to determine limiting failure loads.

Upper bound is where the calculated load is likely higher than the failure load.

Lower bound is where the calculated load is likely lower than the failure load.

3.2 STRESS-STRAIN RESPONSE OF IDEAL MATERIALS

Importance *To conduct analyses we make assumptions on how they would respond to loading based on the stress-strain responses of ideal materials. You need to know the characteristics of these materials and their limitations to judge which one can best represent your soil under the loading conditions imposed by a foundation. In this section, we will study the response of ideal materials to loads that can be used to mimic soil behavior.*

There are two basic classes of ideal materials behavior that are frequently used in analyzing soils under foundation loads. These are *elastic* and *plastic* materials. Within each class there are subclasses. These subclasses can be combined to replicate soil behavior. The mathematical relationships to simulate the stress-strain responses are called *constitutive* relationships. These are the backbone of soil models. We present below the two classes and the common subclasses used in soil models.

3.2.1 Elastic Materials

Perfect elastic materials follow the same load-displacement or stress-strain path on loading and unloading. Their stress-strain relationships can be linear or nonlinear (Fig. 3.1). Isotropic elastic materials show the same stress-strain responses irrespective of the loading direction. Anisotropic elastic materials exhibit different stress-strain responses depending on the direction of loading. Elastic materials obey the principle of superposition; the order in which the load is applied is not important and the total displacement is the sum of the displacements caused by each loading. Hooke's law is used to relate elastic stresses to elastic strains through Young's modulus, also called *elastic modulus*. The elastic response of soils occurs at very small shear strains—less than 0.001%. However, in estimating soil strains in practice, especially in preliminary design, geotechnical engineers often assume elastic behavior even at large strains. At small strains levels, the appropriate elastic modulus to use is the initial elastic modulus (Fig. 3.1). At large strains, the secant or tangent modulus is used.

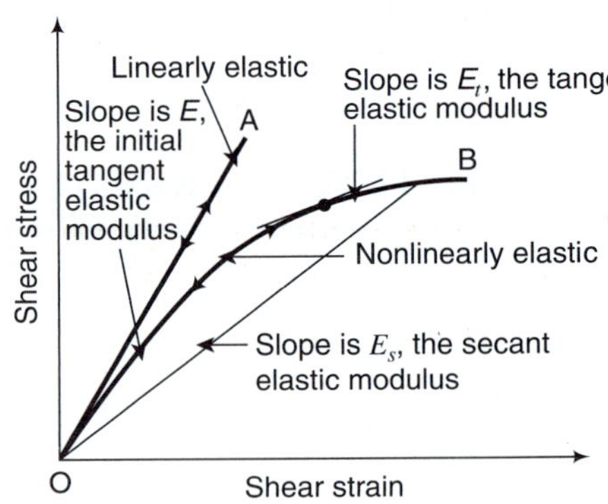

FIGURE 3.1 Shear stress-shear strain response of elastic materials.

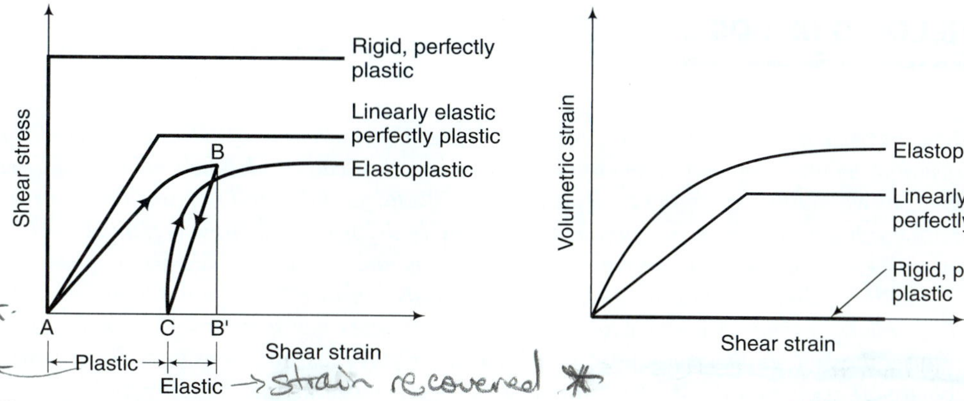

FIGURE 3.2 Shear stress-shear strain responses of ideal plastic and elastoplastic materials.

3.2.2 Plastic Materials

Plastic materials exhibit initial responses that are rigid or elastic or viscoelastic. We will consider rigid–perfectly plastic and elastoplastic materials in this book. Rigid–perfectly plastic materials do not show deformation as loading is increased and then suddenly achieve a perfectly plastic state (a state in which the material undergoes permanent deformation), as shown in Fig. 3.2. A linearly elastic, perfectly plastic material shows an initial linearly elastic response and thereafter reaches a plastic state (Fig. 3.2). If the initial response is nonlinear, the material is called a *nonlinear elastic, perfectly plastic* material. Elastoplastic materials show an initial elastic stress-strain response and then a predominantly plastic stress-strain response (Fig. 3.2).

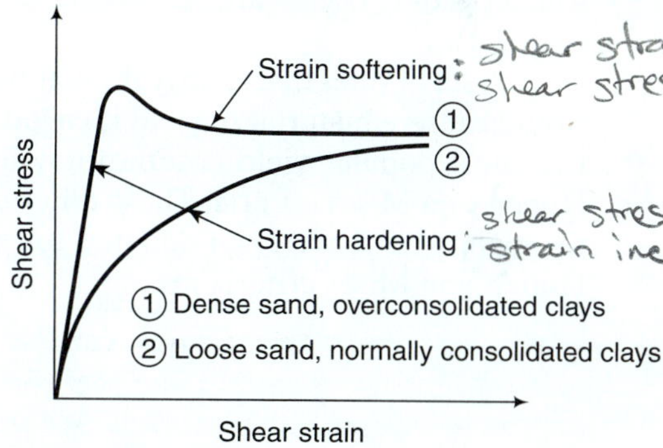

FIGURE 3.3 Strain-hardening and strain-softening.

When an elastoplastic material is unloaded after reaching the plastic state, part of the total strain is recoverable (elastic strain) as shown by B′C in Fig. 3.2, and the other part is the permanent strain (plastic strain) as shown by AC in Fig. 3.2. The stress at which permanent deformation sets in is called the *yield stress*. For most soils, the yield stress is very low (see Section 3.3).

Some soils, for example, loose sand and normally consolidated clays, show a smoothly increasing elastoplastic stress-strain relationship as shown in Fig. 3.3. This type of relationship is described as *strain-hardening*—continuous increase in shear stress with increase in shear strains. Other types of soils, such as dense sand or overconsolidated clays, show a peak shear stress and then the shear stress decreases with shear strain. The latter behavior is termed *strain-softening*. The stress-strain response before the peak shear stress is attained is strain-hardening whereas after peak shear stress the response is strain softening (Fig. 3.3).

Rigid–perfectly plastic material behavior is a popular choice in geotechnical engineering to model soil failure for simplicity.

The* key points *are:

1. ***Perfect elastic materials follow the same load-displacement or stress-strain path on loading and unloading. The stress-strain path could be linear or nonlinear.***
2. ***Plastic materials exhibit initial responses that are rigid or elastic or viscoelastic. Plastic deformations are permanent.***

3.3 YIELDING OF SOILS

Importance *Geotechnical engineers need to make predictions of the amount of deformation that would occur under working loads. If the soil were to behave elastically under the working load, then a simple elastic analysis with only two required soil parameters (elastic modulus and Poisson's ratio, or shear modulus and bulk modulus) can be used to predict the amount of deformation. However, if the working load causes the soil to yield, as in most practical situations, permanent deformations would occur and more involved analysis would be required. It is these permanent deformations that are of particular interest to geotechnical engineers. To predict soil yielding, we have to develop or adopt some rules or criteria that can let us estimate the loads or stresses that will initiate yielding. Any rule we use involves assumptions and you need to be aware of these when you apply that rule to soils. In this section, we will study the fundamentals of soil yielding.*

Yielding of a soil occurs when the elastic limit has been exceeded. Unlike steel, the elastic limit or yield point of soils is very difficult to distinguish, because the shear strains at which yielding occurs are very small, <0.001% for many soils (Fig. 3.4).

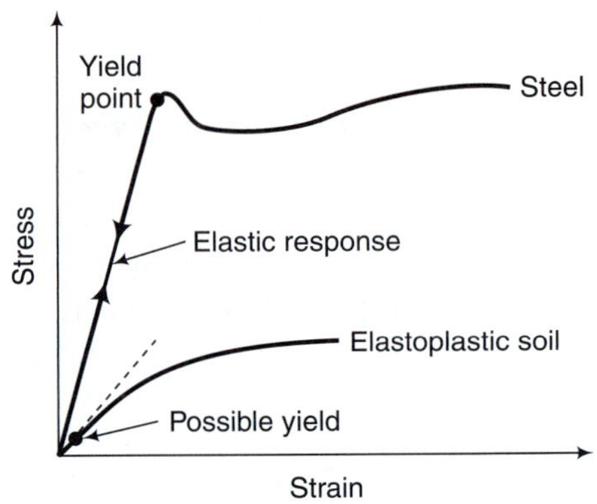

FIGURE 3.4 Yielding of soil compared to steel.

A set of rules that specify the combination of stresses that would cause a material to yield is called a *yield criterion*. The two most popular yield criteria for soils are the Tresca and Huber–von Mises criteria. These criteria have been proposed for metals but have been adopted for soils. The Tresca and Huber–von Mises criteria are

$$\text{Tresca: } \tau_y = \pm\frac{\sigma_1 - \sigma_2}{2}; \quad \tau_y = \pm\frac{\sigma_2 - \sigma_3}{2}; \quad \tau_y = \pm\frac{\sigma_3 - \sigma_1}{2} \tag{3.1}$$

$$\text{Huber–von Mises: } \tau_y = \frac{1}{2\sqrt{2}}\sqrt{[(\sigma_1 - \sigma_2)^2 + (\sigma_2 - \sigma_3)^2 + (\sigma_3 - \sigma_1)^2]} \tag{3.2}$$

where τ_y is the shear stress at yield, and σ_1, σ_2, and σ_3 are principal stresses. The Tresca yield criterion states that yielding would occur at one-half the maximum principal stress difference. The Huber–von Mises criterion states that yielding would occur at one-half the square root of the sum of the squares of one-half the principal stress differences. Tresca yield criterion is more conservative than the Huber-von Mises yield criterion.

The* key points *are:

1. ***Yielding of a soil occurs when the elastic limit has been exceeded.***
2. ***The yield stress for soil occurs at very small strain levels and is often difficult to discern from soil test results.***
3. ***A yield criterion is a set of rules defining the combination of stresses that would cause a material to yield.***
4. ***The Tresca and Huber–von Mises yield criteria, originally developed for metals, are often used in modeling soil yielding.***
5. ***The Tresca yield criterion normally gives conservative predictions compared with the Huber–von Mises criterion***

3.4 FAILURE CRITERIA

Importance *One of the main tasks for geotechnical engineers is to predict (estimate) the loads or stresses that will cause a soil to fail and then design foundations such that only a fraction of these loads or stresses are mobilized. Because of the complexity of soil behavior and the variety of loadings that may be applied to the soils, it is difficult to formulate a single criterion that can accurately predict soil failure. As a result, there are several failure theories (criteria) based on simplifications of soil behavior. For example, some failure criteria are independent of the deformation that occurs in a soil and treat the soil as if it were a rigid body. You need to understand the basis for the popular failure criteria to help you make decision as to which one of them is best for your situation. In this section, we will discuss failure criteria that are relevant to foundation engineering.*

The term *failure* is used very loosely in the literature and has many connotations such as *instability*, *yielding*, *collapse*, and *violation of certain design requirements*. In this book, failure of soils is defined as *the attainment of a constant state of stress without any further change of volume*. This is called *critical state*. Collapse is the sudden decrease in shear strength. The peak shear strength defines incipient collapse. We will discuss failure criteria based on the loose definition but we will indicate, where appropriate, when they strictly refer to collapse or failure as defined in this book. The five common failure criteria for soils are

1. Coulomb
2. Mohr-Coulomb
3. Tresca
4. Taylor
5. Critical state

3.4.1 Coulomb

The Coulomb failure criterion states that failure would occur by sliding on a plane between two rigid bodies. The shear stress at failure is the normal effective stress on the failure plane times the interface friction coefficient. Coulomb's equation for the failure shear strength of a soil, τ_f, is

$$\tau_f = (\sigma'_n)_f \tan\phi' \tag{3.3}$$

where $(\sigma'_n)_f$ is the normal effective stress on the failure plane, and ϕ' is the friction angle. The Coulomb's failure criterion in (σ'_n, τ) space is shown in Fig. 3.5.

The Coulomb failure criterion applies to rigid–perfectly plastic materials (i.e., no deformation occurs prior to failure). This criterion is useful to estimate failure in layered or fissured overconsolidated soils or in soils where a prefailure plane exists.

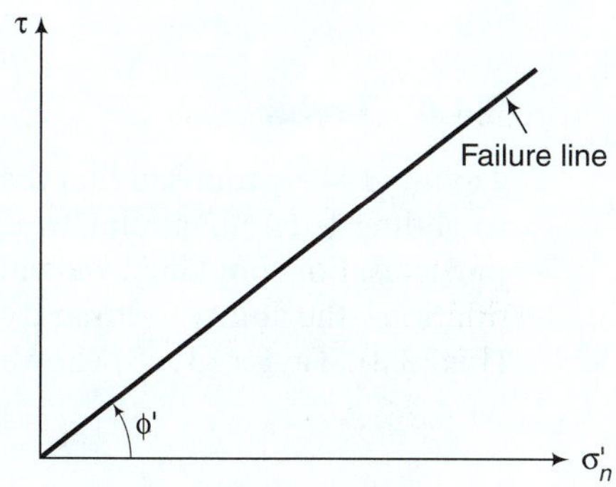

FIGURE 3.5 Coulomb's failure criterion.

3.4.2 Mohr-Coulomb

The Mohr-Coulomb failure criterion states that failure of a soil occurs by sliding on planes of maximum stress obliquity expressed as

$$\left(\frac{\sigma'_1}{\sigma'_3}\right)_{max} = \frac{1+\sin\phi'}{1-\sin\phi'} \quad \text{or} \quad \sin\phi' = \left(\frac{\sigma'_1-\sigma'_3}{\sigma'_1+\sigma'_3}\right)_{max} \tag{3.4}$$

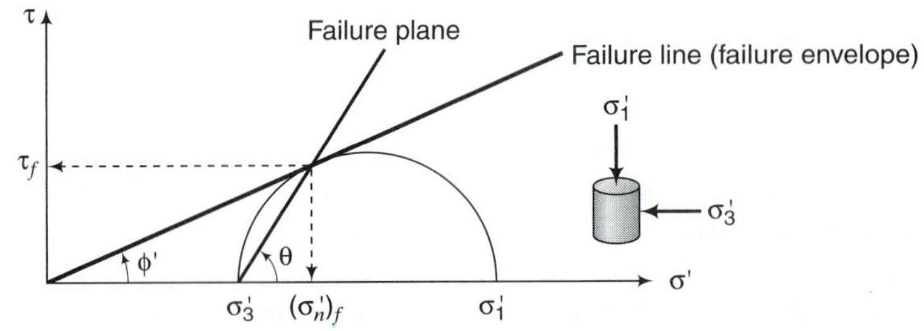

FIGURE 3.6 Mohr-Coulomb failure criterion.

The Mohr-Coulomb failure criterion is shown in Fig. 3.6 and applies to rigid–perfectly plastic materials (i.e., no deformation occurs prior to failure). The failure line (failure envelope) is the line linking the failure stresses. The failure plane is inclined to the plane of the major principal stress (in this case, the horizontal plane) at an angle, $\theta = 45^\circ + \frac{\phi'}{2}$. The Mohr-Coulomb criterion is used to interpret the long-term failure of soils.

3.4.3 Tresca

The Tresca failure criterion is really the Tresca yield criterion that has been adopted as a failure criterion. The shear stress to cause failure according to Tresca's criterion is given as

$$\tau_{max} = \left(\frac{\sigma_1 - \sigma_3}{2}\right)_{max} \quad \text{or} \quad \tau_{max} = \left(\frac{\sigma_2 - \sigma_3}{2}\right)_{max} \quad \text{or}$$
$$\tau_{max} = \left(\frac{\sigma_3 - \sigma_1}{2}\right)_{max} \tag{3.5}$$

On a Mohr's circle of stress, Tresca's failure criterion is represented by the radius of the failure stress circle (Fig. 3.7) and applies to rigid–perfectly plastic materials. Tresca's failure criterion is used to determine the undrained shear strength of soils, s_u, that is used for short-term analysis of geosystems. In soils, we generally write Tresca's failure criterion as

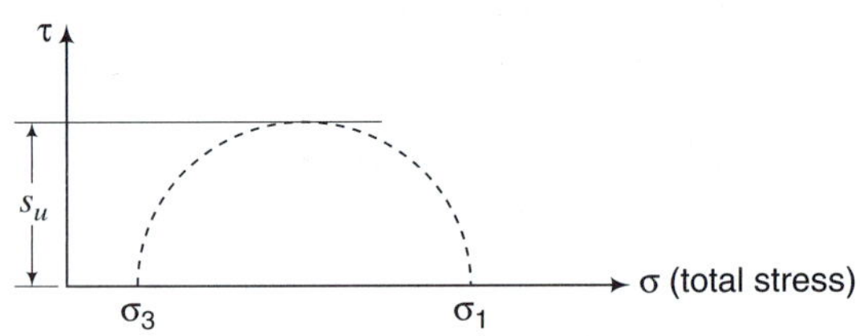

FIGURE 3.7 Tresca's failure criterion interpreted using Mohr's circle of stress.

$$s_u = \left(\frac{\sigma_1 - \sigma_3}{2}\right)_{max} \tag{3.6}$$

3.4.4 Taylor

Taylor (1948) proposed that the shear strength of soil consists of two components. One component is due to sliding friction, similar to Coulomb's failure criterion, and the other is due to the interlocking of particles. For simplicity, we will assume here that the interlocking component can be represented by soil dilation—the ability to expand on shearing. Soil dilation will be represented by the peak dilation angle, α_p (Fig. 3.8). Taylor (1948) showed, with reference to Fig. 3.8, that

$$\left(\frac{\tau}{\sigma'_z}\right)_{max} + \left(\frac{\Delta\varepsilon_z}{\Delta\gamma}\right) = \left(\frac{\tau}{\sigma'_z}\right)_{cs} \tag{3.7}$$

where the subscript cs denotes critical state or failure and the subscript max denotes the maximum or peak, $\Delta\varepsilon_z$ is the vertical strain increase (negative for upward movement), and $\Delta\gamma$ is the shear strain increase. Equation (3.7) can be written as

$$\tan\phi'_p = \tan\phi'_{cs} + \tan\alpha_p \tag{3.8}$$

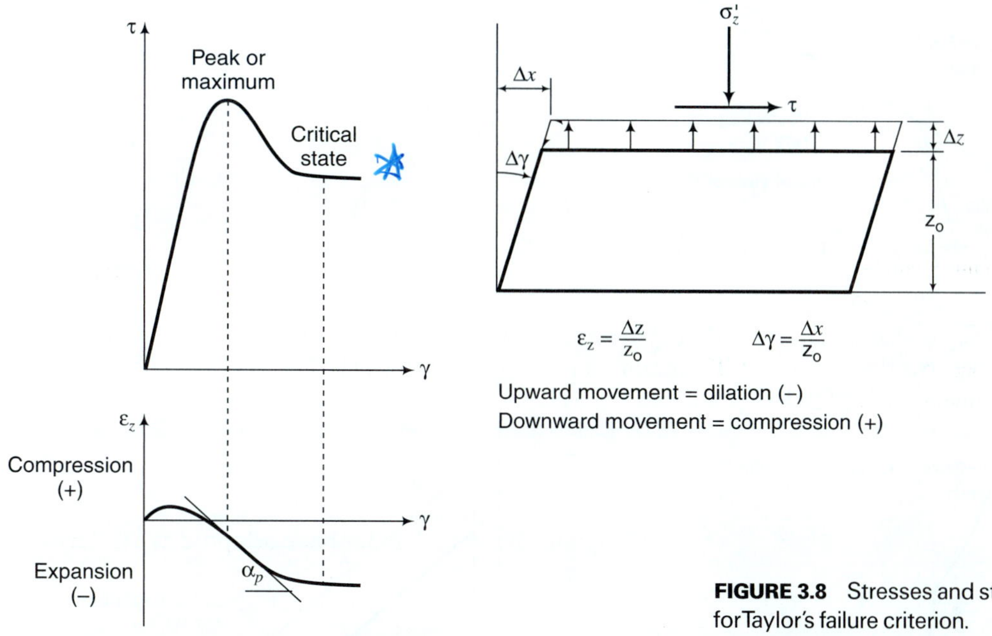

FIGURE 3.8 Stresses and strains considered for Taylor's failure criterion.

where the peak dilation angle is $\alpha_p = \left|\tan^{-1}\left(\frac{\Delta\varepsilon_z}{\Delta\gamma}\right)_p\right|$. Taylor's failure criterion does not require the assumption of any physical mechanism of failure, such as a plane of sliding. It can be applied at every stage of loading for soils that are homogeneous and deform under plane strain conditions.

3.4.5 Critical State

> Access a spreadsheet at http://www.wiley.com/college/budhu. Click Foundations and Earth Structures, Chapter 3, criticalstate.xls to calculate yield and failure stresses based on critical state.

The *critical state failure* criterion states that failure would occur when a constant stress state at constant volume is attained under increasing shearing strains. The critical state failure criterion includes the volumetric response of the soil in defining failure similar to Taylor's failure criterion. According to the critical state criterion, the failure stress state is insufficient to guarantee failure; the soil structure must also be loose enough. The soil mass is treated as an elastoplastic material. The failure state is defined by a straight line (critical state line, CSL) in (p', q) space (Fig. 3.9a), given as

$$q_f = Mp'_f \tag{3.9}$$

where q_f is the deviatoric stress at failure (similar to τ_f), M is a friction constant (similar to $\tan\phi'_{cs}$), and p'_f is the mean effective stress at failure (similar to σ'_n). For compression, $M = M_C$ and for extension $M = M_E$. For triaxial compression (axisymmetric compression),

$$M = M_C = \frac{6\sin\phi'_{cs}}{3 - \sin\phi'_{cs}} \tag{3.10}$$

and for triaxial extension (axisymmetric extension),

$$M = M_E = \frac{6\sin\phi'_{cs}}{3 + \sin\phi'_{cs}} \tag{3.11}$$

An important point to note is that while the friction angle, ϕ'_{cs}, is the same for compression and extension, the slope of the critical state line in (p', q) space is not equal. Therefore, the failure deviatoric

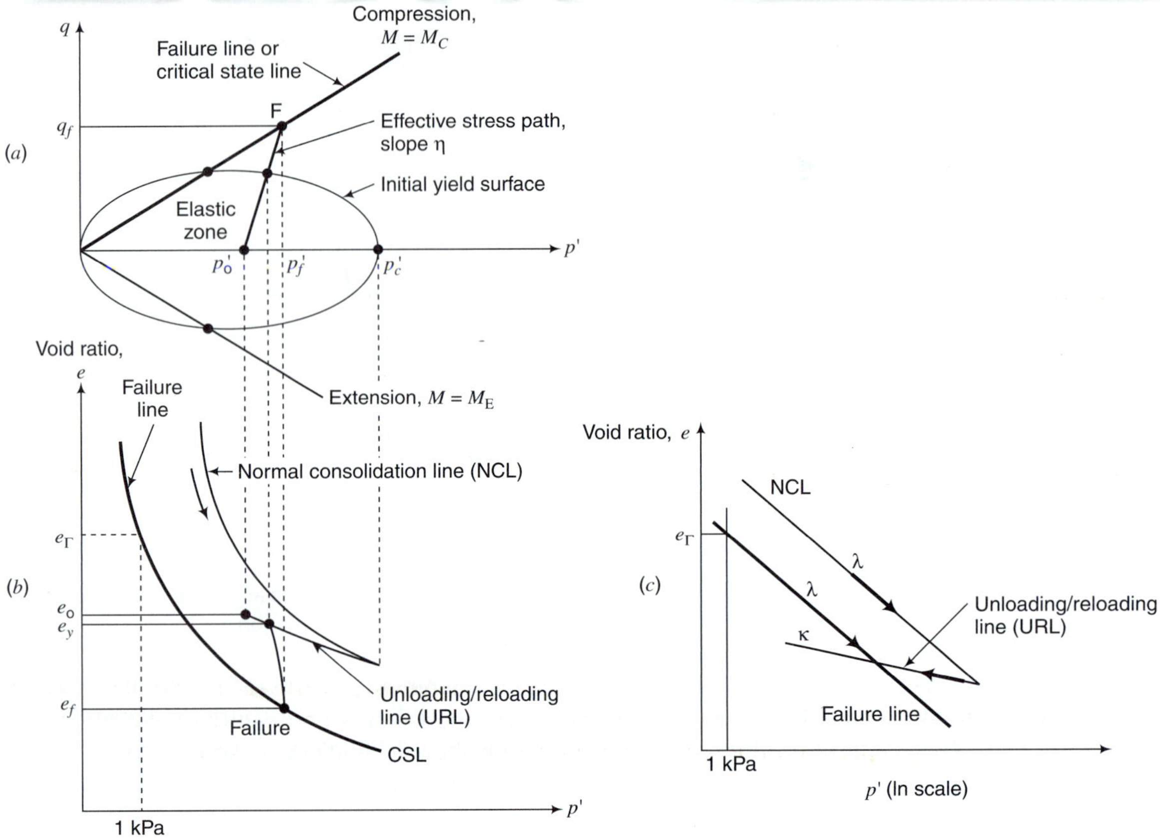

FIGURE 3.9 Critical state failure criterion.

stress in compression and extension is different. Since $M_E < M_C$, the failure deviatoric stress of a soil in extension is lower than that for the same soil in compression.

The failure line in (p', e) space (Figs. 3.9b,c) is represented by

$$e_f = e_\Gamma - \lambda \ln p'_f \tag{3.12}$$

where e_f is the void ratio at failure (critical state), e_Γ is the void ratio on the critical state line when $p' = 1$ kPa, and λ is the compression index (similar to C_c; see Chapter 4) obtained from the slope of the normal consolidation line. The value of e_Γ depends on the units chosen for the p' scale. In this book, we will use kPa for the units of p'. In a plot of p' (ln scale) versus void ratio (arithmetic scale), the failure line is a straight line of slope λ (Fig. 3.9c). The void ratio e_Γ at $p = 1$ kPa is,

$$e_\Gamma = e_o + (\lambda - \kappa)\ln\frac{p'_c}{2} + \kappa \ln p'_o \tag{3.13}$$

where e_o is the initial void ratio, p'_c is the preconsolidation mean effective stress (similar to OCR), p'_o is the initial mean effective stress, and κ is the recompression index (similar to C_r). The initial preconsolidation mean effective stress defines the size of the initial elastic zone (Fig. 3.9). Stress states within the elastic zone cause the soil to behave elastically while stress states outside of the elastic zone cause the soil to behave elastoplastically.

Every point on the failure line in (p', q) space has a corresponding point on the failure line (p', e) space. In reality, the failure surface is a single surface in (p', q, e) space. However, in problem solving, it is easier to sketch and visualize failure according to critical state by using the three illustrated plots in Fig. 3.9.

To use the critical state failure criterion, you must also know p'_o, p'_c, e_o (or w), ϕ'_{cs}, λ, and κ. All of these are obtained from conventional soil tests. The relationships between λ and C_c and κ and C_r are

$$\lambda = \frac{C_c}{2.3} \tag{3.14}$$

and

$$\kappa = \frac{C_r}{2.3} \tag{3.15}$$

Failure occurs when the effective stress path reaches the failure line. For isotropically consolidated soils under undrained loading condition (no change in volume), $e_o = e_f$, and Eq. (3.12) gives

$$p'_f = \exp\left(\frac{e_\Gamma - e_o}{\lambda}\right) \tag{3.16}$$

Since $q_f = Mp'_f$, then

$$q_f = M\exp\left(\frac{e_\Gamma - e_o}{\lambda}\right) \tag{3.17}$$

The undrained shear strength for normally consolidated ($R_o = \frac{p'_c}{p'_o} = 1$) and lightly consolidated soils ($1 < R_o \leq 2$) is

$$s_u = \frac{q_f}{2} = \frac{M}{2}\exp\left(\frac{e_\Gamma - e_o}{\lambda}\right) \tag{3.18}$$

For heavily overconsolidated soils,

$$s_u = \frac{M}{2}\,p'_o\sqrt{R_o - 1}; \quad R_o > 2 \tag{3.19}$$

For a given soil, M, λ, and e_Γ are constants and the only variable in Eq. (3.18) is the initial void ratio. Therefore, the undrained shear strength of a saturated soil depends only on its initial void ratio.

For drained isotropically consolidated loading condition, the effective and total stress paths are the same. Failure occurs when the effective stress path intersects the failure line. The failure stresses are

$$p'_f = \frac{\eta p'_o}{\eta - M} \tag{3.20}$$

$$q_f = \frac{\eta M p'_o}{\eta - M} \tag{3.21}$$

where the slope of the effective stress path is $\frac{q}{p'} = \eta$. For conventional triaxial test, $\eta = 3$.

The critical state failure criterion can be applied to all soils but particularly to soft, normally consolidated clays. It accounts for (a) the initial stress history of the soil, (b) soil deformations, and (c) stress path dependency. All of these are features of the behavior of real soils.

The key points are:

1. ***The Coulomb failure criterion states that failure would occur by sliding on a plane between two rigid bodies. This criterion can be applied to layered or fissured overconsolidated soils or to soils where a prefailure plane exists.***
2. ***The Mohr-Coulomb failure criterion states that failure of a soil occurs by sliding on planes of maximum stress obliquity. The criterion is applied to interpret the long-term failure of soils.***
3. ***The Tresca criterion states that failure would occur at one-half the maximum principal stress difference. This criterion is used to interpret the undrained shear strength of fine-grained soils.***
4. ***Taylor's failure criterion states that the shear strength of the soil is the sum of the sliding frictional strength and strength due to interlocking of the particles.***
5. ***The critical state criterion states that a given soil will fail at a constant stress state and at a constant volume. According to this criterion, the maximum stress state is not sufficient to define failure; the soil must also be loose enough. This criterion can be applied to all soils under***

different drainage and loading conditions. It is particularly suitable for soft, normally consolidated clays.

6. ***All of the criteria presented, except the critical state criterion and Taylor's failure criterion, apply to rigid–perfectly plastic materials. They do not account for such soil behavioral features as stress history and stress path dependency. The critical state failure criterion accounts not only for these features but also soil deformations. In the critical state failure criterion, the soil mass is treated as an elastoplastic material.***

EXAMPLE 3.1 *Failure Stresses According to Tresca Criterion*

A cylindrical soil of diameter 50 mm (2 in) and length 100 mm (4 in) was axially loaded. The maximum load sustained is 196.3 N (44 lb). Calculate the failure shear stress according to Tresca's criterion.

Strategy The test done is known as an *unconfined compression* test. In this case, the axial stress is the major principal stress and the intermediate and minor principal stresses are zero. U.S. units will be used for the solution of this problem.

Solution 3.1

Step 1: Calculate the axial stress.

$$P = 44\,\text{lb}; \quad \text{Area, A} = \pi \times \frac{2^2}{4} = 3.14\,\text{in}^2$$

$$\sigma_1 = \frac{P}{A} = \frac{44}{3.14} = 14\,\text{psi} = 2016\,\text{psf}$$

$$\sigma_2 = \sigma_3 = 0$$

Step 2: Apply Tresca's failure criterion.

$$\text{Tresca:} \quad \tau_f = s_u = \frac{\sigma_1 - \sigma_3}{2} = \frac{2016 - 0}{2} = 1008\,\text{psf} \approx 1\,\text{ksf}$$

■

EXAMPLE 3.2 *Back Calculation of the Friction Angle Using Mohr-Coulomb Failure Criterion*

A tank foundation was loaded incrementally and the soil failed. For a soil element under the center of the tank, the lateral total stress is estimated to be 0.4 times the initial vertical total stress. If the excess porewater pressure is 0.2 times the stress difference (deviatoric stress), estimate the friction angle using the Mohr-Coulomb criterion.

Strategy The Mohr-Coulomb failure criterion is based on effective stresses. You need to calculate the major and minor principal effective stresses in terms of the total initial stress and then apply the Mohr-Coulomb failure criterion. SI units will be used for the solution of this problem.

Solution 3.2

Step 1: Calculate the major and minor principal stresses.

$$\sigma_3 = 0.4\,\sigma_1$$
$$\Delta u = 0.2(\sigma_1 - \sigma_3) = 0.2(\sigma_1 - 0.4\,\sigma_1) = 0.12\,\sigma_1$$
$$\sigma_3' = 0.4\,\sigma_1 - 0.12\,\sigma_1 = 0.28\,\sigma_1$$
$$\sigma_1' = \sigma_1 - 0.12\,\sigma_1 = 0.88\,\sigma_1$$

Step 2: Apply the Mohr-Coulomb failure criterion.

$$\sin\phi' = \frac{\sigma_1' - \sigma_3'}{\sigma_1' + \sigma_3'}$$

$$\phi' = \sin^{-1}\left(\frac{0.88\sigma_1 - 0.28\sigma_1}{0.88\sigma_1 + 0.28\sigma_1}\right) = \sin^{-1}\left(\frac{0.6\sigma_1}{1.16\sigma_1}\right) = 31.1°$$

■

EXAMPLE 3.3 *Prediction of Failure Stresses Using Critical State Failure Criterion*

A saturated soil element under the center of a large square foundation has a void ratio of 1.0 and a compression index, $\lambda = 0.3$. (a) Calculate the mean effective and deviatoric stresses for failure under undrained condition using the critical state model and assuming the soil was normally consolidated. (b) Predict the undrained shear strength of the soil. The frictional constant $M = 1.0$ and $e_\Gamma = 2.2$.

Strategy Since you are dealing with an undrained condition, Eq. (3.18) applies. As a reminder, under undrained conditions in a saturated soil, the void ratio (or soil volume) does not change.

Solution 3.3

Step 1: Calculate p_f'.

$$p_f' = \exp\left(\frac{e_\Gamma - e_o}{\lambda}\right) = \exp\left(\frac{2.2 - 1.0}{0.3}\right) = 54.6\,\text{kPa}$$

Step 2: Calculate q_f.

$$q_f = Mp_f' = 1.0 \times 54.6 = 54.6\,\text{kPa}$$

Step 3: Calculate the undrained shear strength.

$$s_u = \frac{q_f}{2} = \frac{54.6}{2} = 27.3\,\text{kPa}$$

■

3.5 LIMIT EQUILIBRIUM

Importance *Limit equilibrium is the most popular method of analysis in geotechnical engineering. It is the basis for many of the equations that you will use in foundation analyses and design (Chapters 7 through 10). You need to understand the fundamentals of this method and its limitations.*

The *limit equilibrium method* is an iterative process in which you speculate on possible failure mechanisms and then use the equilibrium equations for each mechanism to determine the collapse or failure load. The least load to cause failure is the *ultimate limit state.* Field and experimental observations often help to narrow the range of possible failure mechanisms. Equilibrium is satisfied in only a limited way. For example, a solution of a problem may be obtained only by satisfying moment equilibrium, disregarding force equilibrium and assuming a failure criterion such as Coulomb's law. The soil is assumed to be a homogeneous, rigid–perfectly plastic material.

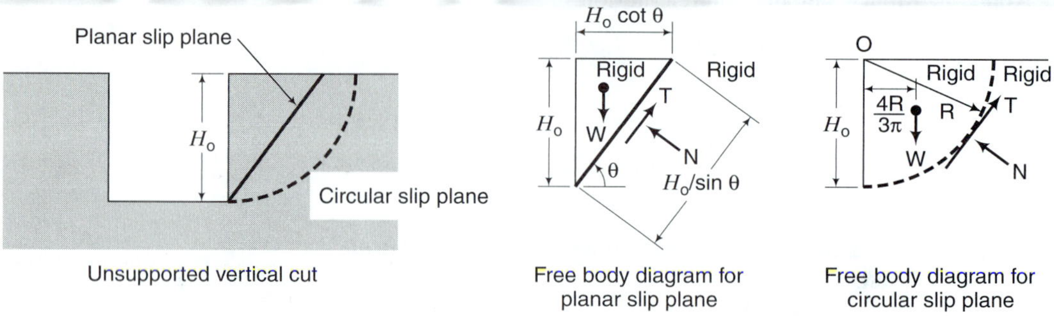

FIGURE 3.10 Limit equilibrium applied to a vertical cut.

Let's assume that we wish to make a vertical cut in a stiff clay (Fig. 3.10) to construct a foundation and we would like to have the cut unsupported for a short time (undrained condition). We are interested in the maximum height of the cut that we can make without any bracing system. Let's use the limit equilibrium method to find the maximum depth of the cut. We can speculate that the vertical cut could possibly fail by sliding of a wedge of soil or by sliding of a quarter-circular mass of soil. These are only two of several plausible mechanisms. We can assume that each failed mass of soil is a homogeneous, rigid body sliding on a slip plane consistent with Coulomb's law. The next step is to draw a free body diagram of the failed mass and use the equilibrium equations to solve for the maximum depth. First, we will consider the planar slip surface (Case 1). The soil resistance per unit length along the slip plane is the undrained shear strength (s_u) times the length of the slip plane (i.e., $T = s_u \frac{H_o}{\sin\theta}$). Recall that we are considering short-term loading so that the appropriate shear strength parameter is s_u. The weight of the wedge of soil is $W = \frac{1}{2}\gamma H_o^2 \cot\theta$.

Equilibrium of the forces along the plane gives

$$T = W \sin\theta \tag{3.22}$$

Substituting for T and W in Eq. (3.22) gives

$$s_u \frac{H_o}{\sin\theta} = \frac{1}{2}\gamma H_o^2 \cot\theta \sin\theta$$

Simplifying, we get

$$H_o = \frac{4 s_u}{\gamma \sin 2\theta} \tag{3.23}$$

For the critical height, we differentiate H_o with respect to θ and set the result equal to zero.

$$\frac{dH_o}{d\theta} = \frac{4 s_u}{\gamma}[-2\cot(2\theta)\csc(2\theta)] = 0$$

The solution is $\theta = \theta_{cr} = 45°$, which on substitution into Eq. (3.23) gives

$$(H_o)_{critical} = \frac{4 s_u}{\gamma} \tag{3.24}$$

Now, let us consider another possible mechanism—the quarter-circular slip surface (Case 2). The radius $R = H_o$ and the centroid for a quarter-circular area is $\frac{4R}{3\pi}$ from the vertical axis through the center. The weight of the circular quadrant is $W = \gamma \frac{\pi H_o^2}{4}$ and the shear resistance along the arc of the quadrant is

$$T = s_u \frac{\pi H_o}{2} \tag{3.25}$$

Taking moments about O, we get

$$W\frac{4H_o}{3\pi} = TH_o \tag{3.26}$$

By substitution Eq. (3.25) into Eq. (3.26) and simplification, we get

$$(H_o)_{\text{critical}} = \frac{1.5\pi s_u}{\gamma} \tag{3.27}$$

Thus Case 2 gives a greater critical height than Case 1. But we are unsure that Case 1 is the lowest critical height. We have to investigate other plausible failure mechanisms and then use the least critical depth.

The limit equilibrium method gives the ultimate load for a particular failure mechanism. It provides no information on soil displacements. Conventionally, soil failure and soil displacements are treated separately in foundation design. However, methods such as finite difference and finite element with appropriate soil models (e.g., critical state soil model) allow for the integration of soil failure and soil displacements.

The key points are:

1. ***The limit equilibrium method requires that you speculate on a possible failure mechanism and then use the equilibrium equations to solve for the collapse load.***
2. ***Not all the equations of equilibrium are necessarily satisfied in the limit equilibrium method.***
3. ***The soil is assumed to be a homogeneous, rigid–perfectly plastic material.***
4. ***The limit equilibrium method gives the ultimate load for a particular failure mechanism but provides no information on soil displacements.***

EXAMPLE 3.4 *Stability of a Slope Near a Canal Using Limit Equilibrium*

An excavation for a canal, 4 m (12 ft) deep, is being considered next to a long foundation carrying a load of 2000 kN/m (137 kips/ft). A potential planar failure surface AB is shown in Fig. E3.4a. Use the limit equilibrium method to determine the factor of safety for the foundation for the given failure surface.

Strategy You are given the potential failure surface. Your next step is to draw a free body diagram and then use the equilibrium equations along the failure plane to determine the collapse load. SI units will be used for the solution of this problem.

Solution 3.4

Step 1: Draw the free body diagram. See Fig. E3.4b.

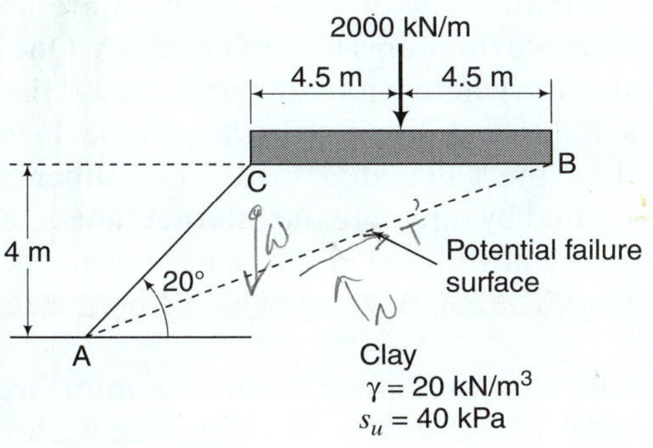

FIGURE E3.4a

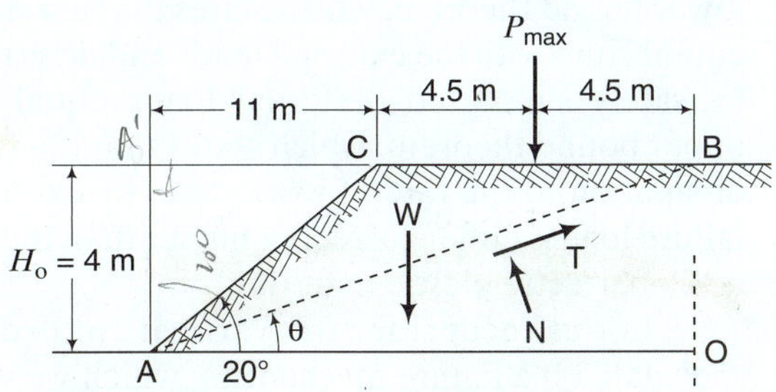

FIGURE E3.4b

Step 2: Calculate the inclination of the failure plane.

$$\mathrm{AO} = 9 + \frac{4}{\tan 20^\circ} = 20\,\mathrm{m}$$

$$\theta = \tan^{-1}\left(\frac{4}{20}\right) = 11.3^\circ$$

Step 3: Calculate the weight of soil wedge ABC.

$$W = \frac{1}{2} \times (\mathrm{BC}) \times H_o \times \gamma = \frac{1}{2} \times 9 \times 4 \times 20 = 360\,\mathrm{kN/m}$$

Step 4: Calculate P_{max} using the force equilibrium equation.

Total vertical disturbing force $= (P_{max} + W)$

The disturbing shear force is $T = (P_{max} + W)\sin\theta$

The soil resistance along AB is

$$R = s_u \times \mathrm{AB} = 40 \times \frac{4}{\sin(11.3^\circ)} = 816.6\,\mathrm{kN/m}$$

For equilibrium, $T = R$,

$$\therefore (P_{max} + W)\sin\theta = 816.6$$

$$P_{max} = \frac{816.6}{\sin(11.3^\circ)} - 360 = 3807.2\,\mathrm{kN/m}$$

Step 5: Determine factor of safety.

Factor of safety, $FS = \dfrac{P_{max}}{P} = \dfrac{3807.2}{2000} = 1.9.$

■

3.6 LIMIT ANALYSIS

Importance *The limit analysis is another popular method of analysis in geotechnical engineering. It has two theorems that give the lower and upper limits of the collapse load. This method gives you a possible range within which lies the true collapse load.*

In *limit analysis*, an idealized stress-strain relationship and failure law or yield condition are used to calculate the maximum load that a soil can support. There are two theorems for limit analysis. One is the lower bound theorem, which states that if a stress distribution anywhere within the soil is found that is in equilibrium with the external loads and does not violate the failure condition, then the external loads can be safely supported—external loads equal to or less than the calculated loads. The other is the upper bound theorem, which states that if a failure load is found by equating the internal rate of energy dissipation to the rate of work done by external forces for any postulated failure mechanism, then the failure load is either correct or higher than the true failure load. The soil is assumed to be a homogeneous, rigid–perfectly plastic material.

Let us reconsider the vertical cut problem. Consider first the upper bound theorem. We first speculate on a failure mechanism, which we will say is a wedge of soil sliding on a slip plane as shown in Fig. 3.11. This failure mechanism is the same as that for Case 1 in Fig. 3.10. Let the velocity be V at an angle

ζ to the failure plane as shown in Fig. 3.11. Then the rate of external work is the weight of the soil multiplied by the vertical component of the velocity,

$$\text{External work} = \frac{1}{2}\gamma\, H_o^2 \cot\theta V \cos(90 - \theta + \zeta) \tag{3.28}$$

The rate of internal work is the shear resistance along the slip plane multiplied by the component of the velocity along the slip plane, that is,

$$\text{Internal work} = s_u \frac{H_o}{\sin\theta} V\cos(\zeta) \tag{3.29}$$

Putting the external work equal to the internal work, we get

$$\frac{1}{2}\gamma H_o^2 \cot\theta V\cos(90 - \theta + \zeta) = s_u \frac{H_o}{\sin\theta} V\cos(\zeta) \tag{3.30}$$

By simplification, we get

$$H_o = \frac{2s_u\cos(\zeta)}{\gamma \cot\theta \cos(90 - \theta - \zeta)} \tag{3.31}$$

By minimization, we get

$$H_o = (H_o)_{\text{critical}} = \frac{4s_u}{\gamma} \tag{3.32}$$

which is identical to the limit equilibrium solution. This means that the limit equilibrium solution is either correct or is an upper bound solution.

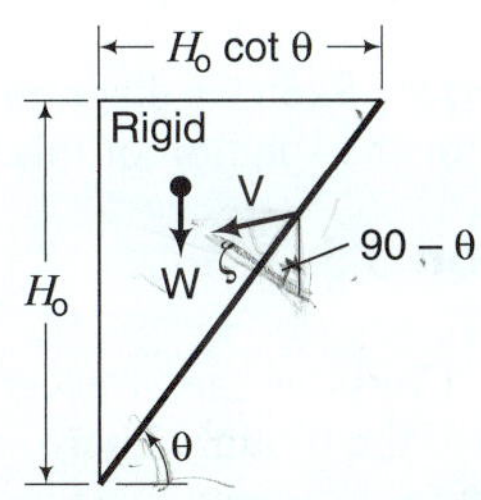

FIGURE 3.11 Failure mechanism for upper bound limit analysis.

The lower bound solution utilizes the stress state of the material and a failure law. Consider the stresses on an element at the base near the vertical wall of the cut. The lateral total stress is zero, while the vertical total stress is γH_o. We can draw Mohr's circle based on these stresses as shown in Fig. 3.12b. If these stresses cause the soil to fail, then the undrained shear strength is

$$s_u = \frac{\gamma H_o}{2} \tag{3.33}$$

Solving for H_o, we get

$$H_o = (H_o)_{\text{critical}} = \frac{2s_u}{\gamma} \tag{3.34}$$

This is the lower bound solution—the correct answer is greater than or equal to this solution. The possible correct answer lies within the range $\frac{2s_u}{\gamma} \leq (H_o)_{\text{critical}} \leq \frac{4s_u}{\gamma}$. In practice, the heights of excavations in stiff overconsolidated clays are designed not to exceed $\frac{3.8s_u}{\gamma}$. The two bounds in limit analysis are particularly useful because they define the range in which the correct solution lies.

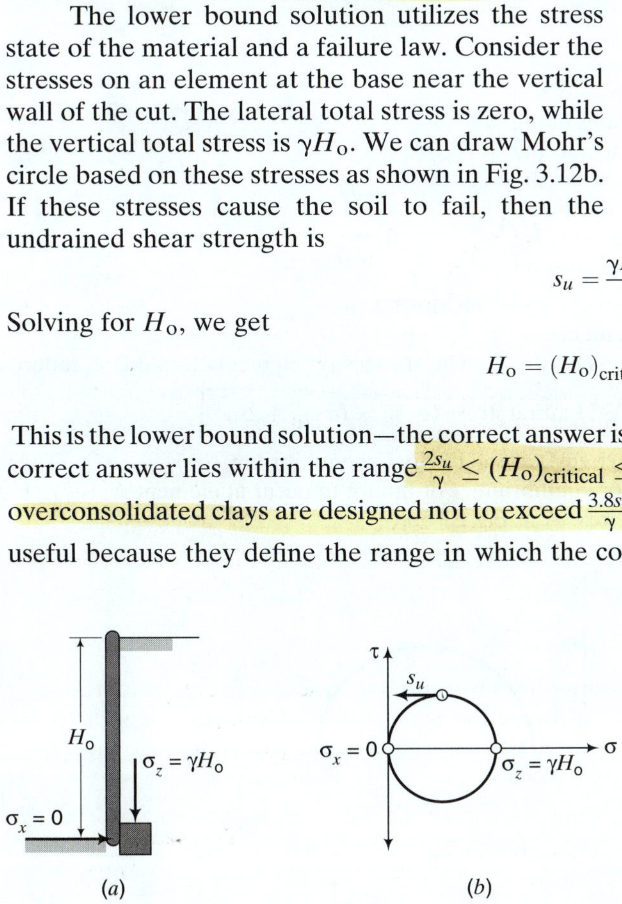

FIGURE 3.12 Stresses and stress state for an element at the base of the cut.

The key points are:

1. ***The limit analysis utilizes two theorems: the upper bound and the lower bound theorems.***
2. ***The upper bound theorem states that if a failure load is found by equating the internal rate of energy dissipation to the rate of work done by external forces for any postulated failure mechanism, then the failure load is either correct or higher than the true failure load.***
3. ***The lower bound theorem states that if a stress distribution anywhere within the soil is found that is in equilibrium with the external loads and does not violate the failure condition, then the external loads can be safely supported.***
4. ***The soil is assumed to be a homogeneous, rigid–perfectly plastic material.***
5. ***The limit analysis gives a range in which the true failure load lies but provides no information on soil displacements.***

EXAMPLE 3.5 *Failure Load for a Long Foundation Using Limit Analysis*

A long concrete foundation of width B rests on the surface of a saturated clay with an undrained shear strength, s_u. Use the limit analysis to find a lower bound solution for the vertical failure load per meter (ft) length of foundation. If $s_u = 100$ kPa (2 ksf) and B = 1 m (3 ft), determine the vertical failure load.

Strategy Sketch a diagram representing the problem and then determine the stresses in the soil. U.S. units will be used for the solution of this problem.

Solution 3.5

Step 1: Consider the stresses on the soil below and soil adjacent to the footing. There will be two stress states in the soil: one for element A (Fig. E3.5a), directly under the footing, and one for elements B and C, adjacent to the footing. The tendency of the footing to move downward will cause the lateral stresses in elements B and C to increase to values that may exceed the vertical stress. The stresses on the elements are as shown in Fig. E3.5a.

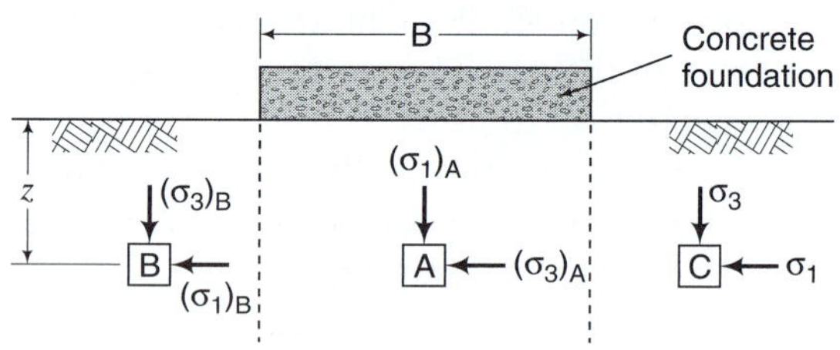

FIGURE E3.5a

Step 2: Determine the stress state of the soil elements.
If failure were to occur, then the failure shear stress is $\tau_f = s_u$. The stresses on elements B and C at failure are:

$$\text{Vertical stress} = (\sigma_3)_B; \quad \text{Lateral stress } (\sigma_1)_B = (\sigma_3)_B + 2s_u$$

The Mohr's circle for elements B and C is shown as circle B in Fig. E3.5b. For equilibrium, $(\sigma_3)_A = (\sigma_1)_B$. The lateral stress on element A is now $(\sigma_3)_A = (\sigma_3)_B + 2s_u$ for equilibrium. For failure to occur at element A, $(\sigma_1)_A$ must be equal to $(\sigma_3)_A + 2s_u$ as shown by Mohr's circle, circle A in Fig E3.5b.

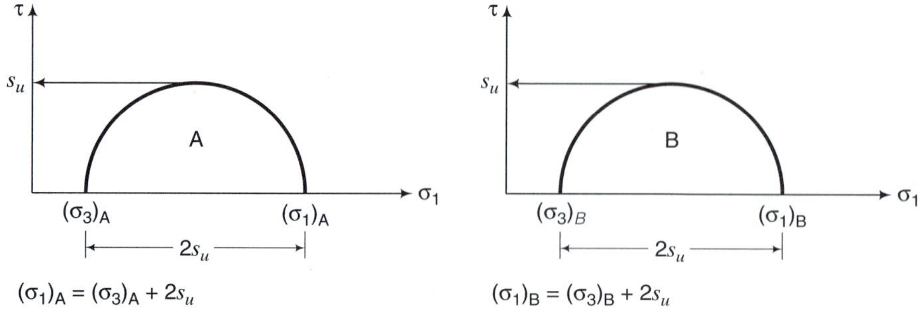

FIGURE E3.5b

Now,

$$(\sigma_1)_A = (\sigma_3)_A + 2s_u = (\sigma_3)_B + 4s_u \quad \text{and} \quad (\sigma_3)_B = \gamma z$$
$$\therefore (\sigma_1)_A = \gamma z + 4s_u$$

The vertical failure load is $P_u = (\sigma_1)_A$ A, where A is the area of the foundation. The area per unit length is $(B \times 1) = B$ and

$$P_u = B(\gamma z + 4s_u)$$

Step 3: Find the vertical failure load.

At the surface, $z = 0$, the failure stress is $(\sigma_1)_A = 4s_u$. The ultimate load/unit width $= P_u = (\sigma_1)_A B = 4Bs_u$; with $B = 3$ ft and $s_u = 2$ ksf,

$$P_u = 4 \times 3 \times 2 = 24 \text{ kips/ft}$$

■

3.7 NUMERICAL METHODS

Importance *Numerical methods are invaluable in foundation engineering as geosystems are becoming more complex and beyond the capability of limit equilibrium and limit analysis. Many commercial programs using numerical methods are now available. You need to understand (1) the basic formulation, (2) the interpretation of the results, and (3) the pitfalls of using these programs. Numerical methods allow you to explore the stresses and displacements based on range of anticipated loading history. Thus, they are particularly useful in sequential construction. The interpretation of results from any numerical method requires familiarity with the method and experience in geotechnical practice. In this section, a brief introduction to numerical methods is presented.*

There are three popular numerical methods to analyze geosystems. One is the finite difference method (FDM) that you may have encountered in your mathematics course. The others are the finite element method (FEM) and the boundary element method (BEM). The FEM has become the premier numerical tool for the analysis of a wide range of problems in engineering. The FDM and FEM require the discretization of the domain of the problem. The BEM requires only the discretization of the boundaries of the problem. The FDM and FEM are methods to solve differential equations, whereas BEM is used to solve integral equations. The theory and methodology behind these numerical methods are described in many textbooks and only an overview is presented here.

Numerical methods require a soil model—simulation of the stress-strain or constitutive relationship and other relationships (e.g., temperature–time relationship)—to describe the general response of the soil to external and internal perturbations. Geotechnical engineers are particularly interested in the deformations and stresses in the soil from applied loading history to safely and economically design foundations on or within it. Soil models are the heart of numerical simulation. We will not discuss the details of the various soil models here. Rather, a general description of soil model construction is presented.

3.7.1 Soil Models

The requirements of a good model for natural systems (Hawkins, 1988) are:

1. It must accurately describe a large class of observations with few arbitrary elements.
2. It must make definite predictions about the results of future observations.

There is no unique soil model. This is not surprising since soil is a natural material that is affected by the environment in which it originates and exists. Even for steel, a man-made material, which we can control in the manufacturing process, there is plethora of models available to predict its response. Many soil models are based on treating soils as a continuum in which smooth changes of stresses and strains occur. Soil is, however, a discrete or particulate material in which discontinuity is to be expected.

Soil models are sometimes called *constitutive* models and are generally categorized into two groups-elastic models and plastic models. Elastic models treat soils as either linear or nonlinear elastic materials that obey Hooke's law. Plastic models have the following basic characteristics:

1. Yield criterion
2. Relationship between plastic strain increment and stress (flow rule)
3. Relationship between displacement and strain (compatibility)

All soil models must contain a criterion for failure. The failure criteria discussed in Section 3.4 are the common ones used and the soil models are titled using the names of these criteria. For example, a Mohr-Coulomb model can simulate a soil that initially behaves linearly elastic and then fails when the maximum stress obliquity is achieved. The use of any soil model depends on the geological condition, the importance of the structure, and the applied loads. Success in using soil models requires high-quality data and experience. A summary of the popular soil models and their common usage is given in Table 3.1.

The basic formulations of FDM, FEM, and BEM are now briefly described. More details are available in textbooks devoted to one or more of these numerical methods.

3.7.2 Finite Difference Method (FDM)

The FDM is a numerical technique to solve differential equations by replacing them with a set of algebraic equations. For example, the axial strain due to a displacement u in the x-direction is

$$\varepsilon_{zz} = \frac{\partial u}{\partial x} \tag{3.35}$$

TABLE 3.1 Summary of Simple Soil Models and Their Possible Usage

Model	Soil simulated as	Geotechnical interest	Soil type best suited	Soil parameters required
Elastic	Elastic	First approximation of soil behavior that does not include failure.	All soils	Shear modulus (G) and bulk modulus (K) or Young's modulus (E) and Poisson's ratio (ν)
Mohr-Coulomb	Elastic–rigid plastic	Long-term failure	Heavily over-consolidated clays and dense sands	ϕ'_p or ϕ'_{cs} and α_p, G and K or E and ν
Tresca	Elastic–rigid plastic	Short-term failure	Clays	s_u, G and K or E and ν
Critical state	Elastoplastic	General soil responses to different loadings	Soft clays and loose sands	Initial stresses: p'_o, q_o, and p'_c or R_o Soil parameters: λ, κ, e_o or w and ϕ'_{cs}

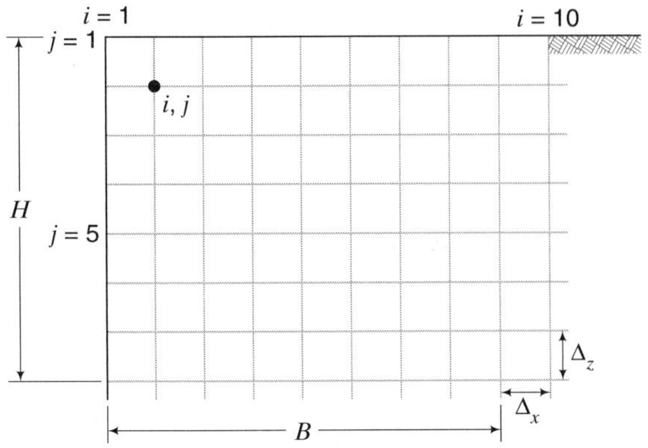

FIGURE 3.13 Discretization of a soil domain for FDM.

We can transform this differential equation into difference form using Taylor's series giving

$$\frac{\partial u}{\partial x} = \frac{u_{j+1} - u_j}{\Delta x} + (\text{truncation error}) \tag{3.36}$$

where j is a discrete point within the domain and Δx is a small distance between the discrete points. There are essentially three ways to transform the differential equation into difference form. Equation (3.36) is called the *forward difference method*. The two other methods are the *backward difference method* and the *central difference method*. The backward difference method is

$$\frac{\partial u}{\partial x} = \frac{u_j - u_{j-1}}{\Delta x} + (\text{Truncation error}) \tag{3.37}$$

and the central difference method is

$$\frac{\partial u}{\partial x} = \frac{u_{j+1} - u_{j-1}}{2\Delta x} + (\text{Truncation error}) \tag{3.38}$$

The adoption of one of these methods depends on the problem that is to be solved.

The soil and structure domains for two-dimensional problems are discretized into a grid of i columns and j rows (Fig. 3.13). The solution of the set of algebraic equations (the transformed differential equation) describing the soil model is found at each node (i, j) of the grid. Truncation errors are reduced by making the grid spacings (Δx and Δz) small.

3.7.3 Finite Element Method (FEM)

The finite element method (FEM) has become the premier numerical tool for the analysis of a wide range of problems in engineering science. Like the finite difference method (FDM), the FEM requires the discretization of the domain of the problem. The basic steps involved in solving a boundary-valued problem using FEM are as follows:

1. The domain is divided into a finite number of elements (Fig. 3.14). Elements are usually clustered in areas where stresses or strains are expected to change rapidly. For two-dimensional problems, triangular and/or quadrilateral elements are often used. The corners of elements are called *nodes* or *nodal points*. Irregular boundaries are easily incorporated by approximating them by discrete line segments or by using mathematical functions.
2. Determine the form of distribution of the unknown quantity. For example, if displacement is the unknown quantity, we can write a mathematical function to describe how the displacement should

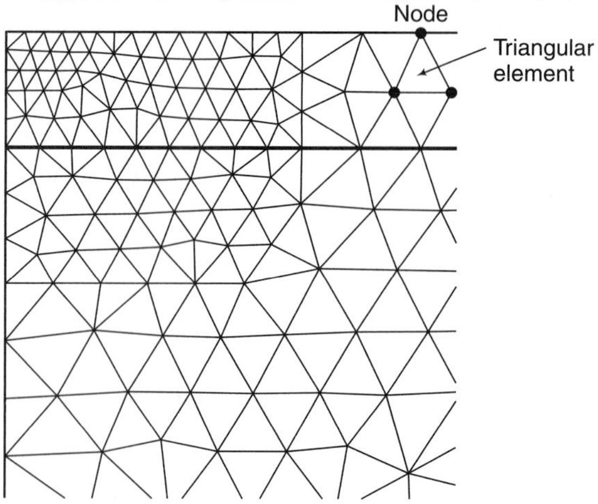

FIGURE 3.14 Discretization of a two-layer soil domain into triangular elements for FEM.

vary across the element using the nodal points as stakes. Polynomial interpolation functions are popular mathematical functions used in FEM. If u is the displacement function, we can write a polynomial interpolation function as $u = A_1 u_2 + A_2 u_2 + \cdots + A_n u_n$, where $u_1, u_2, \ldots u_n$ are the unknown nodal displacements and $A_1, A_2 \ldots A_n$ are interpolation constants.

3. Formulate the response of each element by expressing the unknown quantity as a function of the nodal points and its position inside the element. If displacements are the desired quantity, the relationship can be written as

$$\{u\} = [N]\{d\} \tag{3.39}$$

where d is a vector list of all the nodal displacements and N is a shape function characteristic of the type of element (triangular or quadrilateral) used.

4. The general form of the equations for the response of an element is:

$$\{f\} = [k]\{l\} \tag{3.40}$$

where k is a matrix of the element property, l is a vector list of the unknown quantity at the nodes, and $\{f\}$ is a vector list of the applied loads (perturbation).

5. Formulate the global response equations by assembling the equations for each element for the entire body. The form of the global equations for a two-dimensional stress problem is:

$$\{F\} = [K]\{L\} \tag{3.41}$$

where $[K]$ is the global stiffness matrix, $\{L\}$ is a vector list of the unknown displacements at nodal points within the body, and $\{F\}$ is a vector list of applied loads.

6. Solve the global equation imposing the boundary conditions for the unknown quantity.
7. Calculate other desired quantities, (e.g., stresses) and interpret the results.

3.7.4 Boundary Element Method (BEM)

BEM is used to solve integral rather than differential equations. The differential equation of a problem is first converted into integral equations and then into algebraic equations. In BEM, only the boundaries of the domain have to be discretized, resulting in a smaller number of equations compared to FEM and FDM. The BEM has been used in solving geotechnical problems, particularly pile foundation, but it is less popular than FEM and FDM.

3.7.5 Using Numerical Methods in Foundation Design

Numerical methods are useful tools to explore the possible effects of various parameters such as loads, loading history, soil stiffness, soil friction, undrained shear strength, and sequential construction on foundation design. The accuracy of the methods (FDM, FEM, and BEM) depends on the user's familiarity with the fundamentals of the chosen method, the quality of the input data, and the soil model. Geotechnical data is often limited and is sometimes of suspicious quality. If the input data is incorrect or of poor quality, you should not expect reasonable results ("garbage in, garbage out"). The steps that you should take in using numerical methods are as follows:

1. Get a good understanding of the foundation system to be analyzed. Is the system complex? If it is, how can it be simplified? What are the possible implications of the simplification?
2. Define objectives and scope of the analysis to be performed. Are you interested in possible failure mechanisms? Do you want to get an estimate of the displacements? Do you want to study the effects of possible variations of soil parameters or loading?
3. Collect all available data such as the type of structure, geometry of structure, loads, geological condition of the site, soil properties, and possible construction methods. What data is available? How was it obtained?
4. Evaluate the quality of the data. Is the data high quality? Does it fall within known ranges based on experience?
5. Conduct trial runs using simple models such as a linear elastic model with the available data or data that is consistent with engineering practice and judgment. Where is the soil heavily stressed? What are the displacements? Are they reasonable? You may also calculate the collapse loads, if appropriate, using either the limit equilibrium or limit analysis or empirical methods to give you ballpark reference values.
6. Evaluate the results of the trial runs with different meshes (different sizes and shapes), different soil models, and soil parameters. What are the effects of mesh size, mesh density, soil parameters, and soil models on the results? Is there a mesh density beyond which the results does not change significantly?
7. Conduct runs that closely match the physical system based on the evaluation of the results of the trial runs. Do the results from the simulation of the physical system differ significantly from the trial runs? What are the displacement patterns and values? Does the soil fail? Is the displacement excessive?
8. Evaluate, summarize, and present results. You should produce clear, concise interpretation of the results with appropriate values (e.g., maximum vertical displacement) and selective graphics depicting areas of interest such as areas of high stresses. You should also report the input parameters and any special modeling approach that you have taken.

Some of the areas of special concern in using numerical methods are:

1. ***Extent of boundaries of the system to be modeled.*** You have to select some region of the structure/soil domain to model to minimize the influence of the boundaries and the number of equations to solve. For example, consider the modeling a circular footing of diameter D on a two layered soil with the top layer of thickness, z, and the bottom layer of infinite thickness (Fig. 3.15). The question is what width and thickness of the bottom layer to consider? Assuming the load is uniform, you may exploit symmetry and consider only one-half the width of the soil domain. The lateral extent of the soil should be roughly 10 diameters from the line of symmetry if you are interested in displacements, and may be smaller ($\approx$5 diameters) if your interest is in the collapse load. The thickness of the

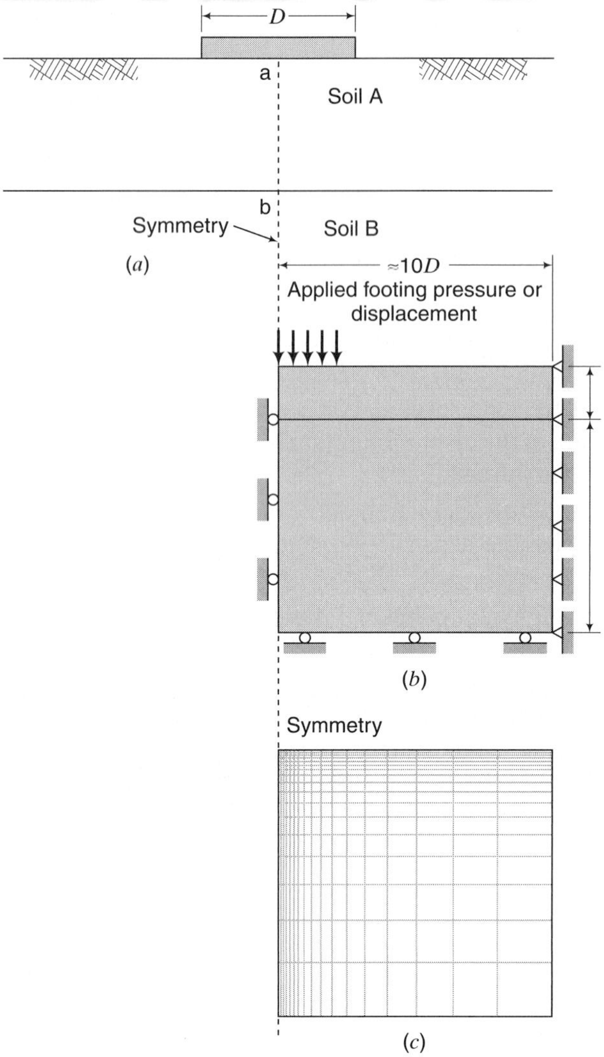

FIGURE 3.15 (*a*) Foundation on a two-layer soil. (*b*) Boundary conditions. (*c*) A graded mesh.

bottom layer should extend past the depth at which the stress increase from the footing is 10%, assuming the soil is elastic. You can usually use a thickness of bottom layer of about 10 diameters in your initial trial run.

2. ***Mesh size.*** In your initial run, you should use a coarse mesh or grid with quadrilateral elements. You can then vary the mesh size or density with higher mesh density near areas of high stress concentration or strain gradients. For example, for the circular footing, we can have a high mesh density under and near vicinity of the footing as shown in Fig. 3.15c. The aspect ratio (length to width) of the grid or mesh should generally be less than 5 to 1 in areas of high strain gradients.
3. ***Boundary conditions.*** You need to impose boundary conditions to the mesh or grid. These boundary conditions are intended to simulate the constraints in the physical system. For example, for the circular foundation, boundary ab (Fig. 3.15a) is an axis of symmetry. Soil elements along this boundary should be prevented from moving in the x-direction. We can simulate this by a roller (Fig. 3.15b). This is accomplished by fixing the grid points or nodal points along ab in the x direction.
4. ***Increment size.*** The size of the increment of loads or displacements or stress depends on the soil model. For a linear elastic model, the increment size is not as important as in a plasticity model. You

should select small increments of displacements or stresses when you are using plasticity models. You should conduct trial runs starting with large increment size, say 1% of the total applied load, displacements, or stresses. Then adjust this accordingly. It may require thousands of increments of loads or displacements or stresses to simulate a foundation system accurately.

5. ***Choice of soil model.*** You should always start with the simplest soil model—a linear, elastic soil model. After evaluating the results, and taking into consideration the objectives of the modeling effort and the quality of the input data, you can use more sophisticated models.
6. ***Decide whether the results are reasonable.*** How do we decide if the results are reasonable? Reasonableness is normally based on experience with similar systems. In the absence of significant experience, you can conduct calculations using simple methods such as limit equilibrium, or limit analysis, or using a linear elastic model. These results will give some indication of the magnitude of the collapse load or the magnitude of the displacements to expect.

The* key points *are:

1. ***Numerical methods are useful tools to simulate geosystems.***
2. ***A variety of soil models are available for use in describing the response of soils to loading. None of these models can accurately describe all the facets of soil behavior.***
3. ***The choice of a model depends on the geological condition, the importance of the structure, and the loads that would be applied.***
4. ***Elastic models can provide a first approximation of the possible response of a soil and set the stage for using more involved soil models.***
5. ***You need to gain familiarity with the numerical method of choice and understand its limitations.***
6. ***Experience is needed to successfully use the results of numerical methods in design.***

3.8 OBSERVATIONAL METHODS

Importance *Soils are complex, natural materials for which none of the methods described previously can be expected to reliably predict the real response of the soils to the anticipated loadings. Consequently, observations of the performance of actual geosystems are invaluable in building a reservoir of knowledge to help you make judgments on the performance of future geosystems. In this section, we will briefly discuss the importance of observational methods and how they can be used with analytical/numerical methods to make decisions on the safe design of geosystems.*

There is no substitute for experience in successful geotechnical engineering practice. In fact, some geotechnical engineers will proportion a foundation relying on prior experience rather than using the results of mathematical analyses. Several of the equations that you will use in design calculations (Chapters 7 through 10) have modification factors that were obtained from observations of previously built geosystems.

The observational method is neither an analytical nor a numerical method. It is the evaluation of the performance of real geosystems to build a reservoir of knowledge to evaluate the predictions from analytical and numerical methods and to build empirical or semi-empirical methods. Empirical equations are based only on observations. Semi-empirical equations are combinations of analytical/numerical methods with observations.

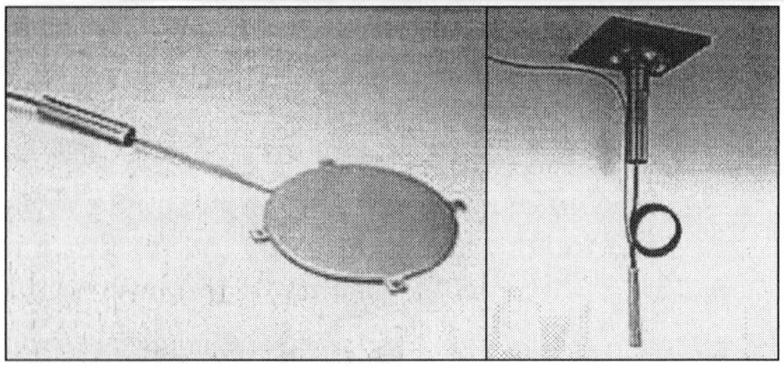

(*a*) Earth pressure cell (*b*) Settlement probe

FIGURE 3.16 An earth pressure cell and a settlement probe. (Source: http://www.geokon.com.)

Wherever possible you should instrument your structure and observe its performance during construction and during its design life, and compare the observations with the original design methods. For example, in a shallow foundation one may bury stress cells under the foundation and install displacement measurement sensors to monitor the stresses induced in the soil and the settlement of the footing. There are a variety of pressure cells, load cells, and displacement measurement devices. An example of a pressure cell and a settlement probe is shown in Fig. 3.16. The earth pressure cell shown here consists of two thin circular plates welded together at the edges and the cavity between the plates is filled with oil. Earth pressure causes the plates to move toward each other, compressing the oil. The change in oil pressure is measured by a vibrating pressure transducer. The settlement probe shown in Fig. 3.16b is used in measuring the surface settlements of embankments and fills.

The basic procedure for the observational method is as follows:

1. Decide on the objective or objectives of the observation. For example, your objective may be to obtain loads and displacements data to assess the performance of the geosystem.
2. Determine locations for the measurements. These may be affected by clearance requirements of the structure. Numerical methods may help you determine the possible locations by conducting simulations of the loading conditions on the geotechnical system.
3. Determine the type of instrumentation or sensors to be used.
4. Determine the method of data acquisition (collecting, transmitting, and storing the data) and security issues.
5. Determine the method of interpretation of the data and presentation of the results.

The* key points *are:

1. ***Observational method is an important part of foundation engineering.***
2. ***You have to use a combination of observational, analytical, and numerical methods to successfully design geosystems.***

3.9 PHYSICAL SOIL MODELS

Importance *A physical model is a scaled version of the real system. These models are useful to (1) study how the real system may perform under simulated conditions, (2) study the behavior and performance of new types of structures, different methods of construction, and new materials, and (3) complement mathematical models. In this section, we briefly describe two popular tools used in physical soil modeling.*

FIGURE 3.17 A geotechnical centrifuge. (Source: http://nees.rpi.edu/IMG/jpg/centr1.jpg.)

Physical models are made from materials that can reasonably simulate the real soil and equipment or systems that can apply simulated loadings and record the results for interpretation. Sometimes real soils are used. A centrifuge and a shake table are two popular pieces of equipment used in geotechnical engineering. The centrifuge is a mechanical device that imposes centrifugal force by rapidly spinning a bucket containing the soil. Centrifuging can drain the fluid from a wet object very quickly. The spin cycle of a modern washing machine is an example of a centrifuge in action. The centrifugal force imposed on a soil can be several hundred times as great as gravity. The ability to apply large gravitational forces and quickly drain a wet soil makes a centrifuge a valuable tool to conduct model studies on soils, especially fine-grained soils. A centrifuge used in geotechnical engineering is shown in Fig 3.17.

Centrifuges in geotechnical engineering are playing important roles in modeling the performance of soil structures, in determining failure mechanisms in soils, and in the verification of numerical models. They are particularly useful in studying geosystems such as slopes, dams, embankments, tunnels, offshore structures, consolidation of dredged fill, bearing capacity of foundations, stability of retaining walls, dynamic liquefaction, flow slides, explosive cratering, contaminant migration, frost heave, and sea ice mechanics under static and dynamic loads.

When designing experiments for the centrifuge, you need to determine the size of the model, the stress history, and the acceleration to be imposed. The key is to reproduce in a small-scale experiment the stress, strain, and flow conditions that the soil will experience in the prototype. Let us assume that we want to test a 1 m^3 container of soil of unit weight (mass density $\times$ g) 20 kN/m^3 in a centrifuge at 100 g. The initial vertical stress at the base of the container is 20 kN/m^2. But at 100 g, the vertical stress imposed at the base will be 100 times greater. This corresponds to testing a soil of unit weight 20 kN/m^3 at a depth 100 m below the surface. The duration of an experiment depends on the time required for the soil to achieve the desired stress history. If n is the ratio of the length of the prototype to the length of the model, the dissipation of porewater pressure is n^2. For problems where permeability plays a principal role, such as in dynamic liquefaction and flow slides, the permeability in a centrifuge model must be reduced by a factor n to accurately simulate the prototype.

A shake table (Fig. 3.18) is a slab of material (steel or concrete) that is configured to apply multiple degrees of motion to a structure to study its response. A scaled soil model is constructed in a container that is then placed on the shake table. The desired motion is imposed and the performance of the soil and any structure placed on or within it is recorded. Shake tables are particularly useful to study the behavior of scaled models of geostructures subjected to cyclical loadings, such as in a seismic event. The major disadvantage of the shake table is that it is practically impossible to simulate the in-situ stresses of real soils.

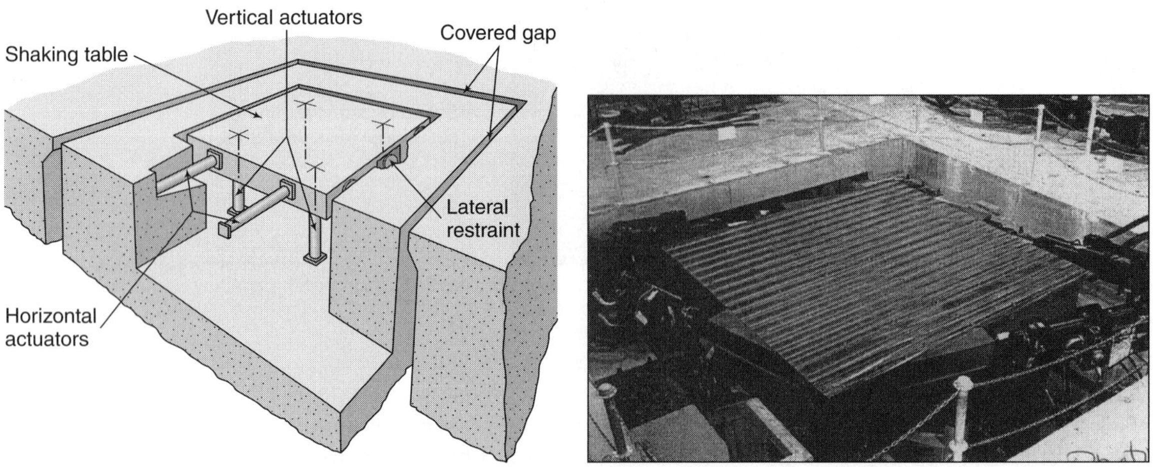

FIGURE 3.18 A shake table. (Source: http://nees.buffalo.edu/seesl/DescriptionOfLaboratoryEquipment.html.)

The* key points *are:

1. ***Physical models are scaled models of the real structure.***
2. ***Physical models are very useful to study the performance of new geostructures and for verification of mathematical models.***
3. ***Centrifuges and shake tables are the two most popular tools for conducting physical soil modeling.***

3.10 SUMMARY

We have briefly discussed yielding, failure criteria, and two popular analytical methods—limit equilibrium and limit analysis—for determining the failure (collapse) loads of soils assuming that they are homogeneous, rigid–perfectly plastic materials. These are simple methods that provide no information on soil deformation. Numerical methods such as FEM and FDM can provide a whole range of soil responses, but their successful use requires significant experience with and understanding of soil models, modeling and simulation methodologies, high-quality soil data, and observations of real geosystems.

Practical Example

EXAMPLE 3.6 ***Forensic Geotechnical Engineering Using Critical State Failure Criterion***

A circular foundation was designed using the undrained shear strength from laboratory tests. The soil profile consisted of 10.2 m (33 ft) of soft, blue clay followed by a deep deposit of heavily overconsolidated gray clay. The geotechnical engineer used a factor of safety of 2.5 (usual range of factor of safety is 2 to 3). The foundation failed just before the construction was fully completed. Reexamination of the data revealed that the average water contents of the samples in the shear tests were about 3.6% lower than the samples taken at about the same depth in the soft, blue clay and used in consolidations tests. The shear test samples were improperly stored and tests were conducted one month after arrival in the laboratory. The average compression index (C_c) of the soft soil was 0.23 and this was judged to be a good representation of the soil compressibility. The original soil samples were destroyed. Select one of the failure criteria described in this chapter and use it to explain the failure. The bearing capacity of soils for short-term condition (undrained condition) is proportional to the undrained shear strength.

Strategy Since the foundation failed just before construction was completed and the soil is a clay, it is best to assume undrained condition. None of the failure criteria, except the critical state failure criterion, is capable of providing an explanation to this forensic geotechnical engineering problem because they do not consider the initial condition of the soil. The data available and given above is sparse. However, the critical state theory is powerful in explaining soil behavior with few available parameters.

Solution 3.6

Step 1: Select a failure criterion.
Select the critical state failure criterion because, unlike the other models, it considers the initial state (in this case, the initial water content) of the soil.

Step 2: Manipulate Eq. (3.18) to compare the two water content conditions.
Let the undrained shear strength of the sample in the shear test (sample A) be $(s_u)_A$ and that of the sample predicted for the consolidation sample (sample B) be $(s_u)_B$, assuming that the water content of the consolidation sample accurately represents the water content of the soil in the field.
The shear strength, according to the critical state failure criterion, is

$$s_u = \frac{q_f}{2} = \frac{M}{2}\exp\left(\frac{e_\Gamma - e_o}{\lambda}\right)$$

The ratio of the undrained shear strength of the two samples is

$$= \frac{(s_u)_B}{(s_u)_A} = \frac{\exp(e_\Gamma - e_c)_B/\lambda}{\exp(e_\Gamma - e_c)_A/\lambda} = \exp\frac{(e_o)_A - (e_o)_B}{\lambda} = \exp\frac{G_s[(w_o)_A - (w_o)_B]}{\lambda} = \exp\frac{G_s\Delta w}{\lambda}$$

Step 3: Calculate the ratio of the undrained shear strength for the two water contents.

$$\lambda = \frac{C_c}{2.3} = \frac{0.23}{2.3} = 0.1. \quad \text{Assume } G_s = 2.7.$$

$$\frac{(s_u)_B}{(s_u)_A} = \exp\frac{Gs\Delta w}{\lambda} = \exp\frac{2.7 \times -0.036}{0.1} = 0.38 = 38\%$$

Step 4: Provide explanation.
The predicted undrained shear strength of the soil, if the water content were 3.6% higher, is 38% of that of the samples tested. A factor of safety of 2.5 is equivalent to multiplying $(s_u)_A$ by 1/2.5 = 0.4, which is greater than 0.38. Consequently, the bearing capacity of the soil was exceeded because the undrained shear strength used in the original design was higher than the intact soil because of the loss of moisture (drying out).

■

EXERCISES

3.1 A soil element 2 m (6 ft) under the center of a circular tank foundation, 4 m (12 ft) in diameter, is subjected to axisymmetric loading. The soil parameters are $\phi'_{cs}=30°$ and $\gamma_{sat}=19.8\,kN/m^3$ (126 pcf). Groundwater is at ground level. If the lateral effective stress increase is 0.3 times the vertical effective stress increase, calculate the increase in the vertical stress that would cause the soil element to fail using the Mohr-Coulomb failure criterion. Assume all stresses are principal stresses. How high can you fill the tank with water before failure becomes imminent? Assume the dead load of the tank is about 4 kPa (75 psf). [*Hint:* Determine the surface stress to induce the calculated vertical stress increase (see Chapter 2), and then use that value to determine the height of water.]

3.2 An unconfined compression test was conducted on a cylindrical clay soil sample, 50 mm (2 in.) in diameter and 100 mm (4 in.) high. The maximum load recorded was 206 N

(46 lb). Using Tresca's failure criterion, determine the undrained shear strength.

3.3 The initial water content of a normally consolidated, saturated clay soil is 67%. The compression index C_c is 0.58. If $e_\Gamma = 3.0$ and $M = 1$, predict the undrained shear strength of the clay using the critical state failure criterion. Make a plot of predicted undrained shear strength for water contents ranging from 40% to 80% for this soil. If the clay were allowed to dry out and lose 2% water, how much change in undrained shear strength would occur assuming that the soil remains saturated?

3.4 The preconsolidation mean effective stress of a soil element under the center of a tank foundation is 120 kPa (2.5 ksf). The current mean effective stress is 100 kPa (2 ksf). When filling the tank, the axial stress increases and the radial stress is approximately constant. Determine the following: (a) the overconsolidation ratio with respect to the mean effective stresses; (b) the failure stresses under drained condition. The soil parameters are: $\lambda = 0.3$, $\kappa = 0.08$, $\phi'_{cs} = 27°$, $\nu' = 0.3$, and $e_o = 1.4$. (c) Plot the stress-strain response for drained condition. You may use the spreadsheet at http://www.wiley.com/college/budhu. Click Foundations and Earth Structures, Chapter 3, criticalstate.xls.

3.5 Describe the differences between the limit equilibrium method and limit analysis. Does either limit equilibrium method or limit analysis account for soil deformation?

3.6 During the construction of a foundation, it became necessary to temporarily support an excavation using a vertical wall with props as shown in Fig. P3.6. The soil is classified according to USCS as SP.

(a) Use the limit equilibrium method to determine the horizontal force per m (ft) length of wall without the props.

(b) Install the props and then find the force per m (ft) length of wall in each prop using statics.

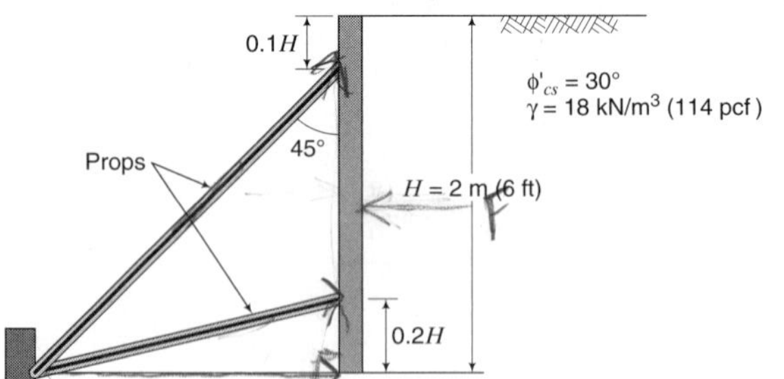

FIGURE P3.6

3.7 For the excavation in a clay soil shown in Fig. P3.7, calculate the upper bound collapse load (limit analysis) in terms of the ratio $\frac{s_u}{\gamma H}$. If $\gamma = 20\,\text{kN/m}^3$ (127 pcf) and $H_o = 3\,\text{m}$ (10 ft), what is the minimum undrained shear strength required to prevent collapse?

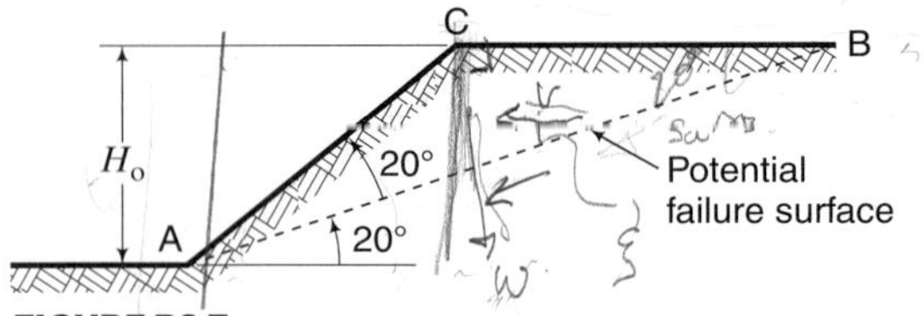

FIGURE P3.7

3.8 An excavation with a vertical face for a foundation is required in a stiff saturated clay with $\gamma = 20\,\text{kN/m}^3$ (127 pcf) and $s_u = 80\,\text{kPa}$ (1.7 ksf). Calculate the depths of the excavation based on the lower bound and upper bound theorems of limit analysis. What maximum depth of excavation would you recommend?

3.9 Determine the uniformly distributed load that will cause collapse of the long foundation shown on Fig. P3.9. The potential planar failure surface is AB. Use the limit equilibrium method.

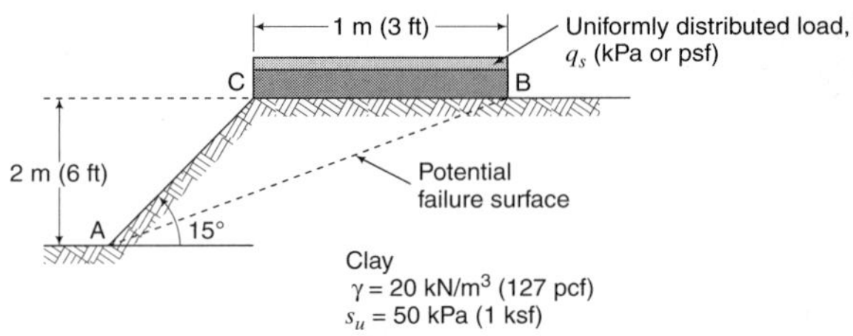

FIGURE P3.9

3.10 Repeat Problem 3.9 except that the potential failure surface is a circular arc with center at C and radius CA.

3.11 Conduct a literature survey of the different types of sensors that are used in foundation monitoring.

3.12 Conduct a literature survey on the instrumentation and data acquisition system used for monitoring (a) an embankment and (b) an excavation for a foundation.

3.13 Conduct a literature survey on the analysis of a long foundation using a numerical method. Identify the mesh and soil parameters used, and the results obtained.

CHAPTER 4

SETTLEMENT, STRENGTH, AND STIFFNESS PARAMETERS FROM LABORATORY TESTS FOR FOUNDATION ANALYSIS

4.0 INTRODUCTION

To estimate bearing capacity, settlement, and stability of geotechnical structures, you need to obtain soil parameters using laboratory and in situ tests. The soil parameters of importance for foundation analysis are as follows.

- ***Settlement parameters:*** compression indexes and overconsolidation ratio
- ***Shear strength parameters:*** undrained shear strength for fine-grained soils and friction angles for all soils
- ***Stiffness parameters:*** shear modulus and bulk modulus, or the elastic modulus and Poisson's ratio

High-quality laboratory tests are required to obtain reasonably accurate values of strength, stiffness, and settlement parameters. Many of these tests, especially shear strength and settlement tests, are time consuming. Consequently, geotechnical engineers have attempted to build correlation among quickly obtained physical soil parameters, shear strength, stiffness, and settlement parameters.

In this chapter, we will discuss the interpretation of laboratory tests and present some of the correlations among physical soil parameters, shear strength, stiffness, and settlement used in geotechnical practice. Experience is required to use these correlations successfully in practice. They are best used for preliminary analysis and design. The focus of this chapter is on settlement, strength, and stiffness. We will interpret the strength parameters based on the failure criteria presented in Chapter 3. When you complete this chapter, you should be able to:

- Interpret laboratory test results to extract design parameters for foundation analysis.
- Understand the implications of the interpretations.
- Estimate settlement, strength, and stiffness parameters from simple, physical soil tests.

4.1 DEFINTIONS OF KEY TERMS

Consolidation is the time-dependent settlement of soils resulting from the expulsion of water from the soil pores.

Excess porewater pressure **(Δu)** is the porewater pressure in excess of the equilibrium porewater pressure.

Drainage path **(H_{dr})** is the longest vertical path that a water particle will take to reach the drainage surface.

Preconsolidation stress or past maximum effective stress (σ'_{zc}) is the maximum vertical effective stress that a soil was subjected to in the past.

Primary consolidation is the change in volume of a fine-grained soil caused by the expulsion of water from the voids and the transfer of load from the excess porewater pressure to the soil grains.

Secondary compression is the change in volume of a fine-grained soil caused by the adjustment of the soil fabric (internal structure) after primary consolidation has been completed.

Shear strength of a soil is the maximum internal resistance to applied shearing forces.

Friction angle **(ϕ')** is a measure of the average friction among soil grains.

Cementation or cohesion **(c_o)** is a measure of the forces that cement soil grains.

Undrained shear strength **(s_u)** is the shear strength of a soil when sheared at constant volume.

Critical state is a stress state reached in a soil when continuous shearing occurs at a constant shear stress to mean effective stress ratio at constant volume.

Dilation is a measure of the change in volume of a soil when it is distorted by shearing.

Dilation angle **(α)** is a measure of the average dilation.

Deformation is the distortion or change in shape of a body.

Settlement is the vertical displacement of a body.

Stiffness is a measure of the resistance to deformation.

4.2 CONSOLIDATION AND SETTLEMENT PARAMETERS

Importance *The prediction of settlement is an important task in foundation analysis and design. The settlements of fine-grained soils are time-dependent. Settlement in fine-grained soils can occur many years (sometimes over 100 years) after the structure is completed. Differential settlement is of particular importance because it causes rotation of the structure and can impose severe stresses in structural members, resulting in cracks and structural separations. In this section, a brief review of one-dimensional consolidation is presented along with the desired settlement parameters and empirical correlations.*

4.2.1 Basic Concept

Consolidation is the densification of a fine-grained soil by expelling water from the voids. Consolidation can occur through the weight of the soil (self-consolidation) or through applied stresses. When a stress is applied to a saturated fine-grained soil, it is assumed to be instantaneously resisted by the porewater. The change in pressure in the porewater is called excess porewater pressure, Δu. With the passage of time (t) and drainage of the excess porewater, the soil volume decreases (soil settles) as shown in Fig. 4.1 for a soil that drains from the top and bottom surfaces only. As time approaches infinity, the excess porewater approaches zero. The difference between the applied stress and the excess porewater pressure at any time t is the amount of stress transferred to the soil solids as effective stress.

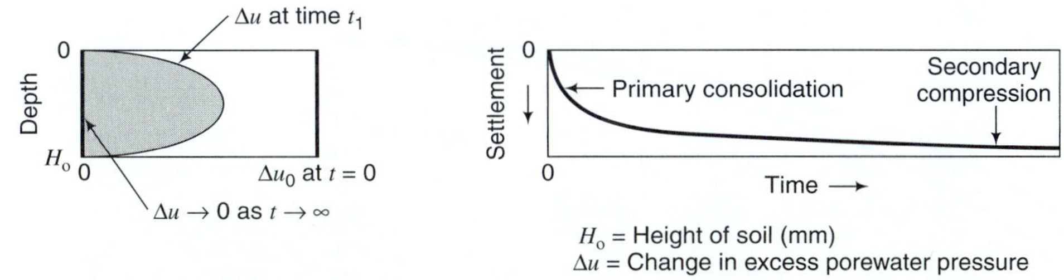

FIGURE 4.1 Settlement and excess porewater distribution with time for double drainage.

Consolidation is a time-dependent process unlike soil compaction, which can be completed quickly. Recall, compaction is soil densification by removing air from the soil mass. Consolidation can occur over several or even hundreds of years. The amount of time required for a certain desired degree of consolidation—desired amount of settlement divided by the total estimated settlement as time approaches infinity—depends on the length of the drainage path and the hydraulic conductivity of the soil. The length of the drainage path (H_{dr}) for single drainage is the thickness of the soil layer; for double drainage it is one-half the thickness of the soil layer. If a homogeneous soil of thickness H_o is drained from only one boundary (single drainage), the time for a certain degree of consolidation would be four times that if drainage of excess porewater were to occur from two opposite boundaries (double drainage).

4.2.2 One-Dimensional Consolidation Theory

Terzaghi (1925) developed a one-dimensional consolidation theory to interpret soil consolidation. Considering one-dimensional flow, the governing consolidation equation is

$$\frac{\partial u}{\partial t} = C_v \frac{\partial^2 u}{\partial z^2} \tag{4.1}$$

where u is excess porewater pressure, t is time, z is location of a point within the soil, and C_v is the coefficient of consolidation. The solution of the governing equation for the average degree of consolidation, U, is

$$U = 1 - \sum_{m=o}^{\infty} \frac{2}{M^2} \exp(-M^2 T_v) \tag{4.2}$$

where $M = \frac{\pi}{2}(2m+1)$ and m is a positive integer with values from 0 to ∞ and

$$T_v = \frac{C_v t}{H_{dr}^2} \tag{4.3}$$

where T_v is known as the time factor; it is a dimensionless term. A plot of the variation of U and T_v for a uniform and a triangular initial distribution of excess porewater pressure is shown in Fig. 4.2. A convenient set of equations for double drainage, found by curve fitting Fig. 4.2, is

$$T_v = \frac{\pi}{4}\left(\frac{U}{100}\right)^2 \quad \text{for} \quad U < 60\% \tag{4.4}$$

and

$$T_v = 1.781 - 0.933 \log(100 - U) \quad \text{for} \quad U \geq 60\% \tag{4.5}$$

The time factor (T_v) provides a useful expression to estimate the settlement in the field from the results of a laboratory consolidation test. If two layers of the same clay have the same degree of consolidation, then their time factors and coefficients of consolidation are the same. Hence,

$$T_v = \frac{(C_v t)_{\text{lab}}}{(H_{dr}^2)_{\text{lab}}} = \frac{(C_v t)_{\text{field}}}{(H_{dr}^2)_{\text{field}}} \tag{4.6}$$

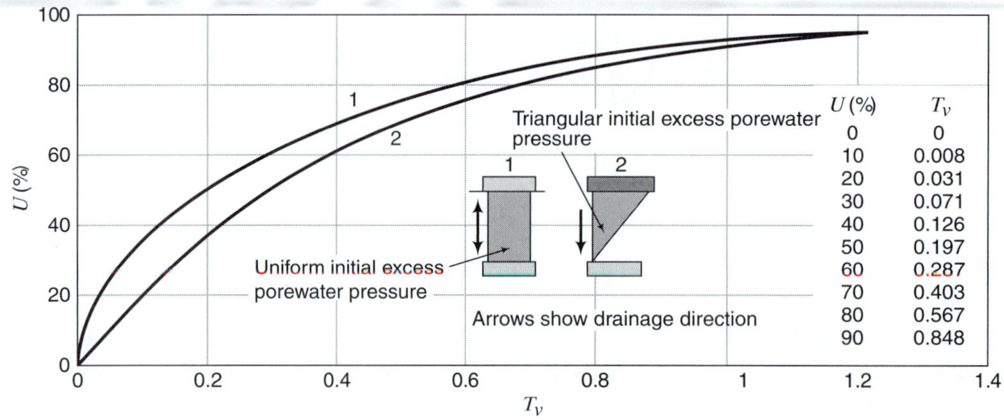

FIGURE 4.2 Relationship between time factor and average degree of consolidation for a uniform and a triangular distribution of initial excess porewater pressure.

and, by simplification,

$$\frac{t_{\text{field}}}{t_{\text{lab}}} = \frac{(H_{dr}^2)_{\text{field}}}{(H_{dr}^2)_{\text{lab}}} \tag{4.7}$$

4.2.3 Consolidation Settlement Parameters

To estimate consolidation settlement, various parameters are required. These are found from a one-dimensional consolidation test, called the *oedometer test* (ASTM D2435). The key settlement parameters are compression index, C_c, recompression index, C_r, secondary compression index, C_α, modulus of volume compressibility, m_v, coefficient of consolidation, C_v, and overconsolidation ratio, σ'_{zc}. The hydraulic conductivity, k, can also be calculated from the test data.

In an oedometer test, a vertical stress is applied to a saturated soil sample contained within a ring. The top and bottom faces of the sample are the drainage boundaries. The settlement is measured at convenient time intervals until the change in settlement (Δz) becomes very small. The applied vertical stress is then doubled and the settlement-time data is again recorded. Further vertical stresses are applied until the desired stress level is achieved. The soil sample can also be unloaded incrementally and settlement-time data recorded.

The void ratio or volumetric strain at the end of consolidation for each increment of applied vertical stress can be plotted as shown in Fig. 4.3. The most popular plot is Fig. 4.3b, where the void ratio is plotted against the applied vertical stress (log scale). Since all the excess porewater pressure under a given stress increment is assumed to dissipate when $\Delta t \to \infty$, the applied vertical stress becomes the vertical effective stress in the soil.

The slope AB of the approximately straight part of the curve in Fig. 4.3b is the compression index, C_c, expressed as

$$C_c = -\frac{e_2 - e_1}{\log\dfrac{(\sigma'_z)_2}{(\sigma'_z)_1}} = \frac{|\Delta e|}{\log\dfrac{(\sigma'_z)_2}{(\sigma'_z)_1}} \quad \text{(no units)} \tag{4.8}$$

where the subscripts 1 and 2 denote two convenient selected points, usually such that the ratio $\frac{(\sigma'_z)_2}{(\sigma'_z)_1}$ is a multiple of 10. The line ABE is known as the *normal consolidation line* (NCL); it represents the initial consolidation of the soil as it was formed. The slope of the line CF approximating the average slope of the unloading curve BC and the reloading curve CD is the recompression index, C_r,

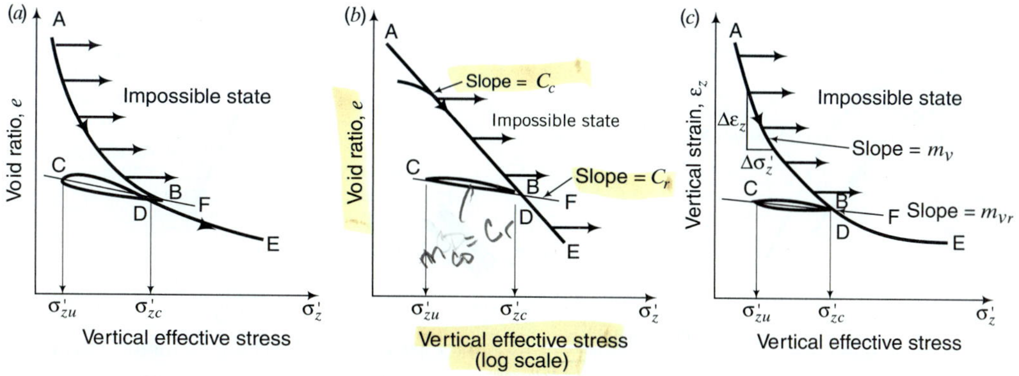

FIGURE 4.3 Three plots of settlement data from soil consolidation.

expressed as

$$C_r = -\frac{e_2 - e_1}{\log \dfrac{(\sigma'_z)_2}{(\sigma'_z)_1}} = \frac{|\Delta e_{zr}|}{\log \dfrac{(\sigma'_z)_2}{(\sigma'_z)_1}} \quad \text{(no units)} \tag{4.9}$$

The line CF is called the *unloading-reloading line* (URL); it represents changes in the stress history of the soil after its initial formation. It is traditionally assumed the soil states on CF will produce elastic response until the vertical effective stress at B is exceeded. Also soil states along CF indicate that the soil has been subjected to stress states greater than its current stress state (i.e., the soil is overconsolidated).

From Fig. 4.3c, we can determine the modulus of volume compressibility, m_v, as

$$m_v = \frac{|\Delta\varepsilon_z|}{\Delta\sigma'_z} \left(\frac{\text{m}^2}{\text{kN}}\right) \tag{4.10}$$

and the modulus of volume recompressibility, m_{vr}, expressed as

$$m_{vr} = -\frac{(\varepsilon_z)_2 - (\varepsilon_z)_1}{(\sigma'_z)_2 - (\sigma'_z)_1} = \frac{|\Delta\varepsilon_{zr}|}{(\Delta\sigma'_z)_r} \left(\frac{\text{m}^2}{\text{kN}}\right) \tag{4.11}$$

where $\Delta\varepsilon_{zr}$ is the change in vertical strain and $(\Delta\sigma'_z)_r$ is the corresponding change in vertical effective stress along the line CF.

In calculating consolidation settlement either C_c and C_r or m_v and m_{vr} are required. The vertical effective stress difference, should not exceed 100 kPa in determining either m_v or m_{vr} because of the nonlinearity of the response shown in Fig. 4.3c.

The secondary compression index, C_α can be found by making a plot as shown in Fig. 4.4. You should note that Fig. 4.4 is for a single applied vertical stress. The value of C_α usually varies with the magnitude of the vertical effective stress and other factors such as load increment ratio.

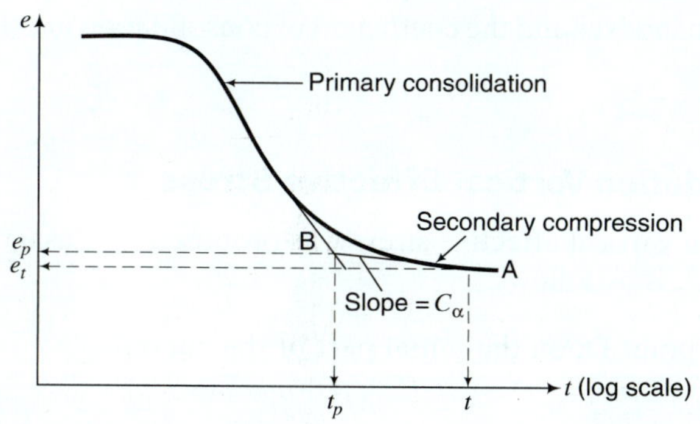

FIGURE 4.4 Determination of secondary compression.

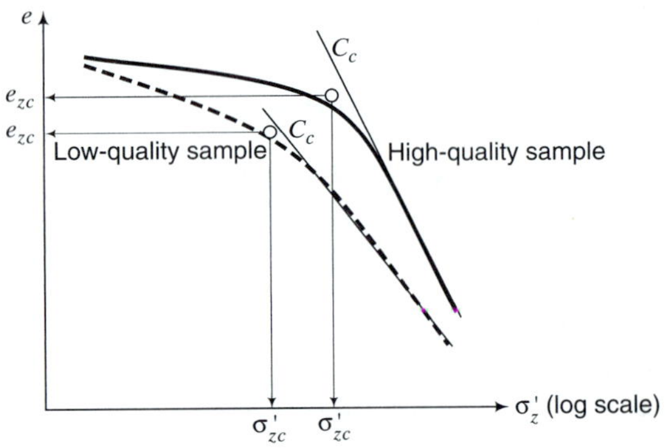

FIGURE 4.5 Effects of sampling disturbance on the compression index and preconsolidation (past maximum) vertical effective stress.

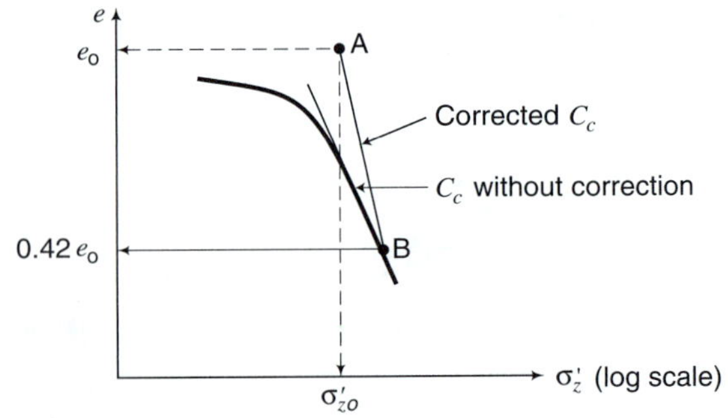

FIGURE 4.6 Schmertmann's method to correct C_c for soil disturbances.

The history of loading of a soil is locked in its fabric. We can get, for example, an estimate of the past maximum vertical effective stress, called the *preconsolidation vertical effective stress*, that the soil has experienced. If the current effective stress is equal to the preconsolidation vertical effective stress, the soil is normally consolidated. If the current vertical effective stress is less than the preconsolidation vertical effective stress, the soil is overconsolidated. Soils states along the NCL constitute a normally consolidated soil while soil states along BC constitute a preconsolidated or overconsoildated soil.

The overconsolidation ratio (OCR) is defined as the ratio of the preconsolidation (past maximum) vertical effective stress, σ'_{zc}, to the current vertical effective stress or overburden pressure, σ'_{zo},

$$\text{OCR} = \frac{\sigma'_{zc}}{\sigma'_{zo}} \tag{4.12}$$

For normally consolidated soils, OCR = 1. For lightly overconsolidated soils, $1 < \text{OCR} \leq 2$, and for heavily overconsolidated soils, $\text{OCR} > 2$.

All the settlement parameters are susceptible to soil disturbances and various corrections have been proposed. Sampling disturbances usually cause the underestimation of the compression index, which leads to the underestimation of the consolidation settlement (Fig. 4.5).

Schmertmann (1953) suggested a correction to the laboratory void ratio–vertical effective stress (log scale) curve to obtain a more representative in-situ value of C_c. His method is as follows:

1. Locate a point A at coordinate (σ'_{zo}, e_o).
2. Draw a horizontal line at $0.42e_o$ to intersect the laboratory e vs. $\log \sigma'_z$ curve at point B as shown in Fig. 4.6.
3. The slope of the line AB is the corrected value for C_c.

Typical ranges of values for the compression indexes and the coefficient of consolidation are shown in Appendix B (Tables B4.1 to B4.4).

4.2.4 Determination of the Preconsolidation Vertical Effective Stress

One method of determining the preconsolidation vertical effective stress was proposed by Casagrande (1936). The procedure, with reference to Fig. 4.7, is as follows:

1. Identify the point of maximum curvature, point D, on the initial part of the curve.
2. Draw a horizontal line through D.

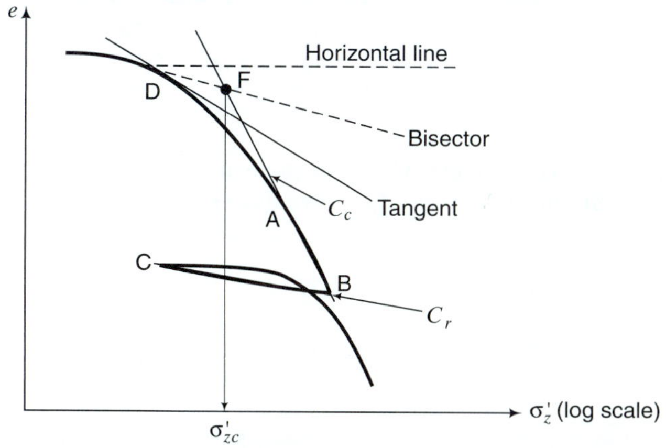

FIGURE 4.7 Determination of the preconsolidation effective stress using Casagrande's method.

3. Draw a tangent to the curve at D.
4. Bisect the angle formed by the tangent and the horizontal line at D.
5. Extend backward the straight portion of the curve (the normal consolidation line), BA, to intersect the bisector line at F.
6. The abscissa of F is the preconsolidation effective stress, σ'_{zc}.

The overconsolidation ratio is also affected by sampling disturbances as shown in Fig. 4.5.

Another method (Becker et. al., 1987) is the strain energy method using a constant rate of strain consolidation test. The procedure is as follows

1. Compute the change in work per unit volume for each strain increment from

$$\Delta W = \left(\frac{\sigma'_{zi} + \sigma'_{zf}}{2}\right)\left(\frac{\Delta H_f - \Delta H_i}{H_o}\right) \quad \text{or} \quad \left(\frac{\Delta\sigma'_{zi} + \Delta\sigma'_{zf}}{2}\right)\left(\frac{\Delta e_f - \Delta e_i}{1 + e_o}\right)$$

   where the subscript i and f denote beginning and end of increment respectively.
2. Plot ΔW (ordinate) versus σ'_{zf} (abscissa).
3. The curve linking the data on the plot should show two distinct slopes. Extend the flatter slope (recompression line) to intersect the steeper slope (the normal consolidation line).
4. The vertical effective stress corresponding to the intersection of the two slopes is σ'_{zc}. The strain energy method is best suited for stiff clays.

EXAMPLE 4.1 *Determination of C_c*

A soil profile consists of a 5 m (16 ft) thick sand layer of unit weight 17 kN/m^3 (106 pcf) followed by a deep deposit of clay. A sample of the clay was taken at a depth of 7 m (23 ft) and tested in an oedometer. The results of the test are shown in the table below. Groundwater level is at 5 m (16 ft) below the ground surface and the initial water content of the sample was 32.5%. The specific gravity was 2.65. Determine the uncorrected and the correct values of C_c, and C_r.

σ'_z (kPa)	15	30	60	120	240	480	240	120	60	120	240	480	960
e	0.84	0.83	0.82	0.75	0.62	0.47	0.47	0.47	0.49	0.48	0.48	0.46	0.32

1 kPa = 20.9 psf.

Strategy Plot the results as e vs. σ'_z (log scale), find C_c (uncorrected) from the slope of the straight portion of the curve. Follow Schmertmann's method to correct C_c. SI units will be used to solve this example problem.

Solution 4.1

Step 1: Plot the results in e vs. σ'_z (log scale).
See Fig. E4.1a.

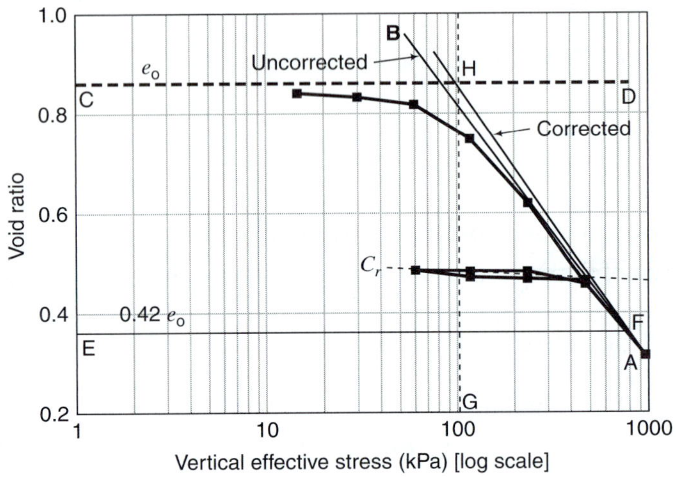

FIGURE E4.1a

Step 2: Calculate the initial void ratio and initial vertical effective stress.

Clay

$$e_o = wG_s = 0.325 \times 2.65 = 0.86$$

$$\gamma' = \frac{G_s - 1}{1 + e_o}\gamma_w = \frac{2.65 - 1}{1 + 0.86} \times 9.8 = 8.7\,\text{kN/m}^3$$

$$\sigma'_{zo} = \gamma_{sand} \times 5 + \gamma' \times 2 = 17 \times 5 + 8.7 \times 2 = 102.4\,\text{kPa}$$

Step 3: Calculate uncorrected C_c.

From the plot, draw a straight line through the normal consolidation part of the results as shown by AB and find the slope of AB,

$$C_c = \frac{0.81 - 0.31}{\log\left(\frac{1000}{100}\right)} = 0.5$$

Step 4: Calculate corrected C_c.

Draw a horizontal line through $e_o = 0.86$ (line CD)
Draw a horizontal line through $0.42e_o = 0.36$ (line EF)
Draw a vertical line through $\sigma'_{zo} = 102.4$ kPa (GH)
Draw a line from F to H. The slope of FH is the corrected C_c,

$$C_c = \frac{0.87 - 0.31}{\log\left(\frac{1000}{100}\right)} = 0.56$$

Step 5: Determine C_r.

Draw an average slope for the unloading-reloading curves. Select vertical effective stresses of 100 kPa and 1000 kPa on this slope for convenience [log (1000/100) = log 10 = 1] and find the corresponding void ratio from the graph.

$$C_r = \frac{0.49 - 0.455}{\log(\frac{1000}{100})} = 0.035$$

Step 6: Determine σ'_{zc}.

Follow the procedure in Section 4.2.4. See plot E4.1b.

$$\sigma'_{zc} = 125\ \text{kPa}$$

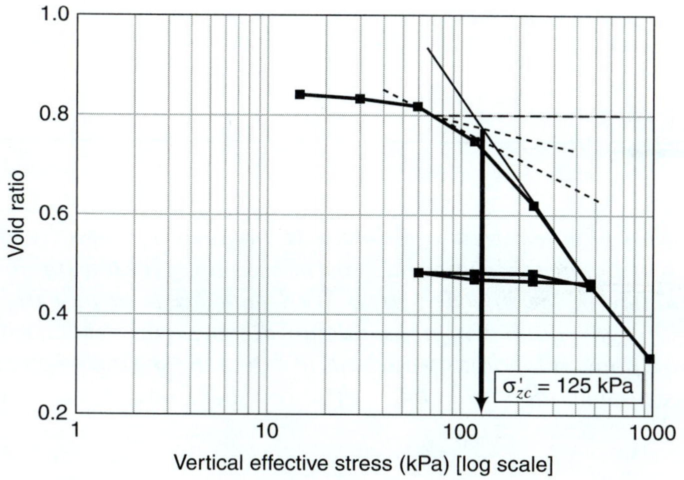

FIGURE E4.1b ■

EXAMPLE 4.2 *Effects of Drainage Boundaries on Time for Consolidation*

A sample, 75 mm (3 in.) in diameter and 20 mm (0.75 in.) high, taken from a clay layer 10 m (32 ft) thick, was tested in an oedometer with drainage at the upper and lower boundaries. It took the laboratory sample 15 minutes to reach 50 percent consolidation. Determine the following.

(a) If the clay layer in the field has the same drainage condition as the laboratory sample, calculate how long it will take the 10 m (32 ft) clay layer to achieve 50% consolidation.

(b) How much more time would it take the 10 m (32 ft) clay layer to achieve 50% consolidation if drainage existed on only one boundary?

Strategy For part (a) there is double drainage in the field and the lab, so the drainage path is one-half the soil thickness. For part (b), there is single drainage in the field, so the drainage path is equal to the soil thickness. U.S. units will be used to solve this example problem.

Solution 4.2

(a)

Step 1: Calculate the drainage path.

$$(H_{dr})_{\text{lab}} = \frac{0.75}{2} = 0.375\ \text{in.} = 0.031\ \text{ft}; \quad (H_{dr})_{\text{field}} = \frac{32}{2} = 16\ \text{ft}$$

Step 2: Calculate field time.

$$t_{\text{field}} = \frac{t_{\text{lab}}(H_{dr}^2)_{\text{field}}}{(H_{dr}^2)_{\text{lab}}} = \frac{15 \times 16^2}{0.031^2} \cong 400 \times 10^4 \text{min} = 7.6 \text{ years}$$

(b)

Step 1: Calculate the drainage path.

$$(H_{dr})_{\text{lab}} = \frac{0.75}{2} = 0.375 \text{ in.} = 0.031 \text{ ft}; \quad (H_{dr})_{\text{field}} = 32 \text{ ft}$$

Step 2: Calculate field time.

$$t_{\text{field}} = \frac{t_{\text{lab}}(H_{dr}^2)_{\text{field}}}{(H_{dr}^2)_{\text{lab}}} = \frac{15 \times 32^2}{0.031^2} = 1600 \times 10^4 \text{ min} = 30.4 \text{ years}$$

You should note that if drainage exists on only one boundary rather than both boundaries of the clay layer, the time taken for a given percent consolidation in the field is four times longer. ■

4.3 SHEAR STRENGTH PARAMETERS

Importance *The shear strength of a soil is required to estimate its load-bearing capacity. This strength depends on several parameters, including the loading path, the intergranular friction, the initial stress state of the soil, and the magnitude of the effective stress. Test data needs to be carefully interpreted to extract the desired shear strength parameters for foundation design. In this section a brief review of the shear strength of soils, some popular laboratory methods to determine shear strength, and the interpretation of test data from three popular laboratory tests—the direct shear, the triaxial, and the direct simple shear—are presented.*

4.3.1 Stress-Strain Response

We are going to summarize the important features of the responses of these two groups of soils when subjected to a constant normal effective stress and increasing shear strain. One group is uncemented soil, the other is cemented. Uncemented soils are soils in which the surface forces between the soil grains are very small, for example, clean sand. Cemented soils are soils in which the soil grains are chemically bonded by cementing agents such as calcium carbonate and by molecular forces. The magnitude of the latter is generally small for practical purposes. An example of a cemented soil is caliche. We will also touch on partially saturated soils but only briefly because partially saturated soils have complexities beyond the objectives of this textbook.

We will consider the shear strain versus the shear stress, the shear strain versus the volumetric strain, and the shear strain versus the void ratio responses of soils as illustrated in Fig. 4.8. These responses are for homogeneous (ideal) soils. The vagaries of natural soils often cause them to disobey our human idealizations. However, we need to idealize complex problems to find solutions. Consequently, the following summary of soil responses should be regarded as a framework for interpreting soil behavior, realizing that real soils may not abide by it.

Group 1—Uncemented Soils Type I soils—loose sands, normally consolidated and lightly overconsolidated clays (OCR ≤ 2)—are observed to:

- Show gradual increase in shear stresses as the shear strain increases (strain hardens) until an approximately constant shear stress, which we will call the critical state shear stress, τ_{cs}, is attained (Fig. 4.8a).

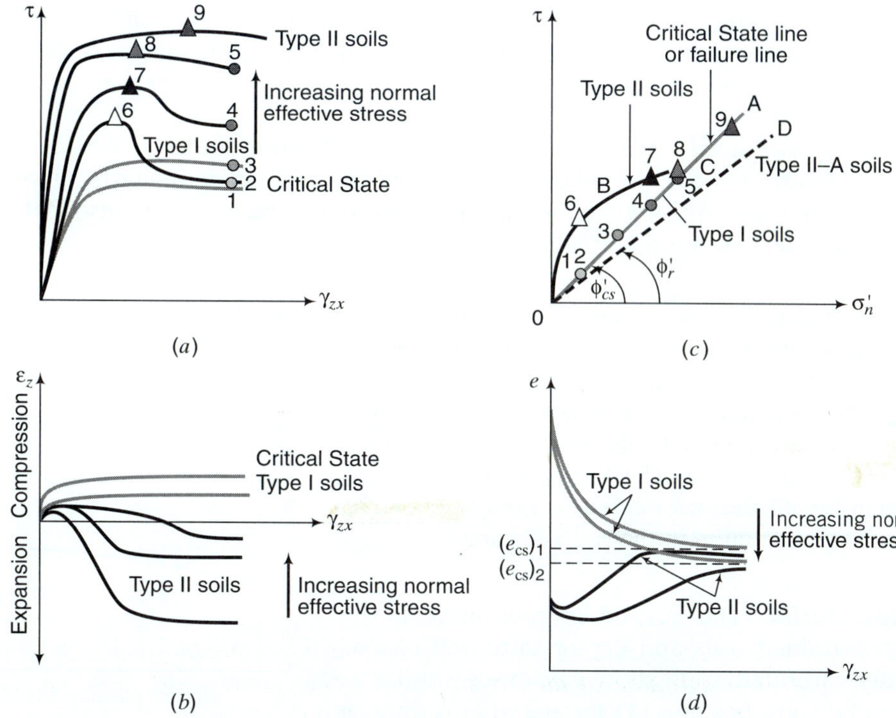

FIGURE 4.8 Response of soils to shearing.

- Compress, that is, they become denser (Figs. 4.8b,d) until a constant void ratio, which we will call the critical void ratio, e_{cs}, is reached (Fig. 4.8d).

Type II soils—dense sands and heavily overconsolidated clays (OCR > 2)—are observed to:

- Show a rapid increase in shear stress reaching a peak value, τ_p, at low shear strains (compared to Type I soils) and then show a decrease in shear stress with increasing shear strain (stain softens), ultimately attaining a critical state shear stress (Fig. 4.8a). The strain softening response generally results from localized failure zones called *shear bands*. The thickness of shear bands was observed from laboratory tests to be about 10 to 15 grain diameters. The soil mass within a shear band undergoes intense shearing while the soil masses above and below it behave as rigid bodies. The development of shear bands depends on the boundary conditions imposed on the soil samples.

  When a shear band develops in some types of overconsolidated clays, the soil grains become oriented parallel to the direction of the shear band, causing the final shear stress of these clays to decrease below the critical state shear stress. We will call this type of soil *Type II-A*, and the final shear stress attained the residual shear stress, τ_r. Type I soils at very low normal effective stress can also exhibit a peak shear stress during shearing.
- Compress initially (attributed to particle adjustment) and then expand, that is, they become looser (Figs. 4.8b,d) until a critical void ratio (the same void ratio as Type I soils) is attained. The ability of a soil to expand is conventionally called *dilation*.

The critical state shear stress is reached for all soils when no further change in volume occurs under continued shearing. We will use the term *critical state* to define the stress state reached by a soil when no further change in the ratio of shear stress to normal effective stress and also no further change in volume occurs under continuous shearing.

When the normal effective stress is increased for Type I soils, the amount of compression and the magnitude of the critical state shear stress will increase (Fig. 4.8a,b). For Type II soils, the peak shear stress tends to disappear, the critical shear stress increases, and the change in volume expansion (positive dilation) decreases (Fig. 4.8a,b).

The initial state of the soil dictates the response of the soil to shearing forces. For example, two overconsolidated soils with different overconsolidation ratios but the same mineralogical composition would exhibit different peak shear stresses and volume expansion. The higher overconsolidated soil usually gives a higher peak shear strength and greater volume expansion.

Group 2—Cemented Soils The soil grains of some soils are cemented (chemically bonded). These soils possess shear strength even when the normal effective stress is zero. Cemented soils behave much like Type II soils except they have an initial shear strength under zero normal effective stress. In a plot of peak shear stress versus normal effective stress, an intercept shear stress, c_o, called cohesion, referred to in this textbook as *cementation*, is observed (Fig. 4.9). You should not rely on c_o in design, because with the application of small shear strains (<0.001%) any shear strength due to cementation in the soil will vanish.

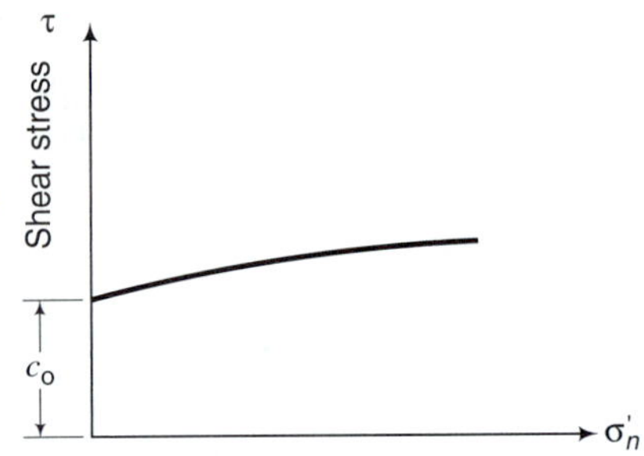

FIGURE 4.9 Peak shear stress envelope for cemented soils.

Partially Saturated Soils The responses presented so far relate to saturated fine-grained soils and dry or saturated coarse-grained soils. Partially saturated soils show responses similar to Type II soils or cemented soils, because (1) the initial porewater is negative and (2) negative excess porewater pressures develop on the application of shearing. The net effect is that the normal effective stress increases. When test data from partially saturated soils is plotted as shear stress versus normal effective stress, an apparent cohesion similar to c_o in Fig. 4.9 normally shows up. You need to be careful in using this apparent cohesion in design as this apparent strength will vanish with time as the excess porewater dissipates.

4.3.2 Interpretation of Shear Strength Parameters from Popular shear Tests

We will consider three popular tests to determine the shear strength parameters for soils. The shear strength parameters are the undrained shear strength, s_u, for fine-grained soils, and the peak and critical state friction angle, ϕ'_p and ϕ'_{cs}, for all soils. We will use the failure criteria discussed in Chapter 3 to interpret the test data and consider the following conditions.

1. ***Drained condition.*** Excess porewater pressures in the soil are dissipated quickly. Drained condition applies to long-term loading—during the life of the geosystem. The drained condition is also called *effective stress condition.* The analysis based on effective stresses is called *effective stress analysis* (ESA). The shear strength parameters are ϕ'_p and ϕ'_{cs}.
2. ***Undrained condition.*** Excess porewater pressures do not dissipate quickly. Undrained condition applies to short-term loading—during construction and shortly afterward. It is also called *total stress condition.* The analysis based on total stresses is called *total stress analysis* (TSA). The shear strength parameter is the undrained shear strength, s_u. For static loading, undrained condition applies only to fine-grained soils. For dynamic loading such as in seismic loading, even clean coarse-grained soils cannot dissipate the excess porewater quickly enough. Although this condition is strictly undrained, an ESA should be used for analyses because these soils rely strictly on frictional strength.

4.3.2.1 Simple Pouring Test to Determine the Friction Angle for Clean Coarse-Grained Soils The critical state friction angle, ϕ'_{cs}, for a clean coarse-grained soil can be found by pouring the soil into a loose heap on a horizontal surface and measuring the slope angle of the heap relative to the horizontal. This angle is sometimes called angle of repose, but it closely approximates ϕ'_{cs}.

★ Shear Box Test ★

4.3.2.2 Direct Shear Test—ASTM D3080 A popular apparatus to determine the shear strength parameters is the *shear box*. This test is useful when a soil mass is likely to fail along a thin zone under plane strain conditions. The shear box (Fig. 4.10) consists of a horizontally split, open metal box. Soil is placed in the box, and one-half of the box is moved relative to the other half. Failure is thereby constrained along a thin zone of soil on the horizontal plane (AB). Serrated or grooved metal plates are placed at the top and bottom faces of the soil to generate the shearing force.

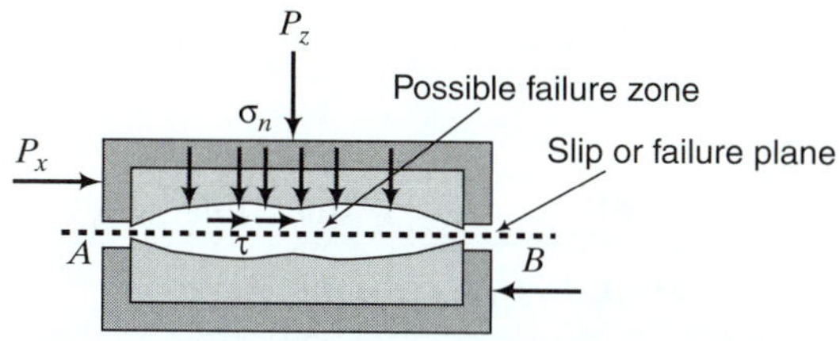

FIGURE 4.10 Shear box.

In this test, the normal force (P_z), the shear (horizontal) force (P_x), and the vertical and horizontal displacements are normally measured. The normal force for each test is constant. Typical results from the direct shear tests are shown in Fig. 4.11. The shear force–horizontal displacement and the vertical displacement–horizontal displacement plots are shown in Figs. 4.11a,b for a single applied vertical stress. If the peak shear forces $(P_x)_p$ and the critical state shear forces $(P_x)_{cs}$ are plotted on a graph of shear force versus normal force for several applied vertical stresses, usually 3, we will normally get a curve OBA (Fig. 4.11c) representing the peak condition and an approximate straight line OA representing the critical state condition.

Coulomb's failure criterion is used to interpret the direct shear test results. From the plot shown as Fig. 4.11c, we can obtain the critical state friction angle (ϕ'_{cs}) and the peak friction angle (ϕ'_p), or you can calculate these friction angles as follows.

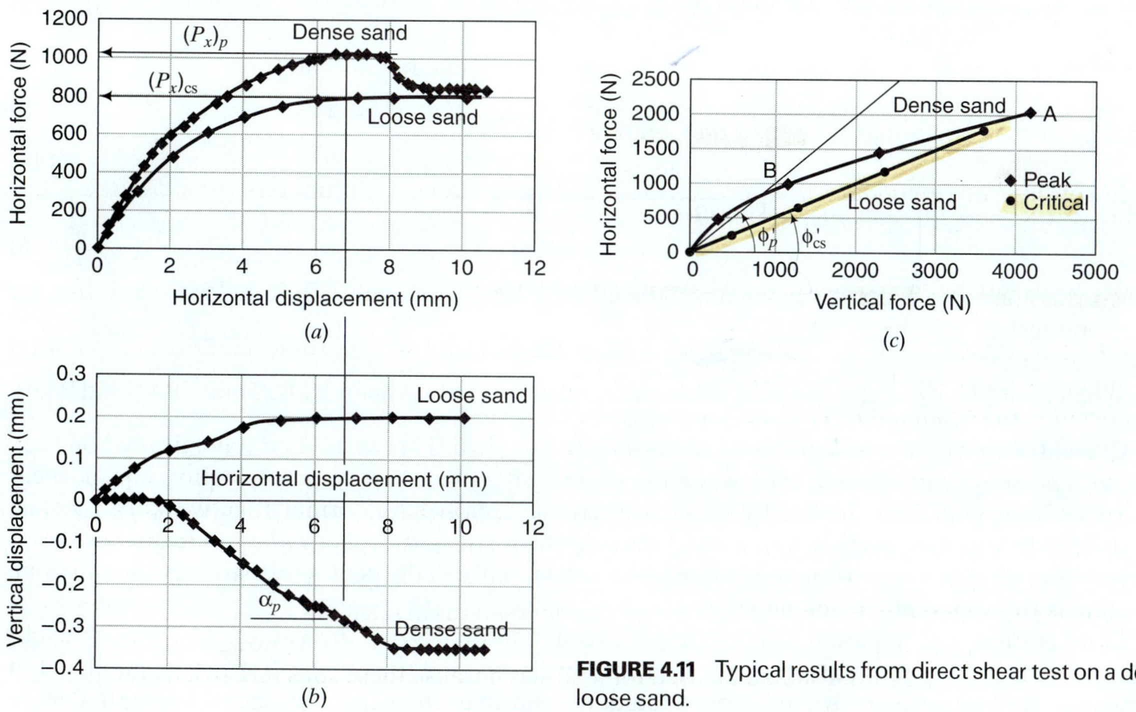

FIGURE 4.11 Typical results from direct shear test on a dense and loose sand.

Uncemented Soils

Critical State Friction Angle, ϕ'_{cs}

1. Extract the critical state horizontal forces from the test data (Fig. 4.11a).
2. Prepare a graph of horizontal force versus vertical force (Fig. 4.11c) and plot the extracted forces.
3. Draw an approximate straight line from the origin linking the critical state points as shown in Fig. 4.11c.
4. The slope of the straight line gives ϕ'_{cs}
5. Alternatively, you can calculate ϕ'_{cs} from

$$\phi'_{cs} = \tan^{-1}\frac{(P_x)_{cs}}{P_z} \tag{4.13}$$

where $(P_x)_{cs}$ is the critical state shear force and P_z is the corresponding normal force.

Peak Friction Angle, ϕ'_p

The peak friction angle is not constant. It varies with the magnitude of the normal force. Thus, the peak friction angle can be only determined for a particular normal force.

1. Extract the peak horizontal force from the test data (Fig. 4.11a).
2. Prepare a graph of horizontal force versus vertical force (Fig. 4.11c) and plot the extracted forces.
3. At a desired normal force, draw a vertical line that intersects the curve OBA (Fig. 4.11c).
4. Draw a line from the origin to the intersection of the vertical line and the curve OBA, for example, point B in Fig. 4.11c.
5. The slope of the line is ϕ'_p
6. Alternatively, you can calculate ϕ'_p for each normal load from

$$\phi'_p = \tan^{-1}\frac{(P_x)_p}{P_z} \tag{4.14}$$

where $(P_x)_p$ is the peak shear force and P_z is the corresponding normal force.

The peak effective friction angle, based on Coulomb's law, is

$$\phi'_p = \phi'_{cs} + \alpha_p \tag{4.15}$$

where α_p is the peak dilation angle. Therefore,

$$\alpha_p = \phi'_p - \phi'_{cs} \tag{4.16}$$

The peak dilation angle decreases as the normal load increases. Test results (Bolton, 1986) show that for plane strain tests:

$$\phi'_p = \phi'_{cs} + 0.8\alpha_p \tag{4.17}$$

Typical ranges of friction angles and peak dilation angles are given in Appendix B (Tables B4.5 and B4.6).

Cemented Soils The soil grains of some soils (e.g., caliche) are cemented (chemically bonded). These soils possess shear strength even when the normal effective stress is zero. In a plot of peak shear stress (peak horizontal force divided by the cross-sectional area) versus normal effective stress (normal force divided by the cross-sectional area), an intercept shear stress, c_o, called *cohesion* (cementation) is observed (Fig. 4.12). An approximate straight line is drawn through the peak shear stresses and the slope of the line is the apparent friction angle, ξ.

Cementation and apparent friction angle are not mobilized at the same shear strain levels. Cementation is predominant at small shear strain levels ($\gamma < 0.001\%$) while friction is mobilized at intermediate-to-large shear strains. You should be cautious in using the shear strength due to

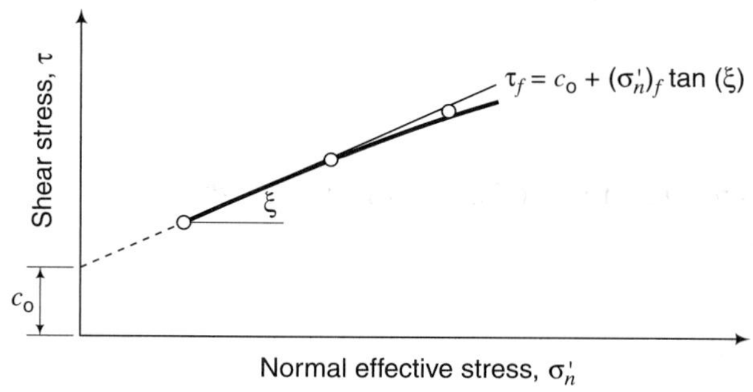

FIGURE 4.12 Interpretation of cementation (cohesion) and apparent friction angle for cemented soils using direct shear test data.

cementation, because in most practical geotechnical problems the shear strains mobilized are generally within an intermediate range of shear strains ($0.001\% \leq \gamma \leq 1\%$). It is best to use the critical state friction angle.

Procedure

1. Extract the peak horizontal forces from the test data.
2. Prepare a graph of the peak horizontal stresses versus vertical effective stresses (normal effective stresses).
3. Draw the best fit line through the data points.
4. The intercept gives the cementation (cohesion) value (c_o) and the slope of the line gives the apparent friction angle, ξ.
5. The critical state friction angle can be found using the procedure for uncemented soils.

EXAMPLE 4.3 *Determination of Friction Angle from Direct Shear Test Results*

The peak and critical state shear forces at different normal forces from tests on an uncemented sand in a direct shear box test are shown in Table E4.3.

TABLE E4.3

			Shear force (lb)	
Soil type	**Test number**	**Vertical load (lb)**	**Peak**	**Critical state**
Sand	Test 1	100	98	55
	Test 2	200	175	125
	Test 3	300	210	183
	Test 4	400	248	248
	Test 5	500	308	308

1 lb = 4.46 N.

Determine the following:

(a) ϕ'_{cs}.

(b) ϕ'_p at a vertical load of 200 lb.

(c) Estimate the dilation angle at a vertical load of 200 lb.

Strategy Follow the procedures outlined in Section 4.3.2.2 for uncemented soils.

Solution 4.3

Step 1: Plot a graph of the normal load versus failure shear force for each sample. See Fig. E4.3

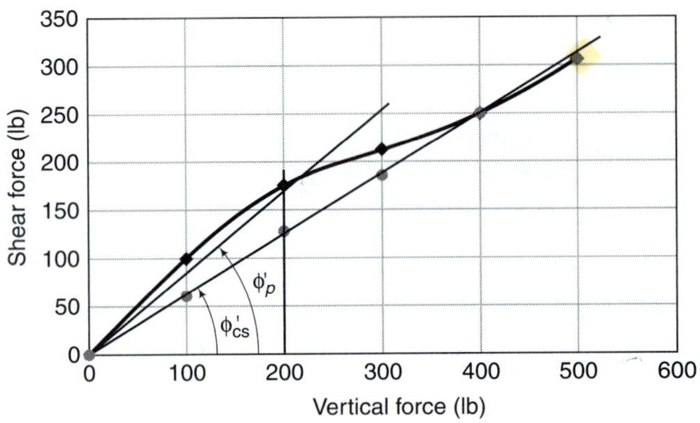

FIGURE E4.3

Step 2: Extract ϕ'_{cs}.

Draw a straight line through the critical state shear forces. The measured slope angle of the line is $\phi'_{cs} = 31.5°$. By calculation,

$$\phi'_{cs} = \tan^{-1}\frac{(P_x)_{cs}}{P_z} = \tan^{-1}\frac{308}{500} = 31.6°$$

Step 3: Determine ϕ'_p.

Draw a vertical line from a normal force of 200 lb to intersect the curve linking the peak shear forces. Draw a line from the origin through this point. The slope angle of this line is $\phi'_p = 41°$.
By calculation,

$$\phi'_p = \tan^{-1}\frac{(P_x)_p}{P_z} = \tan^{-1}\frac{175}{200} = 41.2°$$

Step 4: Estimate α_p.

$$\alpha_p = \phi'_p - \phi'_{cs}$$
$$(\alpha)_{200\,\text{lb}} = 41.2° - 31.5° = 9.7°$$

■

4.3.2.3 *Triaxial Tests* A versatile and widely used apparatus to determine the shear strength parameters and the mechanical behavior of soils is the triaxial apparatus. In this test, two stresses—the vertical and radial stresses—can be controlled independently. A cylindrical sample of soil, usually with a length-to-diameter ratio of 2, is subjected to either controlled increases in axial stresses (σ_a *or* $\sigma_1 - \sigma_3$), also called *deviatoric stress* or *plunger stress*, or axial displacements and radial stresses (σ_r *or* σ_3), also called *confining pressure* or *cell pressure*. The sample is laterally confined by a membrane and radial stresses are applied by pressurizing water in a chamber (Fig. 4.13). The axial stresses are applied by loading a plunger. If the axial stress is greater than the radial stress, the soil is compressed vertically and the test is called *triaxial compression*. If the radial stress is greater than the axial stress, the soil is compressed laterally and the test is called *triaxial extension*. A variety of axisymmetric tests with different drainage conditions can be conducted in the triaxial apparatus. Some of the popular ones are briefly described on the following pages.

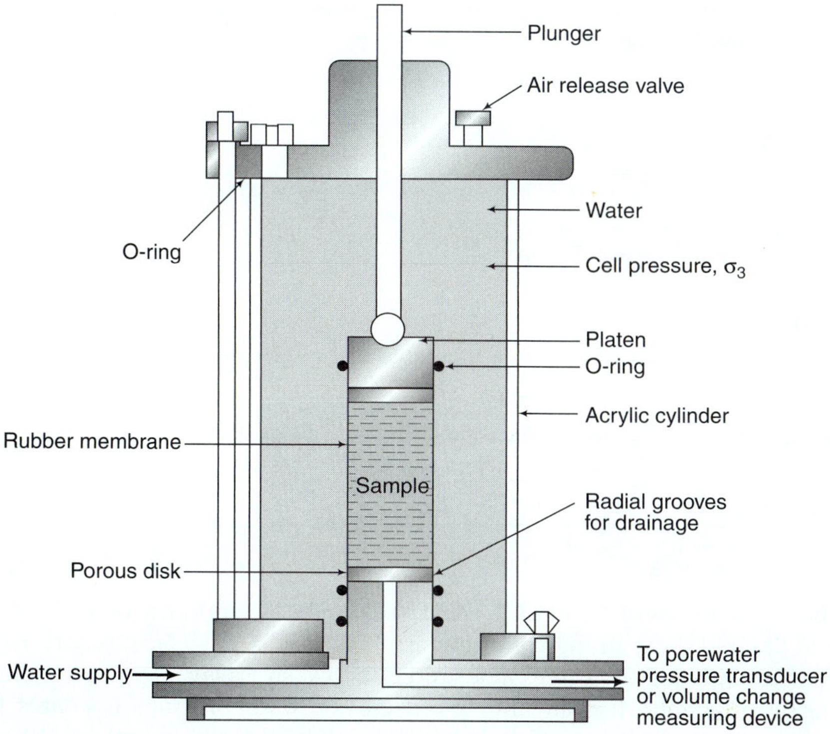

FIGURE 4.13 A triaxial apparatus.

$\sigma_3 = 0$

Unconfined Compression (UC) Test—ASTM D2166 The purpose of this test is to determine the undrained shear strength of saturated clays quickly. In the UC test, no radial stress is applied to the sample ($\sigma_3 = 0$). The plunger load, P_z is increased rapidly until the soil sample fails (i.e., cannot support any additional load). The loading is applied quickly so that the porewater cannot drain from the soil; the sample is sheared at constant volume. Tresca's failure criterion is conventionally used to interpret the test data. The critical state failure criterion can also be used and gives the same result as Tresca's. The undrained shear strength is

$$s_u = \frac{(P_z)_f}{2A} = \frac{(\sigma_1 - \sigma_3)_f}{2} \tag{4.18}$$

where the cross-sectional area, $A = \frac{A_o}{1-\varepsilon_1}$, A_o is the initial cross-sectional area, axial strain, $\varepsilon_1 = \frac{\Delta z}{L}$, Δz is the axial displacement, L is the original length of the soil sample, and the subscript f denotes failure.

The results from UC tests are used to:

- Estimate the short-term bearing capacity of fine-grained soils for foundations.
- Estimate the short-term stability of slopes.
- Compare the shear strengths of soils from a site to establish soil strength variability quickly and cost-effectively (the UC test is cheaper to perform than other triaxial tests).
- Determine the stress-strain characteristics under unconfined, fast (undrained) loading conditions.

EXAMPLE 4.4 *Determination of s_u from UC Test*

A UC test was carried out on a saturated clay sample. The maximum axial load was 97 N (22 lb) and the vertical displacement was 0.6 mm (0.024 in.). The size of the sample was 38 mm (1.5 in.) diameter × 76 mm (3 in.) long. Determine the undrained shear strength.

Strategy The test is a UC test, $\sigma_3 = 0$ and $(\sigma_a)_f = (\sigma_1)_f$ = failure axial stress. You can find s_u by calculating one-half the failure axial stress. U.S. units will be used to solve this example problem.

Solution 4.4

Step 1: Determine area at failure.

$$A_o = \frac{\pi \times D_o^2}{4} = \frac{\pi \times 1.5^2}{4} = 1.77\,\text{in.}^2, \quad \varepsilon_1 = \frac{\Delta z}{z_o} = \frac{0.024}{3} = 0.008$$

$$A = \frac{A_o}{1 - \varepsilon_1} = \frac{1.77}{1 - 0.008} = 1.78\,\text{in.}^2$$

Step 2: Determine the axial stress at failure.

The axial stress is equal to the major principal stress because the radial stress is zero.

$$(\sigma_a)_f = (\sigma_1)_f = \frac{P_z}{A} = \frac{22}{1.78} = 12.4\,\text{psi} = 1786\,\text{psf}$$

Step 3: Calculate s_u.

$$s_u = \frac{P_z}{2A} = \frac{(\sigma_1)_f - (\sigma_3)_f}{2} = \frac{1786}{2} = 893\,\text{psf}$$

■

Consolidated Undrained Compression Test (CU)—ASTM D2850 The purpose of a CU test is to determine the undrained and drained shear strength parameters (s_u, ϕ'_{cs}, ϕ'_p). A CU test is performed in two stages. The first stage is consolidating the soil to a desired effective stress level by pressurizing the water in the cell and allowing the soil sample to drain until the excess porewater pressure dissipates. In the second stage, the pressure in the cell (cell pressure) is kept constant and additional axial loads or displacements are added until the soil sample fails. Drainage of the excess porewater is not permitted but is measured. It is customary to perform a minimum of three tests at different cell pressures.

The CU test is the most popular triaxial test for fine-grained soils. Both the undrained shear strength (s_u) (total stress condition) and the friction angles (ϕ'_{cs}, ϕ'_p) (effective stress condition) can be determined. The Mohr-Coulomb failure criterion is conventionally used to interpret the test data from the triaxial test for the effective stress condition. The critical state failure criterion can also be used. Recall that the Mohr-Coulomb criterion defines failure when the maximum principal effective stress ratio, called *maximum effective stress obliquity*, $\left(\frac{\sigma'_1}{\sigma'_3}\right)_{max}$, is achieved and not when the maximum shear stress, $\left(\frac{\sigma'_1 - \sigma'_3}{2}\right)_{max}$, is achieved. The latter is Tresca's failure criterion. The failure shear stress is then less than the maximum shear stress. Tresca's failure criterion is used to determine failure under total stress condition. Plots of the data from a CU test are shown in Fig. 4.14.

When porewater pressures are measured in CU tests, you can draw two sets of Mohr's circle at failure as depicted in Fig. 4.14. One represents total stress condition (dotted circles) and the other, effective stress condition (solid circles). For each test, the Mohr's circle representing the total stresses has the same size as the Mohr's circle representing the effective stresses but the excess porewater pressure separates them horizontally. The Mohr's circle of effective stresses is shifted to the right if the excess porewater pressure at failure is negative and to the left if the excess porewater pressure is positive. Each Mohr's circle of total stress is associated with a particular value of undrained shear strength, s_u, because each test has a different initial water content (or void ratio) resulting from the different applied cell (confining) pressure during the first stage. For dense sand and overconsolidated clays, you can draw two groups of Mohr circles—one for peak stress and the other for critical state.

Procedure to Extract Shear Strength Parameters s_u, ϕ'_p, and ϕ'_{cs} from CU Tests on Uncemented Soils

1. Plot the axial stress versus axial strain. The axial stress is also called the deviatoric stress, $\sigma'_1 - \sigma'_3 = \sigma_1 - \sigma_3$, and the axial strain is the axial displacement divided by the original sample length.

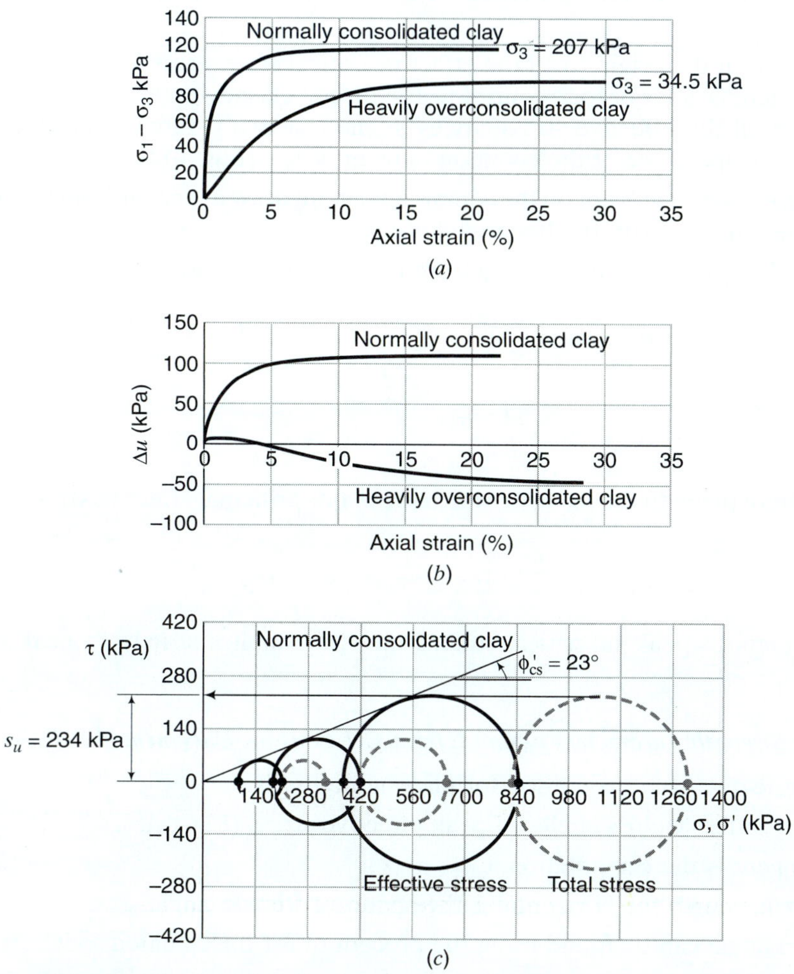

FIGURE 4.14 Triaxial CU tests on clays. (From *The Measurement of Soil Properties in the Triaxial Test*, by Bishop and Henkel, Edward Arnold, 1962.)

2. Plot a companion graph of excess porewater pressure versus axial strain.
3. Identify and extract the peak and critical axial stresses for each test at different cell pressures on the same soil. Dense sand and overconsolidated clays under high normal effective stresses or loose sand and normally consolidated clays do not exhibit a peak axial stress and only the critical axial stress can be extracted. In load control tests, you cannot obtain meaningful data beyond the peak axial stress.
4. Extract the excess porewater pressures at peak and critical state.
5. Calculate the effective minor and major principal stresses by subtracting the related excess porewater pressures. You should note that the axial stress is $\sigma_a = q = \sigma_1 - \sigma_3$, so $\sigma_1 = \sigma_a + \sigma_3$ and $\sigma'_1 = \sigma_a + \sigma_3 - \Delta u = \sigma_a + \sigma'_3$.
6. Plot Mohr's circle for both total and effective stresses.
7. The radius of each Mohr's total stress circle gives the undrained shear strength of the soil at the corresponding confining pressure or initial void ratio. In a graphical procedure, you draw horizontal lines from the top of each Mohr's circle to intersect the ordinate. The intercept gives the undrained

shear strength for each total stress circle. Alternatively, you can calculate the undrained shear strength from Eq. (4.18).

8. Draw a common tangent from the origin to all the effective critical state stress circles. The slope angle of the common tangent is ϕ'_{cs}. For load control tests, you can check whether you can draw a common tangent through all the effective stress circles or the last two if three tests have been conducted. If you can, the slope angle of the common tangent is ϕ'_{cs}. (Fig. 4.14).
9. If an antecedent effective stress circle is not on the common tangent, draw a tangent from the origin to it. The angle the tangent makes with the horizontal is ϕ'_p.
10. Alternatively, you can calculate the friction angles from the effective stresses using

$$\phi'_p = \sin^{-1}\frac{(\sigma' - \sigma'_3)_p}{(\sigma' + \sigma'_3)_p} \tag{4.19}$$

and

$$\phi'_{cs} = \sin^{-1}\frac{(\sigma' - \sigma'_3)_{cs}}{(\sigma' + \sigma'_3)_{cs}} \tag{4.20}$$

The inclination of the failure plane to the plane on which the major principal effective stress acts, θ, in the Mohr's circle is

$$\theta_i = \frac{\pi}{4} + \frac{\phi'_i}{2} \tag{4.21}$$

where the subscript i refers to peak or critical state. Eq. (4.21) applies strictly to peak stress condition.

Procedure to Extract Shear Strength Parameters c_o and ξ from CU Tests on Uncemented Soils

1. Follow steps 1 to 6 for effective stress condition for uncemented soils.
2. Draw a common tangent to the Mohr's circles of peak effective stress (Fig. 4.15).
3. The intercept of this tangent is the cementation, c_o.
4. The angle the tangent makes with the horizontal is the apparent friction angle, ξ.
5. The critical state friction angle can be found using the procedure for uncemented soils.

The Mohr-Coulomb failure expression for cemented soils is

$$\left(\frac{\sigma'_1}{\sigma'_3}\right)_{max} = \frac{1 + \sin\xi\left(1 + \dfrac{2c_o}{\sigma'_3}\cot\xi\right)}{1 - \sin\xi} \tag{4.22}$$

Sometimes cementation (cohesion) is mistakenly identified in soils. Consider a saturated clay that is being unloaded very quickly. The unloading will cause negative excess porewater pressures to develop within the clay mass that pull on the soil grains to bring them closer together. This is analogous to applying an isotropic confining stress to the clay. Consequently, the clay exhibits a shear strength (because it is a

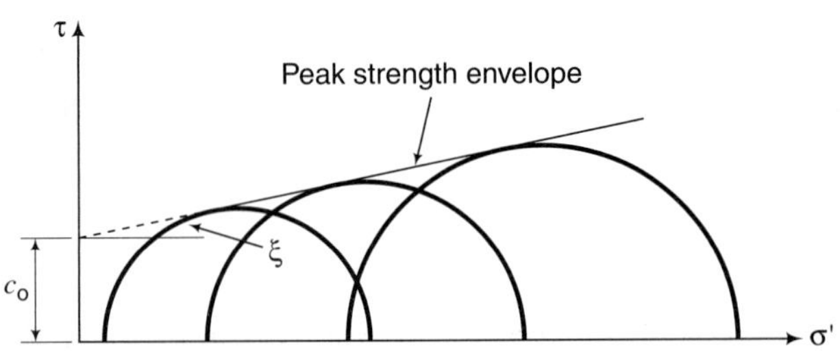

FIGURE 4.15 Mohr's circles for cemented soils.

frictional material) that is often called *cohesion*. But, this is misleading because it is an apparent shear strength that disappears once the excess porewater pressures drain from the soil. The negative excess porewater pressures are exploited in making temporary vertical, unsupported cuts in overconsolidated clays. You need to ensure that the time for the vertical cut to be left unsupported is short because as the negative excess porewater pressures dissipate with time, the effective stresses reduce and failure could occur.

The results from CU tests are used to analyze the stability of slopes, foundations, retaining walls, excavations, and other earthworks.

EXAMPLE 4.5 *Determination of s_u, ϕ'_p, and ϕ'_{cs} from CU test*

A CU test was conducted on a saturated overconsolidated clay by isotropically consolidating the soil using a cell pressure of 150 kPa (3135 psf) and then incrementally applying loads on the plunger while keeping the cell pressure constant. The axial stresses at peak and critical state were 266 kPa (5559 psf) and 148 kPa (3093 psf) respectively. The corresponding excess porewater pressures were −10 kPa (209 psf) and 28 kPa (585 psf). Determine the shear strength parameters, s_u, ϕ'_p, and ϕ'_{cs}.

Strategy You can calculate the effective strength parameters, ϕ'_p and ϕ'_{cs}, by using the Mohr-Coulomb failure criterion equations or you can determine them from plotting Mohr's circle. The undrained shear strength parameter, s_u, is found using Tresca's criterion. Remember that the axial stress is not the major principal stress σ_1 but $(\sigma'_1 - \sigma'_3) = (\sigma_1 - \sigma_3)$. SI units will be used to solve this example problem.

Solution 4.5

Step 1: Calculate the effective stresses at failure.

The subscripts p and cs denote peak and critical state, respectively.

$$(\sigma_1)_p = (\sigma_a)_p + \sigma_3 = 266 + 150 = 416\,\text{kPa}$$
$$(\sigma_1)_{cs} = (\sigma_a)_{cs} + \sigma_3 = 148 + 150 = 298\,\text{kPa}$$

$$(\sigma'_1)_p = (\sigma_1)_p - \Delta u_p = 416 - (-10) = 426\,\text{kPa}$$
$$(\sigma'_1)_{cs} = (\sigma_1)_{cs} - \Delta u_p = 298 - (28) = 270\,\text{kPa}$$

$$\sigma_3 = 150\,\text{kPa}, (\sigma'_3)_p = \sigma_3 - \Delta u_p = 150 - (-10) = 160\,\text{kPa}$$
$$(\sigma'_3)_{cs} = \sigma_3 - \Delta u_{cs} = 150 - (28) = 122\,\text{kPa}$$

Step 2: Determine ϕ'_p and ϕ'_{cs}.

$$\sin\phi'_p = \frac{(\sigma'_1 - \sigma'_3)_p}{(\sigma'_1 + \sigma'_3)_p} = \frac{266}{426 + 160} = 0.45; \quad \phi'_p = 26.7^\circ$$
$$\sin\phi'_{cs} = \frac{(\sigma'_1 - \sigma'_3)_{cs}}{(\sigma'_1 + \sigma'_3)_{cs}} = \frac{148}{278 + 122} = 0.37; \quad \phi'_{cs} = 21.7^\circ$$

Step 3: Determine the undrained shear strength.

$$(s_u)_p = \frac{(\sigma_1)_p - (\sigma_3)_p}{2} = \frac{(\sigma_a)_p}{2} = \frac{266}{2} = 133\,\text{kPa}$$
$$(s_u)_{cs} = \frac{(\sigma_a)_{cs}}{2} = \frac{148}{2} = 74\,\text{kPa}$$

■

Consolidated Drained Compression Test (CD) The purpose of a CD test is to determine the drained shear strength parameters ϕ'_{cs}, ϕ'_p, α_p, τ_p, and τ_{cs} for long-term loading of a soil mass. A CD test

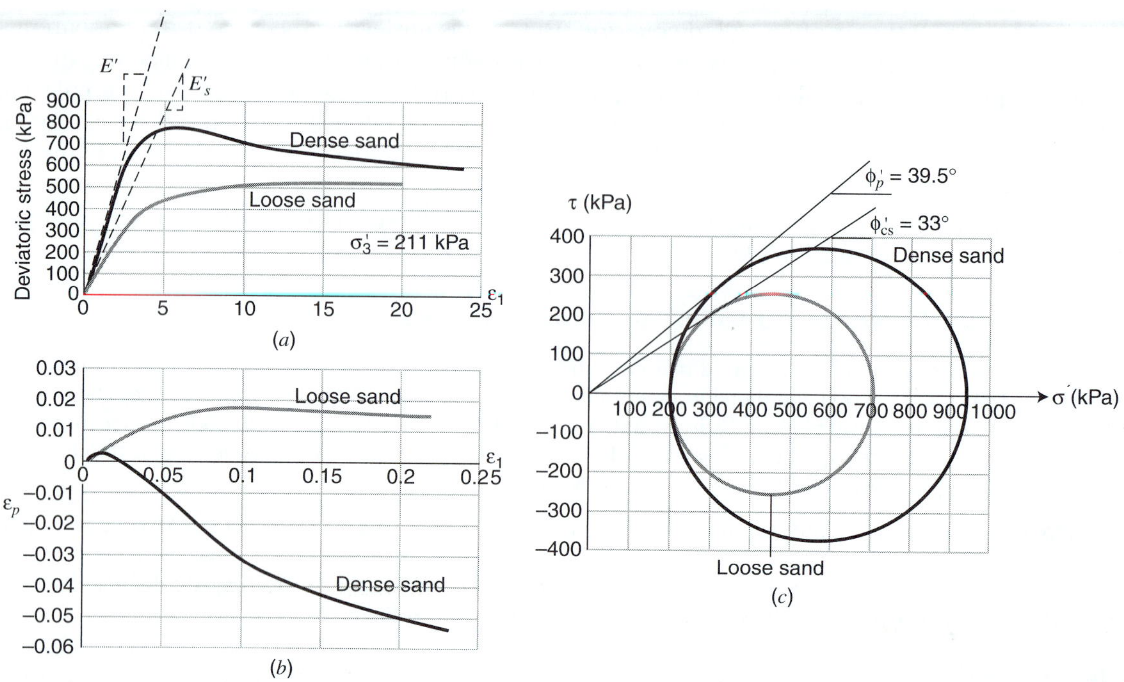

FIGURE 4.16 Triaxial CD tests on dense and loose sands. (From *The Measurement of Soil Properties in the Triaxial Test*, by Bishop and Henkel, Edward Arnold, 1962.)

is performed in two stages similar to the CU test except that in the second stage the soil is allowed to drain. The displacement rate (or strain rate) used must be slow enough to allow the porewater pressure to dissipate. Because the permeability of fine-grained soils is much lower than coarse-grained soils, the displacement rate for testing fine-grained soils is much lower than coarse-grained soils.

The Mohr-Coulomb failure criterion is used to interpret the test data. Typical results from CD tests are shown in Fig. 4.16. The results of CD tests are used to determine the long-term stability of slopes, foundations, retaining walls, excavations and other earthworks. The CD test is not very popular because it takes several days to perform a single test.

4.3.2.4 Direct Simple Shear Tests The direct simple shear test is a plane strain test in which the principal stresses rotate during the test. Two types of tests can be performed in the direct simple shear apparatus (Fig. 4.17). One is the constant load (CL) test in which the vertical load is held constant (drained condition). The other is the constant volume (CV) test in which the vertical load is adjusted to maintain constant volume (undrained condition). The vertical force, the shear (horizontal) force, the vertical displacement and the horizontal displacement are measured. The results from a direct

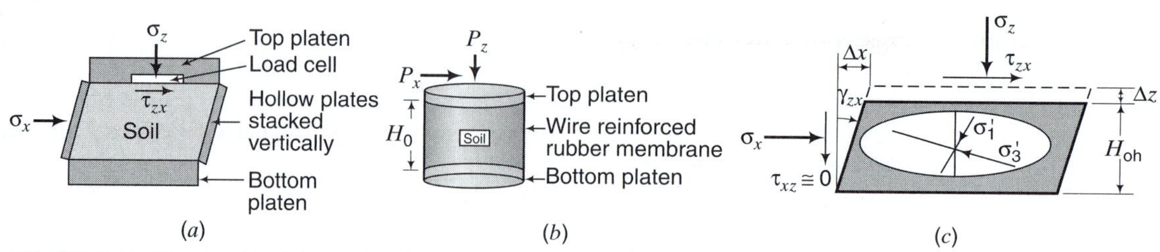

FIGURE 4.17 Two types of direct simple shear apparatuses: (a) rectangular cross-section, (b) circular cross-section, and (c) stresses imposed on samples.

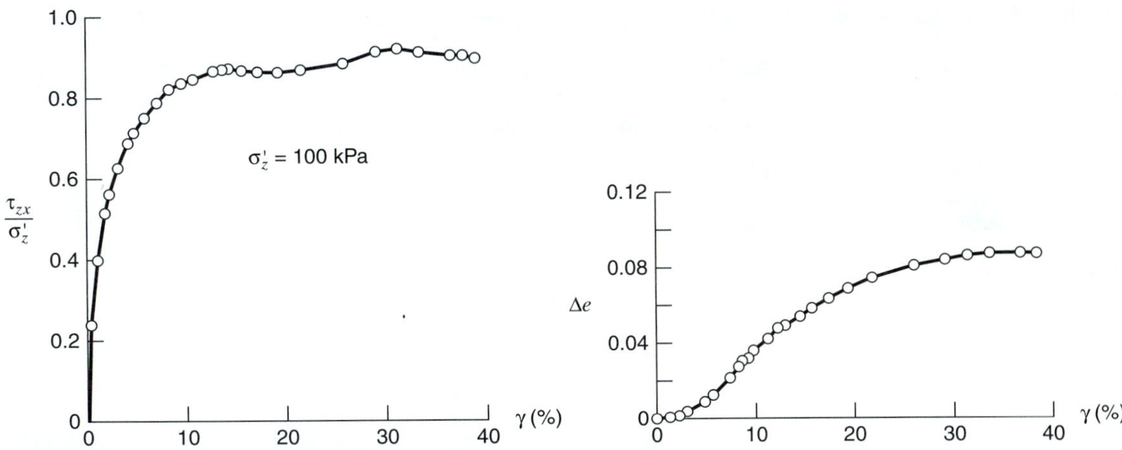

FIGURE 4.18 Direct simple shear results on a sand. (Budhu, 1983.)

simple shear CL test are shown in Fig. 4.18. The shear stress, τ_{zx}, is the horizontal load divided by the horizontal area, the vertical effective stress, σ'_z, is the vertical load divided by the horizontal area, the shear strain is the horizontal displacement (Δx) divided by the original sample height, H_o, and the vertical strain (ε_z) is the vertical displacement (Δz) divided by the original sample height. Compression is taken as positive.

For the CL test, Taylor's failure criterion is used to interpret the test data to give the friction and dilation angles. At the critical state, the volume change is zero. That is, $\frac{\Delta\varepsilon_z}{\Delta\gamma} = 0$ and Eq. (3.7) becomes

$$\phi'_{cs} = \tan^{-1}\left(\frac{\tau_{zx}}{\sigma'_z}\right)_{cs} \tag{4.23}$$

At peak stress condition, the peak dilation angle is $\alpha_p = \left|\tan^{-1}\left(\frac{\Delta\varepsilon_z}{\Delta\gamma}\right)_p\right|$ and Eq. (3.8) can be written as

$$\tan\phi'_p = \tan\phi'_{cs} + \tan\alpha_p \tag{4.24}$$

For the CV test

$$s_u = (\tau_{zx})_f \tag{4.25}$$

and

$$\Delta u = -\Delta\sigma_z \tag{4.26}$$

where $\Delta\sigma_z$ is the change in vertical stress required to maintain constant volume and the subscript f denotes failure. For soils that show both peak and critical state values, s_u would be different. If the soil shows a tendency to expand (negative excess porewater pressure), then additional vertical stress has to be applied to maintain constant volume. The reverse occurs for soils that show tendencies to compress. Thus positive changes in vertical stress are equivalent to negative excess porewater pressures and negative changes in vertical stress (reduction of vertical stress) are equivalent to positive changes in excess porewater pressures. Preferably excess porewater pressures should be measured because the vertical stress changes are not uniform throughout the soil sample.

The direct simple shear test is used to determine the shear strength (and shear stiffness) of soils under plane strain conditions (most prevalent condition in the field) but particularly to simulate the effects of seismic loadings. The direct simple shear test is less popular than either the direct shear or triaxial test because it is less available.

EXAMPLE 4.6 ***Determination of ϕ'_p, ϕ'_{cs} and α_p from Direct Simple Shear CL Test Data***

The peak shear stress in a CL direct simple shear test is 70 kPa (1463 psf) at a vertical effective stress of 100 kPa (2090 psf). If the change in volumetric and shear strains at the peak shear stress are −0.17% and 2% respectively, calculate the peak friction angle, the dilation angle, and the critical state friction angle.

Strategy Use Eqs. (4.23) and (4.24). U.S. units will be used to solve this example problem.

Solution 4.6

Step 1: Calculate peak friction angle.

$$\phi'_p = \tan^{-1}\left(\frac{\tau_{zx}}{\sigma'_z}\right)_p = \tan^{-1}\left(\frac{1463}{2090}\right) = 35^\circ$$

Step 2: Calculate dilation angle.

$$\alpha_p = \left|\tan^{-1}\left(\frac{\Delta\varepsilon_z}{\Delta\gamma}\right)\right| = \left|\tan^{-1}\left(\frac{[-0.0017]}{0.02}\right)\right| = 4.9^\circ$$

Step 3: Calculate critical state friction angle.

$$\begin{aligned}\phi'_{cs} &= \tan^{-1}(\tan\phi'_p - \tan\alpha_p)\\ &= \tan^{-1}(\tan 35^\circ - \tan 4.9^\circ) = 31.6^\circ\end{aligned}$$

■

The key points are:

1. ***The shear strength parameters are conventionally determined using direct shear or triaxial tests. Direct simple shear is used for special cases such as for seismic loadings.***
2. ***The type of test used depends on what parameters are required and what stress conditions need to be simulated.***
3. ***Different failure criteria are applied depending on the test conducted. For the direct shear tests, the Coulomb's failure criterion is used; for triaxial drained tests, the Mohr-Coulomb failure and Tresca's criteria are used; for direct simple shear tests, the Taylor's failure criterion is used.***

4.4 EXCESS POREWATER PRESSURE UNDER AXISYMMETRIC UNDRAINED LOADING

Importance *Knowledge of the excess porewater pressure is needed to evaluate the effective stress in the soil and to make an estimate of the time required for certain desired settlement to occur. Excess porewater pressure changes in soils are due to the changes in mean total and deviatoric stresses. From critical state theory, you can predict the excess porewater pressure for different stress paths. In the case of axisymmetric loading, a method proposed by Skempton (1954) is often used. In this section, Skempton's method is presented.*

Skempton (1954) proposed the following equation to determine the porewater pressure under axisymmetric conditions:

$$\Delta u = B[\Delta\sigma_3 + A(\Delta\sigma_1 - \Delta\sigma_3)] \tag{4.27}$$

where $\Delta\sigma_3$ is the increase in lateral principal stress, $\Delta\sigma_1 - \Delta\sigma_3$ is the deviatoric stress increase, B is a coefficient indicating the level of saturation, and A is an excess porewater pressure coefficient. The coefficient B is 1 for saturated soils and 0 for dry soils. However, B is not directly correlated with saturation except at high values of saturation ($S > 90\%$). The coefficient A was found to be dependent on the overconsolidation ratio (OCR). At failure,

$$A = A_f = \left(\frac{\Delta u_d}{\Delta\sigma_1 - \Delta\sigma_3}\right)_f \tag{4.28}$$

where Δu_d is the change in excess porewater pressure resulting from changes in deviatoric (shear) stresses. Experimental results of A_f presented by Skempton (1954) are shown in Appendix B (Table B4.7) and the variation of A_f with OCR is shown in Fig. 4.19.

The coefficients A and B are referred to as Skempton's porewater pressure coefficients. Eq. (4.27) is useful in estimating excess porewater pressures under axisymmetric condition in the field and in laboratory undrained triaxial tests when measurements of excess porewater pressures were not made.

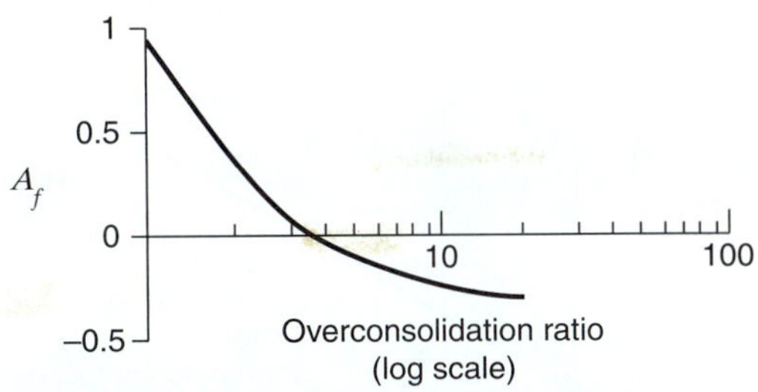

FIGURE 4.19 Variation of A_f with OCR.

The* key points *are:

1. ***Under an axisymmetric loading condition, the porewater pressure can be predicted using Skempton's porewater pressure coefficients, A and B.***
2. ***For a saturated soil, Skempton's porewater pressure coefficient B = 1.***

EXAMPLE 4.7 ***Application of Skempton's Porewater Pressure coefficients.***

The increases in vertical total stress and horizontal total stress below the center of a circular foundation are 10 kPa (209 psf) and 4 kPa (84 psf) respectively. The soil is a saturated overconsolidated clay with OCR = 8. Estimate the change in porewater pressure.

Strategy The soil is saturated, so B = 1. U.S. units will be used in the solution of this example problem.

Solution 4.7

Step 1: Estimate the porewater pressure coefficients.

Soil is saturated. Therefore B = 1.
From Fig. 4.19, with OCR = 8, $A_f = -0.2$.

Step 2: Estimate the change in porewater pressure.

$$\Delta u = \Delta\sigma_3 + A_f(\Delta\sigma_1 - \Delta\sigma_3)$$
$$\Delta\sigma_3 = 84\,\text{psf}, \Delta\sigma_1 = 209\,\text{psf}$$
$$\Delta u = 84 + [-0.2(209 - 84)] = 59\,\text{psf}$$

■

4.5 PRACTICAL IMPLICATIONS OF THE INTERPRETATION OF SHEAR STRENGTH PARAMETERS

Importance *Some of the failure criteria used for soils and how they are applied to interpret laboratory test results were presented in earlier sections. The practical implications of the interpretation of the test results using the three most popular failure criteria (Coulomb, Mohr-Coulomb, and Tresca), which treat soils as rigid bodies, are discussed in this section.*

We are going to define three regions of soil states as illustrated in Fig. 4.20 and consider practical implications of soils in these regions.

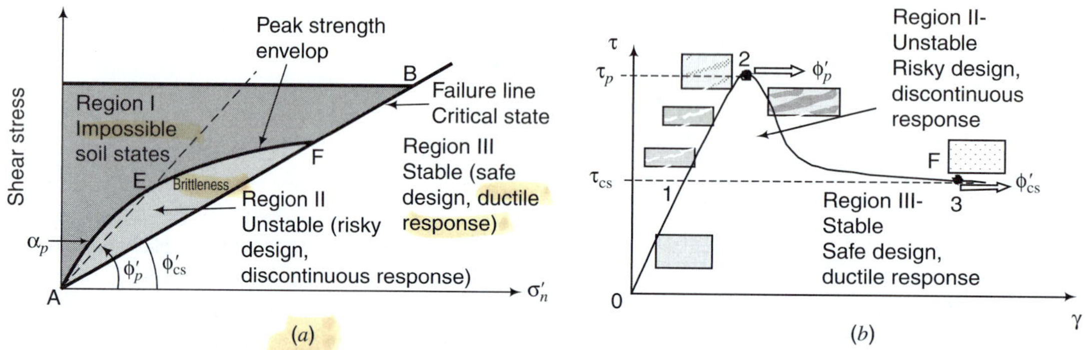

FIGURE 4.20 Interpretation of soil states.

Region I. *Impossible soil states.* A given soil cannot have soil states above the boundary AEFB.

Region II. *Impending instability (risky design).* Soil states within the region AEFA (Fig. 4.20a) or 1-2-3 (Fig. 4.20b) are characteristic of dilating soils that show peak shear strength and are associated with the formation of shear bands. During strain hardening (path 0-1-2, Fig. 4.20b), the grains set up columnar structures parallel to the major principal strain direction. As the peak shear strain is mobilized, these columns buckle and soil grains within them form shear bands as the soil strain softens (gets weaker) (Oda and Kazama, 1998). The shear bands consist of soils that have reached the critical state and are embedded within soil zones with high interlocking stresses due to the particle rearrangement. These shear bands grow as the soil strain softens to the critical state.

Soil stress states within AEFA (Fig. 4.20a) or 1-2-3 (Fig. 4.20b) are analogous to brittle material–type behavior. Brittle material (e.g., cast iron) will fail suddenly. Because of the shear bands that are formed within Region II, the soil permeability increases and if water is available it will migrate to and flow through them. This could lead to catastrophic failures in soil structures such as slopes since the flow of water through the shear bands can trigger flow slides. Design that allows the soil to mobilize stress states that lie within Region II can be classified as a risky design. Soil stress states that lie on the curve AEF can lead to sudden failure. We will call this curve the peak strength envelope (a curve linking the loci of peak shear strengths). The implication is that using ϕ'_p in designing geotechnical structures incurs a greater risk than ϕ'_{cs}.

Region III. *Stable soil states (safe design).* One of your aims as a geotechnical engineer is to design geotechnical systems on the basis that if the failure state were to occur, the soil would not collapse suddenly but will continuously deform under constant load. This is called *ductility*. Soil states that are below the failure line or failure envelop AB (Fig. 4.20a) or 0-1-3 (Fig. 4.20b) would lead to safe design. Soil states on AB are failure states. The implication is that using the critical state friction value, ϕ'_{cs} incurs less risk than ϕ'_p. Also, ϕ'_{cs} is a constant and a fundamental soil parameter while ϕ'_p is not a fundamental soil parameter but depends on the dilation capacity of the soil, which may be suppressed at large normal effective stresses.

A more comprehensive description of soil behavior can be interpreted using the critical state soil model (Schofield, 2001). The author has modified Schofield's conceptual model by introducing some additional regions of soil behavior (see Fig. 4.21). The line OC of slope $q/p' = 3$ delineates the boundary for no-tension in the soil. No soil states can exist in the area bounded by the deviatoric stress, q, axis and the line $q/p' = 3$. The line OB is the failure (critical state) line for compression and the line OG is the failure line for extension. You should note that the failure line for extension is shallower than that for compression, so a soil in extension will have a lower shear strength. The following interpretation of soil states refers to compression stresses. The soil states for extension are similar. For compression, effictive stress paths within the following areas: (1) ODBO will cause plastic yielding—ductile response, (2) BEC will lead to brittle failure and shear banding, (3) OCF will cause fissures or soil cracking to occur, and (4) CFE will cause a mix of cracks and shear bands. An effective stress path that reached the line (1) OC will cause rupture (open fissures), (2) CB will cause peak soil response (collapse) for dilative soils, and (3) OB will cause failure (critical state). The interpretation for effective stress paths that cause extension is similar. There is evidence from experiments especially for soil states in the areas BEC and ODBO to support the conceptual model. The critical state model is intended to provide a simple framework to understand soil behavior accounting for soil yielding, stress path dependency, strains, and stress history. The variability of real soils should not be expected to fit neatly into this framework.

The* key points *are:

1. ***Soil states above the peak shear strength boundary are impossible.***
2. ***Soil states within the peak shear strength boundary and the failure line (critical state) are associated with brittle-type soil responses and risky design.***
3. ***Soil states below the failure line lead to ductile response and are safe.***
4. ***The effective friction angle at the critical state, ϕ'_{cs}, is a fundamental soil parameter.***
5. ***The effective friction angle at peak shear stress for dilating soils, ϕ'_p, is not a fundamental soil parameter but depends on the capacity of the soil to dilate. Large normal effective stresses suppress dilation.***
6. ***Coulomb, Mohr-Coulomb, and Tresca failure criteria give information only on the soil shear strength. They do not give any information on the strains at which soil failure occurs.***
7. ***The critical state soil model provides a framework to interpret soil behavior.***

4.6 SOIL STIFFNESS

Importance *Soil stiffnesses are required to estimate settlement and deformation of soils. They are also invariably required as input parameters in numerical models. Various soil stiffnesses are used. You should understand the implications of using these different soil stiffnesses on the prediction of soil settlement and deformations.*

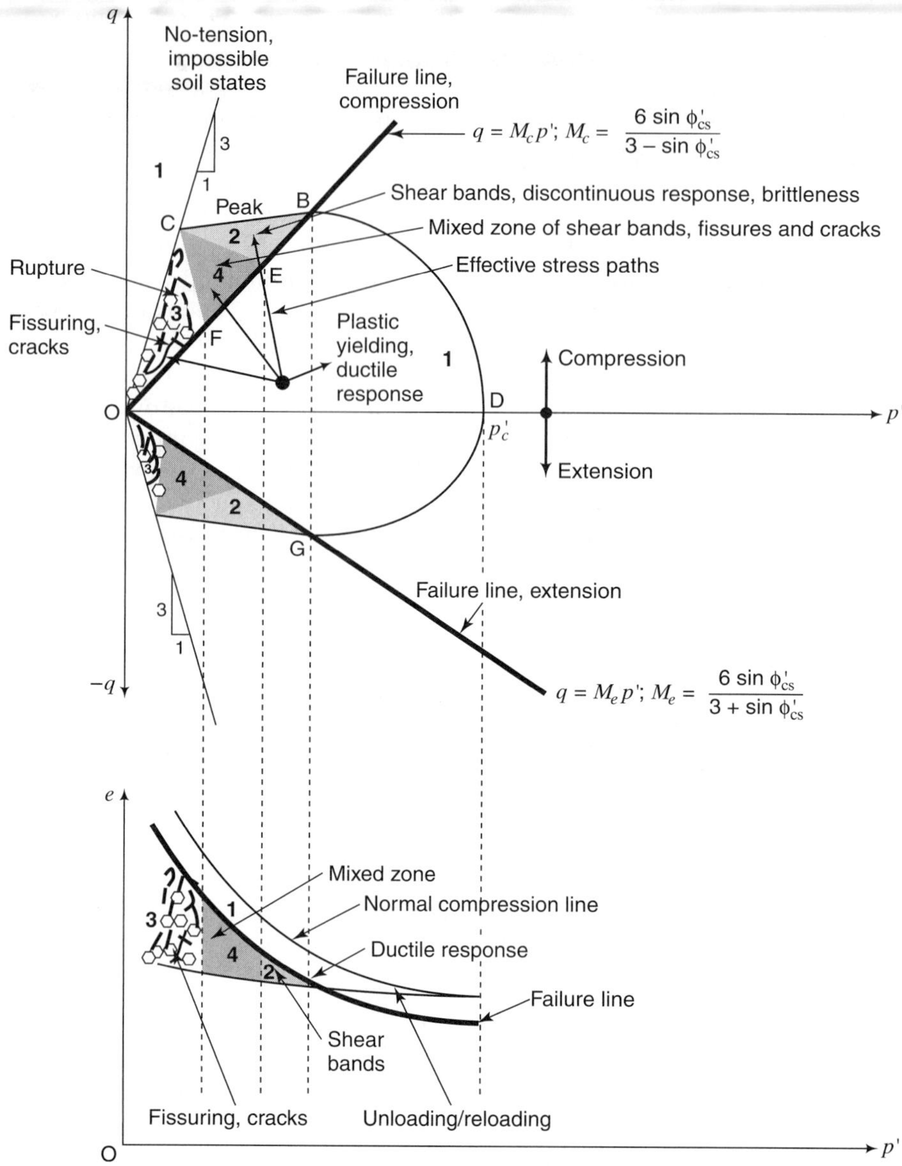

FIGURE 4.21 Interpretation of soil states using critical state theory. (Modified after Schofield, 2001)

The parameters that characterize soil stiffness are the elastic modulus, E, and Poisson's ratio, ν, or the shear modulus, G, and the bulk modulus, K. These moduli are interrelated as follows:

$$G = \frac{E}{2(1 + \nu)} \tag{4.29}$$

$$K = \frac{E}{3(1 - 2\nu)} \tag{4.30}$$

The shear modulus is particularly important because it relates the shearing of the soil to the elastic shear strains as

$$\gamma = \frac{2(1 + \nu)}{E}\tau = \frac{\tau}{G} \tag{4.31}$$

where γ is the shearing strain and τ is the shear stress. The bulk modulus, K, relates the mean effective stress (p') to the volumetric strain (ε_p) as

$$\varepsilon_p = \frac{p'}{K} \tag{4.32}$$

These moduli (E, K, G) strictly refer to linearly elastic materials. The elastic moduli for nonlinear materials (elastic or elastoplastic) such as soils are difficult to obtain from laboratory or in-situ tests because the range of strains for which the soil behaves elastic is very small and often not discernible from the test data. Consequently, the moduli that will be discussed here are approximations assuming that the soil behaves like an elastic material over the range of shear strain considered.

Various shear moduli are used in foundation engineering practice. The popular ones, with reference to Fig. 4.22, are:

1. The initial shear modulus, G_i (also called G_{max}), which is obtained from the slope of the initial part of the shear stress–shear strain plot.
2. The secant shear modulus, G_{sec}, which is obtained from the slope of the line linking the origin to the peak shear stress in the shear stress–shear strain plot. Sometimes the secant shear modulus is taken at other values of shear strain or shear stress such as at one-third the maximum deviatoric stress for mats and piled-raft foundations (Randolph, 1977).
3. The shear modulus, G_{50}, that is obtained from the slope of the line linking the origin to one-half the maximum shear stress.

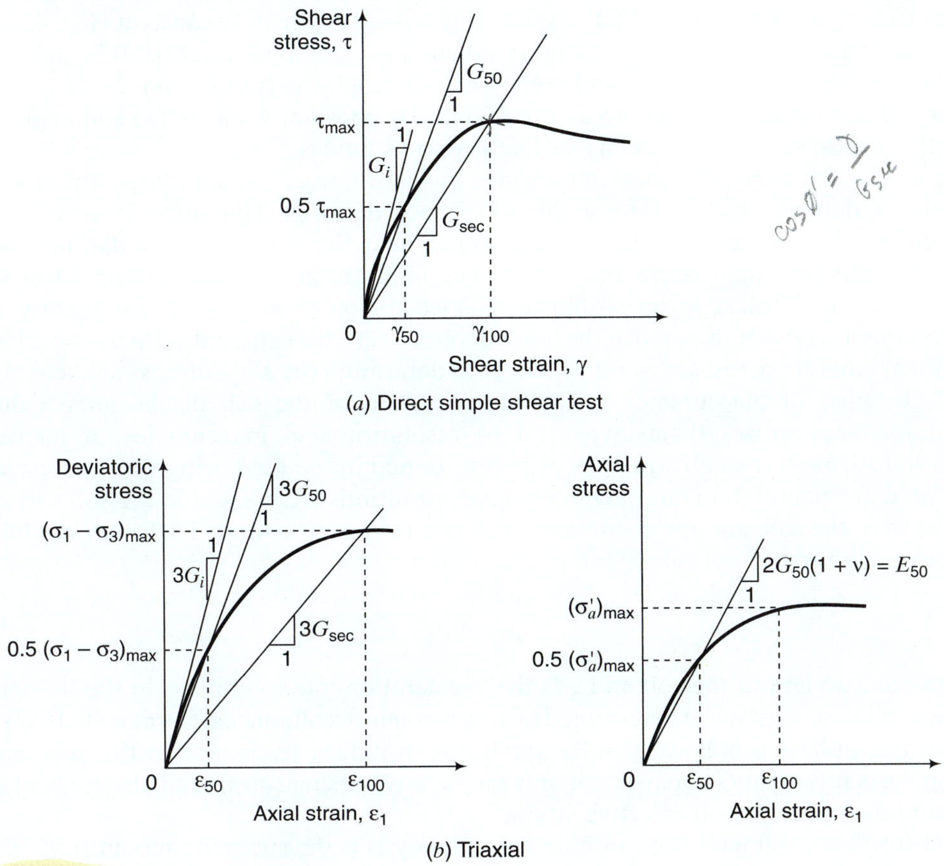

FIGURE 4.22 Illustration of shear moduli.

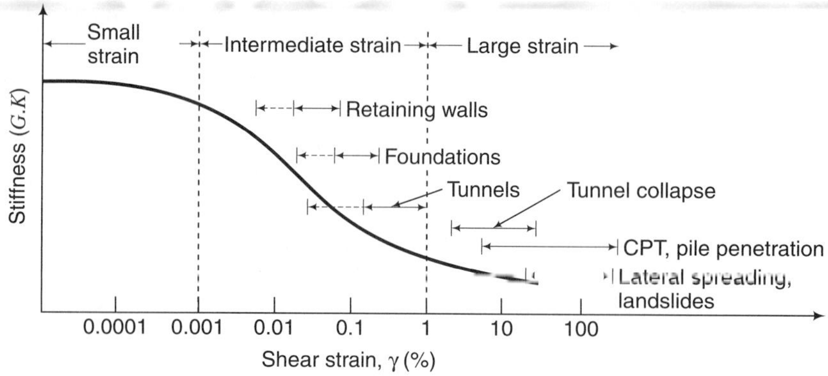

FIGURE 4.23 Soil stiffness at various shear strain ranges for practical problems. (Modified after Mair, 1993.)

When drained condition (effective stress analysis) is considered, the elastic moduli must be obtained from effective stresses. Typical values for ν, E, and G are given in Appendix B (Tables B4.8 and B4.9). For undrained condition, Poisson's ratio is 0.5. A value slightly lower than 0.5 is used in numerical analysis to avoid numerical errors. The effective stress parameters are denoted by the prime symbol.

Soil stiffness is influenced by the amount of shear strains applied to the soil. Increases in shear strains tend to lead to decreases in G and E while increases in volumetric strains lead to decreases in K. The net effect is that the soil stiffness decreases with increasing strains as illustrated in Fig. 4.23.

It is customary to identify three regions of soil stiffness based on the level of applied shear strains. At small shear strains (γ or ε_d usually $<0.001\%$), the soil stiffness is approximately constant (Fig. 4.23) and the soil behaves like a linearly elastic material. At intermediate shear strains between 0.001% and 1%, the soil stiffness decreases significantly and the soil behavior is elastoplastic (nonlinear). At large strains ($\gamma > 1\%$), the soil stiffness decreases slowly to an approximately constant value as the soil approaches critical state. At the critical state, the soil behaves like a viscous fluid.

For most practical problems, the shear strains are in the intermediate and large strain ranges. However, the shear strain distribution within the soil is not uniform. The shear strains decrease with distance away from a structure and local shear strains near the edge of a foundation slab, for example, can be much greater than below the center. The implication of a nonuniform shear strain distribution is that the soil stiffness varies within the loaded region of the soil. Consequently, large settlements and failures are usually initiated in the loaded soil region where the soil stiffness is the lowest.

In conventional laboratory tests, it is not practical to determine the soil stiffness at shear strains less than 0.001% because of inaccuracies in the measurement of the soil displacements due to displacements of the apparatuses themselves and to resolution and inaccuracies of measuring instruments. The soil stiffness at small strains is best determined in the field using wave propagation techniques presented in Chapter 2. In one such technique, vibrations are created at the soil surface or at a prescribed depth in the soil, and the shear wave velocity (V_s) is measured. The shear modulus at small strains is calculated from

$$G = \frac{\gamma V_s^2}{g} \tag{4.33}$$

where γ is the bulk unit weight of the soil and g is the acceleration due to gravity. In the laboratory, the shear modulus at small strains can be determined using a resonance column test (Drnevich, 1967). The resonance column test utilizes a hollow cylinder apparatus to induce resonance of the soil sample. Resonance column tests show that G depends not only on the level of shear strain but also on void ratio, overconsolidation ratio, and the mean effective stress.

Another soil stiffness parameter used in foundation analysis is the subgrade modulus, k_s, that is obtained from a plate load test (see Chapter 5) as the load at failure divided by the vertical settlement. The relation between k_s and elastic modulus (Vesic, 1961) is

$$k_s = \frac{0.65}{B}\left(\frac{E_s B^4}{E_f I_f}\right)^{\frac{1}{12}} \frac{E_s}{1-\nu^2} \tag{4.34}$$

where B is the foundation width, E_f is the elastic modulus of the foundation, and I_f is the second moment of area of the foundation.

The prediction of the amount of soil deformation that occurs is dependent on the choice of the soil stiffness parameter, which requires experience. Two key criteria are the stresses imposed by the loads rather than the magnitude of the loads and the soil type. You should recall from Chapter 2 that the stresses imposed in the soil by a uniform structural load on a foundation are nonuniformly distributed. Consequently, any soil stiffness parameter selected will apply for an averaged stress condition. The following broad recommendation is made in lieu of experience. For lightly stressed soils, where the estimated shear strains are likely to be small ($<0.001\%$), G_{max} should be used. For heavily stressed soils (near to the failure state), G_{sec} should be used; G_{50} should be used for intermediate stress levels. The type of soil is an important consideration in selecting G. For a given stress level, a soft soil will deform much more than a stiff soil and either G_{50} or G_{sec} should be used.

EXAMPLE 4.8 *Determination of G, K, and E from Direct Simple Shear Test Data*

Determine G_{max}, G_{50}, and G_{sec} from the simple shear test results shown in Fig. E4.8. If Poisson's ratio is 0.3, estimate the corresponding bulk moduli (K) and E_{50}.

Strategy Follow the illustration shown in Fig. 4.22 to determine G.

Solution 4.8

Step 1: Determine G_{max}, G_{50}, and G_{sec}.

Estimate the maximum shear stress ratio by drawing a horizontal line as shown in Fig. E4.7. Draw a vertical line from the maximum shear stress ratio to intersect the shear strain axis to get γ_{100}.

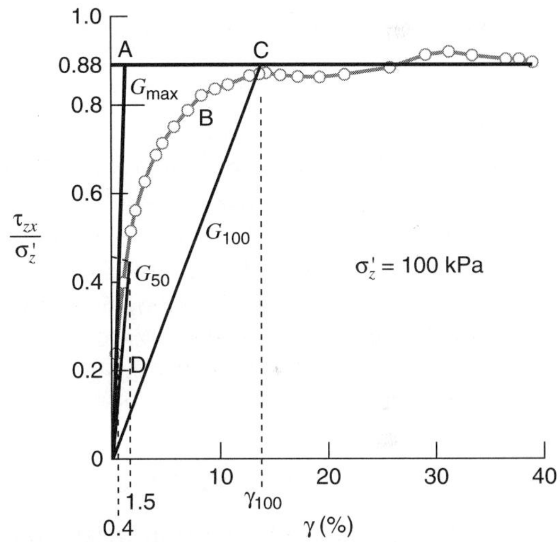

FIGURE E4.8

$$\left(\frac{\tau_{zx}}{\sigma'_z}\right)_{max} = 0.88; \quad \sigma'_z = 100\,\text{kPa}; \quad (\tau_{zx})_{max} = 0.88 \times 100 = 88\,\text{kPa}; \quad \gamma_{100} = 13\%$$

$$\text{one half maximum shear stress} = 0.5\left(\frac{\tau_{zx}}{\sigma'_z}\right)_{max} = 0.44; \quad \tau_{zx} = 0.44 \times 100 = 44\,\text{kPa}, \quad \gamma_{50} = 1.5\%$$

Draw a tangent to the initial portion of the shear stress ratio–shear strain curve as illustrated by OA. Identify the coordinates of a convenient point on the line OA. At D the coordinate is (0.2, 0.4).

$$\text{Slope OA} = G_{max} = \frac{\frac{\tau_{zx}}{\sigma'_z} \times \sigma'_z}{\gamma} = \frac{0.2 \times 100}{0.004} = 5000\,\text{kPa}$$

$$\text{Slope OC} = G_{sec} = \frac{(\tau_{zx})_{max}}{\gamma_{100}} = \frac{88}{0.13} = 677\,\text{kPa}$$

$$\text{Slope OB} = G_{50} = \frac{(\tau_{zx})_{50}}{\gamma_{50}} = \frac{44}{0.015} = 2933\,\text{kPa}$$

Step 2: Estimate K.

From Eqs. (4.29) and (4.30),

$$\frac{K}{G} = \frac{2(1+\nu)}{3(1-2\nu)}$$

$$\therefore \; K = \frac{2(1+\nu)}{3(1-2\nu)}G = \frac{2(1+0.3)}{3(1-2\times 0.3)}G = 2.17G$$

$$K_{max} = 2.17G_{max} = 2.17 \times 5000 = 10850\,\text{kPa}$$

$$K_{sec} = 2.17G_{sec} = 2.17 \times 677 = 1469\,\text{kPa}$$

$$K_{50} = 2.17G_{50} = 2.17 \times 2933 = 6365\,\text{kPa}$$

Step 3: Estimate E_{50}.

From Eq. (4.29),

$$E_{50} = 2G_{50}(1+\nu) = 2 \times 2933(1+0.3) = 7623\,\text{kPa}$$

■

4.7 CORRELATIONS OF SETTLEMENT, STRENGTH, AND STIFFNESS PARAMETERS USING SIMPLE LABORATORY TESTS

Importance *Laboratory tests for settlement and strength parameters take several hours and in most cases days to obtain and are more costly than simple tests to determine physical parameters (Chapter 1). Over the years, geotechnical engineers correlated settlement and strength parameters with these simple tests, particularly the index tests. Many of these correlations are weak (low regression coefficients) and in the case of the strength parameters it is unclear whether peak or critical state parameters were used. In this section, some of the popular correlations are presented. You must be cautious when using these correlations. They are best used for preliminary analysis.*

4.7.1 Settlement Parameters

Some relationships between simple physical laboratory soil tests and consolidation settlement parameters are given in Appendix B (Figs. B4.1 and B4.2, and Tables B4.1 to B4.4 and B4.10).

4.7.2 Shear Strength Parameters

Undrained Shear Strength The undrained shear strength has been correlated with index tests. Some of the correlations are given in Appendix B (Fig. B4.3 and Table B4.11).

Friction Angles Friction angles have been correlated with plasticity index for fine-grained soils and relative density for coarse-grained soils. In many of these correlations, it is unclear whether the friction angle is the peak or critical state value. For relative densities less than 50%, you may assume that $\phi' = \phi'_{cs}$. Typical values of friction angles and dilation angle and correlations with simple tests are given in Appendix B (Tables B4.5 and B4.6, and Figs. B4.4 to B4.6).

4.7.3 Soil Stiffness

A large number of equations have been proposed in the literature for soil stiffness based on correlations with laboratory tests. In this book, only some of the correlations are presented. In most cases, it is unclear whether the correlations refer to the initial or secant modulus and, if the secant modulus was quoted, what stress or shear strain level was used. The correlations presented are assumed to correspond to the secant modulus at peak or maximum shear stress unless otherwise stated. You can use Eqs. (4.29) or (4.30) to calculate the desired modulus if it is not specifically given. Typical values for Poisson's ratio, E and G are shown in Appendix B (Tables B4.8 and B4.9). Correlations are shown in Appendix B (Figs. B4.7 and Table B4.12) Correlations for k_s with USCS is given in Table B4.13.

EXAMPLE 4.9 *Estimation of Settlement and Strength Parameters*

The index limits of normally consolidated saturated clay soil taken at a depth 2 m below the surface are LL = 82%, PL = 28% and the natural water content is 38%. The groundwater is at the surface. Assume $G_s = 2.7$. Estimate

(a) The compression index, C_c
(b) The coefficient of consolidation, C_v
(c) The undrained shear strength
(d) The friction angle
(e) The shear modulus
(f) The bulk modulus assuming $\nu = 0.45$

Strategy Use the data and relate them to the tables and figures given in Appendix B.

Solution 4.9

Step 1: Find PI, LI, and σ'_{zc}.

$$\text{PI} = \text{LL} - \text{PL} = 82 - 28 = 54\%$$
$$\text{LI} = \frac{w - \text{PL}}{\text{PI}} = \frac{38 - 28}{54} = 0.185$$
$$e_o = wG_s = 0.38 \times 2.7 = 1.03$$
$$\gamma_{sat} = \left(\frac{G_s + e_o}{1 + e_o}\right)\gamma_w = \left(\frac{2.7 + 1.03}{1 + 1.03}\right)9.8 = 18.0\,\text{kN/m}^3$$

Step 2: Estimate C_c.
Appendix B (Table B4.2);

$$C_c = 0.009(\text{LL} - 10) = 0.009(82 - 10) = 0.65$$
$$C_c = 0.4(e_o - 0.25) = 0.4(1.03 - 0.25) = 0.31$$
$$C_c = \frac{\text{PI}}{74} = \frac{54}{74} = 0.73$$
$$C_c = 0.00234 \times \text{LL} \times G_s = 0.00234 \times 82 \times 2.7 = 0.52$$

Two of the estimated C_c are close to each other. You can take the average of these as the estimated C_c.

$$C_c = \frac{0.65 + 0.73}{2} = 0.69$$

You should note that the estimated value of C_c is within the range of C_c (0.1 to 0.8) obtained for soils.

Step 3: Estimate C_v.
From Appendix B (Fig. B4.1), with LL = 84% and using the curve for remolded soil (the Atterberg limits are found using remolded soil),

$$C_v = 0.5 \times 10^{-8}\text{m}^2\text{/sec}$$

Step 4: Estimate s_u.

Appendix B (Table B4.11):

$$\left(\frac{s_u}{\sigma'_{zo}}\right)_{nc} = 0.11 + 0.0037\,\text{PI} = 0.11 + 0.0037(54) = 0.31$$

$$\left(\frac{s_u}{p'_o}\right)_{nc} = 0.129 + 0.00435\,\text{PI}$$

Assume $p'_o = \sigma'_{zo}$(i.e., $K_o = 1$, isotropic condition).

$$\left(\frac{s_u}{\sigma'_{zo}}\right) = 0.129 + 0.00435 \times 54 = 0.36$$

Take an average value, $\frac{s_u}{\sigma'_{zo}} = \frac{0.31 + 0.36}{2} = 0.34$.

$$s_u = 0.34\sigma'_{zo} = 0.34 \times 16.8 = 5.7\,\text{kPa}$$

Step 5: Estimate the friction angle.

Since the soil is normally consolidated, $\phi' = \phi'_{cs}$.

Appendix B (Fig. B4.4): PI = 54%, $\phi' = \phi'_{cs} = 21°$.

Step 6: Estimate G_{sec}.

Appendix B (Table B4.12): Use Wroth (1979) expression.

$$\left(\frac{G_{sec}}{s_u}\right) = \frac{2M(1 + e_o)}{\lambda\Lambda(1 - \Lambda)\exp(-\Lambda)}$$

$$\phi'_{cs} = 21°, M = \frac{6 \sin\phi_{cs}}{3 - \sin\phi_{cs}} = \frac{6\sin(21°)}{3 - \sin(21°)} = 0.81$$

Assume $\Lambda = 0.8, \lambda = \frac{C_c}{2.3} = \frac{0.52}{2.3} = 0.23$.

$$\left(\frac{G_{sec}}{s_u}\right) = \frac{2 \times 0.81(1 + 1.03)}{0.23 \times 0.8(1 - 0.8)\exp(-0.8)} = 199$$

$$G_{sec} = 199 \times 5.7 = 1134\,\text{kPa}$$

Step 7: Estimate the bulk modulus.

$$K_{sec} = \frac{2G_{sec}(1 + \nu)}{3(1 - 2\nu)} = \frac{2 \times 1134(1 + 0.4)}{3(1 - 2 \times 0.4)} = 5292\,\text{kPa}$$

■

4.8 DIFFICULT SOILS

Importance *Difficult soils are a genre of soils including fill, rubble, expansive soils, collapsible soils, peat, and sensitive soils. These soils generally have (1) low bearing capacity, (2) high settlement potential, and (3) properties that significantly change with time and environmental conditions. The cost of a foundation system can increase significantly depending on the type and extent of difficult soils at the project site. In this section, a brief introduction to two difficult soils—expansive and collapsible soils—is presented.*

TABLE 4.1 Expansion Potential

EI	Expansion potential
0 to 20	Very low
21 to 50	Low
51 to 90	Medium
91 to 130	High
>130	Very high

Expansive soils expand and contract due to changes in water content. They can cause extensive structural damage through differential settlement (cracking walls, warping, and cracking of floors). The predominant mineral in expansive soils is montmorillonite (smectite). The expansion potential of montmorillonite depends on the exchangeable ion. The three principal exchangeable ions are sodium (Na^{+1}), potassium (K^{+1}), and calcium (Ca^{+2}). Montmorillonite with sodium exchangeable ion has the most expansion potential.

ASTM D4829 and ASTM D4546 provide test methods to determine the expansion potential, which is quantified by the expansion index, EI, given as

$$\mathrm{EI} = 1000\frac{\Delta H}{H_o} \tag{4.35}$$

where ΔH is the change in vertical height and H_o is the initial height of the soil sample. Descriptions of the expansion potential based on EI are given in Table 4.1.

Foundations on soils with $\mathrm{EI} > 20$ must be designed to prevent structural distress from soil expansion. Table 4.2 can give you a quick estimate of the expansion potential based on the fines content (% passing #200 sieve), the Atterberg limits, or SPT.

standard Penetration Test.

TABLE 4.2A Estimation of Potential Volume Changes for Expansive Soils [Source: modified after Chen, 1975]

Percent passing #200 sieve	Liquid limit	Standard penetration resistance	Probable expansion	Swelling pressure	Degree of expansion
	(%)	(blows/ft)	%	(kPa)	
35	60	30	10	100	Very high
60–90	40–60	20–30	3–10	25–99	High
30–60	30–40	10–20	1–5	15–24	Medium
<30	<30	<30	<1	<15	Low

Source: Chen, 1975.

TABLE 4.2B Estimation of Potential Volume Changes for Expansive Soils [Source: Reese and O'Neill, 1988]

Liquid limit	Plastic limit	Suction pressure	Potential swell	Potential swell classification
(%)	(%)	kPa (ksf)	%	
>60	>35	>380 (8)	>1.5	High
50–60	20–30	140 (4)–380 (8)	0.5–1.5	Marginal
<50	10–20	<140 (4)	<0.5	Low

Source: Reese and O'Neill, 1988.

Activity (Skempton, 1953) can also be used to give a quick estimate of the expansion potential. Activity (A) is

$$A = \frac{\text{PI}}{\%\text{ finer than } 2\mu\text{m}} = \frac{\text{PI}}{\%\text{ clay}} \tag{4.36}$$

Expansive soils normally have A values greater than 1.25.

The uplift pressures from expansive soils can be estimated from

$$p_{\text{uplift}} = 4.8(\text{PI} - 10)\text{ kPa} \tag{4.37}$$

$$p_{\text{uplift}} = 100(\text{PI} - 10)\text{ psf} \tag{4.37a}$$

where PI is the plastic index (%). An expansive soil with a plasticity index of 40% will cause an estimated uplift pressure of about 144 kPa (3000 psf).

Collapsible soils are unsaturated and are often found in arid climates, alluvial fans, colluviums, and wind-blown deposits. When water is available to a collapsible soil, it causes the soil structure to collapse. Nonuniform availability of water causes differential settlement and structural damage. Loads can also cause the soil structure to collapse with or without the presence of water.

The collapse potential due to groundwater can be determined using a oedometer. Incremental vertical stresses are imposed on the soil sample followed by flooding. After flooding, a higher vertical stress is added (see ASTM D5333). The collapse potential is the percent change in sample height divided by the original sample height during the period of flooding as illustrated in Fig. 4.24. A description of the collapse potential of a soil is given in Table 4.3.

TABLE 4.3 Description of Collapsible Soils

$\frac{\Delta H}{H_o}$ (%)	Severity
0–1	No concern
1–3	Low
3–5	Moderate
>5	High

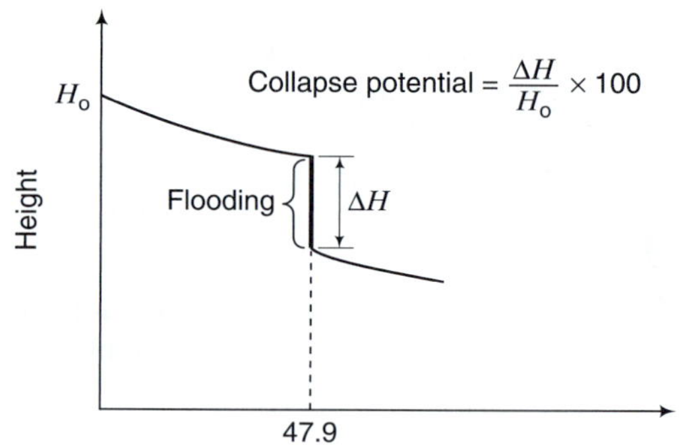

FIGURE 4.24 Collapse potential.

The* key points *are:

1. ***Expansive soils swell in contact with water.***
2. ***Collapsible soils are unsaturated deposits that settle suddenly in contact with water and/or with the application of loads.***
3. ***Expansive and collapsible soils can cause extensive damage to foundations.***

4.9 SUMMARY

In this chapter, we discussed the interpretation of laboratory tests data relevant to foundation design. The key parameters are the settlement parameters (C_c or m_v, C_r or m_{vr}, C_v, C_α, σ'_{zc}), the strength parameters s_u, ϕ'_{cs}, and ϕ'_p, and the deformation parameters G and K. These parameters can be obtained from one-dimensional consolidation, direct shear, triaxial, and simple shear tests. Various correlations have been proposed among these parameters and, in particular, index properties. Most of these correlations are weak and they are best used in preliminary or exploratory analyses.

EXERCISES

4.1 A soil profile consists of a 6 m (20 ft) of dense sand, unit weight, $\gamma = 17\,\text{kN/m}^3$ (106 pcf) followed by a deep deposit of blue clay. The ground water level is at 5 m (15 ft) from the surface. At a depth of 8 m (25 ft), a sample of the clay was extracted and tested in an oedometer. The following results were obtained:

Vertical effective stress—kPa (ksf)	50 (1)	100 (2)	200 (4)	400 (8)	800 (16)	400 (8)	200 (4)	400 (8)	800 (16)	1600 (32)
Void ratio	0.89	0.875	0.82	0.75	0.685	0.68	0.691	0.69	0.67	0.62

Calculate (a) uncorrected C_c, (b) corrected C_c based on Schmertmann's method, (c) the corrected λ, (d) C_r, (e) σ'_{zc}, and (f) OCR. The initial water content of the soil is 34%.

4.2 The following results were obtained from four shear box tests on a sample of sandy clay. The cross section of the shear box was 6 cm (2.5 in.) × 6 cm (2.5 in.).

Normal force N (lb)	1250 (280)	1000 (225)	500 (112)	250 (56)
Shearing force N (lb)	506 (113)	405 (91)	325 (73)	255 (57)

(a) Is the soil dense or loose? Justify your answer.

(b) Which failure law is appropriate to interpret the results of the shear box test?

(c) Determine ϕ'_{cs}.

(d) If the soil sample is dense, determine ϕ'_p at a normal force of 500 N (112 lb). Also calculate the peak dilation angle.

(e) What is the angle of inclination of the failure plane to the horizontal plane in the shear box test?

(f) Are the stresses and strains in the shear box uniform? Justify your answer.

(g) Give one practical example where the results of the shear box may apply.

4.3 A cylindrical sample of clay soil, 50 mm (2 in.) diameter × 100 mm (4 in.) long, was tested under unconfined compression. When the vertical displacement reached 2 mm (0.08 in), the soil failed. The axial force at failure was 270 N (60 lb). Determine the (a) axial strain at failure and (b) undrained shear strength.

4.4 Triaxial CU tests were carried out on two samples of a clay. Each sample was isotropically consolidated before the axial stress was increased. The cell pressure was held constant during shearing. The following results were obtained at maximum deviatoric stress.

Sample no.	$(\sigma_3)_{max}$ kPa (ksf)	$(\sigma_1 - \sigma_3)_{max}$ kPa (ksf)	Δu_{max} kPa (ksf)
I	420 (8.8)	320 (6.7)	205 (4.3)
II	690 (14.4)	365 (7.6)	350 (7.3)

(a) Draw Mohr's circles (total and effective stresses) for each test on the same graph.

(b) Why are the total stress circles not the same size?

(c) Which failure criterion (criteria) is (are) appropriate to interpret the undrained and drained shear strength parameters?

(d) Determine the friction angle(s). Is the soil at critical state? Justify your answer.

(e) Determine the undrained shear strength at a cell pressure of 690 kPa (14.4 ksf).

(f) Determine the shear stress on the failure plane for each sample.

4.5 A standard triaxial CU test was conducted on a soil sample in a triaxial test. The soil was initially consolidated with a cell pressure of 50 kPa (1 ksf), which was held constant during shearing. The results of axial stress and axial strain are as follows:

σ_a (kPa) (ksf)	0	54 (1.1)	108 (2.2)	175 (3.6)	209 (4.4)	249 (5.2)	230 (4.8)	229 (4.78)	229 (4.78)
ε_1	0	0.002	0.0048	0.01	0.02	0.035	0.05	0.075	0.1

Determine (a) s_u, (b) E_{max} and G_{max}, (c) E_{sec} and G_{sec}, and (d) E_{50} and G_{50}.

Assume Poisson's ratio is 0.4.

4.6 A circular tank 5 m (15 ft) in diameter is constructed on a deep deposit of saturated overconsolidated clay with OCR = 10. The weight of the tank and its foundation is 4000 kN (900 kips). The groundwater level is at the surface. The tank will be constructed quickly. Estimate the excess porewater pressure at a depth of 2 m (6 ft) below the center of the tank during construction using Skempton's method.

4.7 The results from a constant volume direct simple shear test on a clay are as follows. Initial vertical stress is $\sigma_{zo} = 100$ kPa (2090 psf); at failure, $(\tau_{zx}) = 62$ kPa(1296 psf), $\Delta\sigma_z = 90$ kPa(1890 psf).

(a) Which failure criterion is appropriate for the direct simple shear test?

(b) Calculate s_u, Δu, and the ratio $\frac{s_u}{\sigma_z'}$.

4.8 The liquid limit and plastic limit of a saturated soil with a natural water content of 32% taken at a depth of 3 m (6 ft) below the ground surface are 73% and 28% respectively. Groundwater level is at the ground surface. Assume $G_s = 2.7$. Estimate the following using the correlations given in Appendix B.

(a) C_c and λ

(b) C_r and κ

(c) OCR

(d) ϕ'

(e) s_u

(f) K_o

(g) G_{sec} and E_{sec}

(h) k_s

(i) Is the compressibility of the soil low, moderate or high?

4.9 The percent swell from an ASTM D4546 expansion potential test of the soil in Problem 4.8 is 14.3%. The percent passing the #4 sieve is 100. → assumption: no gravel

(a) Calculate the expansion index. 14.3 x 1000

(b) Evaluate the expansion potential.

(c) Estimate the uplift pressure.

(d) Assuming that the average dead load of a multistory building is 20 kPa per story, estimate how many stories will be required to balance the uplift pressure.

4.10 One year after occupying her new home, a homeowner observed that the walls had a large number of cracks and sued the builder. The foundation near the landscaped area appeared to be uplifted. The builder delivered the house to the homeowner with no landscaping. After 9 months of owning the home, the homeowner landscaped the area adjacent to the front of the house with grass and shrubs. A watering system was installed but was set at a flow rate three times the absorption rate of the grass and shrubs. The liquid and plastic limits of the soil are 65% and 29% respectively. What is your technical opinion on this matter? Justify your opinion.

CHAPTER 5

SITE CHARACTERIZATION

5.0 INTRODUCTION

Adequate knowledge of ground conditions is very important for analysis, design, and construction of geosystems. Project delays, legal wrangling, soil failures, and cost overrun are often the results of inadequate and inappropriate site investigations. A site investigation must be part of the design process and, while it is important to keep costs down, you should not sacrifice good ground information to save a few dollars to meet minimum cost. In fact, the cost of site investigations for foundation design rarely exceeds 0.5% of project costs.

In this chapter, you will learn about the need for site characterization and how to conduct a soils investigation and specify soil tests. When you complete this chapter, you should be able to:

- Understand the purposes of a site characterization.
- Plan a site investigation.
- Specify laboratory and in-situ tests.

5.1 DEFINITIONS OF KEY TERMS

Groundwater is the free water in a saturated soil.

Shelby tube is a seamless thin-walled sampler of diameter 50 mm (2 in) or 75 mm (3 in) and length 600 mm (2 ft) to 900 mm (3 ft).

Standard sampler, also called *split-spoon sampler*, is a split sampling tube of inside diameter 35 mm (1⅜ in), outside diameter 50 mm (2 in), and length about 456 mm (18 in).

5.2 PURPOSES OF SITE CHARACTERIZATION

Importance *A site characterization program is necessary to provide information for design and construction, and for environmental assessment. In this section, we will list the purposes of site characterization.*

The purposes of a site characterization are:

1. To evaluate the general suitability of the site for the proposed project including environmental impact.
2. To enable an adequate and economical design to be made.
3. To obtain physical and mechanical properties of the soils for design and construction.
4. To obtain groundwater conditions.

5. To disclose and make provision for difficulties that may arise during construction due to ground and other local conditions.
6. To determine suitability of materials for construction.

5.3 PHASES OF SITE CHARACTERIZATION

Importance *In the early stages of a project, the available information is often insufficient to allow a detailed site investigation plan to be made. The extent of the investigations depends on several factors, including the needs of the client and the type of structure. We will discuss the phases of a site characterization program in this section.*

A site characterization program must be developed in phases that depend on the type, size, and importance of the structure, the client's needs, your familiarity with the soils at the site, environmental conditions, and local building codes. Structures that are sensitive to settlement, such as machine foundations and high-use buildings, usually require a thorough site investigation compared to a foundation for a house. A client may wish to take a greater risk than normal to save money and set limits on the type and extent of the site investigation. For this case, you need to consider the professional risks involved. You should not jeopardize the safety and stability of a structure for less expensive but inadequate site characterization. If the geotechnical engineer is familiar with a site, he or she may undertake a very simple soil investigation to confirm his or her experience. Some local building codes have provisions that set out the extent of a site investigation. Regardless of the thoroughness of a site investigation, it is mandatory that a site visit be made to the proposed site.

Five phases of site characterization are discussed below. These phases may be combined or some eliminated depending on the extent of the available information.

Phase I. Collection of available information such as

- ***Maps.*** An assortment of maps giving geology, contours, and elevations, climate, land use, aerial photos, regional seismicity, hydrology, and so forth are constructed using geographical information systems (GIS) and are available on the Internet (e.g., http://www.usgs.gov).
- ***Existing reports.*** Check your company files or associates for any existing information on the site and/or nearby sites.
- ***Newspaper clippings.*** These may provide historical information such as if the site was flooded in the past or a building had existed at the site.
- ***Site plans.*** Giving type of structure, location, size, finish grades, cut and fill, and importance of the structure. Normally, a soil investigation is conducted prior to the development of site plans but, occasionally, they are available.

Phase II. Preliminary reconnaissance or a site visit to provide a general picture of the topography and geology of the site. It is necessary that you take with you on the site visit all the information gathered in Phase I to compare with the current conditions of the site. Your site visit notes should include the following:

- Photographs of the site and its neighborhood.
- Access to the site for workers and equipment.
- Sketch all fences, utility posts, driveways, walkways, drainage systems, and so on.
- Utility services that are available, such as water and electricity.

- Sketch topography and include all existing structures, cuts, fills, ground depression, ponds, and so on.
- The state of any existing building at the site or nearby. Your notes should include exterior and interior cracks, any noticeable tilt, type of construction (e.g., brick or framed stucco building), evidence of frost damages, molds, and any exceptional features.
- Geological features from any exposed area such as a road cut.

Occassionally, a few boreholes may be dug to explore the site.

Phase III. Detailed soils exploration. A soils exploration provides information on the nature of soils at a site, design parameters, groundwater conditions, and possible construction problems. There are several methods of soils exploration. The methods you adopt depend on the extent of the site characterization desired based on the needs of the client, the type and size of the structure, the type and distribution of the structural loads, and local, geological, and environmental conditions. Information about the soils can be obtained by

- Mapping the subsurface (stratigraphy) using nondestructive (geophysical methods) and destructive methods (test pits and boreholes)
- Soil sampling and laboratory tests
- In situ tests

Phase IV. Laboratory testing. The objectives of laboratory tests are:

- To classify the soils.
- To determine soil strength, failure stresses and strains, stress-strain response, permeability, compactability, and settlement parameters. Not all of these may be required for a project.

Phase V. Write a report. The report must contain a clear description of the soils at the site, methods of exploration, soil stratigraphy, in-situ and laboratory test methods and results, and the location of the groundwater. You should include information and/or explanations of any unusual soil, water-bearing stratum, and any soil and groundwater conditions such as frost susceptibility and water-logged areas that may be troublesome during construction.

The **key points** *are:*

1. ***A site investigation must be conducted in phases.***
2. ***It is important that you visit the site.***
3. ***A clear, concise report describing the conditions of the ground, soil stratigraphy, the soil parameters determined, and any potential construction problems must be prepared for the client.***

5.4 MAPPING THE SUBSURFACE USING GEOPHYSICAL METHODS

Geophysical methods (GP) are nondestructive techniques that are used to provide information to characterize soils and rocks and hydrological and environmental conditions at a site. The GP methods are also used for inspection and detection of anomalous conditions such as void spaces beneath foundations. Some of the popular GP methods are:

- Ground penetrating radar
- Seismic surveys
- Electromagnetic survey

5.4.1 Ground Penetrating Radar (GPR)

GPR, also called *georadar,* is a high-resolution, high-frequency (10 MHz to 1000 MHz) electromagnetic wave technique for imaging soils and ground structures. An antenna is used to transmit and recover radar pulses generated by a pulse generator. The returned pulse is then processed to produce images of the soil profile (Fig. 5.1).

The key geotechnical uses are soil profile imaging and location of buried objects. GPR produces continuous-resolution images of the soil profile with very little soil disturbance. GPR is not suitable for highly conductive (>15 milliohms/m) wet clays and silts. GPR resolution decreases with depth.

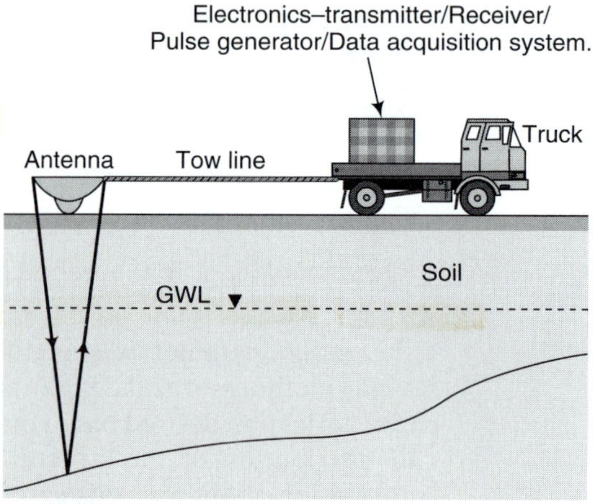

FIGURE 5.1 Ground penetrating radar (GPR).

5.4.2 Seismic Surveys

Seismic investigations utilize the fact that surface waves travel with different velocities through different materials. The subsurface interfaces are determined by recording the magnitude and travel time of the seismic waves (see Chapter 2 for the different types of seismic waves) at a point some distance from the source of the wave.

The velocity of propagation is the most important parameter in the application of seismic methods. P waves always travel fastest through the medium and are first recorded. S waves cannot travel through fluids. Therefore, a saturated soil layer will show the same S wave response as if it were dry. The densities and elastic properties of the geological materials control the velocity of propagation.

For an isotropic elastic medium, the shear wave velocity, V_s, is

$$V_s = \sqrt{\frac{G}{\rho}} \tag{5.1}$$

where G is the shear modulus and ρ is the density of the material.

When a seismic wave encounters a boundary between two elastic media the wave energy is transmitted by *reflection*, *refraction*, and *diffraction*. Seismic reflection and refraction are used in geotechnical site characterization.

Seismic Reflection In seismic reflection tests, the travel times of waves reflected from subsurface interfaces are measured (Fig. 5.2a) by geophones. Geophones are motion-sensitive transducers that convert ground motion to electric signals. The travel times are correlated to depth, size, and shape of the

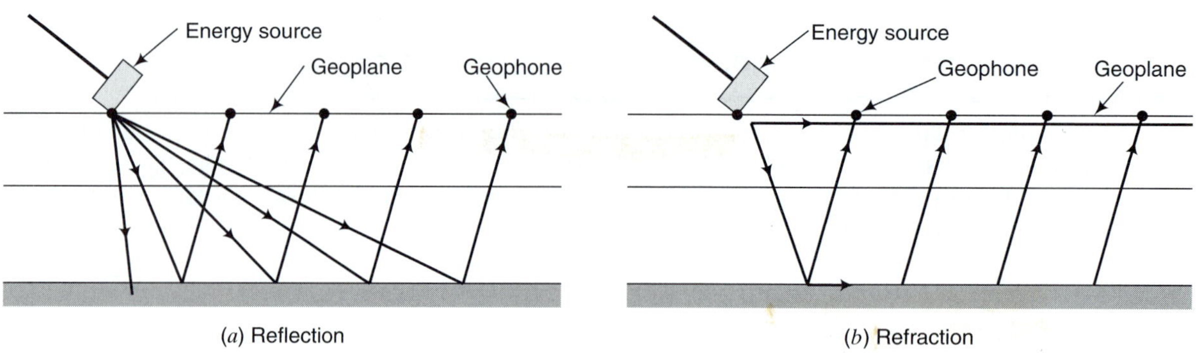

FIGURE 5.2 Transmission of wave energy.

interfaces. The angle of reflection of the waves is a function of the material density contrast. Seismic reflection is used when high resolution of a soil profile is required, especially at large depths (> 50 m).

Seismic Refraction Seismic refraction surveys are very similar to seismic reflection surveys except that refraction waves are measured and the source geophone is placed at a greater distance (Fig. 5.2b). The latter enables the recording of seismic waves that are primarily horizontal rather than vertical. In most refraction surveys, only the initial P waves are recorded. The seismic refraction method is used to determine the depth and thickness of the soil profile and the existence of buried structures.

The type of seismic survey depends on the seismic source and the geophone array. The seismic source used depends on the project costs, convenience, repeatability, depth of investigation, and efficiency. The two most common seismic sources used in field investigations are a sledge hammer and a small explosive.

For shallow depths, a sledge hammer is used; for large depths, a small explosive charge is used. Seismic methods are sensitive to noise and vibration. Various filtering techniques are used to reduce background noise and vibration. A site profile obtained from a seismic survey is shown in Fig. 5.3.

Crosshole Seismic—ASTM D4428 In crosshole seismic tests, the seismic source is located in one borehole and the geophone is located in an adjacent borehole. The velocities of the P and S waves are used

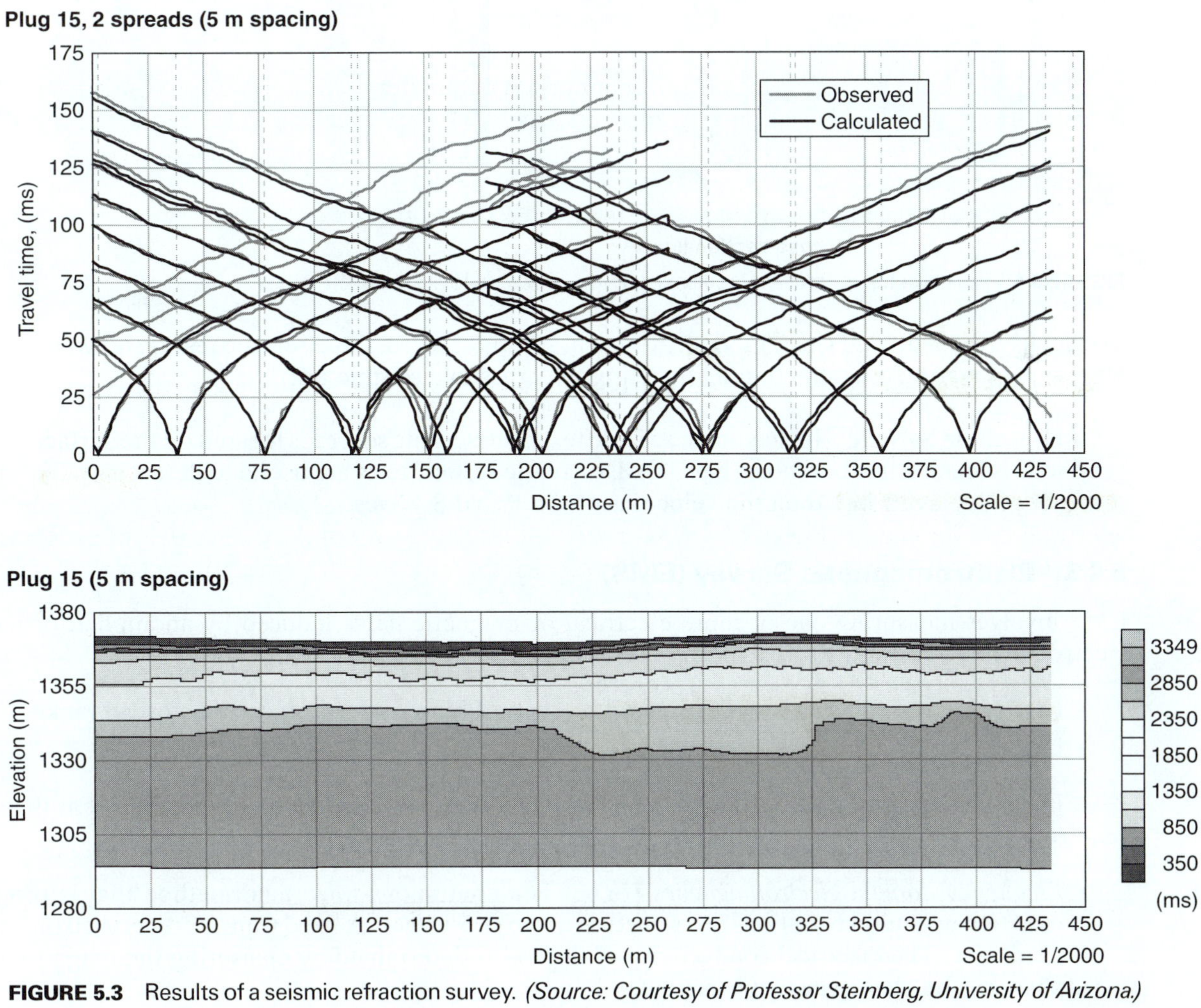

FIGURE 5.3 Results of a seismic refraction survey. *(Source: Courtesy of Professor Steinberg, University of Arizona.)*

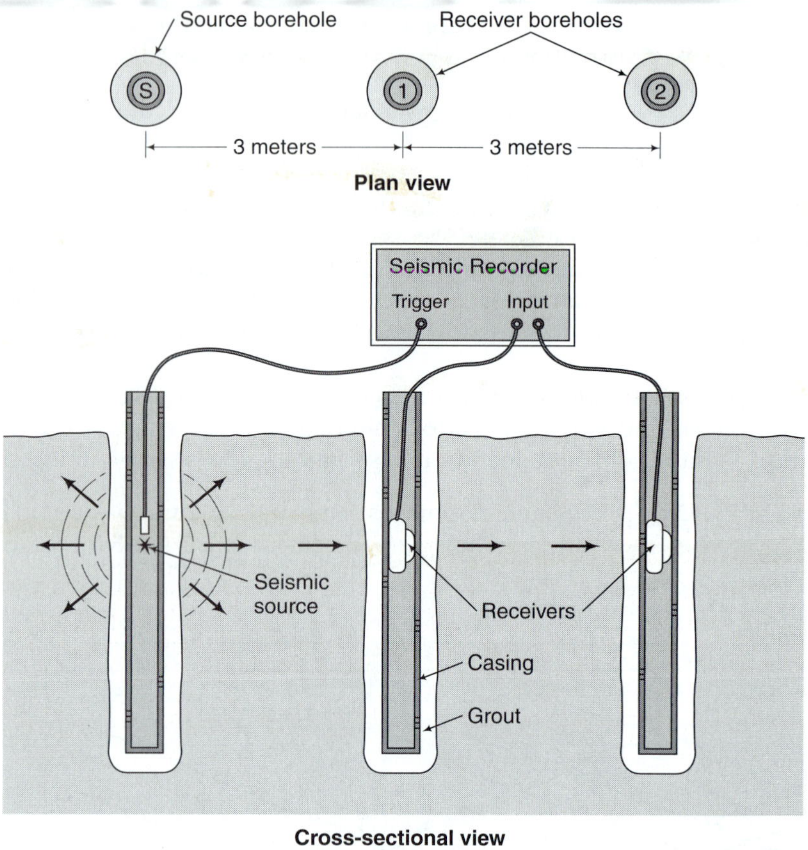

FIGURE 5.4 Typical setup for crosshole seismic survey. *(Source: ASTM D 4428.)*

to calculate the bulk and shear moduli. A typical setup for a crosshole seismic test is shown in Fig. 5.4. Horizontal layering is assumed in interpreting the crosshole seismic test.

Downhole Seismic In downhole seismic tests, the seismic source is located on the surface and the geophone is located in a borehole (Fig. 5.5). Downhole seismic tests are useful to detect layering and the strength of the layers based on the velocities of the P and S waves.

5.4.3 Electromagnetic Survey (EMS)

EMS involves measuring one or more electrical or magnetic fields induced by an applied or natural electromagnetic current. EMS is mostly used to:

- Locate bodies of high electric conductivity such as buried metal pipes and storage tanks.
- Determine the saturation level and type of pore fluids present in soils.

EMS testing is often carried out from an airplane or a helicopter. No ground control at the site is required. The EMS methods used for geotechnical studies are as follows:

- *Frequency domain electromagnetics (FDE)* is a continuous-wave field method that is primarily used to map lateral variations of a structure in the subsurface. A fixed transmitter is used on a single frequency. The electrical conductivity of the soil is determined by measuring the magnitude and phase of the electromagnetic current.

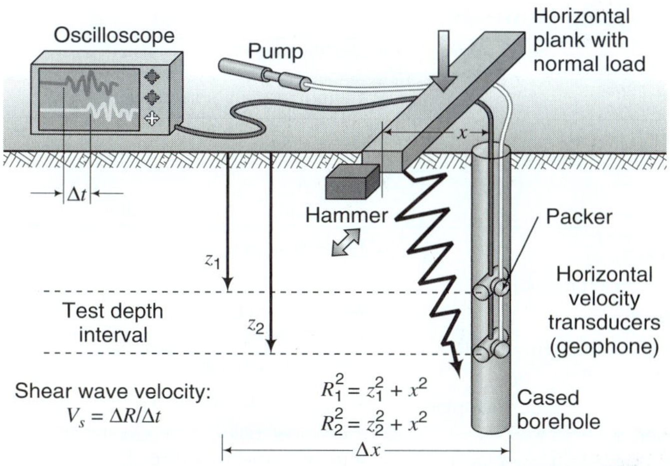

FIGURE 5.5 Downhole seismic survey. *(Source: Professor Paul Mayne, Georgia Tech.)*

- *Very-low-frequency electromagnetics (VLFE)* is also a continuous-wave field method that detects increases in electrical conductivity by measuring the distortion of the very-low-frequency waves. VLFE is very useful in detecting fractures.
- *Time domain electromagnetics (TDE)* measures the rate of decay of pulsating currents over time. TDE is useful in determining the variation of conductivity of soils with depth. You can, for example, obtain variations of soil saturation with depth using TDE.

The* key points *are:

1. ***Nondestructive methods are useful in producing general site data before embarking on a soils exploration using destructive methods such as boreholes.***
2. ***GPR and seismic surveys are very useful in producing soil profile imaging and in detecting buried objects.***
3. ***EMS is useful to detect lateral variation of soil structures and to determine variations in soil saturation.***
4. ***The selection of a geophysical method depends on availability of expertise and equipment, costs, convenience, repeatability, depth of investigation, and efficiency.***

5.5 MAPPING THE SUBSURFACE AND SAMPLING USING DESTRUCTIVE METHODS

The subsurface can also be mapped using methods such as test pits, augering, and drilling. The holes produced by augering and drilling are called *auger holes* and *boreholes* respectively. These methods allow for the extraction of soil samples and the performance of in-situ tests. In this section, we will use the generic term *borings* to include boreholes, auger holes, and test pits. In the planning stages, you should consider the following.

5.5.1 Which Method Is Best for the Project?

The advantages and disadvantages of each of the destructive methods are given in Table 5.1. Test pits and auger holes are best used for lightly loaded structures (one or two-storey buildings) and in preliminary soils exploration for heavy and sensitive structures. Sometimes more than one method has to be used.

TABLE 5.1 Comparison of Subsurface Mapping Methods

Method	Advantages	Disadvantages
Geophysical methods	• Nondestructive • Cost effective for soil profiling and detection of buried objects	• Cannot inspect or sample soils
Test pits A pit is dug either by hand or by a backhoe. 	• Cost effective • Provide detailed information of stratigraphy • Large quantities of disturbed soils are available for testing • Large blocks of undisturbed samples can be carved out from the pits • Field tests can be conducted at the bottom of the pit	• Depth limited to about 6 m (20 ft) in stiff clays • May require side supports in coarse-grained soils and soft clays • Deep pits uneconomical • Excavation below groundwater and into rock difficult and costly • Too many pits may scar site and require backfill soils • Time consuming • Limited to depths above ground water level
Hand augers The auger is rotated by turning and pushing down on the handle bar. 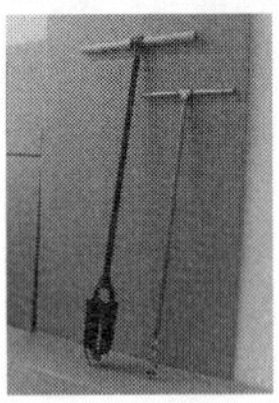	• Cost effective • Not dependent on terrain • Portable • Low headroom required • Used in uncased holes • Groundwater location can be easily identified and measured	• Depth limited to about 3 m (10 ft) to 6 m (20 ft) • Labor intensive • Undisturbed samples can be taken only for soft clay deposit • Cannot be used in rock, stiff clays, dry sand, caliche soils, etc.
Power augers Truck mounted and equipped with continuous flight auger that bores a hole 100 mm to 250 mm in diameter. Augers have solid or hollow stem. 	• Quick • Used in uncased holes • Undisturbed samples can be obtained quite easily • Groundwater location can be identified	• Depth limited to about 15 m (50 ft); At greater depth drilling becomes difficult and expensive. • Site must be accessible to motorized vehicle

Wash boring Water is pumped to bottom of borehole and soil washings are returned to surface. A drill bit is rotated and dropped to produce a chopping action. 	• Can be used in difficult terrain • Low equipment cost • Used in uncased holes	• Depth limited to about 30 m (100 ft) • Slow drilling through stiff clays, and gravels • Difficulty in obtaining accurate location of groundwater level • Undisturbed soil samples cannot be obtained
Rotary drills A drill bit is pushed by weight of drilling equipment and rotated by a motor 	• Quick • Drill through any type of soil or rock • Can drill to depths of 7500 m • Undisturbed samples can be easily recovered	• Expensive equipment • Terrain must be accessible to motorized vehicle • Difficulty in obtaining location of groundwater level • Additional time required for setup and cleanup

5.5.2 Where Should the Borings be Located?

The initial borings should give general geological information about the site that can be used to plan and execute the detailed soils exploration. The initial borings should be located near heavily loaded parts of the structures, special structures, suspected dumpsites, old landslide areas, and ground depression. It is common practice that these boreholes be located no more than a radius of 5 m (15 ft) from the center of the load.

5.5.3 How Many Borings and at What Depths?

It is practically impossible and economically infeasible to completely explore the whole project site. You have to make judgment on the number, location, and depths of borings to provide sufficient information for design and construction. The number and depths of borings should cover the zone of soil that will be affected by the structural loads. There is no fixed rule to follow. In most cases, the number and depths of borings are governed by experience based on the geological character of the ground, the importance of the structure, the structural loads, and the availability of equipment. Building codes and regulatory bodies provide guidelines on the minimum number and depths of borings.

The footprint of a structure (plan area covered by the structure) should be divided using a grid approximately 20 m (60 ft) to 40 m (120 ft) for large areas and boreholes should be located at node points on the grid. A simple numbering system can be devised by labeling one side with numbers and the adjacent side with letters. A borehole at a grid node can be identified as, say, 2C (Fig. 5.6).

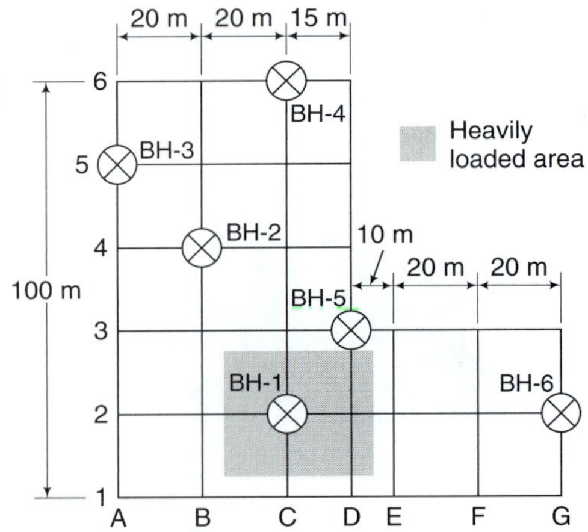

FIGURE 5.6 A preliminary plan of borehole locations at a site.

The number of boreholes should be adequate to detect variations of the soils at the site. If the locations of the loads on the footprint of the structure are known (this is not often the case), you should consider drilling at least one borehole at the location of the heaviest load. As a guide, a minimum of three boreholes should be drilled for a building area of about 250 m^2 (2500 ft^2) and about five for a building area about 1000 m^2 (10, 000 ft^2). Some guidelines on the minimum number of boreholes for buildings and for due diligence in subdivisions are given in Table 5.2. Some guidance on the depth of boreholes is provided in the following:

- In compressible soils such as clays, the borings should penetrate at least between 1 to 3 times the width of the proposed foundation below the depth of embedment or until the stress increment due to the heaviest foundation load is less than 10%, whichever is greater.
- In very stiff clays and dense coarse-grained soils, borings should penetrate 5 m (15 ft) to 6 m (20 ft) to prove that the thickness of the stratum is adequate.
- Borings should penetrate at least 3 m (10 ft) into rock.
- Borings must penetrate below any fills or very soft deposits below the proposed structure.
- The minimum depth of boreholes should be 6 m (20 ft) unless bedrock or very dense material is encountered.

TABLE 5.2 Guidelines for the Minimum Number of Boreholes for Buildings and Subdivisions

Buildings		Subdivisions	
Area—m^2 (ft^2)	No. of boreholes (min.)	Area—m^2 (ft^2)	No. of boreholes (min.)
$<$100 (1000)	2	$<$4000 (40000)	2
250 (2500)	3	8000 (80000)	3
500 (5000)	4	20000 (200000)	4
1000 (10000)	5	40000 (400000)	5
2000 (20000)	6	80000 (800000)	7
5000 (50000)	7	400000 (4000000)	15
6000 (60000)	8		
8000 (80000)	9		
10000 (100000)	10		

5.5.4 What Methods and Procedures Should Be Used to Advance the Borings?

FIGURE 5.7 A test pit dug by a backhoe.

Test pits (Fig. 5.7) can be hand dug if the soil is soft or loose. For stiff or dense soils, machinery such as a backhoe can be used. When the soil is loose, soft or uncemented, the sides of the test pits may require shoring to prevent collapse. The depth of test pits is limited to about 6 m (20 ft) and the width is generally about 400 mm (16 in.—one of the standard sizes of a backhoe bucket) or greater. Stratigraphic information is obtained by visual inspection of the soil at the sides of the excavation. Samples of soils can be extracted for laboratory tests and in-situ tests (Section 5.6) can be performed at the bottom of the excavation.

Auger holes are made using hand or powered augers. Hand augers are limited to depths about 6 m (20 ft) while powered augers can drill to greater depths up to about 15 m (50 ft). Stratigraphic information is obtained by visual inspection of the extracted soil. However, this information is generally reliable only for the first 2 m (6 ft) of extracted soil. At greater depths, the soils tend to intermingle as they make their way up to the surface. Samples of soils can be collected for laboratory tests and in-situ tests can be performed at the bottom of the augerhole. Where there is soft or loose soils, a casing has to be inserted to prevent collapse of the hole. However, auger holes in such soils are avoided and replaced by boreholes.

Boreholes to large depths (greater than 30 m (100 ft)) are made using rotary drills (Fig. 5.8). A borehole is advanced by applying torque to an auger or a drill bit. There are various types of augers and drill bits. The selection depends on the soil type and the purpose of the investigation. A helical screw auger (Fig. 5.8) is the most common type of auger used. For dense soils and rocks, drilling is performed using carbon-tipped (Fig. 5.9) or diamond-tipped drill bits connected to the bottom end of the drill rods.

One type of drilling method employed is wash boring. The drill bit is rotated and dropped to produce a chopping action and water is pumped to the bottom of the borehole through the center of hollow drill rods. Soil washings are returned to the surface. In another method the drill bit is pushed by a hydraulic system and rotated by a motor.

Since the boreholes are discretely placed, the soil stratigraphy has to be inferred from the logs of each borehole. Even with many years of experience, it is difficult for a geotechnical engineer to sketch an

FIGURE 5.8 A truck-mounted rotary drill (left side) and closeup of a drilling in progress (right side).

FIGURE 5.9 A carbon-tipped drill bit (head).

accurate soil stratigraphy using borehole logs, especially if the boreholes are not closely spaced. The soil types between boreholes are speculative. Samples of soils can be extracted for laboratory tests and in situ tests can be performed at the bottom of the boreholes.

5.5.5 How Do We Sample the Soils?

The objective of soil sampling is to obtain soils of satisfactory size with minimum disturbance for observations and laboratory tests. Soil samples are usually obtained by attaching a thin-walled, seamless steel tube of diameter 50 mm (2 in.) to 75 mm (3 in.) and length of 600 mm (2 ft) to 900 mm (3 ft)—called a *Shelby tube* (Fig. 5.10a)—to drill rods and forcing it down into the soil. The tube is carefully withdrawn, hopefully, with the soil inside it. The amount of soil disturbance depends on friction between the soil and the sampling tube, the wall thickness of the sampling tube, the sharpness of the cutting edge, and care and handling of samples during transportation. To minimize friction, the sampling tube should be thin and pushed instead of driven into the ground. Soil disturbance can significantly affect the settlement and strength parameters used for design.

A popular sampler is the "standard" sampler (Figure 5.10b), popularly known as the split-spoon sampler, which has an inside diameter of 35 mm (1⅜ in.), an outside diameter of 50 mm (2 in.), and lengths 456 mm (18 in.) and 762 mm. The sampler has a split barrel that is held together using a screw-on driving shoe at the bottom end and a cap at the upper end. The thicker wall of the standard sampler permits higher driving stresses than the Shelby tube, but does so at the expense of higher levels of soil

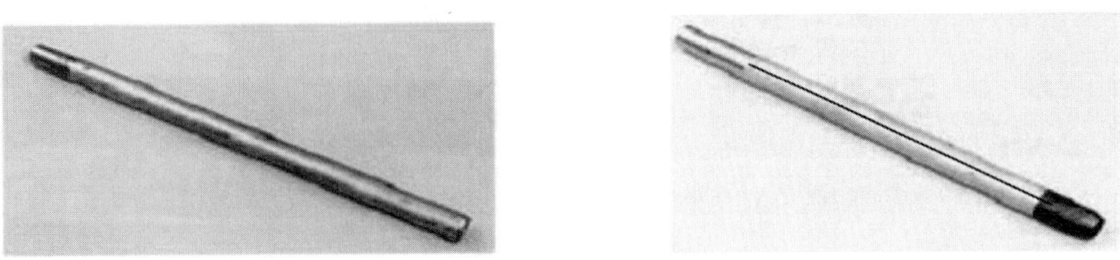

(*a*) (*b*)

FIGURE 5.10 (a) A thin wall tube and (b) a split tube sampler.

disturbance. Split-spoon samples are disturbed. They are used for visual examination and for classification tests.

Hvorslev (1949) conducted a comprehensive study on the sampling disturbances caused by samplers and defined several critical factors such as area ratio, inside clearance ratio, outside clearance ratio, length-to-diameter ratio, method of advancing the sampler, method of sample recovery, and stress relief. One measure of the effects of sampler wall thickness is the recovery ratio, defined as L/H, where L is the length of the sample and H is the distance that the sampler was pushed. Higher wall thickness leads to greater recovery ratio and greater sampling disturbance. Sampling tubes that are in common use have been designed to minimize sampling disturbances.

It is necessary that standard procedures be followed and great care be exercised in sampling soils because sampling disturbances can significantly affect the results of laboratory tests and negatively impact foundation design. For example, if the soil strength parameters from laboratory tests are lower than the actual due to sampling disturbances, then the foundation size is likely to be greater and consequently the cost is higher. If the soils at a site show significant variations, it would be more prudent to map the geological structures carefully rather than to obtain large numbers of samples and laboratory test data.

The sampling instructions must include the type of sample, the number of samples, and locations. Changes in the sampling instructions often occur after the first boring.

The* key points *are:

1. ***For undistributed samples, a thin wall sampler should be carefully pushed into the soil stratum.***
2. ***The standard or split-spoon sampler can sustain high driving stress, but the sample is disturbed.***

5.5.6 What Are the Requirements for Groundwater Observations?

The elevation of the natural groundwater is called the *water table* or *groundwater level* or *phreatic surface*. Many construction failures, court battles, and construction cost overruns have resulted from the nonidentification or nondisclosure of groundwater conditions at a site. The water table invariably fluctuates depending on environmental conditions (e.g., rainfall patterns, winter rains, monsoons, tides, drought), human activities (e.g., pumping groundwater from wells and draw-down during construction), and geological conditions.

Groundwater sometimes gets trapped in a permeable soil layer overlaid by an impervious layer. If a borehole penetrates the impervious layer, the trapped groundwater may break out and shoot up above the ground surface. This is called *artesian condition*. If the water just reaches the ground surface, or stays below, we have a *sub-artesian condition*. It is particularly important to identify and disclose these conditions because failure can occur with possible loss of lives and properties. You should identify not only the groundwater level and any special conditions (e.g., artesian condition), but also the possible range of groundwater-level fluctuations. Groundwater is measured using piezometers or standpipes. Measurements should be made over a period of days to detect any local variability. You should not rely on groundwater measurements from excavations.

The* key points *are:

1. ***It is essential to identify the location of the water table and its variation.***
2. ***Water trapped in a water-bearing stratum beneath impermeable soils can cause considerable hydrostatic pressures and must be identified and disclosed.***

5.5.7 What Types of Tests Should Be Conducted?

The types of tests depend on the type of structure—importance, loads, geometry—the general character of the site, and the design parameters desired. Two classes of tests—in-situ tests and laboratory tests—are used to provide estimates of the physical and mechanical properties of the soils. In situ tests are considered in the next section followed by laboratory tests.

5.6 IN SITU TESTS

Importance *In situ tests are used to identify soils at a site and to provide soil parameters for design. The results are available quickly, often during the tests or soon after, compared with laboratory tests. However, there are a variety of in situ tools available; each has its own advantages and disadvantages and cost implications. As a geotechnical engineer, you would have to decide on the in situ tools to be used and the frequency and location of tests required to provide design and construction information for your project. In this section, we discuss a broad set of in situ tests.*

The most popular in situ devices are

- Standard penetration test (SPT)
- Vane shear test (VST)
- Cone penetrometer test (CPT)
- Flat plate dilatometer (DMT)
- Pressuremeter test (PMT)

The performance and interpretation of the data from these tests are presented in the following.

The data from in situ tests is used to characterize soils and to design foundation. In situ tests are generally cost effective. Most in-situ tests do not measure the strength and deformation properties of the soil directly. Rather, correlation with laboratory test results and experience are used to establish shear strength and deformation parameters, safe design loads, and tolerable settlements. Most of these correlations are weak.

5.6.1 Standard Penetration Test (SPT)—ASTM D1586

Access a spreadsheet at http://www.wiley.com/college/budhu. Click Foundations and Earth Structures, Chapter 5, SPT.xls to correct *N* values.

The standard penetration test (SPT) was developed circa 1927 and is perhaps the most popular in-situ test. The SPT is performed by driving a standard split-spoon sampler into the ground by a drop hammer of mass 63.5 kg (140 lb) falling from a height of 760 mm (30 in.) (Figs. 5.11 and 5.12). The number of blows to drive the sampler at 152 mm (6 in.) intervals into the soil is recorded. The sum of the number of blows for the last two 152 mm (6 in.) intervals is the standard penetration number (*N*). SPTs are normally conducted about every 1.5 m (5 ft).

The word *standard* is a misnomer for the standard penetration test. Different procedures are used in different parts of the world to conduct the test, especially the weight of the hammer and methods of

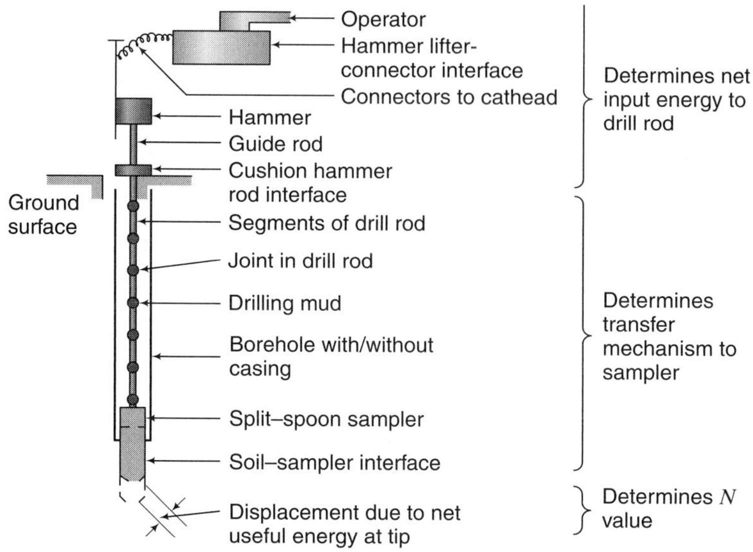

FIGURE 5.11 Details of SPT test.

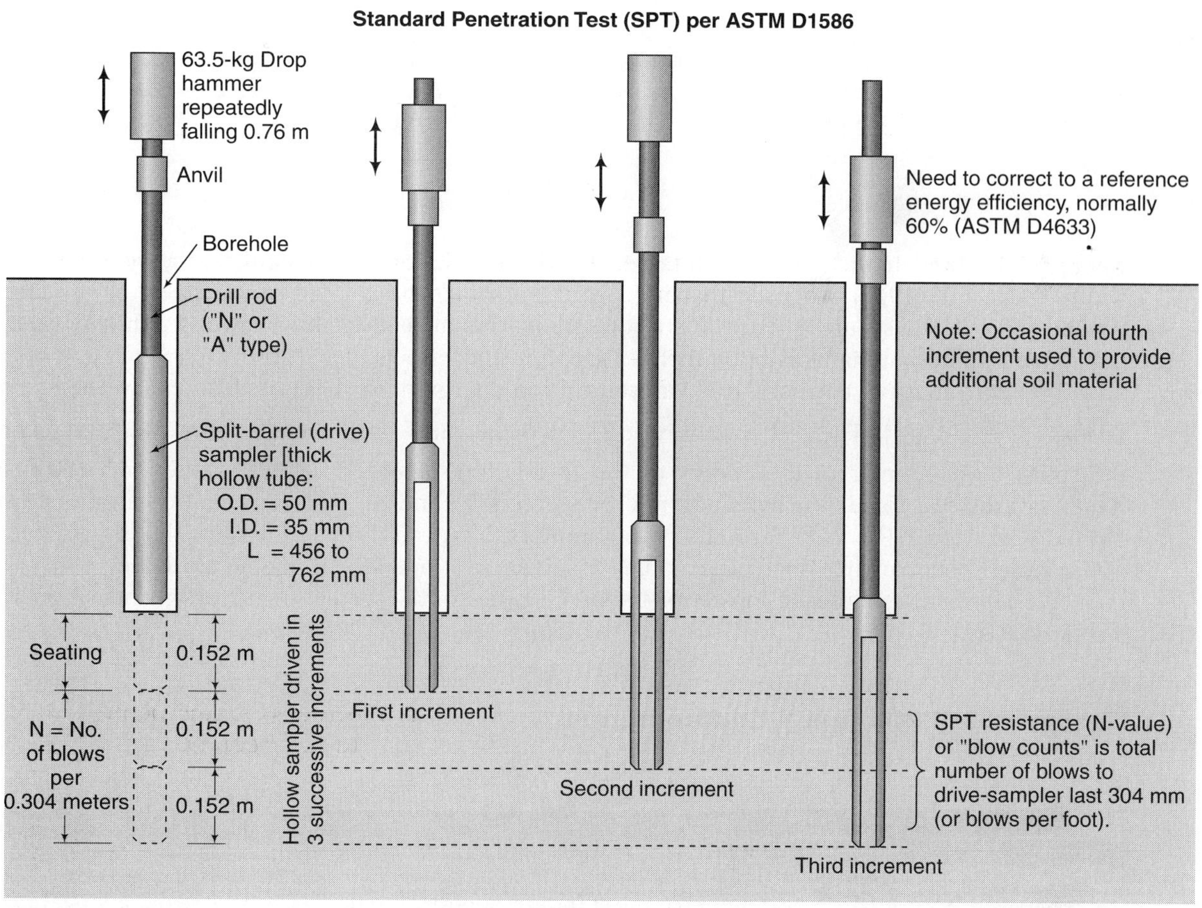

FIGURE 5.12 Driving sequence in an SPT. *(Source: Professor Paul Mayne, Georgia Tech.)*

releasing it. Also different types of anvil, rods, and rod lengths are prevalent. The N values are subjected to errors from various sources such as overdriving, improperly cleaning the hole, loose rod connection, poor drilling procedures (overwashing ahead of casing), and poor drilling method (e.g., full freefall height not achieved). Various corrections are applied to the N values to account for energy losses, overburden pressure, rod length, borehole site, sampler type, and so on. Correction factors for rod lengths, sampler type, and borehole diameter are given in Appendix B, Table B5.1.

Various correction factors for overburden pressure have been suggested by a number of investigators. Only two such suggestions for correcting N values for overburden pressure are included in this text. These are:

$$C_N = \left(\frac{95.8}{\sigma'_z}\right)^{\frac{1}{2}}; \quad C_N \le 2; \quad \sigma'_z\,(\text{kPa})\ (\text{Liao and Whitman, 1985}) \tag{5.2}$$

$$C_N = \left(\frac{2000}{\sigma'_z}\right)^{\frac{1}{2}}; \quad C_N \le 2; \quad \sigma'_z\,(\text{psf}) \tag{5.2a}$$

$$C_N = 0.77\log_{10}\left(\frac{1916}{\sigma'_z}\right); \quad C_N \le 2, \quad \sigma'_z > 24\,\text{kPa}\ (\text{Peck et. al., 1974}) \tag{5.3}$$

$$C_N = 0.77\log_{10}\left(\frac{40}{\sigma'_z}\right); \quad C_N \le 2, \quad \sigma'_z\,(\text{ksf}) > 0.5\,\text{ksf} \tag{5.3a}$$

where C_N is a correction factor for overburden pressure, and σ'_z is the effective overburden pressure in the appropriate units. A further correction factor is imposed on N values if the groundwater level is within a depth B below the base of the footing. The groundwater correction factor is:

$$C_W = \frac{1}{2} + \frac{1}{2}\frac{z}{(D_f + B)} \tag{5.4}$$

where z is the depth to the groundwater table, D_f is the foundation embedment depth, and B is the footing width. Meyerhof (1965) proposed that no correction should be applied to N values for the effects of groundwater as these are already incorporated in the measurement. However, this suggestion is debatable and design manuals often include groundwater effects.

It is customary to correct the N values to a rod energy ratio (ER) of 60%, called the *equipment factor*, $C_E = \left(\frac{\text{ER}(\%)}{60}\right)$. The rod energy ratio (ER) is the ratio (in percent) of the energy delivered to the split-spoon sampler to the energy of the free-falling hammer. Typical values for ER in North America are 45% for donut hammers and between 55% and 83% for automatic hammers. For the United Kingdom, ER = 60% for automatic hammers. You can directly measure the impact energy in a test rather than relying on typical values as given in Appendix B, Table B5.1.

We can write a composite correction factor, C_{RSBNE}, for the correction factors given in Appendix B, Table B5.1 as

$$C_{RSBNE} = C_R C_S C_B C_N C_E \tag{5.5}$$

The corrected N value at 60% hammer energy is

$$N_{1,60} = C_{RSBNE} N \tag{5.6}$$

The N value is used to estimate the relative density, friction angle, bearing capacity, and settlement of foundations in granular soils. Typical correlation among N values, relative density, and ϕ' are given in Appendix B, Table B5.2. You should be cautious in using the correlation in Table B5.2 to determine the mechanical properties of soils and to design foundations because the scatter in data is generally large and the correlation coefficients are low. Experience and judgment are required to successfully use Table B5.2.

The SPT is simple and quick to perform. The equipment is widely available and can penetrate dense materials. SPT results have been correlated with engineering properties of soils, bearing capacity, and settlement of foundations. Most of these correlations are, however, weak. SPT is unreliable for coarse gravel, boulders, soft clays, silts and mixed soils (combinations of boulders, cobbles, clays, and silts). SPT is useful to determine changes in stratigraphy, locating bedrock, and in estimating liquefaction potential of soils. Also, you can inspect the soil in the split-spoon sampler to describe the soil profile and extract disturbed samples for laboratory tests.

EXAMPLE 5.1 *Correction of a Set of N Values*

The blow counts for an SPT test at a depth of 5 m (15 ft) in a coarse-grained soil at every 0.152 m (6 in.) are 8, 12, 15. A donut hammer with ER = 45% and a standard sampler were used in a borehole 152 mm (6 in.) diameter. (1) Determine the N value. (2) Correct the N value for rod length, sampler type, borehole size, and energy ratio of 60%. (3) Make a preliminary description of the compactness of the soil. The rod length measured from the depth of the test to the anvil is 6 m (20 ft). Groundwater level is 10 m (30 ft) below the surface.

Strategy The N value is the sum of the blow counts for the last 0.304 m (12 in.) of penetration. Just add the last two blow counts. U.S. units will be used to solve this example problem.

Solution 5.1

Step 1: Add the last two blow counts.

$$N = 12 + 15 = 27$$

Step 2: Obtain correction factors.

From Appendix B (Table B5.1),

$$L = 20 \text{ ft.}$$

$$C_R = 0.015L + 0.61 = 0.015 \times 20 + 0.61 = 0.91.$$

$C_S = 1.0$ for standard sampler and $C_B = 1.05$ for a borehole of diameter 6 in.

For a donut hammer with ER = 45%, $C_E = \frac{45}{60} = 0.75.$

From Appendix B (Table B5.2), the estimated unit weight is about 119 pcf for $N = 27$.

$$\sigma'_z = 119 \times 15 = 1785 \text{ psf}$$

$$C_N = \left(\frac{2000}{\sigma'_Z}\right)^{\frac{1}{2}} = \left(\frac{2000}{1785}\right)^{\frac{1}{2}} = 1.0 < 2; \quad \text{use } 1.0$$

Step 3: Calculate corrected N value.

$$C_{RSBNE} = C_R C_S C_B C_N C_E = 0.91 \times 1.0 \times 1.05 \times 1.0 \times 0.75 = 0.72$$

$$N_{1,60} = C_{RSBNE}\, N = 0.72 \times 27 = 19$$

Step 4: Use Table B5.2 to describe the compactness.

For $N = 27$, the soil is medium dense. ■

EXAMPLE 5.2 *Correction of N Values from Borehole Logs*

Part of a boring log at site is shown in Fig. E5.2. The SPT tests were conducted using a donut hammer (ER = 45%) and a U.S. sampler without liner. The borehole size was 152 mm (6 in.) and the unit weight was estimated based on the N value. The rod length extends 0.6 m (2 ft) above the surface. Correct the N values to $N_{1,60}$.

BOREHOLE BH - 2

Page 1 of 1

CLIENT	ENGINEER
SITE	PROJECT

Approx. Surface Elev.: 2210.5 ft.

DESCRIPTION	Elev.	DEPTH (FT)	USCS	SAMPLES: NUMBER	SAMPLES: TYPE	SAMPLES: RECOVERY, IN.	SAMPLES: PENETRATION BLOWS / FT	TESTS: MOISTURE, %	TESTS: DRY DENSITY PCF	TESTS: UNCONFINED STRENGTH, PSF
SILTY SAND; poorly graded, fine-grained, dry, light brown			SM		BAG					
loose		5	SM		R	6	15			
very loose		10	SM		R	NR	6			
15.0	2195.5									
SAND; trace gravel and silt, poorly graded, medium dense, dry, light brown		15	SP		SS	15	15			
20.0	2190.5									
SAND; with silt and gravel, poorly graded, medium dense, dry, light brown		20	SP-SM		R	6	26			
25.0	2185.5									
SAND; silty to with silt, with gravel, well graded, dense, dry, light brown		25	SW-SM		SS	8	22			
30.0	2180.5									
SILTY TO CLAYEY SAND; well graded, loose, dry, light brown		30	SC-SM		R	6	22			

Continued Next Page

The stratification lines represent the approximate boundary lines between soil and rock types: in-situ, the transition may be gradual.

Offset: 180 feet Right

WATER LEVEL OBSERVATIONS				
WL	▽ **None WD**	▼ **None AB**	BORING STARTED	
WL	▼	▽	BORING COMPLETED	
WL			RIG	
			APPROVED	JOB #

FIGURE E5.2

Strategy U.S. units will be used in solving this example problem. Prepare a table (preferably using a spreadsheet). One is available at http://www.wiley.com/budhu. Click Foundations and Earth Structures, Chapter 5, SPT.xls); correct the N values using Eq. (5.6). The unit weight should be estimated for a specific depth. The average depth of the SPT is taken as 6 inches above the tip of the sampler.

Solution 5.2

Step 1: Prepare a table or spreadsheet heading as shown in Table E5.2.

$$C_E = \frac{ER}{60} = \frac{45}{60} = 0.75$$

The corrected N values are shown in Table E5.2.

	A	B	C	D	E
1	CORRECTING SPT VALUES			Help	
2	Muni Budhu"Foundations and Earth Retaining Structures", John Wiley & Sons, NY, 2006				
3	Select units	US			
4	Groundwater depth	100	ft		
5	Select sampler type	Standard		C_S	1
6	Borehole size, C_B	1	See table below		
7	Equipment, C_E	0.75	See table below		
8	Rod length from surface to anvil	2	ft		
9	Correct to $N_{1,60}$	Yes			

	Depth	Unit weight	Rod length correction	Vertical total stress	Porewater pressure	Vertical effective stress	C_n Calc.	C_n use	N	$N_{1,60}$
13	ft	pcf	C_R	psf	psf	psf				
14	0	0		0	0	0	0	0	0	0
15	5.5	111	1.00	610.5	0	610.5	1.8	1.8	15	27
16	10.5	95	1.00	1085.5	0	1085.5	1.4	1.4	6	8
17	16	114	0.88	1712.5	0	1712.5	1.1	1.1	15	16
18	21.5	118	0.93	2361.5	0	2361.5	0.9	0.9	26	24
19	26	114	0.94	2874.5	0	2874.5	0.8	0.8	22	18
20	30.5	114	0.95	3387.5	0	3387.5	0.8	0.8	22	17
21			0.00	0	0	0	0.0	0.0		0
22			0.00	0	0	0	0.0	0.0		0
23			0.00	0	0	0	0.0	0.0		0
24			0.00	0	0	0	0.0	0.0		0

	A	B
33	**Borehole diameter:**	C_B
34	65 mm to 115 mm (2.5 in to 4.5 in))	1
35	152 mm (6 in)	1.05
36	200 mm (8 in)	1.15
38	**Equipment:**	C_E
39	Safety Hammer (rope, without Japanese "throw" release)	0.7 to 1.2
40	Donut Hammer (rope, without Japanese "throw" release)	0.5 to 1.0
41	Donut Hammer (rope with Japanese "throw" release)	1.1 to 1.4
42	Automatic-Trip Hammer (Donut or Safety Type)	0.8 to 1.4

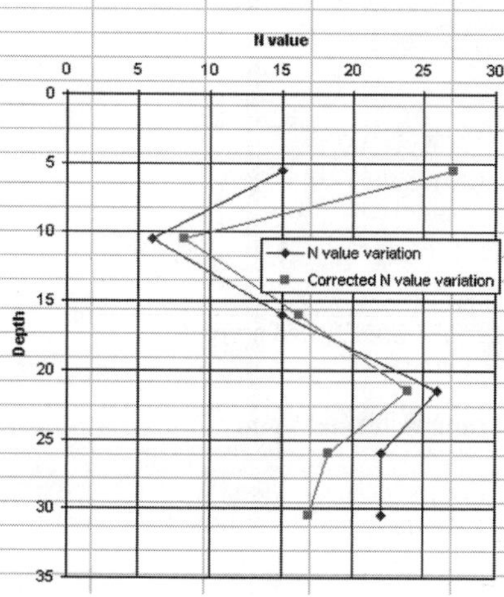

TABLE E5.2

5.6.2 Vane Shear Test (VST)—ASTM D2573

The vane shear device consists of four thin metal blades welded orthogonally (90°) to a rod (Fig. 5.13). The vane is pushed from the bottom of a borehole to the desired depth usually at 1 m (3 ft) intervals. A torque is applied at a rate of 6° per minute by a torque head device (torque meter) located on or above the soil surface and attached to the vane shear rod. The measured peak torque from the vane shear device is converted to undrained shear strength using

$$s_u = \frac{2T_{\max}}{\pi d^3\left(\frac{h}{d}+\frac{1}{3}\right)} \tag{5.7}$$

where $T_{\max}$ is the maximum torque, h is the height of the vane, and d is the diameter of the vane. The ratio h/d is generally 2. The factor 1/3 is based on the assumption of a uniform strain distribution between the top and bottom of the failure surfaces. After the maximum torque is obtained, the shear vane is rotated an additional 8 to 10 revolutions to measure the residual torque, T_r. The ratio of the peak torque to the residual torque is the soil sensitivity, S_t, where

$$S_t = \frac{T_{\max}}{T_r} \tag{5.8}$$

Sensitivity is a measure of the reduction of undrained shear strength due to soil disturbance. The insertion of the vane causes soil remolding. Higher blade thickness ($h/d > 2$) results in greater remolding and lower soil strengths. The blade thickness should not exceed 5% of the vane diameter. The vane shear test is used

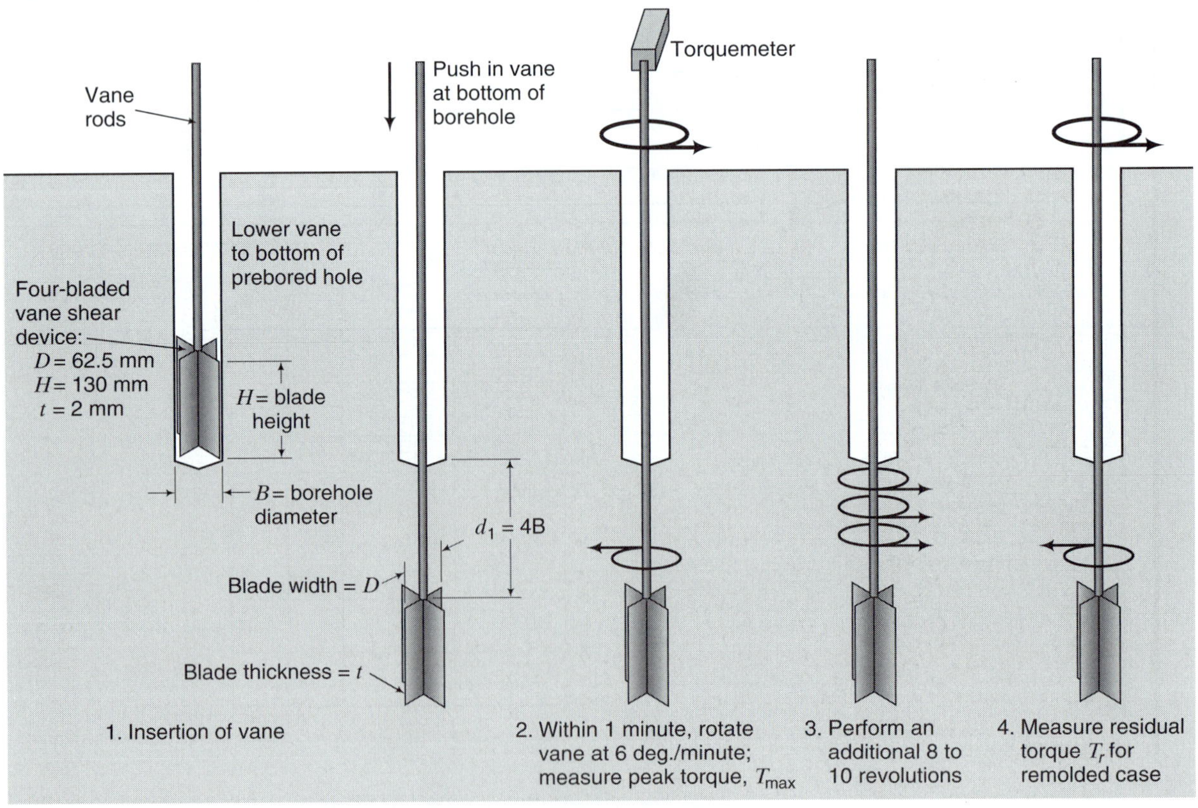

FIGURE 5.13 Vane shear tester. *(Source: Professor Paul Mayne, Georgia Tech.)*.

in soft to medium fine-grained soils. The results of a vane shear test are displayed as undrained or vane shear strength versus depth.

5.6.3 Cone Penetrometer Test (CPT)—ASTM D5778

A basic cone penetrometer (Fig. 5.14a) is a conical probe of base diameter 36 mm (1.4 in.), projected area 10 cm^2 (1.55 $in.^2$), and cone angle 60° that is pushed into the ground at a rate of 2 cm/s (0.79 in/sec). The cone is connected to hollow drill rods, each of length about 1 m (3 ft), or can be connected to standard AW or EW drill rods. An outer sleeve sits inside the drill rod above the cone base. The thrusts required to drive the cone and the sleeve into the ground are measured independently so that the end resistance (or cone resistance) and side friction (or sleeve resistance) may be estimated separately. There are several variants of the basic cone penetrometer. Three of these are the piezocone (uCPT or CPTu), seismic cone (SCPT), and vision cone (VisCPT or VisCPTu). A test using a miniature cone called the Continuous Intrusion Miniature Cone Penetration Test (CIMCPT—Tumay and Kurup, 2001) is also in use. The miniature cone penetrometer has a projected cone area of 2 cm^2 and a friction sleeve surface area of 40 cm^2.

The Piezocone has porous elements inserted into the cone or sleeve to allow for porewater pressure measurements (Fig. 5.14b). The measured porewater pressure depends on the location of the porous elements. The piezocone is a very useful tool for soil profiling and estimation of in situ shear strength, bearing capacity, and consolidation characteristics of soils. The CPT test results are normally recorded electronically (electric cone) by running a cable carrying the signals from a load transducer and porewater pressure transducers through the hollow drill rods and connecting it to a data acquisition system at the surface.

An example of results from a piezocone is shown in Fig. 5.15. The first plot shows the variation of cone or tip resistances (q_c) with depth. The second plot is the variation of sleeve resistance (f_s), with depth. The third plot shows the porewater pressure (u) variation with depth.

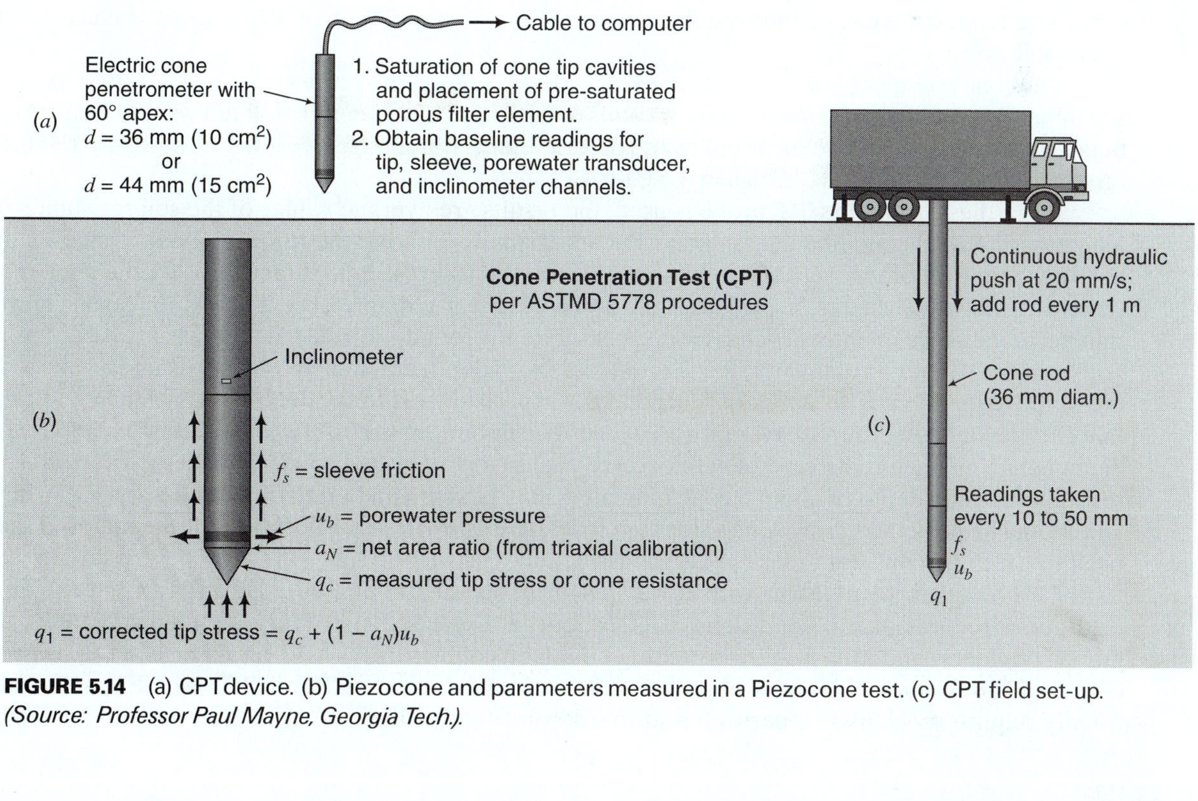

FIGURE 5.14 (a) CPT device. (b) Piezocone and parameters measured in a Piezocone test. (c) CPT field set-up. *(Source: Professor Paul Mayne, Georgia Tech.)*

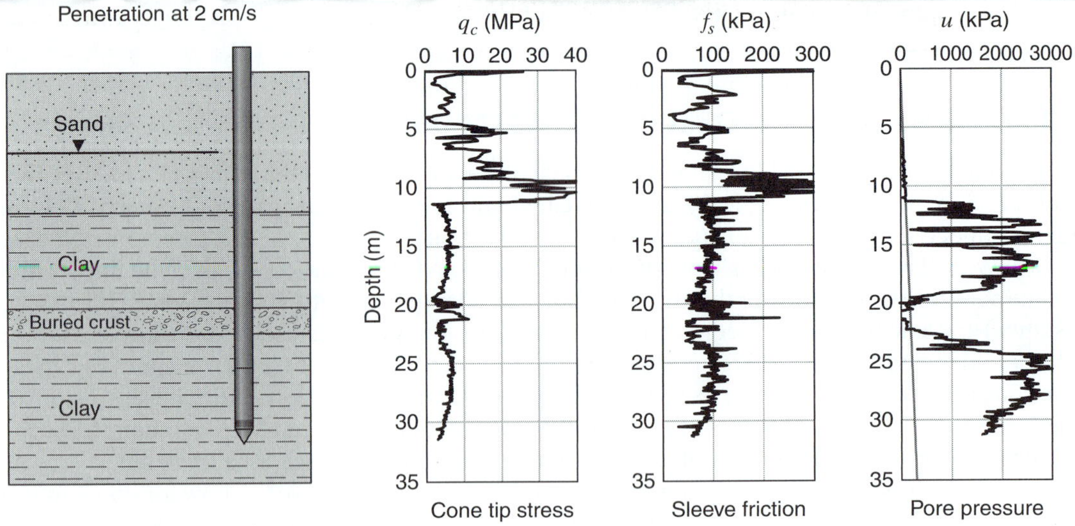

FIGURE 5.15 Piezocone results. *(Source: Professor Paul Mayne, Georgia Tech.)*

In the SCPT, geophones are installed inside the cone. Hammers on the surface are used to produce surface disturbances and the resulting seismic waves are recorded by the geophones (usually three). The recorded data is then analyzed to give damping characteristics and soil strength parameters.

Since no samples are recovered during any of the CPT variants described above, it is not possible to describe the soils from visual examination. The VisCPT (Raschke and Hryciw, 1997) and VisCPTu have miniature cameras installed in the CPT probe that provide continuous images of the soil adjacent to the cone. The soil texture can be inferred through image processing. The VisCPTu can also be used to detect liquefiable soils.

The cone resistance is influenced by several soil variables such as stress level, soil density, stratigraphy, soil mineralogy, soil type, and soil fabric. Results of CPT tests have been correlated with laboratory tests to build empirical relationships for strength and deformation parameters. Investigators have also related CPT results to other in situ tests (particularly SPT).

Regardless of which CPT probe is used, the results are average values of the soil resistance over a length of about 10 cone diameters—about 5 cone diameters above the tip plus about 5 cone diameters below the tip for homogeneous soil. In layered soils, the soil resistances measured by the cone may not represent individual layers, especially thin layers (<5 cone diameters). These thin layers may have significant influence on the soil behavior and consequently foundation design. For practical use, composite graphs of the variations of tip and sleeve resistances with depth are made based on visual inspection of the measured data. There is no standard procedure. One procedure is to divide the soil profile into layers such that each layer has a different average resistance. The minimum size of the layer should be about 10 cone diameters. You need to be careful with excessively large or very low cone resistances. Excessively large resistances over a depth of about 5 cone diameters may be due to a buried object such as a rock fragment and should be neglected. Low cone values are indicative of soft soils or pockets of loose material that you need to carefully consider in design and construction.

CPT is quick to perform with fewer performance errors compared to SPT. CPT can provide continuous records of soil conditions. CPT cannot be used in dense coarse-grained soils (e.g., coarse gravel, boulders) and mixed soils containing boulders, cobbles, clays, and silts. The cone tip is prone to damage from contact with dense objects. The more sophisticated uCPT, SCPT, VisCPT, and VisCPTu usually require specialists to perform and to interpret the results.

EXAMPLE 5.3 ***Interpreting Tip Resistance Variation with Depth from CPT Data***

The results distribution of tip resistance (q_c) from a CPT is shown in Fig. E5.3a. Draw an equivalent composite distribution of q_c with depth. How many distinct layers are present based on the equivalent composite distribution?

Strategy Follow the strategy described in Section 5.6.3.

Solution 5.3

Step 1: Sketch the equivalent composite distribution of q_c.
See Fig. E5.3b for a sketch of composite distribution.

Step 2: Estimate the number of distinct layers.
Number of layers = 9.

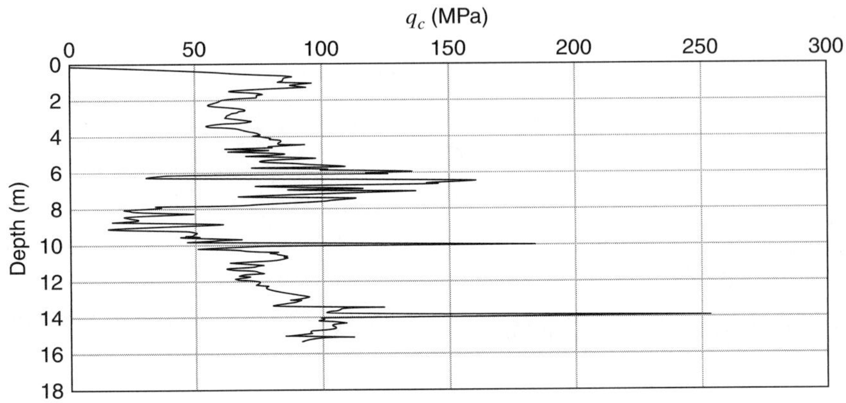

FIGURE E5.3a

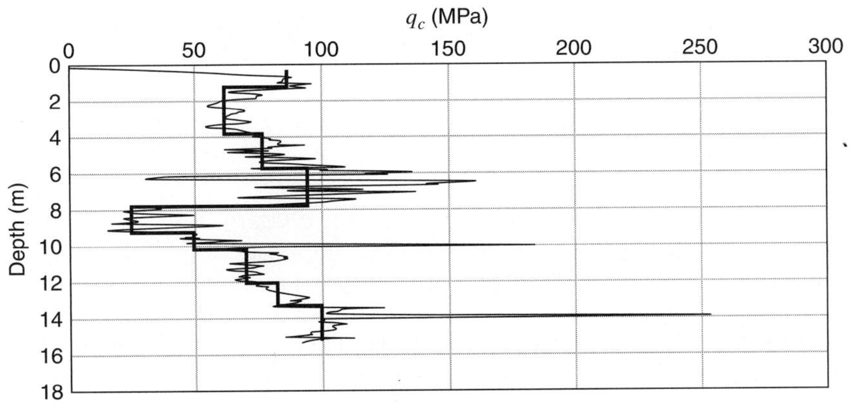

FIGURE E5.3b

■

5.6.4 Flat Plate Dilatometer (DMT)

The flat plate dilatometer consists of a tapered blade 95 mm (3.75 in.) wide and 15 mm (0.6 in.) thick and 240 mm (9.5 in.) long (Fig. 5.16). On the flat face of the dilatometer is a flexible steel membrane 60 mm in diameter that, when inflated, pushes the soil laterally. The blade is attached to drill rods and is pushed into the soil at a rate of 2 cm/s (0.79 in./s) by a drill rig. The pneumatic pressures (1) to bring the membrane flush with the soil surface, (2) to push the soil laterally for a distance of 1.1 mm (0.043 in.), and (3) at which the membrane returns to its original position are recorded.

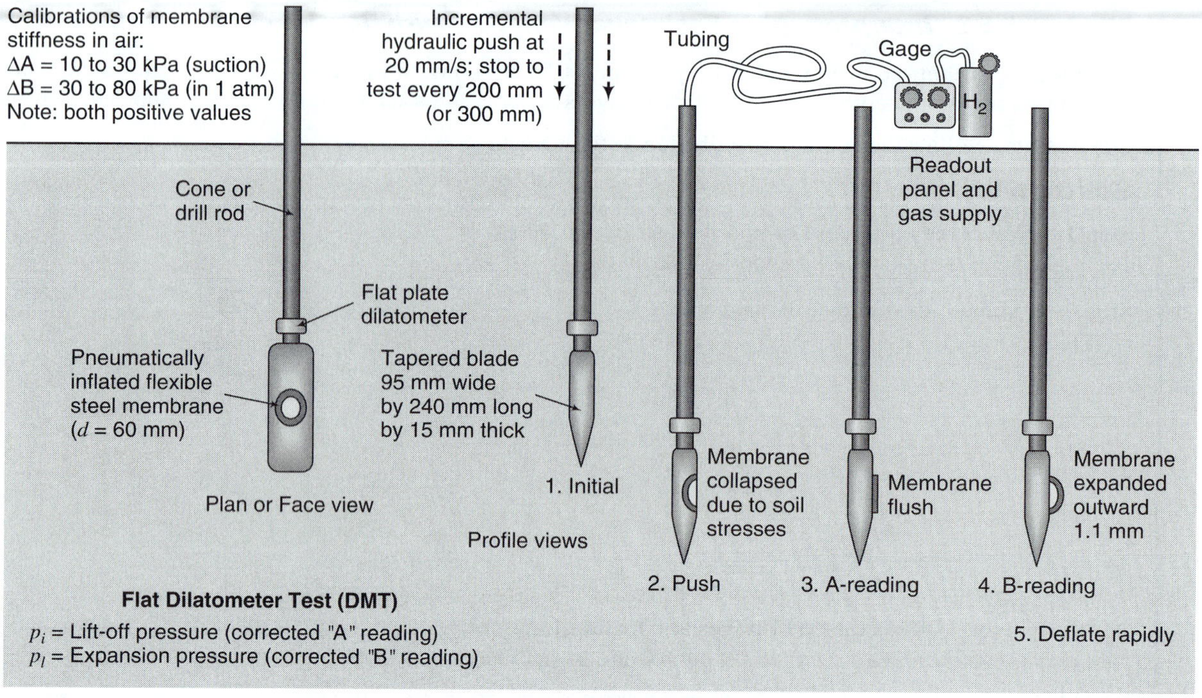

FIGURE 5.16 Flat plate dilatometer. *(Source: Professor Paul Mayne, Georgia Tech.)*

In the dilatometer test, the pressure (p_i) at which the membrane displaces by 0.05 mm (0.002 in.) and the pressure (p_l) at which it displaces 1.1 mm (0.043 in.) are recorded at the membrane center at penetration intervals of 100 mm (4 in.) to 200 mm (8 in.). The parameters of interest are the material index, I_D, the dilatometer modulus, E_D, and the horizontal stress index K_D, where

$$I_D = \frac{p_l - p_i}{p_l - u_o} \tag{5.9}$$

$$K_D = \frac{p_i - u_o}{\sigma'_{zo}} \tag{5.10}$$

$$E_D = \alpha_D(p_l - p_i) \tag{5.11}$$

where u_o is the initial hydrostatic pressure, σ'_{zo} is the initial vertical effective stress at the depth of the test, and α_D is the dilatometer factor, which for a 60 mm (2.36 in.) membrane diameter and 1.1 mm (0.043 in.) displacement is 34.7.

Results from dilatometers have been related to undrained shear strength, lateral earth pressures, overconsolidation ratio, and elastic modulus. Dilatometers cause significant remolding of the soil before the test commences and the results obtained should be used with caution. The dilatometer test is best suited for clays and sands.

5.6.5 Pressuremeter Test (PMT)—ASTM D4719

Menard PMT The Menard-type pressuremeter test (PMT) (Fig. 5.17) consists of a cylindrical probe that is placed at the desired depth and pressure applied to a measuring cell of the probe. The borehole is unlined over the height of the probe. The pressure applied is analogous to the expansion of a cylindrical

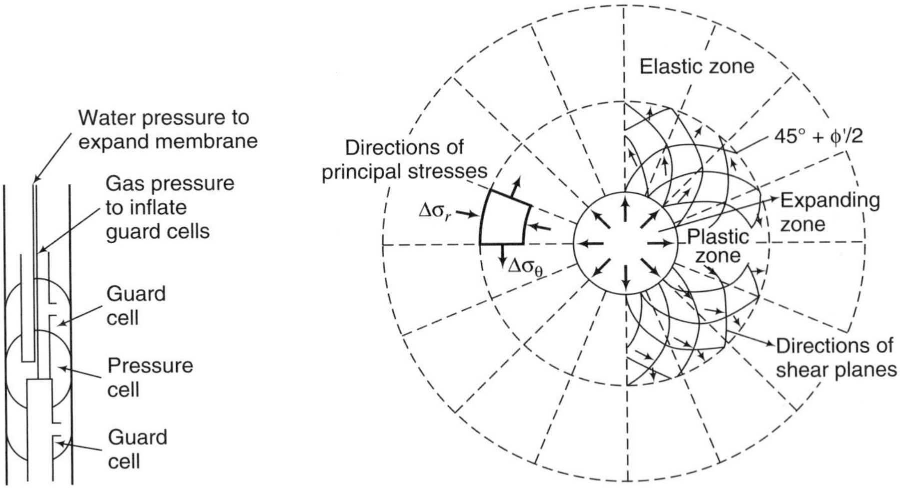

(*a*) Vertical section (*b*) Stresses near probe

FIGURE 5.17 Pressuremeter (PMT).

cavity. The pressure is raised in stages at constant time intervals (usually 1 minute) and volume changes or radial displacements are recorded at each stage. A pressure/volume change curve is then drawn from which the elastic modulus, shear modulus, and shear strength may be estimated.

The data from a PMT is usually plotted as pressure versus volumetric deformation as shown in Fig. 5.18.

The straight section of the curve AB is assumed to be the elastic response of the soil and the shear modulus G is calculated from the test data as

$$G = \left(V_c + \frac{V_o + V_f}{2}\right)\frac{\Delta p}{\Delta V} \tag{5.12}$$

where V_c is the initial volume of the probe before inflation, V_o and V_f are two points on the line AB (Fig. 5.18), $\Delta p = p_f - p_o$, and $\Delta V = V_f - V_o$. Since $G = \frac{E}{2(1+\nu)}$ we can calculate the elastic modulus E if we know the value of Poisson's ratio, ν. For drained test, $\nu \approx 1/3$ and $E \approx 2.67G$.

The Menard PMT is useful for all soils except very soft clays or gravels. One of the disadvantages of the Menard pressuremeter is that it has to be inserted into a predrilled hole and consequently the soil is disturbed. The method of drilling is crucial for accurate pressuremeter results.

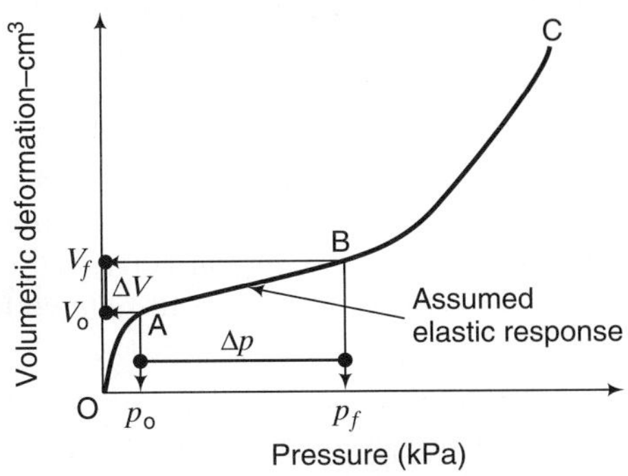

FIGURE 5.18 Illustration of PMTresults.

Self-Boring Pressuremeter Test (SBPMT) The Cambridge Camkometer (SBPMT) (Fig. 5.19) is a self-boring pressuremeter that minimizes soil disturbances. Incremental pressures are applied to radially expand a rubber membrane that is built into the sidewall of the Camkometer and a feeler gauge measures the radial displacement. Thus, the stress-strain response of the soil can be obtained.

PMT and SBPMT provide measurement of horizontal stress, shear modulus, and shear strength. Soil disturbance is small when SBPMT is used. Pressuremeters are more costly than CPT and the flat plate dilatometer and are not widely available. The drainage condition is unknown and consequently this leads to uncertainty in the interpretation of the test data to estimate the shear modulus and shear strength.

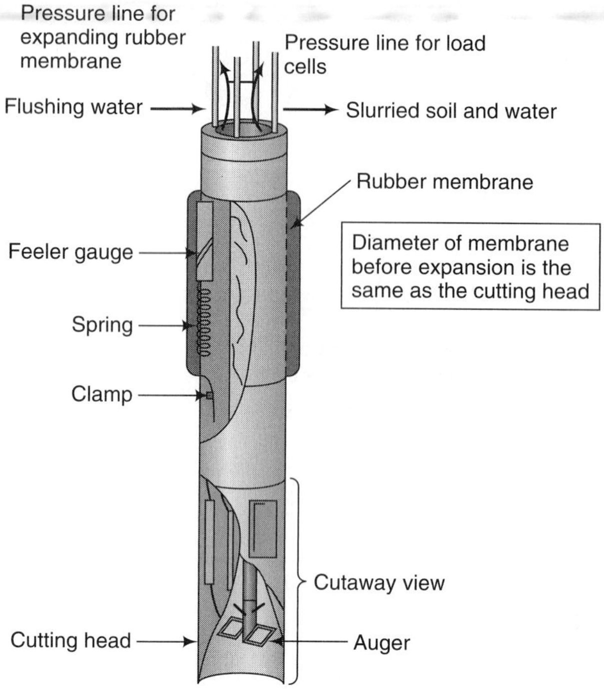

FIGURE 5.19 Schematic of Cambridge camkometer. (Redrawn after Hughes & Wroth, 1972.)

5.6.6 Plate Loading Test (PLT)—ASTM D1194

Plate load tests (Fig. 5.20) are used to estimate bearing capacity and settlement of shallow footings. The plates are made from steel with sizes varying from 152 mm (6 in.) to 900 mm (36 in.). Two common plate sizes used in practice are a square plate of width 300 mm (12 in.) and a circular plate of diameter 300 mm (12 in.). The test is carried out in a pit at the depth of the proposed footing. Loads are applied in increments of not more than 95 kPa (2000 psf) or 10% of the estimated allowable bearing capacity. Each load increment is held until settlement ceases. The final settlement at the end of each loading increment is recorded. Loading is continued until the soil fails or settlements are in excess of 10% of the plate diameter. The maximum load should be at least 1.5 times the estimated allowable bearing capacity.

If the sand were to behave like an elastic material, then the settlement could be calculated from

$$\rho_p = \sigma_{ap} B_p \frac{1-(\nu)^2}{E} I_p \tag{5.13}$$

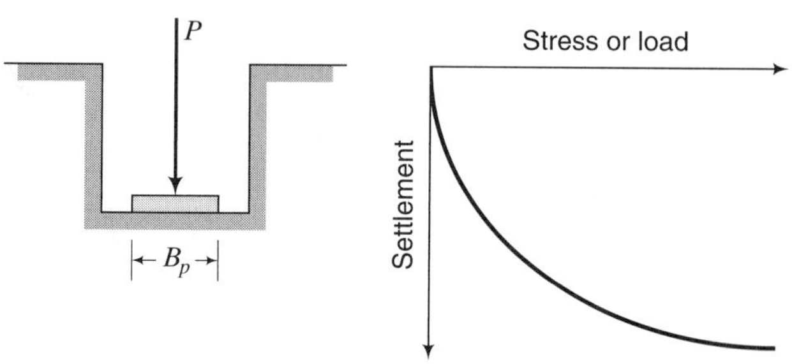

FIGURE 5.20 Plate load test.

where ρ_p is the plate settlement, σ_{ap} is the applied stress, B_p is the width or diameter of the plate, ν is Poisson's ratio, E = elastic modulus, and I_p is an influence factor (0.82 for a rigid plate). The settlement of the real footing (ρ) of width B is related to the plate settlement by

$$\rho = \rho_p \left(\frac{2}{1 + \frac{B_p}{B}} \right)^2 \tag{5.14}$$

In the limit $\frac{B_p}{B} \to \infty$, $\frac{\rho}{\rho_p} \to 4$.

There are several problems associated with the correlating plate load test results and real footing:

- The test is reliable only if the soil layer is thick and homogeneous.
- The depth of soil that is stressed below the plate is significantly lower than the real footing. A weak soil layer below the plate may not affect the test results because it may be at a depth at which the stresses imposed on the weak layer by the plate loads may be insignificant. However, this weak layer can have significant effect on the bearing capacity and settlement of the real footing.
- Local conditions such as a pocket of weak soil near the surface of the plate can affect the test results but these may have no significant effect on the real footing.
- Settlement depends on the size of the plate. Settlement increases with increases in plate size. Bjerrum and Eggestad (1963) found that there is significant scatter in the relationship between plate size and settlement for a given applied stress. Bjerrum and Eggestad (1963) also reported that field evidence indicates that the limit of $\frac{\rho}{\rho_p}$ ranges between 3 and 5 rather than being fixed value of 4.
- Performance of the test is difficult. On excavating sand to make a pit, the soil below the plate invariably becomes looser and this has considerable influence on the test results. Good contact must be achieved between the plate and the soil surface, but this is often difficult. If the plate were just above the groundwater, the results would be affected by negative porewater pressure.
- PLTs are limited to shallow depths. A screw, called a *screw plate*, that consists of a flat pitch auger device and is augered in the ground is used for tests at large depths.

5.6.7 Hydraulic Conductivity Test (HCT)—ASTM D4043

The hydraulic conductivity k in the field is obtained by pumping water at a constant flow rate from a well and measuring the decrease in groundwater level at observation wells (Fig. 5.21). The hydraulic conductivity for a confined acquifer is calculated using Darcy's law and Dupuit's assumption and is given as

$$k = \frac{q \ln\left(\frac{r_2}{r_1}\right)}{\pi(h_2^2 - h_1^2)} \tag{5.15}$$

With measurements of r_1, r_2, h_1, h_2, and q (flow rate of the pump), k can be calculated.

The key points are:

1. ***Various in situ tests are used to determine soil stratigraphy, soil strength, and deformation parameters.***
2. ***Experience is required to select the appropriate tests for a given site.***

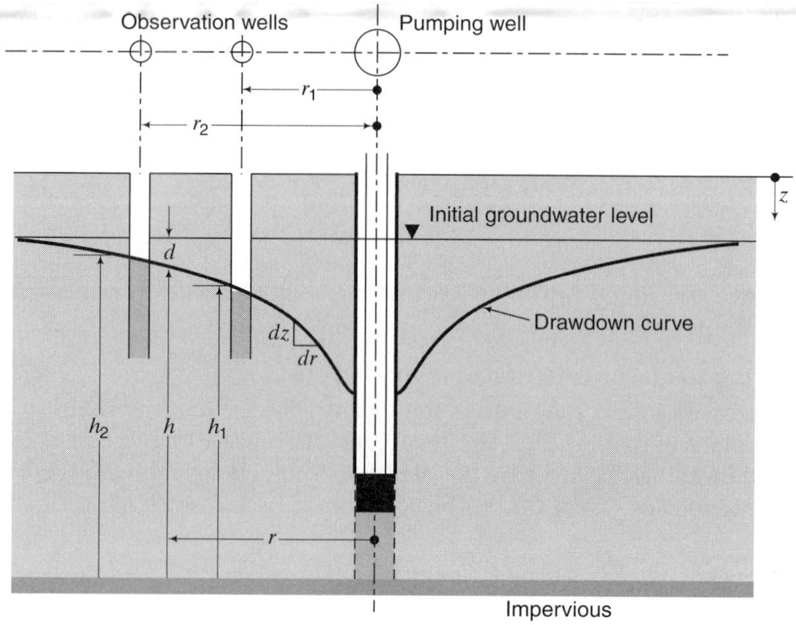

FIGURE 5.21 Pump test.

5.6.8 Comparison of In Situ Tests

A comparison to aid in selecting in situ tests for site characterization is given in Table 5.3.

5.7 LABORATORY TESTS

Importance *Laboratory tests are needed to classify soils and to determine strength, settlement, and stiffness parameters for design and construction. Laboratory tests allow for better control of the test conditions applied to the soil than in situ tests. Laboratory test samples are invariably disturbed and the degree of disturbance can significantly affect the test results. Sufficient care must be taken to reduce testing disturbances. In this section, we will present the types of tests that are commonly specified.*

Laboratory tests can be divided into two classes: Class I tests are tests to determine the physical properties. Class II tests are used to determine the mechanical properties.

Class I: Tests to determine physical properties. The physical properties are related to the following parameters: water content, specific gravity, density, grain size distribution, plasticity index, compaction, and hydraulic conductivity. The tests required are summarized in Table 5.4. The results of these tests are used to classify soils (group soils that are likely to exhibit similar construction characteristics) and to determine compactability and seepage characteristics. Some of the parameters from physical tests (e.g., liquidity index and plasticity index) have been correlated with compressibility and strength parameters.

Class II: Tests to determine mechanical properties. The mechanical properties consist of the stress-strain characteristics and settlement and shear strength parameters. A summary of conventional mechanical tests is shown in Table 5.5. For fine-grained soils, the settlement parameters are C_v, C_c or m_v, C_r, C_α, and σ'_{zc}. These are obtained from one-dimensional consolidation tests. The shear strength parameters are the friction angles (ϕ'_{cs} and ϕ'_p) and the undrained shear strength,

TABLE 5.3 Comparison of In Situ Tests

Comparison Basis	Standard Penetration Test (SPT)	Cone Penetration Test (CPT)	Vane Shear Test (VST)	Pressuremeter Test (PMT and SBPMT)	Flat Dilatometer Test (DMT)
ASTM	D 1586	D 5778	D 2573		
Equipment	Simple, rugged	Complex, rugged	Simple, rugged	Complex, delicate	Simple, rugged
Ease of testing	Easy	Easy	Easy	Complex	Easy
Continuous profile or point values	Point	Continuous	Point	Point	Semi-continuous
Basis for interpretation	Empirical	Empirical, theory	Theory	Empirical, theory	Semi-empirical, theory
Soil types best suited for	Sand	Sand and fine-grained soils	Fine-grained soils	Coarse-grained soils and soft rock	Sand and fine-grained soils
Soils not suited for	Fine-grained	Gravel, dense sand	Coarse-grained soils	Fine-grained (PMT)	Gravel
Suitability in practice	Routine	Routine	Routine	Limited	Routine
Unit cost	Medium	Low	Medium	High	Low
Availability	Excellent	Good	Excellent	Poor (SBPMT), Good (PMT)	Fair
Possible variability in results (COV %)*	15–45	15–85 (MCPT), 5–15 (ECPT)	10–20	15–25 (SBPMT), 10–20 (PMT)	5–15
Geotechnical information provided	• Stratification • Density or compaction • Disturbed samples for lab tests • Inspection of soil for general description and qualitative characterization, e.g. a soil with N value of 30 is likely to be stronger than a soil with an N value of 15	• Continuous resistance (strength) measurement • Porewater pressure measurements if cone fitted with porewater pressure transducers	• Undrained shear strength	• Bearing capacity • Compressibility • Shear and bulk modulus	• Lateral stress coefficient • Over-consolidation ratio • Undrained shear strength (fine-grained soils) • Deformation modulus

* COV—standard deviation/mean, MCPT—manual CPT; ECPT—electronic CPT; SBPMT—self-boring PMT.

SOURCE: Modified from Mitchell (1978), Orchant et al.(1988), Kulhawy and Mayne (1990).

TABLE 5.4 Summary of Laboratory Tests to Determine Physical Properties

Physical Properties	ASTM	Test objective	Parameters determined	Purpose
Specific gravity	D854	To determine the specific gravity of soils.	G_s	To calculate soil density.
Grain size determination	D422 D1140	To determine the grain size distribution.	D_{10}, D_{50} $Cu = \frac{D_{10}}{D_{60}}$ $CC = \frac{D_{30}^2}{D_{10}D_{60}}$	Soil classification.
Water content	D2216	To determine the water content of a soil.	w	Qualitative information on strength and deformation.
Index test	D4318	To determine the water content at which soil changes phases.	PL, LL, PI, SL, LI	Soil classification. Qualitative information on strength and settlement.
Compaction	D698 D1557	To determine the maximum dry density and optimum water content.	$(\gamma_d)_{max}, w_{opt}$	Specification of compaction in the field.
Permeability	D2434	To determine the hydraulic conductivity.	k	Estimation of flow of water and seepage forces. Stability analysis.
Maximum and minimum dry density	D4253 D4254	To determine the maximum and minimum dry density of coarse grained soil.	e_{max} e_{min}	Soil classification.

TABLE 5.5 Summary of Laboratory Mechanical Tests

Test	ASTM	Stress condition	Drainage condition	Soil type	Parameters	Advantages	Disadvantages
Direct shear (DS)	D3080	Plane strain-stress or strain control	Drained	Coarse-grained	ϕ'_{cs}, ϕ'_p, α_p	• Simple • Quick • Commonly available	• Soil fails on predetermined failure plane • Nonuniform stress distribution • Strains cannot be determined
Triaxial (T)	D4767 D2850 D2166 (for UC tests)	Axisymmetric stress or strain control	Drained or undrained	All	ϕ'_{cs}, ϕ'_p, s_u, E and M	• Versatile—two stresses (axial and radial can be controlled independently) • Commonly available	• Principal axes rotate only by 90° instantaneously • Nonuniform stress distribution – reduced by lubricating platen
One-dimensional consolidation	D2435 D4186 D5333 D454	Axisymmetric	Drained	Fine-grained	C_c, C_r, C_α, C_v, σ'_{zc}, m_v or λ, κ, p'_c	• Simple • Readily available	• One-dimensional
Direct simple shear (DSS)		Plane strain	Drained (constant load) or undrained (constant volume)	All	ϕ'_{cs}, ϕ'_p, s_u, G	• Principal axes rotate during test • Closely approximates many field conditions.	• Nonuniform stress and strain distributions • Not readily available

M is critical state frictional constant; m_v is modulus of volume compressability; G is shear modulus.

s_u. These are obtained from conventional laboratory tests such as direct shear, triaxial, and direct simple shear tests.

The stress-strain responses and shear strength of soils are stress path dependent. Therefore tests should be specified that duplicate the stress and strain paths that the structure would experience during construction and during its design life. We also want to know what loads will cause the soil to fail. No single mechanical test is available to subject the soil to the variety of stress paths that are introduced by loading in the field. You will have to make decisions based on the predominant stress path and the availability of test equipment. For example, if simple shear test equipment is not available, then direct shear test will have to be substituted. Stress conditions and the tests that can simulate them based on the conventional laboratory tests for some common foundations are shown in Fig. 5.22.

Sampling disturbances usually cause underestimation of the compression index and alter the shear strength parameters. The result is that the compression index and shear strength parameters are unrepresentative of the intact soil. One technique, called SHANSEP (Ladd and Foot, 1974), for reducing the effects of sampling disturbances on settlement and shear strength parameters is to consolidate the soil sample to about 1.5 to 2 times the estimated preconsolidation pressure followed by swelling (unloading) to the desired stress level. The unloading pressure should cover the range of overconsolidation ratios in the field. The SHANSEP procedure tries to reduce sampling disturbance by reorganizing the soil structure. But, this reorganization may not lead to the same structure as the intact soil. Therefore, the settlement and shear strength parameters do not necessary represent the intact soil's response. The SHANSEP procedure is not recommended for heavily overconsolidated clays (overconsolidation ratio $> \sim 8$) or cemented soils.

The key points are:

1. ***Laboratory tests are used to classify soils, determine their compactability, determine hydraulic conductivity, determine stress-strain response, determine shear strength, and determine settlement characteristics.***
2. ***Care needs to be exercised in testing to reduce sample disturbances.***

5.8 GEOTECHNICAL REPORT

A clear, concise, and accurate report of the site investigation must be prepared. The report should contain at a minimum the following:

1. A document (often a letter) authorizing the investigation.
2. A summary of the work done and recommendations (about one page).
3. Scope of work.
4. Description of the site.
5. Details of the types of investigation conducted, soil and groundwater information including lab and field test results, assumptions and limitations of the investigation, and possible construction difficulties. Soil boring logs (a typical one is shown in Fig. 5.23) are normally used to summarize the soil data.
6. Recommendations.

During soil exploration all pertinent details are recorded and presented in a *boring log*. A typical boring log is shown in Fig. 5.23. Additional information consisting mainly of laboratory and in situ test results is added to complete the boring log.

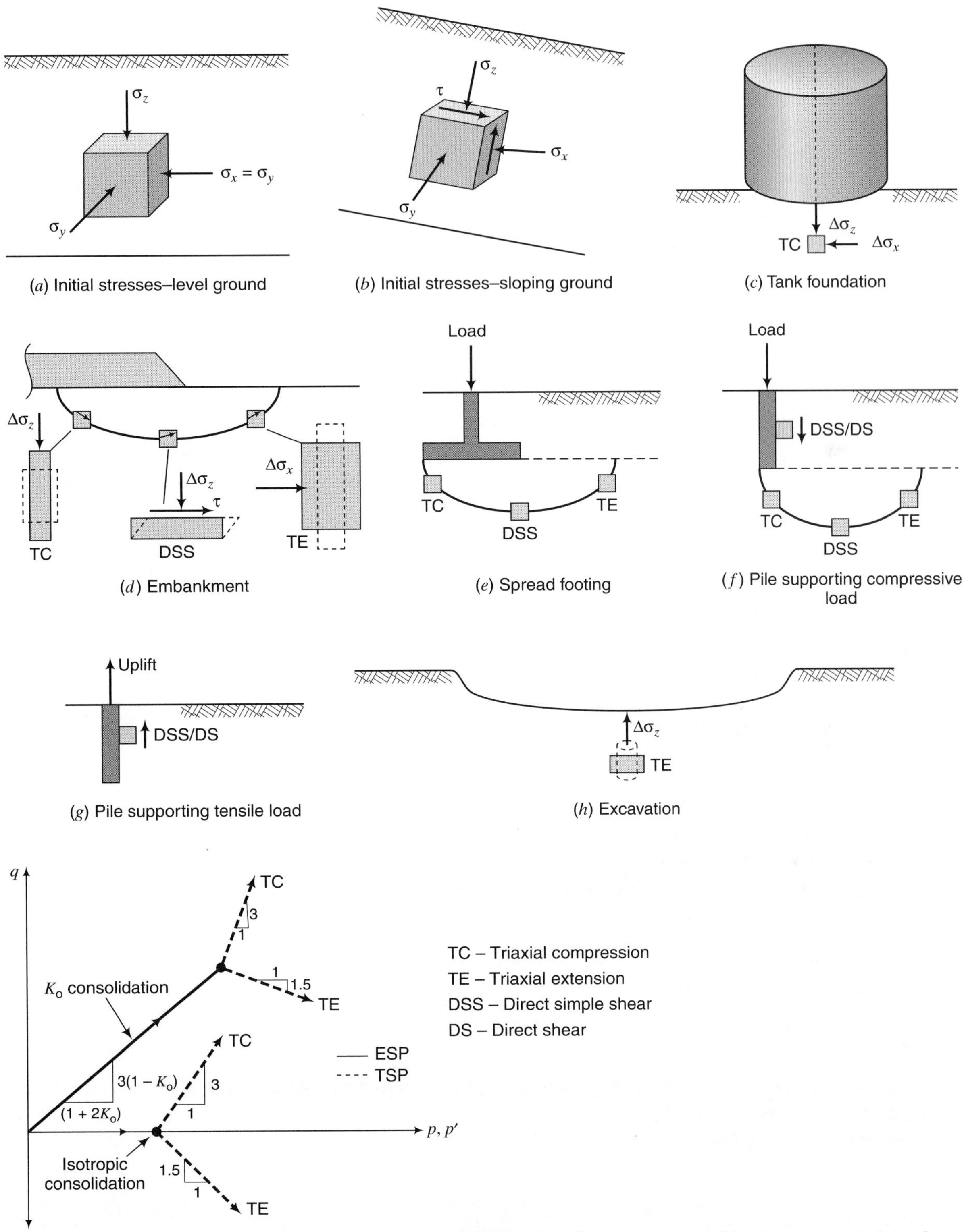

(*i*) Total stress paths in TE and TC triaxial tests.

FIGURE 5.22 Stresses and possible laboratory tests based on stress paths for some common foundations.

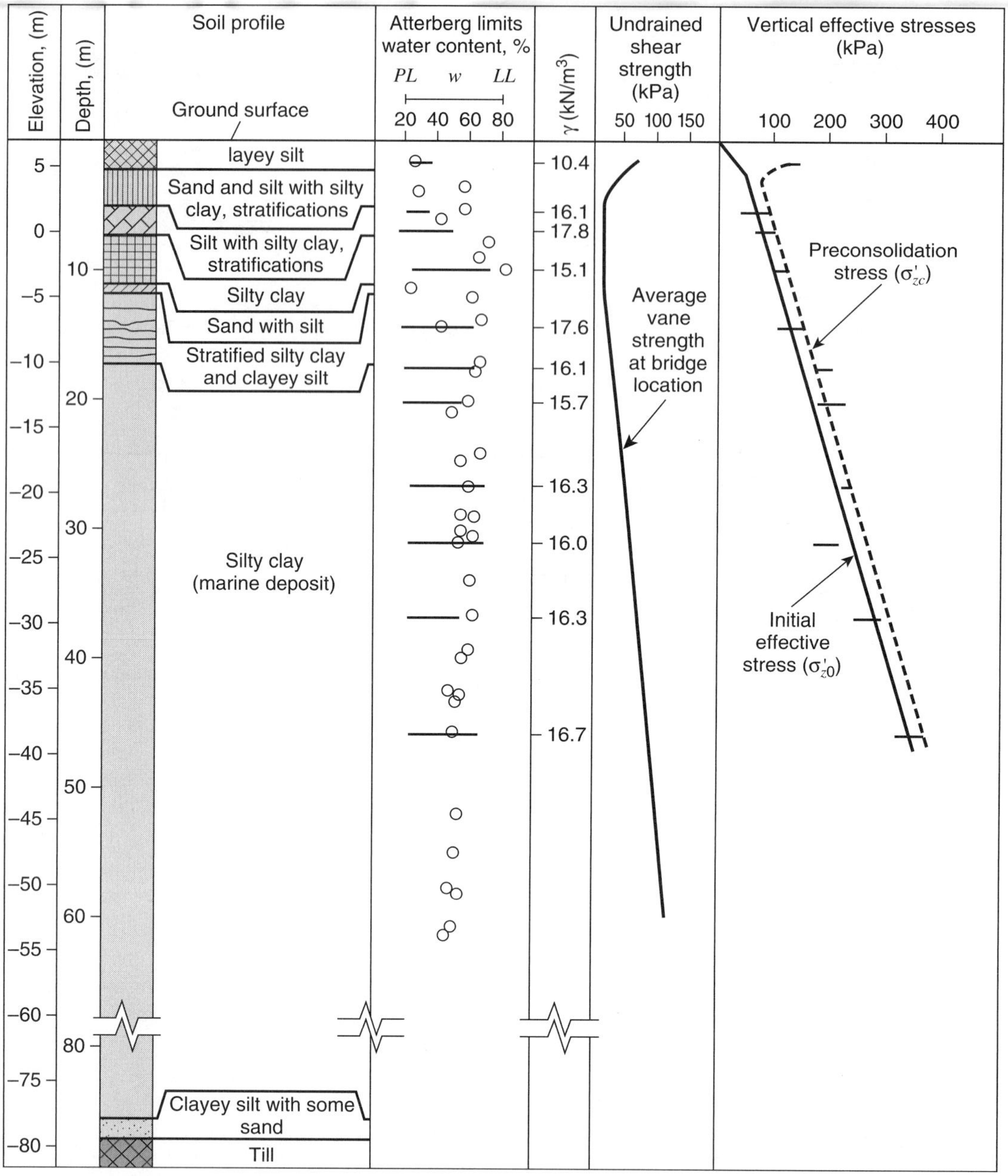

FIGURE 5.23 A borehole log (Redrawn from Blanchet et al., 1980).

5.9 EMPIRICAL RELATIONSHIPS FOR SETTLEMENT AND SHEAR STRENGTH PARAMETERS

Importance *A number of empirical relationships among soil shear strength, settlement and deformation parameters, and in situ tests is available. These relationships are used to make preliminary analysis of the response of foundations.*

Some of these correlations are presented in Appendix B:

1. Unit weight, relative density and friction angle (Appendix B, Table B5.2)
2. Overconsolidation ratio (Appendix B, Table B5.3)
3. Friction angle (Appendix B, Fig. B5.2)
4. Undrained shear strength (Appendix B, Table B5.4)
5. Soil stiffness (Appendix B, Table B5.5)

5.10 LIQUEFACTION POTENTIAL

Importance *Seismic and other dynamic loading can cause soils to liquefy. Dynamic liquefaction of soils is usually catastrophic and is manifested by sand boiling, land subsidence, local flooding, sudden failure of foundation of structures, and uplift or floatation of buried structures. You need to identify soils that have the potential to liquefy and, if they do, you have to take measures to prevent or ameliorate damage. In this section, we discuss semi-empirical methods to identify the potential of a soil to liquefy under a seismic event.*

5.10.1 Basic Concept

During an earthquake, the dynamic load (mass × acceleration) is cyclical and is imposed so quickly that even saturated coarse-grained soils with their excellent drainage characteristics cannot dissipate the induced dynamic excess porewater pressure. Thus, during earthquakes, soils are cyclically loaded under undrained conditions. Cyclical loading reduces the shear strength and stiffness of the soil. For coarse-grained soils, the dynamic excess porewater pressure could cause the effective stress of the soil to reduce to zero, leading to dynamic liquefaction. A structure founded on a soil that dynamically liquefies will fail regardless of its structural integrity, as observed during the 1989 Loma Prieta earthquake in California and the 1964 Niagatta earthquake in Japan (Figure 5.24).

Various forms of dynamic liquefaction such as sand boils and lateral spreading (viscous flow of soils) can occur. Dry coarse-grained soil can also temporarily behave as a viscous fluid during an earthquake but does not flow in the same way as a saturated soil that dynamically liquefies. Dynamic liquefaction of soils can be catastrophic and causes sand boiling, land subsidence, local flooding,

FIGURE 5.24 Collapse of buildings from dynamic liquefaction during (a) the 1989 Loma Prieta earthquake in California, USA and (b) the 1964 Niagatta earthquake in Japan.

sudden failure of foundation of structures, and uplift or flotation of buried structures. In dry soils, earthquake loads can cause reduction in the soil's load-bearing capacity, increase settlement, and trigger failures in retaining walls and slopes.

Consider a soil element at a depth z below the ground in a homogeneous soil mass, and groundwater is at the surface. Under static condition,

$$\sigma'_{zo} = \sigma_{zo} - u_o \tag{5.16}$$

where σ'_{zo} and σ_{zo} are the effective and total vertical stresses respectively, and $u_o = \gamma_w z$ is the initial porewater pressure. During an earthquake event, the porewater pressure increases by Δu_d, where the subscript d denotes dynamic action and the effective stress in the soil reduces to:

$$\sigma'_z = \sigma_{zo} - u_o - \Delta u_d = \sigma_{zo} - (u_o + \Delta u_d) \tag{5.17}$$

If Δu_d increases such that $u_o + \Delta u_d = \sigma_{zo}$, the effective vertical stress in the soil becomes zero and the soil becomes a viscous fluid (i.e., liquefies). In Eq. 5.17, only a one-dimensional stress system was considered. If we were to consider all the stresses, then we could rewrite Eq. (5.17) in terms of stress invariants as

$$p'_o = p - (u_o + \Delta u_d) \tag{5.18}$$

Dynamic liquefaction would occur when $(u_o + \Delta u_d) = p$. When liquefaction occurs, the soil loses its intergranular frictional resistance and is replaced by viscous resistance.

We have considered a homogeneous soil mass and assumed that the earthquake motion induces uniform stresses and strains. Soils are (notoriously) inhomogeneous and earthquake motions do not impose uniform stresses and strains in the soil. The result is that some parts of the soil will exhibit a tendency to compress (positive excess porewater pressure) while other parts will exhibit a tendency to expand (negative excess porewater pressure). The difference in excess porewater pressure will cause a porewater pressure gradient and seepage will occur. An upward component of the seepage force j_z will cause the effective stress to decrease ($\sigma'_z = \sigma_{zo} - u_o - \Delta u_d - j_z z$, where $j_z = i\gamma_w$, i is the hydraulic gradient, and z is the length of the vertical seepage path), while a downward component will cause the effective stress to increase ($\sigma'_z = \sigma_{zo} - u_o - \Delta u_d + j_z z$). The net outcome is that some parts of the soil mass can liquefy before the porewater pressure becomes equal to the total mean stress.

The key points are:

1. ***Vibrations produced by earthquakes cause reduction in shear strength and stiffnesses of soils.***
2. ***Stresses due to an earthquake are imposed on a soil so quickly that even coarse-grained soils cannot dissipate the dynamic excess porewater pressure; the soil mass is loaded under undrained conditions.***
3. ***When the increase in total porewater pressure (initial plus dynamic excess porewater pressure) becomes equal to the total stress on the soil mass, the soil liquefies.***
4. ***Excess porewater pressure gradient will accelerate the potential of a soil to liquefy.***

5.10.2 Evaluation of Liquefaction Potential

Go to http://www.wiley.com/college/budhu. Click Foundations and Earth Structures, Chapter 5, liquefactionpotential.xls for a spreadsheet to calculate liquefaction potential.

The potential of a soil to dynamically liquefy is currently evaluated using laboratory and field tests. The triaxial and simple shear tests are the preferred laboratory tests, while SPT and CPT are popular field tests.

The most widely used method to estimate the liquefaction potential (LP) of sandy soils at a site due to a seismic event is the SPT method proposed by Seed and Idriss (1971) and modified later by several investigators including Youd et al. (2001) and Seed et al. (2003). The procedures based on Seed et al. (2003) for SPT and CPT are as follows:

SPT

1. Correct the N value from SPT to $N_{1,60}$ as described in Section 5.6.1.

2. Correct $N_{1,60}$ for fines content.

$$N_{1,60,\text{cs}} = (1 + 0.004FC)N_{1,60} + 0.05FC; \quad 35\% \geq FC \geq 5\% \tag{5.19}$$

where FC is the fines content (% finer than #200 sieve expressed as an integer e.g. 10.3% = 10); $N_{1,60,\text{cs}}$ is an integer value.

3. Determine the maximum horizontal acceleration, $a_{\max}$, from previous earthquake records at or near the site. Earthquake records are available at http://www.neic.usgs.gov/neis/epic/ and http://www.peer.berkeley.edu/smcat/search.html.

4. Calculate the shear stress ratio, $\text{CSR} = \frac{\bar{\tau}}{\sigma'_z}$, for a magnitude ($M_w$) 7.5 earthquake from the following equation,

$$\text{CSR} = 0.65 \frac{a_{\max}}{g} \frac{\sigma_z}{\sigma'_z} r_d \tag{5.20}$$

where $\bar{\tau}$ is the average shear stress, σ'_z and σ_z are the vertical effective and total stresses, g is acceleration due to gravity (9.8 m/s^2 or 32 ft/s^2), and r_d is a nonlinear shear mass participation factor (Fig. 5.25).

5. Calculate the modified shear stress ratio, CSR*,

$$\text{CSR}^* = \frac{\text{CSR}}{\text{DWF}_m \times K_\sigma} \tag{5.21}$$

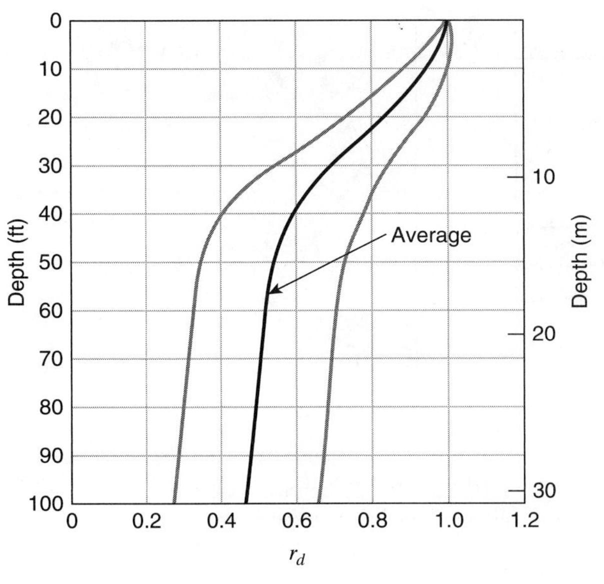

FIGURE 5.25 Shear mass participation factor, r_d. *(Source: Seed et al., 2003.)*

where DWF_m is a magnitude duration weighing factor for the design earthquake (Fig. 5.26) and K_σ is an overburden reduction (Fig. 5.27).

6. Plot the point (CSR^*, $N_{1,60,cs}$) on Fig. 5.28.
7. Identify the probability for liquefaction (P_L).

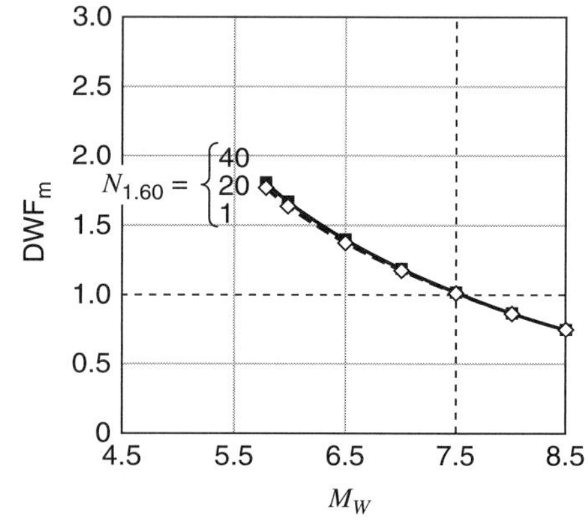

FIGURE 5.26 Magnitude-correlated duration irregular factor. *(Source: Modified after Seed et al., 2003.)*

CPT

1. Calculate CSR^* as given in steps 3 through 5 for SPT.
2. Calculate the normalized cone tip resistance

$$q_{c,1} = q_c \left(\frac{p_{atm}}{\sigma'_z} \right)^c \tag{5.22}$$

where q_c is the cone tip resistance, p_{atm} is atmospheric pressure, and c is an exponent given Fig. 5.29. The factor R_f shown in Fig. 5.29 is the sleeve resistance–tip resistance ratio given as $R_f(\%) = \frac{f_s}{q_c} \times 100$, and f_s is the sleeve resistance. You will have to find $q_{c,1}$ by iteration. For example, if $q_c =$ 10 MPa, $\frac{p_{atm}}{\sigma'_z} = 2$, and $R_f = 1\%$, you can assume a value of c of say 0.35 and calculate $q_{c,1} = 10(2)^{0.35} = 12.7$. In Fig. 5.29, with $R_f = 1\%$ and $c = 0.35$, you get $q_{c,1} = 29$, so we have to assume a higher value of c. Let us assume $c = 0.4$; then the calculated $q_{c,1} = 13.2$. From Fig. 5.29 with $R_f = 1\%$ and $c = 0.4$, $q_{c,1} = 14$, so $c = 0.4$ is good choice.

3. Calculate $q_{c,1,mod}$.

$$q_{c,1,mod} = q_{c,1} + \Delta q_c \tag{5.23}$$

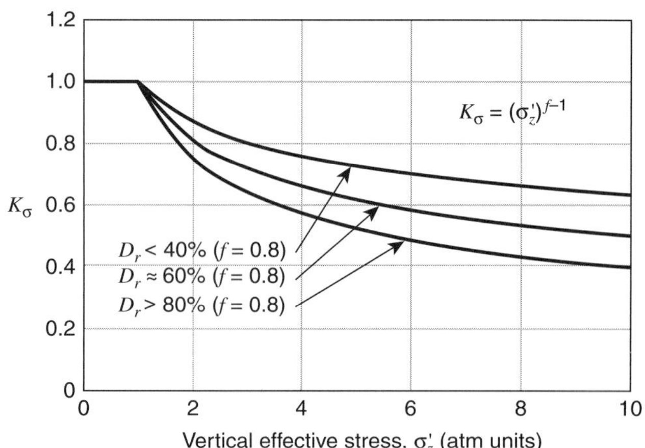

FIGURE 5.27 Overburden correction factor. *(Source: Seed et al., 2003.)*

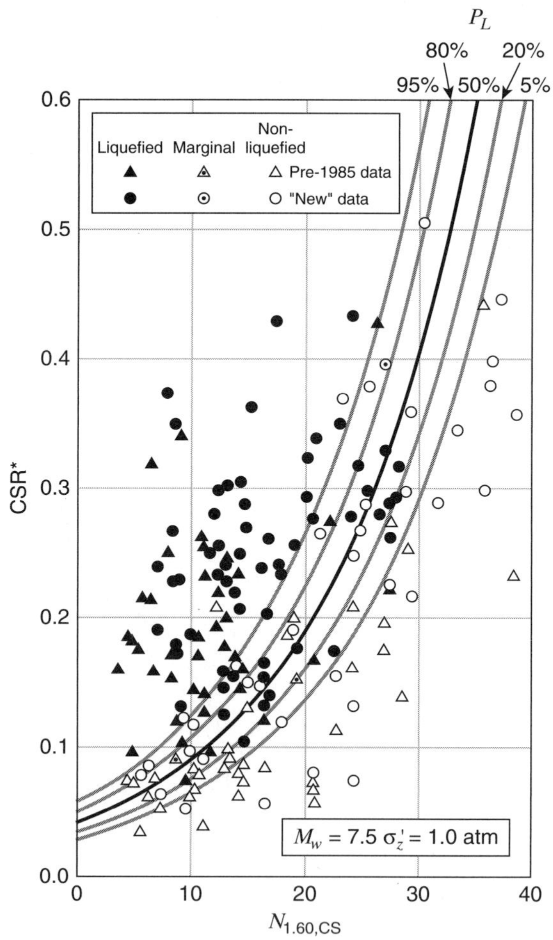

FIGURE 5.28 Recommended "probabilistic" SPT-based liquefaction triggering correlation for $M_w = 7.5$ and $\sigma'_z = 1.0$ atm. *(Source: Seed et al., 2003.)*

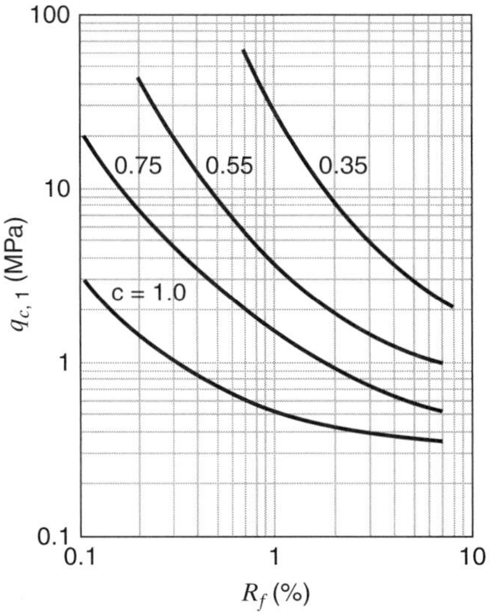

FIGURE 5.29 Exponent *c* for correcting tip resistance. *(Source: Seed et al., 2003.)*

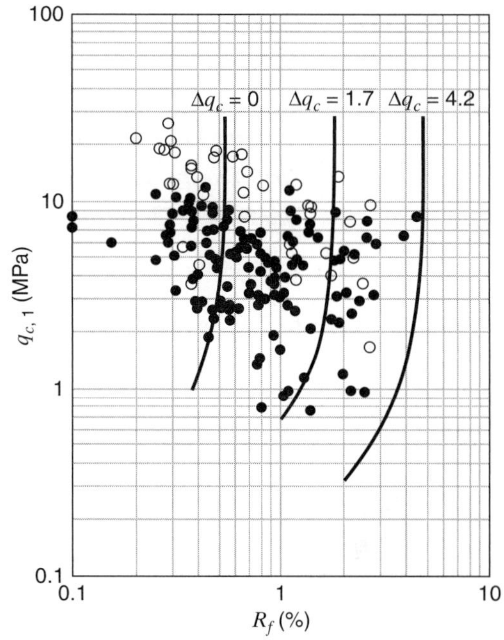

FIGURE 5.30 Modification factor for frictional effects of fines. *(Source: Seed et al., 2003.)*

where Δq_c is a modification factor for the frictional effects of fines content given in Fig. 5.30.

4. Plot the calculated coordinates ($q_{c,1,\text{mod}}$, CSR*) in Fig. 5.31.
5. Identify the probability of liquefaction P_L.

The SPT and CPT methods are merely approximate methods and are intended only as a guide to the possible liquefaction of a soil at a site.

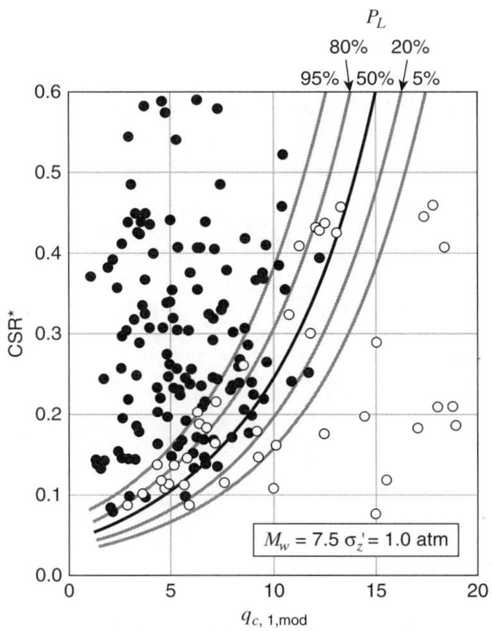

FIGURE 5.31 Probability of liquefaction for CPT ($M_w = 7.5$, $\sigma_z' = 1$ atm). *(Source: Seed et al., 2003.)*

The* key points *are:

1. ***We have a reasonable, albeit not full, understanding of the mechanics of liquefaction.***
2. ***No generally acceptable, reliable, and rational method has evolved to determine whether the soil at a particular site would liquefy.***
3. ***We still do not know how many cycles of a prescribed acceleration would cause a soil to liquefy.***
4. ***The SPT and the CPT methods are only preliminary estimates of possible soil liquefaction at a site.***

EXAMPLE 5.4 *Liquefaction Potential*

The SPT at the center of a soil layer 4 m (13 ft) thick is 8. The soil is classified using the USCS as SW-SM. Groundwater level is at the surface. The following parameters are known: Rod length = 3 m (10 ft), standard sampler, borehole diameter = 100 mm (4 in.), donut hammer without throw release $C_E = 0.6$, $a_{max} = 0.15\,g$, expected earthquake magnitude = 7.5. Estimate the probability of liquefaction.

Strategy Follow the procedure outlined in Section 5.10.2 for SPT. SI units will be used in the solution of this example problem.

Solution 5.4

Step 1: Correct N to $N_{1,60}$.

From Appendix B (Table B5.1),

$$C_R = 0.05L + 0.61$$
$$= 0.05(3) + 0.61 = 0.76$$
$$C_S = 1, \quad C_B = 1, \quad C_E = 0.6$$

From Appendix B (Table B5.2), $N = 8$, $\gamma = 15\,\text{kN/m}^3$.
Center of layer: $\sigma_z = \gamma z = 15 \times 2 = 30$ kPa; $\sigma'_z = \gamma' z = (15 - 9.8) \times 2 = 10.4$ kPa

$$C_N = \left(\frac{95.8}{\sigma'_z}\right)^{\frac{1}{2}} = \left(\frac{95.8}{10.4}\right)^{\frac{1}{2}} = 3.0 > 2; \quad \text{Thus, use 2.0.}$$

$$N_{1,60} = C_R C_S C_B C_N C_E N = 0.76 \times 1 \times 1 \times 2 \times 0.6 \times 8 = 7$$

Step 2: Find $N_{1,60,cs}$

From Fig. 1.3a, FC = 10% for a SW – SM soil
$N_{1,60,cs} = (1 + 0.004\text{FC})N_{1,60} + 0.05\text{FC} = (1 + 0.004 \times 10)7 + 0.05(10) = 7.78$; use 8.

Step 3: Calculate CSR.

$$\text{CSR} = 0.65 \frac{a_{max}}{g} \frac{\sigma_z}{\sigma'_z} r_d$$

From Fig. 5.25, r_d, = 0.83 (use average curve).

$$\text{CSR} = 0.65 \times \frac{0.15g \times 30}{g \times 10.4} \times 0.83 = 0.23$$

Step 4: Calculate CSR*.

From Fig. 5.26, $\text{DWF}_m = 1$.

$$\frac{\sigma'_z}{p_{atm}} = \frac{10.4}{101.3} = 0.103, \quad K_\sigma = 1 \quad \text{(Fig. 5.27)}$$

$$\text{CSR}^* = \frac{\text{CSR}}{\text{DWF}_m \times K_\sigma} = \frac{0.23}{1 \times 1} = 0.23$$

Step 5: Estimate probability of liquefaction
Using the coordinates $(\text{CSR}^*, N_{1,60,cs}) = (0.23, 8)$ and from Fig. 5.28, we get that the probability of liquefaction is about 100%.

■

5.11 SUMMARY

Soils are products of nature that have been subjected to a variety of natural processes over space and time. Thus, a site investigation is necessary to determine the nature of the soils at a proposed site for design and construction. A site investigation needs careful planning and is usually done in phases. The extent of a soils exploration program depends largely on the geology of the site, the type and importance of the structure, the structural loads, and groundwater conditions. A number of tools are available for soil exploration. You need to use judgment as to the types of tools appropriate for a given project, and the types of in-situ and laboratory tests required to characterize the soil and to determine soil strength and deformation parameters.

Practical Example

EXAMPLE 5.5 *Planning a Preliminary Soils Exploration Program*

A two-storey office building 40 m (132 ft) × 100 m (300 ft) is proposed for a site. The soil profile from a geological map at the site and the plan of the building are shown in Fig. E5.5a,b. Discussions with the architect and examination of nearby buildings reveal that the building can be supported by a concrete slab (foundation) of average width about 1.5 m (5 ft) under each column. Plan a preliminary soils exploration program for the site.

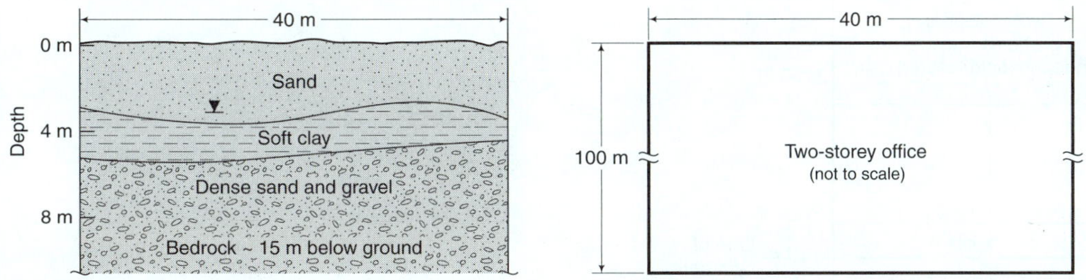

(*a*) Soil Profile (*b*) Plan of building

FIGURE E5.3a,b

Local costs are as follows:

Drilling: Mobilization: \$500.

Sampling and in situ test: SPT: \$60 each sample (including drilling); CPT: \$40 each test; VST: \$30 each test; Shelby sample: \$55 each.

Lab tests: Sieve analysis: \$65 each test; direct shear: \$300 each test; consolidation test: \$400 each test; water content: \$6 each test; Atterberg limits: \$70 each test.

Strategy The geological profile provides a good reference to plan the preliminary soils exploration program. SI units will be used in solving this problem.

Solution 5.5

Step 1: Inspect the geological profile.

Some key points:

- Possibly only three soil layers with sand in the top layer followed by clay above dense sand and gravel.
- Groundwater level is just above clay layer.

Step 2: Decide on the type of soil investigation needed.

Footprint of office building = 40 × 100 = 4000 m^2. From Table 5.2, 6 to 7 boreholes are needed. Use 7 boreholes.

Minimum depth: Based on the effective depth for vertical stress increase, a minimum depth of about $2B = 2 \times 1.5 =$ 3 m is required. However, from Section 5.5.3, 6 m is the minimum depth recommended.

The top layer is sand. Recovery of sand samples is difficult, so in situ tests should be considered. Inspection of Table 5.3 shows that either SPT or CPT can be used. The bottom layer is a sand-gravel mixture for which SPT is better suited than CPT. SPT at 1 m intervals will then be recommended. The clay layer may have an effect on the bearing capacity and settlement. VST can be used to determine the shear strength and Shelby samples of the clay will be taken for consolidation tests. If field inspections of samples show that the clay is homogeneous, then only two samples will be taken and consequently only two consolidation tests will be conducted.

Two sieve analysis tests will be conducted on samples of recovered sand and Atterberg limits will be conducted on the clay samples. If field inspections of samples show that the sand is homogeneous, then only four sieve analysis tests will be conducted. Water content will be conducted on samples from about three boreholes. The above is the minimum requirement.

Step 3: Summarize soil exploration program and estimate cost.

A tentative location plan of the boreholes is shown in Fig. E5.5c. The boreholes will be symmetrical about the centerline of the length of the footprint. Only three boreholes instead of five are shown in Fig. E5.5c because of

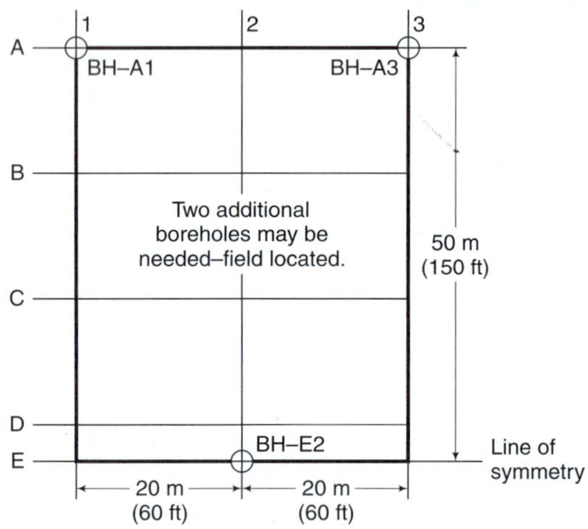

FIGURE E5.5c

symmetry. The location of the two additional boreholes to make a total of seven will depend on observations of the soil as drilling proceeds. It may be possible that these additional boreholes may not be needed if the changes in soil profile and soil types are insignificant.

The costs are shown below. The costs may be lower if five and not seven boreholes are used.

Description	No	Units	Unit cost	Cost
Mobilization			$500.00	$500.00
No. of boreholes	7			
Drilling 7 boreholes @ 6 m deep	42	m		
5 SPT per borehole	35	SPT	$60.00	$2,100.00
1 VST per borehole	7	VST	$30.00	$210.00
Shelby sample	2		$55.00	$110.00
Consolidation tests	2		$400.00	$800.00
Sieve analysis	14		$65.00	$910.00
Water content	21		$6.00	$126.00
Atterberg limits	2		$70.00	$140.00
			Total	$4,896.00

■

EXERCISES

5.1 Select a site within your area for a six-storey office building. Write a short report explaining to a client why a soils exploration program is needed.

5.2 Select an area in your location and prepare a representative soil profile for it.

5.3 Select a site within your area and conduct a Phase I site characterization.

5.4 An earth dam of height 4 m (12 ft) is to be constructed from fill of compacted unit weight 20 kN/m^3 (127 pcf) on a soft clay, 20 m (60 ft) thick, resting on a deep deposit of stiff, heavily overconsolidated clay. Groundwater level is at 2 m (6 ft) below the surface of the soft clay. Describe and justify the type of in-situ and laboratory tests you would specify.

5.5 The soil profile from a geological map of a project site obtained during a Phase I investigation is shown in Fig. P5.5. For the proposed building shown, plan a preliminary soils exploration program (e.g., number of boreholes, location and depth of boreholes, type of tests required). Use the costs given in Example 5.5.

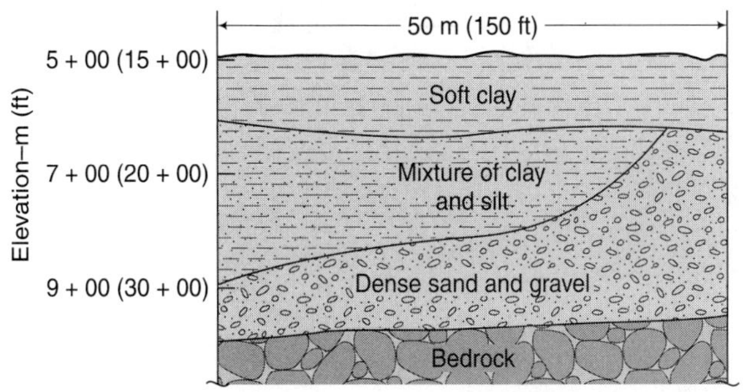

(*a*) Soil profile

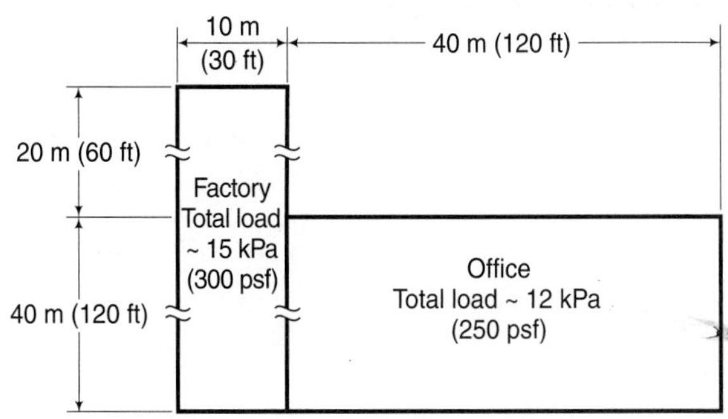

(*b*) Plan

FIGURE P5.5

5.6 A preliminary boring at a site for a four-storey office building 50 m (150 ft) × 150 m (350 ft) is summarized in the table below: The blow counts are SPT values for every 0.152m(6 in).

Depth—m (ft)	Soil description	SPT Depth m (ft)	SPT Blow counts	w (%)	LL (%)	PL (%)
0 to 1.7 (5)	Black organic material					
1.7 to 4.2 (5 to 14)	Silty sand—light brown, very loose to loose	1.7 (5)	2, 1, 4			
		3.0 (9)	4, 3, 6			
	Groundwater at 3.6 m					
4.2 to 8.0 (14 to 25)	Sand—light brown, medium to dense	4.0 (12)	7, 7, 8			
		5.0 (15)	8, 9, 11			
		6.0 (18)	15, 11, 17			
		7.0 (21)	21, 19, 29			
8 to 10 (25 to 30)	Stiff gray clay with yellowish bandings			22	58	31
End						

(a) Plot the boring log.

(b) Is there any soil layer or soil condition that is of concern at this stage of the exploration? Explain your answer.

(c) How many boreholes would you estimate are needed for this site? Justify your answer.

(d) Is the depth of the preliminary borehole satisfactory? Justify your answer.

5.7 SPT test at a site of a sandy soil gave the following results.

Depth—m (ft)	1 (3)	2 (6)	3 (9)	4 (12)	5 (15)	6 (18)	7 (21)	8 (24)	9 (27)
N	4	6	7	4	13	38	44	56	62*

* *Refusal after 0.35 m (14 in) penetration.*

The rod length about the surface is 1 m (3 ft). A donut hammer (ER = 50%) and a standard sampler in a borehole 152 mm (6 in.) diameter were used.

(a) Plot the uncorrected N values versus depth. On this plot, draw an estimated soil profile and describe the soil (e.g., loose, dense, etc.).

(b) Correct the N values to $N_{1,60}$ and plot $N_{1,60}$ with depth on the plot in (a).

(c) Estimate the friction angle and the shear modulus for the soil at a depth of 2.5 m (8 ft).

5.8 The average N value for a soil classified as SP is 20. Estimate

(a) ϕ'

(b) G_{sec}

(c) E_{sec}

Assume $\nu = 0.3$.

5.9 The peak torque measured in a VST is 740 N-m (546 lb ft). The diameter of the vane is 100 mm (4 in.) and its height is 200 mm (8 in.). The residual torque measured is 320 N-m (236 lb ft). Calculate the undrained shear strength and the sensitivity of the soil.

5.10 Results from a CPT test are at http://www.wiley.com/college/budhu. Click Foundations and Earth Structures, Chapter 5, CPT5.10.xls.

(a) Plot the distribution of tip and sleeve resistances, and porewater pressure with depth.

(b) Plot a composite diagram of the tip resistance.

5.11 Estimate the liquefaction potential of the soils shown in Fig. P5.11.

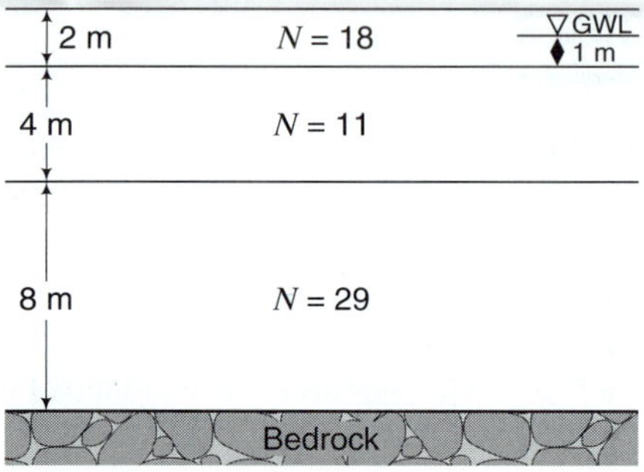

FIGURE P5.11

CHAPTER 6

UNCERTAINTIES IN FOUNDATION DESIGN, PHILOSOPHY AND METHODOLOGIES

6.0 INTRODUCTION

In this chapter, we will discuss some of the key elements of foundation design. An important requirement for successful foundation design is experience and judgment, which cannot be obtained from this or any textbook. Consequently, our discussions on foundation design will be limited to analyses and design guidelines.

All foundation designs involve risk of failure. You need to understand some basic statistical and probability analysis to evaluate the probability of failure of your design. In this chapter, a brief review of pertinent statistical and probability analysis is presented.

We are going to refer to geosystems in this chapter. A *geosystem* or a *geotechnical system* is a single structure or multiple integrated components that transmit loads from a structure to the soil. Examples of geosystems are foundations for buildings and bridges, tunnels, retaining walls, and so forth.

It is common to ask when doing your first design, "Where do I start?" The content of this chapter is intended to help you answer this question. When you complete this chapter, you should be able to:

- Understand some of the uncertainties in foundation design.
- Understand the geotechnical design process.
- Understand design methodologies—allowable stress design (ASD) and load and resistance factor design (LRFD).

6.1 DEFINITIONS OF KEY TERMS

Ultimate limit state (ULS) delineates the strength of a geosystem or component that should not be exceeded by any conceivable loading during its design life.

Serviceability limit state (SLS) delineates a limiting deformation (displacement, rotation, and settlement) of a geosystem or component that, if exceeded, will impair its function.

Failure is the inability of a geosystem either as a whole or in part to perform its intended design functions.

6.2 UNCERTAINTIES IN FOUNDATION DESIGN

Importance *There is nothing certain about a foundation design. Every aspect of the design process involves uncertainties and, consequently, risk of failure. As a geotechnical engineer, you need to understand where major uncertainties occur and how to reduce them to design a safe and economical foundation. In this section, we will briefly discuss some of these uncertainties and review some pertinent statistical and probability measures.*

Uncertainties in foundation design come from a number of sources, but mainly from (1) uncertainty in the magnitude and direction of the loads; (2) natural or inherent uncertainty from soil formation; (3) epistemic uncertainty, which is the lack of knowledge or possession of incomplete or insufficient information; and (4) human errors. Human errors such as incorrect calculations and lack of experience can be eliminated by quality control and the passage of time (experience), and will not be further discussed here.

Geosystems are subjected not only to structural loads but also earth pressures, seepage forces, and hydrostatic pressures (Chapter 2). The uncertainties in structural loads have been well researched in structural engineering, and building codes include factors and load combinations that allow you to account for these uncertainties using reliability methods. However, because of epistemic uncertainty we have not been able to apply reliability methods to earth pressures, seepage forces, and hydrostatic pressures with the same level of confidence as for structural loads.

Soils are natural materials that vary temporally and spatially. The uncertainties in soil formations and structural arrangements are natural uncertainties and cannot be reduced or eliminated. We are stuck with the soil that nature has given us to work with. We then have to develop methods that would allow us to build safe and economical geosystems on and within soil with a certain amount of risk. The amount of risk depends on the type of geosystem, the requirements of the clients, and the amount and quality of available information.

Epistemic uncertainty can be reduced or eliminated based on the amount and quality of the information that is available. For soils, epistemic uncertainty comes from a lack of knowledge about (1) soil parameters, (2) methods of calculations or soil modeling, and (3) construction methods. The uncertainties in the construction method can be substantially reduced through quality control and monitoring, and will not be discussed further. The uncertainties in soil parameters (particularly shear strength, deformation, and settlement parameters) arise from sampling and testing methods (measurement uncertainty), insufficient number of soil samples (data set inadequacies), and methods of interpreting the results (modeling uncertainty).

Consider the data from a CPT test as shown in Fig. 6.1. If we are going to construct a foundation resting 0.75 m below the soil surface, we would like to know if the soil has enough strength to resist the anticipated applied foundation loads safely and that the settlement is tolerable. We may ask, what depth of soil will be effective in resisting the loads? What cone tip resistance (q_c) value should be used? Does the cone stress the soil in the same way as the foundation would? What method of calculations should be used and how accurate are these calculations to predict the load that a foundation of a certain size can safely support? You know from Chapter 2 that if we were to assume that the soil is an elastic solid, then at a depth below $2B$ (B is the foundation width) the soil will have to support about less than 10% of the surface stress. But is the soil elastic? No, it is not. Is it elastoplastic? No, it is not. These material types are idealizations and it is difficult to quantify how accurately they represent soil behavior over the many range of soil behaviors encountered in a single geosystem.

Let's go ahead and assume an elastic soil and take the depth $2B$ as our effective depth. Let us try a square foundation of width 2 m. Then the effective depth is 4 m. Inspection of Fig. 6.1 reveals that q_c varies considerably over the effective depth. Are these values of q_c correct? If not, how much is the error? Do q_c

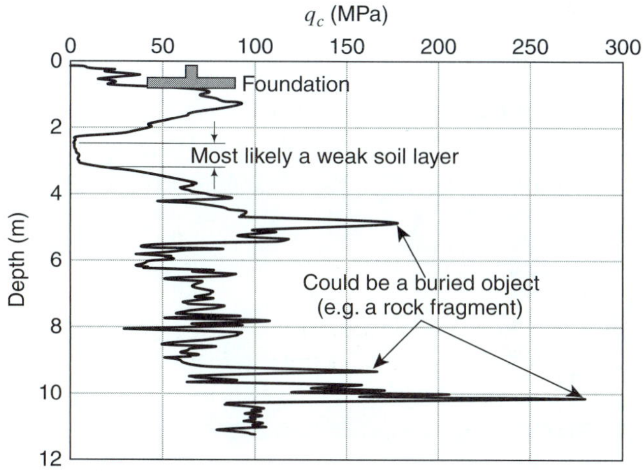

FIGURE 6.1 Tip resistance in a CPT.

values accurately represent the resistance that the soil will offer to the loads? No, they do not. Are there enough borings with CPT data? Normally, the answer is no. But, this is the data we have on hand to design the foundation safely and economically. The common practice is to inspect the data, eliminate suspicious data based on personal judgment, and then use an average value based on inspection or a statistical mean. However, given the variability of the data over the effective depth, using an average value raises some additional issues. For example, the top part of the soil profile just below the base of the foundation within the effective depth seems to be stronger than the bottom part. Is the average value an accurate value? We do not know. It is quite likely that the lower part of soil profile, being much weaker, may control the design, using the paradigm that a chain is only as strong as its weakest link or drawing from our experience in a similar situation. We may divide the soil profile within the effective depth into two layers using personal judgment, and then find the average value in each layer. We now have to select a method of calculating the safe foundation load and displacement. All the methods (Chapters 7–10) of doing so are uncertain. But, we often do not know the extent of the uncertainty.

The foregoing discussion illustrates some of the difficulties that a geotechnical engineer must face in designing foundations. The decisions made are all personal based on experience. However, our desire is to have some rational method of quantifying the uncertainties so that we can design with a higher level of confidence and provide our clients with a safe and economical foundation. How do we deal with uncertainties in a rational method? Rational methods of dealing with uncertainties are based on probability methods. In probability methods, the parameters (dependent and independent variables) describing a process that is well or not well defined are treated as random variables. In contrast, there is the deterministic method, which assumes that the process is known and the parameters that described it are well defined. The deterministic method is popular in foundation design. One of the reasons is that in geotechnical engineering, we have a paucity of data to conduct probability analyses. However, there are methods such as Bayesian theory that provide a framework for using insufficient data sets in probability analysis. Regardless of what methods we choose to determine uncertainties, we still have to combine them with our experience to make decisions on foundation design; natural systems are more complex than our human descriptions of them.

Probabilistic methods make use of the arithmetic mean (called *mean* hereafter), $\bar{X}$, and standard deviation or variance in assessing uncertainties. The mean or average value is obtained by adding up all your data values and then dividing them by the number of data items. Data generally contains low and high values. Your highs and lows could get more extreme as your data set gets larger. The difference in the highest and lowest values gives the range. The most common way to describe the range of variation of your data set from the mean is through the standard deviation, commonly denoted by $\bar{\mu}$. The standard deviation is the square root of the variance, normally denoted as $\bar{\mu}^2$. The variance is obtained by

TABLE 6.1 Typical COV for Soil Parameters (modified after Duncan, 2000)

Measured or interpreted parameter value	Coefficient of Variation, COV (%)
Unit weight, γ	3 to 7%
Buoyant or effective unit weight, γ_b	0 to 10%
Peak friction angle, ϕ'_p	2 to 13%
Critical state friction angle, ϕ'_{cs}	0 to 3%
Undrained shear strength, s_u	13 to 40%
Undrained strength ratio (s_u/σ'_z)	5 to 15%
Compression index, C_c	10 to 37%
Preconsolidation stress, σ'_{zc}	10 to 35%
Hydraulic conductivity of saturated clay, k	68 to 90%
Hydraulic conductivity of partly-saturated clay, k	130 to 240%
Coefficient of consolidation, C_v	33 to 68%
Standard penetration blowcount, N	15 to 45%
Electric cone penetration test, q_c	5 to 15%
Mechanical cone penetration test, q_c	15 to 37%
Vane shear test undrained strength, s_uVST	10 to 20%

subtracting the mean from each data item, squaring the results, adding them up, and then dividing by the number of items less 1. If you take the square root of the variance you get the standard deviation. The standard deviation tells you how tightly the data items are clustered around the mean. A large standard deviation indicates that the data points are far from the mean. Another measure of variation is the coefficient of variation (COV), which is the standard deviation divided by the mean. The smaller the COV, the lower the variation of the data from the mean. Typical COV for soil parameters are shown in Table 6.1.

The mutual dependency between two variables can be found using regression analysis. The coefficient of regression, denoted as *R-squared*, is a measure of the goodness of the fit. An R-squared of 1 means a perfect fit. You need to be careful that although you may get a high correlation coefficient (R-squared $>$ 0.9) between two variables, the relationship between the two variables may not be consistent with physical reality. Thus, when you are building relationships between or among variables, you need to consider the physics (or mechanics) of the problem.

In probability methods, we often assume that the data are normally distributed. If the data set has no negative values, a lognormal distribution can be assumed. These distributions are often represented in a graph of probability distribution function versus standard deviation. The probability density function is any function that is everywhere nonnegative and has an integral from $-\infty$ to $+\infty$ equal to 1 (i.e., the area under the distribution curve is equal to 1). For a data set that has a normal distribution, 68.3% of the values are within one standard deviation away from the mean, 95.4% of the values are within two standard deviations, and 99.7% lie within three standard deviations (Fig. 6.2).

When the data items are tightly bunched together and the probability distribution curve (a bell-shaped curve) is steep, the standard deviation is small (low level of uncertainty). When the data items are spread apart, the bell-shaped curve is relatively flat, the standard deviation is large (high level of uncertainty).

For soils, we are often interested in the relationship between two variables such as the variation of undrained shear strength with depth or the relationship between load (Q) and soil resistance (R). Let us consider the latter. The probability distribution curves for load and resistance are shown in Fig. 6.3. The probability of failure is the size of the intersecting area of the two probability distributions. The larger the intersecting area, the larger the probability of failure. In the deterministic method, we define a factor of safety (FS) as the resistance divided by the load. The probability of failure of a system can increase while

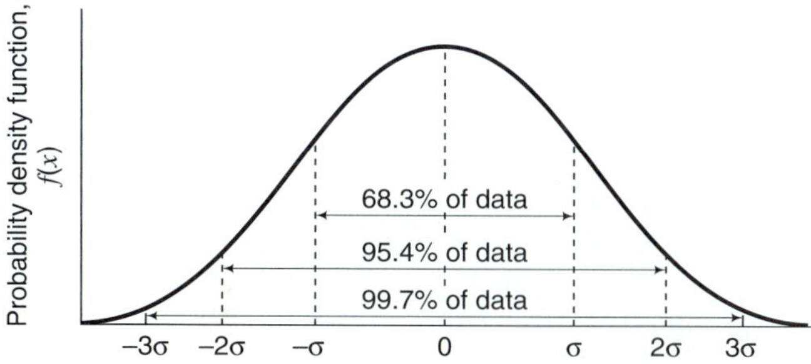

FIGURE 6.2 Probability distribution for normally distributed variables.

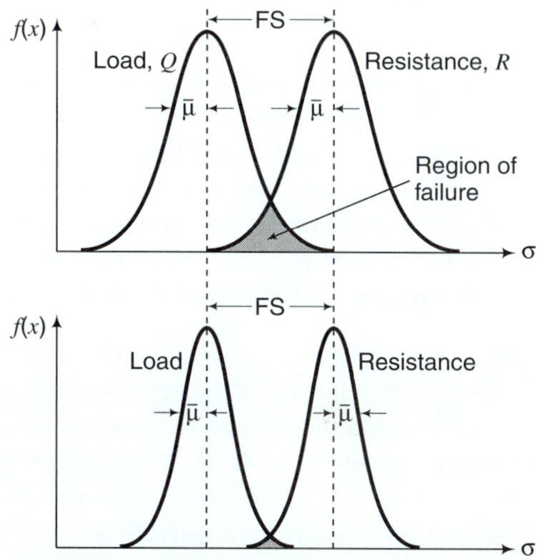

FIGURE 6.3 Probability distributions for load and resistance, and comparison with factor of safety.

the factor of safety remains the same, as shown in Fig. 6.3. The probability of failure (P_f) is defined as

$$P_f = p_p(R < Q) \tag{6.1}$$

where p_p is probability, R is resistance, Q is the unfactored load, and the subscript f denotes failure. Probability of failure is characterized by a parameter called the *reliability index*, β, which is the number of standard deviations between the mean and the origin. The reliability index for a normal distribution of a variable is

$$\beta = \frac{\bar{X} - 1}{\bar{\mu}} \tag{6.2}$$

The reliability index for a lognormal distribution of a variable is

$$\beta_L = \frac{\ln\left(\dfrac{\bar{X}}{\sqrt{1 + (COV)^2}}\right)}{\sqrt{\ln[1 + (COV)^2]}} \tag{6.3}$$

where β_L is the lognormal reliability index.

The relationship between the reliability index and the probability of failure for lognormal distribution of a variable is

$$P_f \cong 460 \exp(-4.3\beta_L) \tag{6.4}$$

Small changes in β_L result in large changes in the probability of failure because of the natural logarithm relationship given in Eq. (6.3). For example, if $\beta_L = 3$, $P_f = 1.15 \times 10^{-3}$. But if β_L increases only by 0.5 (i.e., $\beta_L = 3.5$), $P_f = 1.34 \times 10^{-4}$, which is a decrease in the probability of failure by a factor about 9. The probability of failure for most shallow foundations (Chapters 7 and 9) is within the range 1×10^{-2} to 1×10^{-4} (β_L within the range 2.5 to 3.5). For deep foundations (Chapter 8), the probability of failure is within the range 1×10^{-1} to 1×10^{-2} (β_L within the range 2.0 to 2.5).

Regardless of which analysis you use in design, there is going to be a difference between the predicted and measured value. The ratio of the measured value, X_m, to the predicted value, X_p, is called the *bias factor*, λ_b, expressed as

$$\lambda_b = \frac{X_m}{X_p} \tag{6.5}$$

Ideally, λ_b should be close to 1. If $\lambda_b < 1$, the design is nonconservative; if $\lambda_b > 1$, the design is conservative. Both the loads and the resistance have bias factors. We can write the mean load and mean resistance as

$$\bar{Q} = \lambda_P Q \quad \text{and} \quad \bar{R} = \lambda_R R \tag{6.6}$$

where λ_P and λ_R are the bias factors for unfactored loads, Q, and resistance, R, respectively.

The reliability index for the lognormal distribution of variables—unfactored load (Q) and resistance (R)—is given as

$$\beta_L = \frac{\ln\left[\frac{\bar{R}}{\bar{Q}}\sqrt{\frac{1+\mathrm{COV}_Q^2}{1+\mathrm{COV}_R^2}}\right]}{\sqrt{\ln\left[(1+\mathrm{COV}_Q^2)(1+\mathrm{COV}_R^2)\right]}} \tag{6.7}$$

or

$$\beta_L = \frac{\ln\left[\frac{\lambda_R R}{\lambda_P Q}\sqrt{\frac{1+\mathrm{COV}_Q^2}{1+\mathrm{COV}_R^2}}\right]}{\sqrt{\ln\left[(1+\mathrm{COV}_Q^2)(1+\mathrm{COV}_R^2)\right]}} \tag{6.8}$$

where COV is the coefficient of variation and the subscripts Q and R denote unfactored load and resistance respectively.

There are several commercial computer programs, packages, and spreadsheets (Excel is an example) that can perform probability calculations using your data set. For example, Excel has a function NORMDIST ( ) that returns the normal distribution for a given mean and standard deviation. Regardless of the sophistication of the computer programs that use, you still need to interpret the results and understand their implications before applying them to your project. Statistics and probability can be easily misused.

The key points are:

1. ***Foundation design consists of many uncertainties. One of the goals of a geotechnical engineer is to reduce uncertainties for safety and economics.***
2. ***Uncertainties due to natural processes such as soil formation cannot be eliminated.***
3. ***Epistemic uncertainties can be reduced and in some instances eliminated through sufficient high-quality information and assessment.***
4. ***There are several statistical measures that you can use to quantify uncertainties, but you should use them in conjunction with engineering experience and physical reality.***

EXAMPLE 6.1 *Statistical Calculations for Triaxial Test Data*

A series of triaxial tests were conducted on samples taken at approximately the same depth in a silty clay layer. The undrained shear strengths of 12 tests are shown in the table.

Test	1	2	3	4	5	6	7	8	9	10	11	12
s_u—kPa (psf)	42 (878)	39 (815)	37 (773)	35 (731)	37 (773)	48 (1003)	41 (857)	37 (773)	38 (794)	39 (815)	45 (941)	33 (690)

(a) Calculate the mean, standard deviation, and the coefficient of variation for the test data.

(b) Remove the highest and lowest undrained shear strengths and recalculate the mean, standard deviation, and coefficient of variation. Does the variation from the mean improve?

(c) What value of s_u would you recommend for a foundation design?

Strategy It is best to use a spreadsheet to do the calculations. You can use Excel or similar software. SI units will be used for the solution of this problem.

Solution 6.1

Access the spreadsheet for this example at http://www.wiley.com/college/budhu. Click Foundations and Earth structures, Chapter 6, example6.1.xls.

Step 1: Input the data in a spreadsheet.
The data in Excel is shown in Table E6.1.

Step 2: Use the statistical tools to calculate the required parameters.
For the mean use the statistical function AVERAGE; for standard deviation use the statistical function STDEVA. For our case, (see Table E6.1) cell B14 = AVERAGE (B2:B13), cell B15 = STDEVA (B2:B13), and cell B16 = B15/B14.

Step 3: Recalculate mean, standard deviation, and COV.
Remove the highest value (48) and the lowest value (33). The recalculation of the mean, standard deviation, and COV is shown in column C of Table E6.1.

Step 4: Evaluate the results.
The mean value is about the same for both data sets. However, for the second data set (column C), the standard deviation and COV are lower than the first data set, indicating lower variability from the mean. You need to be cautious on how you interpret the statistical measures because it is not a good practice to remove data from the data set even if the variation from the mean is reduced. You need to carefully evaluate the test results before removing any of them because the lowest shear strength in your case might control the design if it were deemed to be an accurate representation of the soil at the location in which the test was done.

TABLE E6.1

	A	B	C
1	Test	s_u (kPa)	s_u (kPa)
2	1	42	42
3	2	39	39
4	3	37	37
5	4	35	37
6	5	37	37
7	6	48	41
8	7	41	37
9	8	37	38
10	9	38	39
11	10	39	45
12	11	45	
13	12	33	
14	Mean	39.3	39.2
15	Standard deviation	4.2	2.7
16	COV	0.11	0.07
17	Reliability index		53.3

Step 5: Decide on s_u for design.
The mean value of 39 kPa is a good measure of the undrained shear strength for this case ($\beta_L = 53$). ■

EXAMPLE 6.2 *Calculation of the Reliability Index and Probability of Failure of a Foundation*

The mean predicted load on a foundation is 1500 kN (337.2 kips) with a COV of 0.15. The mean measured resistance is 1600 kN (359.7 kips) with a COV of 0.1. Determine the reliability index and the probability of failure. Assume lognormal distributions of the load and resistance.

Strategy Use Eqs. (6.3) and (6.5). SI units will be used for the solution of this example problem.

Solution 6.2

Step 1: Calculate reliability index.

$$\beta_L = \frac{\ln\left[\frac{\bar{R}}{\bar{Q}}\sqrt{\frac{1+\text{COV}_Q^2}{1+\text{COV}_R^2}}\right]}{\sqrt{\ln\left[\left(1+\text{COV}_Q^2\right)\left(1+\text{COV}_R^2\right)\right]}} = \frac{\ln\left[\frac{1600}{1500}\sqrt{\frac{1+0.15^2}{1+0.1^2}}\right]}{\sqrt{\ln\left[\left(1+0.15^2\right)\left(1+0.1^2\right)\right]}} = 2.2$$

Step 2: Calculate the probability of failure.

$$P_f \cong 460\exp(-4.3\beta_L) = 460\exp(-4.3 \times 2.2) = 3.6 \times 10^{-2}$$ ■

6.3 DESIGN PROCESS

Design is an iterative process that requires the integration of

1. The needs of the client
2. Analyses
3. Experience and judgment
4. Economics
5. Constructability
6. Safety
7. Environmental concerns

The design process commences with an understanding of the needs of the client (an individual, corporation, government, etc.) referred to as *conceptualization*. You will then proceed to use your technical knowledge to provide a safe, constructable, and economical system to satisfy the requirements of the client and minimize negative environmental impact. Safety during construction and during the life of a geosystem is a very important design consideration. You should never compromise safety for economics in design. You will be involved in only a limited part of the design process in this book. You do not know the needs of the client, you generally have little or no experience, and you are unlikely to know costing, constructability, and safety issues. In this textbook, you will learn only to calculate the geometry of simple geosystems based on defined conditions. In the next section, we will discuss design issues related to analyses.

6.4 LIMIT STATES

Importance *Design in the context of this book means the dimensioning of a geosystem to satisfy stability and serviceability requirements. What are stability and serviceability requirements? These requirements are called* limit states. *A limit state is reached when a system starts to respond in an undesirable manner or fails to meet the desired design function. As a geotechnical engineer you are required to ensure that the system you are designing will not reach a limit state under any anticipated loading or environmental conditions. There are two limit states that are the backbone of geotechnical design criteria; both criteria must be satisfied. These are described in this section.*

6.4.1 Ultimate-Limit State

The ultimate limit state (stability requirement) delineates the strength of a geosystem or component that should not be exceeded by any conceivable loading during its design life. At the ultimate limit state (ULS), the geosystem is likely to become unstable, resulting in structural damage (local or global—the latter is called *collapse*).

In most geosystems, various modes of instability are possible and each of these modes constitutes an ultimate limit state. For example, a retaining wall could fail by sliding or by rotation or by load bearing inadequacy of the soil or by a mass failure event wherein the wall is part of a large mass of soil that slips as in a slope failure or a landslide. Each of these—called *mode of failure* or *failure mode*—must be checked to ensure that the ultimate limit state is not reached under the anticipated loading and environmental conditions. The ultimate limit state does not account for the strains or displacements required to reach instability. These are stipulated in the serviceability limit state, which we will discuss next.

6.4.2 Serviceability Limit State

The serviceability limit state (serviceability requirement) delineates a limiting deformation (displacement, rotation, and settlement) of a geosystem or component that, if exceeded, will impair its function. What constitutes exceedance of serviceability limit states (SLSs)? Here are a few markers: unsightly cracks, differential settlement, excessive total settlement, intolerable vibrations, intolerable lateral displacements, and rotations.

6.4.3 Limit State Provisions

Codes (e.g., International Building Code, Eurocode, National Building Code of Canada, and the American Concrete Institute) usually provide guidelines on limit states. These codes establish standards and present the best current practice based on practical experience on existing systems. Codes tend to simplify analyses to allow for day-to-day calculations of common systems. Codes state how a system or component should function but do not provide, in most cases, the procedure and methodology to achieve the desired function.

Your design must satisfy *both* limit states. You cannot satisfy one or the other. These limit states are also called *design criteria*.

The* key points *are:

1. ***Foundation design is an iterative process that involves the integration of the needs of the client, analyses, experience and judgment, economics, constructability, safety, and environmental impacts.***
2. ***Ultimate limit state delineates a strength that if exceeded will cause the geosystem or component to fail.***
3. ***Serviceability limit state delineates the deformation of a geosystem or component that if exceeded will impair its intended function.***
4. ***Foundation design requires the satisfaction of both ultimate and serviceability limit states.***
5. ***Codes establish standards of practice (but not necessarily the procedure or methodology) to accomplish a safe, constructable, and economical geosystem.***

6.5 DESIGN METHODS

Importance *There are two design methods used in foundation analysis. One is the allowable stress design (ASD) that has been conventionally used in geotechnical practice. The other is load and resistance factor design (LRFD), which is gradually replacing ASD. You need to understand the implications and differences between these methods. In this section, we will briefly discuss these methods.*

6.5.1 Allowable Stress Design to Satisfy ULS

TABLE 6.2 Typical Values of Factor of Safety

Foundation/earth structures	Factor of safety
Foundations—bearing capacity	2 to 3 (normally 3)
Retaining walls	1.5 to 2
Earthworks	1.3 to 1.5
Seepage—uplift	1.5 to 2
Piping	2 to 3
Slopes	1.25 to 1.75

In *allowable stress design* (ASD), the ultimate resistance of a structure is calculated and then divided by a factor greater than 1—called the *factor of safety*—to give the allowable load or stress,

$$Q_a = \frac{Q_{\text{ult}}}{\text{FS}} \tag{6.9}$$

where Q_a and Q_{ult} are the allowable and ultimate load respectively, and FS is the factor of safety. Typical factors of safety for foundations and earth structures are listed in Table 6.2.

The key points for ASD are:

1. ***In ASD, the loads and resistances are deterministic; that is, the dead loads, live loads, earthquake loads, and so on and soil parameters during the design life of a system are assumed to be known a priori and any variability is ignored.***
2. ***All loads are treated equally and are lumped together.***
3. ***Inaccuracies and risks of the analysis are not considered explicitly.***
4. ***The various levels of risk associated with different structures and their components are not considered explicitly.***
5. ***The FS has no fundamental basis; it is based on experience and judgment of the performance of existing structures.***

In conducting any geosystem analysis, you need to combine loads in a way that will produce the largest negative effect on the system. Codes usually recommend load combinations for the worst-case scenario. For allowable stress design, IBC (2006) recommends the following load combinations:

I. $Q_{\text{DL}} + Q_{\text{FL}}$

II. $Q_{\text{DL}} + Q_{\text{HL}} + Q_{\text{FL}} + Q_{\text{LL}} + Q_{\text{TL}}$

III. $Q_{\text{DL}} + Q_{\text{HL}} + Q_{\text{FL}} + (Q_{\text{RoL}} \text{ or } Q_{\text{SL}} \text{ or } Q_{\text{RL}})$

IV. $Q_{\text{DL}} + Q_{\text{HL}} + Q_{\text{FL}} + 0.75(Q_{\text{LL}} + Q_{\text{TL}}) + 0.75(Q_{\text{RoL}} \text{ or } Q_{\text{SL}} \text{ or } Q_{\text{RL}})$

V. $Q_{\text{DL}} + Q_{\text{HL}} + Q_{\text{FL}} + (Q_{\text{WL}} \text{ or } 0.7Q_{\text{EL}})$

VI. $Q_{\text{DL}} + Q_{\text{HL}} + Q_{\text{FL}} + 0.75(Q_{\text{WL}} \text{ or } 0.7Q_{\text{EL}}) + 0.75Q_{\text{LL}} + 0.75(Q_{\text{RoL}} \text{ or } Q_{\text{SL}} \text{ or } Q_{\text{RL}})$

VII. $0.6Q_{\text{DL}} + Q_{\text{WL}} + Q_{\text{HL}}$

VIII. $0.6Q_{\text{DL}} + 0.7Q_{\text{EL}} + Q_{\text{HL}}$

where Q_{DL} is dead load, Q_{LL} is live load, Q_{RoL} is the roof load, Q_{SL} is snow load, Q_{WL} is wind load, Q_{EL} is earthquake load, Q_{HL} is lateral loads due to earth pressures, groundwater pressures, or pressure from bulk materials, Q_{FL} is loads from fluids with well-defined pressures and maximum heights, Q_{TL} is loads due to temperature changes, and Q_{RL} is rain load. These load combinations apply strictly to structural components and you need to be very careful how you apply them to soils. Structural components are analyzed either as linearly elastic materials or as linearly elastic–rigid plastic materials. Soils are nonlinear, anisotropic, stress-path, and stress history–dependent materials for which the magnitude and

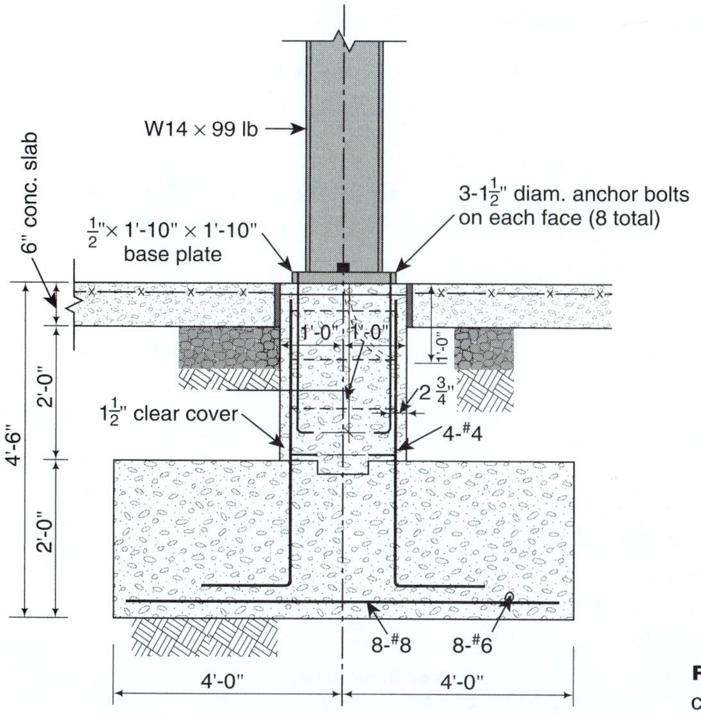

FIGURE 6.4 Steel column to foundation connection.

direction of the loadings are very important. The positive sign in the load combination expressions should be interpreted as load components acting simultaneously and not arithmetic addition.

When a structural design is performed, the reactions at supports that rest on the foundations are calculated. These supports may not be at ground surface. For shallow foundations (Chapter 7), structural loads at the supports have to be transferred to the base of the foundations. In the case of pile foundations (Chapter 8), the structural loads have to be transferred to the pile head at the ground surface. For example, a connection detail for a steel column to the foundation is shown in Fig. 6.4. The loads at the bottom of the column consist of a vertical-centric load, a clockwise moment, and a horizontal load, as shown in Fig. 6.5a. The equivalent loads at the base of the footing for foundation design are as shown in Fig. 6.5b. You should be familiar with the relevant codes as there are exceptions or alternative loads that are recommended.

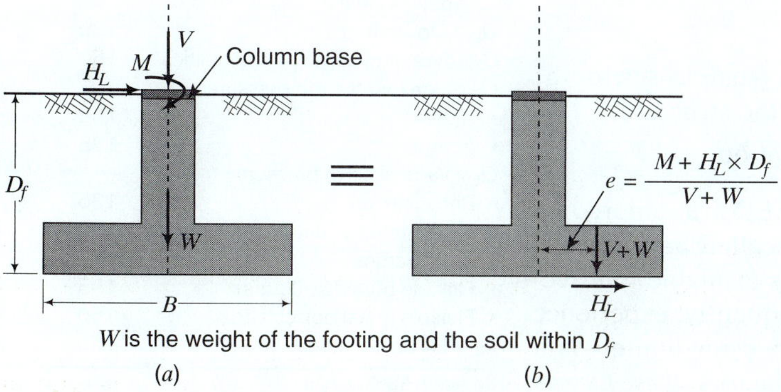

FIGURE 6.5 (a) Loads at base of column. (b) Equivalent loads at base of foundation.

REM 490

Only have H_L:
$M = F \times d$
$F = V + W$
$M = H_L \times D_f$
$d = e$
$e = \frac{H_L \times D_f}{V + W}$

6.5.2 Load and Resistance Factor Design to Satisfy ULS

In the load and resistance factor design (LRFD), reliability theory is used to evaluate the uncertainties of different loads and soil resistances. Each type of load is treated separately and factors are applied to each based on its uncertainties. The ultimate resistance provided by the soil is modified based on the uncertainties of the soil parameters (each parameter is treated separately), the method of sampling and testing, the extent of the soil investigation, the method of analysis, the amount of risk associated with the structure, consequence of failure, and construction practice. LRFD is mathematically defined as

$$\Sigma \varphi_R R_{ult} \geq \Sigma \rho_i Q_i \tag{6.10}$$

where φ_R is a performance or resistance factor, R_{ult} is the ultimate soil resistance, ρ is the load factor for load type *i*, and Q is the magnitude of the nominal load (estimated actual load) for load type *i*. Various load and resistance factors are suggested in codes. Table 6.3 shows a comparison of load factors in IBC (2006), Canadian Foundation Engineering Manual (CFEM), and Eurocode 7 (1993).

The American Association of State Highway and Transportation Officials (AASHTO) recommend a different set of load factors as shown in Table 6.4 AASHTO (2004) also considers a modification factor to the load factors in Table 6.4 to account for ductility, redundancy, and operational importance.

The loads are combined in specific ways to produce the worst-case scenario. IBC (2006) recommends the following load combinations:

TABLE 6.3 Load Factors

Loads	IBC (2006)	CFEM	Eurocode 7
Dead (Q_{DL})	1.4	1.25	1.1
	0.9	0.8	0.9
Live (Q_{LL})	1.6	1.5	1.5
	0		
Fluid (Q_{FL})	1.4	1.25	1.0
	0	0.8	1.0
Earthquake (Q_{EL})	1.0		
Wind (Q_{WL})	1.6		
	0.8		
Lateral loads (Q_{HL})	1.6		
	0		

Top values are maximums and lower values are minimums.

I. $1.4(Q_{DL} + Q_{FL})$

II. $1.2(Q_{DL} + Q_{FL}\, Q_{TL}) + 1.6(Q_{LL} + Q_{HL}) + 0.5(Q_{RoL} \text{ or } Q_{SL} \text{ or } Q_{RL})$

III. $1.2Q_{DL} + 1.6\,(Q_{RoL} \text{ or } Q_{SL} \text{ or } Q_{RL}) + f_1(Q_{LL} \text{ or } 0.8Q_{WL})$

IV. $1.2Q_{DL} + 1.6Q_{WL} + f_1 Q_{LL} + 0.5\,(Q_{RoL} \text{ or } Q_{SL} \text{ or } Q_{WL})$

V. $1.2Q_{DL} + 1.0Q_{EL} + f_1 Q_{LL} + f_2 Q_{SL}$

VI. $0.9Q_{DL} + (1.0Q_{EL} \text{ or } 1.6Q_{HL})$

where:

f_1 = 1 for floors in places of public assembly for live loads in excess of 4.8 kPa and for parking garage live load
= 0.5 for other live loads

f_2 = 0.7 for roof configurations (such as sawtooth) that do not shed snow off the structure
= 0.5 for other roof configuration.

TABLE 6.4 Load Factors

Type of loads	Load factor Maximum	Minimum
Q_{LL}: Live loads	1.75	1.35
Q_{DL}: Component and attachments (dead loads)	1.25	0.90
Q_{DD}: Downdrag	1.80	0.45
Q_{DW}: Wearing surfaces and utilities	1.50	0.65
Q_{EL}: Horizontal earth pressure		
• Active	1.50	0.90
• At rest	1.35	0.90
Q_{EV}: Vertical earth pressure		
• Retaining structure	1.35	1.00
• Rigid buried structure	1.30	0.90
• Rigid frames	1.35	0.90
• Flexible buried structures	1.95	0.90
• Flexible metal box culverts	1.50	0.90
Q_{ES}: Earth surcharge	1.50	0.75

SOURCE: AASHTO, 2004.

Current issues concerning LRFD are centered on the resistance factors (sometimes called *performance factors*) because insufficient data is available to effectively use reliability theory. Consequently, experience and judgment and reliability theory using limited data sets are used to propose resistance factors (*these factors*

TABLE 6.5 Rotation Limits (Curvature) of Structures

Relative rotation δ/L^*	Type of structure
1/100	Danger limit for statically determinate structures and retaining walls
1/150	Safe limit for statically determinate structures with flexible cladding and retaining wall
	Danger limit for open steel and reinforced concrete frames, offshore platforms, steel storage tanks, and tilt of high, rigid structures
1/250	Safe limit for open steel and reinforced concrete frames, offshore platforms, steel storage tanks, and tilt of high, rigid structures
	Danger limit for panel walls of frame buildings and tilt of bridge abutments
	Tilting of high-rise building may become visible
1/300	Limit when difficulties with overhead cranes are to be expected
1/500	Panel walls of frame buildings and tilt of bridge abutments
1/1000	Sagging of unreinforced load-bearing walls
1/2000	Hogging of unreinforced load bearing walls

*δ is the differential settlement along the length L.
SOURCE: Canadian Foundation Engineering Manual.

will be given in the relevant section of this textbook). As more quality field and laboratory test data becomes available, the performance factors may change. Also, in reliability theory one presumes that the correct or expected value is known a priori. This is not the case for soils.

An alternative method to LRFD called the *factored strength method* (FSM) or *partial factor method* (PFM) is used in some countries. In FSM or PFM, a factor (<1) is applied to the soil strength parameters rather than the calculated soil resistance.

6.5.3 ASD and LRFD to Satisfy SLS

In calculating movements of geosystems to satisfy the serviceability limit state, a load factor of 1 is used for all types of loads irrespective of whether you are using ASD or LRFD. Meyerhof (1995) suggested a tentative serviceability limit state for rotation of structures (Table 6.5).

EXAMPLE 6.3 ***Calculation of the Worst-Case Loads Based on ASD and LRFD for Ultimate and Serviceability Limit States***

The unfactored dead load, live load, and horizontal load at the base of a column shown in Fig. 6.4 are 250 kN (56 kips), 180 kN (40 kips), and 22 kN (5 kips) respectively. Calculate the loads based on ASD and LRFD for ultimate and serviceability limit states at the base of the square foundation. Use IBC (2006) load factors. Neglect the effects of lateral soil pressure on the foundation. The unit weights of the concrete and soil are 24 kN/m^3 (150 pcf) and 18 kN/m^3 (115 pcf).

Strategy You need to be careful in calculating the equivalent loads. You finally want to get the resultant load, usually the resultant vertical load and eccentricity and the resultant horizontal load. U.S. units will be used to solve this example problem.

Solution 6.3

Step 1: Calculate the deal load of the foundation, the soil above it, and the concrete slab.
Footing base $= [(8 \times 8 \times 2) \times 150 = 19200$ lb.
Pedestal $= (2 \times 2 \times 2.5) \times 150 = 1500$ lb.
Soil $= [(8 \times 8 \times 2) - (2 \times 2 \times 2)] \times 115 = 13800$ lb.
Concrete slab $= [(8 \times 8 \times 0.5) - (2 \times 2 \times 0.5)] \times 150 = 4500$ lb.
Total $= W = 19200 + 1500 + 13800 + 4500 = 39000 = 39$ kips.

Step 2: Calculate the loads at the base of the foundation using ASD for ultimate limit state.
Unfactored deal load, $Q_{DL} = 56 + 39 = 95$ kips.
Unfactored vertical load, $V = Q_{DL} + Q_{LL} = 95 + 40 = 135$ kips.
Unfactored moment at base, $M_b = Q_{HL} \times D_f = 5 \times 4.5 = 22.5$ kips ft (clockwise).

The eccentricity, $e = \dfrac{M_b}{V} = \dfrac{22.5}{135} = 0.17$ ft.

The maximum eccentricity will occur when only the dead load is acting.

$$e_{max} = \frac{M_b}{Q_{DL}} = \frac{22.5}{95} = 0.24 \text{ ft}$$

The loads at the base using Fig. 6.5b are $V = 135$ kips, $H_L = Q_{HL} = 5$ kips, $e = 0.17$ ft ($e_{max} = 0.24$ ft with dead load only).

Step 3: Calculate the loads at the base of the foundation using LRFD for ultimate limit state.
From load factors in Table 6.2,
Factored vertical load, $V_f = 1.2Q_{DL} + 1.6Q_{LL} = 1.2(95) + 1.6(40) = 178 \text{ kips} > 1.4\,Q_{DL}(133 \text{ kips})$.
Factored lateral load, $(Q_{HL})_f = 1.6\,Q_{HL} = 1.6(5) = 8$ kips.
Factored moment at base, $(M_b)_f = 1.6\,Q_{HL} \times D_f = 8 \times 4.5 = 36$ kips ft (clockwise).

Factored eccentricity, $e_f = \dfrac{(M_b)_f}{V_f} = \dfrac{36}{178} = 0.2$ ft.

The maximum eccentricity will occur when only the minimum factored dead load is acting.

$$\text{Factored } e_{max} = \frac{M_b}{0.9Q_{DL}} = \frac{36}{85.5} = 0.42 \text{ ft.}$$

The factored loads at the base using Fig. 6.5b are $V = V_f = 178$ kips, $H_L = (Q_{HL})_f = 8$ kips, $e = e_f = 0.2$ ft (factored $e_{max} = 0.42$ ft with dead load only).

Step 4: Calculate design load for serviceability limit state.
The design load for both ASD and LRFD is the same. No load factors are applied to the nominal loads. The loads at the base are $V = 135$ kips, $Q_{HL} = 5$ kips, $e = 0.17$ ft ($e_{max} = 0.24$ ft with dead load only). ■

6.6 WHICH DESIGN METHOD SHOULD BE USED?

ASD is called *conventional design* because it has been the method in use over the last 50 years in geotechnical practice. LRFD is a popular choice in structural engineering. LRFD is growing in popularity in geotechnical design mainly because it is desirable that design methods in structures and geotechnical engineering be consistent, coherent, and compatible. On many large projects, a structural engineer would give the structural loads to the geotechnical engineer and consequently the major concern of the geotechnical engineer is with the resistance factors, earth loads, and water pressures. The resistance factors in LRFD, although intended to consider the uncertainties in the analysis, soil properties, and methods of obtaining them, construction practice, and so on, have been altered to match the design obtained by ASD using factors of safety consistent with good engineering practice and experience. Consequently, designs obtained using ASD and LRFD are likely to be similar. The performance factor for a given factor of safety can be calculated using Eq. (6.11):

$$\varphi_R = \frac{\Sigma \rho_i Q_i}{\text{FS}\Sigma Q_i} \tag{6.11}$$

As LRFD matures, it is likely to become the preferable method of design because it attempts to rationally address the many uncertainties of materials, construction, and analysis.

Both design methods, ASD and LRFD, are strength-based methods; that is, the stresses or loads applied on the structural system must not exceed the strength or load carrying capacity (collapse load) of the soil. Displacements require a separate analysis to check that the serviceability limit state is not exceeded. Serviceability limit states rather than ultimate limit states govern the design of many geosystems (e.g., shallow foundations and retaining walls).

A displacement-based rather than a strength-based design approach is better suited for geosystem design. However, a displacement-based design requires fundamental knowledge of soil–structure interactions, accurate modeling of the stress-strain behavior of soils, accurate measurement of soil properties, and careful documentation of actual displacements of the geosystem. Research along these lines is ongoing and has not matured for adoption in routine engineering design and codes. Numerical methods such as the finite difference method or the finite element method are suitable for displacement-based design, but they require good knowledge of the numerical method employed, accurate input parameters (especially soil properties), accurate soil model, and experience to evaluate the results. Normally, equations based on limit equilibrium are used to determine the ultimate limit load in the preliminary design stage, followed by numerical methods if the importance and risk associated with the structure warrant involved analyses.

The* key points *are:

1. ***The allowable stress design method does not consider the uncertainty in loads and soil parameters explicitly. A factor of safety (an arbitrary, subjective value based on experience) is applied to the ultimate resistance to reduce the potential for failure.***
2. ***The load and resistance factor design method is based on consideration of the uncertainties of the loads and the soil resistance.***
3. ***LRFD is calibrated against ASD using factors of safety consistent with good engineering practice. Therefore, LRFD and ASD with the appropriate factor of safety should produce the same design.***

EXAMPLE 6.4 ***Calculation of Performance Factor***

For the vertical loads given in Example 6.3, calculate the equivalent resistance factor to be used in LRFD for FS = 2 used in ASD.

Strategy This is a straightforward application of Eq. (6.11).

Solution 6.4

Step 1: Calculate the resistance factor.

$$\Psi_R = \frac{\Sigma \rho_i Q_i}{\text{FS}\Sigma Q_i} = \frac{178}{2(135)} = 0.66$$

■

6.7 HOW DO I START?

How do I start a design? This is a question that you are likely to ask yourself or your instructor or your employer on your first design project. Here is a list to help you with this question:

1. Gather as much information as possible about the project, such as its location, purpose, loads, importance, nearby buildings, environmental impact, cost estimates, and so forth.
2. Make a site visit and dig a few test pits or shallow borings.

3. Carry out preliminary analyses using available geotechnical data and the presumptive soil parameters in this textbook or those in codes or technical manuals. You should consider a range of soil parameters that you believe to characterize your soil. For example, if the soil at the site is a soft clay and you have water content data only from the test pits, you can use empirical equations from Chapters 4 and 5 to estimate values of undrained shear strength and friction angle. When you complete the preliminary analysis, you should be able to narrow your choice of geosystems and appreciate the sensitivity of the geosystem to changes in soil values. If a structure similar to the one that you are designing exists at the site or nearby, you can conduct back-analyses to get preliminary soil parameters and inspect the structure for any visible problems such as cracks, differential settlement, and so forth. This would help you to formulate the extent of the soil investigation and the type of geosystem that might be required.
4. Plan and execute a soil investigation (Chapter 5). Your preliminary analyses should help you to decide on the soil parameters to obtain and the extent of the soil investigation. You are the one who has to decide on the soil parameters to use in a design. This is a crucial step, because the ultimate and serviceability limit states depend directly on the soil strength (friction angle and undrained shear strength) and deformation (elastic and shear moduli) parameters. Poor assessment of the soil parameters could result in (a) an uneconomical system; (b) failure—ultimate limit state reached or exceeded; (c) the structure not satisfying its design function—serviceability limit state exceeded; and (d) loss of property and lives.
5. Design the geosystem to comply with the appropriate code of practice, for example, IBC (2006). Remember that in executing your design, you should consider how the system would be constructed.
6. You should not regard any value from calculations as an absolute value. Instead, any calculated value such as settlement should be regarded as an estimated expected value. You should examine *what-if* scenarios, especially the effects of the soil parameters and groundwater conditions. It is good practice to consider a range of estimated expected values based on experience. In lieu of experience, consider the effects of the possible range of variations of input soil parameters. For example, if the friction angle were in doubt, how would the design change (type of system, geometry, constructability, and costs) if, say, the error is $\pm 5°$?

You are encouraged to read the excellent article on geosystem design by Casagrande (1960).

6.8 SUMMARY

In this chapter, you were introduced to foundation design philosophy and foundation design methods. Design is an iterative process that requires, among other criteria, experience and judgment. In foundation design, both limit states—ultimate and serviceability—must be satisfied. Design calculations give only estimated values.

EXERCISES

Some of the questions below require that you consult additional reading material in your library or elsewhere. Your instructor may provide references.

6.1 Define and discuss the two limit states.

6.2 What are the differences between LRFD and ASD?

6.3 Are all load combinations in codes the same? Locate two codes and compare the load factors and load combinations. Discuss how these may be used in geosystem design.

6.4 Research and write an essay on the failure of a geosystem. Your essay should contain discussions on the design process.

6.5 The peak friction angles, ϕ'_p, measured in a sand layer 2 m (6 ft) below the surface in 10 tests are given below:

Test	1	2	3	4	5	6	7	8	9	10
ϕ'_p	32	37	31	36	31	29	35	32	36	28

Calculate the mean, standard deviation, and coefficient of variation of ϕ'_p. What value of ϕ'_p would you use in a design of a foundation? What concerns, if any, do you have about the data with regard to the ϕ'_p selected for the design? What range of values will you consider for design?

6.6 The predicted average load on a set of 10 foundations on a particular soil type is 500 kN (112 kips) with $\bar{\mu} =$ 35 kN(7.9 kips). The measured loads are 809 kN (182 kips), 748 kN (168 kips), 695 kN (156 kips), 700 kN (157 kips), 902 kN (203 kips), 895 kN (201 kips), 752 kN (169 kips), 637 kN (143 kips), 741 kN (167 kips), and 697 kN (156 kips). Calculate the reliability index and the probability of failure.

6.7 The unfactored loads at the base of a column located on a pile cap are shown in Fig. P6.7. Calculate the loads for ASD and LRFD on the pile at ground surface using IBC (2006). The unit weight of concrete is 24 kN/m^3 (150 pcf).

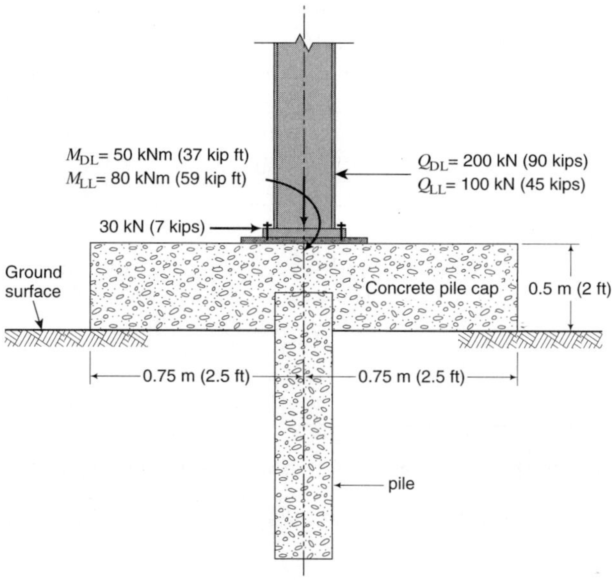

FIGURE P6.7

6.8 The unfactored loads on a retaining wall are shown in Fig. P6.8. Calculate the loads for ASD and LRFD at the base of the wall using IBC (2006). The unit weight of concrete is 24 kN/m^3 (150 pcf). Draw a diagram of the base showing the resultant load and its location.

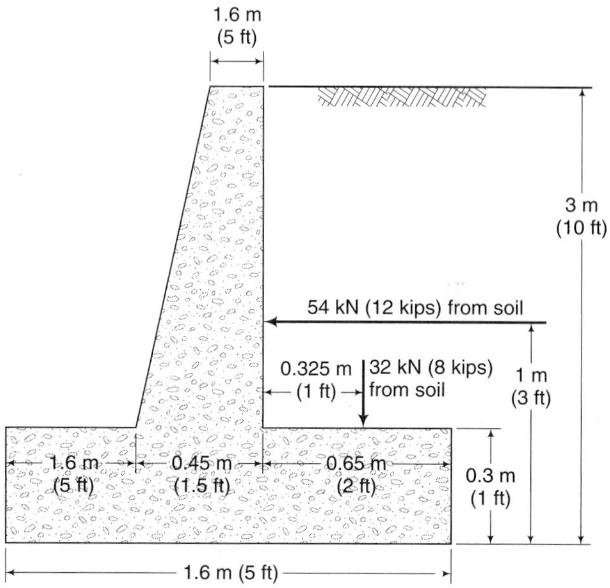

FIGURE P6.8

6.9 The unfactored dead, live, and earthquake loads on the foundation shown in Fig. 6.4 are 180 kN (41 kips), 120 kN (27 kips), and 50 kN (13.5 kips) respectively. Calculate the loads for ASD and LRFD at the base of the wall using IBC (2006). Assume the foundation is part of the support system for an office building. If a factor of safety of 2 is used in the ASD, calculate the equivalent performance factor to be used in LRFD.

6.10 The length and width of the foundation spanning the footprint of a structure are 40 m(125 ft) × 20 m(60 ft). The structure is to be constructed using reinforced concrete. Estimate the limiting tolerable differential settlement (serviceability limit state) of the foundation.

CHAPTER 7

ANALYSIS AND DESIGN OF SHALLOW FOUNDATIONS

7.0 INTRODUCTION

Loads from a structure are transferred to the soil through a foundation. A foundation itself is a structure, often constructed from concrete, steel, or wood. A geotechnical engineer must ensure that a foundation satisfies the following two stability conditions:

1. The foundation must not collapse or become unstable under any conceivable loading.
2. Settlement of the structure must be within tolerable limits.

Both requirements must be satisfied. In addition, the foundation must be constructable and be economical.

In this chapter, we will consider methods—mostly simplified methods that have been used in design practice—to determine the geometry of a shallow foundation to prevent soil failure and to limit settlement to a tolerable value. Often, it is settlement that governs the design of shallow foundations. When you complete this chapter, you should be able to:

- Calculate the safe bearing capacity of soils.
- Estimate the settlement of shallow foundations.
- Estimate the size of shallow foundations to satisfy both ultimate and serviceability limit states.

7.1 DEFINITIONS OF KEY TERMS

Foundation is a structure (usually a slab of concrete) that transmits loads to the underlying soils.

Embedment depth **(D_f)** is the depth below the finished grade (finished ground surface) where the base of the foundation rests.

Shallow foundation is one in which the ratio of the embedment depth, D_f, to the minimum plan dimension, which is usually the width, B, is less or equal 2.5 (i.e., $\frac{D_f}{B} \le 2.5$).

Ultimate net bearing capacity **(q_u)** is the maximum net foundation pressure (pressure in excess of overburden pressure) that the soil can support.

Ultimate gross bearing capacity **(q_{ult})** is the maximum foundation pressure that the soil can support.

Allowable bearing capacity or ***safe bearing capacity*** **(q_a)** is the working pressure that would ensure a margin of safety against collapse of the structure from soil shear failure. The allowable bearing capacity is usually a fraction of the ultimate net bearing capacity.

Ultimate limit state defines a limiting load or stress that should not be exceeded under any conceivable or anticipated loading during the design life of a geotechnical system.

Serviceability-limit state defines a limiting deformation or settlement of a foundation that, if exceeded, will impair the function of the structure that it supports.

Factor of safety or ***safety factor*** **(FS)** is the ratio of the ultimate net bearing capacity to the allowable bearing capacity or to the applied maximum vertical stress. In geotechnical engineering, a factor of safety between 2 and 3 is used (3 most often used) to calculate the allowable bearing capacity.

7.2 TYPES OF SHALLOW FOUNDATIONS

There are several types of shallow foundations. Some are shown in Fig. 7.1. An isolated foundation, also called a *spread* or *individual foundation*, transfers individual column loads directly to the soil (Figs 7.1a,b). If a single spread foundation overlaps with another spread foundation, the two can be combined to form a *combined foundation* (Fig. 7.1c). When there are a large number of spread foundations required for a structure, it may be more economical to use a mat or raft foundation. We will consider mat or raft foundations separately in Chapter 9. Walls are normally supported using a strip foundation (Fig. 7.1d,e). A strip foundation is one in which the width (B) to length (L) ratio is small ($B/L \gg 10$). On sloping ground, a foundation can rest on top of the slope, on the slope, or can be stepped as illustrated in Fig. 7.1f,g,h.

Choice of the type of foundation is based on experience and is governed by:

1. Geological conditions (e.g., topography, soil type, soil layering, and groundwater conditions). For example, an undulating topography may require stepped foundations; soft soils may require a mat foundation (Chapter 9) or deep foundations (piles, Chapter 8).
2. Loading conditions (e.g., the type of loading, such as vertical loads, moments, horizontal loads, and eccentric loads) and magnitudes of the loads. For example, shallow foundations cannot support large moments or large horizontal forces.
3. Construction methods (e.g., local construction methods; type of equipment required and availability; labor requirements, such as skilled or unskilled, and availability).
4. Local codes and regulations.

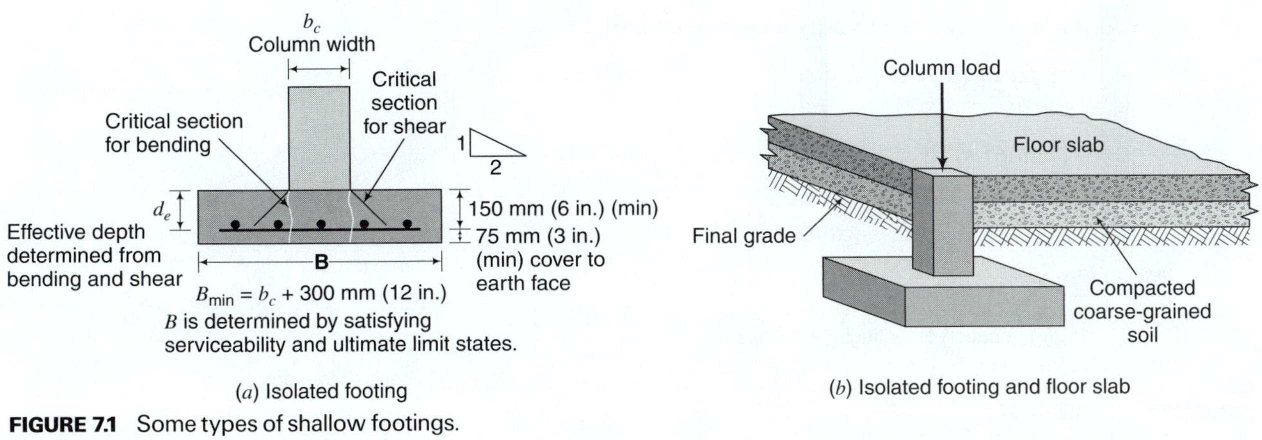

FIGURE 7.1 Some types of shallow footings.

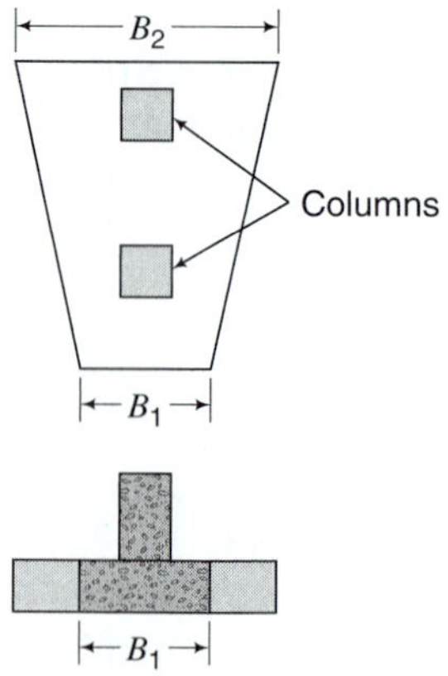

(c) Combined footing

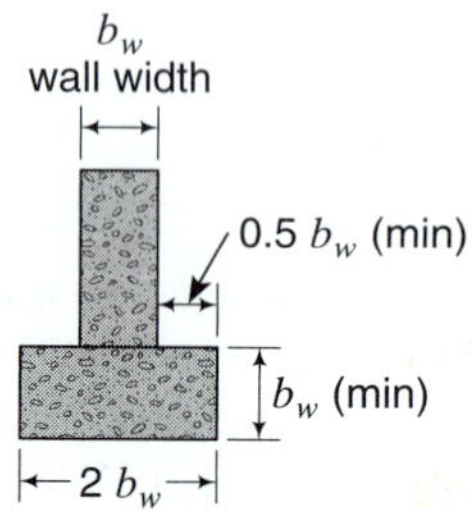

(d) Strip footing

(e) A strip footing under construction

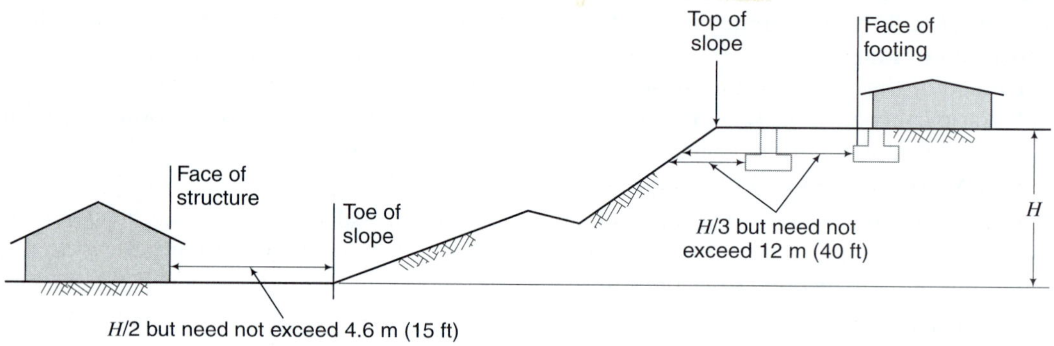

(f) Location of footing near slopes (IBC 2006)
The maximum slope angle for which this figure applies is 45°

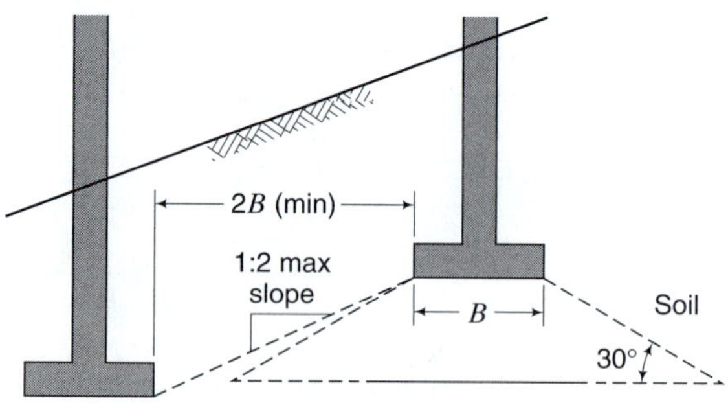

(g) Location of footings on a slope

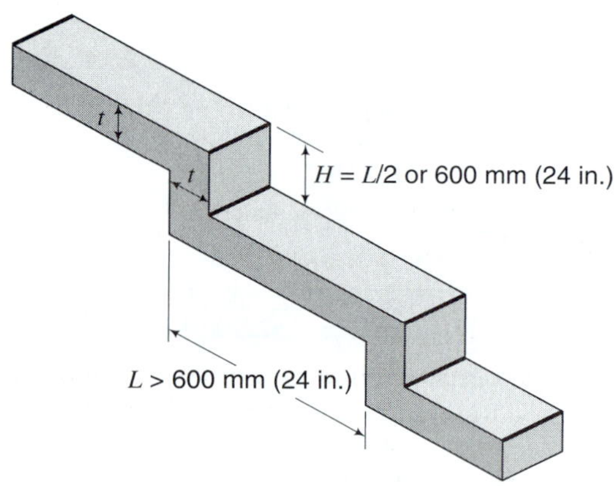

(h) Stepped footings

FIGURE 7.1 (Continued)

5. Environmental conditions such as rainfall, noise, traffic, and disposal of site-excavated material.
6. Impact on neighboring structures. For example, isolated foundations may have to be combined because they would otherwise infringe on adjacent property.
7. Needs of owner such as desired construction completion time and costs.

7.3 BEARING CAPACITY OF SHALLOW FOUNDATIONS

Access a spreadsheet at http://www.wiley.com/college/budhu. Click Foundations and Earth Structure, Chapter 7, bc.xls to calculate bearing capacity factors and to estimate the geometry of shallow foundations to satisfy ultimate limit state. Select the worksheet "homogeneous soil."

7.3.1 General Bearing Capacity for Homogeneous Soils

Importance *A large number of bearing capacity equations are available to estimate the bearing capacity of soils. All generally give different values of the bearing capacity for the same soil conditions. It is important that you understand the assumptions and limitations of these equations for shallow foundation design. In this section, only a few equations have been selected based on their widespread use in practice.*

The theoretical bearing capacity equations in common use in design practice were obtained using the limit equilibrium method (Chapter 3). Prandtl (1920) showed theoretically that a wedge of material is trapped below a rigid plate when it is subjected to concentric vertical loads. Terzaghi (1943) applied Prandtl's theory to a strip foundation with the assumption that the soil is a semi-infinite, homogeneous, isotropic, weightless rigid-plastic material. Based on Prandtl's theory, failure of the foundation occurs by a wedge of soil below the foundation pushes its way downward into the soil. The failure mechanism postulated by Terzaghi (1943) and modified by Vesic (1973) is shown in Fig. 7.2a.

The zone ABC is a rigid wedge of soil trapped beneath the foundation. Zone ABD is fan-like with radial slip lines stopping on a logarithmic spiral slip line. The other zone, ADE, consists of slip lines oriented at angles of $45° + \frac{\phi'_p}{2}$ and $45° - \frac{\phi'_p}{2}$ to the horizontal and vertical planes, respectively. The zone ADE is called the *Rankine passive zone.* The surfaces, AB and AD, are frictional sliding surfaces and their actions are similar to rough walls being pushed into the soil. The pressure exerted is called passive earth pressure.

The shape of the surface BD depends on the ratio $\frac{\gamma B}{\sigma_{zo}}$, where γ is the unit weight of the soil and σ_{zo} is the overburden pressure. For a weightless soil, $\frac{\gamma B}{\sigma_{zo}} \rightarrow 0$, BD is a logarithmic spiral; for $\gamma B = 0$, BD is a circle; for $\gamma B \neq 0$, BD is between a circle and a logarithmic spiral. For clean, coarse-grained soils, BD is always a circle (Vesic, 1973).

For dense, coarse-grained soils or heavily overconsolidated fine-grained soils, the failure planes are expected to reach the ground surface and the failure mode (collapse mechanism) is called *general shear failure* mode. For loose coarse-grained soils or lightly overconsolidated fine-grained soils, the failure planes, if they develop, are expected to lie within the soil layer below the base of the foundation and extend laterally. This failure mode is called *local shear* (Fig. 7.2b). For very loose coarse-grained soils, the failure surfaces may be confined to the surfaces of the rigid wedge. This type of failure mode is termed

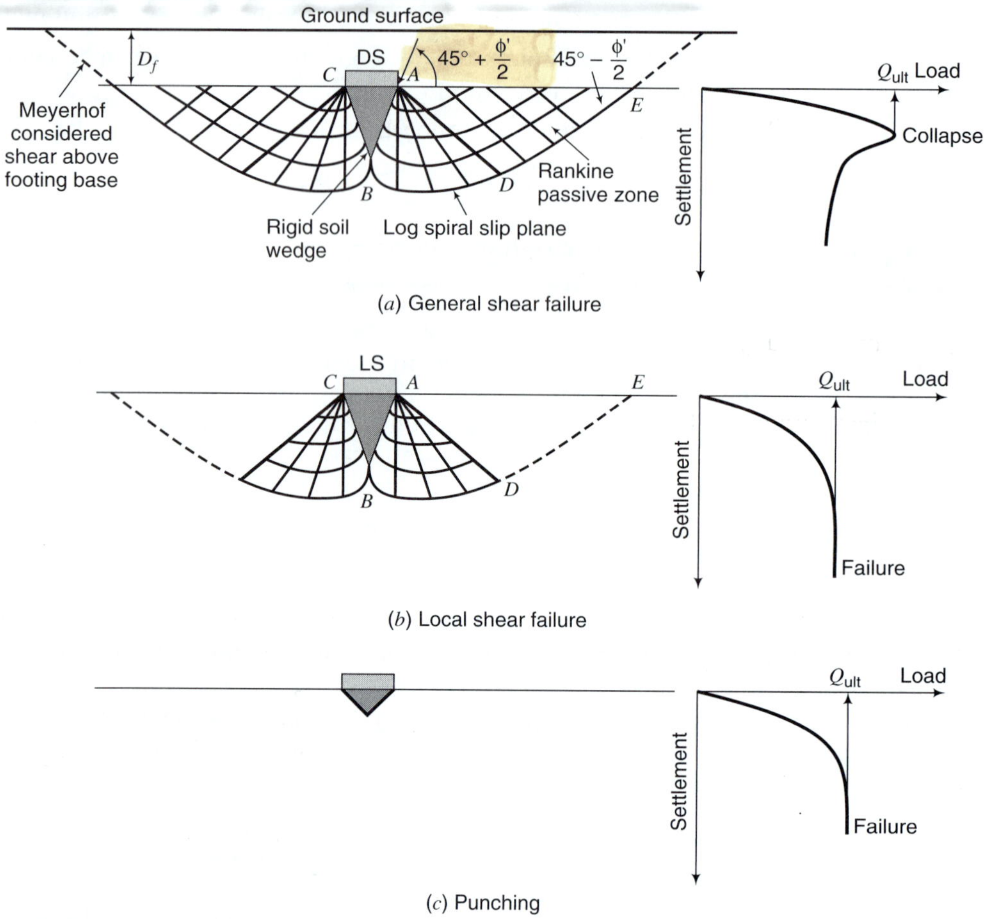

FIGURE 7.2 Failure mechanism.

punching shear (Fig. 7.2c). For coarse-grained soils, each of these collapse mechanisms depends on the relative density (D_r) of the soil and the depth of embedment (Fig. 7.3). For most shallow foundations, general shear failure tends to occur in coarse-grained soils with $D_r \geq 70\%$ and local shear failure when $30\% \leq D_r < 70\%$.

The settlement to mobilize general shear failure under a surface foundation resting on a saturated fine-grained soil is 3% to 7% of the foundation width and about 15% for deeply embedded foundations. For coarse- grained soils, the corresponding settlements are 5% to 15% for surface foundations and 25% for deeply embedded foundations.

A number of bearing capacity equations, based on limiting equilibrium (e.g., Terzaghi, 1943; Meyerhof, 1963; Hansen, 1970; and Vesic, 1973) have been proposed. The ultimate net bearing capacity from all these equations is a function of the plane strain strength parameters (peak friction angle, ϕ'_p, or undrained shear strength, s_u), soil unit weight,

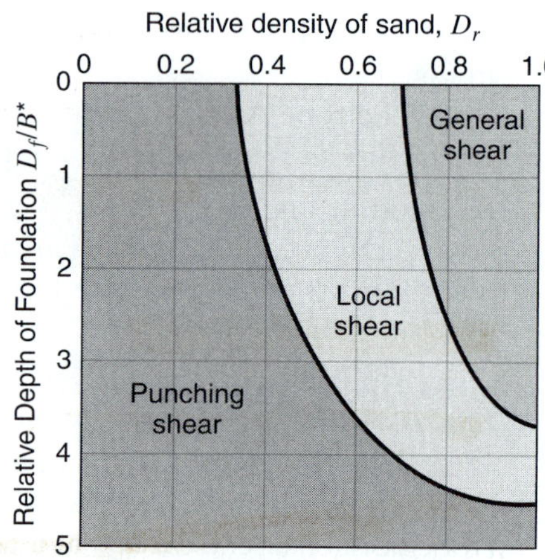

$B^* = B$ for a square or circular footing
$B^* = 2\,BL/(B + L)$ for a rectangular footing

FIGURE 7.3 Collapse mechanism from model tests based on relative density and embedment depth. (Source: Vesic, 1973.)

geometric factors, and compressibility. We can write the ultimate net bearing capacity as

$$q_u = f(\text{strength } [\phi'_p, s_u], \text{unit weight, geometry, compressibility}) \tag{7.1}$$

We will consider two limiting conditions, short-term and long-term, separately rather than lumping them into a single equation. By doing so, we can clearly distinguish which shear strength parameters and method of analysis to use for each condition. Short-term condition requires a *total stress analysis* (TSA). TSA is applicable to fine-grained soils and the shear strength parameter is the undrained shear strength, s_u. Long-term condition requires an *effective stress analysis* (ESA). ESA is applicable to all soils and the shear strength parameter is the peak friction angle, ϕ'_p, from plane strain tests.

Ueno et al. (1998) recommended that the strength parameters should be obtained at a mean effective stress, p', within the range

$$m\gamma B \leq p' \leq n\gamma B \tag{7.2}$$

where, for strip surface foundation, $m = 2$ and $n = 10$, and, for circular surface foundation, $m = 1$ and $n = 15$.

7.3.1.1 Ultimate Net Bearing Capacity

The ultimate net bearing capacity equations[1] are

$$\text{TSA: } q_u = 5.14 s_u s_c d_c i_c b_c g_c r_c \tag{7.3}$$

$$\text{ESA: } q_u = \gamma D_f (N_q - 1) s_q d_q i_q b_q g_q r_q w_q + 0.5\,\gamma B' N_\gamma s_\gamma d_\gamma i_\gamma b_\gamma g_\gamma r_\gamma w_\gamma \tag{7.4}$$

where N_q and N_γ are bearing capacity factors that are functions of ϕ'_p; s_c, s_q, and s_γ are shape factors; d_c, d_q, d_γ are embedment depth factors; i_c, i_q, i_γ are load inclination factors; b_c, b_q, b_γ are base inclination (base tilt) factors; g_c, g_q, g_γ are ground inclination factors; r_c, r_g, r_γ are soil compressibility factors; D_f is the depth of the foundation base from the finished ground surface; and w_q and w_γ are groundwater factors. We will discuss these factors later. In practice, you can use 5 instead of 5.14 in Eq. (7.3).

7.3.1.2 Allowable Bearing Capacity

The allowable bearing capacity for ASD is

$$q_a = \frac{q_u}{\text{FS}} + \gamma D_f \tag{7.5}$$

7.3.1.3 Ultimate Gross Bearing Capacity

The ultimate gross bearing capacity for LRFD is

$$q_{\text{ult}} = q_u + \gamma D_f \tag{7.6}$$

7.3.1.4 Bearing Capacity for Special Cases

Horizontal Surface

Vertical Centric Load Only

$$\text{TSA: } q_u = 5.14 s_u s_c d_c r_c \tag{7.7}$$

$$\text{ESA: } q_u = \gamma D_f (N_q - 1) s_q d_q r_q w_q + 0.5\,\gamma B N_\gamma s_\gamma d_\gamma r_\gamma w_\gamma \tag{7.8}$$

Inclined Centric Load Only

$$\text{TSA: } q_u = 5.14 s_u i_c r_c \tag{7.9}$$

$$\text{ESA: } q_u = \gamma D_f (N_q - 1) i_q r_q w_q + 0.5\,\gamma B N_\gamma i_\gamma r_\gamma w_\gamma \tag{7.10}$$

[1] It is a legacy in foundation engineering to assume that soil is a cohesive-frictional material and to lump cohesion (cementation, c) and friction (ϕ') together by writing a single bearing capacity equation as $q_{\text{ult}} = cN_c + \gamma D_f N_q + 0.5\gamma B' N_\gamma$. The interpretation of soil as a cohesive-frictional material is based on a linearization of the Mohr-Coulomb peak shear strength (failure) envelope. Cohesion and friction are not mobilized at the same shear strain level—c is mobilized at small shear strain levels while ϕ' is mobilized at large shear strains. Thus, as shear strains develop under the foundation during loading, c and ϕ' would not be mobilized simultaneously. In this book, soil is treated as a dilatant-frictional material. Analyses and parameters based on effective stress and total stress are clearly differentiated.

7.3.1.5 *Bearing capacity, geometric, compressibility, and groundwater factors*

Bearing Capacity Factors The bearing capacity factor N_q is

$$N_q = e^{\pi \tan\phi'_p} \tan^2\left(45° + \frac{\phi'_p}{2}\right); \ \phi'_p \text{ in degrees} \tag{7.11}$$

Various equations have been proposed for N_γ in the literature. Among the popular equations are

Caquot and Kerisel (1953)[2]: $N_\gamma = 2(N_q + 1)\tan\phi'_p$; ϕ'_p in degrees (7.12)

Meyerhof (1976): $N_\gamma = (N_q - 1)\tan(1.4\phi'_p)$; ϕ'_p in degrees (7.13)

Davis and Booker (1971)[3] :

smooth foundation, $N_\gamma = 0.0663 \exp(9.3\phi'_p)$; ϕ'_p in radians (7.14)

rough foundation, $N_\gamma = 0.1054 \exp(9.6\phi'_p)$; ϕ'_p in radians (7.15)

Ueno et al. (1998)[4] :

rough foundation, $N_\gamma = 0.477 \exp(6.52\phi'_p)$; ϕ'_p in radians (7.16)

The differences among these popular equations are shown in Fig. 7.4.

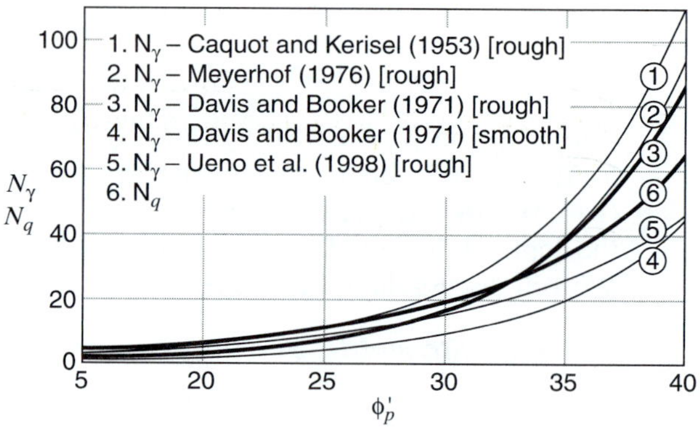

FIGURE 7.4 Bearing capacity factors.

The bearing capacity factor, N_γ, proposed by Davis and Booker (1971) and Ueno et al. (1998), is based on refined plasticity methods that give conservative values compared with analytical solutions based on the limit equilibrium method conducted by Terzaghi (1943), Caquot and Kerisel (1953), and Meyerhof (1976). For ϕ'_p values less than 30°, the differences among the various N_γ factors for rough foundation are not very large for practical purposes. Thus, for $\phi'_p < 30°$ you can use any of the equations for N_γ. However, when $\phi'_p > 30°$, the differences in the various methods increase exponentially. In this book we will primarily use the Booker and Davis (1971) equations.

The crucial and only soil parameter to determine the bearing capacity factors is ϕ'_p. The attainment of ϕ'_p depends on the ability of the soil to dilate, which can be suppressed by large normal effective stresses. Since neither the loads nor the stresses induced by the loads on the soil mass are certain, the use of ϕ'_p is then uncertain. We can write $\phi'_p = \phi'_{cs} + \alpha_p$, where ϕ'_{cs} is the friction angle at critical state (a fundamental soil parameter) and α_p is the peak dilation angle. The uncertainty in ϕ'_p comes mainly from the uncertainty in the value of α_p.

Plane strain conditions have been assumed in developing the theoretical bearing capacity equations. Thus, the values of ϕ'_p should come from plane strain tests such as the direct shear or direct simple shear tests. Triaxial tests on the same soil generally give values of ϕ'_p lower than plane strain tests. Various conversions have been proposed. For practical purposes, the following conversion is sufficiently accurate:

$$(\phi'_p)_{ps} = \frac{9}{8}(\phi'_p)_{tr} \tag{7.17}$$

where the subscripts *ps* and *tr* denote plane strain and triaxial respectively.

Rough foundation denotes a foundation in which full shear is developed at the foundation–soil interface. It is common construction practice to compact a layer of coarse-grained soils for the foundation

[2] Vesic (1973) recommended this equation.

[3] These equations were obtained by curve fitting the theoretical results of Davis and Booker (1971) by Poulos et al. (2002).

[4] This equation was obtained by curve fitting the theoretical results of Ueno et al. (1998) by the author.

to rest on. In this case, N_γ for rough foundation is appropriate. When a foundation rests directly on fine-grained soils, N_γ for smooth foundation should be used.

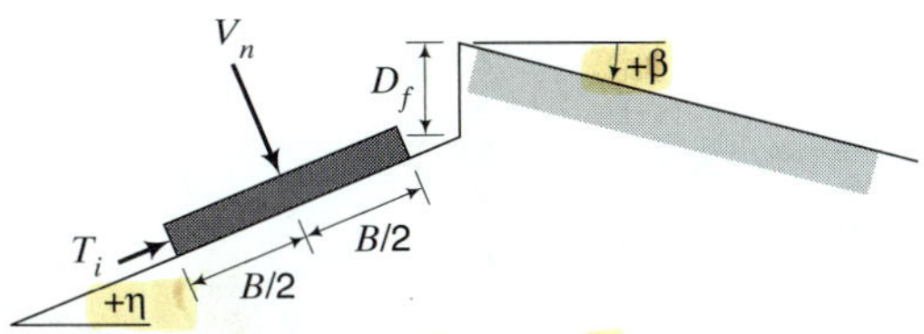

FIGURE 7.5 Footing on a slope.

Geometric Factors Geometric factors account for the shape and slope of the foundation (base tilt), the load inclination and load eccentricity, and the ground slope (Fig. 7.5). Several equations have been proposed for these factors. In this book, we will use a set of factors proposed in the literature that are generally accurate for practical applications (Table 7.1). All the factors listed in Table 7.1 must be regarded as estimates. The bearing capacity equations apply for a single resultant load with normal component, V_n, and shear (parallel to the base of the foundation) component, T. The shear component can be decomposed into a component T_B, parallel to the width, B (the short side), and a component, T_L, parallel to the length, L (the long side). When investigating potential failure along the short side, use $T = T_B$. For failure along the long side, use $T = T_L$. *All loads are assumed to be acting at the base of the foundation.*

When the location of the resultant load (load center) is not coincident with the centroid (center of area) of the foundation, the foundation dimension is theoretically adjusted to align the load center with the centroid. Some possible cases are shown in Fig. 7.6. The effective width B' and effective length L' giving an effective area $A' = B'L'$ must be used in the theoretical bearing capacity equations. The factors given in Table 7.1 are for failure along a plane parallel to the short side, B. For failure along the long side, replace B' and L'.

The normal stresses for eccentrically loaded foundation are assumed to be distributed linearly across the foundation base (Fig. 7.7) with maximum and minimum values calculated from:

Along foundation width:

$$\sigma_{max} = \frac{V_n}{A}\left(1 + \frac{6e_B}{B}\right) \tag{7.18}$$

$$\sigma_{min} = \frac{V_n}{A}\left(1 - \frac{6e_B}{B}\right) \tag{7.19}$$

Along foundation length:

$$\sigma_{max} = \frac{V_n}{A}\left(1 + \frac{6e_L}{L}\right) \tag{7.20}$$

$$\sigma_{min} = \frac{V_n}{A}\left(1 - \frac{6e_L}{L}\right) \tag{7.21}$$

where V_n is the resultant normal load component, and e_B and e_L are eccentricities along the width and length, respectively.

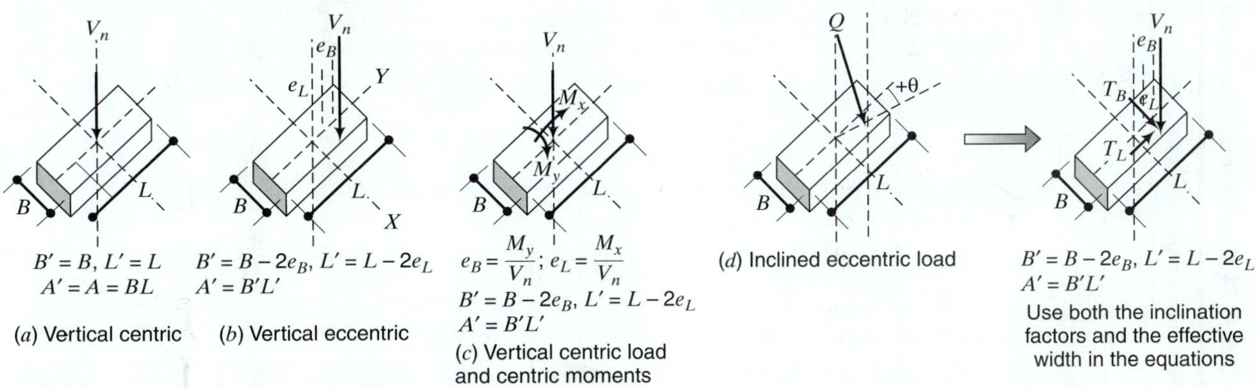

FIGURE 7.6 Some possible load cases.

TABLE 7.1 Geometric Factors for Use in Theoretical Bearing Capacity Equations

Geometric parameters for TSA

s_c	d_c	i_c	b_c	g_c	r_c
$1+0.2\dfrac{B'}{L'}$	$1+0.33\tan^{-1}\left(\dfrac{D_f}{B'}\right)$; for $\dfrac{D_f}{B'}>1$ $1+0.33\left(\dfrac{D_f}{B'}\right)$; for $\dfrac{D_f}{B'}\leq 1$ See note 1	$1-\dfrac{nT}{5.14s_uB'L'}$ See note 2	$\dfrac{\eta^\circ}{147}$ $\beta<\phi'_p$; $\eta^\circ+\beta^\circ<90^\circ$ See Fig. 7.5	$\dfrac{\beta^\circ}{147}$ $\beta<\phi'_p$; $\eta^\circ+\beta^\circ<90^\circ$ See Fig. 7.5	$r_c=0.32+0.12\dfrac{B'}{L'}+0.6\log_{10}I_r$ $I_r=\dfrac{G}{s_u}$, where G is the shear modulus of the soil. See note 3

Geometric parameters for ESA

s_q	d_q	i_q	b_q	g_q	r_q
$1+\dfrac{B'}{L'}\tan\phi'_p$	$1+2\tan\phi'_p(1-\sin\phi'_p)^2\tan^{-1}\left(\dfrac{D_f}{B'}\right)$; for $\dfrac{D_f}{B'}>1$ $1+2\tan\phi'_p(1-\sin\phi'_p)^2\dfrac{D_f}{B'}$; for $\dfrac{D_f}{B'}\leq 1$	$\left(1-\dfrac{T}{V_n}\right)^n$ See note 2	$b_q=b_\gamma$	$(1-\tan\beta)^2$	$r_\gamma=\exp\left[\left(-4.4+0.6\dfrac{B'}{L'}\right)\tan\phi'_p+\dfrac{(3.07\sin\phi'_p)\log_{10}(2I_r)}{1+\sin\phi'_p}\right]$ $I_r=\dfrac{G}{\gamma D_f\tan\phi'_p}$, where G is the shear modulus of the soil or $I_r=0.636\dfrac{D_r}{\sqrt{\gamma D_f}}$, D_r = relative density (%) and γD_f is the overburden pressure (MPa). See note 3
s_γ	**d_γ**	**i_γ**	**b_γ**	**g_γ**	**r_γ**
$1-0.4\dfrac{B'}{L'}$	1	$\left(1-\dfrac{T}{V_n}\right)^{n+1}$ See note 2	$(1-\eta\tan\phi'_p)^2$; η is in radians.	$g_q=g_\gamma$	$r_\gamma=r_q$ See note 3

NOTES:

Note 1: If the shear strength of the soil above the footing is low compared with the soil below the footing, you should set all depth factors to 1.

Note 2: The depth and shape factors for inclined loads should be set to 1. For loading inclined in the direction of the width, B, $\theta=90^\circ$ in Fig. 7.6d, $n=n_B=\left(2+\frac{B'}{L'}\right)/\left(1+\frac{B'}{L'}\right)$. For loading inclined in the direction of the length, L, $\theta=0^\circ$ in Fig. 7.6d, $n=n_L=\left(2+\frac{L'}{B'}\right)/\left(1+\frac{L'}{B'}\right)$. For other loading conditions, $n=n_L\cos^2\theta+n_B\sin^2\theta$. The inclination factors cannot be greater than 1 or less than 0. Also the load inclination angle, $\tan^{-1}\left(\frac{T}{V_n}\right)$, must be less than the critical state friction angle.

Note 3: The critical rigidity index for long-term condition is $I_{rc}=0.5\exp\left[\left(3.30-0.45\frac{B'}{L'}\right)\cot\left(45^\circ-\frac{\phi'_p}{2}\right)\right]$ and for short-term condition is $I_{rc}=0.5\exp\left(3.30-0.45\frac{B'}{L'}\right)$. If $I_r>I_{rc}$, the soil is practically incompressible, general shear failure is most likely to occur and the compressibility factors should be set to 1. Set all compressibility factors to 1 if relative density $D_r\geq 70\%$.

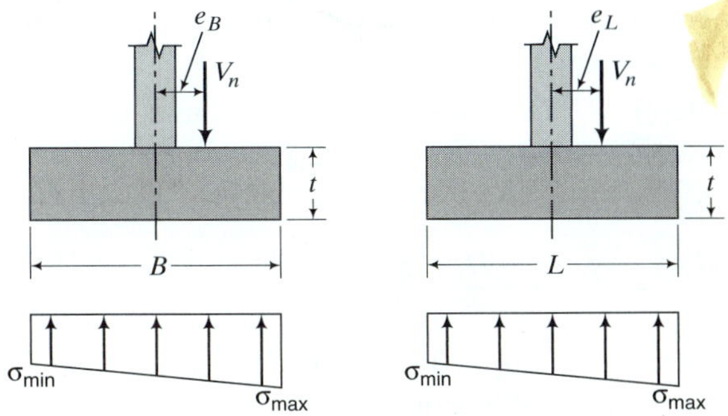

FIGURE 7.7 Assumed vertical stress distribution below an eccentrically loaded footing.

Let us examine σ_{min}. If $e_B = \frac{B}{6}$ or $e_L = \frac{L}{6}$, then $\sigma_{min} = 0$. If, however, $e_B > \frac{B}{6}$ or $e_L > \frac{L}{6}$ then $\sigma_{min} < 0$, and tension develops. Since the tensile strength of soil is zero, part of the foundation will not transmit loads to the soil. You should try to avoid this situation by designing the foundation such that $e_B < \frac{B}{6}$ or $e_L < \frac{L}{6}$.

The ultimate net load is:

$$P_u = q_u B'L' \tag{7.22}$$

For circular foundation, $B = L = D$, where D is diameter. The equivalent foundation area for a circular foundation subjected to an eccentric load is

$$A' = \frac{D^2}{2}\left[\cos^{-1}\frac{2e_c}{D} - \frac{2e_c}{D}\sqrt{1-\left(\frac{2e_c}{D}\right)^2}\right] \tag{7.23}$$

where e_c is the eccentricity. The equivalent width, B', is $2\sqrt{\frac{A'}{\pi}}$.

In calculating the factor of safety, FS, replace q_a in Eq. (7.5) by σ_{max}; that is, FS is

$$\text{FS} = \frac{q_u}{\sigma_{max} - \gamma D_f} \tag{7.24}$$

When the foundation rests on sloping ground, the positive direction of the slopes and the foundation depth are as shown in Fig. 7.5.

Compressibility Factors The compressibility factors (Vesic, 1973) are intended to approximately compensate for soil conditions (relative density and embedment depth) for which a general failure mechanism may not develop. For practically incompressible coarse-grained soils, $I_r > I_{rc}$ (see note 3 in Table 7.1), general shear failure is likely to occur and the compressibility factors should be set to 1. For fine-grained soils, you must know the shear modulus G to calculate the compressibility factor. However, the value of G is not always known from laboratory and field tests for many practical situations. In this case, you can use empirical relationships between s_u and G (see Chapter 4). For coarse-grained soils, use the relative density or G, if known. The compressibility factor is 1 for relative densities in excess of 70%.

Groundwater Factors You have to make some adjustment to the theoretical bearing capacity equations for groundwater conditions. The term γD_f in the bearing capacity equations for an ESA refers to the vertical stress of the soil above the base of the foundation. The last term, γB, refers to the vertical stress of a soil mass of thickness B below the base of the foundation. There are three groundwater situations as shown in Fig. 7.8.

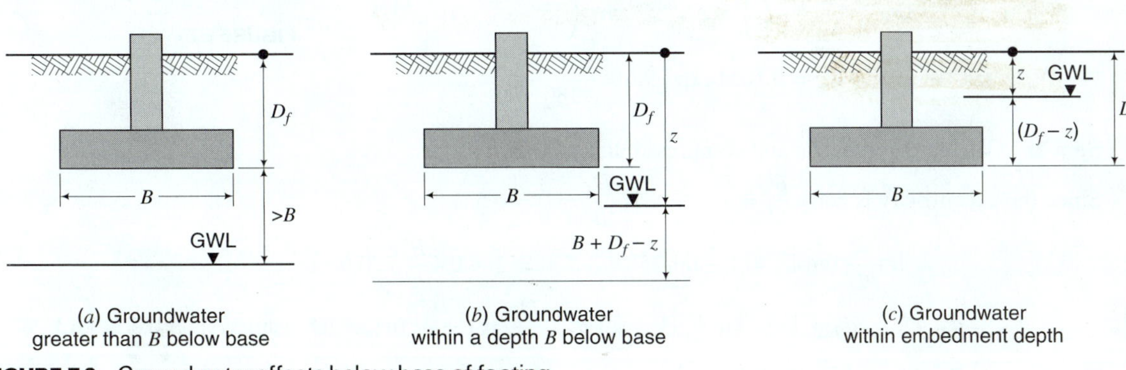

FIGURE 7.8 Groundwater effects below base of footing.

Situation 1: Groundwater level (GWL) at a level B or greater below the base of the foundation (Fig. 7.8a). No correction is required (i.e., $w_q = w_\gamma = 1$).

Situation 2: If the groundwater level is within a depth B below the base of the foundation $D_f < z < (B + D_f)$ (Fig. 7.8b), then the term γB is $\gamma(z - D_f) + \gamma'(D_f + B - z)$ or $\gamma_{sat}(z - D_f) + \gamma'(D_f + B - z)$. The latter equation is used for saturated soils. The term γD_f remains unchanged. The groundwater factors are

$$w_q = 1, \quad w_\gamma = \frac{(z - D_f)}{B} + \frac{\gamma'}{\gamma_{sat}}\left(1 + \frac{D_f}{B} - \frac{z}{B}\right) \tag{7.25}$$

Situation 3: If the groundwater level is within the embedment depth $0 \leq z \leq D_f$ (Fig. 7.8c), then the term γD_f is $\gamma z + \gamma'(D_f - z)$ or $\gamma_{sat} z + \gamma'(D_f - z)$. The term γB becomes $\gamma' B$. The groundwater factors are

$$w_q = \frac{z}{D_f} + \frac{\gamma'}{\gamma_{sat}}\left(1 - \frac{z}{D_f}\right) \text{ and } w_\gamma = \frac{\gamma'}{\gamma_{sat}} \tag{7.26}$$

If the groundwater level is at the surface, $w_q = \frac{\gamma'}{\gamma_{sat}}$, $w_\gamma = \frac{\gamma'}{\gamma_{sat}}$; the ratio $\frac{\gamma'}{\gamma_{sat}} \cong \frac{1}{2}$.

EXAMPLE 7.1 *Determination of Bearing Capacity for a Square Foundation on a Compacted Sand*

A foundation 2 m (6 ft) square is to be located at a depth of 1.0 m (3 ft) below the ground surface in a deep deposit of compacted sand ($\phi'_p = 35°$ from plane strain tests and $\gamma_{sat} = 18\,\text{kN/m}^3$ (114 pcf)) to support a vertical centric load. The groundwater level is 5 m (15 ft) below the ground surface but the soil above the groundwater is saturated. Determine the ultimate net bearing capacity, ultimate gross bearing capacity, the allowable bearing capacity using FS = 3, and the allowable vertical centric load. Use Booker and Davis (1971) bearing capacity factors.

Strategy U.S. units will be used in solving this example problem. It is a good policy to sketch a diagram illustrating the conditions given (See Fig. E7.1.). The groundwater level is located at (15 ft − 3 ft) = 12 ft from the foundation base. That is, the groundwater level is more than $B = 6$ ft below the base. We can neglect the effects of groundwater (i.e., $w_\gamma = w_q = 1$). Since the soil is compacted and no information is given about the relative density, we will assume that general soil failure would occur (i.e., $r_q = r_\gamma = 1$) and rough foundation contact with the soil. Only an ESA is required for coarse-grained soils.

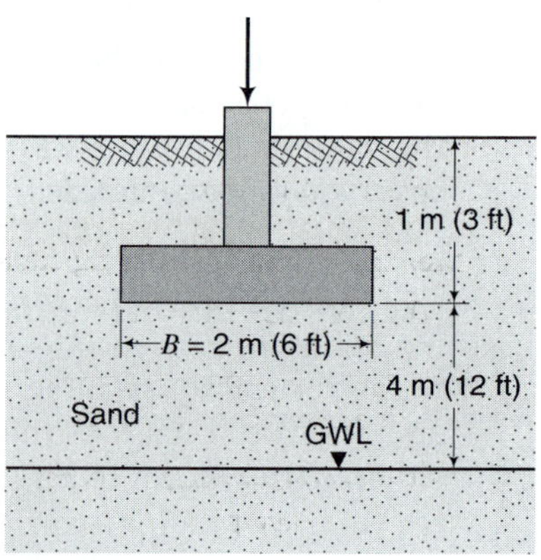

FIGURE E7.1

Solution 7.1

Step 1: Calculate the bearing capacity factors.

$$N_q = e^{\pi \tan \phi'_p} \tan^2\left(45° + \frac{\phi'_p}{2}\right) = e^{\pi \tan 35°} \tan^2\left(45° + \frac{35°}{2}\right) = 33.3$$

$$N_\gamma = 0.1054 \exp 9.6\,\phi'_p = 0.1054 \exp\left(9.6 \times \frac{\pi \times 35}{180}\right) = 37.1$$

Step 2: Calculate geometric and compressibility factors.

Since the eccentricity is zero, $\frac{B'}{L'} = \frac{B}{L} = \frac{6}{6} = 1$; $\frac{D_f}{B'} = \frac{3}{6} = 0.5$.

$$s_q = 1 + \frac{B'}{L'} \tan \phi'_p = 1 + \tan 35° = 1.7 \quad s_\gamma = 1 - 0.4\frac{B'}{L'} = 0.6$$

$$d_q = 1, \quad d_q = 1 + 2 \tan \phi'_p (1 - \sin \phi'_p)^2 \frac{D_f}{B'} = 1 + 2 \tan(35°)(1 - \sin 35°)^2 \times 0.5 = 1.13$$

$$r_q = r_\gamma = 1, \quad w_\gamma = w_q = 1$$

Step 3: Substitute the values in steps 1 and 2 into the long-term bearing capacity equation to estimate q_u.

$$q_u = \gamma D_f(N_q - 1)s_q d_q r_q w_q + 0.5\gamma B' N_\gamma s_\gamma d_\gamma r_\gamma w_\gamma$$
$$= 114 \times 3 \times (33.3 - 1) \times 1.7 \times 1.13 \times 1 \times 1 + 0.5 \times 114 \times 6 \times 37.1 \times 0.6 \times 1 \times 1 \times 1 = 28833 \text{ psf}$$

Step 4: Determine q_{ult}, q_a, and Q_a.

$$q_{\text{ult}} = q_u + \gamma D_f = 28833 + 3 \times 114 = 29175 \text{ psf}$$
$$q_a = \frac{q_u}{FS} + \gamma D_f = \frac{28833}{3} + 3 \times 114 = 9953 \text{ psf}$$
$$Q_a = \text{allowable ultimate centric load}$$
$$Q_a = q_a A' = 9953 \times 6^2 = 358308 \text{ lbs} = 358.3 \text{ kips}$$

■

EXAMPLE 7.2 *Effects of Groundwater Level on Bearing Capacity*

If the groundwater were located at (a) the base and (b) the ground surface in Example 7.1, calculate the ultimate bearing capacity and ultimate gross bearing capacity.

Strategy Groundwater will now affect the bearing capacity. You need only calculate the groundwater factors for the two cases as all other parameters remain the same.

Solution 7.2

Step 1: Calculate groundwater factors.

GWL at base (situation 2): $w_q = 1$, $z = D_f$, $\gamma' = 114 - 62.4 = 51.6$ pcf.

$$w_\gamma = \frac{z - D_f}{B} + \frac{\gamma'}{\gamma_{\text{sat}}}\left(1 + \frac{D_f}{B} - \frac{z}{B}\right)$$
$$= \frac{3 - 3}{2} + \frac{51.6}{114}\left(1 + \frac{3}{6} - \frac{3}{6}\right) = 0.45$$

GWL at surface: $w_q = w_\gamma = \dfrac{\gamma'}{\gamma_{\text{sat}}} = 0.45.$

Step 2: Calculate bearing capacity.

GWL at base: $q_u = 114 \times 3 \times (33.3 - 1) \times 1.7 \times 1.13 \times 1 \times 1 \times 1 + 0.5 \times 114 \times 6 \times 37.1 \times 0.6 \times 1 \times 1 \times 1 \times 0.45$
$= 24646$ psf.

GWL at surface: $q_u = 28833 \times 0.45 = 12975$ psf.

Note that the ultimate net bearing capacity is reduced by 55% if the groundwater rises to the surface. A leaking water pipe or sewage pipe can cause the groundwater level to rise locally. You should take special precaution to drain any potential water (or liquid) leak away from the foundation.

■

EXAMPLE 7.3 *Bearing Capacity of a Foundation with an Inclined Load*

A foundation 1.5 m (4.5 ft) × 3.0 m (10 ft) with a thickness 300 mm (1 ft) is located at a depth of 1.5 m (4.5 ft) below the ground surface in a deep deposit of saturated overconsolidated clay. A steel column is fixed to the top of the foundation and imposes a 500 kN (112.5 kips) load located at the center of the foundation and inclined at an angle of 20° to the vertical along the long side. Triaxial tests with porewater pressure measurements were conducted on the soil and the results are $s_u = 120$ kPa (2500 psf), $\phi'_p = 21.3°$. The groundwater was measured over a period of one week and was stable at a depth of 5.0 m (15 ft) below the surface. Determine the factor of safety for ultimate limit state. The ratio $\frac{G}{s_u} = 150$ and $\gamma_{\text{sat}} = 18\text{ kN/m}^3$ (114 pcf). The foundation base will rest on a 100 mm (4 in.) compacted sand-gravel layer constructed on top of the clay surface.

Strategy Because the soil is fine-grained, calculate the short-term and long-term bearing capacities. The lesser would govern. Since triaxial tests were conducted, you need to convert the friction angle to the equivalent plane strain friction angle. Also, you should check short and long side failure modes. Since the load is inclined, the shape and depth factors are equal to 1. The load is applied on the top of the foundation so you need to transfer it to the base by finding an equivalent load system. The foundation can be assumed to be on a rough base. SI units will be used in solving this example problem.

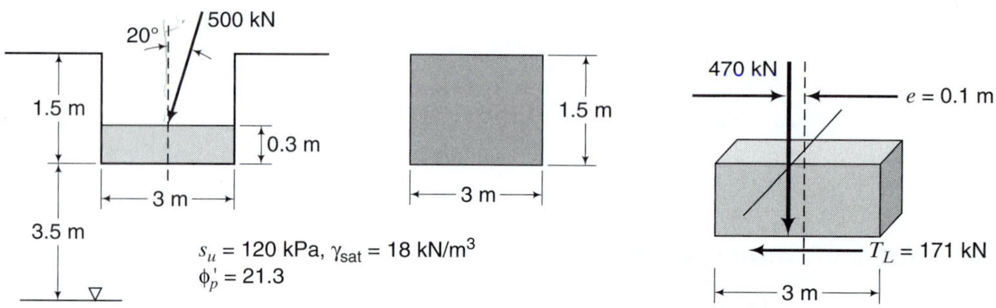

(*a*) Section (*b*) Plan (*c*) Equivalent load

FIGURE E7.3

Solution 7.3

Step 1: Convert $(\phi'_p)_{tr}$ to $(\phi'_p)_{ps}$.

$$(\phi'_p)_{ps} = \frac{9}{8}(\phi'_p)_{tr} = \frac{9}{8} \times 21.3° = 24°$$

Step 2: Find equivalent load system at base.

T_L = horizontal force along long side; T_B = horizontal force along short side
$T_L = 500 \times \sin 20° = 171$ kN; $T_B = 0$; $V_n = 500 \times \cos 20° = 470$ kN
t = base thickness = 0.3 m

Moment $= T_L \times t = 171 \times 0.3 = 51.3$ kN/m.

$$e_L = \frac{M}{V_n} = \frac{51.3}{470} = 0.11\text{ m};\ e_B = 0$$

Move horizontal force to base: $M = F \cdot d = 171 \cdot 0.3$
Get rid of moment by moving vertical force over ⇒
$d = \frac{M}{F} = \frac{(171)(0.3)}{470} = 0.1$

Step 3: Check eccentricity does not exceed $\frac{L}{6}$.

$\frac{L}{6} = \frac{3}{6} = 0.5\text{ m} > e_L$; no tension

Step 4: Calculate maximum vertical stress.

LSF: Long side failure
SSF: Short side failure

$$\text{LSF: } \sigma_{max} = \frac{V_n}{A}\left(1 + \frac{6e_L}{L}\right) = \frac{470}{3 \times 1.5}\left(1 + \frac{6 \times 0.11}{3}\right) = 127\text{ kPa}$$

$$\text{SSF: } \sigma_{max} = \frac{470}{3 \times 1.5} = 104\text{ kPa}$$

Step 5: Calculate geometric factors.

$$B' = B = 1.5\text{ m}, \quad L' = L - 2e_L = 3 - 2(0.11) = 2.8\text{ m}$$

$$A' = B' \times L' = 1.5 \times 2.8 = 4.2\text{ m}^2$$

$$\frac{B'}{L'} = \frac{1.5}{2.8} = 0.54, \quad \frac{L'}{B'} = \frac{2.8}{1.5} = 1.87$$

$$\text{LSF}: n = n_L = \left(\frac{2+\frac{L'}{B'}}{1+\frac{L'}{B'}}\right) = \frac{2+1.87}{1+1.87} = 1.35$$

$$\text{SSF}: n = n_B = \left(\frac{2+\frac{B'}{L'}}{1+\frac{B'}{L'}}\right) = \frac{2+0.54}{1+0.54} = 1.65$$

TSA

$$I_r = \frac{G}{s_u} = 150; s_c = d_c = 1.0 \text{ (for inclined loads)}$$

$$\text{LSF}: I_{rc} = 0.5\exp\left(3.30 - 0.45\frac{L'}{B'}\right) = 0.5\exp(3.30 - 0.45 \times 1.87) = 5.84$$

$$\text{SSF}: I_{rc} = 0.5\exp\left(3.30 - 0.45\frac{B'}{L'}\right) = 0.5\exp(3.30 - 0.45 \times 0.54) = 10.63$$

$I_r > I_{rc}$; soil is practically incompressible and $r_c = 1$

$$\text{LSF}: i_c = 1 - \frac{nT_L}{5.14 s_u B' L'} = 1 - \frac{1.35 \times 171}{5.14 \times 120 \times 1.5 \times 2.8} = 0.91$$

$$\text{SSF}: i_c = 1, \text{ since } T_B = 0$$

ESA

$$\tan\omega = \tan 20^\circ = 0.36$$

$$s_q = s_\gamma = d_\gamma = d_q = 1$$

$$\text{LSF}: i_\gamma = \left(1 - \frac{T_L}{V_n}\right)^{(n+1)} = (1-0.36)^{(1.35+1)} = 0.35$$

$$\text{SSF}: i_\gamma = 1.0, \text{ since } T_B = 0$$

$$\text{LSF}: i_q = \left(1 - \frac{T_L}{V_n}\right)^{n} = (1-0.36)^{1.35} = 0.55$$

$$\text{SSF}: i_q = 1.0, \text{ since } T_B = 0.$$

$$r_\gamma = r_q = 1;\ w_q = w_\gamma = 1 \text{ (groundwater below } B \text{ from base)}$$

Step 6: Calculate bearing capacity factors.

ESA

$$N_q = e^{\pi\tan\phi'_p}\tan^2\left(45 + \frac{\phi'_p}{2}\right) = e^{\pi\tan 24^\circ}\tan^2\left(45 + \frac{24}{2}\right) = 9.6$$

$$N_\gamma = 0.1054\exp(9.6\phi'_p) = 0.1054\exp\left(9.6 \times \frac{24\pi}{180}\right) = 5.9$$

Step 7: Calculate bearing capacity.

TSA

$$\text{LSF}: q_u = 5.14\, s_u i_c r_c = 5.14 \times 120 \times 0.91 \times 1 = 561\,\text{kPa}$$
$$\text{SSF}: q_u = 5.14 \times 120 \times 1 \times 1 = 617\,\text{kPa}$$

ESA

$$\begin{aligned}\text{LSF}: q_u &= \gamma D_f(N_q - 1)i_q r_q w_q + 0.5\gamma L' N_\gamma i_\gamma r_\gamma w_\gamma \\ &= (18 \times 1.5(9.6 - 1) \times 0.55 \times 1 \times 1) + (0.5 \times 18 \times 2.8 \times 5.9 \times 0.35 \times 1 \times 1) \\ &= 180\,\text{kPa}\end{aligned}$$

$$\begin{aligned}\text{SSF}: q_u &= \gamma D_f(N_q - 1)i_q r_q w_q + 0.5\gamma B' N_\gamma i_\gamma r_\gamma w_\gamma \\ &= (18 \times 1.5(9.6 - 1) \times 1 \times 1 \times 1) + (0.5 \times 18 \times 1.5 \times 5.9 \times 1 \times 1 \times 1) \\ &= 381\,\text{kPa}\end{aligned}$$

Step 8: Determine factor of safety.

The minimum bearing capacity occurs for long side failure under long-term conditions.

$$\text{LSF}: \text{FS} = \frac{180}{125 - 1.5 \times 18} = 1.8$$

The factor of safety is well below acceptable values (FS of 2 to 3). The foundation size and/or foundation depth should be increased. ■

EXAMPLE 7.4 *Foundation for a Retaining Wall*

A section of a long concrete retaining wall is shown in Fig. E7.4a. The lateral force per unit length of the wall is 30 kN acting 1 m from the base. The unit weight of concrete is 24 kN/m^3. Determine the factor of safety of the wall foundation. Groundwater is 10 m below the base of the foundation.

Strategy The retaining wall is long, so you can assume the foundation to be a strip foundation. Transfer the load (self-weight of concrete) and horizontal loads to an equivalent force system at the base. SI units will be used in solving this example problem.

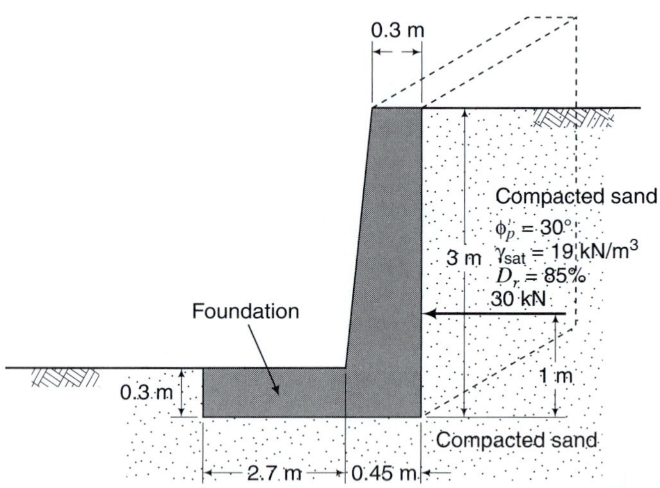

FIGURE E7.4a

Solution 7.4

Step 1: Calculate weight of wall and load eccentricity.

Divide the wall into parts with simple shapes as shown in Fig. E7.4b.

Part	Weight (kN)	Centroid from O (m)	Moment at O (kN/m)—clockwise positive
1	0.3 × 2.7 × 24 = 19.4	3	58.2
2	0.5 × 0.15 × 2.7 × 24 = 4.9	2.78	13.6
3	3.15 × 0.3 × 24 = 22.7	1.58	35.8
	ΣW = 47		
	Moment due to horizontal force = −30 × 1 = −30		
			ΣM = 77.6

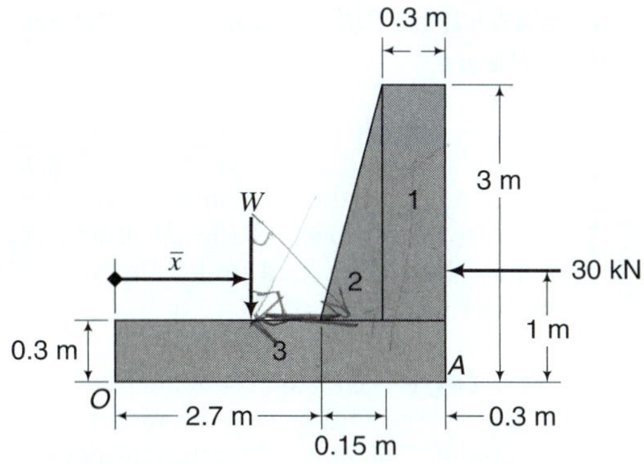

FIGURE E7.4b

All loads are per unit length.

$$\bar{x} = \frac{\Sigma M}{\Sigma W} = \frac{77.6}{47} = 1.65\,\text{m};\quad e_B = \frac{B}{2} - \bar{x} = 1.58 - 1.65 = -0.07\,\text{m}$$

The resultant vertical load is to the right of the centerline as shown in Fig. E7.4c.

$$\frac{B}{6} = \frac{3.15}{6} = 0.53 > 0.07\,\text{m; no tension—}\textit{okay}$$

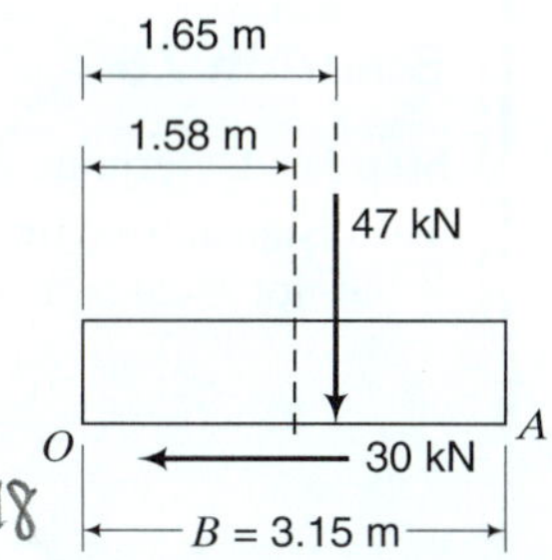

FIGURE E7.4c Equivalent system.

Step 2: Calculate maximum vertical stress.

Maximum vertical stress will occur at A

$$\sigma_{max} = \frac{W}{A}\left(1 + \frac{6e}{B}\right) = \frac{47}{3.15 \times 1}\left(1 + \frac{6 \times 0.07}{3.15}\right) = 16.9\,\text{kPa/m}$$

Step 3: Calculate the equivalent width.

$$B' = B - 2e = 3.15 - 2(0.07) = 3.01\,\text{m}$$

Step 4: Calculate bearing capacity and geometric factors.

You only need to calculate N_γ since the foundation is at the surface.
Assume rough foundation and use Davis and Booker (1971) equation.

$$N_\gamma = 0.1054\exp(9.6\phi'_p) = 0.1054\exp\left(9.6 \times \frac{30}{180} \times \pi\right) = 16$$

The resultant load is inclined to the vertical at an angle, $\theta = \tan^{-1}(30/47) = 32.5°$, $i_\gamma = 0.08$ (spreadsheet bc.xls). Since $D_r = 85\%$, $r_\gamma = 1$.

Step 5: Calculate the net bearing capacity.

$$\text{ESA: } q_u = 0.5\gamma B' N_\gamma i_\gamma = 0.5 \times 18 \times 3.01 \times 16 \times 0.08 = 34.7\,\text{kPa}$$

Step 6: Calculate factor of safety.

$$\text{FS} = \frac{q_u}{\sigma_{max} - \gamma D_f} = \frac{34.7}{16.9 - 0} = 2.0 < 3;\ \textit{not okay}$$

The foundation will not fail but the FS is low. The modes of failure for retaining walls are rotation, translation, and bearing capacity. All modes must be safe. In the case of translation, the maximum friction resistance is $W\tan\phi'_p = 47\tan 30° = 27.1\,\text{kN/m}$ and the factor of safety is $\frac{27.1}{30} = 0.9 < 1.5$; *not okay*. (This value of 1.5 is the conventional factor of safety for sliding; see Chapter 10.) The foundation width is inadequate for safety against translation also. The resultant load inclination cannot be greater than the critical state friction angle unless the horizontal load is resisted by lateral supports such as ties between individual foundations. ■

EXAMPLE 7.5 *Foundation Near a Slope*

A strip foundation 3 m (10 ft) wide is to be located on top of a slope as shown in Fig. E7.5a. Determine (a) the minimum foundation distance from the base of the slope and (b) the allowable centric load under short-term condition for a factor of safety of 3.

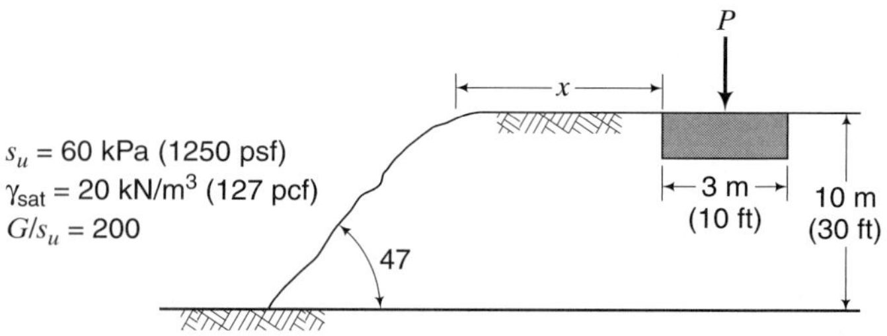

FIGURE E7.5a

Strategy Use the recommendation given in Fig. 7.1f to find the minimum distance from the slope. Calculate the ultimate net bearing capacity and then the allowable centric load. U.S. units will be used in solving this example problem. (See Fig. E7.5b.)

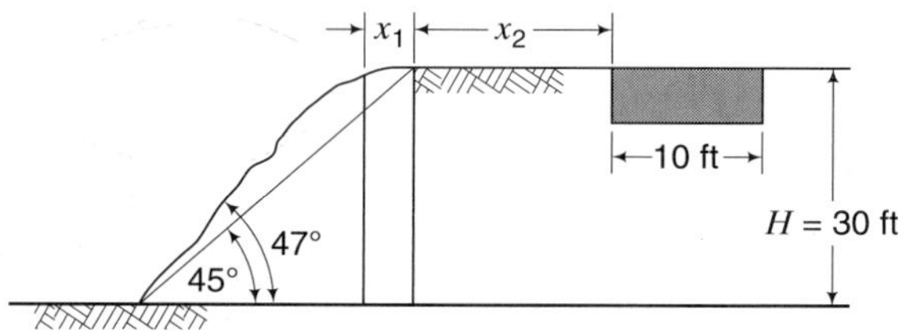

FIGURE E7.5b

Solution 7.5

Step 1: Determine edge distance.

The distance from the outer edge of the foundation $x_2 = \frac{H}{3}$ but not greater than 12 m (40 ft).

$$\frac{H}{3} = \frac{30}{3} = 10 \text{ ft} < 40 \text{ ft}$$

$$\therefore x_2 = 10 \text{ ft(minimum)}$$

$$x_1 = H\left(\frac{1}{\tan 45^\circ} - \frac{1}{\tan 47^\circ}\right) = 30 \times 0.067 = 2 \text{ ft}$$

$$x = x_1 + x_2 = 10 + 2 = 12 \text{ ft}$$

Step 2: Calculate geometric and compressibility factors.

All geometric factors are equal to 1.

$$I_r = \frac{G}{s_u} = 200; \quad \frac{B^1}{L^1} \approx 0 \text{ for strip foundation}; \quad I_{rc} = 0.5\exp\left(3.30 - 0.45\frac{B'}{L'}\right) = 13.6 < I_r \therefore r_c = 1.0$$

Step 3: Calculate ultimate net bearing capacity.

$$q_u = 5.14 s_u r_c = 5.14 \times 1250 \times 1.0 = 643 \text{ psf}$$

Step 4: Calculate allowable load.

$$q_a = \frac{q_u}{FS} = \frac{643}{3} = 214 \text{ psf}$$

$$Q_a = q_a \times B \times 1 = 218 \times 10 \times 1 = 2180 \text{ lb/ft}$$

■

7.3.2 Combined Loading

The general bearing capacity equations [Eqs. (7.3) and (7.4)] do not account for vertical and horizontal loading, and moment applied simultaneously. When vertical load (V_n), horizontal load (T), and moments (M) act on a foundation, a two-dimensional approach is unsatisfactory because the foundation response is three-dimensional. The preferred approach is to consider a failure surface that encloses a safe combination of these loads, as shown in Fig. 7.9. Combinations of loads that lie outside of a failure surface are unsafe. We will consider combined loadings for circular and strip foundations.

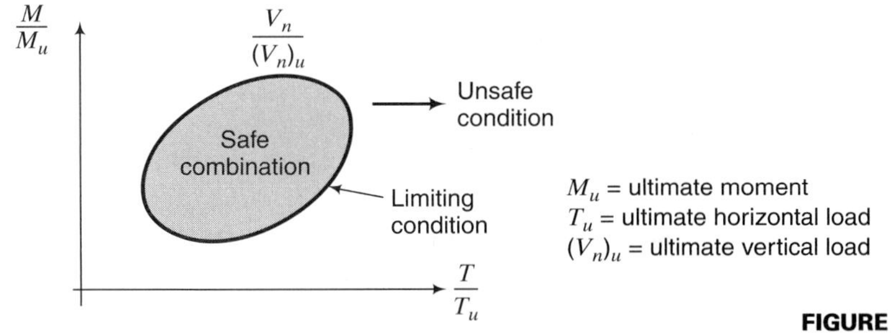

FIGURE 7.9 Load combination.

7.3.2.1 Circular and Rectangular Foundations

> Access a spreadsheet at http://www.wiley.com/college/budhu. Click Foundations and Earth Structures, Chapter 7, bc.xls to calculate bearing capacity of shallow foundations under combined loads. Select the worksheet "clay under combined load."

Taiebat and Carter (2000) proposed a failure surface for combined loading (V_n, M, and T) for a foundation resting on fine-grained soils using the finite element method. The failure surface is approximately represented by

$$\text{TSA: } f = \left(\frac{V_n}{(V_n)_u}\right)^2 + \left[\frac{M}{M_u}\left(1 - \alpha_1 \frac{TM}{T_u|M_u|}\right)\right]^2 + \left(\frac{T}{T_u}\right)^3 - 1 = 0 \tag{7.27}$$

and illustrated in Fig. 7.10. The value of α_1 is 0.3 for most practical problems.

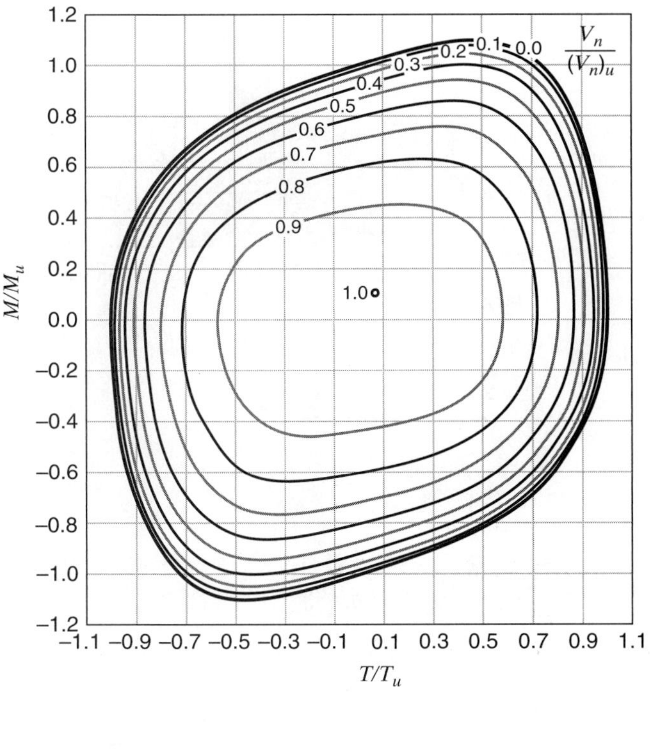

FIGURE 7.10 Approximate failure surface for combined loads (V_n, T, and M). (Source: Taiebat and Carter, 2000.)

The ultimate vertical load capacity, $(V_n)_u$, the ultimate moment capacity, M_u, and the ultimate horizontal load capacity, T_u, are:

$$(V_n)_u = 5.7\,As_u \text{ for smooth foundation} \tag{7.28}$$

$$(V_n)_u = 6\,As_u \text{ for rough foundation} \tag{7.29}$$

$$\text{Vertical load and moment only: } M_u = 0.8\,ADs_u \tag{7.30}$$

$$\text{Horizontal load and moment only: } M_u = 0.89\,ADs_u; \quad T_u = 0.71\,As_u \tag{7.31}$$

You should use factored load values of V_n, M, and T for LRFD. In the case of ASD, multiply the design loads (V_n, M, T) by the factor of safety.

7.3.2.2 Strip Foundation The failure envelope, based on a two-dimensional finite element analysis of a strip foundation resting on a fine-grained soil and subjected to a vertical load and a moment about a plane parallel to the length (Houlsby and Purzin, 1999), is:

$$\text{TSA: } f = 5.14\left(1 - \frac{2M}{V_n B}\right)s_u - \frac{V_n}{A} = 0 \tag{7.32}$$

That is,

$$\frac{V_n}{A} = 5.14\left(1 - \frac{2M}{V_n B}\right)s_u \tag{7.33}$$

where V_n and M are the factored vertical load and moment or the design vertical load and design moment multiplied by the factor of safety.

EXAMPLE 7.6 *Circular Foundation under Combined Loading*

The factored loads and moments on a circular concrete foundation 6 m (18 ft) in diameter are shown in Fig. E7.6. The unit weight of concrete is 24 kN/m^3 (150 pcf). The vertical load includes the weight of the foundation. Groundwater is 10 m (30 ft) below the surface. Determine if the size of the foundation is adequate to prevent bearing capacity failure under short-term condition.

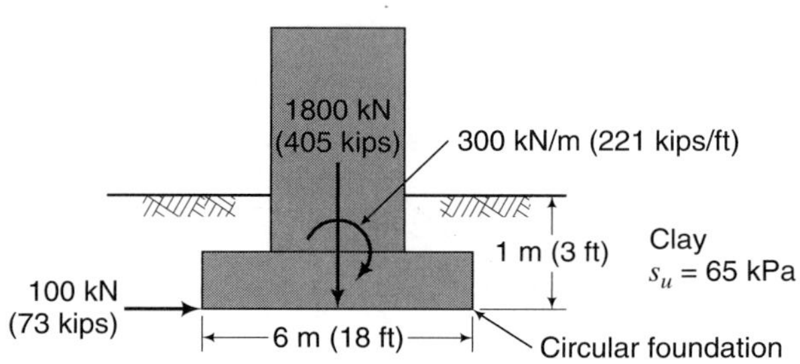

FIGURE E7.6

Strategy Calculate the M_u, T_u, and $(V_n)_u$ and then use Eq. (7.27). If $f < 0$, then foundation is safe. SI units will be used in solving this example problem.

Solution 7.6

Step 1: Find ultimate loads and moments.

$$\text{Area, } A = \frac{\pi D^2}{4} = \frac{\pi \times 6^2}{4} = 28.3\,\text{m}^2$$

$$(V_n)_u = 5.7\,As_u = 5.7 \times 28.3 \times 65 = 10485\,\text{kN}$$

$$M_u = 0.8\,ADs_u = 0.8 \times 28.3 \times 6 \times 65 = 8830\,\text{kN}\cdot\text{m}$$

$$T_u = 0.71\,As_u = 0.71 \times 28.3 \times 65 = 1306\,\text{kN}$$

Step 2: Substitute values into Eq. (7.27).

$$f = \left(\frac{V_n}{(V_n)_u}\right)^2 + \left\{\frac{M}{M_u}\left(1 - \alpha_1 \frac{TM}{T_u|M_u|}\right)\right\}^2 + \left|\left(\frac{T}{T_u}\right)^3\right| - 1 = 0$$

Assume $\alpha_1 = 0.3$

$$f = \left(\frac{1800}{10485}\right)^2 + \left\{\frac{300}{8830}\left(1 - 0.3\frac{100 \times 300}{1306|8830|}\right)\right\}^2 + \left|\left(\frac{100}{1306}\right)^3\right| - 1 = -0.97$$

Foundation safe: ($f < 0$). ■

7.4 LAYERED SOILS

Importance *The methods discussed in Section 7.3 apply strictly to homogeneous soils. However, soils are invariably mixed and deposited in layers. In this section, we discuss how to deal with layered soils.*

Shallow foundations on layered soils are difficult to analyze. The basic problem lies in determining and defining the soil properties for layered (stratified) soils and the mechanism of failure. A geotechnical engineer can estimate the bearing capacity for simple two-layered systems using equations obtained from limit equilibrium or limit analyses or results from numerical methods, or adopt a set of practical guidelines. For multiple-layered soils, it is best to use numerical tools. Numerical methods such as FEM or FDM are very useful in determining not only the bearing capacity of foundations on stratified soils but also the foundation settlements. Accurate solutions depend on experience in discretization of the foundation system, selection of an appropriate soil model, quality of soil data, and method of solution. It is customary in FEM to use higher-order elements such as 8-noded quadrilateral elements with reduced 2×2 integration scheme for accurate results. Consider a shallow foundation resting on a two-layered soil profile as shown in Fig. 7.11.

According to Prandtl's failure mechanism, if the thickness, H_{cr}, of the soil below the foundation in the top layer is greater than

$$H_{\text{crit}} = \frac{B}{2\cos\left(45^\circ + \frac{\phi'_p}{2}\right)} \exp[A \tan \phi'_p] \tag{7.34}$$

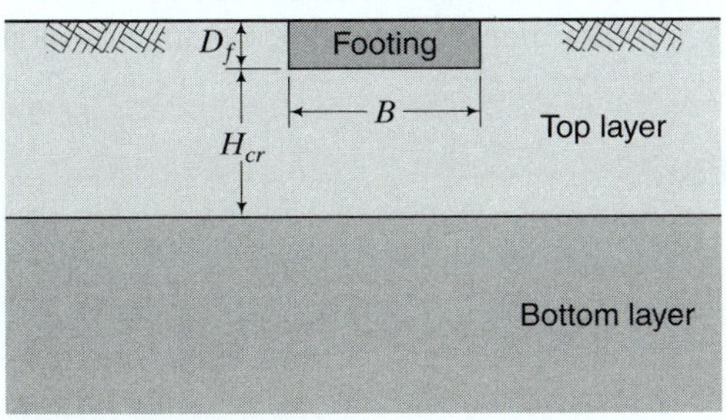

FIGURE 7.11 Footing on a two-layered soil.

where $A = (45^\circ - \frac{\phi'_p}{2})$ radians, the failure surface will be confined in the top layer and it is sufficiently accurate to calculate the bearing capacity based on the properties of the soil in the top layer. Otherwise, the failure surface would be influenced by the bottom layer and may extend into it. For example, a shallow foundation resting on a thin stiff crust followed by a layer of soft soil could punch through the crust, causing sudden failure. For fine-grained soils, the bottom layer would theoretically not influence the bearing capacity if $\frac{H_{cr}}{B} > 0.707$.

An alternative equation for H_{cr} is

$$H_{\text{crit}} = \frac{3B \ln \frac{(q_u)_{\text{top}}}{(q_u)_{\text{bot}}}}{2\left(1 + \frac{B}{L}\right)} \tag{7.35}$$

where $(q_u)_{\text{top}}$ is the ultimate net bearing capacity of the top layer assuming it is infinitely thick and $(q_u)_{\text{bot}}$ is the ultimate gross bearing capacity of the same size foundation resting on the bottom layer; both q_u values are calculated with the depth factor equal to 1.

We consider next a few cases of practical interest. These cases are intended for hand calculations and to serve as a reference for numerical methods.

7.4.1 Two-Layered Fine-Grained Soils

Access a spreadsheet at http://www.wiley.com/college/budhu. Click Foundations and Earth Structures, Chapter 7, bc.xls to calculate bearing capacity of shallow foundations on two-layered fine-grained soils. Select the worksheet "two layers of clay."

The ultimate net bearing capacity for foundations resting on the surface of a two-layered fine-grained soil (Fig. 7.12) can be written as

$$q_u = N_m (s_u)_{\text{top}} \tag{7.36}$$

where N_m is a modified bearing capacity factor and $(s_u)_{\text{top}}$ is the undrained shear strength of the top layer. We consider two cases.

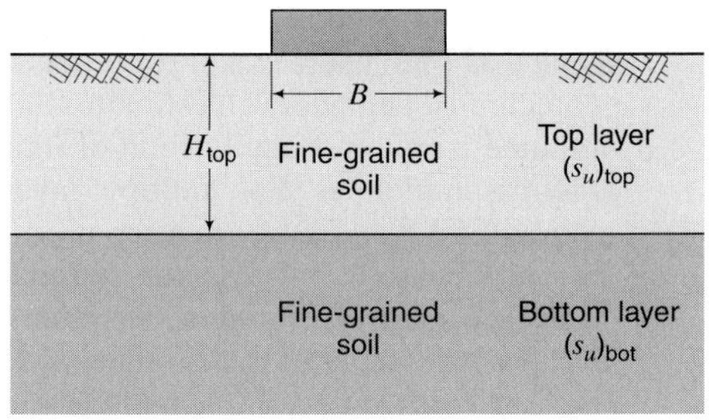

FIGURE 7.12 Two-layered fine-grained soil.

Case 1: Top Layer Stiffer than Bottom Layer Based on experimental results, Brown and Meyerhof (1969) suggested that for a strip foundation resting on a two-layered fine-grained soil with the top layer much stronger than the bottom layer, the modified bearing capacity factor is

$$\text{TSA: } N_m = 1.5\left(\frac{H_{\text{top}}}{B}\right) + 5.14\left[\frac{(s_u)_{\text{bot}}}{(s_u)_{\text{top}}}\right] \tag{7.37}$$

and for a circular foundation

$$\text{TSA: } N_m = 1.5\left(\frac{H_{\text{top}}}{B}\right) + 6.05\left[\frac{(s_u)_{\text{bot}}}{(s_u)_{\text{top}}}\right] \tag{7.38}$$

Case 2: Top Layer Weaker than Bottom Layer Vesic (1970) proposed values of N_m for a top layer overlying a stronger bottom layer. Values of N_m for rectangular, square, and circular foundations are shown in Table 7.2.

7.4.2 Coarse-Grained Soil Over Fine-Grained Soil

Access a spreadsheet at http://www.wiley.com/college/budhu. Click Foundations and Earth Structures, Chapter 7, bc.xls to calculate bearing capacity of shallow foundations on coarse-grained soil over a fine-grained soil. Select the worksheet "sand over clay."

TABLE 7.2 N_m Values for Top Weak Layer Over Bottom Strong Layer of Fine-Grained Soil

Long rectangular footing ($L/B \leq 5$)

$\frac{(s_u)_{bot}}{(s_u)_{top}}$ / B/H	2	4	6	8	10	20	∞
1.0	5.14	5.14	5.14	5.14	5.14	5.14	5.14
1.5	5.14	5.31	5.45	5.59	5.70	6.14	7.71
2	5.14	5.43	5.69	5.92	6.13	6.95	10.28
3	5.14	5.59	6.00	6.38	6.74	8.16	15.42
4	5.14	5.69	6.21	6.69	7.14	9.02	20.56
5	5.14	5.76	6.35	6.90	7.42	9.66	25.70
10	5.14	5.93	6.69	7.43	8.14	11.40	51.40
∞	5.14	6.14	7.14	8.14	9.14	14.14	∞

Square or circular footing ($L/B = 1$)

$\frac{(s_u)_{bot}}{(s_u)_{top}}$ / B/H	4	8	12	16	20	40	∞
1	6.17	6.17	6.17	6.17	6.17	6.17	6.17
1.5	6.17	6.34	6.49	6.63	6.76	7.25	9.25
2	6.17	6.46	6.73	6.98	7.20	8.10	12.34
3	6.17	6.63	7.05	7.45	7.82	9.36	18.51
4	6.17	6.73	7.26	7.75	8.23	10.24	24.68
5	6.17	6.80	7.40	7.97	8.51	10.88	30.85
10	6.17	6.96	7.74	8.49	9.22	12.58	61.70
∞	6.17	7.17	8.17	9.17	10.17	15.17	∞

SOURCE: Vesic, 1970. *H* is thickness of top weak layer.

Strip Foundation Using a failure mechanism (Fig. 7.13) from centrifuge model tests and the limit equilibrium method, Okimura et al. (1998) proposed the following equation for coarse-grained soils overlying fine-grained soils:

$$q_u = \left(1 + \frac{H}{B}\tan\varpi\right)(5.14s_u + \gamma D_f) + \left(\frac{K_p \sin(\phi'_p - \varpi)}{\cos\phi'_p \cos\varpi}\right)\left(\frac{H}{B}\gamma D_f\right) - \gamma D_f\left(1 + \frac{H}{B}\tan\varpi\right) \tag{7.39}$$

where

$$\varpi = \tan^{-1}\left[\frac{m_1 - m_2(1 + \sin^2\phi'_p)}{m_2 \cos\phi'_p \sin\phi'_p + 1}\right] \tag{7.40}$$

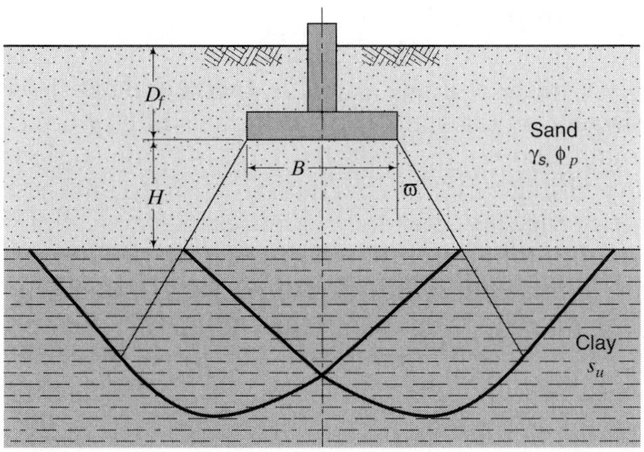

FIGURE 7.13 Mechanism assumed by Okimura et al. (1998).

$$m_1 = 5.14s_u\left(1 + \frac{1}{\lambda_c}\frac{H}{B} + \frac{\lambda_p}{\lambda_c}\right); \quad m_2 = \frac{m_1 - \sqrt{(m_1)^2 - \cos^2\phi'_p[(m_1)^2 + 1]}}{\cos^2\phi'_p}; \tag{7.41}$$

$$\lambda_p = \tfrac{D_f}{B} \text{ and } \lambda_c = \frac{5.14s_u}{\gamma B}.$$

Circular Foundation Okimura et al. (1998) proposed the following equation for circular foundation.

$$\begin{aligned} q_u = {} & \left(1 + 2\frac{H}{B}\tan\varpi\right)^2 (6.17s_u + \gamma\{D_f + H\}) + \left(\frac{4K_p\sin(\phi'_p - \varpi)}{\cos\phi'_p\cos\varpi}\right) \\ & \times \left[\frac{H}{B}\left(\gamma D_f + \frac{\gamma H}{2}\right) + \gamma D_f\tan\varpi\left(\frac{H}{B}\right)^2 + \frac{2}{3}\gamma H\tan\varpi\left(\frac{H}{B}\right)^2\right] \\ & - \frac{\gamma H}{3}\left[4\left(\frac{H}{B}\right)^2\tan^2\varpi + 6\frac{H}{B}\tan\varpi + 3\right] \end{aligned} \tag{7.42}$$

The equations are satisfactory provided $\lambda_p < 4.8$ and $\lambda_c > 26$. You must also calculate the bearing capacity for the sand as a single layer, assuming it is infinitely thick, and the lesser bearing capacity value should be used.

7.4.3 Practical Guidelines for Layered Soils

Numerical methods, particularly the FEM and the FDM, are suitable to analyze layered soils. In the absence of these methods, you can resort to some practical guidelines for three common cases, a soft clay over a stiff clay, a stiff clay over a soft clay, and thinly stratified soils.

Soft Clay Over Stiff Clay In general, shallow (individual or isolated) foundations on soft clays should be avoided, except lightly loaded structures such as houses and one-story buildings. You should investigate removing the soft clay and replacing it with compacted fills. An inexperienced geotechnical engineer should calculate the bearing capacity using the methods described previously before making a decision to replace the soft clay.

Stiff Clay Over Soft Clay The bearing capacity for this case is the smaller value of:

1. Treating the stiff clay as if the soft clay layer does not exist, or
2. Assuming that the foundation punches through the stiff clay and is supported on the soft clay.

The bearing capacity is the sum of the shear required to punch through a vertical plane in the stiff clay and the bearing capacity of the soft clay layer. Only a fraction, about $\frac{2}{3}$ to $\frac{1}{2}$, of the undrained shear strength should be used in computing the shear resistance on the vertical plane in the stiff clay layer.

Another method is to place an imaginary foundation on the soft clay layer with dimensions $(B + t_{sc}) \times (L + t_{sc})$, where B and L are the real width and length of the foundation and t_{sc} is the thickness of the stiff layer, and then calculate the ultimate bearing capacity using the bearing capacity equations.

Stratified Soils In stratified soils with weak layers, especially if the layers are dipping, mat or deep foundations should be used. If either mat or deep foundations are uneconomical, then the bearing capacity can be calculated by using the shear strength parameters for the weakest layer. Alternatively, harmonic mean values for s_u and ϕ' can be calculated and then these values can used to calculate the bearing capacity.

7.5 SOIL STRENGTH VARYING WITH DEPTH

> Access a spreadsheet at http://www.wiley.com/college/budhu. Click Foundations and Earth Structures, Chapter 7, bc.xls to calculate bearing capacity of shallow foundations on coarse-grained soil over a fine-grained soil. Select the worksheets "clay with linear strength" and "clay with crust."

Davis and Booker (1973) used the method of stress characteristics to derive the bearing capacity of a strip foundation resting on a fine-grained soil (1) with shear strength varying linearly with depth and (2) with a crust of uniform shear strength followed by a linear variation and $(s_u)_o$ is the undrained shear strength at the surface of strength with depth. The ultimate net bearing capacity is given as

$$q_u = F\left[5.14(s_u)_o + \frac{mB}{4}\right] \tag{7.43}$$

where F (Fig. 7.14) is a function of the roughness of the foundation and the strength ratio $\frac{mB}{(s_u)_o}$ or $\frac{(s_u)_o}{mB}$, m is the slope of the shear strength versus depth variation and $(s_u)_o$ is the undrained shear strength at the surface. For circular foundation the ultimate bearing capacity can be approximated as

$$q_u = F\left[6(s_u)_o + \frac{mB}{4}\right] \tag{7.44}$$

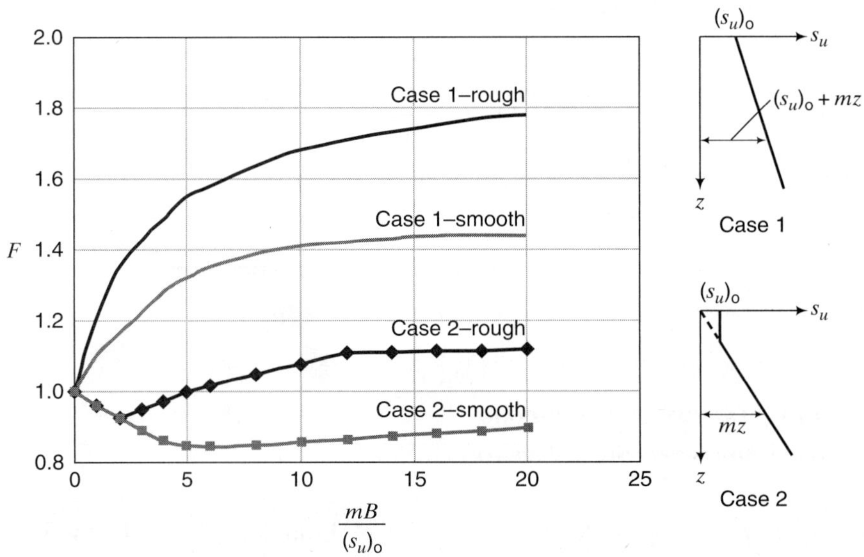

FIGURE 7.14 Values of F for footing on a soil with (a) case 1—undrained shear strength varying linearly with depth and (b) case 2—a top crust of uniform undrained shear strength followed by a linear variation of undrained shear strength with depth (replotted from Davis and Booker, 1973).

EXAMPLE 7.7 *Foundation on a Stiff Clay Above a Soft Clay*

A soil profile consists of a top crust 3 m (10 ft) thick followed by a very thick [>10 m (30 ft)] layer. The undrained shear strength of the crust is 90 kPa (1800 psf) and 30 kPa (600 psf) for the soft layer. A strip foundation, 2 m (6 ft) wide, is located at 1 m (3 ft) in the crust. Estimate the ultimate net bearing capacity.

Strategy This is a two-layer system with the top layer stiffer than the bottom layer. Use Eqs. (7.36) and (7.37). U.S. units will be used in solving this example problem.

Solution 7.7

Step 1: Calculate N_m.

$$N_m = 1.5\left(\frac{H_{top}}{B}\right) + 5.14\left[\frac{(s_u)_{bot}}{(s_u)_{top}}\right]$$

$$H_{top} = 10\,\text{ft} - 3\,\text{ft} = 7\,\text{ft}$$

$$N_m = 1.5\left(\frac{7}{6}\right) + 5.14\left(\frac{600}{1800}\right) = 3.5$$

Step 2: Calculate the net ultimate bearing capacity.

$$q_u = N_m (s_u)_{top} = 3.5 \times 1800 = 6300\,\text{psf}$$

■

7.6 PRESUMPTIVE ALLOWABLE BEARING CAPACITY

Building codes usually provide recommended values of bearing capacity for local conditions. You can use these values for preliminary design but you should check these values using soil test data and the analytical methods described previously. Table 7.3 shows the allowable bearing capacity values for general soil types (IBC, 2006).

TABLE 7.3 Allowable Bearing Capacities and Friction Coefficient

	[1]Allowable bearing capacity, q_a—kPa (psf)	Lateral[2] bearing capacity—kPa (psf)	Friction coefficient
Sandy gravel/gravel (GW, GP)	144 (3000)	9.6 (200)	0.35
Sand, silty sand, clayey sand, silty gravel (SW, SP, SM, SC, GM, GC)	96 (2000)	7.2 (150)	0.35
Clay, sandy clay, silty clay, and clayey silt (CL, ML, MH, CH)	72 (1500)	4.8 (100)	Lateral resistance 6.2 kPa (130 psf)

[1]Increase by 1/3 for load combinations that include wind and/or seismic loads.
[2]Values are for grade level. Increase by 1/3 for load combinations that include wind and/or seismic loads.
SOURCE: IBC 2006.

7.7 SETTLEMENT OF SHALLOW FOUNDATIONS

Access a spreadsheet at http://www.wiley.com/college/budhu. Click Foundations and Earth Structures, Chapter 7, bc.xls to calculate settlement of shallow foundations. Select the worksheet "settlement."

Importance *It is practically impossible to prevent settlement of foundations. At least, elastic settlement will occur. Your task as a geotechnical engineer is to prevent the foundation system from reaching a serviceability limit state. The design of most shallow foundations is controlled by serviceability limit state rather than by ultimate limit state. However, both limit states must be satisfied. In this section, we will deal with settlement of shallow foundations.*

The settlement of shallow foundations is divided into three time segments: immediate or elastic settlement, primary consolidation settlement, and secondary compression (creep). We will discuss each of these in the following.

7.7.1 Immediate Settlement

We can use the theory of elasticity to determine the immediate or elastic settlement of shallow foundations (Chapter 2). But the elastic equations do not account for shape of the foundation and the depth of embedment. An embedded foundation has the following effects in comparison with a surface foundation.

- Soil stiffness generally increases with depth, so the foundation loads will be transmitted to a stiffer soil than the surface soil. This can result in smaller settlement.
- Normal stresses from the soil above the foundation have been shown (Eden, 1974; Gazetas and Stokoe, 1991) to reduce the settlement by providing increased confinement on the deforming half space. This is called the *trench effect* or *embedment effect.*
- Part of the load on the foundation is transmitted to the side walls depending on the amount of shear resistance mobilized at the soil–wall interface. The accommodation of part of the load by side resistance reduces the vertical settlement. This is called the *side–wall contact effect.*

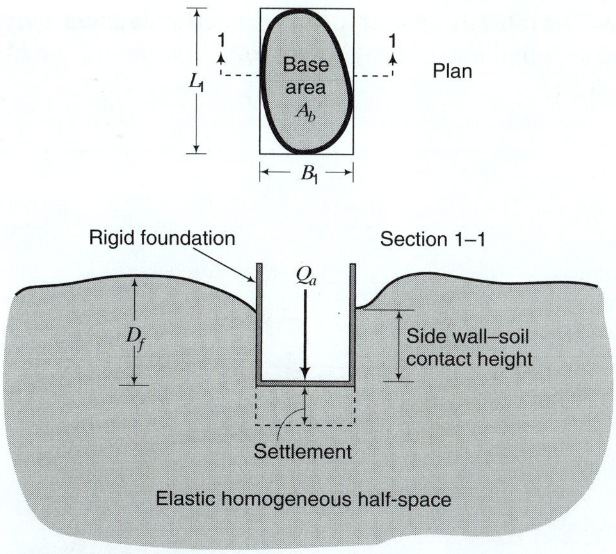

FIGURE 7.15 Elastic settlement of shallow footings—geometry (after Gazetas et al., 1985).

Gazetas et al. (1985) considered an arbitrarily shaped rigid foundation embedded in a deep homogeneous soil (Fig. 7.15) and proposed the following equation for the elastic settlement:

$$\rho_e = \frac{2Q_a}{EL_1}\left(1 - \nu_u^2\right)\mu_s\mu_{emb}\mu_{wall} \qquad (7.45)$$

where Q_a is allowable (design) vertical load, E is the elastic modulus ($E = E_u$ for fine-grained soils, $E = E'$ for coarse-grained soils; the subscript u denotes undrained load condition), L_1 is the length of a circumscribed rectangle, ν_u = Poisson's ratio for undrained condition, and μ_s, μ_{emb}, and μ_{wall} are shape, embedment, and wall adhesion factors given as:

$$\mu_s = 0.45\left(\frac{A_b}{{L_1}^2}\right)^{-0.38} \qquad (7.46)$$

$$\mu_{emb} = 1 - 0.08\frac{D_f}{B_1}\left\{1 + \frac{4}{3}\left(\frac{A_b}{L_1^2}\right)\right\} \qquad (7.47)$$

$$\mu_{wall} = 1 - 0.16\left(\frac{A_w}{A_b}\right)^{0.54} \qquad (7.48)$$

where A_b is the actual area of the base of the foundation, A_w is the actual area of wall in contact with the embedded portion of the foundation, and B_1 is the width of a circumscribed rectangle. The values of the dimensionless shape parameter, $\frac{A_b}{L_1^2}$, for common foundation geometry are shown in Table 7.4.

TABLE 7.4 Values of $\frac{A_b}{L_1^2}$ for common shapes of footings

Footing shape	$\frac{A_b}{L_1^2}$
Square	1
Rectangle	B/L
Circle	0.785
Strip	0

The equations proposed by Gazetas et al. (1985) apply to a foundation of arbitrary shape on a deep homogeneous soil. There is no clear definition of what *deep* signifies. The author suggests that Gazetas et al. (1985) equations be used when the thickness of the soil layer just beneath the foundation base is such that 90% of the applied stresses are distributed within it.

The accuracy of settlement predictions from elastic equation depends particularly on the accuracy of the elastic modulus. It is common laboratory practice to determine a secant E_u from undrained triaxial tests or unconfined compression tests at a deviatoric stress equal to one-half the maximum shear strength. However, for immediate settlement it is better to determine E_u over the range of deviatoric stress pertaining to the problem. In addition, the elastic modulus is strongly dependent on depth and Eq. (7.45) is cast in terms of a single value E_u. One possible solution is to divide the soil into sublayers and use a weighted harmonic mean value of E.

The full wall resistance will be mobilized only if sufficient settlement occurs. It is difficult to ascertain the quality of the soil–wall adhesion. Consequently, you should be cautious in relying on the reduction of settlement resulting from the wall factor. If wall friction and embedment are neglected, then $\mu_{wall} = 1$ and $\mu_{emb} = 1$. Equation (7.45) strictly applies to fine-grained soils under short-term loading. For long-term loading in fine-grained soils and for coarse-grained soils, you should use E' and ν' (~0.3 to 0.35 for most coarse-grained soils).

EXAMPLE 7.8 *Immediate Settlement of a Rectangular Foundation*

Determine the immediate settlement of a rectangular foundation 4 m (12 ft) × 6 m (18 ft) embedded in a deep deposit of homogeneous clay as shown in Fig. E7.8. Assume wall friction is negligible.

Strategy You have sufficient information to directly apply Eq. (7.45). U.S. units will be used in solving this example problem.

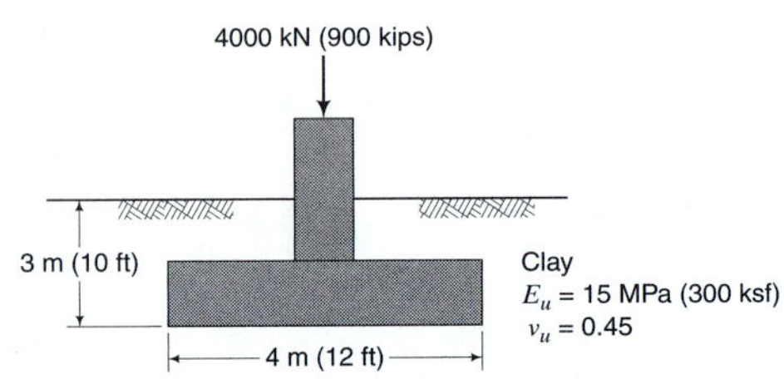

FIGURE E7.8

Solution 7.8

Step 1: Determine geometric parameters.

$$L_1 = 18\,\text{ft},\ B_1 = 12\,\text{ft};\ A_b = 12 \times 18 = 216\,\text{ft}^2;$$

Step 2: Calculate the shape and embedment factors.

$$\frac{A_b}{L_1^2} = 0.67;\ \mu_s = 0.45\left(\frac{A_b}{L_1^2}\right)^{-0.38} = 0.45(0.67)^{-0.38} = 0.52$$

$$\mu_{emb} = 1 - 0.08\frac{D_f}{B_1}\left\{1 + \frac{4}{3}\left(\frac{A_b}{L_1^2}\right)\right\} = 1 - 0.08 \times \frac{10}{12}\left\{1 + \frac{4}{3}(0.67)\right\} = 0.92$$

Step 3: Calculate immediate settlement.

$$\rho_e = \frac{2Q_a}{E_u L_1}\left(1 - \nu_u^2\right)\mu_s \mu_{emb} \mu_{wall} = \frac{2 \times 900}{300 \times 18}(1 - 0.45^2) \times 0.52 \times 0.92 \times 1 = 0.13\,\text{ft} = 1.6\,\text{in.}$$

■

EXAMPLE 7.9 *Immediate Settlement of an Irregular Foundation*

Determine the immediate settlement of the foundation shown in Fig. E7.9. The undrained elastic modulus varies with depth as shown in the figure and $\nu_u = 0.45$.

Strategy You have to determine the length L and width B of a circumscribed rectangle. The undrained elastic modulus varies with depth, so you need to consider the average value of E_u for each of the layers and then find the harmonic mean. You also need to find the shape parameter $\frac{A_b}{L^2}$.

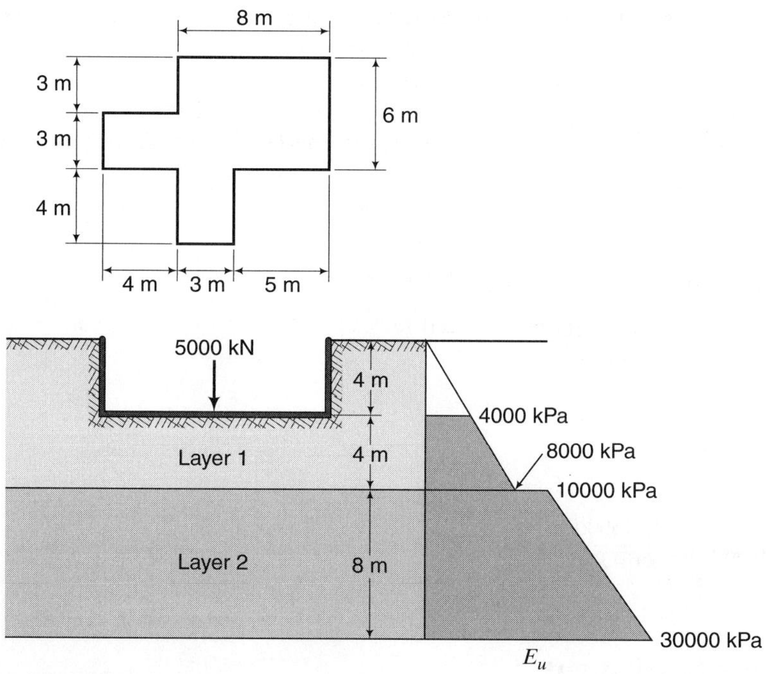

FIGURE E7.9

Solution 7.9

Step 1: Length and width of circumscribed rectangle.

$$L_1 = 8 + 4 = 12\,\text{m}; \quad B_1 = 3 + 3 + 4 = 10\,\text{m}$$

Step 2: Determine E_u.

Layer 1

$$E_u \text{ at base level} = \frac{4}{8} \times 8000 = 4000\,\text{kPa}, \; E_u \text{ at bottom of layer} = 8000\,\text{kPa}$$

$$(E_u)_{avg} = \frac{4000 + 8000}{2} = 6000\,\text{kPa}$$

Layer 2

$$E_u \text{ at top of layer} = 10{,}000\,\text{kPa},\ E_u \text{ at bottom of layer} = 30{,}000\,\text{kPa}$$

$$(E_u)_{avg} = \frac{10{,}000 + 30{,}000}{2} = 20{,}000\,\text{kPa}$$

Step 3: Find weighted harmonic mean E_u.

$$E_u = \frac{2(6000) + 1(20{,}000)}{3} = 10{,}667\,\text{kPa}$$

Step 4: Find shape parameter $\frac{A_b}{L_1^2}$.

$$A_b = (3 \times 4) + (8 \times 6) + (3 \times 4) = 72\,\text{m}^2, \quad \frac{A_b}{L_1^2} = \frac{72}{12^2} = 0.5$$

Step 5: Find shape, embedment, and wall factors.

$$\mu_s = 0.45(0.5)^{-0.38} = 0.59, \quad \mu_{emb} = 1 - 0.08\frac{4}{10}\left(1 + \frac{4}{3} \times 0.5\right) = 0.94$$

$$A_w = \text{perimeter} \times \text{depth} = (3 + 4 + 3 + 4 + 4 + 3 + 4 + 5 + 6 + 8) \times 4 = 176\,\text{m}$$

$$\frac{A_w}{A_b} = \frac{172}{72} = 2.44, \quad \mu_{wall} = 1 - 0.16(2.44)^{0.54} = 0.74$$

Step 6: Calculate immediate settlement.

$$\rho_e = \frac{2Q_a}{E_u L_1}\left(1 - \nu_u^2\right)\mu_s\mu_{emb}\mu_{wall} = \frac{2 \times 5000}{10667 \times 12}(1 - 0.45^2) \times 0.59 \times 0.94 \times 0.74 = 0.026\,\text{m} = 26\,\text{mm}.$$

■

7.7.2 Primary Consolidation Settlement

Primary consolidation settlement for fine-grained soils is normally estimated from parameters deduced from the one-dimensional consolidation test. In using the parameters from the one-dimensional consolidation test, you are assuming (1) vertical uniform strain only (lateral strain is zero), (2) no settlement occurs from shear stresses, (3) saturated soil, (4) the initial excess porewater pressure is equal to the change in applied stress at the instant the load is applied, and (5) excess porewater pressures are dissipated only vertically. In practice, lateral strains are significant except when the ratio of the layer thickness to lateral dimension of the loaded area is small (approaches zero). Shear stresses also cause settlement and excess porewater pressures can dissipate not only in the vertical but in any direction.

The initial excess porewater pressure is equal to the change in vertical stress at the instant the vertical stress is applied is possible only if the lateral stresses are equal to the vertical stresses. If the lateral strains are zero, then under undrained conditions (at the instant the load is applied), the vertical settlement is zero.

Normally Consolidated Clays The primary consolidation settlement for normally consolidated soils (OCR = 1) occurs solely along the compression line (NCL) and is given as

$$\rho_{pc} = H\frac{\Delta e}{1+e_o} = \frac{H}{1+e_o}C_c \log\frac{\sigma'_{fin}}{\sigma'_{zo}}; \quad \text{OCR} = 1 \tag{7.49}$$

where H is the thickness of the soil layer, $\Delta e = C_c \log\frac{\sigma'_{fin}}{\sigma'_{zo}}$, C_c is the compression index, $e_o(e_o = wG_s)$ is the initial void load ratio, $\sigma'_{fin} = \sigma'_{zo} + \Delta\sigma_z$, σ'_{zc} is the preconsolidation vertical effective stress, and $\Delta\sigma_z$ is the increase in vertical stress at the center of the layer.

Overconsolidated Clays If the soil were overconsolidated, we would have to consider two cases depending on the magnitude of $\Delta\sigma_z$ (Fig. 7.16). In the first case (Fig. 7.16a), the increase in $\Delta\sigma_z$ is such that $\sigma'_{fin} = \sigma'_{zo} + \Delta\sigma_z$ is less than the preconsolidation vertical effective stress, σ'_{zc}. In this case, consolidation occurs along the *unloading reloading line* (URL), and

$$\rho_{pc} = \frac{H}{1+e_o}C_r \log\frac{\sigma'_{fin}}{\sigma'_{zo}}; \quad \sigma'_{fin} < \sigma'_{zc} \tag{7.50}$$

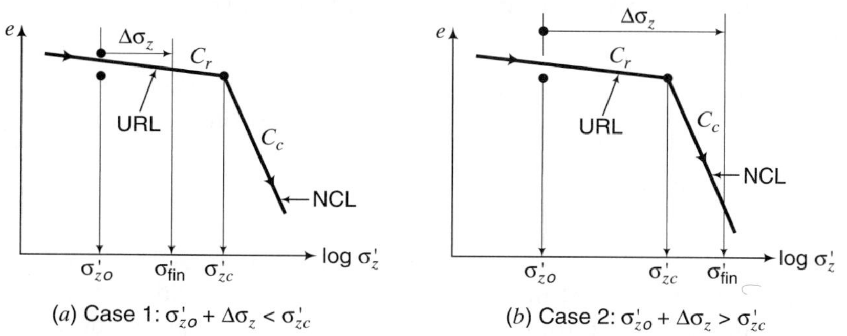

FIGURE 7.16 Two cases to consider for calculating primary consolidation settlement of overconsolidated soils.

In case 2 (Fig. 7.16b), the increase in $\Delta\sigma_z$ is such that $\sigma'_{fin} = \sigma'_{zo} + \Delta\sigma'_z$ is greater than σ'_{zc}. In this case, we have to consider two components of settlement—one along the URL and the other along the NCL. The primary consolidation for this case is

$$\rho_{pc} = \frac{H}{1+e_o}\left(C_r \log\frac{\sigma'_{zc}}{\sigma'_{zo}} + C_c \log\frac{\sigma'_{fin}}{\sigma'_{zc}}\right); \quad \sigma'_{fin} > \sigma'_{zc} \tag{7.51}$$

or

$$\rho_{pc} = \frac{H}{1+e_o}\left\{C_r \log(\text{OCR}) + C_c \log\frac{\sigma'_{fin}}{\sigma'_{zc}}\right\}; \quad \sigma'_{fin} > \sigma'_{zc} \tag{7.52}$$

Primary Consolidation Settlement Using m_v You can also calculate the primary consolidation settlement using the modulus of compressibility, m_v. However, unlike C_c, which is constant, m_v varies with vertical effective stress levels. You should compute an average value of m_v over the stress range σ'_{zo}

to σ'_{fin}. The vertical effective stress difference over which an m_v value is calculated should not exceed 100 kPa because of nonlinearity. If $\sigma'_{\text{fin}} - \sigma'_{zo} > 100$ kPa you can find the values of m_v for each difference of vertical effective stress less than or equal to 100 kPa from σ'_{zo} to σ'_{fin}. You can then take the harmonic mean value of m_v. The primary consolidation settlement, using m_v, is

$$\rho_{pc} = Hm_v\Delta\sigma_z \tag{7.53}$$

The advantage of using Eq. (7.53) is that m_v is readily determined from settlement data in consolidation tests. You do not have to calculate void ratio changes from the test data as required for C_c. Replace m_v by m_{vr} for overconsolidated fine-grained soils (OCR > 1).

7.7.3 Secondary Compression (Creep)

The secondary compression is estimated from

$$\rho_{sc} = \frac{H}{(1+e_p)} C_\alpha \log\left(\frac{t}{t_p}\right) \tag{7.54}$$

where (t_p, e_p) is the coordinate at the intersection of the tangents to the primary consolidation and secondary compression parts of the void ratio verses logarithm of time curve, C_α is secondary compression index (Chapter 4), and t is time after primary consolidation. Overconsolidated soils do not creep significantly but creep settlements in normally consolidated and sensitive soils can be significant.

7.7.4 Modification to One-dimensional Consolidation Settlement to Account for Lateral Stresses

Skempton and Bjerrum (1957) proposed a method to modify the one-dimensional consolidation equation to account for lateral stresses but not lateral strains. They proposed the following equation.

$$(\rho_{pc})_{\text{SB}} = \int_{\text{O}}^{H} m_v \Delta u_{\text{o}}\, dz \tag{7.55}$$

where Δu_{o} is the initial porewater pressure. Skempton and Bjerrum (1957) suggested that neglecting lateral strains could lead to an error of up to 20% in the estimation of the consolidation settlement. Skempton's porewater pressure parameter, $B = 1$, for saturated soils and under triaxial conditions (axial symmetry) is:

$$\Delta u = \Delta\sigma_1\left[A + \frac{\Delta\sigma_3}{\Delta\sigma_1}(1-A)\right] \tag{7.56}$$

By substituting Eq. (7.56) into Eq. (7.55) we get

$$(\rho_{pc})_{\text{SB}} = \int_{\text{O}}^{H} m_v \Delta\sigma_1\left[A + \frac{\Delta\sigma_3}{\Delta\sigma_1}(1-A)\right] dz = \Sigma(m_v\Delta\sigma_z H)\mu_{\text{SB}} = \Sigma\rho_{pc}\mu_{\text{SB}} \tag{7.57}$$

where ρ_{pc} is the one-dimensional primary consolidation settlement, $\mu_{\text{SB}} = A + \alpha_{\text{SB}}(1-A)$ is a settlement coefficient to account for the effects of the lateral stresses, and $\alpha_{\text{SB}} = \frac{\int \Delta\sigma_3 dz}{\int \Delta\sigma_1 dz}$. The values of μ_{SB} typically vary from 0.6 to 1.0 for soft clays and from 0.3 to 0.8 for overconsolidated clays. Values of μ_{SB} for circular and strip foundations are shown in Fig 7.17. For square or rectangular foundations use an equivalent circular foundation of diameter $D = 2\sqrt{\frac{A}{\pi}}$, where A is area of the rectangle or the square. Equation (7.57) applies only to situations where axial symmetry occurs, such as under the center of a circular foundation.

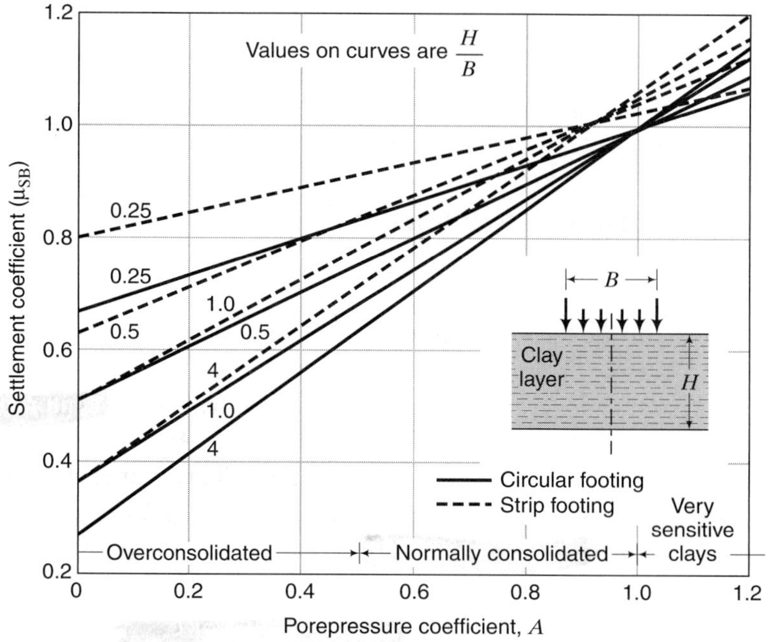

FIGURE 7.17 Values of μ_{SB} for circular and strip footing (redrawn after Scott, 1963).

7.7.5 Time Rate of Settlement

The time rate of settlement for one-dimensional consolidation can be calculated from

$$t = \frac{T_v H_{dr}^2}{C_v} \tag{7.58}$$

where t is time, H_{dr} is the length of the drainage path ($H_{dr} = \frac{H}{2}$ for double drainage, $H_{dr} = H$ for single drainage), C_v is the coefficient of consolidation, and T_v is time factor that can be calculated for double drainage condition, in the case of uniform distribution of initial excess porewater pressures, from

$$T_v = \frac{\pi}{4}\left(\frac{U}{100}\right)^2 \text{ for } U < 60\% \tag{7.59}$$

and

$$T_v = 1.781 - 0.933\log(100 - U) \text{ for } U \geq 60\% \tag{7.60}$$

where $U = \dfrac{\text{Settlement at time, } t}{\text{Total primary consolidation settlement}}$ is the average degree of consolidation.

7.7.6 Thick Layers

For better accuracy, when dealing with thick layers, $H > 2$ m (6 ft), you should divide the soil layer into sublayers (about 2 to 5 sublayers) and find the settlement for each sublayer. Add up the settlement of each sublayer to find the total primary consolidation settlement. You must remember that the value of H in the primary consolidation equations is the thickness of the sublayer. An alternative method is to use a harmonic mean value of the vertical stress increase for the sublayers in the equations for primary consolidation settlement. The harmonic mean stress increase is

$$\Delta\sigma_z = \frac{n(\Delta\sigma_z)_1 + (n-1)(\Delta\sigma_z)_2 + (n-2)(\Delta\sigma_z)_3 + \cdots + (\Delta\sigma_z)_n}{n + (n-1) + (n-2) + \cdots + 1} \tag{7.61}$$

where n is the number of sublayers and the subscripts $1, 2, \ldots$ mean the first (top) layer, the second layer, and so on. The advantage of using the harmonic mean is that the settlement is skewed in favor of the upper part of the soil layer. You should recall from Chapter 2 that the increase in vertical stress decreases with depth. Therefore, the primary consolidation settlement of the upper portion of a soil layer can be expected to be more than the lower portion because it is subjected to higher vertical stress increases.

7.7.7 Procedure to Calculate Consolidation Settlement

The procedure to calculate consolidation settlement is as follows:

1. If the soil layer is thick, $H > 2\,\text{m}(6\,\text{ft})$, divide the soil layer into convenient sublayers of say 1 m (3 ft) thickness or less.
2. Calculate the initial vertical effective stress (σ'_{zo}) and the initial void ratio (e_o) at the center of the soil layer or each sublayer.
3. Calculate the applied vertical stress increase $(\Delta\sigma_z)$ at the center of the soil layer or each sublayer using the appropriate method in Chapter 2.
4. Calculate the final vertical effective stress $\sigma'_{\text{fin}} = \sigma'_{zo} + \Delta\sigma_z$.
5. Calculate the elastic settlement using Eq. (7.45).
6. Calculate the primary consolidation settlement using the corrected value of C_c.
   - If the soil is normally consolidated (OCR = 1), the primary consolidation settlement is

$$\rho_{pc} = \frac{H}{1+e_o} C_c \log \frac{\sigma'_{\text{fin}}}{\sigma'_{zo}}$$

   - If the soil is overconsolidated and $\sigma'_{\text{fin}} < \sigma'_{zc}$, the primary consolidation settlement is

$$\rho_{pc} = \frac{H}{1+e_o} C_r \log \frac{\sigma'_{\text{fin}}}{\sigma'_{zo}}$$

   - If the soil is overconsolidated and $\sigma'_{\text{fin}} > \sigma'_{zc}$, the primary consolidation settlement is

$$\rho_{pc} = \frac{H}{1+e_o} \left\{ C_r \log(\text{OCR}) + C_c \log \frac{\sigma'_{\text{fin}}}{\sigma'_{zc}} \right\}$$

   - Alternatively, you can calculate the primary consolidation settlement using $\rho_{pc} = m_v \Delta\sigma_z H$ for the vertical stress increases within the normal consolidation line (OCR = 1) and $\rho_{pc} = m_{vr} \Delta\sigma_z H$ for vertical stress increases within the unloading/reloading line (OCR > 1).
7. To account for lateral stresses, extract the modifying factor μ_{SB} from Fig. 7.17 and correct the primary consolidation settlement as $(\rho_{pc})_{SB} = \mu_{SB}\rho_{pc}$.
8. For normally consolidated (OCR = 1) or sensitive soils, calculate the secondary compression or creep settlement using Eq. (7.54).
9. Add the elastic settlement, the corrected primary consolidation settlement, and the secondary compression to get the total settlement.

EXAMPLE 7.10 ***Primary Consolidation Settlement***

The soil profile at a site for a proposed office building consists of a layer of fine sand 11 m (34 ft) thick above a layer of soft normally consolidated clay 2 m (6 ft) thick (Fig. E7.10). Below the soft clay is a deposit of coarse sand. The groundwater table was observed at 3 m (10 ft) below ground level. The void ratio of the sand is 0.76 and the water content of the clay is 43%. One of the individual square foundations, 4 m (12 ft) wide, will apply a vertical stress of 50 kPa (1000 psf) at the middle of the clay layer. Considering the clay as single layer, (1) estimate the primary consolidation settlement of the clay without the Skempton and Bjerrum (1957) correction; (2) estimate the time for 50% primary consolidation settlement. Assume the soil above the water table to be saturated, $C_c = 0.3$, $C_v = 0.05\ \text{m}^2/\text{year}$ ($0.5\ \text{ft}^2/\text{year}$), and $G_s = 2.7$.

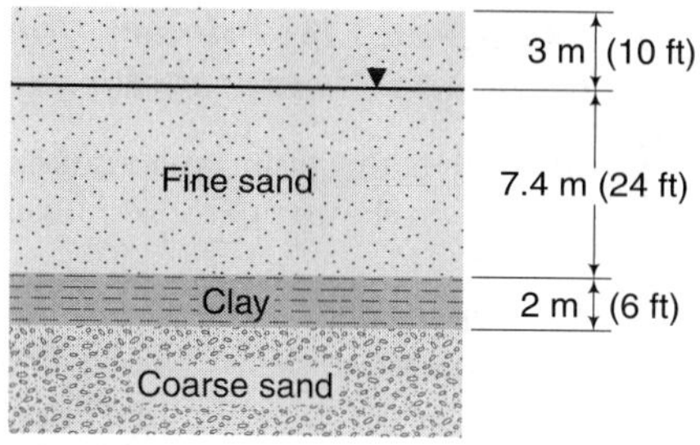

FIGURE E7.10

Strategy U.S. units will be used in solving this example problem. You should write down what is given or known and draw a diagram of the soil profile. In this problem, you are given the stratigraphy, groundwater level, stress increase, and the following soil parameters and soil condition: e for sand = 0.76; w for clay = 43%; sand is saturated, normally consolidated saturated clay, $H = 6$ ft, $\Delta\sigma_z = 1000$ psf, $C_c = 0.3$, $G_s = 2.7$. Since you are given a normally consolidated clay (OCR = 1), use Eq. (7.49). Because the clay is sandwiched between two coarse-grained soils, you can assume double drainage.

Solution 7.10

Step 1: Calculate σ'_{zo} and e_o at the center of the clay layer.

$$\text{Sand}: \gamma_{sat} = \left(\frac{G_s + e}{1+e}\right)\gamma_w = \left(\frac{2.7 + 0.76}{1 + 0.76}\right)62.4 = 122.7\ \text{pcf}$$

$$\gamma_{sat} - \gamma_w = 122.7 - 62.4 = 60.3\ \text{pcf}$$

$$\text{Clay}: e_o = wG_s = 2.7 \times 0.43 = 1.16;\ \gamma' = \left(\frac{G_s - 1}{1+e}\right)\gamma_w = \left(\frac{2.7 - 1}{1 + 1.16}\right)62.4 = 49.1\ \text{pcf}$$

The vertical effective stress at the mid-depth of the clay layer is

$$\sigma'_{zo} = (122.7 \times 10) + (60.3 \times 24) + (49.1 \times 3) = 2822\ \text{psf}$$

Step 2: Calculate the increase of vertical stress at the mid-depth of the clay layer. You do not need to calculate $\Delta\sigma_z$ for this problem. It is given as $\Delta\sigma_z = 1000$ psf.

Step 3: Calculate σ'_{fin}.

$$\sigma'_{fin} = \sigma'_{zo} + \Delta\sigma_z = 2822 + 1000 = 3822\ \text{psf}$$

Step 4: Calculate the primary consolidation settlement.

$$\rho_{pc} = \frac{H}{1 + e_o} C_c \log\frac{\sigma'_{fin}}{\sigma'_{zo}} = \frac{6}{1 + 1.16} \times 0.3 \log\frac{3822}{2822} = 0.11\ \text{ft} = 1.3\ \text{in.}$$

Step 5: Calculate time for 50% primary consolidation settlement.
The average degree of consolidation (% primary consolidation) = 50% < 60%.

$$T_v = \frac{\pi}{4}\left(\frac{U}{100}\right)^2 \text{ for } U < 60\% = \frac{\pi}{4}\left(\frac{50}{100}\right)^2 = 0.196$$

Since the top and bottom of the clay layer consist of coarse-grained soils, we will assume they are clean and free-draining. Thus, $H_{dr} = 6/2 = 3$ ft.

$$t = \frac{T_v H_{dr}^2}{C_v} = \frac{0.196 \times 3^2}{0.5} = 3.53\ \text{years}$$

If one of the drainage boundaries becomes blocked, the time for the same degree of consolidation would quadruple because H_{dr} would now be 6 ft.

■

EXAMPLE 7.11 *Primary Consolidation Settlement Using the Skempton-Bjerrum Method*

Determine the primary consolidation settlement under the square foundation shown in Fig. E7.11 using the Skempton-Bjerrum method.

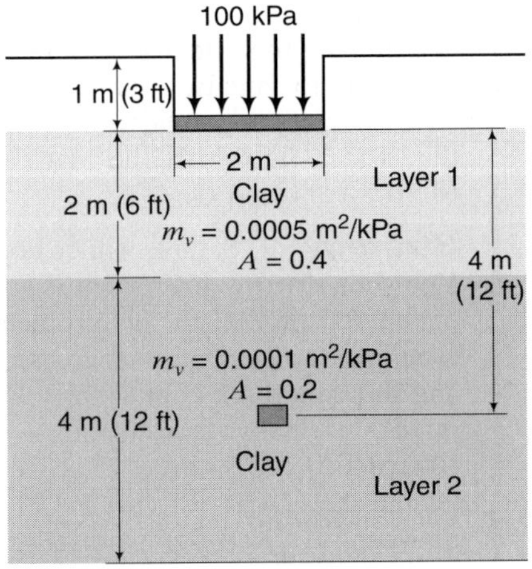

FIGURE E7.11

Strategy You have to first calculate the one-dimensional primary consolidation settlement and then determine μ_{SB} from Fig. 7.17. Treat each soil layer separately and then add the results. SI units will be used in solving this example problem.

Solution 7.11

Step 1: Calculate applied stress increase at the center of each layer below the center of the foundation.
From the program STRESS,

$$\text{Layer 1: } B = \frac{2}{2} = 1\,\text{m}, \quad L = \frac{2}{2} = 1\,\text{m}, \quad z = 1\,\text{m};$$

$$q_s = 100\,\text{kPa}, \quad \Delta\sigma_z = 70\,\text{kPa}.$$

$$\text{Layer 2: } B = \frac{2}{2} = 1\,\text{m}, \quad L = \frac{2}{2} = 1\,\text{m}, \quad z = 4\,\text{m};$$

$$q_s = 100\,\text{kPa}, \quad \Delta\sigma_z = 10.8\,\text{kPa}.$$

Step 2: Determine μ_{SB} for each layer.

$$\text{Area of base} = 2 \times 2 = 4\,\text{m}^2$$

$$\text{Equivalent diameter of circular foundation, } D = 2\sqrt{\tfrac{4}{\pi}} = 2.26\,\text{m}$$

$$\text{Layer 1: } \frac{H}{B} = \frac{2}{2.26} = 0.88, \quad A = 0.4; \quad \text{from Fig. 7.17,} \quad \mu_{SB} = 0.6.$$

$$\text{Layer 2: } \frac{H}{B} = \frac{4}{2.26} = 1.76, \quad A = 0.2; \quad \text{from Fig. 7.17,} \quad \mu_{SB} = 0.43.$$

Step 3: Calculate primary consolidation settlement.

$$(\rho_{pc})_{SB} = \Sigma m_v \Delta\sigma_z H \mu_{SB} = (0.0005 \times 70 \times 2 \times 0.6) + (0.0001 \times 10.8 \times 4 \times 0.43) = 0.044\,\text{m} = 44\,\text{mm}.$$

■

7.8 DETERMINATION OF BEARING CAPACITY AND SETTLEMENT USING DATA FROM IN SITU TESTS

Importance *It is difficult to obtain undisturbed samples of coarse-grained soils for testing in the laboratory. Consequently, the allowable bearing capacity and settlement of foundations on coarse-grained soils are often based on empirical methods using test data from in-situ tests. Various empirical equations have been proposed. Their use requires significant experience with foundation design on soils similar to those from which the empirical equations were obtained. Only a few of these equations are presented here.*

Access a spreadsheet at http://www.wiley.com/college/budhu. Click Foundations and Earth Structures, Chapter 7, bc.xls to calculate bearing capacity and settlement from SPT and CPT data. Select worksheet "SPT" or "CPT."

7.8.1 SPT

There are various relationships between the bearing capacity of coarse-grained soils and N values. Two such relationships for a vertical centric load are presented here.

AASHTO (2004) recommended that

$$q_{ult} = 32N_1B\left(C_{w1} + C_{w2}\frac{D_f}{B}\right)(\text{kPa}); \quad B(\text{m}); \quad D_f(\text{m}) \tag{7.62}$$

$$q_{ult} = 200N_1B\left(C_{w1} + C_{w2}\frac{D_f}{B}\right)(\text{psf}); \quad B(\text{ft}); \quad D_f(\text{ft}) \tag{7.62a}$$

TABLE 7.5 Groundwater Factors

Depth of groundwater below finished grade	C_{w1}	C_{w2}
0	0.5	0.5
D_f	0.5	1.0
1.5B + D_f	1.0	1.0

SOURCE: AASHTO, 2004.

where N_1 is the SPT value corrected for overburden pressure only, B is the foundation width, and C_{w1} and C_{w2} are groundwater factors given in Table 7.5. In practice, the average N_1 for a depth up to B or up to $1.5B$ below the foundation base is used in the above expression. The term $\frac{D_f}{B}$ is equal to zero if the soil above the base of the foundation is weaker than the soil below the base. The total settlement is expected to be less than 25 mm (1 in.). Normally, it is differential settlement that is the key concern. Meyerhof (1965) suggested that the settlement under a shallow foundation can be estimated from SPT value using

$$\rho(\text{mm}) = \frac{2q_a}{N_1}; \quad B \le 1.2 \text{ m}; \quad q_a(\text{kPa}) \tag{7.63}$$

$$\rho(\text{mm}) = \frac{3.2q_a}{N_1}\left(1 + \frac{1}{3.28B}\right)^{-2}; \quad B > 1.2 \text{ m}; \quad q_a(\text{kPa}) \tag{7.64}$$

$$\rho(\text{in.}) = \frac{4q_a}{N_1}; \quad B \le 4 \text{ ft}; \quad q_a(\text{ksf}) \tag{7.63a}$$

$$\rho(\text{in.}) = \frac{6q_a}{N_1}\left(1 + \frac{1}{B}\right)^{-2}; \quad B > 4 \text{ ft}; \quad q_a(\text{ksf}) \tag{7.64a}$$

where q_a is the allowable bearing capacity.

By estimating the settlement under each shallow foundation, you can determine the differential settlement.

Burland and Burbridge (1985) did a statistical analysis of settlement records from 200 foundations located in sand and in gravel. They proposed the following equation for the settlement of a foundation in a normally consolidated sand at the end of construction.

$$\rho = f_s f_1 q_a B^{0.7} I_c; \quad q_s(\text{kPa}), \quad B(\text{m}) \tag{7.65}$$

where ρ = settlement (mm), f_s = shape factor $= \left(\frac{1.25\frac{L}{B}}{\frac{L}{B}+0.25}\right)^2$, $f_1 = \frac{H_o}{z_1}\left(2 - \frac{H_o}{z_1}\right)$ is a correction factor if the thickness (H_o) of the sand stratum below the foundation base is less than an influence depth z_1, q_a is the allowable bearing capacity or applied stress (kPa), B and L are the width and length of the foundation (m) respectively,

$$I_c = \text{compressibility index} = \frac{1.71}{N^{1.4}} \tag{7.66}$$

and N is the uncorrected N value. The influence depth is the depth below the foundation that will influence the settlement and bearing capacity. If N increases with depth or N is approximately constant, the influence depth is taken as $z_1(\text{m}) = B^{0.763}$. If N tends to decrease with depth, the influence depth is $z_1 = 2B$.

If the sand is overconsolidated,

$$\rho = f_1 f_s\left(q_a - \frac{2}{3}\sigma'_{zc}\right)B^{0.7}I_c; \quad \text{if } q_a > \sigma'_{zc}(\text{kPa}),\ B\,(\text{m}) \tag{7.67}$$

$$\rho = q_a B^{0.7}\frac{I_c}{3}, \quad \text{if } q_a < \sigma'_{zc} \tag{7.68}$$

For a desired settlement, we can estimate allowable bearing capacity by rearranging Eq. (7.65) to give

Normally consolidated sand:

$$q_a = \frac{\rho}{f_1 f_s B^{0.7} I_c}; \quad q_a\,(\text{kPa}),\ B\,(\text{m}),\ \rho\,(\text{mm}) \tag{7.69}$$

Overconsolidated sand:

$$q_a = \frac{\rho}{f_1 f_s B^{0.7} I_c} + \frac{2}{3}\sigma'_{zc}; \quad q_a > \sigma'_{zc}; \quad q_a(\text{kPa}),\ B\,(\text{m}),\ \rho\,(\text{mm}),\ \sigma'_{zc}\,(\text{kPa}) \tag{7.70}$$

$$q_a = \frac{3\rho}{B^{0.7} I_c}; \quad q_a < \sigma'_{zc} \tag{7.71}$$

Overconsolidation is difficult to determine for coarse-grained soils. As an estimate, if $D_r \geq 85\%$, assume the soil is overconsolidated, else assume the soil is normally consolidated.

The procedure for the Burland-Burbridge method is as follows:

1. Determine the influence depth z_1.
2. Find the average N value within the depth z_1 below the foundation.
3. Calculate I_c from Eq. (7.66).
4. Determine ρ from the appropriate equation [Eq. (7.65) or Eq. (7.67) or Eq. (7.68)], or if ρ is specified, you can determine q_a.

For a given set of SPT data, these empirical methods give different bearing capacity values. Adequate site information, experience, and good engineering judgment are needed to successfully use these methods. For inclined and eccentric loads, you can modify the allowable bearing capacity using the appropriate factors given in Table 7.1 or in design manuals.

EXAMPLE 7.12 *Bearing Capacity Using SPT Data*

The SPT results at various depths in a soil are shown in Table E7.12a. The borehole was 100 mm (4 in.) in diameter and a standard sampler was used. Determine the allowable bearing capacity for a foundation 2 m (6 ft) wide and located at 1 m (3 ft) below the surface. The groundwater level is 10 m (30 ft) below the ground surface. The tolerable settlement should not exceed 25 mm (1 in.).

TABLE E7.12a

Depth m (ft)	*N* blows/0.304 m
0.6 (2)	25
1.2 (4)	28
1.8 (6)	33
2.4 (8)	29
3.0 (10)	28
3.6 (12)	29
4.2 (14)	31
4.8 (16)	35
5.4 (18)	41

Strategy U.S. units will be used in solving this example problem. The N_1 value to use is the average value for a depth $1.5\,B = 9$ ft below the foundation (i.e., the average N_1 value from a depth = 3 ft to a depth = 12 ft.) The N_1 values above the base are neglected. The unit weight is not given so we have to estimate this based on the description and the N_1 values. The groundwater level is deep and its effects can be neglected.

Solution 7.12

Step 1: Determine q_a.

Use a spreadsheet to do the calculation as shown in Table E7.12b. This spreadsheet is available at http://www.wiley.com/college/budhu. Click Foundations and Earth Structures, Chapter 7, bc.xls. Select worksheet "SPT."

TABLE 7.12b Bearing Capacity of Shallow Footings Using SPT Data

Select units q_a	US Eq. (7.62a)/FS 18400	Eq. (7.69) 22850	psf		Help
p_a	1	in	C_{w1}	1	
Width of footing	6	ft	C_{w2}	1	
Depth of footing	3	ft	f_s	1.00	
Length of footing	6	ft	I_c	0.015	
Groundwater	30	ft	z_1	5.20	ft
FS	3		H_o	15	ft
Variation in N	0		f_1	1	
Consider groundwater	Yes		C_S	1	
Relative density, D_r	85				
Preconsolidation stress		psf			
Sampler	Standard				
Borehole size, C_B	1				
Equipment, C_E	0.8				
Rod length from surface to anvil	0.8	ft			
Correct to $N_{1,60}$	No				

B = 6ft
Df = 3ft

Depth ft	Unit weight pcf	Rod length correction C_R	Vertical stress psf	Porewater pressure psf	Vertical effective stress psf	C_n calc.	C_n use	N	N_1
0	0		0	0	0	0	0	0	0
2	118	1	236	0	236	2.9	2.0	25	50
4	121	1	478	0	478	2.0	2.0	28	56
5	121	1	599	0	599	1.8	1.8	28	51
6	122	1	721	0	722	1.7	1.7	33	55
8	121	1	963	0	964	1.4	1.4	29	42
10	121	1	1205	0	1206	1.3	1.3	28	36
12	121	1	1447	0	1448	1.2	1.2	29	34
14	122	0.832	1691	0	1692	1.1	1.1	31	34
16	123	0.862	1937	0	1938	1.0	1.0	35	36
18	123	0.892	2183	0	2184	1.0	1.0	41	39
							Avg	29	46

The average N_1 value from 4 ft to 12 ft is 46. This soil can be classified as very dense (Chapter 4) with a relative density > 85%. As a conservative estimate, we will neglect preconsolidation stress as this is unknown.

Step 2: Decide on the value of q_a.

Evaluate the results. The estimated values for q_a are 18.4 ksf using the AASHTO (2004) method and 22.9 ksf using the Burland and Burbridge (1985) method. The question is which value to use. The answer depends solely on experience.

Without experience, you can use the least value (conservative) or take an average value or use the equation specified in your design manual. For example, if you are designing using the AASHTO (2004) design manual, then you would use 18.4 ksf. This example illustrates how different the predictions can be from empirical relationships using SPT. The standard deviation for the N value is 2 while that of N_1 is 10. A geotechnical engineer may then decide to use the N value rather than N_1. ■

7.8.2 CPT

Schmertmann (1970) and Schmertmann et al. (1978) proposed a methodology to determine settlement from the quasi-static cone test data for sands. They assumed that the sand is a linearly elastic material and only stress changes within depths of $2B$ for axisymmetric conditions and $4B$ for plane strain conditions influence the settlement. Settlement is calculated by integrating the vertical strains,

$$\rho = \int \varepsilon_z \, dz \tag{7.72}$$

The equation proposed by Schmertmann et al. (1978) is:

$$\rho = \frac{c_D c_t}{\beta} \sigma_{net} \sum_{i=1}^{n} \frac{(I_{co})_i}{(q_c)_i} \Delta z_i \tag{7.73}$$

where c_D = depth factor $= 1 - 0.5\frac{\sigma'_{zo}}{\sigma_{net}} \geq 0.5$, c_t = creep factor $= 1.0 + A\log_{10}\left|\frac{t}{0.1}\right|$, β is cone factor [$\beta = 2.5$ for square foundations (axisymmetric conditions, $L = B$), $\beta = 3.5$ for strip foundation (plane strain) conditions, $\frac{L}{B} > 10$], σ_{net} = the net foundation pressure (applied vertical stress minus soil pressure above the base of foundation), σ'_{zo} = initial vertical effective stress in kPa at the depth of the foundation, t = time in years ($t \geq 0.1$), A = empirical factor taken as 0.2, Δz_i = thickness of the ith layer, $(I_{co})_i$ = influence factor of the ith layer given as:

Axisymmetric: $L = B$

$$I_{co} = 0.1 + 2(I_{cp} - 0.1) \text{ for } \frac{z}{B} \leq \frac{1}{2} \tag{7.74}$$

$$I_{co} = I_{cp}\left[1 - \frac{2}{3}\left(\frac{z}{B} - \frac{1}{2}\right)\right] \text{ for } 2 \geq \frac{z}{B} > \frac{1}{2} \tag{7.75}$$

Plane strain: $\frac{L}{B} > 10$

$$I_{co} = 0.2 + (I_{cp} - 0.2)\frac{z}{B} \text{ for } \frac{z}{B} \leq 1 \tag{7.76}$$

$$I_{co} = I_{cp}\left[1 - \frac{1}{3}\left(\frac{z}{B} - 1\right)\right] \text{ for } 4 \geq \frac{z}{B} > 1 \tag{7.77}$$

$$\text{where } I_{cp} = 0.5 + 0.1\sqrt{\frac{\sigma_{net}}{\sigma'_{zp}}} \tag{7.78}$$

$(q_c)_i$ is the cone tip resistance for the ith layer, σ'_{zp} is original vertical effective stress at the depth of $\frac{B}{2}$ for axisymmetric condition and B for plane strain, and n is the number of sublayers.

The procedure to determine the settlement from cone data is as follows:

1. Divide the soil below the foundation into a number of sublayers. For square foundations, the total depth of the sublayers is $2B$ and a reasonable number of sublayers is 4. For strip foundations, the total depth is $4B$ and a reasonable number of sublayers is 7. The number of sublayers can be dictated by the layering based on the distribution of q_c.
2. Estimate the average value of $(q_c)_i$ for each sublayer from the field data of q_c versus depth.
3. Find I_{co} at the center of each sublayer.
4. Find allowable bearing capacity from $q_a = \frac{\Sigma q_c \Delta z}{\Sigma \Delta z}$.
5. Estimate ρ using Eq. (7.73).

EXAMPLE 7.13 *Bearing Capacity and Settlement Using CPT Data*

The end bearing resistance in a CPT at a site is shown in Fig. E7.13a. A square foundation 3 m (10 ft) wide imposing an applied stress of 217 kPa (4.5 ksf) is to be located 1 m (3 ft) below ground level at this site. Determine the ultimate bearing capacity and the settlement of the foundation one year after construction. The bulk unit weight of the sand is 17 kN/m^3 (108 pcf). Groundwater level is 8 m (25 ft) below the ground surface.

Strategy For a square foundation, the influence depth is $2B$. You need to divide this depth into sublayers and then use Eq. (7.73). SI units will be used in solving this example problem.

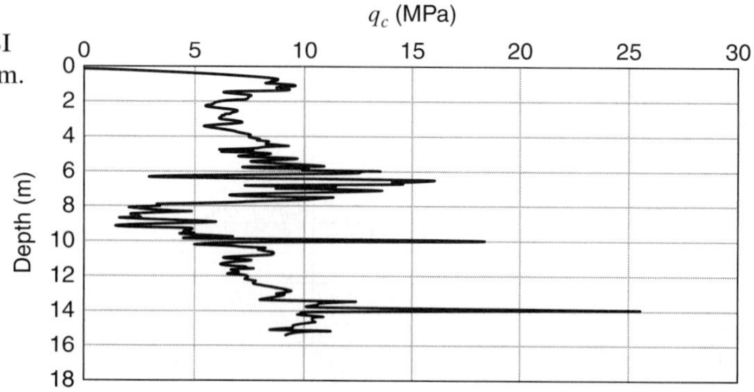

FIGURE E7.13a

Solution 7.13

Step 1: Divide influence depth into layers.
Influence depth $= 2B = 6\,\text{m}$.

Step 2: Visually inspect the data, disregard any excessively large value, and identify sublayers. See Fig. E7.13b for sublayering of the first 6 m of data. For each sublayer, use the average value of q_c. (See Table E7.13.)

Use a spreadsheet to do the calculation as shown in Table E7.12b. This spreadsheet is available at http://www.wiley.com/college/budhu. Click Foundations and Earth Structures, Chapter 7, bc.xls. Select worksheet "CPT."

$$\sigma'_{zp} = 17 \times \frac{3}{2} = 25.5\,\text{kPa}; \quad I_{cp} = 0.5 + 0.1\sqrt{\frac{200}{25.5}} = 0.78$$

$$\sigma'_z = 1 \times 17 = 17\,\text{kPa}$$

$$\sigma_{\text{net}} = 217\,\text{kPa} - 1 \times 17 = 200\,\text{kPa}$$

$$c_D = 1 - 0.5 \times \frac{17}{200} = 0.96,$$

$$c_t = 1 + 0.2\log_{10}\left|\frac{1}{0.1}\right| = 1.2$$

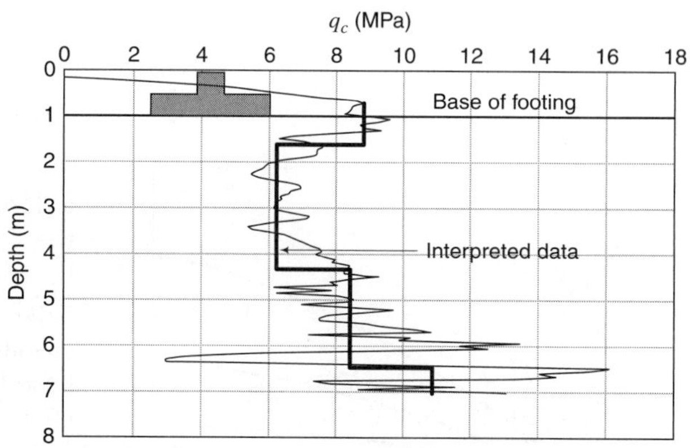

FIGURE E7.13b

Step 3: Calculate the bearing capacity.
(see worksheet "CPT" in spreadsheet bc.xls for summation values)

$$q_u = \frac{\Sigma q_c \Delta z}{\Sigma \Delta z} = \frac{42.15}{6} = 7\,\text{MPa}$$

Step 4: Calculate the settlement.
(see worksheet "CPT" in spreadsheet bc.xls for summation values)

$$\rho = \frac{c_D c_t}{\beta}\sigma_{\text{net}}\sum_{i=1}^{n}\frac{(I_{co})_i}{(q_c)_i}\Delta z = \frac{0.96 \times 1.2}{2.5} \times 200 \times 0.29 \approx 26\,\text{mm}$$

■

	A	B	C	D	E	F	G	H
1	**Settlement of shallow footings using cone penetrometer results**							
2	Condition	1	Axisymmetric			HELP		
3	Settlement		26.2	mm				
4	Bearing capacity		7.0	MPa				
5								
6	q	217	kPa		c_D	0.96		
7	D_f	1	m		c_t	1.20		
8	γ	17	kN/m³					
9	q_{net}	200	kPa					
10	B	3	m					
11	L	3	m					
12	I_{cp}	0.78						
13	t	1	yr					
14	**Layer**	z	$\Delta z'$	z/B	I_{co}	q_c	$(I_{co}/q_c)\Delta z$	$q_c\Delta z$
15	**m**	m	m			MPa		MPa.m
16								
17	1	0.75	0.75	0.25	0.44	5.80	0.06	4.35
18	2	2.25	1.50	0.75	0.65	7.00	0.14	10.50
19	3	3.75	1.50	1.25	0.39	8.50	0.07	12.75
20	4	5.25	1.50	1.75	0.13	9.70	0.02	14.55
21						Sum	0.29	42.15
22								

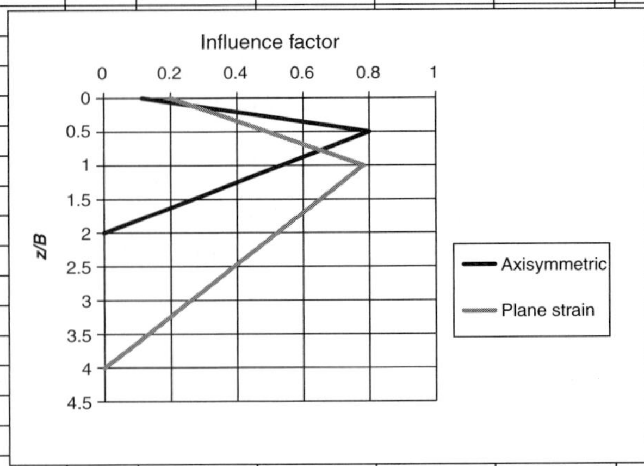

TABLE E7.13

7.8.3 Pressuremeter

The ultimate bearing capacity from the PMT (Briaud, 1992) is

$$q_{\text{ult}} = K_g p_{LE}^* + \sigma_{zo} \tag{7.79}$$

The allowable bearing capacity is

$$q_a = \frac{K_g p_{LE}^*}{\text{FS}} + \sigma_{zo} \tag{7.80}$$

where $p_{LE}^* = p_L - p_{OH}$ is the equivalent limit pressure within a depth 1.5B above and 1.5B below the foundation level, B is the foundation width, p_L is the limit pressure, p_{OH} is the horizontal pressure at the foundation level, σ_{zo} is the vertical total stress at the foundation level, FS is a factor of safety (usually 3), and K_g a pressuremeter bearing capacity factor based on the geometry of the foundation and the soil type (Fig. 7.18). For foundations on soils of varying strength the equivalent limit pressure is

$$p_{LE}^* = \sqrt{p_{l1} p_{l2} p_{l3}} \tag{7.81}$$

where p_{l1}, p_{l2}, p_{l3} are the net limit pressures (total limit pressure—the horizontal pressure at depth of test) at B above the foundation level, at the foundation level, and B below the foundation level, respectively. The equivalent depth of embedment, D_f^e, is

$$D_f^e = \frac{1}{p_{LE}^*} \sum_{i=1}^{n} \Delta z_i p_{Li}^* \tag{7.82}$$

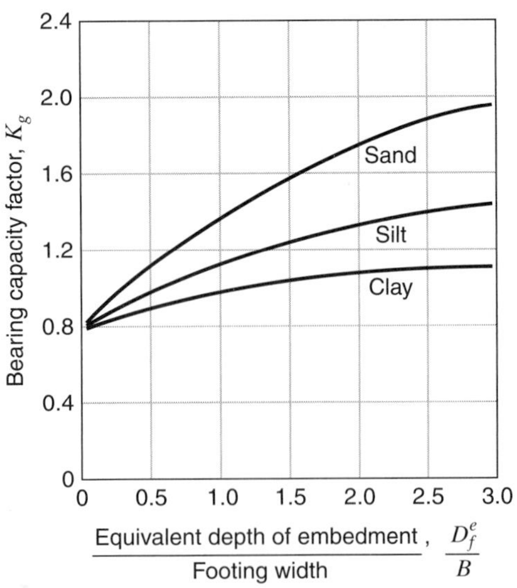

FIGURE 7.18 The PMT bearing capacity factor, kg (Source: Briaud, 1992.)

TABLE 7.6 Menard's Rheological Factors

Soil type	Peat		Clay		Silt		Sand		Sand and gravel	
	E/p_L^*	α_r	E/p_L^*	α_r	E/p_L^*	α_r	E/p_L^*	α_r	E/p_L^*	α_r
Overconsolidated	*	*	>16	1	>14	2/3	>12	1/2	>10	1/3
Normally consolidated	For all values	1	9–16	2/3	8–14	1/2	7–12	1/3	6–10	1/4
Weathered and/or remoulded	*	*	7–9	1/2	*	1/2	*	1/3	*	1/4
Rock	* Extremely fractured $\alpha_r = 1/3$		Other $\alpha_r = 1/2$				Slightly fractured or extremely weathered $\alpha_r = 2/3$			

Note: * Values are unknown.

where Δz_i are the discrete thicknesses from where the pressuremeter test data is extracted and p_{Li}^* are the net limit pressures between ground surface and the foundation level.

The total settlement, ρ, is given as

$$\rho = q_a\left[\frac{2}{9E_d}B_o\left(\lambda_d\frac{B}{B_o}\right)^{\alpha_r} + \frac{\alpha_r}{E_c}\lambda_c B\right] \tag{7.83}$$

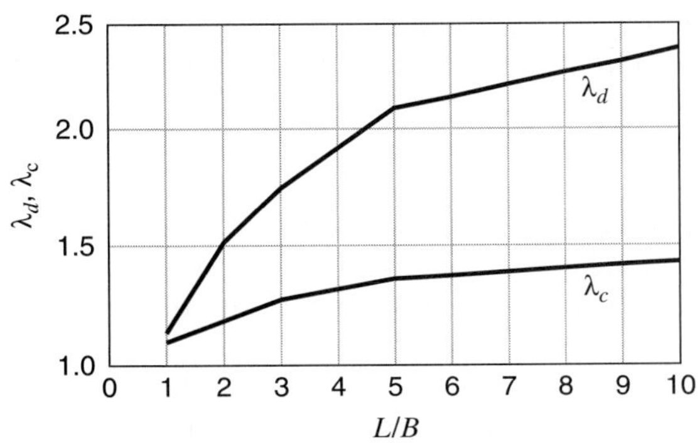

FIGURE 7.19 Menard's shape factors for PMT. (SOURCE: Briaud, 1992.)

where q_a is the allowable bearing capacity or design vertical stress, E_c is the pressuremeter modulus under the first layer below the foundation, E_d is the equivalent harmonic mean pressuremeter modulus for 16 layers, each B/2 thick, below the foundation, B is the foundation width, B_o is a reference width of 60 cm, α_r is a rheological factor (Table 7.6), and λ_c and λ_d are shape factors (Fig. 7.19). The design of shallow foundations using pressuremeter results requires experience with interpreting the data from the tests and in using them in design practice.

7.8.4 Plate Load Test

The settlement from plate load tests is calculated using

$$\rho = \rho_p\left(\frac{2}{1+\frac{B_p}{B}}\right)^2 \tag{7.84}$$

where ρ_p is the plate settlement (assuming the soil is elastic).

If the sand were to behave like an elastic material, then the settlement can be calculated from

$$\rho_p = q_a B_p \frac{1-(\nu')^2}{E'} I_p \tag{7.85}$$

where q_a is the allowable bearing capacity or σ_{ap} is the applied stress, B_p is the width or diameter of the plate, ν' is Poisson's ratio, E' = elastic modulus, and I_p is an influence factor (0.82 for a rigid plate).

In the limit $\frac{B_p}{B} \to \infty$, $\frac{\rho}{\rho_p} \to 4$.

7.9 SEISMIC BEARING CAPACITY AND SETTLEMENT OF SHALLOW FOUNDATIONS

Access a spreadsheet at http://www.wiley.com/college/budhu. Click Foundations and Earth Structures, Chapter 7, bc.xls to calculate seismic bearing capacity factors and to estimate the geometry of shallow foundations to satisfy a seismic event. Select the worksheet "homogeneous soil."

Importance *A seismic event can significantly reduce the bearing capacity of soils and cause a foundation to exceed its serviceability limit state. In nonliquefiable soils, you need to check that the static bearing capacity is not degraded to a level that will cause the structure to fail and that seismic settlement does not impair the function of the structure. In this section, we will discuss seismic bearing capacity and settlement.*

Richards et al. (1990) considered the effects of seismic forces on the bearing capacity and settlement of a strip foundation on a dry coarse-grained soil using limit equilibrium with a wedge-type failure mechanism. They showed that seismic forces can reduce the bearing capacity of shallow foundations and can cause excessive settlement. Further, a clean, dry coarse-grained soil can behave like a viscous fluid at certain levels of acceleration. They used the term *fluidization* to describe this fluid-like state. Two states of fluidization were described. One state is the initial fluidization state of the soil when slip planes are initiated in the soil mass below the foundation. Soil movement (rigid body movement) will then occur on these planes when a certain level of acceleration is reached but the soil mass as a whole does not fluidize. The other fluidization state is general fluidization, in which the whole mass of soil behaves as a viscous fluid.

What are the practical implications of fluidization? For a soil element below the groundwater level, initial dynamic liquefaction is possible for a single pulse of acceleration that brings the soil to the initial fluidization state. Dynamic liquefaction is a special form of fluidization in which displacements of a saturated soil in the presence of groundwater are unbounded. Dynamic liquefaction of soils can be catastrophic and cause boiling, land subsidence, local flooding, sudden failure of foundation of structures, and uplift or floatation of buried structures. In dry coarse-grained soils, initial fluidization will result in the reduction in the soil's bearing capacity and will cause foundation settlement that may exceed the serviceability limit state. It can also trigger failures in retaining walls and slopes. Unlike dynamic liquefaction, the seismic settlement of foundations on dry coarse-grained soils is bounded. At general fluidization state, foundations can settle excessively, buried structures can rise, and retaining structures and slopes can collapse.

The initial fluidization acceleration ratio is

$$\left(\frac{k_h^*}{1-k_v^*}\right)_i = \tan\Omega_o = \frac{\sin\phi'}{2}\{(\sin\phi'-3)(\sin\phi'-1)\}^{\frac{1}{2}} \tag{7.86}$$

where k_v and k_h are the vertical and horizontal acceleration coefficient respectively, the superscript* denotes critical, the subscript, i, denotes initial, and ϕ' is a generic friction angle of the soil (discussed later in this section). The acceleration ratio at general fluidization is

$$\left(\frac{k_h^*}{1-k_v^*}\right)_g = \tan\phi' \tag{7.87}$$

where the subscript g denotes general.

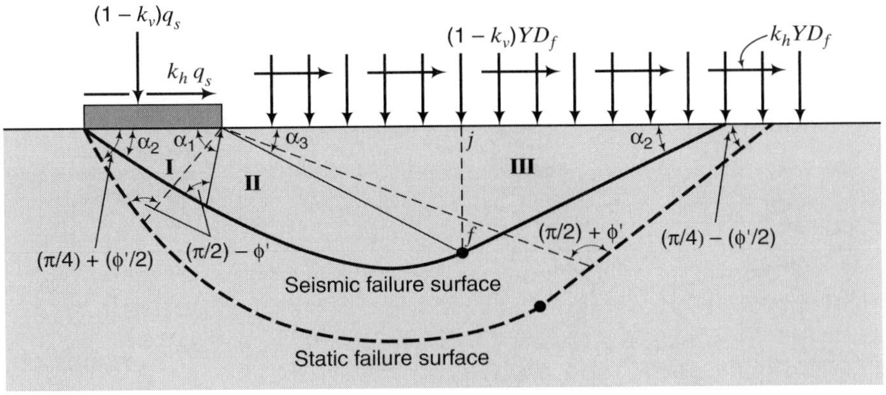

$$\alpha_3 = \frac{\pi}{4} - \frac{\phi'}{2} + \alpha_p$$

$$\tan\alpha_p = \frac{2\tan\theta}{K_{pE} - 1 + [(K_{pE} - 1)^2 + 4\tan^2\theta]^{\frac{1}{2}}}$$

$$K_{pE} = \left(\frac{\sin^2\phi' + 1}{\cos^2\phi'}\right) + \left[\left(\frac{\sin^2\phi' + 1}{\cos^2\phi'}\right)^2 - \frac{4\tan^2\theta}{\cos^2\phi'} - 1\right]^{\frac{1}{2}}$$

q_s = stress at base of footing.

FIGURE 7.20 Static and seismic failure surfaces. (Source: Budhu and Al-Karni, 1993.)

Budhu and Al-karni (1993) used a Prandtl-type mechanism (Fig. 7.20) and the limit equilibrium method to establish seismic bearing capacity coefficients. The theoretical seismic bearing capacity coefficients for different generic values of ϕ' were curve fitted for practical use. These seismic bearing capacity coefficients for a horizontal acceleration, $k_h g$, and a vertical acceleration, $k_v g$, where g is the acceleration due to gravity are

$$N_{qE} = (1 - k_v)N_q \exp(-\beta_q) \tag{7.88}$$

where

$$\beta_q = \frac{5.3\,k_h^{1.2}}{1 - k_v} \tag{7.89}$$

$$N_{\gamma E} = \left(1 - \frac{2}{3}k_v\right)N_\gamma \exp(-\beta_\gamma) \tag{7.90}$$

where

$$\beta_\gamma = \frac{9k_h^{1.1}}{1 - k_v} \tag{7.91}$$

The ultimate seismic bearing capacity for clean, coarse-grained soils is

$$q_{uE} = \gamma D_f N_{qE} s_q + 0.5\gamma B N_{\gamma E} s_\gamma \tag{7.92}$$

The ultimate seismic bearing capacity for finegrained soils is

$$q_{uE} = 5.14\,N_{cE} s_u s_c \tag{7.93}$$

where

$$N_{cE} = \exp(-\beta_c); \quad \beta_c = 4.3k_h^{\left(1+\frac{s_u}{\gamma H_E}\right)} \tag{7.94}$$

and H_E is the depth of the seismic failure surface from the finished grade given as

$$H_E = \frac{0.5B}{\cos\alpha_3}\exp^{\frac{\pi}{2}\tan\phi'} + D_f \tag{7.95}$$

The angle, α_3, is described in Fig. 7.20.

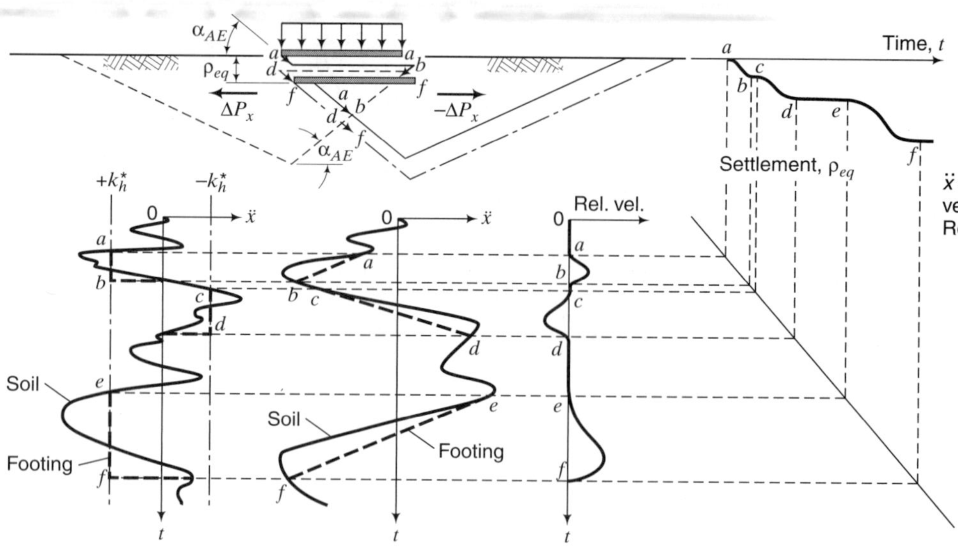

FIGURE 7.21 Ratchet-type settlement after initial fluidization. (Source: Richards et al., 1993.)

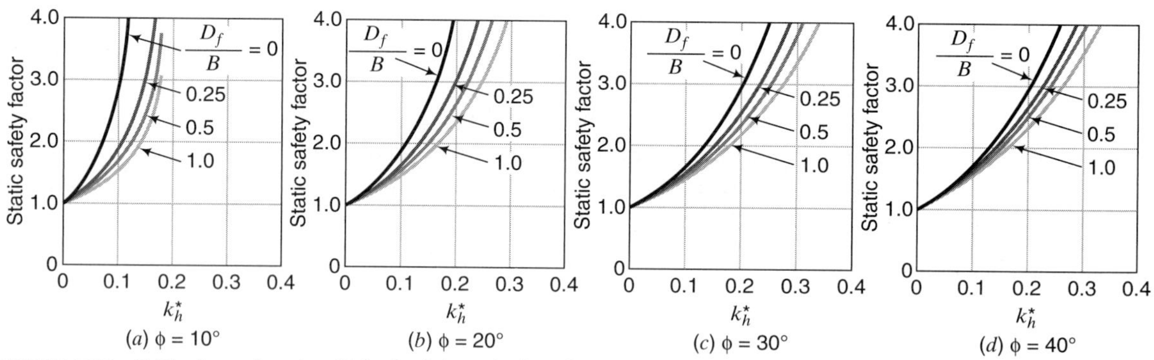

FIGURE 7.22 Critical acceleration k_h^* for incipient shallow foundation settlement. (Source: Richards et al., 1993.)

If the soil above the foundation base is weaker (about one-half) than the soil below it, you should neglect D_f in Eq. (7.95).

Seismic settlement will occur in a ratchet-type fashion as described by Richards et al. (1993) and shown in Fig. 7.21. Every time the acceleration is greater than the initial fluidization acceleration (Fig. 7.22), the foundation will move. It will cease to move when the acceleration is less than the initial fluidization acceleration.

The seismic settlement can be estimated for two way shear using a sliding block analysis from

$$\rho_{eq} = 0.174 \frac{V^2}{Ag} \left| \frac{k_h^*}{A} \right|^{-4} \tan \alpha_{AE} \tag{7.96}$$

where V is the peak design velocity, A is the peak design acceleration coefficient, g is acceleration due to gravity, α_{AE} is the inclination of the slip plane to the horizontal plane due to a seismic event given by

$$\alpha_{AE} = a + \tan^{-1} \left\{ \frac{\sqrt{(1 + \tan^2 a)[1 + \tan(\delta + \theta)\cot a]} - \tan a}{1 + \tan(\delta + \theta)(\tan a + \cot a)} \right\} \tag{7.97}$$

where $\theta = \tan^{-1}\left(\frac{k_h}{1-k_v}\right)$, $a = \phi' - \theta$, and δ is the interface friction angle taken as $\frac{\phi'}{2}$. The values of $\tan(\alpha_{AE})$ for various values of ϕ' are shown in Fig. 7.23.

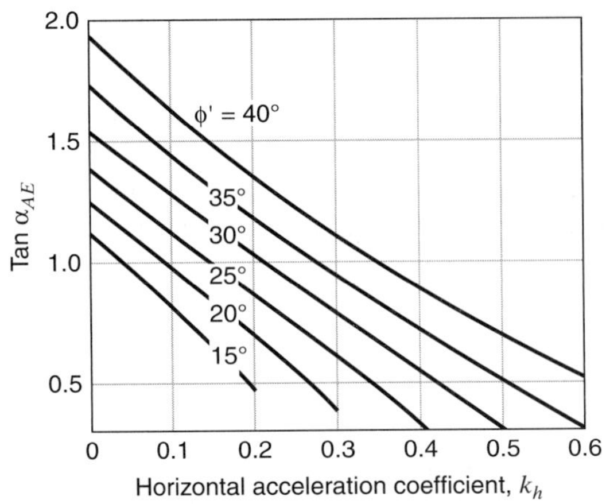

FIGURE 7.23 Tangents of Coulomb wedge for seismic settlement for $\delta = \phi'/2$. (Source: Richards et al., 1993.)

The seismic bearing capacity equations are based on a pseudostatic analysis (i.e., it is time independent) and does not account for the seismic shear strains that are responsible for soil deformation and reduction in soil strength. These equations should be used cautiously. You should use a conservative value of friction angle in the seismic bearing capacity equations. The critical state friction value is recommended, or you may take an average of ϕ'_{cs} and ϕ'_p.

The procedure to determine the seismic bearing capacity and settlement of shallow foundations is as follows:

1. Obtain the peak design acceleration and peak design velocity from the response spectra for the site using your local codes or IBC (2006 or later).
2. Conduct a static bearing capacity analysis.
3. Determine the seismic bearing capacity using the appropriate equation.
4. Calculate the factor of safety for seismic loading. If FS > 1, the foundation will not fail. A factor of safety of 1.25 to 1.5 under seismic loading is satisfactory.
5. Determine the critical acceleration for initial fluidization.
6. Check if the critical acceleration from item 5 is less than the design acceleration. If it is, the foundation will displace each time the design acceleration exceeds the critical acceleration. If not, seismic displacement will not occur.

Since the design of most shallow foundations is controlled by settlement, you can evaluate the desired static factor of safety for a prescribed allowable seismic settlement as follows.

1. Guess a value of tan α_{AE} usually 1.0.
2. Calculate the uni-directional seismic settlement, i.e., one-half the prescribed allowable seismic settlement.
3. Calculate k_h^* from Eq. (7.96).
4. Check initial guess of tan α_{AE} using Fig. 7.23. If it is nearly correct, go to the next step. If not, guess a different value and repeat.
5. Obtain the desired static factor of safety from Fig. 7.22.

EXAMPLE 7.14 *Seismic Bearing Capacity and Settlement of a Shallow Foundation*

A square foundation 2.5 m (8 ft) wide is founded 1 m (3 ft) below the ground surface in a dense coarse-grained soil. Laboratory direct shear tests on samples of the soil compacted to a relative density of 85% gave $\phi'_{cs} = 30°$, $\phi'_p = 35°$ at a vertical stress of 125 kPa (2.6 ksf), and $\gamma = 18$ kN/m^3 (113 pcf). Groundwater level is 15 m (50 ft) below the ground surface. The dead load including the weight of the foundation is 900 kN (200 kips) and the live load is 1100 kN (245 kips). The peak acceleration and peak velocity for 5% damping at the site are 0.2 g and 0.5 m/s (1.6 ft/s) respectively.

(a) Determine the factor of safety against bearing capacity failure under (1) static load and (2) seismic load.

(b) Estimate whether initial fluidization will occur.

(c) Would seismic settlement occur?

(d) If the average friction angle is 30°, and $A = 0.3$, estimate the static factor of safety for an allowable seismic settlement of 25 mm (uni-directional settlement).

Strategy SI units will be used in solving this example problem. The vertical stress at which the laboratory test was conducted for the ϕ'_p reported is 125 kPa. If we consider a soil element at a depth of $0.5B$ to represent the soil within the failure zone, the vertical effective stress is $(1 + \frac{2.5}{2}) \times 18 = 40.5$ kPa, and assuming a 2:1 approximate distribution of stress (Chapter 2), the increase in vertical stress at this depth is $\frac{2000}{(2.5+1.25)^2} = 142$ kPa. The total vertical effective stress on the soil element is $142 + 40.5 = 182.5$ kPa, which is higher than the vertical effective stress of 125 kPa at which the soil test was done. Recall from Chapter 4 that ϕ'_p decreases with increasing normal effective stress due to reduced dilatancy. One can surmise then that ϕ'_p as reported may not be mobilized and it is prudent to use a friction value between ϕ'_p and ϕ'_{cs}. Use an average value of $\phi' = (30 + 35)/2 = 32.5°$.

Solution 7.14

Step 1: Calculate static bearing capacity and factor safety.
Use the spreadsheet bc.xls. (See Table E7.14)
Input: $\phi' = 32.5°$, $D_r = 85\%$, $\gamma = 18\,\text{kN/m}^3$, $\gamma_{sat} = 18\,\text{kN/m}^3$, $D_f = 1$ m, $B = 2.5$ m, $D_w = 10$ m, and $V_n = 2000$ kN.
Output: $N_q = 24.6$, $s_q = 1.64$, all other geometric factors = 1. $N_\gamma = 24.4$ (Davis and Booker, 1971), $s_\gamma = 0.6$, all other geometric factors = 1, $q_u = 1100$ kPa, FS = 3.6.
A partial listing from the spreadsheet is shown in Table E7.14

Step 2: Calculate the seismic bearing capacity and factory of safety.

$$\beta_q = \frac{5.3k_h^{1.2}}{1 - k_v} = \frac{5.3 \times 0.2^{1.2}}{1 - 0.05} = 0.81$$

$$N_{qE} = (1 - k_v)N_q \exp(-\beta_q) = (1 - 0.05) \times 24.6 \times \exp(-0.81) = 10.4$$

	A	B	C	D	E	F
1	**THEORETICAL BEARING CAPACITY OF SHALLOW FOUNDATIONS**					
2	**Muni Budhu "Foundation and Earth retaining Structures," John Wiley & Sons, NY, 2006**					
3	**Parameter**	**Value**	**Units**			
4	**1. Soil Properties** Undrained Shear strenght, s_u	0	kPa			
5	Angle of friction, ϕ_p	32.5	°	**FINAL RESULTS**		
6	Relative density	85	%	SHORT SIDE FAILURE		
7	Shear modulus, G	0	kPa	Ultimate bearing capacity, q_{ult}	***1118***	kPa
8	Bulk unit weight, γ_b	18	kN/m³	Net bearing capacity, q_u	***1100***	kPa
9	Saturated unit weight, γ_s	18	kN/m³	Computed Factor of Safety, FS	**3.6**	
10	**2. Proposed Footing Geometry** Depth of embedment, D	1	m	N_q value (rough footing)	24.6	
11	Thickness, T	0	m			
12	Footing length, L, or "d", or "s"	s	m	N_γ values (rough footing)		
13	Footing width, B	2.500	m	Davis and Booker	24.4	
14	Smooth (s) or rough (r)	r	°	Caquot and Kerisel (Vesic)	32.6	
15	**3. Loading Conditions** Applied Vertical Load,	2000	kN	Meyerhof	48.0	
16	H_B	0	kN			
17	M_B	0	kNm	SEISMIC - SHORT SIDE		
18	e_L	0.000	m	Ultimate bearing capacity, q_{uE}	370	kPa
19	H_L	0	kN	Computed Factor of Safety, FS	1.2	
20	M_L	0	kNm	Critical acceleration, k_h*	0.287	
21	e_B	0	m	Would seismic settlement occur?	NO	
22	Depth to Goundwater Table, D_w	10	m			
23	Inclination of load to vertical, ω	0	°			
24	Load inclined along length,L or width, B	B				
25	Inclination of load to longitudinal axis, θ	0	°			
26	Base tilt, η	0	°			
27	Ground Slope, β	0	°			
28	Unit weight of concrete, γ_c	25	kN/m³			
29	**4. Seismic Acc. factors**					
30	Horiz. Accl. Coeff. k_h	0.2				
31	Vertical accl. Coef. k_v	0.05				
32	Friction angle	32.5	°			

FIGURE E7.14

$$\beta_\gamma = \frac{9k_h^{1.1}}{1-k_v} = \frac{9 \times 0.2^{1.1}}{1-0.05} = 1.61$$

$$N_{\gamma E} = \left(1 - \frac{2}{3}k_v\right) N_\gamma \exp(-\beta_\gamma) = \left(1 - \frac{2}{3} \times 0.05\right) \times 24.4 \times \exp(-1.61) = 4.7$$

$$\begin{aligned} q_{uE} &= \gamma D_f N_{qE} s_q + 0.5\gamma B N_{\gamma E} s_\gamma \\ &= 18 \times 1 \times 10.4 \times 1.64 + 0.5 \times 18 \times 2.5 \times 4.7 \times 0.6 \\ &= 370\,\text{kPa} \end{aligned}$$

Applied vertical stress, $\sigma_a = \frac{2000}{2.5^2} = 320\,\text{kPa}$.

$$\text{FS} = \frac{q_{uE}}{\sigma_a - \gamma D_f} = \frac{357}{320 - 18 \times 1} = 1.18$$

Step 3: Determine k_h^* for seismic settlement.

$A = 0.2, \quad V = 0.5\,\text{m/s}$

Estimate k_h^* by linear interpolation using Fig. 7.22c and d.

For FS $= 3.6$ and $\frac{D_f}{B} = \frac{1}{2.5} = 0.4$.

For $\phi' = 30°$, $k_h^* = 0.28$.

For $\phi' = 40°$, $k_h^* = 0.3$.

$$\phi' = 32.5°, \; k_n^* = 0.28 + \frac{(0.3 - 0.28)}{(40 - 30)} \times (32.5 - 30) = 0.285.$$

For practical purposes, $k_h^* = 0.29$.

Since $k_h^* > k_h$, seismic settlement will be small (negligible). Note that Eq. (7.96) should be used only if $k_h > k_h^*$.

Step 4: Determine the static factor of safety for the prescribed allowable seismic settlement.

$A = 0.3, \quad V = 0.5\,\text{m/s}$

Assume a value of tan $\alpha_{AE} = 0.9$

Calculate k_h^* from Eq. (7.96).

$$0.025 = 0.174 \frac{0.5^2}{0.3g} \left| \frac{k_h^*}{0.3} \right|^{-4} 0.9$$

Solving for k_h^* gives $k_h^* = 0.26$.

From Fig. 7.23, $\tan\alpha_{AE} \sim 0.9$ for $k_h^* = 0.26$. Therefore, the initial guess is correct.

From Fig. 7.22, using $k_h^* = 0.26$ and $\frac{D_f}{B} = \frac{1}{2.5} = 0.4$, the required static factor of safety is 3.3. ■

7.10 BEARING CAPACITY AND SETTLEMENT USING NUMERICAL METHODS

Importance *Numerical methods are useful for complex loadings, irregular-shaped foundations, and layered soils, and for exploring the effects of design variables such as the effects of B, s_u, and ϕ'_p. Good quality soil data, an understanding of soil models (Chapter 3), and experience are essential in using numerical methods. In this section, a brief discussion on the use of numerical methods is presented.*

Numerical methods such as the FEM are particularly useful in dealing with layered soils. However, care is required in the implementation of numerical methods to foundations because the results could be inaccurate due to the constraints placed on the relationship between plastic strain increment and stress. This leads to a load-displacement curve that keeps rising rather than reaching failure (a plateau).

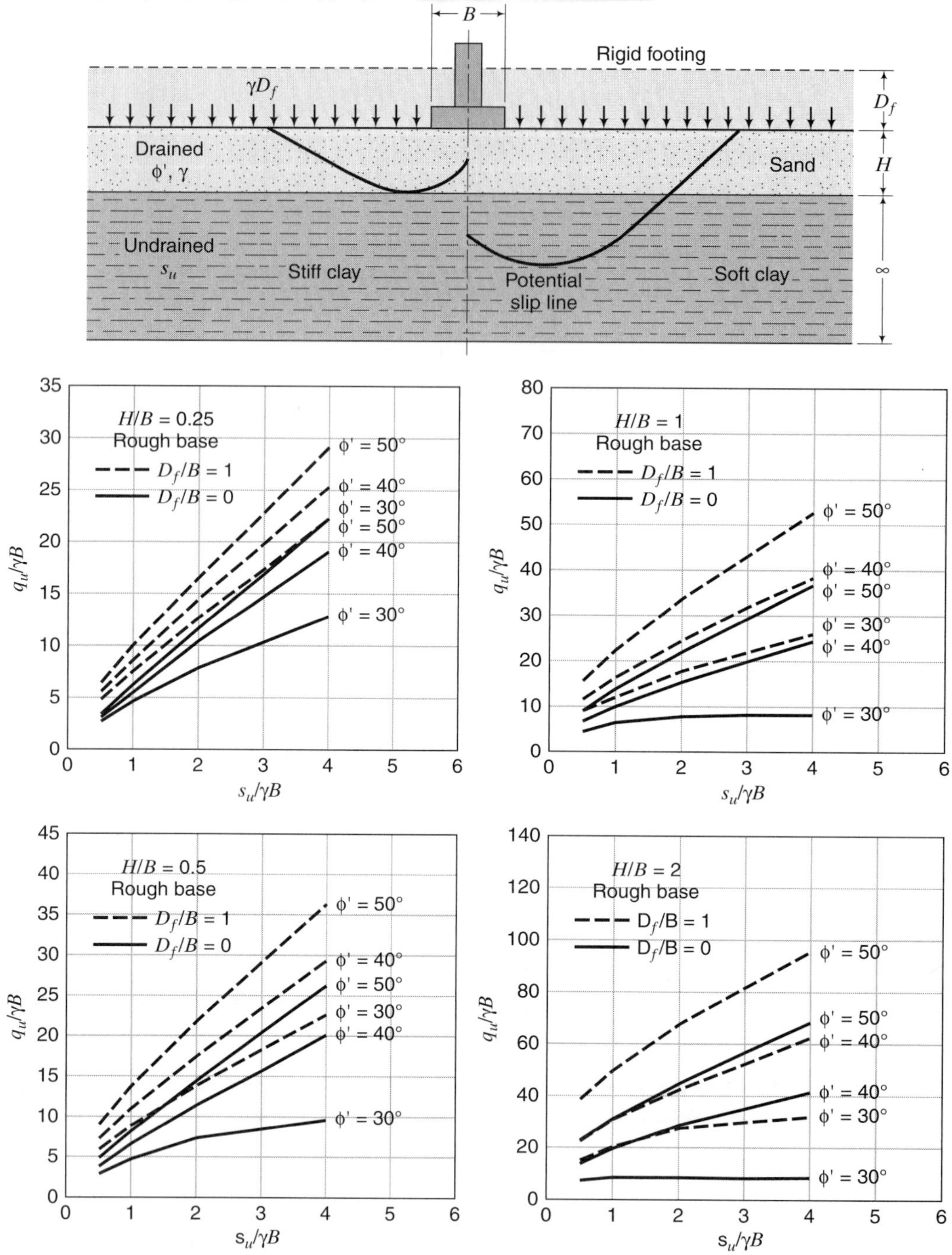

FIGURE 7.24 Bearing capacity of a strip footing in a sand layer underlain by a clay layer (Shiau et al., 2003).

The bearing capacity of a strip foundation resting in a sand layer underlain by a clay layer was investigated using the FEM and the limit theorems by Shiau et al. (2003). Their results for rough foundation base are shown in Fig. 7.24. The effects of foundation base roughness on the bearing capacity are shown in Fig. 7.25. For $\phi'_p < 35°$, the bearing capacity for a foundation on a smooth base is about one-half that of the same foundation on a rough base.

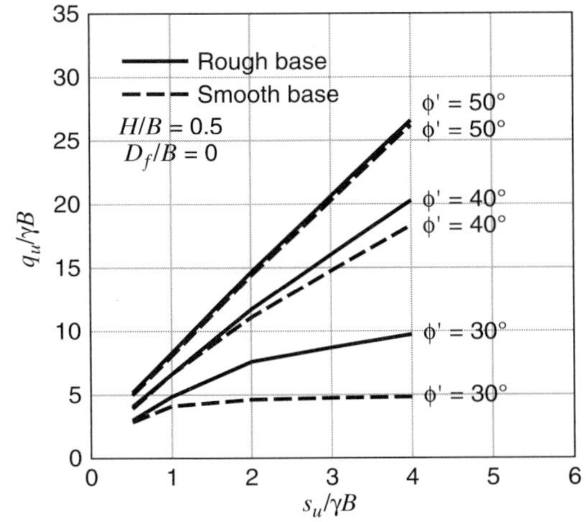

FIGURE 7.25 Effects of base roughness (Shiau et al., 2003).

EXAMPLE 7.15 *Using a Numerical Method to Analyze a Foundation on a Three-Layered Soil*

A strip foundation of width $B = 1$ m (3 ft) rests on the finished grade on a fine sand as shown in Fig. E7.15a. Predict the ultimate load capacity.

Strategy We are going to use a commercial program called FLAC. FLAC uses FDM. You can exploit symmetry and discretize only one-half the soil's domain. The soil domain should extend laterally about $2.5B$ to $5B$, and about $5B$ and greater in depth. You do not have to discretize the foundation. This is a plane strain problem. Symmetry will not be exploited here so that you can see the solution for the whole mesh.

Solution 7.15

Step 1: Discretize the soil domain.

A simple grid 60×14 is initially chosen as shown in Fig. E7.15b.

Step 2: Apply boundary conditions.

The left and right vertical boundaries are fixed in the x direction. The bottom boundary is fixed in the x and z direction.

Step 3: Select a soil model.

The Mohr-Coulomb model is selected since our main interest is on the collapse load.

Step 4: Input soil parameters.

Input the soil parameters as given in Fig. E7.15a.

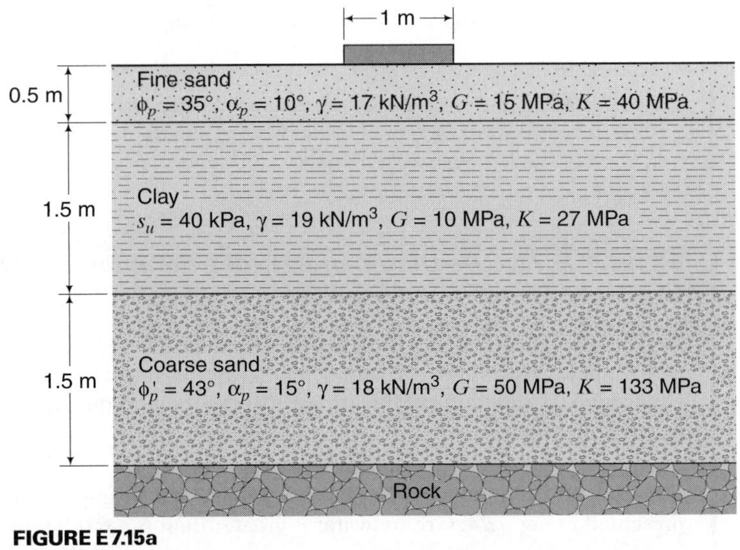

FIGURE E7.15a

Step 5: Apply load condition.

In this problem, we will apply vertical displacement increments. In FLAC, you have to use velocity increments. We select small increments of size -2.5×10^{-5} m/s (the negative sign indicates that the velocity is downward).

Step 6: Run program and interpret results.

The results are shown in Fig. E7.15b,c,d,e.

$$q_u = 283 \text{ kPa}$$

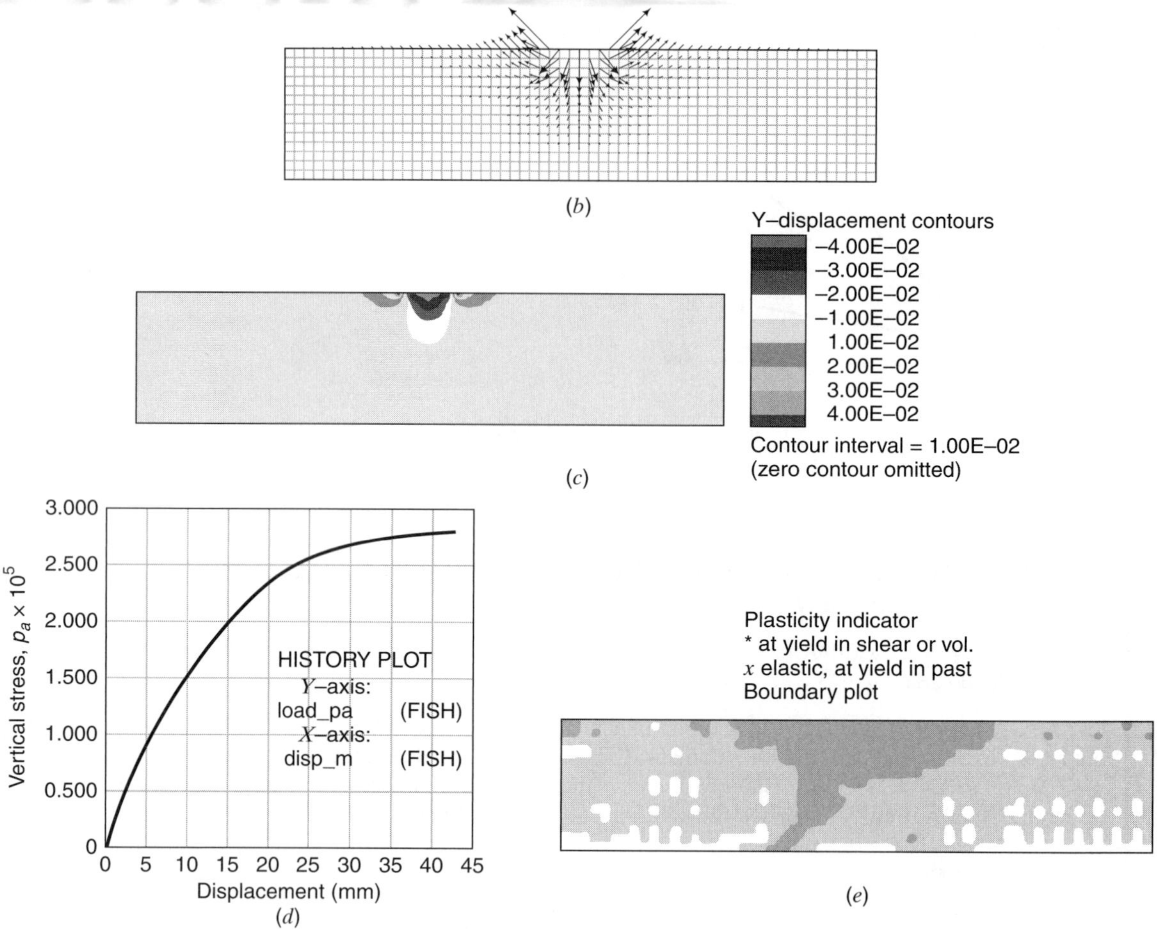

FIGURE E7.15b-e (b) Mesh with results of displacement vectors; (c) Vertical displacement contours; (d) Vertical stress–vertical displacement relationship, $q_u = 283$ kPa; (e) Yielding of soil elements.

■

EXAMPLE 7.16 *Bearing Capacity of a Foundation in Sand Underlain by a Deep Deposit of Stiff Clay*

A strip foundation of width $B = 2$ m (6 ft) is located 1 m (3 ft) below the ground surface in a sand layer 4 m (12 ft) thick with $\gamma = 18.2\ \text{kN/m}^3$ (116 pcf) and $\phi'_p = 35°$. Below the sand is a deep deposit of stiff clay with $s_u = 80$ kPa (1600 psf). Determine the ultimate bearing capacity of the foundation. Groundwater is 6 m (20 ft) below the surface. Assume a rough foundation base.

Strategy U.S. units will be used in solving this example problem. Use the FEM results of Shiau et al. (2003) presented in Fig. 7.24. Groundwater is greater than $B = 6$ ft below the base and will not affect the bearing capacity.

Solution 7.16

Step 1: Calculate parameters to use Fig. 7.24.

$$H = 12 - 3 = 9\,\text{ft}, \frac{H}{B} = \frac{9}{6} = 1.5; \quad \frac{s_u}{\gamma B} = \frac{1600}{116 \times 6} = 2.3; \quad \frac{D_f}{B} = \frac{3}{6} = 0.5$$

Step 2: Extract ultimate bearing capacity from Fig. 7.24.
We will use the plots $\frac{H}{B} = 1$ and $\frac{H}{B} = 2$ for and interpolate for $\frac{H}{B} = 1.5$.
Also we have to interpolate between $\phi'_p = 30°$ and $\phi'_p = 40°$, and between $\frac{D_f}{B} = 1$ and $\frac{D_f}{B} = 0$.

$$\text{For } \phi'_p = 30°, \quad \frac{H}{B} = 1, \quad \frac{q_u}{\gamma B} = 14.$$

$$\text{For } \phi'_p = 40°, \quad \frac{H}{B} = 1, \quad \frac{q_u}{\gamma B} = 24$$

$$\therefore \phi'_p = 35°, \quad \frac{H}{B} = 1, \quad \frac{q_u}{\gamma B} = \frac{14 + 24}{2} = 19$$

$$\text{For } \phi'_p = 30°, \quad \frac{H}{B} = 2, \quad \frac{q_u}{\gamma B} = 20.$$

$$\text{For } \phi'_p = 40°, \quad \frac{H}{B} = 2, \quad \frac{q_u}{\gamma B} = 39$$

$$\therefore \phi'_p = 35°, \quad \frac{H}{B} = 2, \quad \frac{q_u}{\gamma B} = \frac{20 + 39}{2} = 29.5$$

$$\therefore \phi'_p = 35°, \quad \frac{H}{B} = 1.5, \quad \frac{D_f}{B} = 1, \quad \frac{q_u}{\gamma B} = \frac{19 + 29.5}{2} = 24.3$$

$$q_u = 24.3\gamma B = 24.3 \times 116 \times 6 = 16913 \text{ psf} = 16.9 \text{ ksf}$$

■

7.11 DESIGN FOR DUCTILITY

The design methods presented so far in this chapter to satisfy ultimate limit state are based on the assumption of the formation of slip planes between rigid soil bodies (Fig. 7.2) when the peak friction angle is mobilized. The factor of safety (ASD) and the performance factor (LRFD) are then used to provide a margin of safety. The stress states of soils below a footing are complex. At ultimate load, two thumblike regions of soils at critical state extend below the edges of the footing (Fig 7.26).

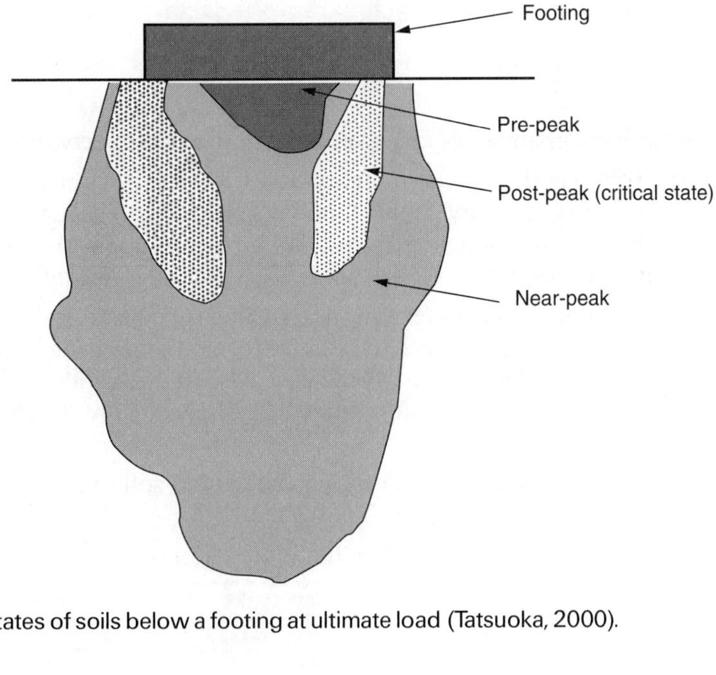

FIGURE 7.26 Stress states of soils below a footing at ultimate load (Tatsuoka, 2000).

An alternative method is to design for *ductility*—the ability to deform without rupture or sudden failure. In ductile design, the critical state friction angle, ϕ'_{cs}, is used. To use the existing equation based on limit analysis or limit equilibrium, it is assumed that as soon as slip planes develop, the soil along the slip planes strain softens to the critical state.

The procedure to design a shallow foundation for ductile response is as follows.

Long-Term Condition:

1. Use the ϕ'_{cs} in Eq. (7.4) for long-term conditions.
2. Apply a factor of safety in the range 1.25 to 1.5 for ASD or a performance factor in the range 0.9 to 0.95 for LRFD.

Short-Term Condition:

1. Calculate the mean effective stress at a depth $\frac{B}{2}$ below the foundation.
2. Correct the undrained shear strength based on the field condition. The water content of the soil in the field must be the same as the water content at which the test was done to determine the undrained shear strength. If not, an estimate can be made using Eq. (3.18) as

$$(s_u)_{\text{field}} = s_u \exp\left[G_s\left(\frac{w - w_{\text{field}}}{\lambda}\right)\right]; \quad R_o \leq 2 \tag{7.98}$$

where $(s_u)_{\text{field}}$ and w_{field} are the estimated undrained shear strength and the existing water content is the field, s_u and w are the currently known values of undrained shear strength and water content of the soil, and $\lambda(= \frac{C_c}{2.3})$ is the compression index. If $R_o > 2$, we can use Eq. (3.19) to estimate s_u as follows:

$$(s_u)_{\text{field}} = s_u \frac{(p'_o)_{\text{field}}}{p'_o}; \quad R_o > 2 \tag{7.99}$$

where s_u and p'_o are the lab test values of undrained shear strength and mean effective stress respectively. A FS in the range 1.25 to 1.5 should be used for soils with $R_o \leq 2$ and in the range 1.5 to 2 should be used for soils with $R_o > 2$.

The advantages of the ductility approach are as follows:

1. The critical state friction angle, ϕ'_{cs}, is a fundamental soil property, unlike ϕ'_p.
2. Elements of the soil mass loaded by a foundation follow various stress paths (Fig. 7.26), most of which cannot be reproduced by laboratory or field tests. The critical state friction angle is independent of stress paths and test methods, so the magnitude is the same regardless of the tests or stress paths that are followed.
3. Only simple tests are required to determine ϕ'_{cs}. Indeed for coarse-grained soils, you can find ϕ'_{cs} by simply measuring the angle of response—the angle formed by pouring coarse-grained soils to form a heap—and measuring the slope angle of the heap. Also, it does not matter whether plane strain or axisymmetric tests are conducted to determine ϕ'_{cs}. Theoretically ϕ'_{cs} should be the same for both types of tests.
4. Even the highest stressed soil elements within the loaded soil mass will be in Region III (ductile region, Fig. 4.20), under working loads.
5. The uncertainties in the soil's capacity to dilate are eliminated.

2. Evaluate the geotechnical data in each soil layer. How many different layers exist? Are the layers horizontal or are there layers that dip? How thin are the layers? How strong (qualitatively) is the different layer? Where is the groundwater level?
3. Use the structural and earth pressure loading to determine the worst-case load combination using ASD or LRFD.
4. Determine the minimum embedment depth from local code requirements and Section 7.12.
5. Check whether the foundation will rest on a soil with suitable strength and thickness below the base. Ideally, the thickness of the soil below the base should be at least 1.5 times the estimated width of the foundation. Adjust the foundation depth if necessary.

The procedure for conventional design of a shallow rectangle foundation using ASD is as follows:

1. Estimate a foundation size. For eccentric loads, check that either $e_B \leq \frac{B}{6}$ or $e_L \leq \frac{L}{6}$. If $e_B > \frac{B}{6}$, increase the width; if $e_L > \frac{L}{6}$, increase the length.
2. Determine settlement using a load factor of 1.
3. Check that settlement is less than or equal to the serviceability limit state. If settlement exceeds serviceability limit state, increase the foundation size and recalculate. If the settlement is much less than the serviceability limit state, then reduce foundation size and recalculate. The maximum tolerable settlement for most shallow foundations is about 25 mm (1 in.). Differential settlement has more stringent requirements. Use Table 6.5 as a guide.
4. For eccentric loads, calculate $B' = B - 2e_B$ and $L' = L - 2e_L$.
5. Calculate the maximum vertical stress, $(\sigma_a)_{\max} = \frac{Q_a}{A}(1 + \Gamma)$, where Q_a is the applied (design) load and $\Gamma = \frac{6e_B}{B}$ (short side failure) and $\Gamma = \frac{6e_L}{L}$ (long side failure), where B and L are the actual width and length.
6. Calculate the ultimate net bearing capacity (q_u) using the bearing capacity equations if lab test results are available or from empirical equations using in-situ test data. For fine-grained soils, calculate the ultimate bearing capacity for both short-term (TSA) and long-term (ESA) conditions. For free-draining coarse-grained soils, calculate the ultimate bearing capacity for long-term (ESA) conditions, which is also applicable to short-term conditions.
7. Calculate the allowable bearing capacity using $q_a = \frac{q_u}{\text{FS}} + \gamma D_f$.
8. Check that $q_a \geq (\sigma_a)_{\max}$.
9. Assess the constructability of the foundation.
10. Design reinforcement (beyond the scope of this textbook).

The performance or resistance factors to determine the bearing capacity of shallow foundations based on LFRD are shown in Table 7.8.

The procedures for the design of a shallow rectangle foundation using LRFD is as follows:

1. Estimate a foundation size. For eccentric loads, check that either $e_B \leq \frac{B}{4}$ or $e_L \leq \frac{L}{4}$. If $e_B > \frac{B}{4}$, increase the width; if $e_L > \frac{L}{4}$, increase the length.
2. Determine settlement using a load factor of 1.
3. Check that settlement is less than or equal to the serviceability limit state. If settlement exceeds serviceability limit state, increase the foundation size and recalculate. If settlement is much less than the serviceability limit state, then reduce foundation size and recalculate. The maximum tolerable settlement for most shallow foundations is about 25 mm (1 in.). Differential settlement has more stringent requirements. Use Table 6.5 as a guide.

TABLE 7.8 Performance Factors for Bearing Capacity Calculations Using LRFD[6]

Bearing capacity	Resistance factor, Ψ_R
ESA: Coarse-grained and fine-grained soils	
ϕ'_{cs} from lab tests	0.95*
ϕ'_p from lab tests	0.8
ϕ'_p from SPT	0.45
ϕ'_p from CPT	0.55
SPT	0.45
CPT	0.55
Plate load test	0.55
TSA: Fine-grained soils	
s_u from lab tests	0.60
Shear vane	0.60
CPT	0.50
FVT	0.60
Plate load test	0.55
Rock	0.6
Sliding	
Precast concrete placed on sand	
ϕ'_{cs} from lab tests	0.95*
ϕ'_p from SPTor CPT	0.90
Concrete cast-in-place	
ϕ'_p from SPTand CPT	0.80

[1]Collected from different sources. You need to check your local code for recommended values.

*Author's recommendation.

4. Calculate P_u using load factors from local codes for different combinations of loads.
5. For eccentric loads, calculate $B' = B - 2e_B$ and $L' = L - 2e_L$.
6. Calculate $(\sigma_u)_{\max} = \frac{Q_u}{A}(1 + \Gamma)$, where Q_u is the applied factored load and $\Gamma = \frac{4e_B}{B}$ (short side failure) and $\Gamma = \frac{4e_L}{L}$ (long side failure), where B and L are the actual width and length.
7. Calculate the ultimate net bearing capacity (q_u) using the bearing capacity equations if lab test results are available or from empirical equations using field test data. For fine-grained soils, calculate the ultimate net bearing capacity for both short-term (TSA) and long-term (ESA) conditions. For free-draining coarse-grained soils, calculate the ultimate net bearing capacity for long-term (ESA) conditions, which is also applicable to short-term conditions.
8. Calculate the gross resistance, $R = \psi_R q_u + \rho_i \gamma D_f$, where γD_f is the effective soil pressure above the foundation base, and ρ_i is the dead load factor for earth pressure (usually 0.9).
9. Check that $R \geq (\sigma_u)_{\max}$.
10. Assess the constructability of the foundation.
11. Design reinforcement (beyond the scope of this textbook).

You can take the following actions if serviceability and ultimate limit states are not satisfied after the first trial:

1. Increase the size of the foundation. This will reduce the applied stress. Note that in TSA, your bearing capacity is independent of foundation size.
2. Consider a mat foundation (Chapter 9) if the isolated foundation becomes too large with the potential for stress overlaps. The mat foundation will spread the load over a larger area than isolated foundations, thereby reducing the applied stress.
3. Consider soil improvement.
4. Consider deep foundations (Chapter 8).

7.14 SUMMARY

In this chapter, we described ways in which you can estimate the bearing capacity of soils and settlement of shallow foundations. The critical criterion for design of most shallow foundations is serviceability limit state. The equations that are used in designing foundations are based on simplifying assumptions and empirical methods. Consequently, good judgment based on past experience is required to successfully use these equations. In the case of coarse-grained soils, empirical methods based on in-situ tests, SPT, and CPT, are often the best choice. When designing a foundation, you must consider several issues such as constructability, economics, and local code requirements.

Practical Examples

EXAMPLE 7.18 *Design of Foundation at the Center of an Office Building*

A vertical column, 0.25 m (0.8 ft) × 0.25 m (0.8 ft) at the center of a two-storey building must transfer a dead load of 500 kN (112.5 kips) and a live load of 1000 kN (225 kips) to the foundation. The soil profile at the site is shown in Fig. E7.18a. Triaxial and one-dimensional consolidation tests were conducted on the stiff clay and the results are summarized in Fig. E7.18a. Groundwater is 8 m (25 ft) below finished grade. The total settlement of the foundation should not exceed 25 mm (1 in.). Minimum foundation depth according to the local code for frost action is 1.0 m (3 ft). Design (geometry only) a shallow foundation using (1) ASD with FS = 3 and (2) LRFD.

Strategy Determine the foundation size to satisfy serviceability limit state and then ultimate limit state. SI units will be used in solving this example problem.

Solution 7.18

Step 1: Assemble design parameters.

Loads

$$Q_{DL} = 500\,\text{kN}, \quad Q_{LL} = 1000\,\text{kN}$$

Soil

Triaxial tests: $\phi'_p = 27°, \phi'_{cs} = 25°,$

$$s_u = 120\,\text{kPa}, \quad E_u = 70\,\text{Mpa}.$$

Consolidation tests: $C_c = 0.12$, $C_r = 0.012$, OCR = 10.

Convert triaxial friction angle to plane strain friction angle.

$$(\phi'_p)_{ps} = \frac{9}{8}(\phi'_{cs})_{tr}$$

Design friction angle values: $\phi'_p = \frac{9}{8} \times 27 \approx 30°$.

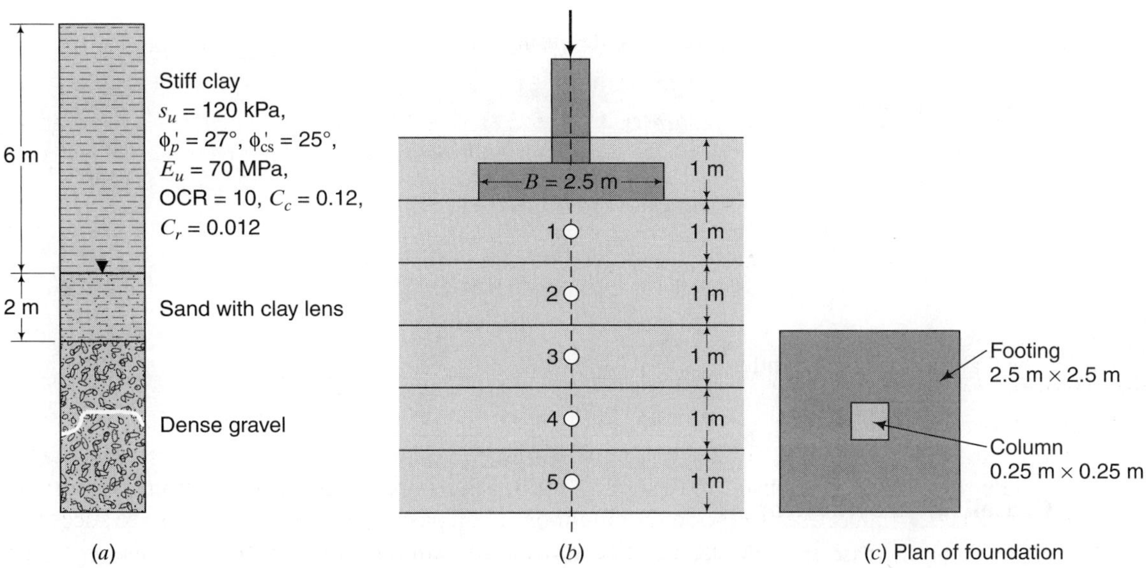

FIGURE E7.18a,b,c

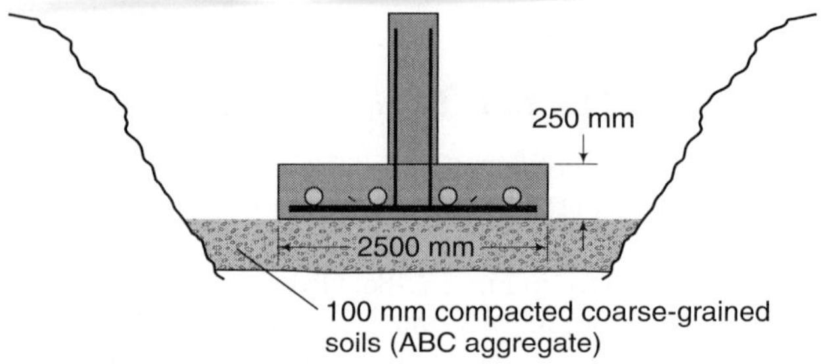

FIGURE E7.18d

Assume $G_s = 2.7$ and $v = 0.35$

$$\gamma_{sat} = 20\text{kN/m}^3, \quad e_o = \frac{G_s - \dfrac{\gamma_{sat}}{\gamma_w}}{\dfrac{\gamma_{sat}}{\gamma_w} - 1} = \frac{2.7 - \frac{20}{9.8}}{\frac{20}{9.8} - 1} = 0.633$$

Concrete $\gamma_c = 24\,\text{kN/m}^3$

Design Code
IBC (2006).

Design Criteria
ASD: FS = 3; $\delta = 25$ mm; $(D_f)_{min} = 1.0$ m.

Step 2: Assume foundation size.
Try a square foundation $B = L = 2.5$ m.
Minimum foundation thickness (Fig. 7.1a) = 0.225 m.
Assume foundation thickness $t_f = 0.25$ m.
Sketch foundation—see Fig. E7.18b,c

Step 3: Calculate weight of foundation.

$$W = B^2 t_f \gamma_c = 2.5^2 \times 0.25 \times 24 = 37.5\,\text{kN}$$

Step 4: Calculate loads using load combinations in IBC (2006).
Total $Q_{DL} = 500 + 37.5 = 537.5$ kN

ASD

$$Q_a = Q_{DL} + Q_{LL} = 537.5 + 1000 = 1537.5\,\text{kN}$$

LRFD

I : $1.4\,Q_{DL} = 1.4 \times 537.5 = 752.5$ kN

II : $1.2\,Q_{DL} + 1.6Q_{LL} = 1.2 \times 537.5 + 1.6 \times 1000 = 2245$ kN

Use $Q_u = 2245$ kN.

Step 5: Calculate settlement.
Settlements for both ASD and LRFD are calculated using a load factor of 1.

Elastic settlement

$$\rho_e = \frac{2Q_a(1 - v^2)}{E_u L_1} \mu_s \mu_{emb} \mu_{wall}$$

$$Q_a = 1537.5\,\text{kN}, \quad v = 0.35$$

$$\mu_s = 0.45\left(\frac{A_b}{L_1^2}\right)^{-0.38} = 0.45(1)^{-0.38} = 0.45$$

Neglect wall and embedment depth ($\mu_{emb} = \mu_{wall} = 1.0$).

$$\rho_e = \frac{2 \times 1537.5(1 - 0.35^2) \times 0.45}{70 \times 2.5} = 6.9\,\text{mm}$$

Consolidation Settlement

Most of the stress increase from the load will be distributed within a depth of $2B$ (see Chapter 2) below the foundation. Assume an effective depth below the foundation of $2.0 \times 2.5 = 5.0$ m.

Divide the effective depth into 5 equal layers of thickness 1 m. Calculate the vertical stress increase and overburden pressure at the center of each layer. You can use the program STRESS to calculate the vertical stress increase.

ASD : Surface stress (base of footing),

$$q_s = \frac{Q_a}{B^2} = \frac{1537.5}{2.5^2} = 246\,\text{kPa}.$$

Set up a spreadsheet to do the calculations as shown in Table E7.18.

TABLE E7.18

Layer	Thickness	$\Delta\sigma_z$	σ'_{zo}	σ'_{fin}	σ'_{zc}	ρ_{pc}
	m	kPa	kPa	kPa	kPa	mm
1	0.5	236	30	266	300	7.0
2	1.5	149	50	199	500	4.4
3	2.5	83	70	153	700	2.5
4	3.5	49	90	139	900	1.4
5	4.5	32	110	142	1100	0.8
					sum	16.1

Calculations for layer 1 are as follows:
From STRESS with $q_s = 246\,\text{kPa}$, $B = L = \frac{2.5}{2}$, $z = 0.5$ and clicking submit 4 times, we get

$$\Delta\sigma_z = 236\,\text{kPa}$$

$$\sigma'_{zo} = 20 \times 1.5 = 30\,\text{kPa}$$

$$\sigma_{\text{fin}} = \sigma'_{zo} + \Delta\sigma_z = 30 + 236 = 266\,\text{kPa}$$

$$\sigma'_{zc} > \sigma'_{fin};\ \text{Use}\ \rho_{pc} = \frac{H}{1+e_o} C_r \log\frac{\sigma'_{fin}}{\sigma'_{zo}}$$

$$\rho_{pc} = \frac{1.0}{1+0.633} \times 0.012 \log\left(\frac{266}{30}\right) = 0.007\,\text{m} = 7\,\text{mm}$$

See Table E7.18 for calculations in other layers.

The total settlement = elastic + consolidation = 6.9 + 16.1 = 23 mm < 25 mm = *okay*.

Step 6: Determine if failure surface will be confined to top layer for bearing capacity calculations.

$$H_{\text{crit}} = \frac{B}{2\cos\left(45^\circ + \frac{\phi'_p}{2}\right)} \exp[A \tan\phi'_p]$$

$$\text{Short term}: H_{\text{crit}} = \frac{B}{2\cos 45^\circ} = \frac{2.5}{1.414} = 1.77\,\text{m} < 5\,\text{m}$$

$$\text{Long term}: A = \left(45^\circ - \frac{\phi'_p}{2}\right)\text{radians} = \left(45^\circ - \frac{27^\circ}{2}\right)\frac{\pi}{180} = 0.55$$

$$H_{\text{crit}} = \frac{B}{2\cos\left(45^\circ + \frac{\phi'_p}{2}\right)} \exp[A \tan\phi'_p] = \frac{2.5}{2\cos(45^\circ + \frac{27^\circ}{2})} \exp[0.55 \tan 27^\circ] = 3.2\,\text{m} < 5\,\text{m}$$

The failure surface will be confined to the stiff clay layer.

Step 7: Check bearing capacity.

Assume rough foundation—a 100 mm layer of sand and gravel will be placed between the foundation base and the clay surface.

Short-Term Condition—TSA

$$q_u = 5.14 s_u d_c r_c$$

$$G = \frac{E_u}{2(1+\nu)} = \frac{70}{20(1+0.35)} = 26\,\text{MPa}$$

$$\frac{G}{s_u} = \frac{26 \times 10^3}{120} = 217 > 200$$

$$\therefore r_c = 1$$

Neglect effects of embedment depth; $d_c = 1$.

$$s_c = 1 + 0.2\frac{B'}{L'} = 1 + 0.2 = 1.2$$

$$q_u = 5.14 \times 120 \times 1.2 \times 1 \times 1 = 740\,\text{kPa}$$

$$\text{FS} = \frac{740}{246 - 1 \times 20} = 3.3$$

LRFD

Table 7.8: For s_u from lab tests, $\Psi_R = 0.6$.

$$R = \Psi_R q_u + \rho_i \gamma D_f = 0.6 \times 740 + 0.9 \times 20 \times 1 = 462\,\text{kPa}$$

$$(\sigma_u)_{\max} = \frac{Q_u}{BL} = \frac{2245}{2.5^2} = 359\,\text{kPa} < R; \quad \therefore \textit{okay}$$

Long-Term Condition—ESA

$$q_u = \gamma D_f(N_q - 1)s_q d_q r_q w_q + 0.5\gamma B N_\gamma s_\gamma d_\gamma r_\gamma w_\gamma$$

$$d_q;\ r_q = 1; w_q = 1; d_\gamma = 1, r_\gamma = 1, w_\gamma = 1$$

$$\phi'_p = 30^\circ \text{ (after conversion to plane strain condition)}$$

$$s_q = 1 + \frac{B'}{L'}\tan\phi'_p = 1 + 1\tan 30^\circ = 1.58$$

$$s_\gamma = 1 - 0.4\frac{B'}{L'} = 0.6$$

$$N_q = e^{\pi\tan\phi'_p}\tan^2\left(45 + \frac{\phi'_p}{2}\right) = e^{\pi\tan 30}\tan^2\left(45 + \frac{30}{2}\right) = 18.4$$

Davis and Booker (1971): $N_\gamma = 0.1054\exp\left(9.6 \times \frac{30 \times \pi}{180}\right) = 16.1.$

$$q_u = 20 \times 1(18.4 - 1) \times 1.58 + 0.5 \times 20 \times 2.5 \times 16.1 \times 0.6 = 791\,\text{kPa}$$

$$\text{FS} = \frac{791}{246 - 1 \times 20} = 3.5 > 3$$

LRFD

Table 7.8: For ϕ_p from lab test, $\Psi_R = 0.45$.

$$R = \Psi_R q_u + \rho_i \gamma D_f = 0.45 \times 791 + 0.9 \times 20 \times 1 = 374\,\text{kPa} > (\sigma_u)_{\max}\ (=359\,\text{kPa}); \quad \therefore \textit{okay}$$

Settlement governs the design. Long-term condition is more critical than short-term condition. See sketch of final foundation (Fig. E7.18d). ■

EXAMPLE 7.19 ***Sizing Foundation for a Building Using SPT Data***

The column layout for part of a building is shown in Fig. E7.19a. Three boreholes are located within this part of the building and the SPT data is shown in Figs. E7.19b,c,d. Determine satisfactory sizes for the foundations under columns $B2$ and $B4$ using ASD. The loads on these columns are given in Fig. E7.19a. The design criteria based on local codes are: maximum settlement = 25 min (1 in), minimum depth of foundation = 300 mm (12 in.), and the finished grade elevation is 780 m. Groundwater was not reached in any borehole.

Strategy It is best to draw an estimated soil profile based on the three borehole data items. It is customary to size the foundation under the largest load and then use the maximum vertical stress under this foundation to size the other foundations. In order for the settlement of the two foundations to be the same, the maximum vertical stress imposed in the soil should be the same. It is best to locate the foundations on the same layer, if possible. SI units will be used in solving this example problem.

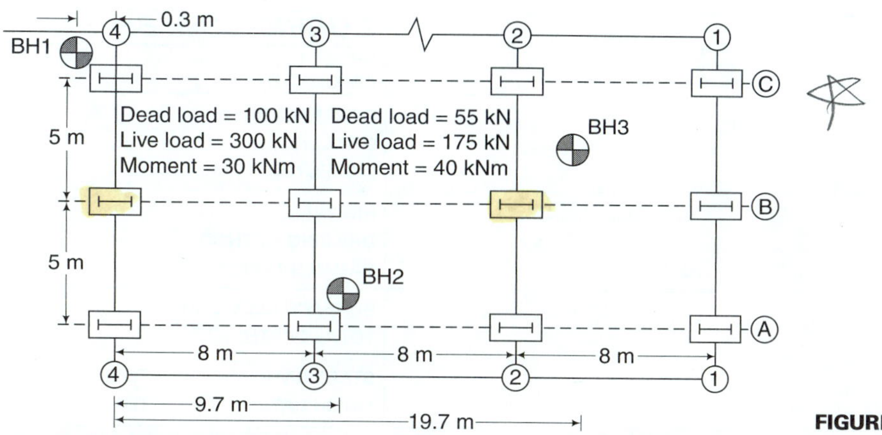

FIGURE E7.19a

Solution 7.19

Step 1: Draw a composite soil profile based on nearby boreholes.

1. Map boreholes in a graph of elevation versus distance using column line ④–④ as a reference line (see Fig. E7.19e).
2. Inspect soil type and N values in each borehole.
3. Sketch lines linking similar soil types. These lines are speculative since it is possible that the soil type could change or even that a void could exist within the distance between boreholes.
4. Divide each soil type, if necessary, into different density groups (dense, medium, loose) based on N values and soil description (see Fig. E7.19e). Assign values of unit weight based on Table B5.2 (Appendix B).

Step 2: Decide on a possible layer for the location.

Layer 2 with $N = 33$ (dense soil) at an elevation of 778.5 m appears to be a good location.

Step 3: Assume a foundation size for B4.

Assume a square foundation 1.2 m (4 ft) × 1.2 m (4 ft).

Step 4: Correct N values.

Use an average depth of $\frac{1.5B}{2}$ below the foundation $\left(\text{i.e.,} \frac{1.5 \times 1.2}{2} = 0.9\,\text{m; i.e., elevation } 777.6\text{ m}\right)$.

Original ground level at B4: elevation 781.5 m.

$$\sigma'_{zo} = \gamma z = 18(781.5 - 780.5) + 20(780.5 - 777.6) = 76\,\text{kPa}$$

$$c_N = \left(\frac{95.8}{\sigma'_{zo}}\right)^{\frac{1}{2}} = \left(\frac{95.8}{76}\right)^{\frac{1}{2}} = 1.12 < 2;\ okay$$

$$N_1 = c_N N = 1.12 \times 33 = 37$$

Step 5: Estimate q_u.

$$q_u = 32 N_1 B = 32 \times 37 \times 1.2 = 1421\,\text{kPa}$$

Step 6: Calculate maximum imposed vertical stress.

$$Q_{DL} = 100\,\text{kN}, \quad Q_{LL} = 300\,\text{kN}; \quad \text{vertical total load (ASD)}, \quad V_n = 400\,\text{kN}$$

$$M = \text{moment} = 30\,\text{kN/m}$$

$$eccentricity,\ e = \frac{30}{400} = 0.075\,\text{m}$$

$$\frac{B}{6} = \frac{1.2}{6} = 0.2 > e;\ \therefore\ \text{no tension}$$

$$(\sigma_a)_{\max} = \frac{V_n}{A}\left(1 + \frac{6e}{B}\right) = \frac{400}{1.2 \times 1.2}\left(1 + \frac{6 \times 0.075}{1.2}\right) = 382\,\text{kPa}$$

	BORING LOG: BH-1
	SHEET 1 of 1 PROJECT #:
PROJECT: CLIENT: CLIENT PROJECT:	CONTRACTOR: DRILLER: INSPECTOR:
COORDINATES N: E: LOCATION: Tucson REF. ALIGNMENT: STATION: OFFSET:	RIG TYPE: DRILLING METHOD: HAMMER TYPE:
COMMENTS:	SURFACE ELEV. : 781.5 TOTAL DEPTH: 23 START DATE: TIME: FINISH DATE: TIME:

	Casing	Split Spoon	Ring Sampler	Shelby Tube	Cuttings	Core Barrel
Type/Symbol		S	R	U	CU	C
I.D.		34.93mm	63.50mm		–	
O.D.		50.80mm	76.20mm		–	
Length		457.20mm	457.20mm		–	
Hammer WT	622.75 N		Drill Rod Size			
Hammer Fall	762mm		I.D. (O.D.)			

GROUNDWATER DATA					
Date	Time	Water Depth (m)	Casing Depth (m)	Hole Depth (m)	Symbol

DEPTH BELOW SURFACE (m)	ELEVATION (m)	TYPE	NUMBER	SYMBOL	DEPTH FROM (m)	DEPTH TO (m)	BLOWS 0-152.4 mm	BLOWS 152.4-304.8-mm	BLOWS 304.8-457.2mm	N-VALUE	RECOVERY (m)	VISUAL MATERIAL CLASSIFICATION AND REMARKS	MOISTURE, %	DRY DENSITY	SAMPLES SENT TO LAB
	780	S	S1		0.5	0.957	4	5	9	14	.457	Silty sand with gravel (SM)			
												Poorly graded sand with gravel and boulders (SP)			
5	775	S	S2		6	6.457	20	21	17	38	.457				
10	770	S	S3		10	10.457	12	14	10	24	.457	Lean Clay, hard, dark brown to gray (CL)			
15	765	S	S4		14	14.457	25	30	41	71	.229	Well graded sand, very dense (SW)			
20	760														
25												EOB @ 23m. No ground water encountered.			

FIGURE E7.19b

	BORING LOG: BH-2
	SHEET 1 of 1 **PROJECT #:**
PROJECT: **CLIENT:** **CLIENT PROJECT**	**CONTRACTOR:** **DRILLER:** **INSPECTOR:**
COORDINATES N: **REF. ALIGNMENT:** **E:** **STATION:** **LOCATION:** Tucson **OFFSET:**	**RIG TYPE:** **DRILLING METHOD:** **HAMMER TYPE:**
COMMENTS:	**SURFACE ELEV.:** 781 **TOTAL DEPTH:** 20 **START DATE:** **TIME:** **FINISH DATE:** **TIME:**

	Casing	Split Spoon	Ring Sampler	Shelby Tube	Cuttings	Core Barrel
Type/Symbol		S	R	U	CU	C
I.D.		34.93mm	63.50mm		–	
O.D.		50.80mm	76.20mm		–	
Length		457.20mm	457.20mm		–	
Hammer WT	622.75 N		Drill Rod Size			
Hammer Fall	762mm		I.D. (O.D.)			

GROUNDWATER DATA					
Date	Time	Water Depth (m)	Casing Depth (m)	Hole Depth (m)	Symbol

DEPTH BELOW SURFACE (m)	ELEVATION (m)	GRAPHIC	SOIL SAMPLE TYPE	NUMBER	SYMBOL	DEPTH (m) FROM	DEPTH (m) TO	BLOWS 0-152.4 mm	BLOWS 152.4-304.8-mm	BLOWS 304.8-457.2mm	N-VALUE	RECOVERY (m)	VISUAL MATERIAL CLASSIFICATION AND REMARKS	MOISTURE, %	DRY DENSITY	SAMPLES SENT TO LAB
	780		S	S1		1	1.457	6	9	10	19	.457	Silty sand with gravel (SM)			
			S	S2		2	2.457	11	12	10	22	.457	Pooly graded sand with trace of gravel (SP)			
5	775															
			S	S3		7	7.457	20	28	20	48	.457	Becomes fine to coarse			
10			S	S4		9	9.457	10	13	15	28	.457	Lean Clay, hard, dark brown to gray (CL)			
	770															
15			S	S5		14	14.457	25	35	31	66	.229	Well graded sand, very dense			
	765															
20																
	760												EOB @ 20m. No ground water encountered.			
25																

FIGURE E7.19c

	BORING LOG: BH-3
	SHEET 1 of 1 **PROJECT #:**
PROJECT: **CLIENT:** **CLIENT PROJECT:**	**CONTRACTOR:** **DRILLER:** **INSPECTOR:**
COORDINATES N: **REF. ALIGNMENT:** **E:** **STATION:** **LOCATION:** Tucson **OFFSET:**	**RIG TYPE:** **DRILLING METHOD:** **HAMMER TYPE:**
COMMENTS:	**SURFACE ELEV.:** 782 **TOTAL DEPTH:** 21 **START DATE:** **TIME:** **FINISH DATE:** **TIME:**

	Casing	Split Spoon	Ring Sampler	Shelby Tube	Cuttings	Core Barrel
Type/Symbol		S	R	U	CU	C
I.D.		34.93mm	63.50mm		–	
O.D.		50.80mm	76.20mm		–	
Length		457.20mm	457.20mm		–	
Hammer WT.	622.75 N		Drill Rod Size			
Hammer Fall	762mm		I.D. (O.D.)			

GROUNDWATER DATA					
Date	Time	Water Depth (m)	Casing Depth (m)	Hole Depth (m)	Symbol

DEPTH BELOW SURFACE (m)	ELEVATION (m)	GRAPHIC	SOIL SAMPLE TYPE	NUMBER	SYMBOL	DEPTH (m) FROM	DEPTH (m) TO	BLOWS 0-152.4 mm	BLOWS 152.4-304.8-mm	BLOWS 304.8-457.2-mm	N-VALUE	RECOVERY (m)	VISUAL MATERIAL CLASSIFICATION AND REMARKS	MOISTURE, %	DRY DENSITY	SAMPLES SENT TO LAB
													Silty sand with gravel (SM)			
	780															
			S	S1		2.5	2.957	5	7	9	16	.457				
													Poorly graded sand with trace of gravel (SP)			
5																
	775															
			S	S2		7.5	7.957	14	15	18	33	.457	Becomes fine to coarse			
10													Lean Clay, hard, dark brown to gray (CL)			
	770		S	S3		11	11.457	13	14	12	26	.457				
													Well graded sand, very dense (SW)			
15			S	S4		15	15.457	15	23	16	39	.457	Becomes brown to light brown			
	765															
			S	S5		18	18.229	25	36	42	78	.229				
20																
	760												EOB @ 21m. No ground water encountered.			
25																

FIGURE E7.19d

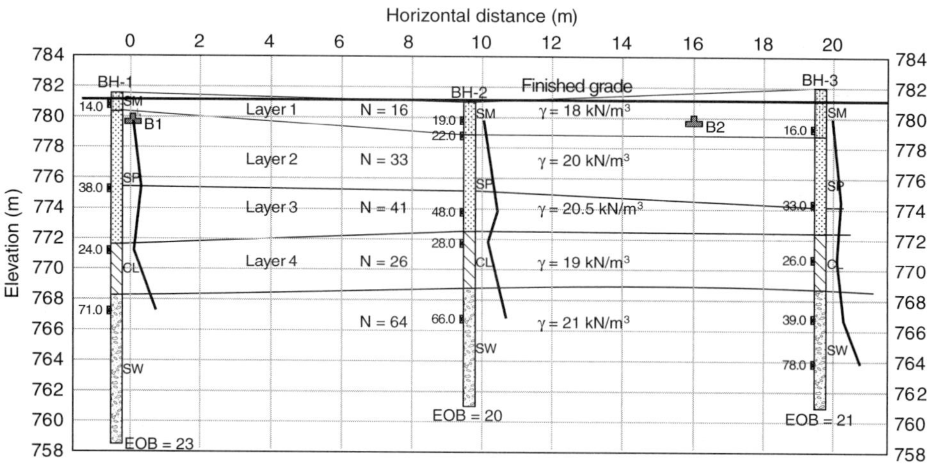

FIGURE E7.19e

	A	B	C	D	E	F	G	H
1	Bearing capacity from SPT		Help					
2	ρ_a	25	mm	Dead load	300	kN		
3	Finished grade elevation	780	m	Live load	100	kN		
4	Elevation of base of footing	778.5	m	Moment	30.00	kN.m		
5	Depth of footing	1.5	m	Total load	400.00	kN		
6	Width of footing	1.15	m	e	0.08		no tension	
7	Length of footing	1.15	m	σ_{max}	421			
8	Groundwater	100	m					
9	FS	3.26						
10								
11				Vertical				
12	Elevation	Thickness	Unit weight	effective stress	C_N	N	N_1	q_{ult}
13	m	m	kN/m³	kPa	use			kPa
14	781.5	0	0	0.0	0	0	0	0
15	780.5	1	18	18.0	2.0	16	32	1178
16	777.6375	2.8625	20	75.2	1.1	33	37	1370

(f) Footing *B4*

	A	B	C	D	E	F	G	H
1	Bearing capacity from SPT				Help			
2	ρ_a	25	mm	Dead load	55	kN		
3	Finished grade elevation	780	m	Live load	175	kN		
4	Elevation of base of footing	778.5	m	Moment	40.00	kN.m		
5	Depth of footing	1.5	m	Total load	230.00	kN		
6	Width of footing	1.05	m	e	0.17		no tension	
7	Length of footing	1.05	m	σ_{max}	416			
8	Groundwater	100	m					
9	FS	3.12						
10								
11				Vertical				
12	Elevation	Thickness	Unit weight	effective stress	C_N	N	N_1	q_{ult}
13	m	m	3	kPa	use			kPa
14	781.5	0	0	0.0	0	0	0	0
15	778.6	2.9	18	52.2	1.4	16	22	728
16	777.7	0.89	20	70.0	1.2	33	39	1298

(g) Footing *B2*

FIGURE E7.19f,g

Step 7: Check factor of safety.

$$\text{FS} = \frac{q_u}{(\sigma_a)_{\max}} = \frac{1421}{382} = 3.7$$

Step 8: Calculate size of foundation B2.

$$V_n = 55 + 175 = 230\,\text{kN}; \quad \text{moment} = 40\,\text{kN/m}$$

$$e = \frac{40}{230} = 0.17\,\text{m}$$

$$B_{\min} \text{ for no tension} = 6e = 6 \times 0.17 = 1.02\,\text{m}$$

Assume a square foundation with size to produce a similar maximum vertical stress on the soil as foundation B4.

$$(\sigma_a)_{\max} = \frac{V_n}{A}\left(1 + \frac{6e}{B}\right) = 382$$

$$382 = \frac{230}{B^2}\left(1 + \frac{6 \times 0.17}{B}\right)$$

$$B = 1.07\,\text{m}$$

Use $B = 1.1$ m.

Step 9: Assess design.
The foundations are greater than the minimum size for individual foundations (Fig. 7.1a) and there will be no overlap. You can investigate a more economical size of foundation by changing the size of the foundations to achieve an FS closer to 3. This is best done using a spreadsheet as shown in Figs. E7.19f and E7.19g. For practical purposes, the foundation sizes are $B4 = 1.2\,\text{m} \times 1.2\,\text{m}$ and $B2 = 1.1\,\text{m} \times 1.1\,\text{m}$. ■

EXERCISES

Use Davis and Booker (1971) expression for N_γ and IBC (2006) unless otherwise noted.

7.1 A foundation for a long concrete block wall of thickness 0.2 m (8 in.) will be founded on a dense sand (ϕ'_p from direct shear test = 32°; $D_r = 85\%$ and $\gamma_{sat} = 17.8\,\text{kN/m}^3$ (113 pcf)). The wall will transmit a vertical live load load of 60 kN/m (4000 lb/ft) in addition to the estimated foundation dead load of 7.5 kN/m (500 lb/ft). The foundation is to be located 0.5 m (1.5 ft) below finished grade. The groundwater level is 5 m (16 ft) below finished grade. Determine the size of the foundation (width) using ASD for FS = 3. Investigate the effects of a rise of the groundwater to (a) the ground surface and (b) the foundation base.

7.2 A square foundation, 3 m (10 ft) wide, is located 1.5 m (5 ft) below the finished grade in dense sand ($D_r = 80\%$). The soil parameters are: ϕ'_p from triaxial tests = 35°, $G_s = 2.65$, $e = 0.3$, and $w = 5\%$. Groundwater level is at an average depth 3.5 m (12 ft) below the base of the foundation but seasonally fluctuates ±1 m (±3 ft). Determine the minimum allowable bearing capacity using FS = 3.

7.3 A water tank 5 m (16 ft) in diameter is to be located on a shallow foundation 6 m (18 ft) in diameter and embedded 0.5 m (1.5 ft) from the finished ground surface. The soil is classified as CL with ϕ'_p from triaxial test= 23°, $\gamma_{sat} =$ 19.2 kN/m^3(122 pcf), s_u=55 kPa(1200 psf), and G=44 MPa (900 ksf). Determine the allowable bearing capacity using an FS of 3. Groundwater is at 2 m (6 ft.) below the surface.

7.4 A column carrying a dead load of 200 kN (45 kips) and a live load of 500 kN (112.5 kips) is to be founded on a rectangular foundation at a depth of 0.5 m (1.5 ft) below the finished ground surface in a deep deposit of dense sand ($D_r = 85\%$). The soil parameters are: ϕ'_p from direct shear tests = 38°, $\gamma = 19.8\,\text{kN/m}^3$(126 pcf), and $S = 70\%$. The groundwater level is at 8 m (25 ft) below the base of the foundation. Determine the size of the foundation to satisfy ultimate limit state based on ASD (FS = 3) and LRFD.

7.5 Repeat Problem 7.4 with a moment of 200 kN/m (147.5 kips/ft) about the Y-axis in addition to the vertical load.

7.6 A square foundation 1 m (3 ft) wide is embedded 2 m (8 ft) in a soil classified as SM. The soil parameters are: ϕ'_p from triaxial test = 28°, $\gamma = 18.5\,\text{kN/m}^3$(18 pcf), $D_r = 60\%$, and $S = 70\%$. Groundwater is at 3 m (10 ft) below finished grade. The foundation is required to carry a load of 1000 kN (225 kips) inclined at 15° to the vertical plane.

(a) Determine whether the foundation size is satisfactory to support the load using ASD and a desired FS of 2.5.

(b) If the foundation size is not satisfactory but cannot be increased and the load and its direction are fixed, recommend a remedial method that will enable the foundation to satisfy ultimate limit state.

(c) The acceleration coefficients for the site are $k_h = 0.1$ and $k_v = 0.02$ Determine whether the foundation size is satisfactory for a seismic event considering the vertical component of the load only.

7.7 Design a square foundation to be embedded 1.0 m (3 ft) in the sand layer at a site with a soil profile as shown in Fig. P7.7 to support a dead load of 100 kN (22.5 kips) and a live load of 150 kN (34 kips). Establish whether the clay layer will influence the bearing capacity. The design criteria are FS = 3 for ASD and maximum tolerable settlement is 25 mm (1 in.).

7.8 The foundation (base) of a long retaining wall is required to support the load (including self-weight of the base and the wall) and moment (this comes from the lateral loads on the wall) shown in Fig. P7.8. Determine the factor of safety against bearing capacity failure. Calculate the elastic settlement at the two corners of the base. Would the settlement of the base be uniform? Discuss your answer.

7.9 The column loads from a building at a site are shown in Fig. P7.9. The results shown for the stiff clay are averaged values obtained from triaxial undrained and one-dimensional consolidation tests. Design a suitable size for the foundation to support the column using LRFD. The maximum tolerable settlement is 20 mm (0.75 in.).

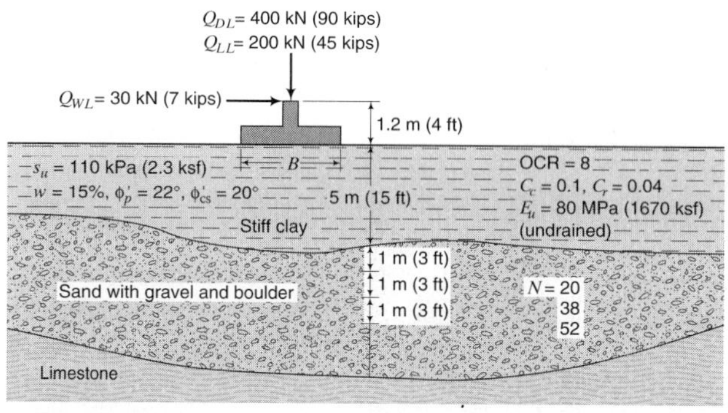

FIGURE P7.9

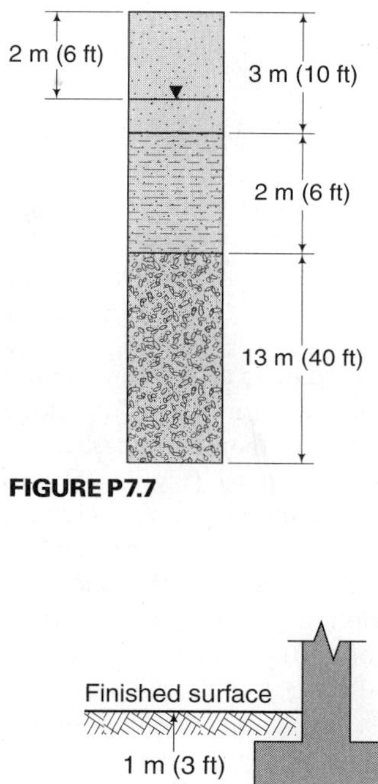

Sand
Triaxial tests: ϕ'_p = 37°, γ = 16.5 kN/m^3 (105 pcf), γ_{sat} = 17 kN/m^3 (108 pcf), E' = 45 MPa (940 ksf), ν' = 0.3

Stiff clay
Triaxial test: s_u = 60 kPa (1250psf), ϕ'_p = 26°, E_u = 40 MPa (835 ksf), ν' = 0.35.
Consolidation tests: OCR = 5, C_c = 0.3, C_r = 0.05.
Physical tests: w = 35%, G_s = 2.7

Dense gravel
N = 52 (average)

FIGURE P7.7

7.8 FS = 26.6, settlement will NOT be uniform; ratio of settlement = 2

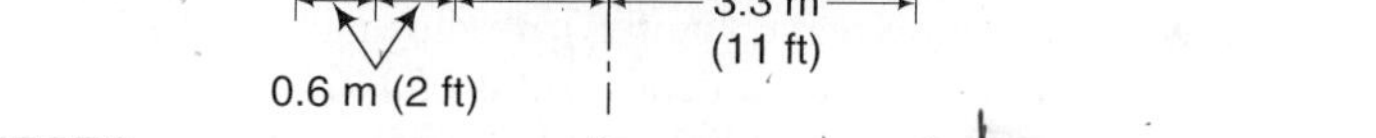

Sand — ESA
Triaxial tests: ϕ'_{cs} = 30°; ϕ'_p = 36°, E' = 50 MPa (1044 ksf)
Physical tests: D_r = 85%, γ = 18 kN/m^3 (114 pcf),
soil-wall base friction = 25° ??
Groundwater: 10 m (32 ft) below finished surface

FIGURE P7.8

7.10 An office building is to be located at a site with a representative soil stratigraphy as shown in Fig. P7.10. The loads from two columns of preliminary size 0.3 m (1 ft) × 0.3 m (1 ft) are also shown in Fig. P7.10. Determine suitable sizes of shallow foundations to safely transmit the loads. The total settlement should not exceed 25 mm (1 in) and the differential settlement should not exceed 10 mm. The finished elevation is 0 + 00. Use both LRFD and ASD with an FS = 3. For LRFD, use the load factors given in Chapter 6. You may use the spreadsheet at http://www.wiley.com/college/budhu. Click Foundations and Earth Structures, Chapter 7, bc.xls.

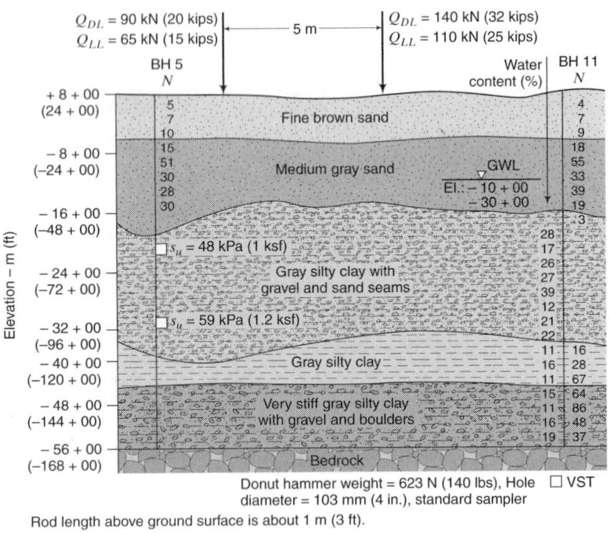

FIGURE P7.10

7.11 The identical office building in Problem 7.10 is to be located at another site. The CPT data for a borehole located between the two columns is at http://www.wiley.com/college/budhu. Click Foundations and Earth Structures, Chapter 7, problem 7.11 CPT data.xls. Determine suitable sizes of shallow foundations to safely transmit the loads using the same design criteria as in Problem 7.10. You may use the spreadsheet at http://www.wiley.com/college/budhu. Click Foundations and Earth Structures, Chapter 7, bc.xls.

7.12 A building is to be supported on a foundation 58 m (190 ft) wide × 75 m(246 ft) long × 3 m(10 ft) thick resting on a very stiff deposit of saturated clay. The foundation is to be located at 3 m (10 ft) below ground level. The average vertical stress at the base of the foundation is 350 kPa (7300 psf). Groundwater level is at 12 m (36 ft) below the surface. Triaxial and one-dimensional consolidation tests gave the following results:

Depth—m (ft)	0.5 (1.5)	6 (20)	25 (80)
s_u—kPa (psf)	58 (1200)	122 (2500)	156 (3200)

$e_o = 0.57$, $C_c = 0.16$, $C_r = 0.035$, $\phi'_p = 28°$, $\phi'_{cs} = 24°$, $E_u =$ 1 MPa(20 ksf), $\nu_u = 0.45$, $E' = 900$ kPa(18 ksf), and $\nu' = 0.3$. The subscript u denotes undrained condition and the superscript prime denotes effective stress condition. Determine the total settlement and the factor of safety against bearing capacity failure.

7.13 Repeat Problem 7.12 but design for ductility with FS = 1.25.

7.14 Design a square foundation to be embedded 1.0 m (3 ft) in the clay crust at a site with a soil profile as shown in Fig. P7.14. The dead load is 120 kN (27 kips) and the live load is 150 kN (34 kips). The design criteria are FS = 3 for ASD and maximum settlement is 25 mm (1 in.). Compare your design using LRFD.

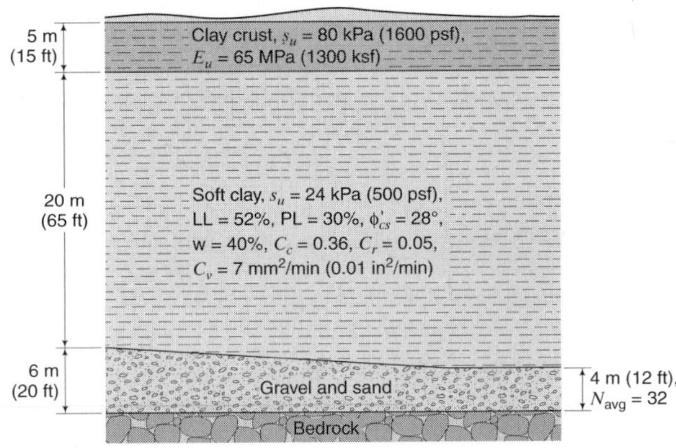

FIGURE P7.14

7.15 A factory building is required in an area where the soil profile from borehole data is as shown in Fig. P7.15a. One part of the factory, isolated from other buildings, has a building with a crane, also called a *gantry* (Fig. P7.15b), to load raw materials. The span of the building is 40 m (120 ft). The area is a marsh with organic matter in the top soil layer. Below this layer are layers of sands and gravels as shown in Fig. P7.15a. The USCS description for each layer is given. The topsoil up to an elevation of −10 m (−30 ft) was removed and replaced with compacted fill. After the fill was compacted at approximately 90% maximum dry unit weight (average), further boreholes were made with SPT conducted to ascertain the level of compaction. The SPT results for two boreholes, BH2 and BH5, near two of the outer columns (column A and column B) are shown in Fig. P7.15c. On each column, the total dead load is 650 kN (150 kips), live load except crane is 1300 kN (300 kips), and the dynamic (live)

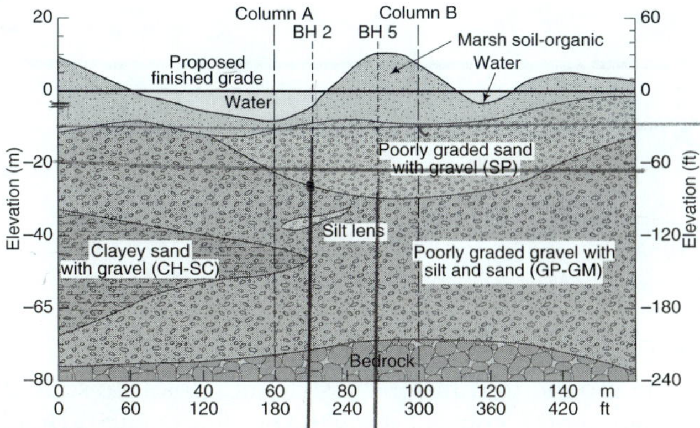

FIGURE P7.15a

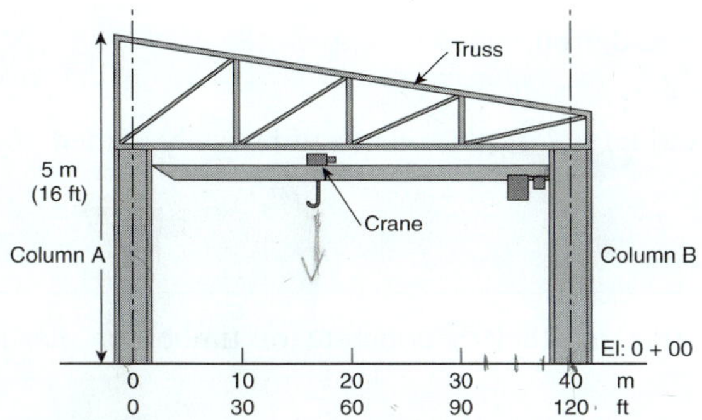

FIGURE P7.15b

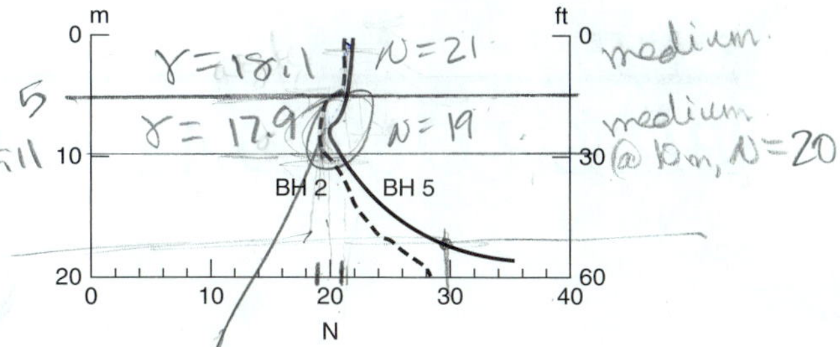

FIGURE P7.15c

load from the crane is 2000 kN (400 kips). Assume that the equivalent static live load from the crane is 125% of the dynamic (live) load. Design suitable shallow foundations for columns A and B. The total settlement is limited to 25 mm (1 in.) and the differential settlement cannot exceed 6 mm (1/4 in.). The minimum embedment depth is 1.5 m (5 ft). Use either ASD with FS = 3 or LRFD with the load factors given in Chapter 6.

Write a report describing and justifying your design. Clearly state all design assumptions, limitations (e.g. uncertainty of design data from the tests conducted) and criteria use. Describe how the foundation should be constructed and provide any necessary information to the contractor. Your report must contain a design summary about one-half page long. All drawings should be neatly drawn by hand or preferably by a CAD program. [Hint: One of the cases you should consider is when the crane is fully loaded and is near each of the columns.]

Footing dimensions
reinforcement

worst case scenario:
vert. loading = crane by one of the columns
moment: when crane is at opposite end of beam.
can you combine the worst case scenarios?
How do you deal w/ the moment?
=> e

CHAPTER 8

PILE FOUNDATIONS

8.0 INTRODUCTION

A pile is a long, slender, structural member installed in the ground to transfer the structural loads to soils at some significant depth below the base of the structure. In this chapter, you will study the bearing capacity (load capacity) and settlement of single and group piles under axial and lateral loads. When you complete this chapter, you should be able to:

- Estimate the allowable load capacity of single and group piles.
- Estimate the settlement of single and group piles.
- Estimate the allowable lateral load capacity and lateral displacements of laterally loaded piles.

8.1 DEFINITIONS OF KEY TERMS

Pile is a long, slender, structural member consisting of steel or concrete or timber or plastic or combinations of these materials.

***Drilled shaft* (*bored pile*)** is a pile that is constructed using concrete in a prebored hole.

Displacement pile is a pile that displaces a large volume of soil. Piles with solid sections are displacement piles.

Nondisplacement pile is a pile that displaces only a small volume of soil relative to its external volume. Steel H-piles and open-ended pipe piles are nondisplacement piles.

***Skin frictional stress* or *shaft frictional stress* or *adhesive stress* (f_s)**, is the frictional or adhesive stress on the shaft of a pile.

***End bearing stress* or *point resistance stress* or *base resistance stress* or *tip resistance stress* (f_b)**, is the stress at the base or tip of a pile.

***Ultimate load capacity* (Q_{ult})** is the maximum load that a pile can sustain before soil shear failure occurs.

***Ultimate group load capacity* $(Q_{ult})_g$** is the maximum load that a group of piles can sustain before soil shear failure occurs.

***Skin friction* or *shaft friction* (Q_f)** is the maximum frictional resistance generated on the shaft of a pile.

***End bearing* or *point resistance* or *base resistance* or *tip resistance* (Q_b)** is the resistance generated at the base or tip of a pile.

End bearing* or *point bearing pile is one that transfers almost all the structural loads to the soil at the base of the pile.

Friction pile is one that transfers almost all the structural loads to the soil by side shear (skin friction or adhesion) along a substantial length of the pile.

Floating pile is a friction pile in which the end bearing resistance is neglected.

The key terms used in pile foundations are illustrated in Fig. 8.1.

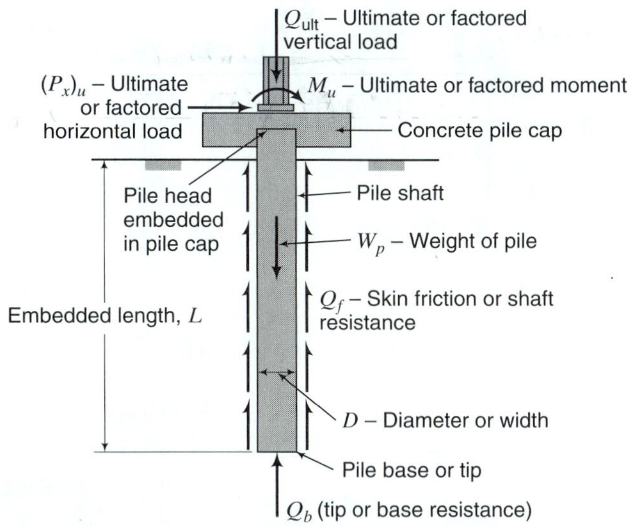

FIGURE 8.1 Pile terminology.

8.2 CONSIDERATIONS FOR USE OF PILE FOUNDATIONS

Pile foundations are used when

- The soil near the surface does not have sufficient bearing capacity to support the structural loads using shallow foundations.
- The estimated settlement of the soil exceeds tolerable limits (i.e., settlement greater than the serviceability limit state).
- Differential settlement due to soil variability or nonuniform structural loads is excessive.
- The structural loads consist of lateral loads and/or uplift forces.
- Excavations to construct a shallow foundation on a firm soil layer are difficult and expensive.

8.3 PILE TYPES

Importance *In designing pile foundations, you have to select a pile based on site conditions, structural loads, and local availability. In this section, some pile types are described. More information on how to select piles is presented in Section 8.16.*

Piles (Fig. 8.2) are made from timber, concrete, steel, or plastic or combinations of these materials (composites). The selection of a material type depends on the magnitude of the loading, the soil type, availability, and the environment in which the pile will be installed (e.g., corrosive environment, marine environment, etc.).

There are several types of concrete piles that are commonly used. These include cast-in-place concrete piles (Fig. 8.2a), precast concrete piles (Fig. 8.2b), and drilled shafts (Fig. 8.2f,g). Cast-in-place concrete piles are formed by driving a cylindrical steel shell into the ground to the desired depth, and then filling the cavity of the shell by fluid concrete. Plain concrete is used when the structural load is only compressive.

Precast concrete piles are fabricated in a construction yard from reinforced or prestressed concrete. They are preferred when the pile length is known in advance. A very popular type of precast concrete piles is the Raymond cylindrical prestressed pile. This pile comes in sections, and lengths up to 75 m (250 ft) can be obtained by stacking the sections. Concrete piles are best suited for high load capacity, long lengths, and for soils in which the environment is corrosive. However, the concrete within the corrosive environment needs to be protected. The disadvantages of precast piles are problems in transporting long piles, cutting, and lengthening.

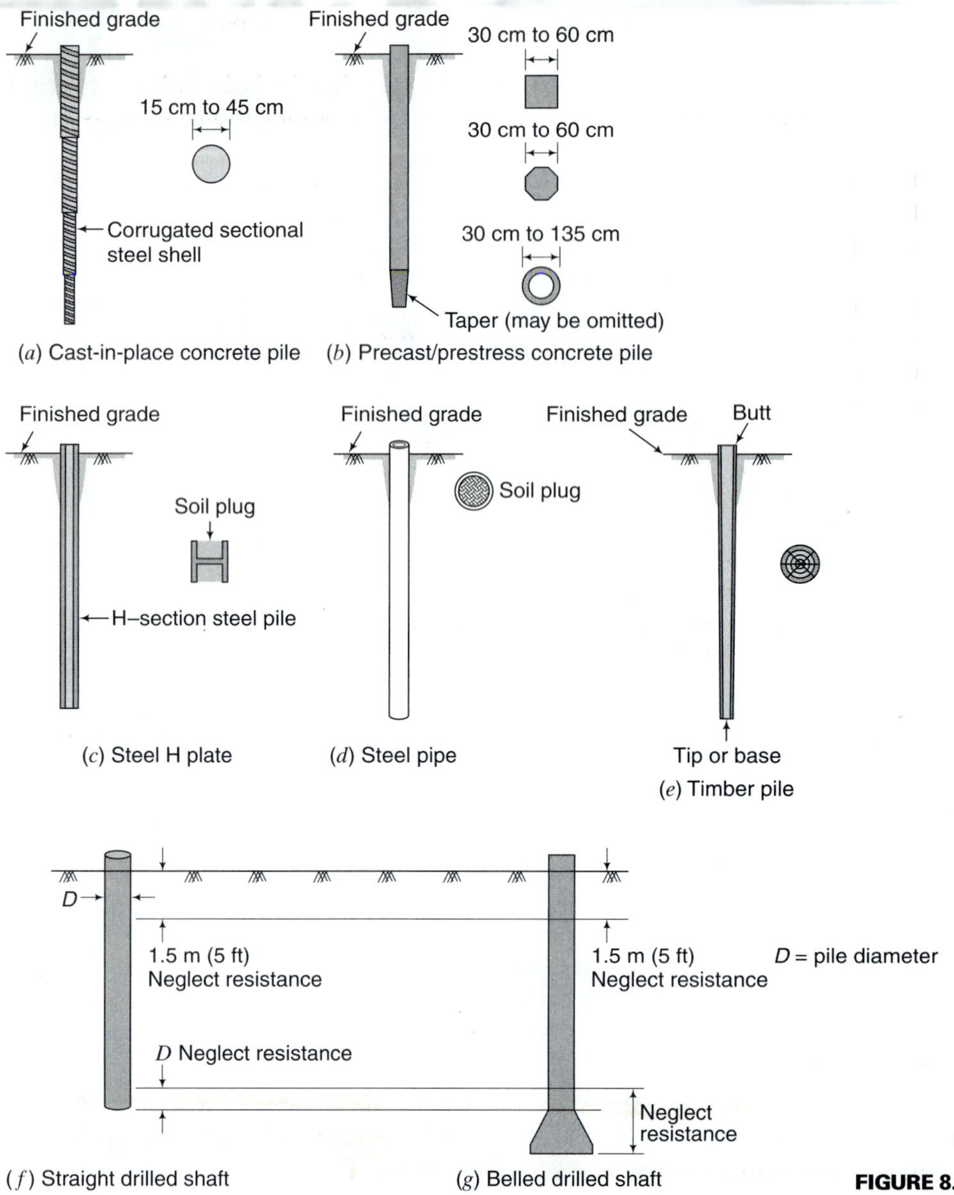

FIGURE 8.2 Pile types.

Steel piles (Fig. 8.2c,d) are best used for any noncorrosive soil type and high load capacity and long pile lengths (up to 70 m or 230 ft). For corrosive environment, steel piles have to be protected by one or more coatings of commercially available anticorrosive materials.

Timber piles (Fig. 8.2e) are best suited for fine-grained soils, low capacity loading (< 250 kN or < 55 kips), and short lengths, and for an environment in which the groundwater level is stable. The lengths of timber piles depend on the types of trees used to harvest the piles, but common lengths are about 12 m (40 ft). Longer lengths can be obtained by splicing several piles. Timber piles are susceptible to termites, marine organisms, and rot within zones exposed to seasonal changes.

Concrete, steel, and timber can be combined to form a composite pile. For example, the portion of a timber pile above groundwater level that is likely to suffer from decay due to termites or rot may be replaced by concrete. Similarly, the portion of a steel pile within a corrosive environment can be covered with concrete or other protective materials.

Plastic piles comprise a variety of composite materials that include polymer composites, PVC, and recycled materials. These piles are used in special applications such as in marine environment.

The cross sections of piles are circular or square or rectangular or octagonal and can vary along the length of the pile (Fig. 8.2). Straight piles have constant cross-sectional areas. Tapered piles have cross-sectional areas that decrease with pile length. Timber piles either are cylindrical or have square cross-sections. Cast-in-place piles (e.g., drilled shafts) can be straight or have a belled bottom (Fig. 8.2f,g). Steel piles come in various shapes and sizes, and include cylindrical, tapered, and H-piles. Steel H-piles are rolled steel sections. Steel pipe piles (Fig. 8.2d) are seamless pipes. Steel piles (H-piles or pipe piles) can be welded to give lengths up to 70 m (250 ft). Pipe piles are usually driven with open ends into the soil. A conical tip is used where the piles have to penetrate boulders and rocks. To increase the load capacity of steel pipe piles, the soil plug (Fig 8.2d) is excavated and replaced by concrete. These piles are called *concrete-filled steel pipe piles*.

Micropiles (also called *minipiles, pin piles, needle piles, or root piles*) are small diameter, 50 mm (2 in.) to 340 mm (13 in.), piles installed as pipes piles (pushed or driven) or grouted (jet or post or pressure) piles. They are particularly useful for sites with (1) low headroom (2) congestion, and (3) restricted access.

A summary of the different piles types is shown in Table 8.1. This table should assist you in pile selection.

The* key points *are:

1. ***Piles are made from timber, steel, concrete, plastic, or combinations of these material types.***
2. ***Piles are available in several standard shapes and lengths.***
3. ***You need to select piles based mainly on site conditions, structural loads, local availability, and method of installation.***

8.4 PILE INSTALLATION

Importance *Piles have to be installed into the ground. The method of installation affects the structural integrity of the piles and strength and deformation characteristics of the soil. Consequently, the load capacity is strongly dependent on pile installation methods. In this section, the popular methods of pile installation are presented.*

Piles can either be driven into the ground (driven piles) or be installed in a predrilled hole (bored piles or drilled shafts). When the installation process displaces only a small amount of soil ($\sim <10\%$), the pile is called a nondisplacement pile; otherwise the pile is a displacement pile. H-piles, open-ended pipe piles, and drilled shafts are nondisplacement piles. Timber piles and other driven piles with solid sections, such as driven precast concrete piles, are displacement piles. Displacement piles tend to cause significantly more soil remolding adjacent to the pile compared with nondisplacement piles. Consequently, the soil strength parameters used in design have to be adjusted based on whether the pile is a displacement or a nondisplacement pile.

TABLE 8.1 Comparison of Different Piles

Pile Type	Section* m (in)	Common Lengths m (ft)	Average Loads kN (kips)	Allowable Stress MPa (ksi)	Allowable driving stress – MPa (ksi)	Advantages	Disadvantages
Cast-in-place concrete	0.15 to 1.5** (6 to 60**)	$\leq$ 35 (115)	600 (135)	4.5 to 8.5 (0.65 to 1.2)	$0.85f'_c$	Can sustain hard driving, resistant to marine organisms, easily inspected, length can be changed easily.	Concrete can arch during placement, can be damaged if adjacent piles are driven before concrete sets.
Precast concrete rebar	0.15 to 0.3 (6 to 12)	$\leq$ 35 (115)	750 (170)	4.5 to 7 (0.65 to 1)	$0.85f'_c$	Economical for specified length, higher capacity than timber.	Cutting and lengthening of piles are expensive, handling is a problem. Shipping long piles is expensive, may crack during driving.
Precast concrete Prestressed	0.15 to 0.6 (6 to 24)	$\leq$ 35 (115)	1000 (225)	$\frac{f'_c}{3} - 0.27f_{pe}$ $(f_{pe})_{min} = 5$ MPa (0.7) $(f_{pe})_{max} = 34.5$ MPa (5)	$0.85f'_c - f_{pe}$	Economical for specified length Less permeable than reinforced concrete. Very good for marine environment.	Cutting and lengthening cannot be performed. Handling is a problem. Shipping long piles is expensive. Compressive strength decreases with increase in prestressing force.
Tapered cylinder	variable	< 60 (200)	2,000 (450)	40 to 70 (5.8 to 10)		Can ship in sections, high capacity, long length.	Cost (expensive).
Steel pipe	0.2 to 1 (8 to 36)	< 35 (115)	900 (200)	59 to 83 (8.5 to 12)	186 (27) – ASTM A252, f_y = 207 (30) GR 1 217 (31.5) ASTM A572, f_y = 241 (35) GR 2 310 (40.5) ASTM A572, f_y = 345 (50) GR 3	High axial and lateral capacity, can take hard driving, easy inspection and handling, length can be changed easily, resistant to deterioration.	Needs treatment for corrosive environment.
Concrete-filled pipe	0.2 to 1 (8 to 36)	< 35 (115)	900 (200)	Concrete: 4.5 to 8.5 (0.65 to 1.2) Steel: 62 to 83 (9 to 12)	186 (27) – ASTM A252, f_y = 207 (30) GR 1 217 (31.5) ASTM A572, f_y = 241 (35) GR 2 310 (40.5) ASTM A572, f_y = 345 (50) GR 3	(Similar to steel pipe pile)	(Similar to steel pipe pile)
Steel H-Pile	Webs: 1 to 3 (36 to 117) Flange: 0.2 to 0.35 (8 to 14)	< 60 (200)	900 (200)	59 to 83 (8.5 to 12)	223 (32.4) – ASTM A36, f_y = 248 (36) 310 (40.5) ASTM A572, f_y = 345 (50)	Nondisplacement pile, can take hard driving, easy handling, high axial and lateral capacity, length can be changed easily.	(Similar to steel pipe pile)
Timber	0.125 to 0.45 (5 to 18)	12 to 35 (40 to 115)	250 (55)	5.5 to 8.5 (0.8 to 1.2)	5.5 to 8.3 (0.8 to 1.2)	Low cost, easy to handle, renewable resource.	Low capacity. Can deteriorate above groundwater if not protected. Cannot take hard driving.
Micropiles	0.05 to 0.3 (2 to 12)	< 25 (80)	1000 (225)	Concrete: 4.5 to 8.5 (0.65 to 1.2) Steel: 59 to 83 (8.5 to 12)	(Similar to concrete and steel pipe pile)	High capacity, low headroom, restricted access, low noise and vibration	

*Diameter or width as appropriate. **The larger sizes (> 0.6 m) refer to drilled shafts.
f'_c = ultimate strength of concrete = 21 MPa, GR = grade, f_{pe} is prestress.

Cast-in-place (CIP) pipe piles can be installed by driving a steel closed-end pipe into the ground up to the desired depth and then either filling the void left as the pipe is removed with concrete or leaving the pipe in place and filling the inside with concrete. A popular method of installing CIP piles such as drilled shafts is to use a continuous flight hollow-stemmed auger (a screw auger with a hollow core). A cap is positioned at the base of the stem and a hole is bored to the desired depth. Concrete is pumped through the hollow core, which then pushes the cap out and fills the void left in the soil as the auger is removed. The auger is removed while rotating in the same direction as the insertion and at a rate that allows the spoils to be expelled and preventing the hole from collapsing. A special tool—an underreamer tool—may be used in stiff soils to widen the base of the hole to increase the end resistance of the pile. A rebar cage, if needed, is then inserted from the surface into the fresh concrete.

Driven piles can be installed using one of the following methods:

- Drop hammers
- Jetting
- Jacking
- Vibration
- Explosion

The key components of drop hammers are the leads and the hammer. The leads are used to align the hammer to strike the pile squarely (Fig. 8.3). Hammers can be simple drop hammers or steam/pneumatic hammers. Two popular types of steam/pneumatic hammers are shown in Fig. 8.4. The single-acting hammer utilizes steam or air to lift the ram and its accessories (cushion, drive caps, etc.). The double-acting hammer is used to increase the number of blows per minute and utilizes steam/air to lift the ram and force it down.

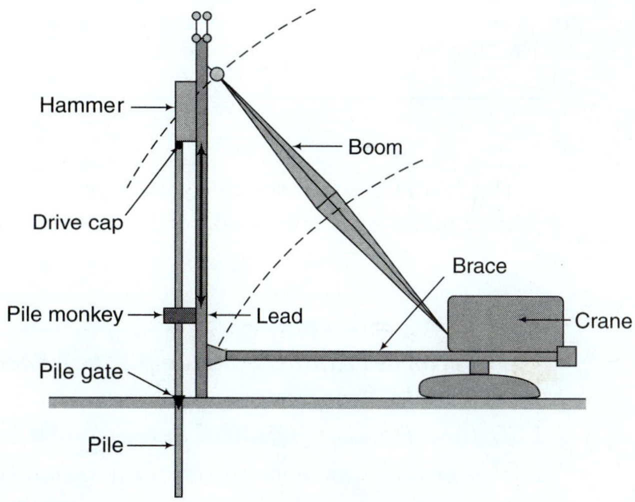

FIGURE 8.3 Key components of pile driving operation.

Piles can be installed using water jets. This technique is best suited for sandy or gravelly soils. They are not suitable for stiff clays or coarse-grained soils containing boulders. Piles in soft clays or loose coarse-grained soils can be jacked into the ground using hydraulic jacks. The soil disturbance around the pile is much less than for driven piles. However, the jacking method is suited only for short or minipiles. Piles in sandy or gravelly soils can be vibrated to the required depth using vibratory hammers. These hammers are either electrically powered or hydraulically powered. They consist of rotating eccentric masses within a housing that is attached to the pile head. Vibratory hammers are generally selected based on experience (Table 8.2). Explosion can be used to loosen compacted or cemented soils or bedrock to install piles.

When a pile is driven into the soil, the force applied must be more than sufficient to fail the soil at the tip and overcome the friction on its sides. The stresses imposed during driving are generally higher than the stresses imposed under static loading. When selecting the pile material, you should consider the short-term driven stresses (Table 8.1) to ensure that the pile load capacity is not compromised during installation.

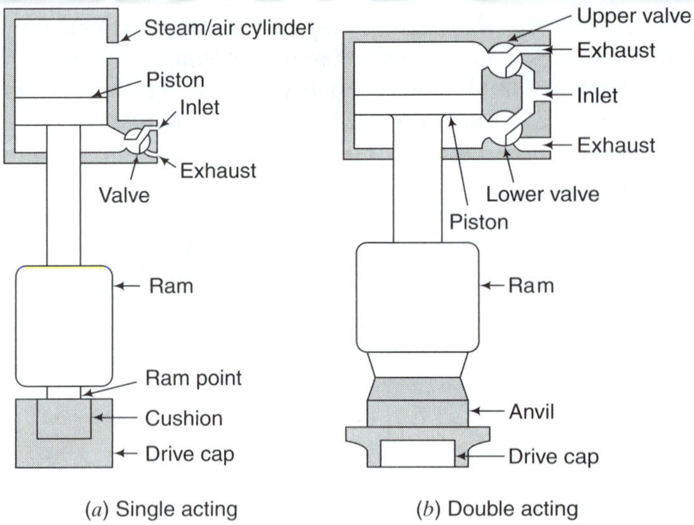

FIGURE 8.4 Two types of pile driving hammers.

TABLE 8.2 Selection of vibratory hammers for coarse-grained soils and soft, low plasticity clays

Pile length m (ft)	Type of vibratory hammer	Eccentric moment N.m (lb.in)	Frequency (RPM)
0 to 10 (0 to 30)	small	135 (1200)	1200–1600
10 to 15 (30 to 50)	medium	270 (2400)	1200–1600
>15	large	540 (4800)	1200–1600

Pile installation changes the soil characteristics and significantly affects the pile load capacity. Care should be taken to minimize soil disturbances during pile installation.

The key points are:

1. ***Piles can be installed using simple drop hammers, but most often they are installed using steam or pneumatic hammers.***
2. ***Drilled shafts are installed in a prebored hole.***
3. ***Pile installation methods can cause structural damage, remold the soil, and reduce the pile load capacity.***

8.5 LOAD CAPACITY OF SINGLE PILES

Importance *The load capacity of piles has been studied extensively. However, no universally acceptable method of determining the load capacity has evolved. The analysis of the load capacity of piles for most practical situations is based on empirical equations, experience, and judgment. For* *special structures and loading conditions (e.g., seismic loads), computer programs based on FDM, FEM, and BEM are used. Analyses of pile load capacity are approximations because it is difficult, if not impossible, to account for the variability of soil types and the differences in the quality of construction practice, especially pile installation. In this section, some broad ideas regarding the determination of pile load capacity are presented.*

In practice, the ultimate load capacity, Q_{ult}, of a pile is divided into two parts. One part is due to friction (Fig. 8.1), called *skin friction* or *shaft resistance*, Q_f, and the other is due to end bearing at the base or tip of the pile, Q_b. If the skin friction is greater than about 80% of the end bearing load capacity, the pile is deemed a friction pile; if the reverse, it is an end bearing pile. If the end bearing is neglected, the pile is called a *floating pile*.

From statics (Fig. 8.1):

$$Q_{\text{ult}} = Q_f + Q_b - W_p \tag{8.1}$$

where W_p is the weight of the pile. In most situations, W_p is small in comparison to Q_{ult} and is neglected. For large diameter (>1 m; >3 ft) and long concrete piles (>8 m; >25 ft), you should consider including W_p. The allowable load capacity based on ASD is expressed as:

$$Q_a = \frac{Q_{\text{ult}}}{\text{FS}} \tag{8.2}$$

where FS is a gross factor of safety usually greater than 2. In the case of LRFD, the condition for satisfying ultimate limit state (ULS) is

$$(\text{combination of factored load for worst case}) \leq (\varphi_f Q_f + \varphi_b Q_b) \tag{8.3}$$

where φ_f and φ_b are the performance factors for skin friction and end bearing respectively.

The distribution of skin frictional stress to the adjacent soil depends on the soil type. In fine-grained soils, the load transfer is nonlinear and decreases with depth as illustrated in Fig. 8.5. As a result, elastic

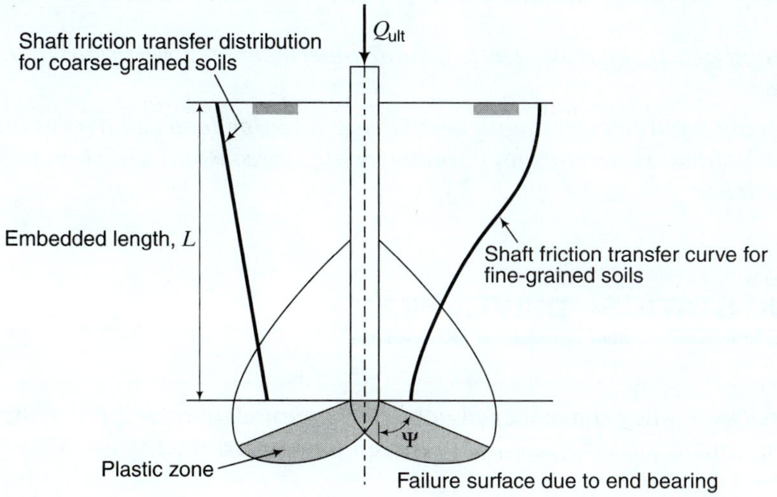

FIGURE 8.5 Load transfer characteristics.

compression of the pile is not uniform; more compression occurs on the top part than on the bottom part of the pile. For coarse-grained soils, the load transfer is approximately linear with depth (higher loads at the top and lower loads at the bottom).

In order to mobilize skin friction and end bearing, some movement of the pile is necessary. Field tests have revealed that to mobilize the full skin friction, a vertical displacement of 2.5 mm (0.1 in.) to 10 mm (0.4 in) is required. The actual vertical displacement depends on the strength of the soil and is independent of the pile length and pile diameter. The full end bearing resistance is mobilized in driven piles when the vertical displacement is about 8% to 10% of the pile tip (base) diameter. Similarly for drilled shafts, a vertical displacement of about 30% of the pile tip diameter is required. Because the full end bearing is not mobilized at the same displacement as the full skin friction—the full skin friction is mobilized about one-tenth the displacement required to mobilize the full end bearing resistance—different factors of safety or performance factors can be applied to skin friction and end bearing.

The full end bearing resistance is mobilized when slip or failure zones similar to shallow foundations are formed (Fig. 8.5). The end bearing resistance can then be calculated by analogy with shallow foundations with consideration given to the overburden soil pressure above the base and the skin friction. The important bearing capacity factor is N_q.

Piles driven into loose, coarse-grained soils tend to densify the adjacent soil. When piles are driven into dense, coarse-grained soils, the soil adjacent to the pile becomes loose. Pile driving usually remolds fine-grained soils near the pile shaft. The implication of pile installation is that the intact shear strength of the soil is changed and you must account for this change in your estimation of the load capacity.

Except for loose, coarse-grained soils, the soil mass adjacent to a pile reaches the critical state or close to it. You should then use critical state shear strength parameters in determining the load capacity of piles. Loose, coarse-grained soils adjacent to driven piles are densified and ϕ' tends to ϕ'_p. The actual value of ϕ'_p depends on the magnitude of the normal effective stress. Since the magnitude of the normal effective stress is uncertain, you should be cautious when using ϕ'_p in estimating the load capacity. In saturated clays, driven piles tend to remold the adjacent soil in an annulus about one or more diameters. It would be prudent to use $\phi' = \phi'_{cs}$ and remolded values of s_u in all cases except for some clays with a predominance of flat particles, where it is advisable to use $\phi' = \phi'_r$.

The* key points *are:

1. ***Pile load capacity depends on the soil type, method of installation, and construction practice.***
2. ***The full skin friction is mobilized at about one-tenth of the displacement required to mobilize the full end bearing resistance.***
3. ***The critical state shear strength parameters should be used in estimating load capacity of piles, except you should use the residual shear strength parameters for overconsolidated clays with a predominance of flat particles.***

8.6 LOAD CAPACITY USING STATICS—DRIVEN PILES

Access a spreadsheet at http://www.wiley.com/college/budhu. Click Foundations and Earth Structures, Chapter 8, pile.xls to calculate the load capacity of driven piles and drilled shafts.

Importance *To get an estimate of the pile load capacity, at least in the preliminary design stages, recourse is made to statics, experience, and to correlations using lab and field test results. In this section, we will examine how statics is applied to obtain an estimate of pile load capacity.*

A number of methods are available to predict Q_f and Q_b. We will deal with two methods:

1. Statics: α and β methods
2. Pile load test

8.6.1 α-Method (Total Stress Analysis)—Short-Term Loading in Fine-Grained Soils

The α-method is based on a total stress analysis (TSA). It should be used to estimate the short-term load capacity of driven piles embedded in fine-grained soils.

Skin or Shaft Resistance In the α-method, a coefficient, α_u, is used to relate the undrained shear strength, s_u, to the adhesive or skin frictional stress (f_s) along the pile shaft. The skin friction, Q_f, over the embedded length of the pile is the product of the adhesive or skin frictional stress, $f_s = \alpha_u s_u$, and the surface area of the shaft (perimeter × embedded length). Thus,

$$Q_f = \sum_{i=1}^{j} (\alpha_u)_i (s_u)_i \times (\text{perimeter})_i \times (\text{length})_i \tag{8.4}$$

where j is the number of soil layers within the embedded length of the pile. For a cylindrical pile of uniform cross section, diameter, D, and embedded length, L, in a homogeneous soil the skin friction is:

$$Q_f = \alpha_u s_u \pi DL \tag{8.5}$$

For H-piles, the perimeter is calculated for the block size (flange width × gross web depth). The value of α_u to use in determining the skin friction of piles is a subject of much debate and testing. Most tests to determine α_u are laboratory tests on model piles installed in a uniform deposit of soil. The major problems with these laboratory tests are:

- It is difficult to scale up the laboratory model test results to real piles.
- The soils in the field are mostly nonuniform compared with carefully prepared uniform soils in the laboratory.
- Pile installation in the field strongly influences α_u, which cannot be accurately duplicated in the laboratory.

Full scale field tests on real piles are preferred, but such tests are expensive and the results may apply only to the site where the tests are performed. The results from cone penetrometers and SPT have been linked to α_u, but these are found from statistical correlations with rather low coefficient of correlation. Randolph and Murphy (1985) recommended that f_s be estimated using the lower of the following two expressions

$$f_s = 0.5\sqrt{s_u \sigma'_{zo}} \tag{8.6}$$

$$f_s = 0.5\, s_u^{0.75} (\sigma'_{zo})^{0.25} \tag{8.7}$$

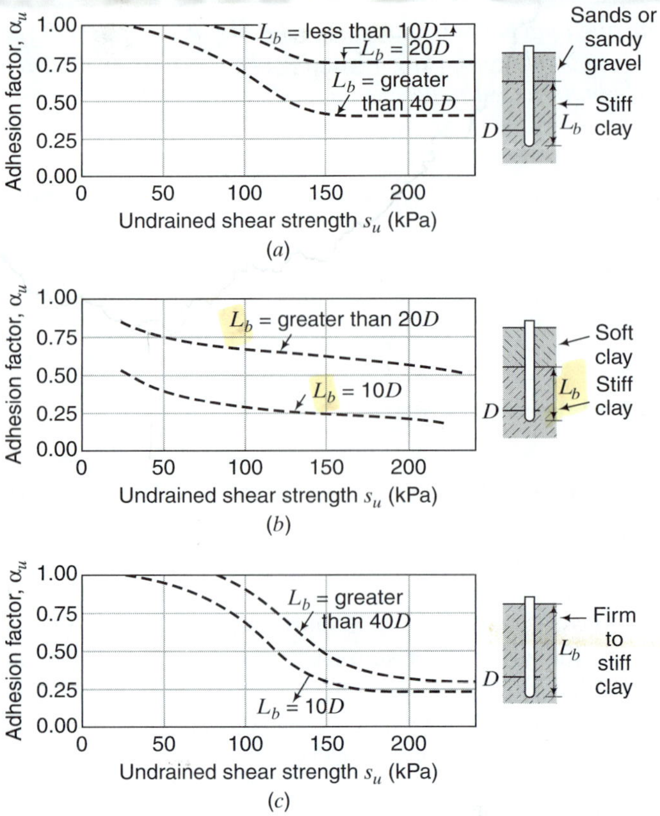

FIGURE 8.6 α_u for fine-grained soils (after Tomlinson, 1987).

where s_u is the undrained shear strength and σ'_{zo} is the vertical effective stress (overburden pressure). The American Petroleum Institute (API, 1993) adopted expressions (8.6) and (8.7). Tomlinson (1987) recommended values of α_u based on field measurements as shown in Fig. 8.6. You should use the critical state undrained shear strength (remolded shear strength) for driven piles because of soil disturbances caused by pile installation.

Shaft friction tend to decrease with embedment depth. An expression to account for embedment was proposed by Kolk and van der Velde (1996) as

$$f_s = 0.55\, s_u^{0.7}\, (\sigma'_{zo})^{0.3} \left(\frac{40}{L/D}\right)^{0.2} \tag{8.8}$$

End Bearing or Base Resistance The end bearing capacity is found by analogy with shallow foundations and is expressed as:

$$Q_b = f_b A_b = N_c (s_u)_b A_b \tag{8.9}$$

where f_b is the base resistance stress, N_c is a bearing capacity coefficient, $(s_u)_b$ is the average undrained shear strength of the soil within two pile diameters below the base, and A_b is the cross-sectional area of the base of the pile. The expressions for N_c are as follows:

$$N_c = 9 \quad \text{for} \quad (s_u)_b > 25\,\text{kPa}(500\,\text{psf}) \tag{8.10}$$

$$N_c = 6 \quad \text{for} \quad (s_u)_b \le 25\,\text{kPa}(500\,\text{psf}) \tag{8.11}$$

8.6.2 β-Method (Effective Stress Analysis)—Long-Term and Short-Term Conditions in Coarse-Grained Soils and Long-Term Condition in Fine-Grained Soils

The β-method is based on an effective stress analysis. It is used to estimate the short-term and long-term pile load capacity in coarse-grained soils and the long-term load capacity of fine-grained soils.

Skin or Shaft Resistance The friction along the pile shaft is found using Coulomb's friction law. The shaft frictional stress is $f_s = \mu\sigma'_x = \sigma'_x \tan\phi'_i$, where μ is the coefficient of friction, σ'_x is the lateral effective stress, and ϕ'_i is the soil–pile interfacial effective friction angle. Typical ranges of interface friction values are given in Table 8.3.

TABLE 8.3 Typical Range of Interfacial Friction Value

Material	steel	concrete	timber
ϕ'_i	$\frac{2}{3}\phi'_{cs}$ to $0.8\phi'_{cs}$	$0.9\phi'_{cs}$ to $1.0\phi'_{cs}$	$0.8\phi'_{cs}$ to $1.0\phi'_{cs}$

The skin friction is expressed as:

$$Q_f = \sum_{i=1}^{j} (\sigma'_x)_i \tan\phi'_i \times (\text{perimeter})_i \times (\text{length})_i \tag{8.12}$$

The lateral effective stress is proportional to the vertical effective stress (σ'_x) by a coefficient K (Chapter 2). Therefore, we can write Eq. (8.12) as

$$Q_f = \sum_{i=1}^{j} K_i(\sigma'_z)_i \tan\phi'_i \times (\text{perimeter})_i \times (\text{length})_i \tag{8.13}$$

We can replace the two coefficients K_i and $\tan\phi'_i$ by a single factor β to yield:

$$Q_f = \sum_{i=1}^{j} \beta_i(\sigma'_z)_i \times (\text{perimeter})_i \times (\text{length})_i \tag{8.14}$$

Several expressions for β have been proposed. In this textbook, we will use the following expressions:

- Fine-grained soils (Burland, 1973)

$$\beta = K\tan\phi'_i = K_o^{oc}\tan\phi'_i = (1 - \sin\phi'_{cs})(OCR)^{0.5}\tan\phi'_i \tag{8.15}$$

- Sands (Burland, 1973)

$$\beta = (1 - \sin\phi'_{cs})\tan\phi'_i \tag{8.16}$$

- Uncemented calcareous sand (Poulos, 1988)

$$\beta = 0.05 \text{ to } 0.1 \tag{8.17}$$

End Bearing or Base Resistance The end bearing capacity is calculated by analogy with the bearing capacity of shallow footings considering only overburden pressure and is determined from

$$Q_b = f_b A_b = N_q(\sigma'_z)_b A_b \tag{8.18}$$

where $f_b = N_q(\sigma'_z)_b$ is the base resistance stress, N_q is a bearing capacity coefficient that is a function of ϕ', $(\sigma'_z)_b$ is the vertical effective stress at the base, and A_b is the cross-sectional area of the base.

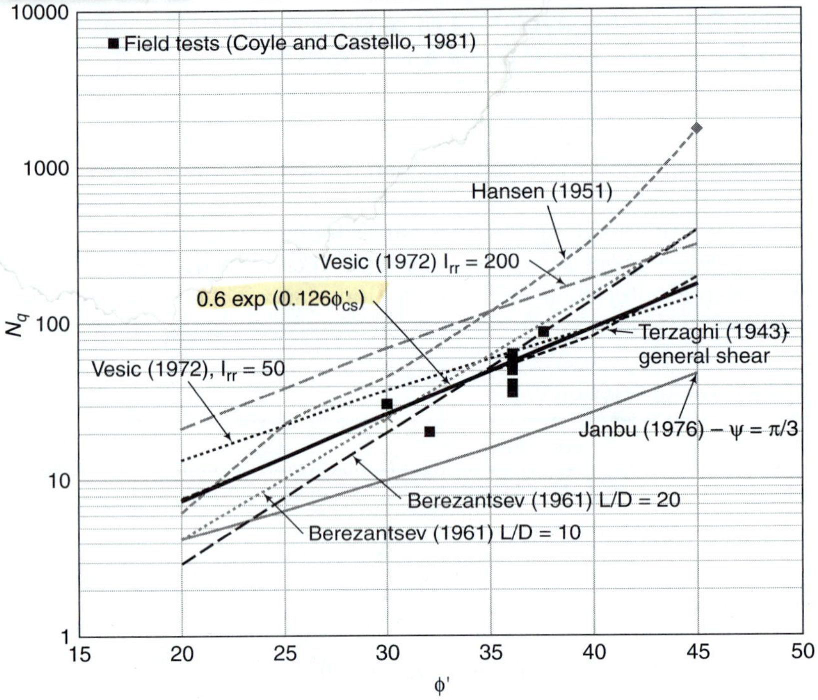

FIGURE 8.7 Comparison of N_q values.

Various recommendations for N_q have been proposed. The popular ones are compared in Fig. 8.7. When using N_q, you need to be aware of the following (modified after Meyerhof, 1976):

1. There are significant differences among the various recommended values of N_q.
2. N_q is applicable only up to a certain critical depth of penetration into the end bearing layer (L_e). The critical depth, L_e, is dependent on the friction angle, the soil compressibility, and method of installation. An approximate equation of the critical depth is

$$\frac{L_e}{D} \leq 0.556 \exp(0.085\phi'_{cs}); \quad 20° \leq \phi'_{cs} \leq 40° \tag{8.19}$$

3. N_q is not applicable when the pile penetrates a compressible soil and the tip is seated in a firm end bearing layer that is located below the critical depth of that layer. For such a situation, the end bearing stress can be estimated from

$$f_b = \frac{(f_b)_l L_e}{10D}; \quad \frac{L_e}{D} \leq 10 \tag{8.20}$$

where $(f_b)_l$ is the end bearing stress of the firm layer.

4. The full skin friction is mobilized at about one-tenth the displacement required for the mobilization of the full end bearing resistance. Thus, the allowable end bearing resistance at the serviceability limit state may not be mobilized if the same factor of safety as for the allowable skin friction is used.
5. It is best to use conservative values for N_q.
6. You should use ϕ'_{cs} instead of ϕ'_p to account for soil disturbances and uncertainties in soil dilation.
7. In general, N_q, is applicable only to homogeneous soils.

8. For sand, the end bearing stress f_b should generally be limited as follows.

$$f_b(\text{kPa}) = 0.17 \exp(0.29\phi'_{cs}); \quad 30° \leq \phi'_{cs} \leq 40° \tag{8.21}$$

$$f_b(\text{ksf}) = 0.00355 \exp(0.29\phi'_{cs}); \quad 30° \leq \phi'_{cs} \leq 40° \tag{8.22a}$$

9. All of the above items are debatable because they are based on simplifying assumptions and experience on certain soil types. The natural vagaries of soil deposits make them theoretically intractable and it is practically impossible to create all inclusive expressions.

A compromise expression for N_q based on some field test results for driven piles is

$$N_q = 0.6 \exp(0.126\phi'_{cs}) \tag{8.23}$$

Equation (8.23) is plotted in Fig. 8.7. The lowest value for N_q in Fig. 8.7 was reported by Janbu (1976) as

$$N_q = (\tan\phi' + \sqrt{1 + \tan^2\phi'})^2 \exp(2\psi \tan\phi') \tag{8.24}$$

where ψ is the plastification angle (Fig. 8.5). The value of ψ varies from $\psi \leq \frac{\pi}{3}$ for soft fine-grained soils to $\psi \leq 0.58\pi$ for dense coarse-grained soils and overconsolidated fine-grained soils. For soft, compressible soils, the pile tip may easily penetrate the soil without causing significantly large plastic zones. For these types of soils, ψ should not exceed $\frac{\pi}{3}$. In dense coarse-grained soils, the plastic zones could be substantial, but it is recommended (Janbu, 1976) that for these soils ψ should not exceed $\frac{\pi}{2}$. Janbu's expression requires $\phi' = \phi'_p$ but it is best to use $\phi' = \phi'_{cs}$ because of soil disturbance.

The* key points *are:

1. ***The α-method is used to estimate the short-term pile load capacity in fine-grained soils.***
2. ***The β-method is used to estimate the short-term and long-term pile load capacity in coarse-grained soils and the long-term pile load capacity in fine-grained soils.***
3. ***The actual values of α_u, β, and N_q are uncertain.***
4. ***The limiting values given are not theoretical limits but are based on experience on certain soil deposits. They do not pertain to all situations.***

8.7 LOAD CAPACITY FOR DRIVEN PILES BASED ON SPT AND CPT RESULTS

Access a spreadsheet at http://www.wiley.com/college/budhu. Click Foundations and Earth Structures, Chapter 8, pile.xls to calculate the load capacity of driven piles using SPT values.

Importance *SPT and CPT are common field tests for which the results are available at the time of the tests. Various correlations, some with very low regression coefficients, have been proposed among results from these tests and pile load capacity. Only a few are given in this textbook.*

8.7.1 SPT (Meyerhof, 1976)

Skin Friction in Sand For driven displacement piles (solid piles), the ultimate skin friction is

$$Q_f = f_s \times \text{perimeter} \times \text{length}; \quad f_s = 1.9N_{60} < 100\,\text{kPa} \tag{8.24}$$

$$Q_f = f_s \times \text{perimeter} \times \text{length}; \quad f_s = 0.04N_{60} < 2\,\text{ksf} \tag{8.24a}$$

For nondisplacement driven piles (H-piles), the ultimate skin friction is

$$Q_f = f_s \times \text{perimeter} \times \text{length}; \; f_s = 0.95N_{60}; f_s < 50\,\text{kPa} \tag{8.25}$$

$$Q_f = f_s \times \text{perimeter} \times \text{length}; \; f_s = 0.02N_{60}; f_s < 1\,\text{ksf} \tag{8.25a}$$

End Resistance in Coarsed Sand and Gravels (Meyerhof, 1976) The ultimate end bearing resistance is

$$Q_b = f_b A_b; \; f_b = CN_{60}(\text{kPa}); \quad C = 38\frac{L_s}{D} \le 380 \tag{8.26}$$

$$Q_b = f_b A_b; \; f_b = CN_{60}(\text{ksf}); \quad C = 0.8\frac{L_s}{D} \le 8 \tag{8.26a}$$

End Resistance in Nonplastic Silts The ultimate end bearing resistance is

$$Q_b = f_b A_b; \; f_b = 300N_{60}(\text{kPa}) \tag{8.27}$$

$$Q_b = f_b A_b; \; f_b = 6N_{60}(\text{ksf}) \tag{8.27a}$$

where N_{60} is the N value with no correction for overburden pressure but corrected for 60% energy; A_b = area of base.

8.7.2 CPT

End Resistance The pile end bearing stress f_b from CPT has been observed to vary from about $0.4q_c$ to $2q_c$, where q_c is the cone tip resistance. However, it is generally assumed that $f_b = q_c$. The end bearing capacity of a single driven pile is estimated from

$$Q_b = q_c A_b; \quad q_c \le 20\,\text{MPa}(400\,\text{ksf}) \tag{8.28}$$

where A_b is the area of the pile base.
For a drilled shaft, Fleming (1992) suggested the following expression for the end bearing stress

$$f_b = \left(\frac{\dfrac{s_b}{D}}{\dfrac{s_b}{D} + \dfrac{0.5q_c}{E_b}} \right) q_c \tag{8.29}$$

where s_b is the base settlement, D is the shaft diameter and E_b is the elastic modulus below the pile base. Both Eq. (8.28) and (8.29) are estimates and their accuracy is unknown.

The determination of q_c from cone penetrometer results is controversial and several methods have been suggested. The differences in the methods come from the influence zone over which the cone values are extracted and the averaging procedures used. One method, proposed by Fleming and Thorburn (1983), is to average cone values over an influence zone of eight pile diameters above the pile base and two pile diameters below the pile base. The average cone resistance value is calculated from

$$q_c = \frac{q_{c1} + q_{c2} + 2q_{c3}}{4} \tag{8.30}$$

where q_{c1} is the arithmetic average of cone resistance values over two pile diameters below the pile base, q_{c2} (is the minimum cone resistance value over two pile diameters below the pile base, and q_{c3} is the arithmetic average of minimum cone resistance values below q_{c2} over eight pile diameters above the pile base.

Another method (Eslami and Fellenius, 1997) is to consider the cone resistance over an influence zone eight pile diameters above the base and four pile diameters below the base—a total of twelve pile diameters—for piles penetrating a weak soil and resting on a dense soil. For piles penetrating a dense soil and resting in a weak soil, the cone resistance is considered over an influence zone of two pile diameters above the pile base. The geometric mean cone resistance is

$$(q_c)_{ag} = \{q_{c1}q_{c2}q_{c3}\cdots q_{cn}\}^{\frac{1}{n}} \tag{8.31}$$

where q_{c1} to q_{cn} are discrete cone resistances over a distance twelve pile diameters or two pile diameters depending on the soil layering, and n is the number of q_c values.

Skin Friction The sleeve resistance, q_{sl}, in the CPT is a measure of the skin or shaft frictional stress. The relative density of the soil and soil compressibility affect the sleeve resistance while the relative density of the soil, method of installation, soil compressibility, pile geometry, and surface roughness affect skin frictional stress on a pile. In the case of fine-grained soils, the cone sleeve friction value is strongly influenced by soil consolidation around the sleeve. Thus, significant differences between short-term and long-term load capacity can be expected using cone penetrometer results. Pile test results show that the short-term load capacity estimated from cone penetrometer results can be as low as 20% of the long-term load capacity. Where the sleeve resistance is measured, either an arithmetic or a geometric mean value over the embedded pile depth is used as the shaft or skin frictional stress. If the cone sleeve resistance is not measured, an estimate can be made from one of several equations proposed in the literature. Some of these are:

- Eslami and Fellenius (1997):

$$f_s = C_s q_{cs} \tag{8.32}$$

where q_{cs} is the cone resistance after adjustments for porewater pressure measured at the cone shoulder, and C_s is a coefficient that depends on soil type as shown in Table 8.4.

TABLE 8.4 Values of C_s

Soil Type	C_s
Soft sensitive soils	0.08
Clay	0.05
Stiff clay and mixture of clay and silt	0.025
Mixture of silt and sand	0.01
Sand	0.004

- Vesic (1977)—coarse-grained soils:

$$f_s = 0.11\exp(-3\tan\phi'_{cs})q_c \tag{8.33}$$

- Jardine et al. (1998)—coarse-grained soil:

$$f_s = \sigma'_{rc}\tan\phi_i \tag{8.34}$$

where σ'_{rc} is the radial effective stress on the shaft and is empirically related to the cone resistance as $\sigma'_{rc} = 0.029\, q_c \left(\frac{\sigma'_{zo}}{p_{atm}}\right)^{0.13} \left(\frac{h}{R^*}\right)^{-0.38}$, σ'_{zo} is the initial vertical effective stress at a depth z, p_{atm} = atmospheric pressure, h = depth from pile base to the depth at which the shaft friction is considered, R^* = pile radius, which for open-ended piles is $R^* = (R_o^2 - R_i^2)^{0.5}$ and for closed-ended piles is $R^* = R$, R_o is outer radius, R_i is inner radius, q_c is the cone resistance, and ϕ_i is the interface friction angle ($\leq \phi'_{cs}$).

- Tumay and Fakhroo (1981)—Stiff clays:

$$f_s = 0.5q_c \tag{8.35}$$

$$Q_f = f_s \times \text{perimeter} \times \text{length} \tag{8.36}$$

The **key points** *are:*

1. ***SPT and CPT results are used to estimate the short-term pile load capacity in soils.***
2. ***The estimation of the pile load capacity from these tests is controversial.***

8.8 LOAD CAPACITY OF DRILLED SHAFTS

Access a spreadsheet at http://www.wiley.com/college/budhu. Click Foundations and Earth Structures, Chapter 8, drilledshaft.xls to calculate the load capacity of drilled shafts.

Importance *The load capacity of drilled shafts is calculated based on semi-empirical equations assuming high-quality construction in stable excavations. In this section, the expressions for estimating the load capacity of drilled shaft are presented.*

8.8.1 α-Method for Fine-Grained Soils (Total Stress Analysis)

Skin Resistance The skin friction, Q_f, over the embedded length of a drilled shaft is calculated using Eq. (8.4). The values of α_u (O'Neill and Reese, 1999) are

$$\alpha_u = 0.55 \quad \text{for} \quad \frac{s_u}{p_{\text{atm}}} \leq 1.5 \tag{8.37}$$

$$\alpha_u = 0.55 - 0.1\left(\frac{s_u}{p_{\text{atm}}} - 1.5\right) \quad \text{for} \quad 1.5 < \frac{s_u}{p_{\text{atm}}} \leq 2.5 \tag{8.38}$$

$$f_s = \alpha_u s_u \leq 380 \text{ kPa} \quad (8000 \text{ psf})$$

where p_{atm} is atmospheric pressure (14.7 psi or 2117 psf or 101.3 kPa), and f_s is the skin frictional stress. The skin friction for the top 1.5 m (5 ft) and the last one-diameter length of the shaft are neglected for drilled shafts in clays because of environmental and construction effects (Figs. 8.2f,g).

End Bearing Resistance The end bearing capacity is calculated using Eq. (8.9). The expressions for N_c are as follows:

$$N_c = 0.63(8.87 - \ln(s_u)_b)(\ln(s_u)_b - 1.35); \quad 25\,\text{kPa} \leq (s_u)_b \leq 190\,\text{kPa}; \quad f_b \leq 4000\,\text{kPa} \tag{8.39}$$

$$N_c = 0.63(11.91 - \ln(s_u)_b)(\ln(s_u)_b - 4.39); \quad 500\,\text{psf} \leq (s_u)_b \leq 4000\,\text{psf}; \quad f_b < 80{,}000\,\text{psf} \tag{8.39a}$$

8.8.2 β-Method for Coarse-Grained and Fine-Grained Soils (Effective Stress Analysis)

Skin Resistance The skin resistance along the shaft is found using Eq. (8.14) where β (Reese and O'Neill, 1988) is given as follows.

Clean Sand:

$$\beta = 1.5 - 0.245\sqrt{z}; \quad 1.2 \geq \beta \geq 0.25; \quad z = \text{depth (m)}, \quad N_{60} \geq 15 \tag{8.40}$$

$$\beta = \frac{N_{60}}{15}(1.5 - 0.245\sqrt{z}); \quad 1.2 \geq \beta \geq 0.25; \quad z = \text{depth (m)}, \quad N_{60} < 15 \tag{8.41}$$

$$\beta = 1.5 - 0.135\sqrt{z}; \quad 1.2 \geq \beta \geq 0.25; \quad z = \text{depth (ft)}, \quad N_{60} \geq 15 \tag{8.40a}$$

$$\beta = \frac{N_{60}}{15}(1.5 - 0.135\sqrt{z}); \quad 1.2 \geq \beta \geq 0.25; \quad z = \text{depth (ft)}, \quad N_{60} < 15 \tag{8.41a}$$

$$\beta = 0.25 \quad \text{for} \quad z > 26\,\text{m (86 ft)} \tag{8.42}$$

N_{60} is the SPT value without correction for overburden pressure but corrected for 60% energy.

$$f_s = \beta\sigma'_z \leq 200\,\text{kPa (4200 psf)} \tag{8.43}$$

Gravels and Sandy Gravels (GW and GP):

$$\beta = 2.0 - 0.15z^{0.75}; \quad 1.8 \geq \beta \geq 0.25; \quad z = \text{depth (m)} \tag{8.44}$$

$$\beta = 2.0 - 0.0615z^{0.75}; \quad 1.8 \geq \beta \geq 0.25; \quad z = \text{depth (ft)} \tag{8.44a}$$

$$\beta = 0.25 \quad \text{for} \quad z > 26\,\text{m (86 ft)} \tag{8.45}$$

$$f_s = \beta\sigma'_z \leq 200\,\text{kPa (4200 psf)}$$

Fine-Grained Soils:

Use Eq. (8.15).

End Bearing Resistance The end bearing resistance is calculated using Eq. (8.18). The end bearing stress is related to SPT values as follows.

Coarse-Grained Soils:

$$f_b(\text{kPa}) = 57.5N_{60} \leq 2900\,\text{kPa}; \quad L \geq 10\,D\,\text{(m)} \tag{8.46}$$

$$f_b(\text{kPa}) = 5.75LN_{60} \leq 290L\,\text{kPa}; \quad L < 10\,D\,\text{(m)} \tag{8.47}$$

$$f_b(\text{ksf}) = 1.2N_{60} \leq 60\,\text{ksf}; \quad L \geq 10D \tag{8.46a}$$

$$f_b(\text{ksf}) = 0.036LN_{60} \leq 1.82L\,\text{ksf}; \quad L < 10D \tag{8.47a}$$

Fine-Grained Soils:

Use Eq. (8.18).

The expressions for β for drilled shafts have no fundamental basis. They were obtained from back calculations of load tests in coarse-grained cemented soils. In these types of soils and for pile diameters about 1 m (3 ft), the load capacity predictions are likely to be satisfactory. Pile test results from drilled shafts in clean sands show that the predicted load capacity using the expressions presented is always conservative. In addition, observations reveal that at the base of the excavation the sand normally becomes loose, while just below the base of the drilled shaft the sand becomes denser. It is best to consider reducing or neglecting the frictional resistance over one diameter above the base of the drilled shaft. You should specify one or more load tests (Section 8.14) to calibrate the predicted load capacity for the project site. The expression, Eq. (8.46), to calculate the end bearing stress is based on a settlement of the drilled shaft of 5% of the diameter and for an embedment depth from the ground surface greater than 10 diameters. For embedment depth less than 10 diameters, a prorated value of f_b is assumed as given by Eq. (8.47).

The key points are:

1. ***The load capacity of drilled shafts calculated from the expressions given in this section is based on observations in particular soil types, mostly cemented sand.***
2. ***You should verify the accuracy of the predicted load capacity from pile load tests (Section 8.14) or other methods described in Section 8.16.***

EXAMPLE 8.1 *Load Capacity of a Timber Pile in a Homogeneous Clay*

A cylindrical timber pile of diameter 457 mm (18 in.) is driven to a depth of 11 m (40 ft) into homogeneous normally consolidated clay. The soil parameters are: $s_u = 40\,\text{kPa}(835\,\text{psf})$, $\phi'_{cs} = 28°$, $\gamma_{sat} = 18\,\text{kN/m}^3(110\,\text{pcf})$. Groundwater level is at the surface. Estimate the allowable pile load capacity for a factor of safety of 2. Is the pile a friction pile?

Strategy The solution is a straightforward application of the static pile load capacity equations. Since the soil is fine-grained, you should check both the short-term and long-term pile load capacity. U.S. units will be used for this solution.

Solution 8.1

Step 1: Calculate geometric and other parameters for driven piles.

$$L = 40\text{ ft; diameter} = 1.5\,\text{ft, perimeter} = \pi D = \pi \times 1.5 = 1.5\pi\,\text{ft}, \; A_b = \frac{\pi D^2}{4} = \frac{\pi \times 1.5^2}{4} = 0.563\,\pi\text{ft}^2$$

$$\gamma' = 110 - 62.4 = 47.6\,\text{pcf}$$

At center, $\sigma'_{zo} = 47.6 \times \frac{40}{2} = 952\,\text{psf}$

$$\text{Eq. (8.6)}: f_s = 0.5\sqrt{s_u \sigma'_{zo}} = 0.5\sqrt{835 \times 952} = 446\,\text{psf}$$

$$\text{Eq. (8.7)}: f_s = 0.5 s_u^{0.25}(\sigma'_{zo})^{0.25} = 0.5 \times 835^{0.75} \times 952^{0.25} = 431\,\text{psf}$$

Use $f_s = 431\,\text{psf}$

For OCR = 1 (normally consolidated soil), $\beta = (1 - \sin 28°)(1)^{0.5}\tan 28° = 0.28$.

Step 2: Calculate Q_a using TSA.

$$Q_f = f_s(\pi D)L = 431 \times 1.5\pi \times 40 \approx 81.3\,\text{kips}$$

$$Q_b = 9 s_u A_b = 9 \times 835 \times 0.563\pi \approx 13.3\,\text{kips}$$

$$Q_{ult} = Q_f + Q_p = 81.3 + 13.3 \approx 94.6\,\text{kips}$$

$$\frac{Q_f}{Q_{ult}} = \frac{81.3}{94.6} = 0.86 > 0.8.$$

Therefore, based on TSA, we have a friction pile.

$$Q_a = \frac{Q_{ult}}{FS} = \frac{94.6}{2} = 47.3\,\text{kips}$$

Step 3: Calculate Q_a using an ESA.

$$Q_f = \beta\sigma'_z(\pi \text{DL})$$

Use an average value of σ'_z over the embedded length,

$$Q_f = 0.28 \times 952 \times 1.5\pi \times 40 = 50252\,\text{lb} = 50.3\,\text{kips}$$

End bearing

$$\text{Eq. (8.23)}: N_q = 0.6\exp(0.126\phi'_{cs}) = 0.6\exp(0.126 \times 28) = 20.4$$

$$Q_b = N_q(\sigma'_z)_b A_b = 20.4 \times (47.6 \times 40) \times 0.563\pi = 68709\,\text{lb} = 68.7\,\text{kips}$$

If you use Janbu's equation, $N_q = \left(\tan\phi'_{cs} + \sqrt{1+\tan^2\phi_{cs'}}\right)^2 \exp(2\psi\tan\phi'_{cs})$, you will get for $\phi'_{cs} = 28°$ and $\psi = \frac{\pi}{3}$ (soft clay), $N_q = \{\tan 28° + \sqrt{1+\tan^2(28°)}\}^2 \exp\left(\frac{2\pi}{3}\tan 28°\right) = 9.5$.

This gives a 53% decrease in end bearing capacity and illustrates the uncertainties in estimating the end bearing resistance.

$$Q_{ult} = Q_f + Q_b = 50.3 + 68.7 = 119\,\text{kips}$$

$$\frac{Q_f}{Q_{ult}} = \frac{50.3}{119} = 0.42 = 42\%$$

Therefore, based on ESA, we have a combination of friction and end bearing pile.

$$Q_a = \frac{Q_{ult}}{2} = \frac{119}{2} = 59.5\,\text{kips}$$

The load capacity from an ESA is greater than that from a TSA. Therefore, use the pile load capacity from the TSA (i.e., $Q_a = 47.3$ kips). ■

use lower one

EXAMPLE 8.2 *Load Capacity of a Square Concrete Pile in a Stratified Soil*

A square concrete pile is required to support a load of 300 kN (67 kips) with a factor of safety of 2. The soil stratification consists of 5 m (15 ft) of sand ($\gamma_{sat} = 15\,\text{kN/m}^3$(94 pcf), $\phi'_{cs} = 30°$) followed by a deep deposit of overconsolidated clay ($s_u = 60\,\text{kPa}$(1250 psf), $\phi'_{cs} = 24°$, OCR $= 4$, $\gamma_{sat} = 20\,\text{kN/m}^3$(125 pcf)). Groundwater level is at 2 m (6 ft) below the ground surface. You can select among driven piles of cross-sections of 0.3 m (1 ft) × 0.3 m (1 ft), 0.45 m (1.5 ft) × 0.45 m (1.5 ft), and 0.6 m (2 ft) × 0.6 m (2 ft) to support the load. Investigate the length required for each.

Strategy You are given Q_a and FS, so you can calculate $Q_{ult} = Q_a \times \text{FS}$. Decide on a base layer. The overconsolidated clay is an obvious choice. You should consider both short-term and long-term conditions and select the larger length. Use the spreadsheet pile.xls and select "driven pile." SI units will be used in solving this example problem.

Solution 8.2

Step 1: Determine ultimate load.

$$Q_a = 300\,\text{kN}; \quad Q_{ult} = Q_a \times \text{FS} = 300 \times 2 = 600\,\text{kN}$$

Step 2: Draw a sketch of the soil profile and pile, and calculate the required geometric and other parameters. See sketch in Fig. E8.2a.

Sand: $\beta = (1 - \sin 30°)\tan 30° = 0.29$;
$\gamma' = 15 - 9.8 = 5.2\,\text{kN/m}^3$

Stiff clay: $s_u > 50\,\text{kPa}$, use Fig. 8.6. Since we do not know the embedded length, we will make an estimate of the adhesion factor using Fig. 8.6a. Use $\alpha_u = 0.9$. Assume $\phi' = \phi'_i = \phi'_{cs}$, we get $\beta = (1 - \sin\phi'_{cs})(\text{OCR})^{\frac{1}{2}}\tan\phi'_{cs} = (1 - \sin 24°)(4)^{\frac{1}{2}}\tan 24° = 0.53$

For the 0.3 m × 0.3 m pile:
Perimeter $= 4B = 4 \times 0.3 = 1.2$ m;
$A_b = 0.3^2 = 0.09\,\text{m}^2$
Eq. (8.23):
$N_q = 0.6\exp(0.126\phi'_{cs}) = 0.6\exp(0.126 \times 24) = 12.3$
$N_c = 9$

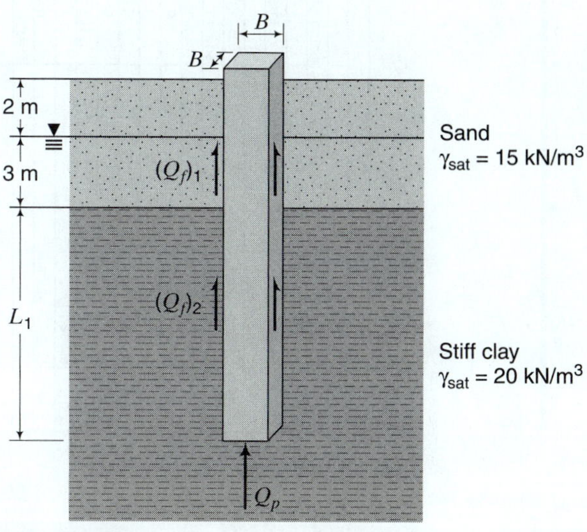

FIGURE E8.2

Step 2: Calculate required length using a TSA.

TABLE E8.2a,b

	A	B	C	D	E	f'	G	H	W	X	Y	Z	AA	AB
1	Muni Budhu "Foundations and Earth Retaining Structures," John Wiley & Sons, 2006.													
2	DRIVEN PILES													
3	Straight, prismatic pile as an individual pile or in a group													
4	Select units	SI		γ_w	9.81	kN/m3								
5	Design load	300	kN	FS	2									
6	Pile section shape	square		Group	no									
7	Adhesion factor, $?_u$	Tomlinson		Spacing	4	m								
8	Enter adhesion factor, su	0.9		Matrix	3									
9	Groundwater	2												
10	End bearing layer	5	m			diameter or width			SINGLE PILE					
11	N_c	9			0.3	0.45	0.6	m	DIAMETER OR WIDTH					
12	ω_p clay	0.5	rad	Per	1.2	1.8	2.4	m	0.3					
13	ω_p sand	0.4		Area	0.09	0.2	0.36	m2	FRICTION		END BEARING		Quit	Quit
14	Use N values	n							TSA	ESA	TSA	ESA	TSA	ESA
15	layer	Depth	thickness	Soil Type	Unit weight		s_u	OCR	Q_f	Q_f	Q_b	Q_b		
16		m	m		kN/m3	deg	kPa		kN	kN	kN	kN	kN	kN
17	1	0	0		0	0		0	0.0	0.0	0	0	0	0
18	groundwater	2	2	s	15	30		0	10.4	10.4	0	0	10.4	10.4
19	2	5	3	s	15	30		0	49.7	49.7	0	0	49.7	49.7
20		6	1	c	20	24	60	4	114.5	81.8	48.6	61.9	163.1	143.7
21		7	1	c	20	24	60	4	179.3	120.4	48.6	73.3	227.9	193.6
22		8	1	c	20	24	60	4	244.1	165.4	48.6	84.6	292.7	250
23	4	9	1	c	20	24	60	4	308.9	216.9	48.6	95.9	357.5	312.8
24		10	1	c	20	24	60	4	373.7	274.8	48.6	107.2	422.3	382.1
25		11	1	c	20	24	60	4	438.5	339.3	48.6	118.6	487.1	457.8
26		12	1	c	20	24	60	4	503.3	410.1	48.6	129.9	551.9	540
27		13	1	c	20	24	60	4	568.1	487.5	48.6	141.2	616.7	628.7
28		14	1	c	20	24	60	4	632.9	571.3	48.6	152.5	681.5	723.8
29	5	15	1	c	20	24	60	4	697.7	661.5	48.6	163.8	746.3	825.4
30		16	1	c	20	24	60	4	762.5	758.2	48.6	175.2	811.1	933.4
31		17	1	c	20	24	60	4	827.3	861.4	48.6	186.5	875.9	1047.9
32		18	1	c	20	24	60	4	892.1	971.1	48.6	197.8	940.7	1168.9
33		19	1	c	20	24	60	4	956.9	1087.2	48.6	209.1	1005.5	1296.3
34		20	1	c	20	24	60	4	1021.7	1209.7	48.6	220.4	1070.3	1430.1

Sand

We will treat the sand above the groundwater as a separate layer.
At center of layer above groundwater: $\sigma'_{zo} = 1 \times \gamma_{sat} = 15\,\text{kPa}$
At center of layer below groundwater: $\sigma'_{zo} = 2\gamma_{sat} + 1.5\gamma' = 2 \times 15 + 1.5 \times 5.2 = 37.8\,\text{kPa}$

$$Q_f = \sum_{i=1}^{j} (\beta)_i(\sigma'_z)_i \times (\text{perimeter})_i \times (\text{length})_i = 0.29(15 \times 2 + 37.8 \times 3) \times 1.2 = 49.9\,\text{kN}$$

For the sand layer TSA and ESA are the same.

Stiff Clay

$$Q_f = \alpha_u s_u \times \text{perimeter} \times \text{length} = 0.9 \times 60 \times 1.2 \times L_1 = 64.8\,L_1\,\text{kN}$$
$$Q_b = 9 s_u A_b = 9 \times 60 \times 0.09 = 48.6\,\text{kN}$$

Ultimate Load Capacity—Short Term

$$Q_{ult} = (Q_f)_{sand} + (Q_f)_{stiff\ clay} + Q_b$$

Therefore

$$600 = 49.9 + 64.8\,L_1 + 48.6$$

Solving, we get

$$L_1 = 7.7\,\text{m}$$
$$\text{Total length required} = 7.7 + 5 = 12.7\,\text{m}.$$

Step 3: Calculate required length using an ESA.
The calculations are done using pile.xls. Portions of the spreadsheet with the input and output data are shown in Tables E8.2a,b.
The summary of the results is as follows:

Pile section	0.3 m × 0.3 m	0.45 m × 0.45 m	0.6 m × 0.6 m
Length required (m)	13	10	8

■

EXAMPLE 8.3 *Load Capacity of a Drilled Shaft*

Determine the load capacity of a drilled shaft 1 m (3 ft) in diameter and length 16 m (50 ft) in the soil shown in Fig. E8.3. The factor of safety required is 2.

Strategy Apply the appropriate equation for drilled shafts. The solution will be solved using U.S. units.

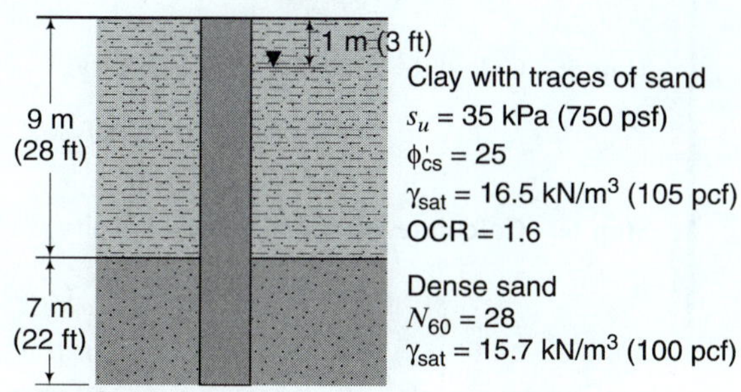

FIGURE E8.3

Solution 8.3

Step 1: Calculate required geometric and other parameters.

$$L = 50\,\text{ft},\ D = 3\,\text{ft},\ \text{perimeter} = \pi D = \pi \times 3 = 9.43\,\text{ft}, A_b = \frac{\pi D^2}{4} = \frac{\pi \times 3^2}{4} = 7.07\,\text{ft}^2$$
$$\gamma' = 105 - 62.4 = 42.6\,\text{pcf}$$

Clay:

$$\frac{s_u}{p_{atm}} = \frac{750}{2117} = 0.35 < 1.5; \quad \alpha_u = 0.55$$

$$\beta = (1 - \sin\phi'_{cs})\tan\phi'_{cs}(\text{OCR})^{1/2} = (1 - \sin 25°)(\tan 25°)(1.6)^{1/2} = 0.34$$

Sand: Depth to center of sand layer $= 28 + 22/2 = 39$ ft

$$\beta = 1.5 - 0.135\sqrt{z} = 1.5 - 0.135\sqrt{39} = 0.66 < 1.2$$

Clay:

Short term: Neglect the resistance for the first 5 ft below the surface due to installation effects.

$$Q_f = \sum_{i=1}^{j}(\alpha_u)_i(s_u)_i \times (\text{perimeter})_i \times (\text{length})_i = 0.55 \times 0.750 \times 9.43 \times (28 - 5) = 89.5 \text{ kips}$$

Long term: σ'_{zo} at center of clay (5 ft to 28 ft) is $(3 \times 105) + 42.6 \times 2 + 42.6 \times \left(\frac{28-5}{2}\right) = 890.1$ psf.

$$f_s = \beta\sigma'_z = 0.34 \times 890.1 = 302.6 \text{ psf}$$

$$(Q_f)_{clay} = \beta\sigma'_z \times \text{perimeter} \times \text{length} = 302.6 \times 9.43 \times (28 - 5) = 65.6 \text{ kips}$$

Therefore, long-term governs design since the frictional resistance under long term condition is smaller than short term condition.

Step 4: Calculate skin friction and end bearing for sand.

Sand:

$$\sigma'_{zo} \text{ at center of sand} = (105 \times 3) + 42.6 \times 25 + (100 - 62.4)\frac{22}{2} = 1793.6 \text{ psf} = 1.79 \text{ ksf}$$

$$f_s = \beta\sigma'_{zo} = 0.66 \times 1.79 = 1.18 \text{ ksf} < 4.2 \text{ ksf}$$

Neglect friction on last one diameter of shaft due to construction effects.

$$(Q_f)_{sand} = f_s \times \text{perimeter} \times \text{length} = 1.18 \times 9.43 \times (22 - 3) = 211.4 \text{ kips}$$

$$L/D = 50/3 = 16.7 > 10\,D; \quad \text{use Eq. (8.46a)}$$

$$f_b = 1.2N_{60} = 1.2 \times N_{60} = 1.2 \times 28 = 33.6 \text{ ksf} < 60 \text{ ksf}$$

$$(Q_b)_{sand} = A_b f_b = 33.6 \times 7.07 = 237.6 \text{ kips}$$

Step 5: Calculate ultimate load capacity.

$$Q_{ult} = (Q_f)_{clay} + (Q_f)_{sand} + (Q_b)_{sand} = 65.6 + 211.4 + 237.6 = 514.6 \text{ kips}$$

Step 6: Calculate allowable load capacity.

$$Q_a = \frac{Q_{ult}}{\text{FS}} = \frac{514.6}{2} = 257 \text{ kips}$$

■

EXAMPLE 8.4 *Load Capacity of a Driven H Pile*

An H-pile (HP 14 × 89 lb, flange width = 14.7 in., gross web depth = 13.8 in.) is driven into a sand deposit to a depth of 30 ft. The SPT results are shown in Table E8.4. Estimate the allowable load capacity for a factor of safety of 2.

Strategy Use Eqs. (8.25a) and (8.26a). You can divide the sand deposit into layers based on the range of N values given in Table E8.4. Here we take a simple approach by averaging the N values.

TABLE E8.4

Depth (ft)	3	10	15	20	25	30	35	40
N_{60} (blows/ft)	22	18	25	20	30	36	39	45

Solution 8.4

Step 1: Determine average value of N over the depth.

$$(N_{60})_{av} = \frac{22 + 18 + 25 + 20 + 30 + 36}{6} = 25$$

Step 2: Determine load capacity.

$$\text{Perimeter} = 2(13.8 + 14.7) = 57\,\text{in} = 4.75\,\text{ft.};\ \text{Area} = 13.8 \times 14.7 = 202.9\,\text{in}^2 = 1.41\,\text{ft}^2$$

$$Q_f = f_s \times \text{perimeter} \times \text{length};\ f_s = 0.02N_{60} = 0.02 \times 25 = 0.5 < 1\,\text{ksf}\,(\textit{okay})$$

$$Q_f = 0.5 \times 4.57 \times 30 = 68.6\,\text{kips}$$

$$Q_b = f_b A_b;\ f_b = CN_{60}(\text{ksf});\quad C = 0.8\frac{L_s}{D} = 0.8\frac{30 \times 12}{13.8} = 20.9 > 8;\quad \text{use } C = 8$$

$$Q_b = 8 \times 36 \times 1.41 = 406.1\,\text{kips}$$

$$Q_{\text{ult}} = Q_f + Q_b = 68.6 + 406.1 = 474.7\,\text{kips}$$

$$Q_a = \frac{Q_{\text{ult}}}{\text{FS}} = \frac{474.7}{2} \approx 237\,\text{kips}$$

■

8.9 UPLIFT RESISTANCE

Importance *Piles are not only subjected to compressive forces but also to uplift forces from applied loads and moments. Two expressions to estimate uplift load capacity are presented in this section.*

The uplift resistance and compressive resistance are approximately the same for a pile in fine-grained soils. For piles in sands, Nicola and Randolph (1993) proposed the following expression for the uplift skin frictional stress (f_{s_u}).

$$\frac{f_{s_u}}{f_s} = \left[1 - 0.2\log_{10}\frac{100}{\left(\frac{L}{D}\right)}\right]\left[1 - 8\eta + 25\eta^2\right];\quad \frac{L}{D} \leq 100 \tag{8.48}$$

where f_s is the skin frictional stress for compressive loads and η is a compressibility factor given as

$$\eta = \nu \tan\phi_i'\left(\frac{L}{D}\right)\left(\frac{G_{\text{avg}}}{E_p}\right) \tag{8.49}$$

and ν is the Poisson's ratio of the pile, G_{avg} is the average shear modulus of the soil over the length of the pile, ϕ_i' is the soil–pile interfacial friction angle (Table 8.3), and E_p is Young's modulus of the pile. For $\frac{L}{D} > 100, \frac{f_{s_u}}{f_s} = 1$.

$$(Q_{\text{ult}})_{\text{uplift}} = f_{s_u} \times \text{perimeter} \times \text{length} \tag{8.50}$$

A quick estimate of the uplift force can be obtained by applying a reduction factor to the skin frictional stress f_s for compressive loads as

$$(Q_{\text{ult}})_{\text{uplift}} = K_t f_s \times \text{perimeter} \times \text{length} \tag{8.51}$$

where K_t is 0.5 to 0.7 (sand and silt) and 0.7 to 1.0 (clay).

8.10 PILES SUBJECTED TO NEGATIVE SKIN FRICTION

Importance *Piles located in settling soil layers (e.g., soft clays, fills) are subjected to negative skin friction called* downdrag. *The settlement of the soil layer causes the friction forces to act in the same direction as the vertical loading on the pile. Rather than providing resistance, the negative friction imposes additional loads on the pile. The net effect is that the pile load capacity is reduced and pile settlement increases.*

The allowable load capacity for negative skin friction is given, with reference to Fig. 8.8, as

$$Q_a = \frac{Q_b + Q_f}{\text{FS}} - Q_{nf} \tag{8.52}$$

For soft, normally consolidated soils, the negative skin friction is usually calculated over one-half its thickness. Negative skin friction should be computed for long-term conditions (i.e., you should use an ESA).

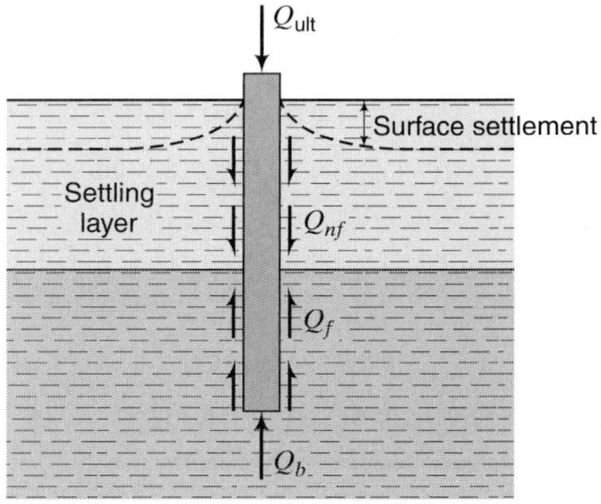

FIGURE 8.8 Negative skin friction.

EXAMPLE 8.5 *Pile Load Capacity with Negative Skin Friction*

Determine the allowable load capacity of the 0.45 m (1.5 ft) diameter pile shown in Fig. E8.5. The fill is recent and was dumped at the site. To eliminate negative skin friction, a steel shell is proposed around the pile within the fill. The groundwater level is at 1 m (3 ft) below the fill in the soft clay but is expected to rise to the surface. A factor of safety of 2 is required.

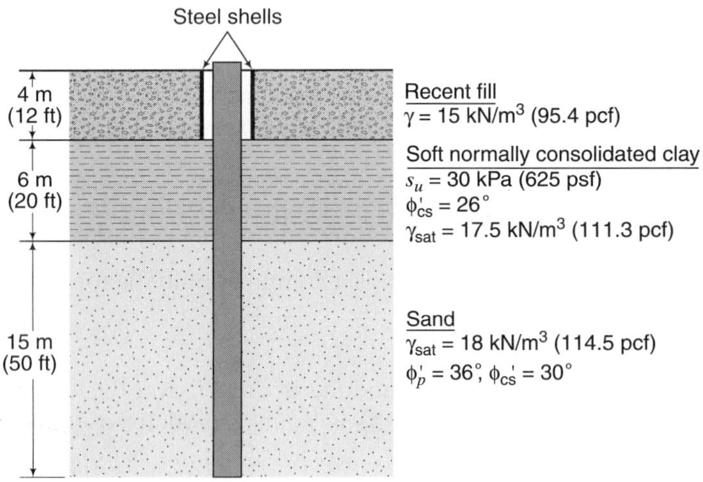

FIGURE E8.5

Strategy Under the load of the fill, the soft clay will settle, dragging the pile down. Therefore, we have to consider negative skin friction imposed by the soft clay. You should use an ESA. Since the pile would likely loosen the dense sand, you should use ϕ'_{cs} rather than ϕ'_p. SI units will be used in solving this example problem.

Solution 8.5

Step 1: Determine β and N_q and other relevant parameters.

$$\frac{L}{D}=\frac{25}{0.45}=55.6,\ \text{perimeter}=\pi D=0.45\pi=1.41\ \text{m},\ A_b=\frac{\pi D^2}{4}=\frac{\pi\times 0.45^2}{4}=0.159\ \text{m}^2$$

$$\gamma'_{\text{fill}}=(15-9.8)=5.2\ \text{kN/m}^3$$

Clay

$$\gamma'_{\text{clay}}=(17.5-9.8)=7.7\ \text{kN/m}^3$$

At center of the top half of the clay layer, $\sigma'_{zo}=5.2\times 4+7.7\times 1.5=32.4\ \text{kPa}$.

$$\beta=(1-\sin\phi'_{cs})\tan\phi'_{cs}(\text{OCR})^{0.5}=(1-\sin 26^\circ)\tan(26^\circ)(1)^{0.5}=0.27$$

Sand

$$\gamma'_{\text{sand}}=(18-9.8)=8.2\ \text{kN/m}^3$$

Assume that the sand is normally consolidated (i.e., OCR = 1).

$$\beta=(1-\sin\phi'_{cs})\tan\phi'_{cs}(\text{OCR})^{0.5}=(1-\sin 30^\circ)\tan(30^\circ)=0.29$$

$$N_q=0.6\exp(0.126\phi'_{cs})=26$$

Step 2: Calculate negative skin friction for clay layer.

Assume the groundwater will rise to the surface and the top one-half of the soft clay will be subjected to negative skin friction.

$$Q_{nf}=\beta\sigma'_z\times\text{perimeter}\times\text{length}=0.27\times 32.4\times 1.41\times 3=37\ \text{kN}$$

Step 3: Calculate skin friction of lower half of soft clay layer.

At center of lower half of clay: $\sigma'_{zo}=5.2\times 4+7.7\times 4.5=55.5\ \text{kPa}$.

Step 4: Calculate skin friction for clay.

$$Q_f=\beta\sigma'_z\times\text{perimeter}\times\text{length}=0.27\times 55.5\times 1.41\times 3=63.4\ \text{kN}$$

Step 5 Calculate skin friction and end bearing in sand.

At center of sand layer: $\sigma'_{zo}=5.2\times 4+7.7\times 6+8.2\times 7.5=128.5\ \text{kPa}$.

$$(Q_f)_{\text{sand}}=0.29\times 128.5\times 1.41\times 15=788\ \text{kN}$$

$$(\sigma'_{zo})_b=5.2\times 4+7.7\times 6+8.2\times 15=190\ \text{kPa}$$

$$Q_b=N_q(\sigma'_{zo})_b A_b=26\times 190\times 0.159=785.5\ \text{kN}$$

Step 6: Calculate allowable load capacity.

$$Q_a=\frac{(Q_f)_{\text{clay}}+(Q_f)_{\text{sand}}+(Q_b)_{\text{sand}}}{\text{FS}}-Q_{nf}=\frac{63.4+788+785.5}{2}-37=781.5\ \text{kN}$$

■

8.11 GROUP PILES

Importance *Piles are rarely used singly. They are often clustered into groups. The group load capacity may not be equal to the single pile load capacity multiplied by the number of piles. In this section, we describe how to estimate the group load capacity.*

In most practical situations, piles are used in groups. They are arranged in geometric patterns (squares, rectangles, circles, and octagons) at a spacing, s, (center to center distance) not less than $3D$ (where D is the diameter or width of the pile). For pile spacing less than $3D$, the stress distribution adjacent to each pile overlaps, significantly reducing the pile load capacity.

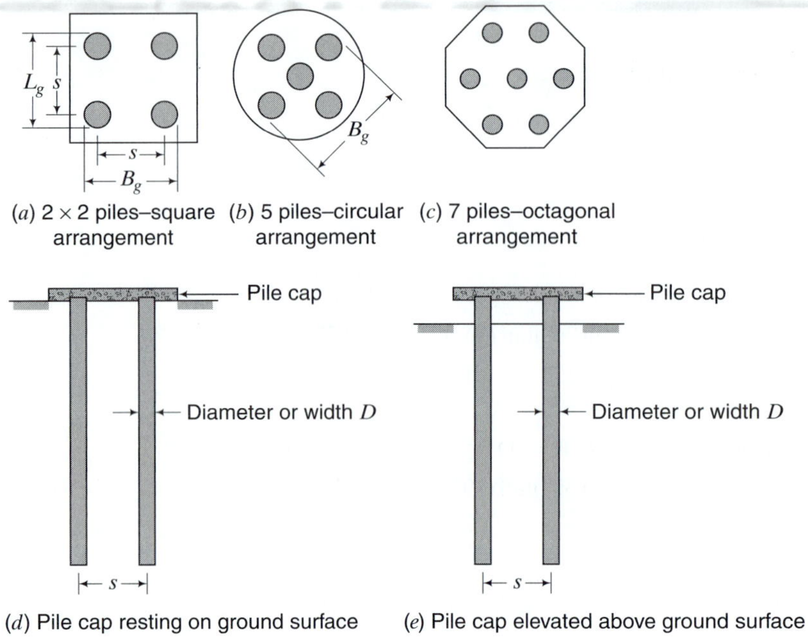

FIGURE 8.9 Pile groups.

The piles in a group are connected at their heads by a concrete cap (pile cap), which may or may not be in contact with the ground (Fig. 8.9). If the pile cap is in contact with the ground, part of the load will be transferred directly to the soil through the pile cap.

The load capacity for a pile group is not necessarily the load capacity of a single pile multiplied by the number of piles. In fine-grained soils, the outer piles tend to carry more loads than the piles in the center of the group. In coarse-grained soils, the piles in the center carry more loads than the outer piles.

The ratio of the load capacity of a pile group, $(Q_{ult})_g$, to the total load capacity of the piles acting as single piles (nQ_{ult}) is called the efficiency factor, η_e

$$\eta_e = \frac{(Q_{ult})_g}{nQ_{ult}} \tag{8.53}$$

where n is the number of piles in the group and Q_{ult} is the ultimate load capacity of a single pile. The efficiency factor is usually less than 1. However, piles driven into a loose coarse-grained soil tend to densify the soil around the piles and η_e could exceed 1.

Two modes of soil failure are normally investigated to determine the load capacity of a pile group. One mode, called block failure (Fig. 8.10), may occur when the spacing of the piles is small enough to cause the pile group to fail as a unit. The group load capacity for block failure mode is:

ESA

$$(Q_{ult})_{gb} = \sum_{i=1}^{j} \{\beta_i(\sigma'_z)_i \times (\text{perimeter})_{igb} \times L_i\} + N_q(\sigma'_z)_b(A_b)_{gb} \tag{8.54}$$

TSA

$$(Q_{ult})_{gb} = \sum_{i=1}^{j} \{(\alpha_u)_i(s_u)_i \times (\text{perimeter})_{igb} \times L_i\} + N_c(s_u)_b(A_b)_{gb} \tag{8.55}$$

where the subscript gb denotes block mode of failure for the group.

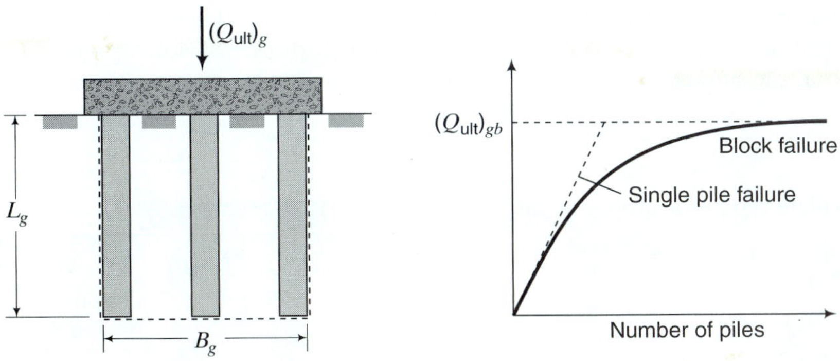

FIGURE 8.10 Block and single pile failure mode.

The other failure mode is failure of individual piles, called *single pile failure mode* or *punching failure mode*. The key assumption in the single pile failure mode is that each pile mobilizes its full load capacity. Thus, the group load capacity is:

$$(Q_{ult})_g = nQ_{ult} \tag{8.56}$$

The group efficiency in fine-grained soils is defined as:

$$\eta_e = (Q_{ult})_{gb}[(Q_{ult})^2_{gb} + (nQ_{ult})^2]^{-\frac{1}{2}} \tag{8.57}$$

The values of ψ to use in determining N_q in Janbu's expression (Eq. 8.24) depends on $\frac{s}{D}$ ratio and the effective angle of friction. Janbu (1976) showed that:

$$\frac{s}{D} = 1 + 2\sin\psi(\tan\phi' + \sqrt{1+\tan^2\phi'})\exp(\psi\tan\phi') \tag{8.58}$$

The value of ψ is not significantly affected by $\frac{s}{D} \le 2.5$.

Generally, single pile failure mode governs for driven piles in loose sand with spacing greater than $3D$. The main reason is that the sand adjacent to the piles compacts, forming an overlapping compaction zone, increasing the skin friction on individual piles. In the case of drilled shafts in homogeneous sand, block failure mode seems to predominate because the skin friction on individual piles reduces within the group due to overlapping shear zones.

The* key points *are:

1. ***The ultimate load capacity of a pile group is not necessarily the ultimate load capacity of a single pile multiplied by the number of piles in the group.***
2. ***A pile group can fail either by the group failing as a single unit (called block failure mode) or as individual piles (called single pile failure mode).***

EXAMPLE 8.6 *Pile Group Load Capacity in a Two-Layer Soil*

A pile group of 9 driven piles, 0.45 m (18 in.) diameter, is arranged in a 3×3 matrix at a spacing of 1.2 m (4 ft). The piles penetrate a medium silty clay [$s_u = 20$ kPa(420 psf), $\phi'_{cs} = 30°$, $\gamma = 17$ kN/m^3(108 pcf), $\gamma_{sat} = 18$ kN/m^3(115 pcf), OCR = 2] of thickness 8 m (26 ft) and are embedded 2 m (6.5 ft) in a stiff gray clay ($s_u = 86$ kPa(1800 psf), $\phi'_{cs} = 28°$, $\gamma_{sat} = 18.5$ kN/m^3(118 pcf), OCR = 5). Calculate the group allowable load capacity for a factor of safety of 2. Groundwater is at 2 m (6.5 ft) below the surface but can rise to the surface due to seasonal changes.

Strategy Use a sketch to illustrate the problem. You need to calculate the ultimate load capacity assuming (1) block failure mode and (2) single pile failure mode. The lesser value is the ultimate pile group load capacity. The worst case is when the groundwater is at the surface. U.S. units will be used to solve this example problem.

Solution 8.6

Step 1: Draw a sketch and calculate geometric properties and other parameters.
See Fig. E8.6.

$$\text{Single pile: } D = 1.5\,\text{ft; perimeter} = \pi D = \pi \times 1.5 = 4.71\,\text{ft;}$$

$$A_b = \frac{\pi D^2}{4} = \pi \times \frac{1.5^2}{4} = 1.77\,\text{ft}^2$$

$$\text{Group: perimeter} = 4(2s + D) = 4\{2(4) + 1.5\} = 38\,\text{ft; base area} = (A_b)g = (2s + D)^2 = 9.5^2 = 90.25\,\text{ft}^2$$

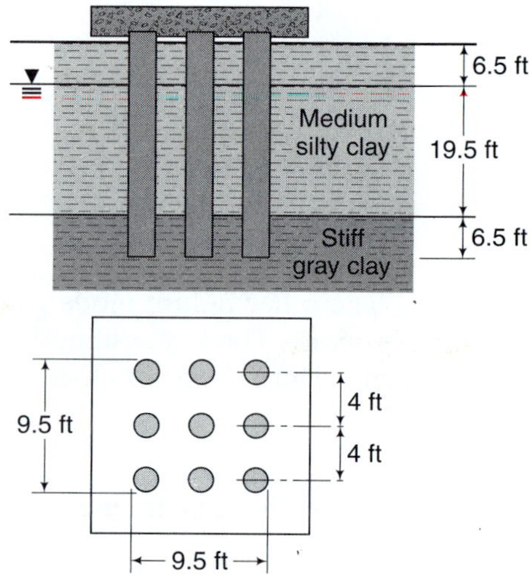

FIGURE E8.6

Assume groundwater is at the surface.

Medium clay:

$$\gamma' = (115 - 62.4) = 52.6\,\text{pcf} = 0.0526\,\text{kcf}$$

Center of layer:

$$\sigma'_{zo} = \left(0.0526 \times \frac{26}{2}\right) = 0.684\,\text{ksf}$$

$$\text{Eq.(8.6): } f_s = 0.5\sqrt{0.42 \times 0.684} = 0.27\,\text{ksf}$$

$$\text{Eq.(8.7): } f_s = 0.5\, s_u{}^{0.75}(\sigma'_{zo})^{0.25} = 0.5 \times 0.42^{0.75} \times 0.684^{0.25} = 0.24\,\text{ksf use } f_s = 0.24\,\text{ksf}$$

$$\beta = (1 - \sin\phi'_{cs})\tan\phi'_{cs}(\text{OCR})^{1/2} = (1 - \sin 30°)\tan(30°)(2)^{1/2} = 0.41$$

Stiff clay:

$$\gamma' = (118 - 62.4) = 55.6\,\text{pcf} = 0.0556\,\text{kcf}$$

Center of layer:

$$\sigma'_{zo} = (0.0526 \times 26) + \left(0.0556 \times \frac{6.5}{2}\right) = 1.548\,\text{ksf}$$

$$\text{Eq.(8.6): } f_s = 0.5\sqrt{1.8 \times 1.548} = 0.83\,\text{ksf}$$

$$\text{Eq.(8.7): } f_s = 0.5 \times 1.8^{0.75} \times 1.548^{0.25} = 0.87\,\text{ksf use } f_s = 0.83\,\text{ksf}$$

$$\beta = (1 - \sin 28°)\tan(28°)(5)^{1/2} = 0.63$$

Step 2: Calculate ultimate load capacity using TSA.
Block failure (short term)

$$Q_f = f_s \times \text{perimeter} \times \text{length}$$

$$(Q_f)_{\text{med. clay}} = 0.24 \times 38 \times 26 = 237.1\,\text{kips}$$

$$(Q_f)_{\text{stiff clay}} = 0.83 \times 38 \times 6.5 = 205\,\text{kips}$$

Stiff clay—end bearing

$$(Q_p)_{\text{stiff clay}} = 9_{s_u}(A_b)_g = 9 \times 1.8 \times 90.25 = 1462\,\text{kips}$$

Block failure mode load capacity

$$(Q_u)_{gb} = (Q_f)_{\text{soft clay}} + (Q_f)_{\text{stiff clay}} + (Q_p)_{\text{stiff clay}}$$

$$(Q_u)_{gb} = 237.1 + 205 + 1462 = 1904.1\,\text{kips} = 1904\,\text{kips}$$

TSA—Single pile failure mode You can use proportion since only the parameters and area change.

Soft clay—skin friction

$$(Q_f)_{\text{soft clay}} = 237.1 \times \frac{4.71}{38} = 29.4\,\text{kips}$$

Stiff clay—skin friction

$$(Q_f)_{\text{stiff clay}} = 205 \times \frac{4.71}{38} = 25.4\,\text{kips}$$

Stiff clay—end bearing

$$(Q_p)_{\text{stiff clay}} = 1462 \times \frac{1.77}{90.25} = 28.7\,\text{kips}$$

Single pile failure mode group load capacity

$$Q_u = n[(Q_f)_{\text{soft clay}} + (Q_f)_{\text{stiff clay}} + (Q_p)_{\text{stiff clay}}]; \quad n = 3 \times 3 = 9\,\text{piles}$$

$$Q_{\text{ult}} = 9(29.4 + 25.4 + 28.7) = 751.5\,\text{kips} = 752\,\text{kips}$$

Step 3: Calculate the ultimate load capacity using an ESA.

ESA—Block Failure Mode

Skin friction

$Q_f = \beta\sigma'_z \times (\text{perimeter})_{gb} \times L$; (σ'_z) is calculated at mid-depth of the layer.

$$\sigma'_z = \sigma'_{zo}$$

Medium clay: $Q_f = 0.41 \times 0.684 \times 38 \times 26 = 277\,\text{kips}$

Stiff clay: $Q_f = 0.63 \times 1.548 \times 38 \times 6.5 = 241\,\text{kips}$

End bearing—stiff clay

$$Q_b = N_q(\sigma'_z)_b(A_b)_{gb}$$

$$N_q = 0.6\exp(0.126\phi'_{cs}) = 20.4$$

$$(\sigma'_z)_b = (0.0526 \times 26) + 0.0556 \times 6.5 = 1.73\,\text{ksf}$$

$$Q_b = 20.4 \times 1.73 \times 90.25 = 3185\,\text{kips}$$

Group load capacity

$$(Q_{\text{ult}})_{gb} = 277 + 241 + 3185 = 3703\,\text{kips}$$

ESA—Single Pile Failure Mode

Skin friction

Medium clay: $Q_f = 277 \times \dfrac{4.71}{38} = 34.3\,\text{kips}$

Stiff clay: $Q_f = 241 \times \dfrac{4.71}{38} = 29.9\,\text{kips}$

End bearing—stiff clay

$$Q_p = 3703 \times \frac{1.77}{90.25} = 72.6$$

Group load capacity

$$(Q_{\text{ult}})_g = nQ_{\text{ult}} = 9(34.3 + 29.9 + 72.6) = 1231 \text{ kips}$$

	Load capacity (kips)	
Analysis	**Block mode**	**Single pile mode**
TSA	1904	752
ESA	3703	1231

Step 4: Decide which failure mode and conditions govern.

Lowest ultimate load capacity is 752 kips for a TSA of single pile mode failure.

$$Q_a = \frac{752}{2} \; 376 \text{ kips}$$

■

8.12 COMBINED AXIAL LOAD AND MOMENTS

Importance *Rarely are piles subjected to vertical centric loads. They are normally subjected to combined loads that can induce both uplift and compressive forces. In this section, we briefly describe the effects of combined loadings.*

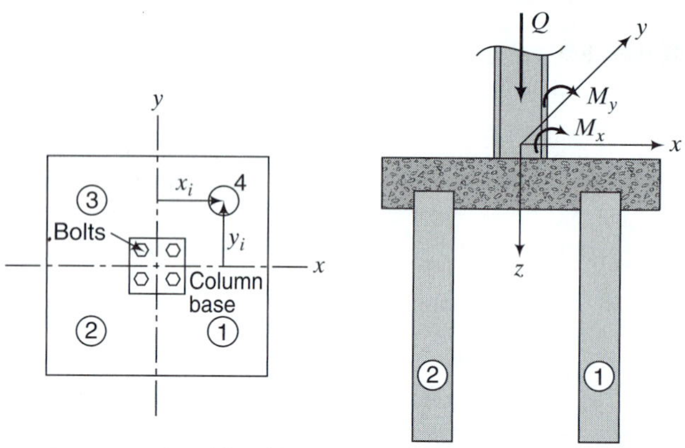

FIGURE 8.11 Combined loads and axial moments on group piles.

Pile groups are the foundation of choice to support combined axial and lateral loads and moments. Lateral loads will be considered in Section 8.15. A pile within a group may be subjected to compression or uplift depending on the direction of the applied moments (Fig. 8.11). In Fig. 8.11, pile 1 will be subjected to uplift while pile 4 will be subjected to compression from the applied moments.

The stresses imposed on a pile, i, within a group (Fig. 8.11) are

$$\text{Axial stress} = f_a = \frac{Q_a}{n \times A_p} \tag{8.59}$$

$$\text{Bending stress due to } M_y = f_{by} = \frac{M_y x_i}{I_y} \tag{8.60}$$

$$\text{Bending stress due to } M_x = f_{bx} = \frac{M_x y_i}{I_x} \tag{8.61}$$

where Q_a is the total allowable load or design load, n is the number of piles, A_p is the cross-sectional area of the pile, I_y and I_x are the second moment of area (moment of inertia) about the y and x axes respectively, and x_i and y_i are the distances from the y and x (centroidal) axes to the pile.

The combined axial and bending compression must satisfy the relationship

$$\left| \frac{f_a}{F_a} + \frac{f_{bx}}{F_b} + \frac{f_{by}}{F_b} \right| \leq 1.0 \tag{8.62}$$

where F_a and F_b are the allowable axial and bending stress of the pile material.

8.13 SETTLEMENT OF PILES

Importance *Although piles are used in situations where the serviceability limit state in other types of foundations is exceeded, it is still necessary to estimate the pile settlement under working loads to check that the serviceability limit state is satisfactory. The prediction of pile settlement is complicated and, for practical purposes, semi-empirical equations are used. In this section, the estimation of pile settlement is discussed.*

8.13.1 Elastic Settlement

Single Driven Pile The elastic settlement of a single pile depends on the relative stiffness of the pile and the soil, the length to diameter ratio of the pile, the relative stiffness of the soil at the base and the soil over the pile length, and the distribution of elastic modulus of the soil along the pile length. The relative stiffness of the pile to the soil is $K_{ps} = \frac{E_p}{E_s}$, where E_p is the elastic modulus of an equivalent solid cylindrical cross-section of the pile and E_s is elastic modulus of the soil. The elastic modulus of the soil at the base or tip of the pile will be denoted by E_{sb}. Usually, the secant elastic modulus is used in design practice.

Various analyses have been proposed to calculate the settlement of single piles and pile groups. Poulos (1989) provided an excellent discussion on the various numerical procedures to calculate settlement of piles. The settlement consists of three components: skin friction, end bearing, and elastic shortening.

Skin friction tends to deform the soil near the shaft analogous to simple shear strain (Fig. 8.12), and the shear strain, γ_{zx}, is:

$$\gamma_{zx} = \frac{\tau}{G} = \frac{f_s}{G} \tag{8.63}$$

where G is the shear modulus of the soil and f_s is the skin frictional stress. The shear strains can be integrated over the pile length to give the elastic settlement (ρ_{es}) as,

$$\rho_{es} = \frac{1}{G(z)} \int_o^L \tau(z)dz \tag{8.64}$$

where (z) means that the parameter in front of it varies with depth. To solve Eq. (8.64), we need to know how G and τ vary with depth, but this we do not know. Therefore, we have to speculate on their variations and then solve Eq. (8.64). A further complication arises in that the boundary conditions for a pile problem are complex. Therefore, we have to solve Eq. (8.64) using numerical procedures. For example, we can assume for a cylindrical pile that

$$\tau(z) = \int F(z, i)\tau(i)di \tag{8.65}$$

where $F(z, i)$ is a stress function and $\tau(i)$ is the shear stress at an ordinate, i. We can use finite element or boundary element to solve equations Eqs. (8.64) and (8.65). Stress functions for a point load within a half space were discussed in Chapter 2.

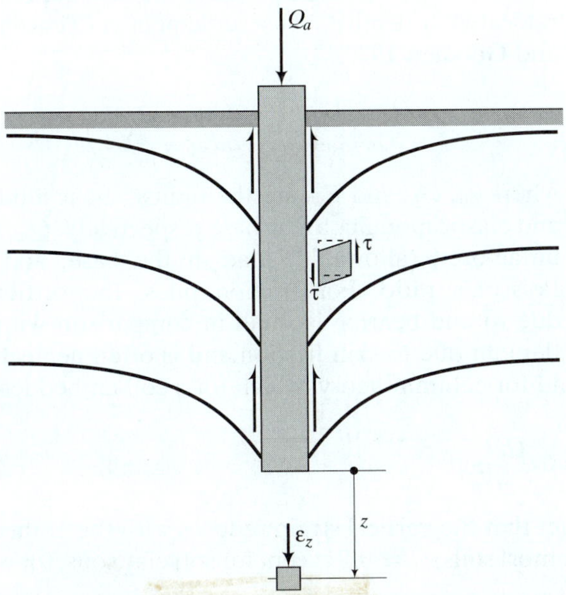

FIGURE 8.12 Shear deformation from skin friction and compression from end bearing.

For a homogeneous soil (E_s is constant with depth), a solution of Eq. (8.64) using Eq. (8.65) for the elastic settlement of a single pile is:

allowable skin friction.

$$\rho_{es} = \frac{Q_{af}}{E_s L} I \tag{8.66}$$

where I is an influence factor and Q_{af} is the unfactored (allowable) load transferred as skin friction. An approximate equation for I is:

$$I = 0.5 + \log\left(\frac{L}{D}\right) \tag{8.67}$$

Soft soils tend to have elastic moduli that vary linearly with depth,

$$E_s = mz \tag{8.68}$$

where m is the rate of increase of E_s with depth (units: kPa/m or psf/ft). For soft soils, the elastic settlement is:

$$(\rho_{es})_{so} = \frac{Q_{af}}{mL^2} I_{so} \tag{8.69}$$

where

$$I_{so} = 2.0 \log\left(\frac{L}{D}\right) \tag{8.70}$$

Poulos (1989) developed another solution for the elastic settlement. He showed that ρ_{es} for a floating pile can be determined from

$$\rho_{es} = \frac{Q_{af}}{E_s D} I_\rho \tag{8.71}$$

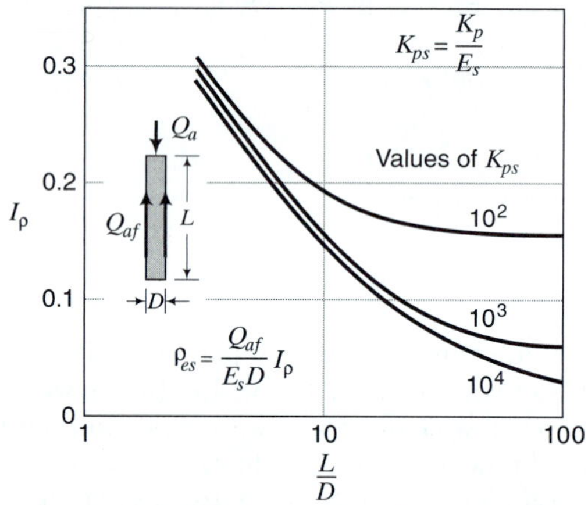

FIGURE 8.13 Influence factor for vertical settlement of a single floating pile (after Poulos, 1989).

where I_ρ is an influence factor that depends on the $\frac{L}{D}$ ratio and K_{ps} as shown in Fig. 8.13.

The soil mass under a pile is subjected to compression and shearing from end bearing. The elastic settlement for end bearing piles is calculated by assuming the pile base is a rigid punch on the surface of the soil relocated at depth L. This settlement is (Timoshenko and Goodier, 1970):

$$\rho_{be} = \frac{Q_{ab}}{r_b G_b}\left(\frac{1-\nu}{4}\right) = \frac{Q_{ab}}{r_b E_{sb}}\left(\frac{1-\nu^2}{2}\right) \tag{8.72}$$

where r_b, G_b, and E_{sb} are the radius, shear modulus, and elastic modulus at the base respectively, Q_{ab} is the unfactored (allowable) load at the base, and ν is Poisson's ratio. For friction piles, the settlement due to end bearing is small in comparison with settlement due to skin friction and is often neglected.

The elastic shortening of the pile (ρ_p) is found for column theory, which for a soil embedded in a homogeneous soil is:

$$\rho_p = C\frac{Q_a L}{E_p A_p} \tag{8.73}$$

where C is a reduction factor to account for the fact that the vertical strain reduces with the embedded length of the pile and Q_a is the unfactored load. For most soils, $C \approx 0.5$ except for soft clay soils, for which $C \approx 0.7$. Elastic shortening is significant only for slender piles, $\left(K_{ps} = \frac{E_p}{E_s} < 500\right)$.

Field observations indicate that at design loads, the elastic settlement of a single driven pile is within the range 0.9 to 1.25% of the pile diameter. Hull (1987) found, from numerical analyses, that embedding a pile beyond a certain critical length does not reduce the settlement. The critical length normalized to the diameter (width) is:

$$\frac{L_c}{D} = \sqrt{\frac{\pi E_p A_p}{E_s D^2}} = \sqrt{\frac{\pi K_{ps} A_p}{D^2}} \tag{8.74}$$

where L_c is the critical length and A_p is the area of the cross-section of the pile.

Since the skin friction is mobilized at about only one-tenth the displacement to mobilize the end bearing resistance (Section 8.5), it is prudent to assume that settlement from end bearing load will occur only after the full skin friction has been mobilized. Also, pile compression will induce skin friction mostly in the top part of the pile. For friction or floating piles, you should neglect pile compression.

Group Driven Piles The settlement of a group of piles is generally not equivalent to a single pile at the same average load level because of pile interaction. The smaller the spacing, the greater the interaction and the larger the settlement. The settlement effect of a single pile within a group is influenced by the geometry of the group. There are various methods suggested to estimate the settlement of group piles. These include interaction factor method (IFM), equivalent raft method (ERM), single pile–group pile factor (SPGM), equivalent pier method (EPM), empirical equations from SPT and CPT tests, and numerical methods. We will consider only IFM, SPGM, and empirical equations from SPT and CPT tests.

A group efficiency (η) of 70–100% is often assumed in calculating the unfactored load to be supported by individual piles in a group. The lowest group efficiency value is used when only a simple soils investigation was carried out and quality control is minimal. The highest group efficiency value is used when a detailed soils investigation was conducted, high quality control including pile integrity checks will be instituted, and critical state parameters were used in estimating the pile load capacity. The unfactored load per pile in a group of n piles is the total unfactored load/$(n \times \eta)$. For example, if the total unfactored load on a group is 10 MN (2248 kips) and the number of piles is 10, then the unfactored load per pile using $\eta = 80\%$ is $10/(10 \times 0.8) = 1.25$ MN(281 kips).

- ***IFM.*** The settlement of a pile, i, in a group of n piles under an average working load in the jth pile, Q_j, is given as

$$p_i = \rho_1 \sum_{j=1}^{n} Q_j \alpha_{ij} \tag{8.75}$$

where ρ_1 is the settlement of a single pile under a unit load, and α_{ij} is the interaction factor between pile i and pile $j\,(i = j = 1$, i.e., $\alpha_{11} = \alpha_{22} = \alpha_{33} = 1)$. For a homogeneous, infinite soil layer the interaction factor is shown in Fig. 8.14.

- ***SPGM.*** For convenience, pile group settlement is related to a single pile settlement through a group settlement factor R_s defined as:

$$R_s = \frac{\text{Settlement of group}}{\text{Settlement of single pile at the same average load}} \tag{8.76}$$

An empirical relationship between R_s and n (Fleming et al. 1985) is

$$R_s = n^{\Phi} \tag{8.77}$$

where the exponent Φ is between 0.4 to 0.6.

- ***Empirical equations from SPT and CPT tests.*** The settlement for group piles in coarse-grained soils from SPT and CPT can be estimated from:

$$\text{SPT}: \rho_{es} = \frac{0.91 q_s I \sqrt{B_g}}{N_{cor}} \text{mm [units: } q_s \text{ in kPa}, B_g \text{ in m]} \tag{8.78}$$

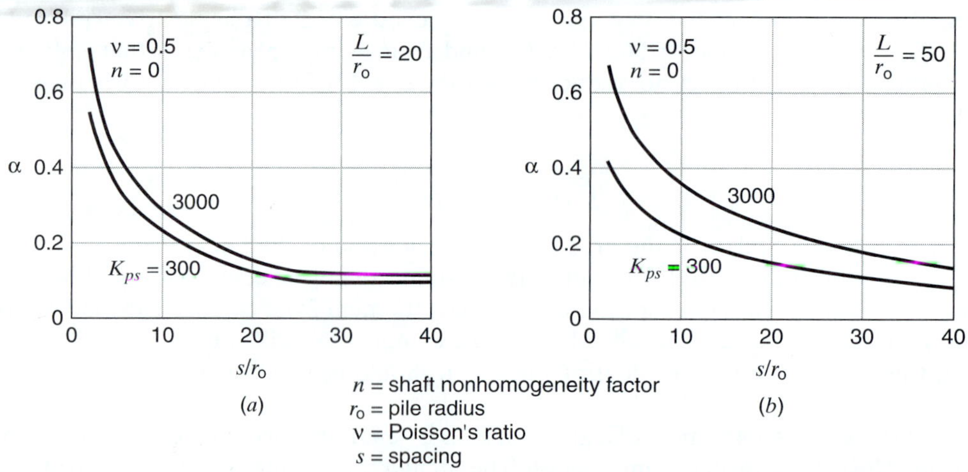

FIGURE 8.14 Interaction factors for a homogeneous infinite soil layer (after Guo and Randolph, 1999).

$$\text{SPT: } \rho_{es} = \frac{q_s I \sqrt{B_g}}{N_{cor}} \text{in. [units: } q_s \text{ in ksf, } B_g \text{ in ft]} \tag{8.78a}$$

$$\text{CPT: } \rho_{es} = \frac{41 q_s I B_g}{q_c} \text{mm [units: } q_s \text{ in kPa, } q_c \text{ in kPa, } B_g \text{ in m]} \tag{8.79}$$

$$\text{CPT: } \rho_{es} = \frac{q_s I B_g}{2q_c} \text{in. [units: } q_s \text{ in ksf, } q_c \text{ in ksf, } B_g \text{ in ft]} \tag{8.79a}$$

where $I = 1 - 0.08\frac{L}{B_g} \geq 0.5$, L is the embedded length of the pile, N_{cor} is the average N value over $\frac{2L}{3}$ corrected for overburden pressure, $q_s = \frac{Q_a}{B_g \times L_g}$ is the average allowable (unfactored) stress of the pile group, B_g and L_g are the width and cross-sectional length of the pile group (Fig. 8.15), and q_c is the arithmetic average of the cone resistance over two pile diameters below the cone tip.

8.13.2 Settlement of Drilled Shafts

The settlement of drilled shafts, based on field data from instrumented piles, 0.6 m (24 in.) to 0.9 m (36 in.) diameter, by Reese and O'Neill (1988), is shown in Fig. 8.16. The full friction for fine-grained soils is mobilized at settlements between 0.2% and 0.8% of the pile diameter while the full end bearing is mobilized at settlements between 2% and 5% of the pile diameter. For coarse-grained soils, the full friction is mobilized at settlements between 0.1% and 1% of the pile diameter, but end bearing continues to increase with increasing settlement—no plateau was observed.

8.13.3 Consolidation Settlement Under a Pile Group

Sometimes, a pile group may be embedded above a soft clay layer and transfer sufficient load to it (soft clay) to cause consolidation settlement. To estimate the consolidation settlement, the full design load is assumed to act at a depth of $\frac{2}{3}L$ and is then distributed in the ratio of 2:1 (vertical: horizontal). The increase in vertical stress at a depth z in the soft clay layer, shown in Fig. 8.15, is:

$$\Delta\sigma_z = \frac{Q_a}{(B_g + z)(L_g + z)} \tag{8.80}$$

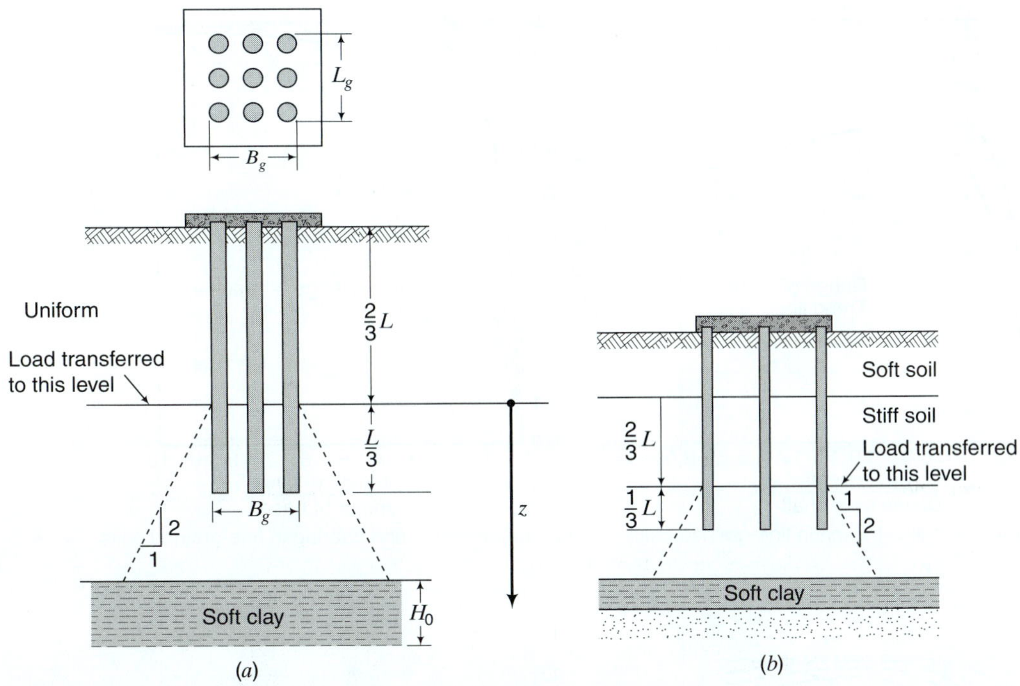

FIGURE 8.15 Assumed load distribution for calculating settlement of a pile group.

where Q_a is the group allowable load, B_g is the width of the group, L_g is the length of the group, and z is the depth from the load transfer point on the pile group to the location in the clay layer at which the increase in vertical stress is desired. The primary consolidation settlement is then calculated using the procedure described in Chapter 4.

8.13.4 Procedure to Estimate Settlement of Single and Group Piles

The procedure to estimate the settlement of single piles and pile groups is:

Driven Piles:

1. Obtain the required parameters Q_{af}, Q_{ab}, E_s, E_p, and consolidation parameters.
2. For friction or floating piles, calculate the elastic settlement, ρ_{es}, for a single pile using Eq. (8.66) or Eq. (8.69) or Eq. (8.71). For end bearing piles, calculate the elastic settlement using Eq. (8.72).
3. For friction–end bearing piles, calculate the elastic settlement, ρ_{es}, for a single pile using Eq. (8.66) or Eq. (8.69) or Eq. (8.71). If the design (allowable) load, Q_a, is less than or equal to the ultimate skin friction (Q_f), then Q_{af} is equal to Q_a. If Q_a is greater than the ultimate skin friction, then Q_{af} is Q_f. The balance of the unfactored load ($Q_a - Q_f$) is Q_{ab}. Calculate the end bearing settlement using Eq. (8.72).
4. For end bearing piles and piles that extend over 2 m (6 ft) above the ground surface, calculate elastic shortening using Eq. (8.73).
5. Calculate the total elastic settlement by adding the settlements due to friction and end bearing.

For group piles, skip step 5 and continue as follows for SPGM.

6. Calculate R_s (Eq. 8.77) using an estimated value of Φ. A value of $\Phi = 0.5$ is a good first approximation.
7. Calculate the group elastic settlement due to skin friction from $(\rho_{es})_g = R_s \rho_{es}$

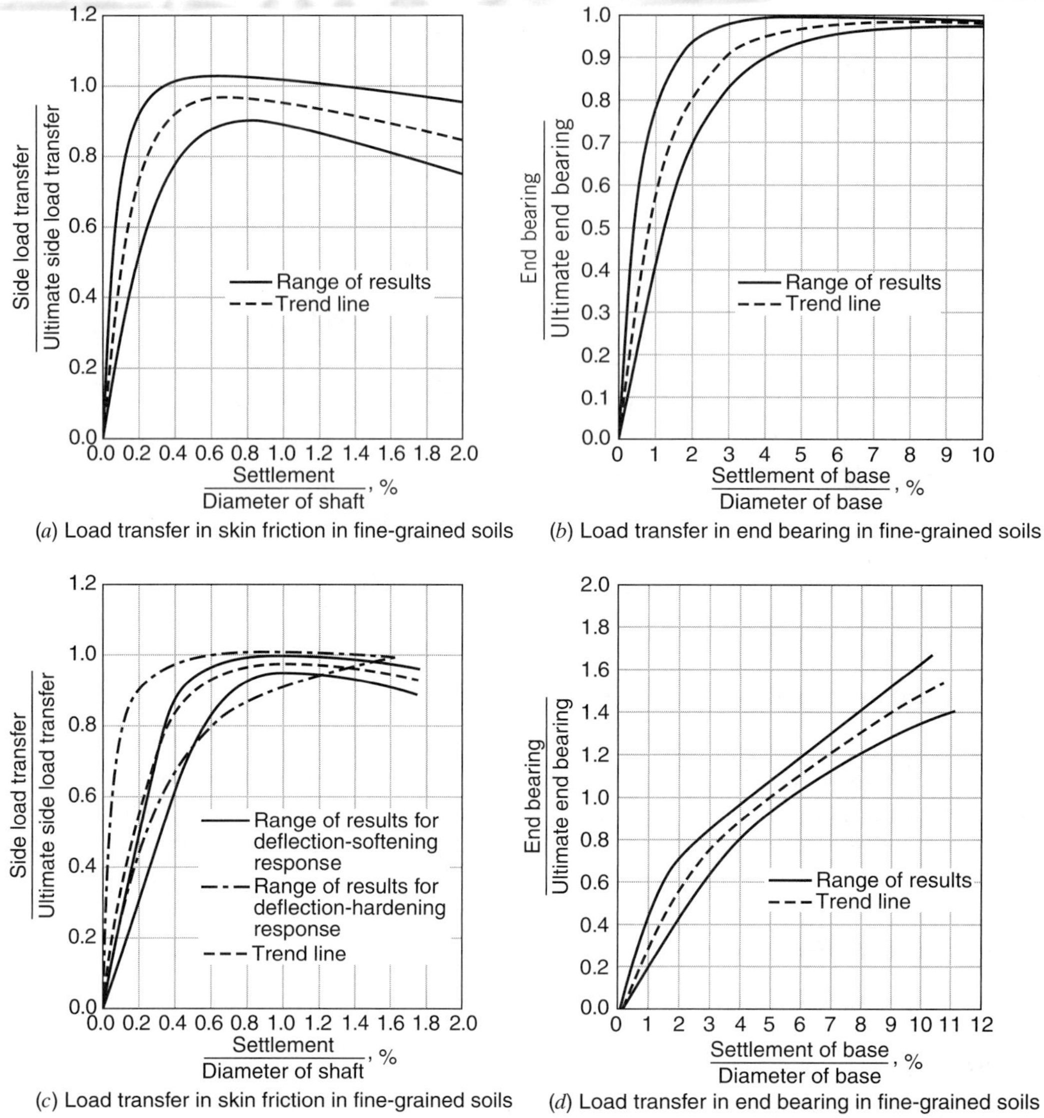

(*a*) Load transfer in skin friction in fine-grained soils (*b*) Load transfer in end bearing in fine-grained soils

(*c*) Load transfer in skin friction in fine-grained soils (*d*) Load transfer in end bearing in fine-grained soils

FIGURE 8.16 Load transfer-settlement curves for drilled shafts. (Source: Reese and O'Neill, 1988.)

8. Calculate consolidation settlement, ρ_{pc}, if necessary.

9. Calculate total group settlement.

Drilled Shafts:

1. Assume a trial settlement, ρ.

2. Calculate the settlement as a percentage of the diameter.

3. Extract the side load transfer/ultimate side load transfer ratio, R_1, from Fig. 8.16a—fine-grained soils—or Fig. 8.16c—coarse-grained soils.

4. Calculate the side load transfer as $R_1 \times Q_f$.

5. Calculate the elastic shortening, ρ_p, using Eq. (8.73) with $C = 1$ and Q_a as the average working (allowable) load.

6. Calculate the base settlement, $\rho_b = \rho - \rho_p$.
7. Calculate the base settlement as a percentage of the diameter.
8. Extract the end bearing/ultimate ratio, R_2, from Fig. 8.16b—fine-grained soils—or Fig. 8.16d—coarse-grained soils.
9. Calculate the end bearing as $R_2 \times Q_b$.
10. Add the end bearing and side load transfer to give Q_1. If $Q_1 \cong Q_a$, then the assumed settlement is the estimated settlement. If not, repeat the procedure.
11. Calculate consolidation settlement, ρ_{pc}, if necessary.
12. Calculate the total settlement.

The key points are:

1. ***Settlement of piles is determined using numerical analyses assuming the soil is an elastic material.***
2. ***The equations for pile settlement will give only an estimate of the expected settlement.***
3. ***The settlement for drilled shafts is found by iteration using load-transfer settlement curves from load tests.***

EXAMPLE 8.7 *Elastic Settlement of a Single Pile*

Determine the settlement of the pile in Example 8.1 using $E_p = 10 \times 10^6$ kPa(0.2×10^6 ksf), $E_s = 5000$ kPa, (105 ksf), and $\nu = 0.4$.

Strategy This is a straightforward application of the procedure in Section 8.13.4. SI units will be used in solving this example.

Solution 8.7

Step 1: Determine influence factor.

$$\frac{L}{D} = \frac{10}{0.457} \approx 22$$

$$\text{Eq. (8.67)}: I = 0.5 + \log\left(\frac{L}{D}\right) = 0.5 + \log(22) = 1.84$$

$$K_{ps} = \frac{E_p}{E_s} = \frac{10 \times 10^6}{5000} = 2000$$

From Fig. 8.13, $I_\rho = 0.09$ for $K_{ps} = 2000$ and $\frac{L}{D} = 22$.

Step 2: Calculate elastic settlement.

$$Q_f = 50.3 \text{ kips} = 224 \text{ kN}, \quad Q_a = 59.5 \text{ kips} = 265 \text{ kN}; Q_a > Q_f$$

$$\text{Eq. (8.66)}: \rho_{es} = \frac{Q_{af}}{E_s L} I = \frac{224}{5000 \times 10} \times 1.84 = 0.0082 \text{ m} = 8.2 \text{ mm}$$

$$\text{Eq. (8.71)}: \rho_{es} = \frac{Q_{af}}{E_s D} I_\rho = \frac{224}{5000 \times 0.457} \times 0.09 = 0.0088 \text{ m} = 8.8 \text{ mm}$$

Difference in two solutions = 0.6 mm. Use $\rho_{es} = 9$ mm.

$$\text{Eq. (8.72): } \rho_{be} = \frac{Q_{ab}}{r_b E_{sb}}\left(\frac{1-\nu^2}{2}\right) = \frac{(265-224)}{(0.457/2)\times 5000}\left(\frac{1-0.4^2}{2}\right) = 0.015\,\text{m} = 15\,\text{mm}$$

Step 3: Calculate total settlement.

$$\rho_{et} = \rho_{es} + \rho_{be} = 9 + 15 = 24\,\text{mm}$$

$$\frac{\rho_{et}}{D} = \frac{24}{457}\times 100 = 5\%$$

■

EXAMPLE 8.8 *Bearing capacity and settlement of a pile group.*

A 3 × 3 pile group consisting of driven concrete piles of diameter 0.4 m (1.3 ft) supports a load of 3 MN (674 kips) (Fig. E8.8). (1) Determine the factor of safety of the pile group. (2) Calculate the total settlement of the pile group.

Strategy Because the sand is dense, driving of the piles will likely loosen it around the pile. You should use ϕ'_{cs} in calculating the pile load capacity. SI unit will be used in the solution of this problem.

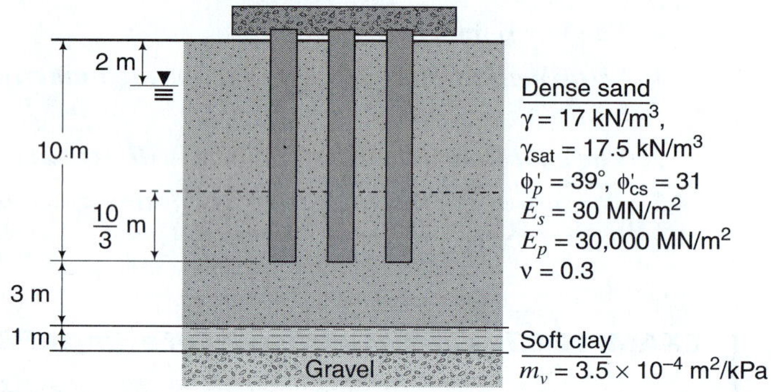

FIGURE E8.8

Solution 8.8

Step 1: Determine geometric parameters, β and N_q.

$$D = 0.4\,\text{m},\quad \frac{L}{D} = \frac{10}{0.4} = 25;\quad n = 9\,\text{piles},\quad s = 1.2\,\text{m};\quad K_{ps} = \frac{E_p}{E_s} = \frac{30\times 10^6}{30000} = 1000$$

$$\text{Single pile: Perimeter} = \pi D = \pi \times 0.4 = 1.26\,\text{m},\quad A_b = \frac{\pi D^2}{4} = \frac{\pi\times 0.4^2}{4} = 0.126\,\text{m}^2$$

$$\text{Group piles: } L_g = B_g = 2s + D = (2.4\times 1) + 0.4 = 2.8\,\text{m};\ \text{perimeter for group} = 2.8\times 4 = 11.2\,\text{m},$$

$$(A_b)_g = 2.8^2 = 7.84\,\text{m}^2$$

$$\beta = (1-\sin\phi'_{cs})\tan\phi'_i = (1-\sin 31°)\tan 31° = 0.29$$

$$\text{Eq. (8.23): } N_q = 0.6\exp(0.126\phi'_{cs}) = 0.6\exp(0.126\times 31) = 29.8$$

Center of sand layer within the embedment length of the pile: $\sigma'_{zo} = 2\times 17 + 3(17.5 - 9.8) = 57.1$ kPa.
At base: $(\sigma'_z)_b = 2\times 17 + 8(17.5 - 9.8) = 95.6$ kPa.

Step 2: Determine single pile failure mode load capacity.

$$\text{Skin friction: } Q_f = \beta\sigma'_z \times \text{perimeter}\times\text{length} = 0.29\times 57.1\times 1.26\times 10 = 208.6\,\text{kN}$$

$$\text{End bearing: } Q_b = N_q(\sigma'_z)_b A_b = 29.8\times 95.6\times 0.126 = 359\,\text{kN}$$

$$\text{Ultimate group load capacity: } Q_{\text{ult}} = 9(Q_f + Q_b) = 9(208.6 + 359) = 5108\,\text{kN}$$

$$\%\ \text{friction} = \frac{208.6}{208.6 + 359}\times 100 = 37\%$$

Step 3: Determine block failure mode load capacity.
Use proportion.

$$\text{Skin friction: } (Q_f)_g = \frac{208.6}{1.26}\times 11.2 = 1854\,\text{kN}$$

$$\text{End bearing: } (Q_b)_g = \frac{359}{0.126} \times 7.84 = 22337.8\,\text{kN}$$

$$\text{Ultimate load capacity: } (Q_{ult})_g = (Q_f)_g + (Q_b)_g = 1854 + 22337.8 = 24191.8\,\text{kN}$$

Step 4: Calculate factor of safety.

Single pile failure mode governs.

$$\therefore \text{FS} = \frac{5108}{3000} \approx 2$$

Step 5: Calculate elastic settlement of pile.

Assume a pile group efficiency of 90%; then the load to be carried by each pile is $Q_a = \frac{3000}{9 \times 0.9} = 370\,\text{kN}$.

$$Q_f = 208.6\,\text{kN} \cong 209\,\text{kN}$$

Since $Q_a > Q_f$, follow Section 8.11.

$$\text{Eq. (8.67): For } \frac{L}{D} = 25,\ I_\rho = 0.5 + \log(25) = 1.9$$

$$\text{Eq. (8.66): } \rho_{es} = \frac{Q_{af}}{E_s L} I = \frac{209}{30000 \times 10} \times 1.9 = 0.0013\,\text{m} = 1.3\,\text{mm}$$

$$\text{Eq. (8.72): } \rho_{be} = \frac{Q_{ab}}{r_b E_{sb}} \left(\frac{1 - \nu^2}{2}\right) = \frac{(370 - 209)}{0.2 \times 30000} \left(\frac{1 - 0.3^2}{2}\right) = 0.012\,\text{m} = 12\,\text{mm}$$

Neglect elastic shortening of the pile since $K_{ps} > 500$.
Assume $\Phi = 0.5$.

$$R_s = n^{\Phi} = 9^{0.5} = 3$$

$$(\rho_{es})_g = 1.3 \times 3 = 3.9\,\text{mm}$$

Step 6: Calculate consolidation settlement.

The design load is transferred to $\frac{2}{3}L$ from surface. Distance of load transfer plan to center of clay is

$$\frac{10}{3} + 3 + \frac{1}{2} = 6.83\,\text{m}$$

$$\Delta\sigma_z = \frac{Q_{ag}}{(B_g + z)^2} = \frac{3000}{(2.8 + 6.83)^2} = 32.3\,\text{kPa}$$

$$\rho_{pc} = m_v \text{H} \Delta\sigma_z = 3.5 \times 10^{-4} \times 1 \times 32.3 = 113 \times 10^{-4}\,\text{m} = 11.3\,\text{mm}$$

Step 7: Compute total settlement.

$$\rho_t = (\rho_{es})_g + \rho_{be} + \rho_{pc} = 3.9 + 12 + 11.3 = 27.2\,\text{mm} \cong 27\,\text{mm}$$ ■

8.14 PILE LOAD TEST

Importance *A pile load test is another method to determine the pile load capacity. Generally, the reliability of pile load tests at a site is greater than with other methods for estimating the pile load capacity. The results of pile load tests are often used to check the estimated pile load capacity from other methods such as statics. In this section, we will discuss pile load tests.*

The purposes of a pile load test are:

- To determine the load capacity of a single pile
- To verify load capacity estimated from statics

- To determine the settlement of a single pile at working loads
- To obtain information on load transfer in skin friction and end bearing
- To verify pile length
- To check structural integrity of the pile, especially the effects of the proposed installation method
- To confirm contract price

There are four types of test loading directions.

1. Compression test
2. Uplift test
3. Lateral load test
4. Torsion load test

In a typical compression loaded pile test, two reaction piles are installed followed by the test pile in between. A reaction beam, usually a steel I beam, is attached to the reaction piles as shown in Fig. 8.17. A hydraulic jack is placed between the reaction beam and the pile head. A settlement sensor (a dial gauge or a linear displacement transducer) is positioned near the pile head to measure the vertical displacement.

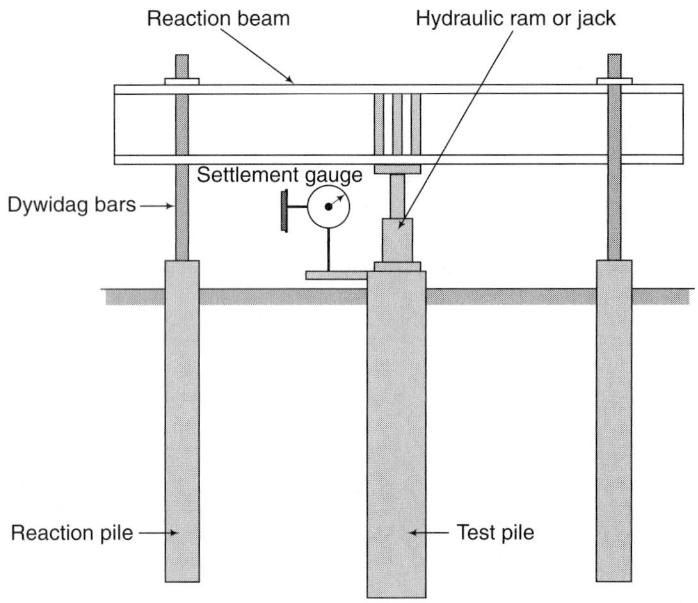

FIGURE 8.17 A pile load test setup.

Tests can be carried out as monotonic loading or as cyclical loading (loading–unloading cycles). In cyclical loading tests, the permanent settlement, also called *permanent pile set*, can be extracted from the results. In Fig. 8.18a, the pile was loaded (OB), unloaded (BC), and then reloaded (CD). During loading, the initial straight portion of the curve (OA) is assumed to be the elastic response. The settlement CB′ is the elastic settlement while the settlement OC is the permanent (plastic) settlement. In most cases, the shaft friction is mobilized initially and as the full frictional resistance is approached, the end bearing begins to be mobilized (Fig. 8.18b). Thus, full friction and full end bearing are not mobilized at the same settlement as discussed in Section 8.5.

Skin friction and end bearing can be measured using load cells such as the Osterberg cell (O-cell) as shown in Fig. 8.19. The O-cell measures both the skin friction (upward resistance) and the end bearing (downward resistance). In the case of laterally loaded piles, strain gages are attached to the test pile at various depths to deduce the soil resistance and bending moments.

The ultimate (failure) load is not always well defined. The load-settlement curve, (a), in Fig. 8.20 shows a well-defined ultimate load while curve (b) does not. To obtain the ultimate load from curve (b) various empirical procedures have been suggested. One simple method is to find the intersection

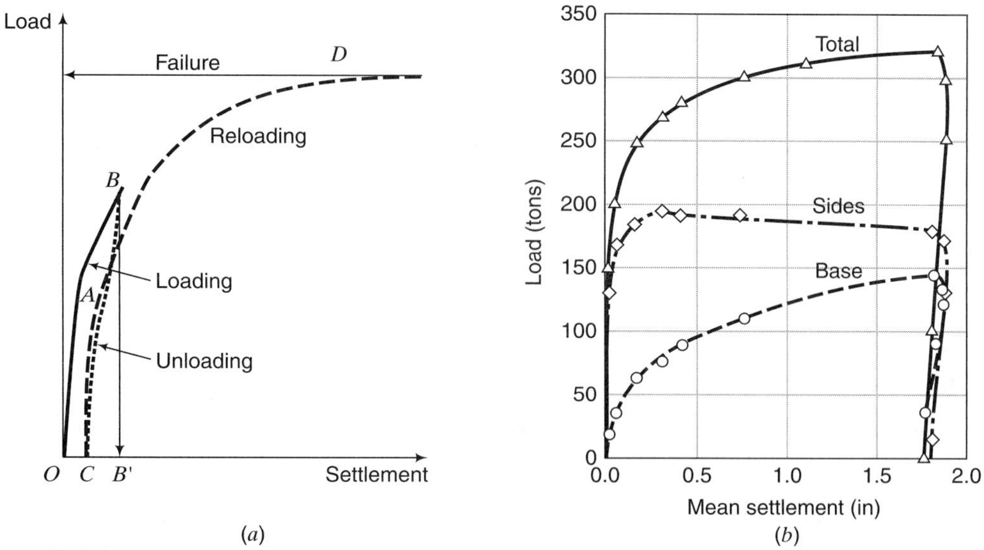

FIGURE 8.18 (a) Schematic pile load test results. (b) Load settlement curves for a drilled shaft. (Source: O'Neill and Reese, 1970.)

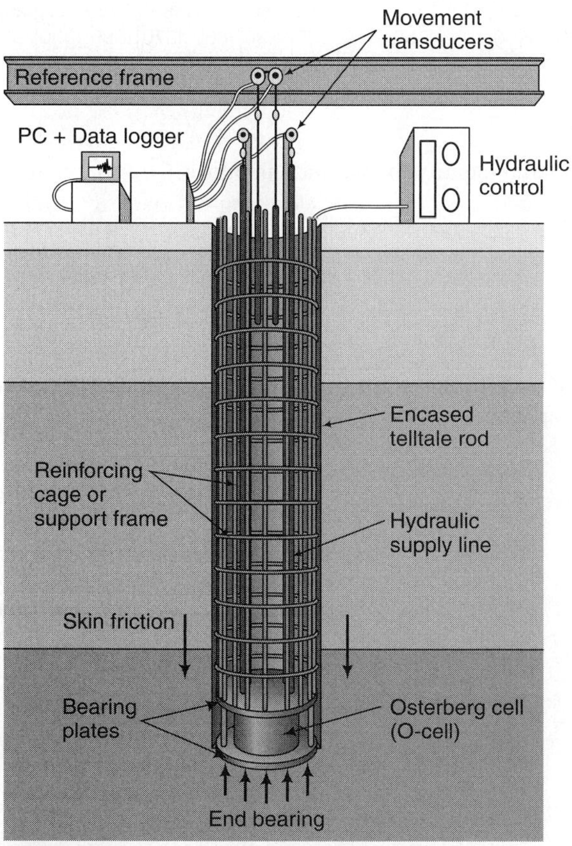

FIGURE 8.19 An O-cell installed in an instrumented test pile. (Source: http://www.loadtest.co.uk/Loadtest%20Ltd/the%20osterberg%20cell_files/ocell.htm.)

of the tangents of the two parts of the curve. The value at the ordinate of the intersection (C in Fig. 8.20) is Q_{ult}. The allowable load capacity is found by dividing the ultimate load by a factor of safety, usually 2. An alternative criterion is to determine the allowable pile load capacity for a desired serviceability limit state (e.g., a settlement of 10% of the pile diameter). The settlement at the allowable (working) load capacity is readily determined from the load-settlement plot (Fig. 8.20).

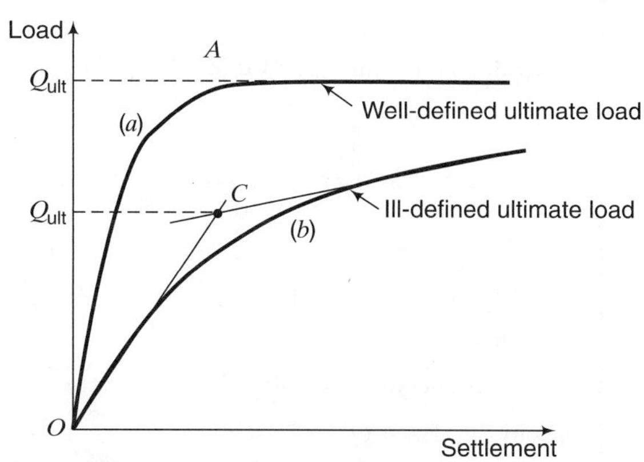

FIGURE 8.20 Interpretation of failure from load-settlement data.

The* key points *are:

1. ***A pile load test provides the load capacity and the settlement of a pile at the working load at a particular location in a job site.***
2. ***Various criteria and techniques are used to determine the allowable load capacity from pile load tests.***
3. ***The tests require accurate measurements of loads and displacements, and careful installation.***

EXAMPLE 8.9 *Interpretation of a Pile Load Test*

The results of a load test on a 0.45 m diameter pile are shown in Table E8.9. Determine (1) the ultimate pile load capacity, (2) the allowable load for a factor of safety of 2, (3) the allowable load capacity at 10% pile displacement.

TABLE E8.9

Displacement (mm)	0.0	1.3	2.5	5.1	7.6	10.2	12.7	15.2	17.8	20.3	22.9	25.4	27.9	30.5	33.0	35.6	38.1	40.6	43.2	45.7	47.0
Load (kN)	0	200	350	670	870	1070	1250	1400	1500	1600	1700	1750	1780	1810	1830	1860	1870	1890	1890	1900	1905

Strategy Plot a graph of displacement versus load and identify the ultimate pile load.

Solution 8.9

Step 1: Plot displacement-load graph.

See Fig. E8.9.

Step 2: Determine ultimate pile load capacity.

The failure load is ill-defined. Locate the intersection of the tangents at the beginning and the end of the curve. The ordinate of this intersection is the ultimate pile load capacity.

$$Q_{\text{ult}} = 1780\,\text{kN}$$

Step 3: Determine allowable pile load capacity.

$$\text{FS} = 2$$

$$Q_a = \frac{Q_{\text{ult}}}{\text{FS}} = \frac{1780}{2} = 890\,\text{kN}$$

Step 3: Determine pile load capacity at 10% pile diameter.

$$\text{Displacement} = 450 \times 0.01 = 4.5\,\text{mm}$$

From Fig. E8.9, $Q_a = 510\,\text{kN}$.

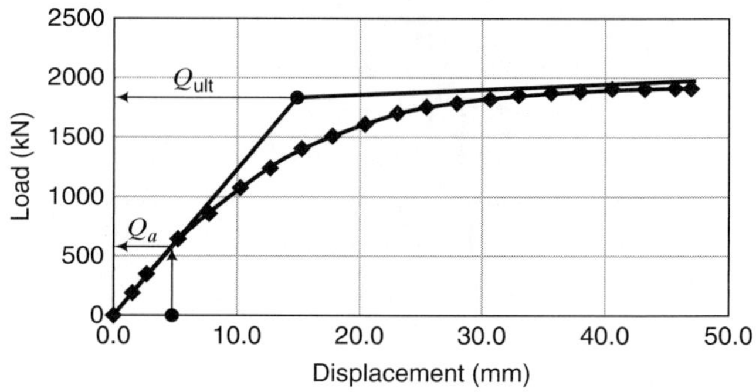

FIGURE E8.9

■

8.15 LATERALLY LOADED PILES

Access a spreadsheet at http://www.wiley.com/college/budhu. Click Foundations and Earth Structures, Chapter 8, p-y.xls to conduct a *p-y* analysis and an interactive computer program apiles based on a continuum approach to estimate the load-displacement response of a singly laterally loaded pile.

Importance *Structures founded on piles are often subjected to lateral loads and moments in addition to vertical loads. Lateral loads may come from wind, traffic, seismic events, waves, docking ships, and earth pressures. Moments may come from the eccentricity of the vertical force, fixity of the superstructure to the piles, and the location of the lateral forces on the pile with reference to the ground surface. Laterally loaded piles, particularly group piles, are difficult to analyze mainly because of the complexity of the soil–structure (pile) interaction. In this section, a few methods to estimate the lateral displacements, rotations, and bending moment are discussed.*

8.15.1 Basic Concept

In designing laterally loaded piles, we need to know the pile deflection, particularly the pile head deflection, to satisfy the serviceability requirement and the bending moments for sizing the pile. The pile head deflection depends on soil type, pile installation method, pile flexibility (or pile stiffness), and loading condition, and on how the pile is attached to the superstructure and pile cap. A pile that is attached to the pile cap such that no rotation occurs is called a *fixed-head pile* (Fig. 8.21a). A pile that is attached to the pile cap such that rotation is unrestricted is called a *free-head pile* (Fig. 8.21b). Piles in general are neither fixed head nor free head. They have undetermined fixity somewhere between free-head and fixed-head conditions. You can think of fixed head and free head as two limiting conditions; piles in practice will respond somewhere within these conditions.

The mechanism of failure depends on the length to diameter or width ratio, soil type, and the fixity of the pile head. Free-head piles tend to fail by rotation. Lateral loads and moments applied to a free-head pile are initially resisted by the soil near ground level. For very small pile deflections, the soil behaves elastically. As the deflection increases, the soil yields and then permanent soil displacement occurs. The soil resistance is shifted to the lower part of the pile as yielding progressively occurs from the top to the bottom of the pile. Fixed-head piles tend to fail by translation.

The response of a pile group to lateral loads and moments is influenced by

- Geometry of the group
- Pile-soil interaction
- Stiffness or flexibility of the piles
- Load conditions
- Individual pile response
- Pile group response resulting from individual pile responses

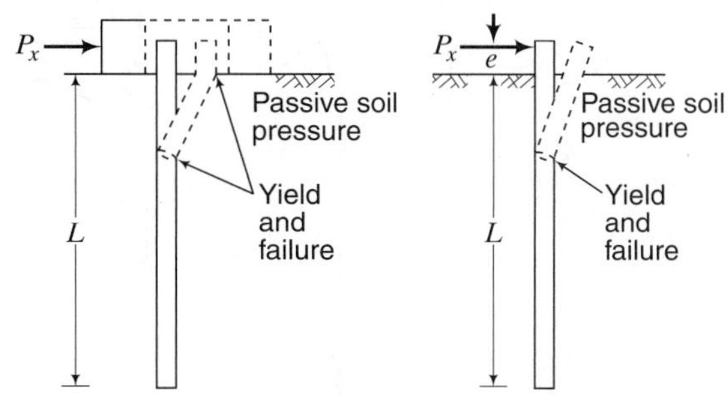

(*a*) (*b*)

FIGURE 8.21 Possible failure modes for long piles ($L/D > 10$) (a) fixed head, (b) free head.

The displacement and rotations are in the directions of the resultant lateral load and resultant moment (Figure 8.22). Outer piles in a group are subjected to

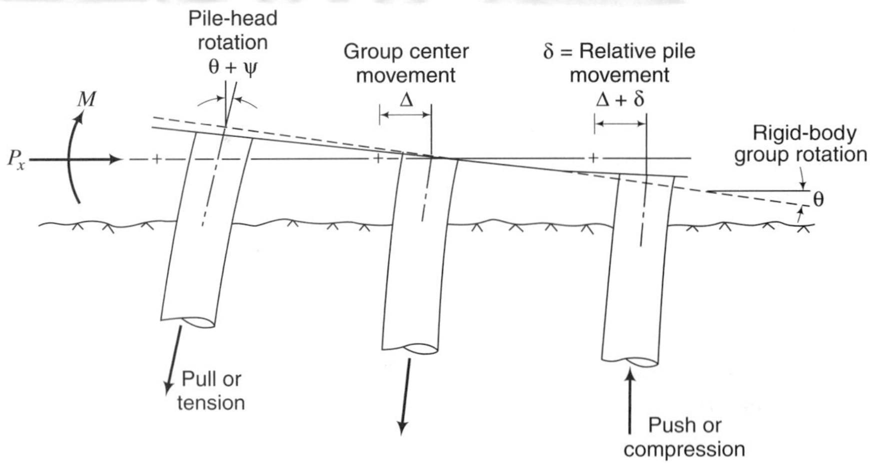

FIGURE 8.22 Group piles subjected to lateral loads and moments.

uplift (pull) and compressive (push) forces while the piles in the center translate at the level of the superstructure connection.

When a pile is subjected to lateral forces and moments, the pile tends to bend or deflect as illustrated in Fig. 8.23. The deflection of the pile causes strains in the soil mass. To satisfy equilibrium, the soil must provide reactions along the length of the pile to balance the applied loads and moments.

Because soil is a nonlinear material, the soil reaction is not linearly related to the pile deflection. Consequently at every point along the length of the pile, a nonlinear relationship between soil resistance (p) and pile deflection (y) exists as illustrated in Fig. 8.23.

8.15.2 Basic Structural Mechanics for Laterally Loaded Piles

Consider an element of thickness, dz, of the pile shown in Fig. 8.23 at a depth z. The free body diagram for this element is shown in Fig. 8.24. We will assume that the pile is symmetrical about the z-axis, the loads lie in the yz plane, the lateral deflection of the pile occurs only along the y-axis (i.e., no out-of-plane deflection), and shear deflection is negligible.

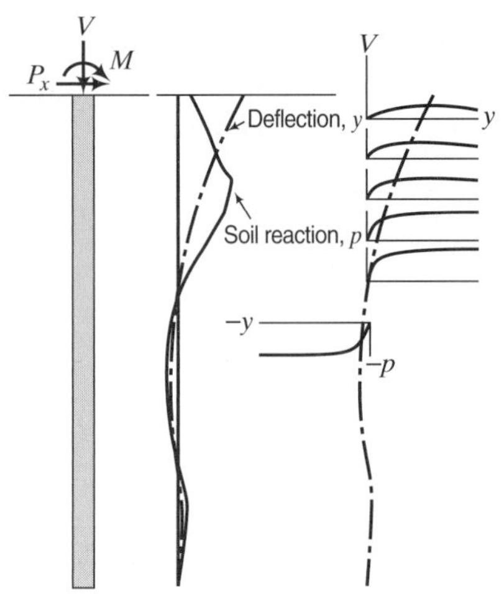

FIGURE 8.23 Pile–soil response to lateral loads and moments.

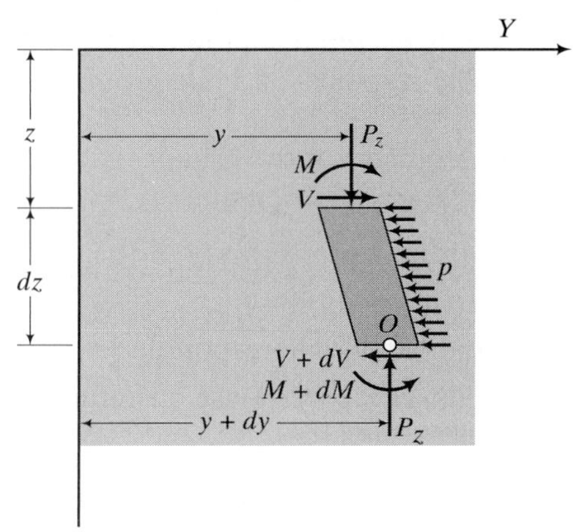

FIGURE 8.24 Free body diagram of a pile section under lateral loads and moments.

Taking moments about O (counterclockwise is positive), we get

$$(M + dM) - M + P_z dy - V dz + p\,dz\frac{dz}{2} = 0 \tag{8.81}$$

Neglecting higher terms (e.g., dz^2) and rearranging Eq. (8.81) gives

$$\frac{dM}{dz} + P_z\frac{dy}{dz} - V = 0 \tag{8.82}$$

Differentiating Eq. (8.82) with respect to z, we get

$$\frac{d^2M}{dz^2} + P_z\frac{d^2y}{dz^2} - \frac{dV}{dz} = 0 \tag{8.83}$$

From basic mechanics, we have the following identities

$$\frac{d^2M}{dz^2} = E_p I_p\frac{d^4y}{dz^4} \tag{8.84}$$

$$\frac{dV}{dz} = p \tag{8.85}$$

$$p = -ky \tag{8.86}$$

where P_z is the vertical load, p is the soil resistance, y is the soil compression (or pile deflection), E_p is the elastic modulus of the pile, k is a soil (stiffness) parameter, and I_p is the second moment of area of the pile. Substituting Eqs. (8.84) to (8.86) into Eq. (8.81) gives the governing equation for a laterally loaded elastic pile as

$$E_p I_p\frac{d^4y}{dz^4} + P_z\frac{d^2y}{dz^2} + ky = 0 \tag{8.87}$$

The governing equation does not include side shear. If the pile above the ground level is subjected to a distributed load (e.g., from water), then the governing equation becomes

$$E_p I_p\frac{d^4y}{dz^4} + P_z\frac{d^2y}{dz^2} + ky - W_l = 0 \tag{8.88}$$

where W_1 is the resultant of the distributed load.

The governing equations can be solved using numerical methods such as FDM or FEM or BEM (Chapter 3). The term ky in the governing equations changes depending on the assumption made regarding the soil response. One can assume an elastic soil or an elastoplastic soil or some other type of load-deformation or stress-strain response. Therefore, the solution of the governing equations depends on the soil model chosen. Three methods of obtaining solutions of the governing equations based on three different types of soil responses are briefly described below.

1. ***Elastic beam on elastic foundation method.*** In this method, the pile is assumed to be an elastic beam that is attached to discrete springs representing the soil (Fig. 8.25). The soil is assumed to be elastic but discontinuous. The soil stiffness parameter k is taken as the lateral subgrade modulus K_h of the soil. The lateral subgrade modulus is the ratio of the horizontal soil reaction per unit area to the lateral soil displacement. The determination of the lateral subgrade modulus is often difficult and expensive. It can be obtained from field tests using instrumented test piles. Empirical relationships are often used to relate results from simple, routine tests such as SPT, cone tests, and/or laboratory

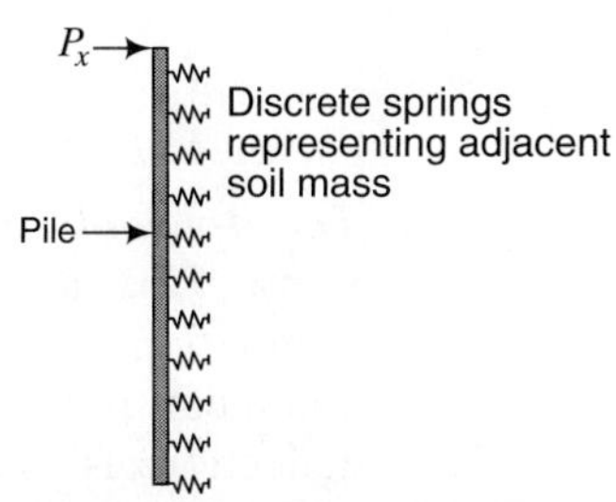

FIGURE 8.25 Simulation of adjacent soil mass as a set of discrete springs.

shear tests to K_h. The value of K_h is assumed to be constant for fine-grained soils and linearly increasing with depth for coarse-grained soils.

The main problem with the elastic beam on elastic foundation approach is that it neglects soil continuity and soil shearing resistance. This method is satisfactory for very small strain levels ($< 0.001\%$). Analytical solution (Matlock and Reese, 1962) for the lateral displacement (ρ_L) and rotation (θ) at ground level for a flexible free-head pile (piles with length longer than the critical length) and constant K_h is

$$\rho_L = \sqrt{2}\frac{P_x}{K_h}\left(\frac{l_c}{4}\right)^{-1} + \frac{M}{K_h}\left(\frac{l_c}{4}\right)^{-2} \tag{8.89}$$

$$\theta = \frac{P_x}{K_h}\left(\frac{l_c}{4}\right)^{-2} + \sqrt{2}\frac{M}{K_h}\left(\frac{l_c}{4}\right)^{-3} \tag{8.90}$$

where $l_c = 4(\frac{E_p I_p}{K_h})^{\frac{1}{4}}$ is the critical length, E_p is the elastic modulus of the pile, I_p is the second moment of area of the pile, P_x is the lateral load, and M is the applied moment.

For K_h varying linearly with depth such that $K_h = m_h z$ where m_h is the slope of the variation of K_h with depth (z), the lateral displacement and rotation are

$$\rho_L = 2.43\frac{P_x}{m_h}\left(\frac{l_c}{4}\right)^{-2} + 1.62\frac{M}{m_h}\left(\frac{l_c}{4}\right)^{-3} \tag{8.91}$$

$$\theta = 1.62\frac{P_x}{m_h}\left(\frac{l_c}{4}\right)^{-3} + 1.73\frac{M}{m_h}\left(\frac{l_c}{4}\right)^{-4} \tag{8.92}$$

$$\text{where } l_c = 4\left(\frac{E_p I_p}{m_h}\right)^{\frac{1}{5}}$$

The accuracy of the lateral displacement depends largely on the accuracy of K_h. But there is no reliable value of K_h. As a first approximation, where K_h varies with depth such that

$$K_h = m_h z \tag{8.93}$$

We can estimate m_h from

Coarse-grained soils

$$m_h = \frac{7}{D}(D_r)^{1.7}; \quad 25\% \le D_r \le 100\%; \quad m_h(\text{kN/m}^3), \quad D(\text{m}) \tag{8.94}$$

$$m_h = \frac{0.045}{D}(D_r)^{1.7}; \quad 25\% \le D_r \le 100\%; \quad m_h(\text{kcf}), \quad D(\text{ft}) \tag{8.94a}$$

Fine-grained soils

$$m_h = \frac{0.32 s_u(s_u + 122)}{D}; \quad s_u(\text{kPa}), \quad m_h(\text{kN/m}^3), \quad D(\text{m}) \tag{8.95}$$

$$m_h = \frac{4.6 s_u(s_u + 2.6)}{D}; \quad s_u(\text{ksf}), \quad m_h(\text{kcf}), \quad D(\text{ft}) \tag{8.95a}$$

where D is the pile diameter or width and D_r is relative density.

2. ***The p-y analysis.*** The *p-y* (soil resistance–pile deflection) method (Matlock and Ripperger, 1956; Reese, Cox, and Koch, 1975 and others) is conceptually similar to the elastic beam on elastic foundation method except that the soil stiffness parameter k is not constant. In the *p-y* method, a nonlinear relationship between the soil resistance and soil displacement (equal to the pile deflection) is assumed along the length of the pile. The nonlinear relationships along the pile length are obtained from results of well-instrumented laterally loaded test piles. Typical *p-y* curves for many soil types are available in software programs such as Com624 (http://www.fhwa.dot.gov/bridge/geosoft.htm#com624p) and various commercial derivatives to conduct preliminary design.

However, it is recommended that laterally loaded instrumented pile tests be conducted whenever such tests are feasible and economical.

Like the elastic beam on elastic foundation approach, the *p-y* method neglects soil continuity and soil shearing resistance. An approximate solution of the governing equations based on the *p-y* approach is at http://www.wiley.com/college/budhu. Click Foundations and Earth Structures, Chapter 8, p-y.xls.

3. ***Continuum analysis.*** In this method of analysis, the soil is treated as a continuous medium with assumptions made on its stress-strain behavior or constitutive relationships. The simplest stress-strain behavior is elastic, described by Hooke's law. The solution gives the load-deformation response of the pile for small strain levels (Poulos, 1971; Randolph, 1981). The effects of soil yielding or pile yielding cannot be obtained from an elastic analysis. To account for soil yielding, the soil can be assumed to be an elastoplastic material (Davies and Budhu, 1986; Budhu and Davies, 1986, 1987).

   In the elasto-(rigid) plastic solution proposed by Davies and Budhu (1986) and Budhu and Davies (1986, 1987), the soil is assumed to behave linearly elastic at small strain levels. Yielding occurs when the normal and shear stresses on the pile reach values determined from a limiting equilibrium analysis. Bearing failure on the front face, shear stresses along the sides, and tension failure along the back face (Fig. 8.26) are incorporated into the analysis. Failure on the bearing face occurs when

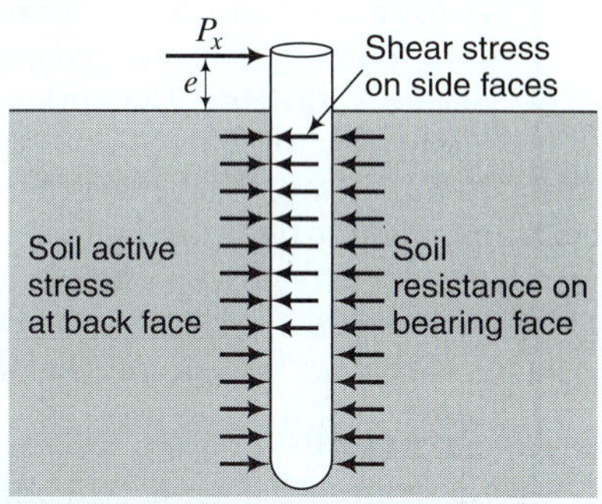

FIGURE 8.26 Soil stresses on a pile segment resulting from lateral loads treating soil as a continuum.

$$\sigma_{bf} = N_c s_u \tag{8.96}$$

where σ_{bf} is the limiting compressive stress, and N_c is the bearing capacity factor that is assumed to vary from 2 at the surface to a constant value of 9 at a depth of three pile diameters and below.

The limiting shear stress at the sides of the pile is

$$\tau_{ss} = \alpha_u s_u \tag{8.97}$$

where the adhesion factor, α_u, is taken as 1 for soft clays. On the back face or tension face, the normal stress cannot exceed the in-situ lateral effective stress,

$$\sigma_{bf} \leq K_{oi}\sigma'_{zo} \tag{8.98}$$

where K_{oi} is the lateral earth pressure coefficient that accounts for pile installation. For soft clays, K_{oi} is usually 1.0.

The important findings from the continuum analysis are as follows:

1. The analysis is valid only for a pile with a length greater than its effective length. The effective length is the length of pile that is effective in resisting the loads and moments. Piles longer than their effective lengths will behave identically. The effective length can be estimated from $L_e = 1.3D\,K_{ps}^{\frac{2}{9}}$.
2. The maximum bending moment in free-head piles would normally occur at depths less than 5 diameters below ground level.
3. Eccentric lateral loads increase the maximum bending moment (regardless of the type of analysis).
4. Fixed-head piles reduce the pile head deflection by about 50%.
5. Laterally loaded piles, in general, fail by yielding of the pile section well before the soil fails.
6. The lower the flexibility of the pile the greater the pile deflection and the greater the maximum bending moment.

Several computer programs are commercially available to analyze laterally loaded piles based on the continuum approach. A computer program based on the Budhu and Davies approach is at http://www.wiley.com/college/budhu. Click Foundations and Earth Structures, Chapter 8, APILES. The default shape in the program is circular. For other shapes, use an equivalent diameter. For example, the equivalent diameter for a square pile of width B is $\frac{2B}{\sqrt{\pi}}$.

EXAMPLE 8.10 *Predicted Response of a Single Laterally Loaded Pile Using the Continuum Approach*

A 460 mm (18 in.) diameter timber pile of total length 9.75 m (32 ft) is embedded to a depth, L, of 9.14 m (30 ft) in a soft, normally consolidated clay deposit of medium plasticity. The strength and elastic modulus of the soil vary linearly with depth. The working stress, σ_w, for timber of this quality is 11 MN/m^2 (230 ksf) and its Young's modulus of elasticity is 10.3 GPa (215064 ksf). A lateral load is to be applied at a height, e, of 690 mm (27 in.) above ground level. Required are: (1) the working load; (2) the lateral deflection at the working load, and (3) the maximum bending moment at the working load. Assume the following: adhesion factor, $\alpha_u = 1$, lateral earth pressure coefficient, $K_{oi} = 1$, effective unit weight, $\gamma' = 9.4\,\text{kN/m}^3$ (60 psf); the slope of the linear variation of undrained shear strength with depth, $c = 2.7\,\text{kPa/m}$ (56.4 psf/ft); and the slope of the linear variation of elastic modulus with depth, $m = 2.7\,\text{MPa/m}$ (56376 psf/ft) that is, $m/c = 1000$. Groundwater level is at ground surface.

Strategy Use the program APILES at http://www.wiley.com/college/budhu. Click Foundations and Earth Structures. Chapter 8, APILES.

Solution 8.10

Step 1: Enter input data interactively using the program APILES.

Step 2: Evaluate results.

The echoed input data and the results are given below.
Predictions of Program APILES – Version 1.0 (1988).
This analysis for APILES was developed by M. Budhu and T. Davies.
This PC version of APILES was written by M. Budhu.
This analysis is valid for piles whose lengths are greater than their effective length.
The program will stop when either the working load is exceeded or the pile head displacement exceeds about one tenth the pile diameter.

TITLE: Example 8.10
Your input data is as follows

PILE DATA
Head type: Free head with eccentricity

Eccentricity = 0.69 m
Length = 9.14 m
Diameter = 0.46 m
Young's Modulus = .10E + 08 kPa
Working stress = .11E + 05 kPa

SOIL DATA
Soil type: Soft clay

Eccentricity unit weight of soul = 9.4000 kN/m^3
Earth pressure coefficient = 1.00
Adhesion factor = 1.00
Young's modulus at surface = .00E + 00 kPa
Rate of increase of Young's modulus with depth = 0.27E + 04 kN/m^3
Undrained shear strength at surface = .00 kPa
Rate of increase of undrained shear strength with dept = 0.27E + 01 kN/m^3

The working load is: 42.54 kN; lateral displacement = 31 mm; maximum bending moment = 103 kN/m
The lateral displacements are computed at the point at which the load is applied.

Load (kN)	Disp (mm)	Moment (kN · m)	Rotation (radians)	Max. B.M. (kN · m)	Depth to max. B.M. (m)
0	0	0	0	0	0
4.50	1.82	3.10	.04	7.45	1.66
4.91	1.99	3.39	.04	8.14	1.66
6.65	2.74	4.59	.06	11.29	1.66
9.40	4.03	6.48	.08	16.51	1.74
12.96	5.85	8.94	.12	23.45	1.84
13.77	6.27	9.50	.13	25.06	1.85
13.88	6.32	9.57	.13	25.25	1.86
14.48	6.65	9.99	.13	26.42	1.86
18.42	8.88	12.71	.18	34.38	1.92
19.62	9.61	13.54	.19	36.98	1.94
20.79	10.35	14.34	.20	39.64	1.96
23.74	12.31	16.38	.24	46.31	1.99
28.67	15.82	19.78	.30	57.83	2.06
30.96	17.61	21.36	.34	63.71	2.10
31.31	17.90	21.61	.34	64.64	2.10
37.83	23.57	26.10	.44	81.49	2.20
39.94	25.62	27.56	.47	87.52	2.24
45.25	31.14	31.22	.57	102.82	2.34
46.58	32.56	32.14	.59	106.68	2.36
50.68	37.58	34.97	.67	119.38	2.45
55.77	44.17	38.48	.77	135.47	2.54

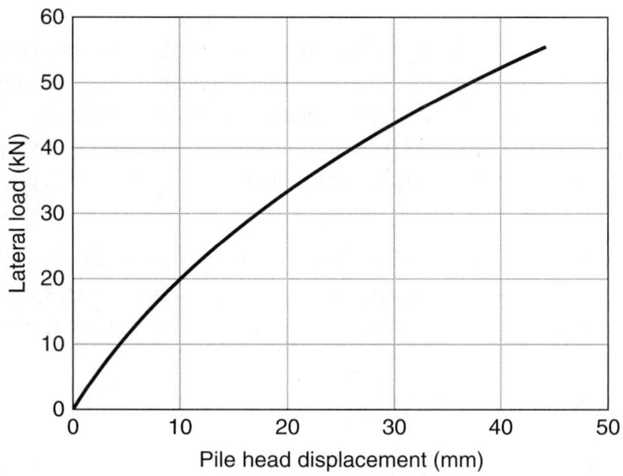

FIGURE E8.10

The bending moments and rotations are computed at ground surface.
The lateral load–lateral displacement curve is shown in Fig. E8.10. ■

8.16 DESIGN ISSUES AND PROCEDURES

Importance *Successful pile design requires experience and understanding of pile installation methods. The pile load capacity, determined by the methods discussed earlier in this chapter, is based on simplifying assumptions and can be regarded only as an estimate. Soil conditions are notoriously spatially variable and as such several design changes—for example, increase in pile length or method of installation—have to be made in the field. Design (office) procedure and some design issues are discussed in this section.*

8.16.1 Evaluate the Need for a Pile Foundation

Evaluate the loading, soil, and environmental conditions at the project site and decide whether pile foundation is suitable (Section 8.2). If it is, review the soil profile and identify a potential end bearing layer. If you have a number of borehole logs, use them to sketch a composite soil profile as discussed in Chapter 5. A suitable end bearing layer may consist of a dense sand ($N_{60} > 15$) or dense gravel ($N_{60} > 15$) or a stiff clay layer ($s_u > 50$ kPa) at least five pile diameters thick. The pile should penetrate at least one pile diameter into the end bearing layer. If the end bearing layer is very deep, you may have to design a floating pile foundation.

8.16.2 Pile Selection

Pile Material Determine the pile material required based on the project site conditions and pile availability. If steel piles or timber piles are readily available in an area where the soil is corrosive and/or the groundwater fluctuates, then decide on the method of protection. Steel piles can be protected from corrosion by one or a combination of corrosion protection paint, electrochemical (cathode) protection, concrete cover, and polyethylene coating. Timber piles can be protected against rotting due to water fluctuation by a concrete cover or by a coating of several commercial products.

Pile Length and Splicing The pile length depends on the magnitude of the loads to be supported, the pile configuration (single or group piles), the method of installation, and the depth of a suitable end bearing layer. Load capacity analysis and/or the depth of the end bearing layer will provide an estimate of the pile length. The pile length should be such as to avoid splicing and cumbersome handling. Splicing is a field operation that increases construction time and costs, and may reduce the structural integrity of the pile. In certain situations (e.g., low headroom) splicing is known before construction, and you must consider such splicing in the design process (including cost estimation) and prepare specifications accordingly.

Steel piles are spliced by welding. Timber piles can be spliced by attaching a steel tube or steel plates to join the two pieces together. Splicing can also be accomplished by using composite piles (e.g., concrete pile attached to a timber pile).

Pile Shape The popular cross-sectional shape is circular because of ready availability, ease of installation, and equal resistance to loads in any direction. The deciding factor on shape, apart from ease of installation and availability, is the second moment of area (moment of inertia). The higher the second moment of area the lower the bending stress.

8.16.3 Pile Spacing

Pile spacing influences the pile load capacity. Some effects of close pile spacing ($< 3D$, $D =$ diameter or width of the pile) for driven piles are

- Reduced load capacity by overstressing the piles through stress interactions
- Soil heaving

- Heaving of existing neighboring piles
- Damage to existing neighboring piles
- Difficulty in pile driving—reduced pile driving clearance requirement

Pile spacing is usually $3D$ to $5D$, where D is the pile diameter or the width. As the pile spacing increases, the pile cap size for a pile group will increase. You should investigate various spacing to determine the least cost.

8.16.4 Pile Handling

Pile handling during transportation and during construction can lead to pile failure. The common failures are bending and spalling. Piles must be lifted at points that will reduce the bending stresses in the pile. One popular method is the two-point pickup method, whereby the two lift points are each located at 20% of the pile length from each end. The bending moment is $\frac{1.5W_pL}{40}$, where W_p is the weight of the pile, L is the pile length, and the factor 1.5 is an impact load factor.

8.16.5 Scour

Piles located in waterways and air-blown deposits are subjected to scour. The effect of scouring is to reduce the embedded length of the pile, possibly leading to pile buckling, failure, and excessive settlement. Scouring can be several meters deep. You should estimate the depth of scour using geohydraulic methods, which are beyond the scope of this textbook.

8.16.6 Pile Installation

Make a decision on the method of installation (e.g., driven or bored). If the soil is granular and contains boulders, then drilled shafts are preferable or you can use driven H steel piles. For fine-grained soils, driven piles are often used. For driven piles, you should select a hammer (Section 8.4) that does not damage the pile and reduce its load capacity. For drilled shafts, the pouring of concrete must be such as to minimize segregation and reduce voids along the pile length.

8.16.7 Pile Verticality and Pile Batter

Straight, vertical piles are preferable to piles with batters. However, soil conditions and driving methods can cause vertical piles to be misaligned. The axial load capacities of misaligned piles are reduced by as much as 30%. Batter piles are used for large lateral loads. The batter should not be greater than $2(V)$ to $1(H)$. The lower the batter ratio (V) to (H), the greater the resistance to lateral loads. A batter of $2.5(V)$ to $1(H)$ is better than $5(V)$ to $1(H)$ to support lateral loads.

8.16.8 Determine Load Capacity Using Parameters that will Lead to a Ductile Response

Consider how the loads will be transferred to the piles and the installation method. The structural loads can be transferred using fixed (piles must support moments), pinned (no moment transferred), or partially fixed (some moments transferred) supports.

Decide on the method of analysis and the design methodology—ASD or LRFD. Combine the loads as required by the governing code (e.g., IBC, 2006) for the chosen design methodology. Calculate the pile load capacity for one or more pile diameters and then decide on the final diameters and configuration—single or group piles. You should use the critical state soil parameters to design for ductile response.

TABLE 8.5 Performance Factors for Pile Load Capacity Using LRFD

Method		Performance factor
1. *Driven piles*		
(a) Compression		
Statics – Skin friction		ψ_f
α-method using remolded (critical state) undrained shear strength.		= 0.7
α-method using estimated undrained shear strength		= 0.35
β Method (ϕ'_{cs})		= 0.75
β Method (ϕ'_p)		= 0.5
End bearing		ψ_b
Fine-grained soils (N_c)		= 0.7
Coarse-grained soils (N_q with ϕ'_{cs})		= 0.7
Coarse-grained soils (N_q with ϕ'_p)		= 0.5
Field tests		$\psi_f = \psi_b$
SPT	N values	= 0.45
CPT	q_c	= 0.55
Load tests		= 0.8
(b) Uplift		= 0.3
2. *Pile driving*		
Pile driving analyzer		= 0.65
Engineering News Record formula		= 0.1
3. *Drilled shafts*		
Fine-grained soils		
Skin friction		= 0.65
End bearing		= 0.55
Coarsed-grained soil		
Skin friction		= 0.5
End bearing		= 0.5

For most situations, the methods of analysis discussed earlier in this chapter are suitable. For complex foundation systems and for seismic loading more involved analysis based on FDM or FEM or BEM may be required to complement the simpler (empirical and semi-empirical methods in this chapter).

The performance factors for LRFD at the time of writing are not standardized and may be different depending on country, federal, and state standards. The performance factors for LRFD for use in this textbook are given in Table 8.5.

The vertical settlements for serviceability limit state are usually less than 25 mm (1 in.) for buildings on pile foundations in coarse-grained soils and less than 50 mm (2 in.) if the pile foundation is in fine-grained soils. The lateral displacements are usually less than 12 mm (0.5 in.). Differential settlements are limited to about 6 mm (1/4 in.), or use Table 6.5 as a guide.

8.16.9 Pile Integrity During and After Installation

The pile installation method coupled with unknown soil conditions can significantly decrease the pile load capacity. Structural problems arise from excessive bending, concrete cracking, and buckling of steel. For drilled shafts, problems occur mainly from improper excavation of the hole during drilling, improper bottom cleaning, improper placement of concrete, poor-quality concrete, improper removal of drill, and casing. You need to check the pile integrity, especially for drilled shafts, during and after installation. Various methods are available. Some of these are as follows:

Pile Driving Formula Driving records can provide useful information on the consistency of a soil at a site. For example, if the number of blows to drive a pile at a certain depth is A blows at one location and B blows at another location at the same site, then the soil stratification is different. You should reexamine your design and the soil boring records and make the necessary adjustments (e.g., increase or decrease the pile length).

Sometimes during installation, a pile moves under its own weight or offers very little resistance to driving, indicating perhaps an unanticipated weak soil layer. You will then have to make field adjustments such as driving the pile deeper. In some situations, this may require splicing the pile. During driving, a pile may reach a refusal state—a state in which the pile does not move further when driven—before achieving the estimated embedment depth. You need to consider whether the hammer is too light or the pile has hit an unknown obstruction or a stiff soil layer exists that was not revealed in the soils investigation.

A number of empirical equations have been proposed to relate the energy delivered by a hammer during pile driving and the pile load capacity. One of the earliest equations is the ENR (*Engineering News Record*) equation given as:

$$Q_{u/t} = \frac{\eta_1 W_R h}{s + C_1} \tag{8.99}$$

where η_1 is an efficiency factor (Table 8.6), W_R is the weight of the ram, h is the height of fall, s is the penetration per blow, and C_1 is a constant. For drop hammers, $C_1 \cong 25$ mm; for steam hammers $C_1 \cong 2.5$ mm. The units of h and s must be the same. The ENR equation was developed for timber piles using drop hammers for installation but is used to estimate or check the ultimate load capacity of all types of piles.

TABLE 8.6 Hammer Efficiency

Hammer type	η_1
Drop hammer	0.75–1.0
Single-acting hammer	0.75–0.85
Double-acting hammer	0.85
Diesel hammer	0.85–1.0

You can compare your estimated ultimate pile load capacity with that calculated from the pile driving formulas. The reliability of the pile driving formula is difficult to ascertain. Experience is required to make a decision on whether to rely on the pile load capacity from the pile driving formula or other methods discussed earlier in this chapter.

Wave Equation When a hammer strikes a pile the shock sets up a stress wave, which moves down the pile at the speed of sound. From elastic theory, the change in force on the pile over an infinitesimal length Δz, assuming the pile to be a rod, is:

$$\frac{\partial F}{\partial z} = E_p A_p \frac{\partial}{\partial z}(\varepsilon_z) = E_p A_p \frac{\partial^2 u}{\partial z^2} \tag{8.100}$$

where $\varepsilon_z = \frac{\partial u}{\partial z}$ is the vertical strain, E_p = Young's modulus of the pile, A_p is the cross-sectional area of the pile, and u is pile displacement. From Newton's second law,

$$\frac{\partial F}{\partial z} = A_p \frac{\gamma}{g} \frac{\partial^2 u}{\partial t^2} \tag{8.101}$$

where g is the acceleration due to gravity and t is time. Setting Eq. (8.100) equal to Eq. (8.101), we get

$$E_p A_p \frac{\partial^2 u}{\partial z^2} = \frac{\gamma}{g} A_p \frac{\partial^2 u}{\partial t^2} \tag{8.102}$$

which, by simplification, leads to

$$\frac{\partial^2 u}{\partial t^2} = V_c^2 \frac{\partial^2 u}{\partial z^2} \tag{8.103}$$

where $V_c = \sqrt{\frac{E_p g}{\gamma}}$ is the vertical wave propagation velocity in the pile.

The solution of Eq. (8.103) is found using appropriate boundary conditions and can be modified to account for soil resistance. Computer programs (e.g., GRPWEAP) have been developed for routine wave-analysis of pile driving operations.

Nondestructive Methods Several nondestructive methods are also available for testing pile integrity. These include the surface reflection method, direct transmission method, crosshole ultrasonic method, and gamma-gamma method:

- ***Surface reflection method (SRM):*** An impact by a hammer at the pile head is used to generate a stress wave at the surface of the pile foundation. The stress waves move down the pile and are reflected when a defect is reached. A geophone records the shaft response. Both the impact force and the wave velocity are measured.
- ***Direct transmission method (DTM):*** A borehole is drilled adjacent to the pile and filled with water. A stress wave is set up through impact by a hammer at the pile head. The stress wave travels down the shaft and through the soil where the compression wave is measured by hydrophones in the adjacent water-filled borehole. The times of arrival of the stress wave are measured. If there is no defect, the first arrival time of the stress wave will vary linearly with depth. If a defect occurs, the first arrival time will be longer. Through this method, the length of the pile can be calculated as the base of the pile will be recorded as a defect.
- ***Cross-hole logging (CSL):*** Two or more steel or PVC tubes are integrated into the pile and a transmitter down one of them sends an ultrasonic signal that is detected by receivers in the others. The transmitter and receiver are raised up or lowered down the tubes to scan the length of the pile. CSL utilizes the first arrival time to detect defects such as concrete voids, poor-quality concrete, and soft soil at the pile base.
- ***Gamma-gamma method (GGM):*** GGM utilizes gamma rays to detect change in density. Gamma rays emitted by a probe are absorbed and scattered by the concrete and steel (if any). A detector is used to count the reflected rays over a fixed interval. Denser material will have lower gamma ray counts than looser material. Gamma ray counts are plotted against depth. Defects show up as changes in the gamma ray counts.

8.16.10 Lateral Squeeze

Lateral squeeze occurs when piles, and also other substructures, are installed in soft compressible soils. These, substructures may be subjected to lateral stresses caused by settlement of the soil. The additional lateral stresses can cause bending of piles and lateral movement of abutments. Lateral squeeze tend to occur when the overburden pressure is greater than three times the undrained shear strength of the soil. The horizontal movement of an abutment is generally about 25% of the vertical settlement. To reduce the effects of lateral squeeze, you may

1. Preload the compressible soil to diminish the amount of settlement.
2. Excave the compressible soil and replace by compacted fill.
3. Use H-piles with high tensile strength.
4. Provide expansion shoes to accommodate the horizontal movement.

8.16.11 Cost

Estimate the cost of each potential pile type considering the expertise and availability of piling contractors. Try to use standard pile diameters and simple shapes (circular, square) in your design as these are cost saving measures. If a simple soils investigation were conducted, then you should anticipate additional costs due to potential problems from unrevealed soil, environmental, and local (e.g., labor) conditions.

The **key points** *are:*

1. ***Estimating the pile load capacity based on statics or other methods is not sufficient for successful pile design. Experience and the ability to make decisions during pile installation are necessary.***
2. ***The integrity of a pile is dependent on handling and the method of installation.***
3. ***Driving records can provide some useful information regarding the character of the soil at a site and help you evaluate the estimated pile load capacity.***

8.17 SUMMARY

Piles are used to support structural loads that cannot be supported on shallow foundations. The predominant types of pile materials are steel, concrete, and timber. The selection of a particular type of pile depends on availability, environmental conditions, pile installation methods, and cost. Pile load capacity cannot be determined accurately because the method of installation invariably changes the soil properties near the pile. We do not know the extent of these changes. The equations for pile load capacities and settlement are, at best, estimates. Load capacities from pile load tests are preferred but these tests are expensive and may be cost effective only for large projects.

Practical Examples

EXAMPLE 8.11 *Design of a Pile Foundation Under a Waterfront Structure*

A wharf structure as shown in Fig. E8.11 is required by a client. The columns are equally spaced on a 3 m (20 ft) grid. Column A will have to support a dead load of 260 kN (60 kips) and a live load of 130 kN (30 kips). Available piles are timber piles 0.45 m (18 in.) diameter and lengths up to 10 m (32 ft), and square precast concrete piles of minimum width 0.3 m (12 in.) and increasing in increments of 25 mm (1 in.). The precast concrete pile lengths can be customized. The elastic moduli of the timber and concrete piles are 13 GPa (2000 ksi) and 21 GPa (3000 ksi) respectively. The allowable driving stresses for timber and precast concrete piles are 1 MPa (0.15 ksi) and 10 MPa (1.5 ksi) respectively. The settlement should not exceed 1% of the pile diameter or width. From experience on this site, it is difficult to drive piles beyond 14 m (45 ft) in the stiff overconsolidated clay. Design the pile foundation for column A. Assume a factor of safety of 2.

Strategy Decide on a pile type and method of installation. Use the spreadsheet pile.xls to do the calculations. The transient mud, very soft clay above the clayey silt, and the clayey silt should be neglected. U.S. units will be used in solving this problem.

Solution 8.11

Step 1: Determine the pile type and end bearing layer.
The end bearing layer is the stiff overconsolidated clay at a depth of about 30 ft under column A. Timber piles will not be suitable because they will need dry rot protection in the tidal variations region and the maximum available length is likely to be less than the required depth. We will select precast concrete piles. The piles will be installed using a pneumatic hammer.

Step 2: Determine the method of design and the ultimate load.
We will use LRFD and IBC (2006).

$$Q_{DL} = 60 \text{ kips}$$

$$Q_{LL} = 30 \text{ kips}$$

$$\text{Factored load} = 1.4Q_{DL} + 1.6Q_{LL} = 1.4(60) + 1.6(30) = 132 \text{ kips}$$

Step 3: Determine ultimate load capacity of a single pile.
Use the spreadsheet pile.xls.

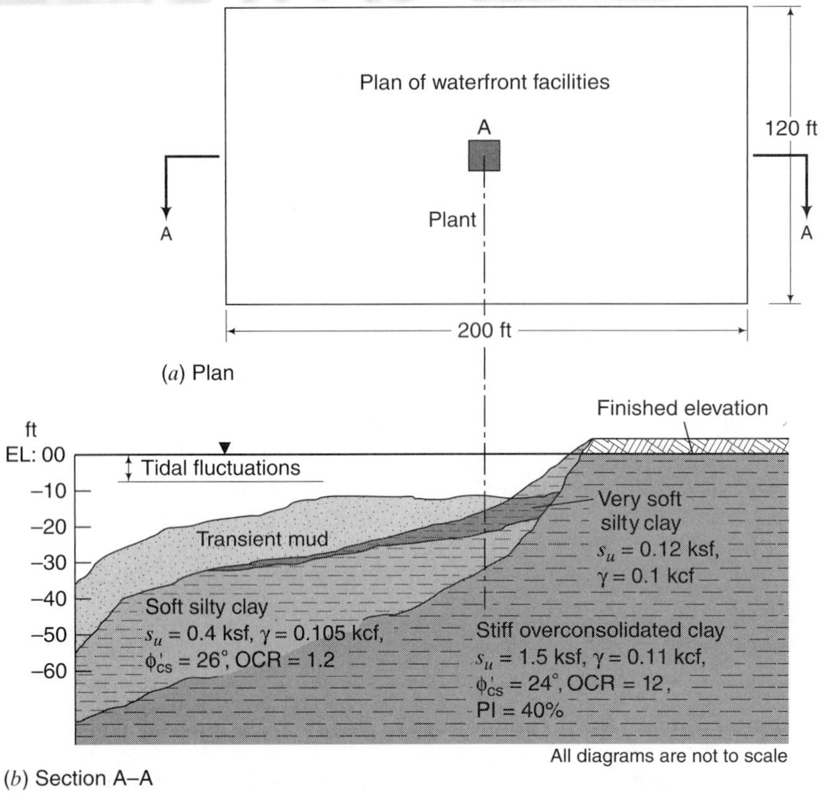

FIGURE E8.11

The input data are shown in Table E8.11. The results of the spreadsheet show that a precast concrete pile of diameter 3 ft and length 45 ft will support a skin friction force of 89 kips (TSA) and 96 kips (ESA), and an end bearing load of 121 kips (TSA) and 103 kips (ESA). The ultimate load capacity for TSA (210 kips) is very close to that of ESA (199 kips). Since piles tend to mobilize the skin friction first and f_s for TSA is less than ESA, we will assume TSA governs.

TABLE E8.11

	A	B	C	D	E	F	G	H
1	Muni Budhu "Foundations and Earth Retaining Structures", John Wiley & Sons, 2006.							
2	**DRIVEN PILES**							
3	**Straight, prismatic pile as an individual pile or in a group**						**Help**	
4	Select units	**US**		γ_w	0.0624	kcf		
5	Design load	132	kips	FS	2			
6	Pile section shape	square		Group	no			
7	Adhesion factor, α_u	API		Spacing	4	ft		
8				Matrix	3			
9	Groundwater	0						
10	End bearing layer	30	ft		diameter or width			
11	N_c	9			2.5	3	3.5	ft
12	ω_p clay	0.5	rad	Perimeter	10.00	12.00	14.00	ft
13	ω_p sand	0.4		Area	6.25	9	12.25	ft2
14	Use N values	n						
15	layer	Depth	thickness	Soil Type	Unit weight	ϕ'	s_u	OCR
16		ft	ft		kcf	deg	ksf	
17	groundwater	0	0		0	0		0
18	transient mud, silt, soft clay	25	25	c	0.0624	0	0	0
19	clayey silt	30	5	c	0.109	26	0.4	1.2
20	stiff clay	44	14	c	0.11	24	2.5	12

Step 4: Check design using LRFD.

$$\varphi_f Q_f + \varphi_b Q_b \geq (\text{combination of factored load for worst case})$$

The performance factors are obtained from Table 8.4 for ESA.

$$0.7(89) + 0.7(121) = 147\,\text{kips} > 132\,\text{kips}.$$

Therefore, okay.

Step 5: Calculate settlement.

Percent of load carried by friction is $\frac{89}{210} \times 100 = 42\%$. Most of this is within the stiff overconsolidated clay.

$$\frac{L}{D} = \frac{45 - 30}{3} = 5$$

Make an estimate of the elastic modulus using Fig. B4.7 (Appendix B). The maximum value of OCR in this figure is 10 but the curves tend to be horizontal after OCR about 8. With OCR = 12 and PI = 40%, $\frac{E_u}{s_u} \cong 150$. Therefore, $E_u = 150 \times 1.5 = 225$ ksf. This value appears to be low in comparison with the range of elastic modulus for stiff clays (Appendix B Table B4.9). We will use the lower value of $E_u \approx 600$ ksf for stiff clays from Appendix B, Table B4.9. Using Eqs. (8.67) and (8.72) to calculate the elastic settlement, we get

$$I = 0.5 + \log(5.7) \approx 1.2$$

$$Q_{af} = \frac{Q_f}{Fs} = \frac{89}{2} = 44.5\,\text{kips}; \quad Q_a = 30 + 60 = 90\,\text{kips}$$

$$\rho_{es} = \frac{Q_{af}}{E_s L} I = \frac{44.5}{600 \times (45 - 30)} \times 1.2 = 0.006\,\text{ft} \approx 0.07\,\text{in.}$$

$$\rho_{be} = \frac{Q_{ab}}{r_b E_{sb}} \left(\frac{1 - \nu^2}{2} \right) = \frac{(90 - 44.5)}{1.5 \times 600} \left(\frac{1 - 0.3^2}{2} \right) = 0.028\,\text{ft} = 0.3\,\text{in.}$$

$$\text{Total settlement} = 0.07 + 0.3 = 0.37\ \text{in.}$$

The settlement is very small, about 1% of the diameter (0.36 in.).

Step 6: Check driveability.

The allowable driving force = allowable driving stress × cross-sectional area

Allowable driving stress for precast concrete piles is 1.5 ksi.

$= 1.5 \times 36^2 = 1944\,\text{kips} > 90\,\text{kips}$; *okay.*

Step 7: Summarize.

Use 3 ft square precast concrete piles of length 45 ft (2 ft freeboard above maximum water level). ■

EXAMPLE 8.12 *Pile Design Using CPT Data*

The results of cone tip and sleeve resistances from a CPT are shown in Fig. E8.12. Estimate the ultimate load capacity of a driven pile of diameter 0.6 m (2 ft) and length 10 m (32 ft).

Strategy It is best to select two or more methods to determine the cone tip and sleeve resistances and then make a decision on the pile load capacity after conducting the calculations. We are going to select the methods proposed by Fleming and Thorburn (1983) and Eslami and Fellenius (1997). SI units were used in solving this example problem.

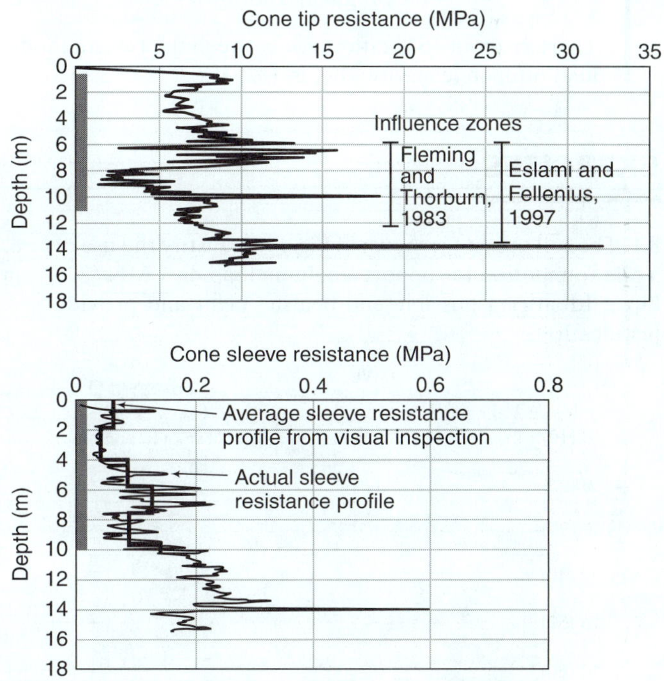

FIGURE E8.12

Solution 8.12

Step 1: Determine the influence zone.

We will assume the pile is penetrating a weak soil and resting on a strong soil since the cone tip resistance is increasing just below the pile base.

$$\text{Eslami and Fellenius (1997): Above base} = 8D = 8 \times 0.6 = 4.8\,\text{m}$$
$$\text{Below base} = 4D = 4 \times 0.6 = 2.4\,\text{m}$$

$$\text{Fleming and Thorburn (1983): Above base} = 8D = 8 \times 0.6 = 4.8\,\text{m}$$
$$\text{Below base} = 2D = 2 \times 0.6 = 1.2\,\text{m}$$

Step 2: Determine cone tip resistance.

Eslami and Fellenius (1997): Geometric mean value, $(q_c)_{ag} = 6.39$ MPa from depth 4.8 m above the base to 2.4 m below the base. Use GEOMEAN function in Excel to find the geometric mean.

$$\text{Fleming and Thorburn (1983): } q_c = \frac{7.5 + 4.77 + 2(3.07)}{4} = 4.6\,\text{MPa}$$

Step 3: Determine average sleeve resistance.

Average sleeve resistance = 0.075 MPa over a depth of 10 m.

Step 4: Calculate load capacity.

$$\text{Area of base, } A_b = \frac{\pi D^2}{4} = \frac{\pi \times 0.6^2}{4} = 0.28\,\text{m}^2$$

$$\text{Eslami and Fellenius (1997): } Q_b = (q_c)_{ag} A_b = 6.39 \times 0.28 = 1.79\,\text{MN}$$

$$\text{Fleming and Thorburn (1983): } Q_b = q_c A_b = 4.6 \times 0.28 = 1.29\,\text{MN}$$

$$\text{Skin friction, } Q_f = f_s \pi D L = 0.075 \times \pi \times 0.6 \times 10 = 1.41\,\text{MN}$$

$$\text{Eslami and Fellenius (1997): } Q_{\text{ult}} = 1.79 + 1.41 = 3.2\,\text{MN}$$

$$\text{Fleming and Thorburn (1983): } Q_{\text{ult}} = 1.29 + 1.41 = 2.7\,\text{MN}$$

There is about 15% difference between the two methods. You can take an average value as an estimate of the pile load ultimate capacity, that is, $(3.2 + 2.7)/2 = 2.95\,\text{MN} \approx 3\,\text{MN}$. ■

EXERCISES

8.1 The soil at a site is shown in Fig. P8.1. Justify the use of a pile foundation to support a column load on 1 MN (225 kips). Identify a possible end bearing layer and provide justification.

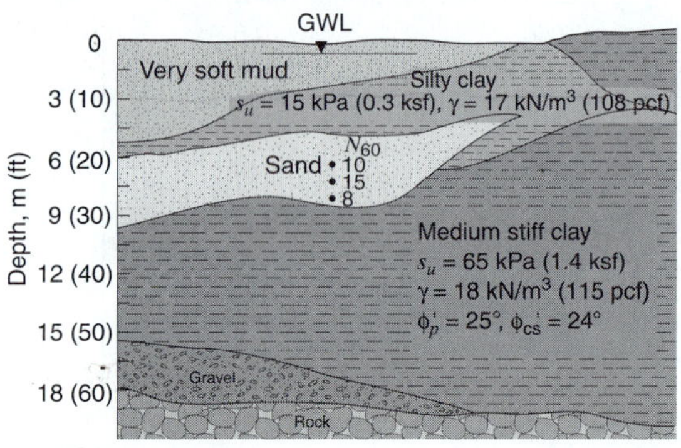

FIGURE P8.1

8.2 The soil profile at a site for a bridge abutment is shown in Fig. P8.2. The load on each pile is 2 MN (450 kips).

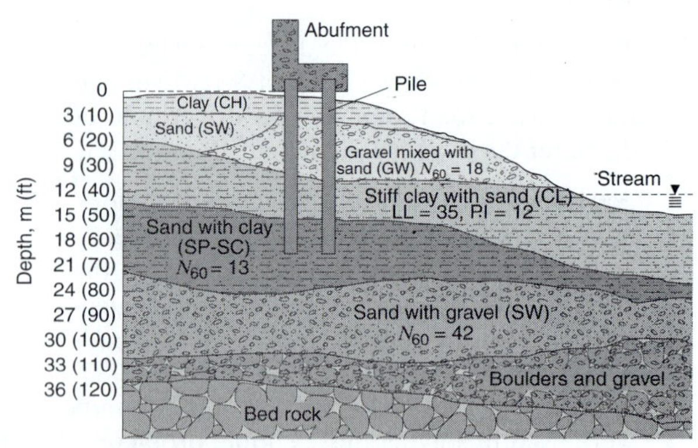

FIGURE P8.2

1. Select a suitable pile type.
2. Would you use driven or drilled shafts?
3. Identify a suitable end bearing layer.

Provide justification for each of your answers.

8.3 Discuss the limitations in predicting pile load capacity.

8.4 A timber pile of diameter of 0.45 m (18 in.) will be driven into the soil shown in Fig. P8.4. Calculate the length required to support the load based on ASD with FS = 2.0.

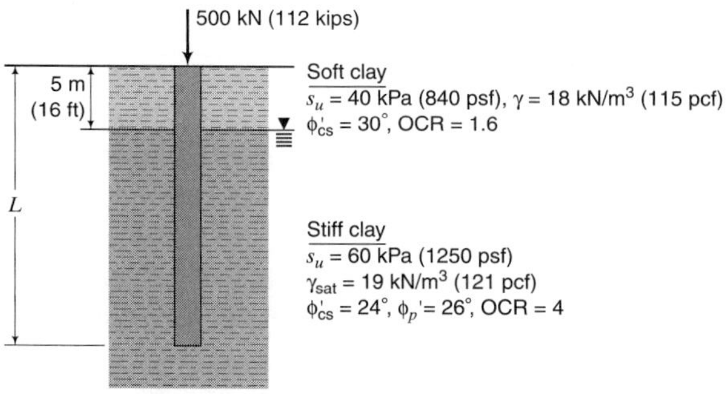

FIGURE P8.4

8.5 Determine the allowable load for a concrete driven pile, 0.6 m (24 in.) diameter, driven 20 m (66 ft) into the soil profile shown in Fig. P8.5. Groundwater is at 2 m (6.5 ft) below the surface but you can assume it will rise to the surface. A factor of safety of 2 is required.

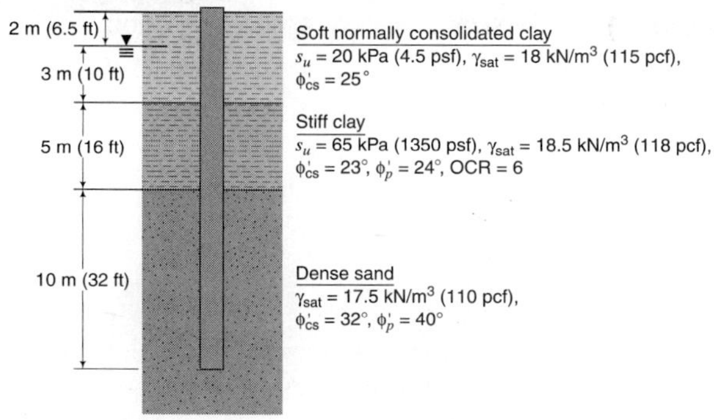

FIGURE P8.5

8.6 A square precast concrete pile of sides 0.45 m (18 in.) is to be driven 15 m (49 ft) into the soil strata shown in Fig. P8.6.

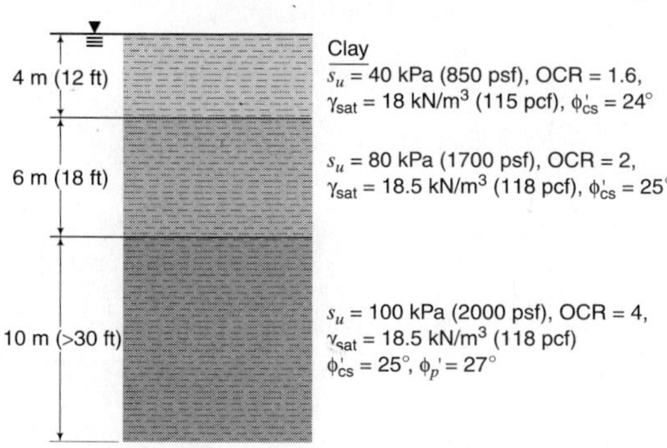

FIGURE P8.6

Estimate the allowable load capacity for a factor of safety of 2. Owing to changes in design requirements, the pile must support 20% more load. Determine the additional embedment depth required.

8.7 Estimate the allowable load capacity of a 1 m (3 ft) diameter drilled shaft embedded 16 m (50 ft) in the soil profile shown in Fig. P8.7. The factor of safety required is 2. The *N* values are blows/ft averaged over the thicknesses of the soil layers.

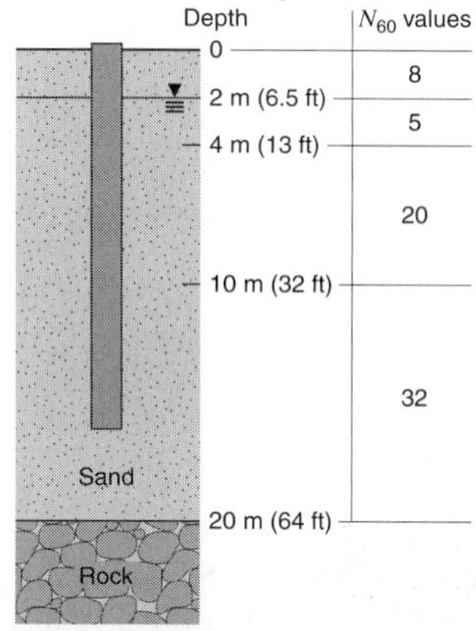

FIGURE P8.7

8.8 A soil profile consists of 3 m (10 ft) of a loose fill of unit weight 18 kN/m^3 (89 pcf) over a fissured overconsolidated clay. The shear strength of the overconsolidated clay increases linearly with depth. At the top of the clay, s_u = 90 kPa (1800 psf), γ_{sat} = 18 kN/m^3 (115 pcf), and at a depth of 20 m (65 ft) [17 m (56 ft) into the clay], s_u = 180 kPa (3600 psf). The angle of friction at critical state is ϕ'_{cs} = 28°. Determine the ultimate load capacity of a driven H-pile ASTM Gr. 50, HP 310 × 110 (HP 12 × 74 lb), flange width = 310 mm (12.2 in.), depth = 308 mm (12.1 in.), embedded 12 m (40 ft) into the clay using LRFD. Groundwater is at 5 m (16 ft) below the ground surface. You may assume that the soil above the groundwater level is saturated and a sleeve will be used over the pile within the loose fill.

8.9 Estimate the allowable load capacity of a 3 × 3 pile group. Each pile has a diameter of 0.6 m (2 ft) and is driven 15 m into a soft clay whose undrained shear strength varies linearly with depth according to $s_u = 0.25\sigma'_z$. The angle of friction at the critical state is 30° and γ_{sat} = 17.5 kN/m^3 (110 pcf). The spacing of the piles is 1.0 m (3 ft). Groundwater level is at ground surface.

8.10 A concrete free head pile 0.6 m × 0.6 m (2 ft × 2 ft) is required to support a working lateral load of 100 kN (23 kips) as shown in Fig. P8.10.

1. What is the value of the eccentricity?
2. Calculate the moment at the ground surface.
3. Calculate K_{ps}.

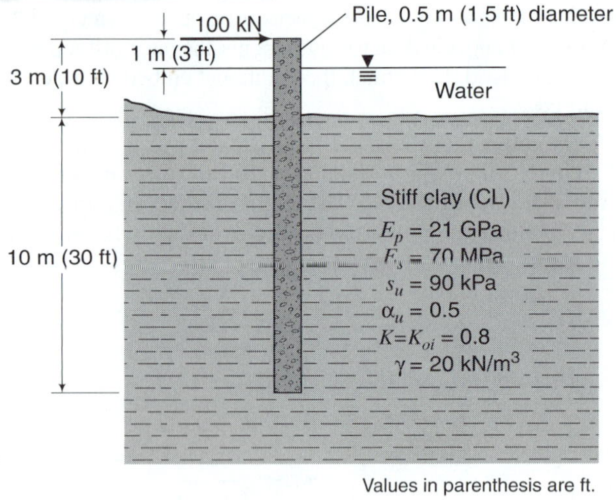

FIGURE P8.10

4. Calculate the equivalent diameter and the equivalent second moment of area.
5. Calculate the effective length.
6. Using the program APILES
   (a) Plot the lateral load–lateral displacement response.
   (b) Identify the lateral displacement at the working load.

8.11 The bridge abutment [3 m (15 ft) wide × 13 m (40 ft) long] in Fig. P8.2 will have to support a dead load of 10 MN (2250 kips) and a live load of 6 MN (1350 kips). With the limited soil parameters given in Fig. P8.2 and a live load impact factor of 1.25, make a preliminary design of a pile group to support the load using LRFD. Specifically, you should determine the following:

1. The type of piles to be used assuming that all types and methods of installation are available
2. The size of the piles (cross section and length)
3. The pile spacing.
4. Pile load capacity.
5. Estimated settlement.
6. Estimated cost using local data.
7. Would the sand layer liquefy under a seismic acceleration of 0.2 g?
8. If the sand layer were to liquefy, how would it affect the design? For example, would there be any frictional resistance within this layer?

You may assume an initial pile group configuration of 16 piles arranged in a 2 × 8 matrix. You may use the correlations in Appendix B to estimate the desired soil parameters and the program pile.xls to do the calculations. Make good-quality drawings showing at least a plan and a section of the pile group with the necessary dimensions. Write a short report, about a page, describing the shortcomings of the design and the proposed method of construction.

CHAPTER 9

MAT FOUNDATIONS

9.0 INTRODUCTION

Mat foundations are similar to spread foundations, which were discussed in Chapter 7. However, the analysis for the design of mat foundations is much more complex than spread foundations. This complexity arises from the unknown stress distribution under the slab. In this chapter, we will discuss practical criteria for the use of mat foundations, the estimation of bearing capacity and settlement, and some of the methodologies that are utilized for analysis. When you complete this chapter, you should be able to:

- Make a decision on whether a mat is the appropriate foundation for the superstructure under consideration and what type of mat is best.
- Estimate the allowable bearing capacity and settlement of the selected mat foundation.
- Understand the popular methods of analyses.
- Understand the limitations of the analyses.
- Carry out simple analyses for lightly loaded structures.

9.1 DEFINITIONS OF KEY TERMS

A ***mat*** foundation is a large slab, usually concrete, for transmitting the structural loads to the underlying soil.

A ***raft*** **(*also floating mat or floating raft*)** is a mat foundation that is located at a depth to compensate for part or all of the loads from the superstructure.

Embedment depth **(D_f)** is the depth below the finished grade (finished ground surface) where the base of the mat rests.

9.2 CONSIDERATIONS FOR USING MAT FOUNDATIONS

A mat foundation is conventionally a concrete slab used when

- Spread footing covers over 50% of the foundation area because of large column loads.
- The soil is soft with a low bearing capacity.
- Variable soil types and/or pockets of soft soils are present within the footprint of the structure.
- Serviceability limit state (tolerable total and differential settlement) and/or ultimate limit state for the structure using spread footings (Chapter 7) are exceeded.
- Hydrostatic uplift resistance is needed.

9.3 TYPES OF MAT FOUNDATIONS

A mat foundation can be located at the surface or buried deep into the soil to compensate for all or part of the applied loads. Mat foundations are described frequently as raft foundations because they act like rafts when part or all of the loads from the superstructure are compensated through embedment. The common types of rafts (Fig. 9.1) are

- ***Slab-on-grade***—a concrete slab placed on the finished ground surface (Fig. 9.1a). This type of slab is used for light column loads such as in residential and one- or two-storey office structures and to control structural distress due to differential settlement. The reinforcement can be conventional rebar or post-tensioned strands (Fig. 9.1b). The latter is often used to control

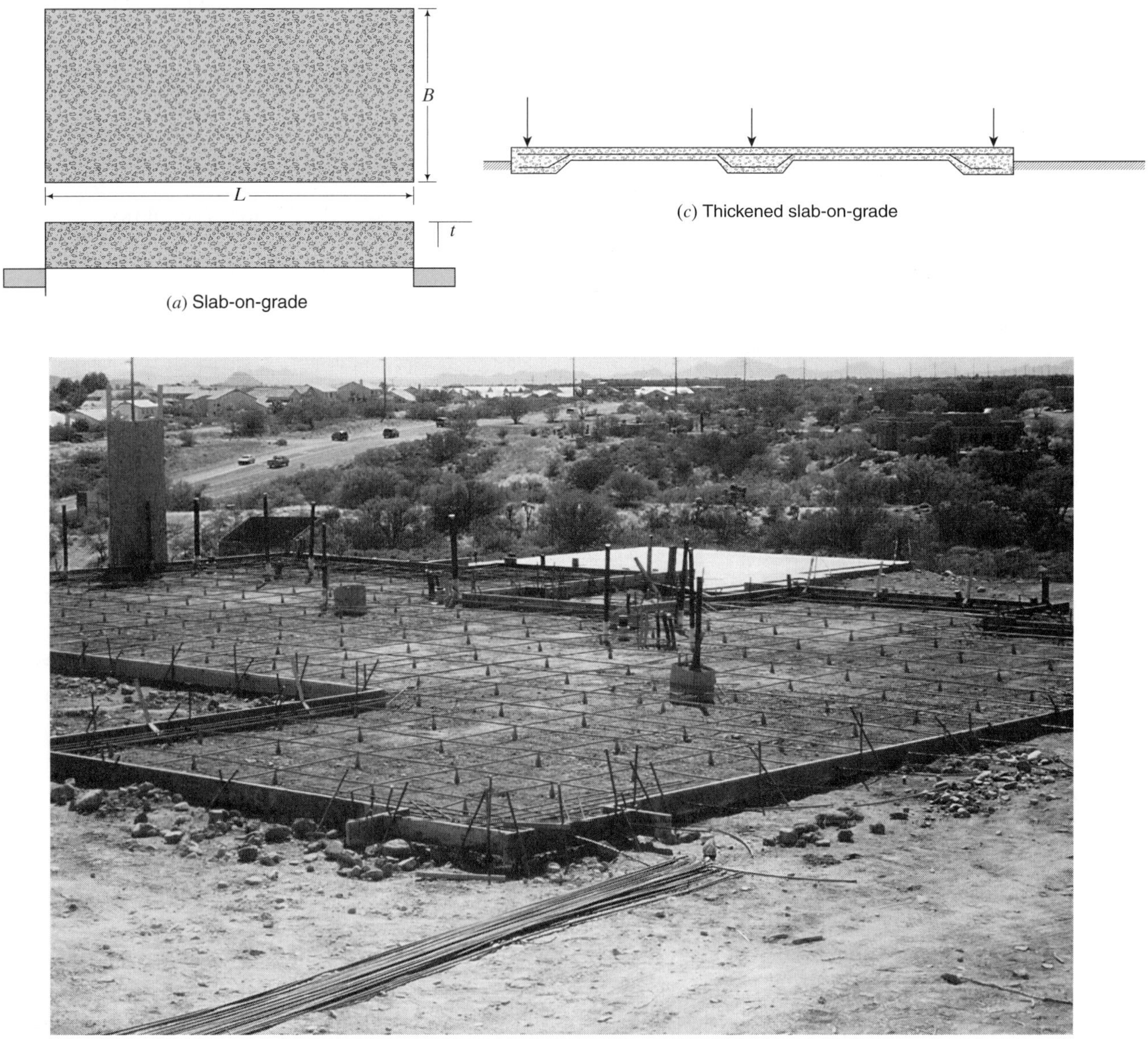

(*a*) Slab-on-grade

(*c*) Thickened slab-on-grade

(*b*) Post tensioned slab-on-grade for a residence

FIGURE 9.1 Some types of mats.

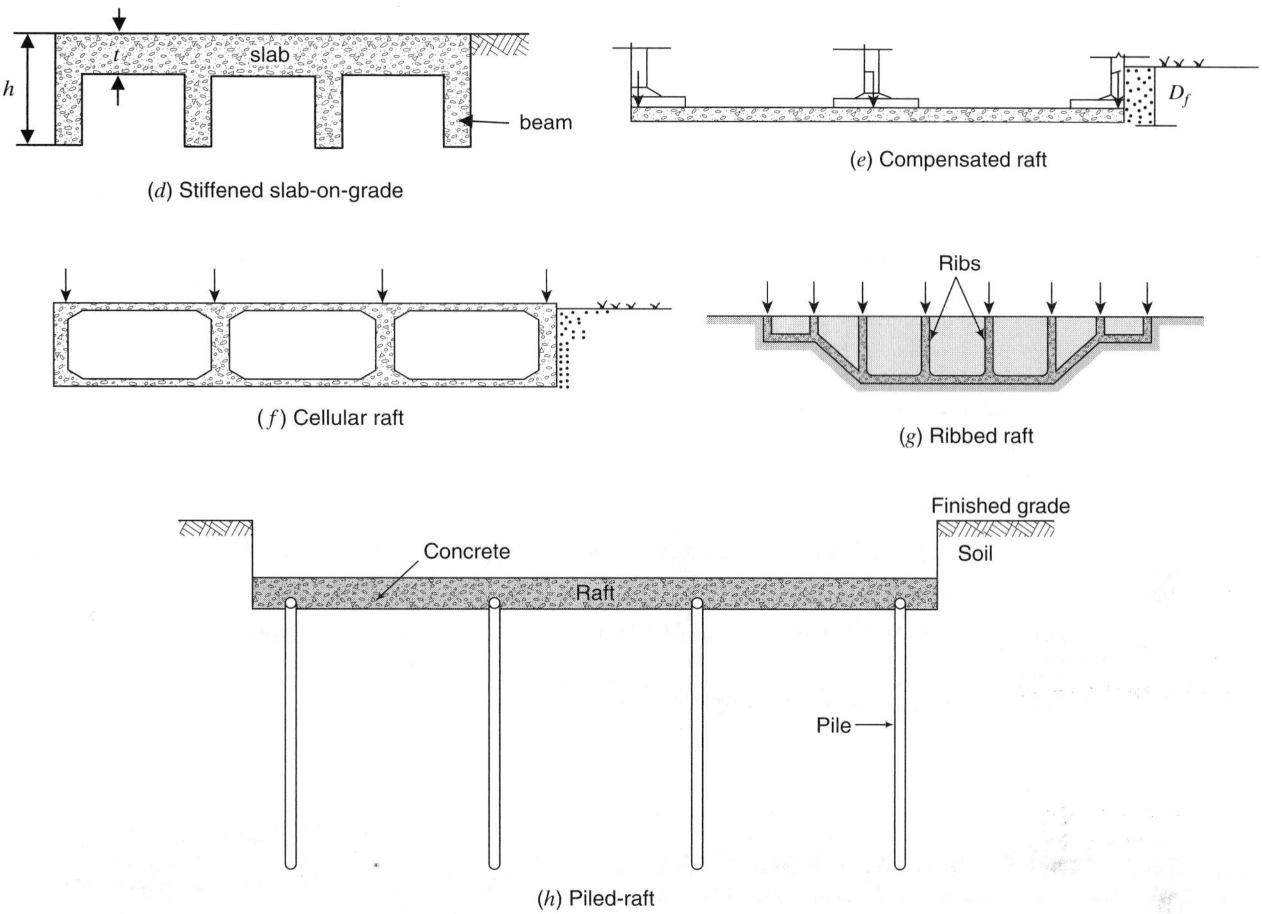

(*d*) Stiffened slab-on-grade

(*e*) Compensated raft

(*f*) Cellular raft

(*g*) Ribbed raft

(*h*) Piled-raft

FIGURE 9.1 *(Continued)*

cracking, reduce mat thickness, and eliminate joints, and when the soil is expansive or collapsible.

- ***Thickened slab-on-grade***—a concrete slab placed on the finished ground surface and thickened under columns and walls for better concrete shearing resistance (Fig. 9.1c). Similar usage as for slab-on-grade but in this case the column or wall loads are larger so that the slab thickness under the columns and walls is increased to provide the needed shear resistance.
- ***Stiffened slab-on-grade***—a concrete slab placed on the finished ground surface with beams to increase the stiffness of the slab (Fig. 9.1d). This type of mat is used particularly in expansive soils.
- ***Compensated raft***—a concrete slab placed in an excavation at a depth such that the overburden pressure removed compensates for all or part of the applied stress (Fig. 9.1e). This type of raft is used for multistorey structures or other structures when the column loads are very large and a deep underground space is not needed. By removing soil mass to compensate for all or part of the applied stress, the potential for shear failure and excessive settlement is lowered.
- ***Cellular raft***—a concrete box-like structure placed in an excavation at a depth such that the overburden pressure removed compensates for all or part of the applied stress (Fig. 9.1f). Cellular rafts are used for very high column loads, for underground space, and as a water barrier when placed below groundwater level.

- ***Ribbed raft***—a concrete slab with ribs aligned with the columns placed in an excavation at a depth such that the overburden pressure removed compensates for all or part of the applied stress (Fig.9.1g). The floor slab is not integrated with the raft. Commonly used for lightly loaded structures (imposed vertical stress <72 kPa [1500 psf]).
- ***Piled raft***—a combination of piles and a concrete slab. The top of the slab can be at the ground surface or in an excavation at a depth such that the overburden pressure removed compensates for part of the applied stress. The balance of the applied loads is carried by the piles (Fig. 9.1h). Piled rafts consist of three components; concrete slab, pile, and soil. They are particularly useful in reducing total and differential settlement. They are very popular in multistorey buildings (skyscrapers).

9.4 DESIGN CONSIDERATIONS

The design of mat foundations requires the following considerations:

- The loads to be supported.
- Selection of the type of mat. Very heavy loads may require a fully or partially compensated raft.
- The sensitivity of the structure and any machines to settlement.
- Determination of the bearing capacity (ultimate limit state) and settlement (serviceability limit state).
- Stability of the excavation for compensated rafts.
- Structural analysis of the mat: stress distribution, shear force, and bending moments.

9.5 PRESSURES ON A MAT FOUNDATION

The pressures on a mat are shown in Fig. 9.2. The applied pressures come from

- Lateral hydrostatic pressures on the side walls.
- Lateral soil pressures on the side walls.
- Uplift pressures on the mat base (slab) from groundwater and soil heaving.
- Seepage pressures on the side walls and the base. Seepage increases the effective stress if the flow gradient has a downward component. The net effect is an increase in the lateral soil pressure. Upward component of the flow gradient reduces the effective stress that can lead to piping or

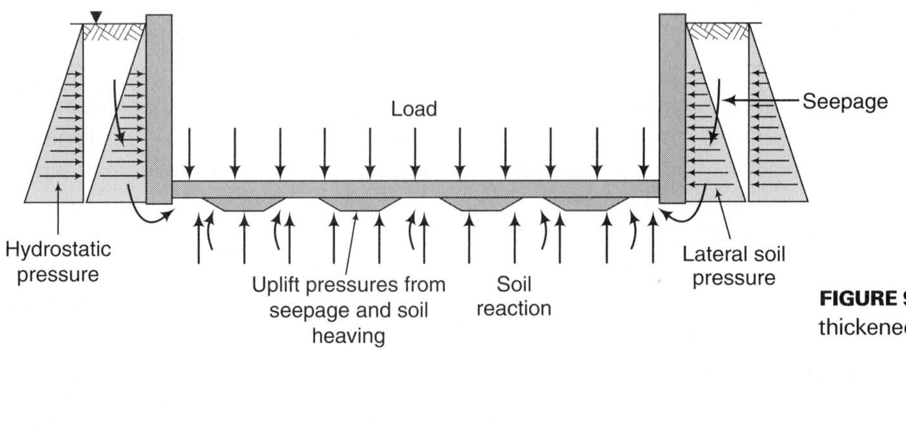

FIGURE 9.2 Pressures acting on a thickened raft buried into soil.

boiling. Piping is "pipe-shaped" erosion that initiates near the edges of the mat base. Boiling is a state of static liquefaction.

- Loss of support. Collapsible soils tend to settle in the presence of water. This settlement may be nonuniform, resulting in loss of support in some sections of the mat base. Poor construction such as an unleveled soil base can also lead to loss of support. The net effect is that certain sections of the soil under the mat will be subjected to higher stresses than anticipated while some sections will be unstressed.
- Applied structural loads.

Not all of these pressures may be present in a given mat. For example, if the groundwater level is at a depth greater than the width of the mat and is stable (i.e., does not fluctuate), hydrostatic and seepage effects will be absent.

EXAMPLE 9.1 *Embedment Depth for a Raft Foundation*

The average vertical stress at the ground surface from a superstructure including the self-weight of the raft is 50 kPa (1045 psf). If the saturated unit weight of the soil is 20 kN/m^3 (125 pcf), determine the depth of embedment of the raft for 100% and 80% compensation. Groundwater is at 10 m (33 ft) below the ground surface.

Strategy Calculate how much soil should be removed so that the total stress of soil removed balances the percent reduction sought. SI units will be used in solving this example problem.

Solution 9.1

Step 1: Calculate the depth of soil required for 100% compensation.
The embedment depth for 100% compensation is

$$D_f = \frac{\text{Average applied stress}}{\text{Unit weight of soil}} = \frac{50}{20} = 2.5 \text{ m}$$

Step 2: Calculate the depth of soil required for 80% compensation.
The embedment depth for 80% compensation is

$$D_f = \frac{\%\,\text{Compensation} \times \text{Average applied stress}}{\text{Unit weight of soil}} = 0.8 \times \frac{50}{20} = 2 \text{ m}$$

■

9.6 BEARING CAPACITY AND SETTLEMENT OF MAT FOUNDATIONS

Importance *Mat foundations must meet ultimate limit state (ULS) and serviceability limit state (SLS) requirements. In this section, we describe how the bearing capacity for ULS and settlement for SLS can be estimated.*

The bearing capacity for mat foundations is calculated in a manner similar to spread footings. The key unknown is how the stresses are distributed under the mat to evaluate the safety margin for shear failure. If the mat is rigid, the settlement is likely to be uniform but the stresses will be distributed nonuniformly. If the mat is flexible, the stresses are likely to be uniformly distributed but the settlement will be nonuniform. The rigidity of the mat is found using Eq. (2.31).

Rigid slabs can undergo large rotations without damage to the superstructure. Flexible slabs can cause significant structural distress due to differential settlements. Superstructure damage is caused by the curvature of the mat, not rigid body rotation. More information on the influence of rigidity is provided in Section 2.4.11.

The estimation of the settlement of the mat is much more complex than for the bearing capacity. The settlement depends on the rigidity of the mat and the structure it supports with respect to the soil, the type of soils, the homogeneity of the soils, groundwater conditions, type of mat, and construction methods. A first approximation for the settlement of mats, essentially flexible mats, is calculated based on Chapter 7, Section 7.7 for spread footings. Tolerable settlement is generally 25 mm (1 in.) to 50 mm (2 in.). The tolerable curvature (δ/L, δ is the maximum differential deflection within a distance L) is 1/150 to 1/360 (in practice, $\delta/L < 250$) for flexible mats, 1/360 to 1/600 for semirigid mats, and 1/600 to 1/1000 for rigid mats.

EXAMPLE 9.2 *Bearing Capacity and Settlement of a Circular Mat on a Stiff Clay*

A foundation is required to support closely spaced columns distributed in a circle. Spread footings cannot be used because the overlap is greater than 50%. The geotechnical engineer decided to try a circular mat foundation of 8 m (26 ft) diameter. The total vertical centric load including the weight of the mat is 3.5 MN (787 kips). The soil is a deep deposit of a stiff clay with $s_u = 100$ kPa (2090 psf), $\phi'_p = 25°$ (plane strain test), $\gamma_{sat} = 19.8$ kN/m^3 (124 pcf), OCR = 10, $C_c = 0.15$, $C_r = 0.02$, $w = 20\%$, $\nu = 0.3$, and $E_s = 90$ MPa (1881 ksf). Groundwater is at the surface. Determine the factor of safety for ULS. Estimate the settlement of the mat assuming that it is located on the finished grade (surface).

Strategy For this mat, you need to consider both short-term and long-term conditions. SI units will be used in solving this example problem.

Solution 9.2

Step 1: Calculate the bearing capacity and geometric factors.

$$N_\gamma = 0.1054 \exp(9.6\phi'_p) = 0.1054 \exp\left(9.6 \times 30\frac{\pi}{180}\right) = 8.6$$

The foundation is at the surface. Set all depth factors to 1. $D = B = L = L' = B' = 8$ m.

$$s_c = 1 + 0.2\frac{B'}{L'} = 1 + 2 = 1.2$$

$$s_r = 1 - 0.4\frac{B'}{L'} = 0.6$$

Step 2: Calculate the short-term bearing capacity.

$$q_u = 5.14 s_u s_c d_c = 5.14 \times 100 \times 1.2 \times 1 = 617 \text{ kPa}$$

Step 3: Calculate the long-term bearing capacity.

$$q_u = 0.5\gamma' B' N_\gamma\, s_\gamma\, d_\gamma = 0.5 \times 10 \times 8 \times 8.6 \times 0.6 \times 1 = 206.4 \text{ kPa}$$

Step 4: Calculate factor of safety.
Long-term condition controls.

$$\sigma_{ap} = \frac{\text{Load}}{\text{Area}} = \frac{3500}{\pi\frac{8^2}{4}} = 69.7 \text{ kPa}$$

$$\text{FS} = \frac{206.4}{69.7} \cong 3$$

Step 5: Calculate elastic settlement.

$$\rho_e = \frac{2Q_a}{EL_1}(1 - \nu_u^2)\mu_s \mu_{emb} \mu_{wall}$$

The mat is at ground surface, so the embedment and wall factors are 1. From Table 7.4,

$$\mu_s = 0.785 \qquad \rho_c = \frac{2Q_a}{EL_1}\left(1 - \nu_u^2\right)\mu_s = \frac{2 \times 3.5}{90 \times 8}(1 - 0.3^2)0.785 = 6.9 \times 10^{-3}\,\text{m} = 6.9\,\text{mm}$$

Step 6: Calculate consolidation settlement.
Assume an effective depth of $1.5D = 1.5 \times 8 = 12\,\text{m}$, where D is the diameter.
Divide the effective depth into 3 layers, 4 m thick, and calculate the increase in vertical stress at the center of each layer. Using the program STRESS, the increases in vertical stress are as follows:

Depth (m)	2	6	10
$\Delta\sigma_z$ (kPa)	63.5	29.6	13.9

Use a weighted harmonic mean vertical stress increase

$$\Delta\sigma_z = \frac{3\,(63.5) + 2(29.6) + 1(13.9)}{3 + 2 + 1} = 43.9\,\text{kPa}$$

Calculate the initial effective vertical stress at the center of the effective depth,

$$\sigma'_{zo} = (19.8 - 9.8) \times 6 = 60\,\text{kPa}$$

Calculate the preconsolidation effective stress,

$$\sigma'_{zc} = \text{OCR} \times \sigma'_{zo} = 10 \times 60 = 600\,\text{kPa}.$$

$$\sigma_{\text{fin}} = \sigma'_{zo} + \Delta\sigma_z = 60 + 43.9 = 10.3.9\,\text{kPa} < 600\,\text{kPa}$$

Calculate primary consolidation settlement,

$$\rho_{pc} = \frac{H}{1 + e_o} C_r \log \frac{\sigma'_{\text{fin}}}{\sigma'_{zo}}, \quad \sigma_{\text{fin}} < \sigma'_{zc}.$$

$$\rho_{pc} = \frac{12000}{1 + 0.54} 0.02 \log \frac{103.9}{60} = 37.2\,\text{mm}$$

Step 7: Calculate total settlement.
Total expected settlement $= 6.9 + 37.2 = 44.1\,\text{mm}$.

■

9.7 STRUCTURAL ANALYSIS OF MAT FOUNDATIONS

Importance *The design of a mat foundation is essentially a structural problem involving the determination of the displacements, bending moments, and shear forces. The key issue is how to model the soil. We discuss some of the assumptions normally made in practice.*

Various assumptions are made in modeling the soil for the analysis of mat foundations. The issue is how the stresses (or soil resistance) are distributed at the base of the mat so that the bending moments and shear forces can be calculated. Some popular assumptions on the distribution of the soil resistance are as follows:

1. Uniform contact pressure
2. Constant subgrade modulus
3. Variable subgrade modulus
4. Treating the soil as an elastic material or an elastoplastic material. The analyses performed are classified into three groups:
   a. Simple analysis—linear contact pressure
   b. Mat on springs—Winkler spring model with constant or variable subgrade modulus
   c. Mat on soil as a continuum—soil treated as an elastic or elastoplastic material

We will discuss some of the implications of these assumptions.

9.7.1 Approximate Analysis—Uniform (Planar) Contact Pressure

This type of analysis is useful as a first approximation when the soil is homogeneous, the slab is semirigid ($0.1 < K_r < 10$; see Section 2.4.11), the columns are approximately equally spaced, and the loads are approximately equal (i.e., the differential settlement between columns is expected to be small). The contact pressure is taken as the average vertical stress on the mat. The mat is divided into a convenient number of strips in each direction with each strip having a line of continuous columns along its centerline (Fig. 9.3). The solution for the bending moment and shear force are then analogous to a beam (strip) on continuous supports (columns). If the soil is compressible—soft fine-grained soils with unfactored applied vertical stress less than 72 kPa (1500 psf)—you can assume two limiting pressure distributions. One is a uniform contact pressure. The other is a linearly varying pressure with a starting value of zero at the center of the slab and a maximum value of twice the uniform contact pressure at the edge. The design should be based on which one of the pressure distributions gives the worse case.

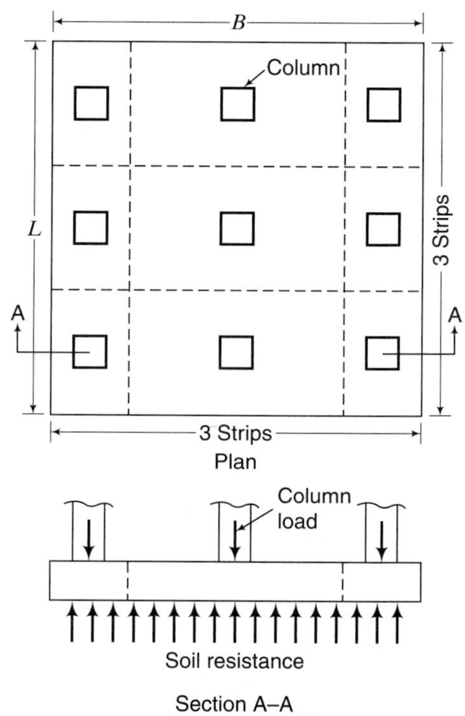

FIGURE 9.3 Approximate strip analysis for slab-on-grade.

9.7.2 Mat on Springs—Winkler Spring Model

In this model, the mat is treated either as several strips (beams) or as a thin plate and the soil is represented by discrete springs located at arbitrarily selected nodes along the strips or plate (Winkler spring model, Fig. 9.4). The governing equation when the mat is a thin plate is

$$D\left(\frac{\partial^4 \rho}{\partial x^4} + \frac{2\partial^4 \rho}{\partial x^2 \partial y^2} + \frac{\partial^4 \rho}{\partial y^4}\right) + k_s \rho = 0 \tag{9.1}$$

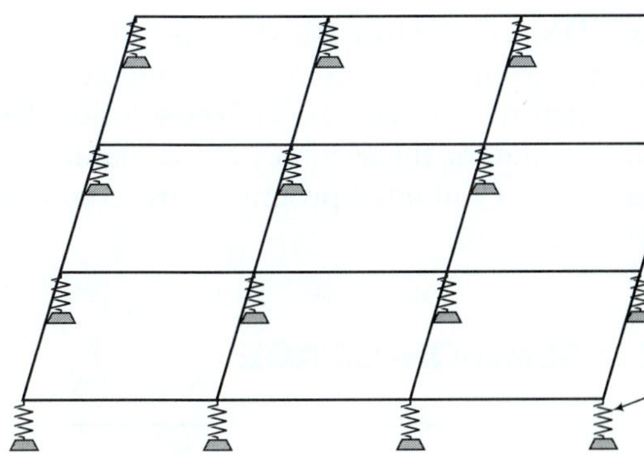

FIGURE 9.4 Winker spring model idealization.

where D is the flexural rigidity of the mat given as

$$D = \frac{Et^3}{12(1 - \nu^2)} \tag{9.2}$$

and ρ, E, ν, and t are the deflection, elastic modulus, Poisson's ratio, and thickness of the mat respectively and k_s is the spring stiffness or subgrade modulus (units: FL^{-3}) of the soil (see Appendix B Table B 4.13). Computer programs are commercially available to solve Eq. (9.1) using FDM.

A constant subgrade modulus is assumed when the soil is homogeneous, the slab is semirigid or flexible, the columns are approximately equally spaced, and the loads are approximately equal (i.e., the differential settlement between columns is expected to be small). A variable subgrade modulus is used when the soil is layered, the slab is flexible, the columns are unequally spaced, and the loads are variable.

The Winkler spring model is best used for coarse-grained soils. The limitations in using the solutions of Eq. 9.1 representing soil as discrete springs are as follows:

- Interactions between adjacent soil elements are neglected.
- If the mat is uniformly loaded and rests on a homogeneous soil, the Winkler spring model will predict uniform displacement and zero bending moment, whereas the mat will bend nonuniformly and incur significant bending moments. For such a situation, you should use values of k_s that are higher near the edges of the slab compared with the center.
- The solution gives vertical displacements for vertical loads, whereas substantial horizontal displacements can also occur.
- The solution is applicable to a single soil layer (i.e., the model cannot effectively deal with soil layering).
- The subgrade modulus, which is obtained from plate loading tests, is not constant but depends on the plate size, the location where the test is performed, and the depth of the soil layer.
- The number of springs to represent the soil is unknown and is arbitrarily set by the analyst based on experience.

9.7.3 Mat on Soil as a Continuum

In this analysis, the mat is treated as a structural element—a thin plate (thickness small compared with the other dimensions)—and the soil as an elastic or an elastoplastic medium. The governing equation is

$$D\left(\frac{\partial^4\rho}{\partial x^4} + \frac{2\partial^4\rho}{\partial x^2\partial y^2} + \frac{\partial^4\rho}{\partial y^4}\right) = q_s + \frac{P_z}{\partial x\,\partial y} \tag{9.3}$$

where q_s is the uniformly distributed vertical load (including the self-weight of the mat) and P_z is the column loads. The settlement of the mat is assumed to be small in comparison with its thickness.

The solution of the governing equation is found using FDM or FEM or BEM. This type of analysis can simulate complex soil conditions such as soil layering, groundwater conditions, and construction sequencing and it is often used for sensitive, high-valued structures. The accuracy of the solution depends on the experience of the analyst, the accuracy of the soil parameters, the accuracy of the material model representing the soil, and the boundary conditions imposed. Computer programs are commercially available to solve mat problems treating soil as a continuum.

9.8 APPROXIMATE ANALYSIS FOR STIFFENED SLAB-ON-GRADE ON EXPANSIVE AND COLLAPSIBLE SOILS

Importance *Expansive and collapsible soils can cause extensive damage to properties. Stiffened slab-on-grade is one of the foundation types often used for lightly loaded structures—applied stress less than 72 kPa (1500 psf). Approximate solutions have been developed based on experience in certain geographic areas. These are briefly discussed in this section.*

Access a spreadsheet at http://www.wiley.com/college/budhu. Click Foundations and Earth Structures, Chapter 9, mat.xls. to analyze lightly loaded structures using an approximate method.

Stiffened slab-on-grade (Fig. 9.1d) is often used in expansive and collapsible soils. There are three popular design methodologies:

1. ***BRAB*** (British Research Advisory Board, 1968)—a semi-empirical method in which the mat is supported as shown in Fig. 9.5. The center heave and edge lift pertain to expansive soils while the center settlement and edge settlement pertain to collapsible or compressible soils. To calculate the maximum bending moment, shear force, and settlement, the mat is then treated as a simple beam with an effective support length, CL, where C is support factor (<1).

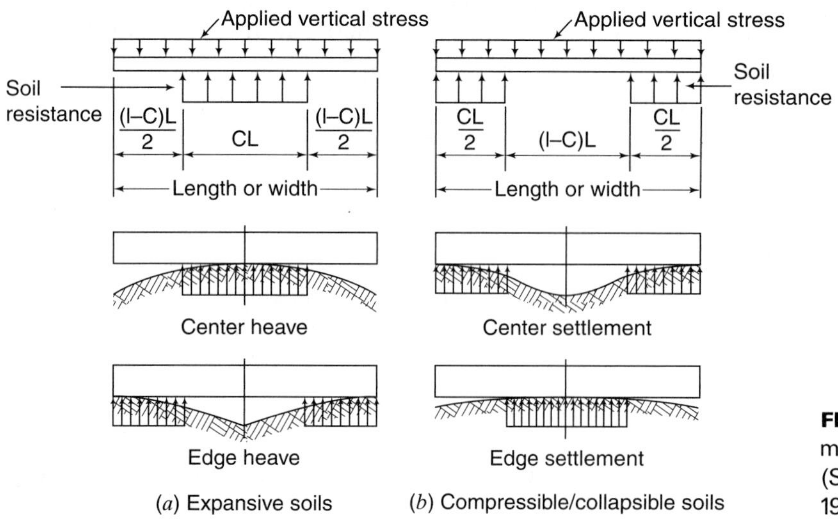

FIGURE 9.5 BRAB design methods.
(Source: Modified after BRAB, 1968.)

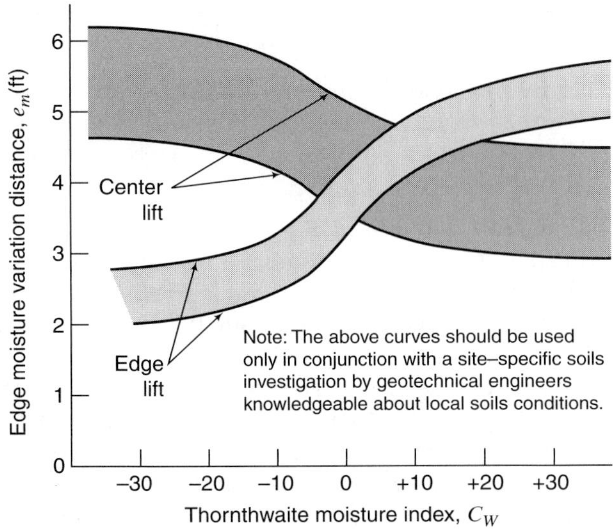

FIGURE 9.6 Estimation of edge and center lift moisture variation distance. (Source: PTI, 1996.)

2. ***WRI/CSRI***—similar to BRAB but the maximum bending moment, shear, and settlement are calculated using a cantilever formulation with a modified length different from BRAB.
3. ***PTI*** (Post-Tension Concrete Institute)—based on equations obtained by curve fitting FEM results of a slab on an elastic soil.

The PTI method is similar in approach to BRAB except that the key (and controversial) design parameters are the edge moisture variation distance (e_m) and the differential soil movement (y_m). An estimate of the edge moisture variation distance can be obtained from Fig. 9.6 and an estimate of the differential soil movement can be inferred from tables in the worksheet—design parameters—at http://www.wiley.com/college/budhu. Click Foundations and Earth Structures, Chapter 9, mat.xls. The moment and shear force are found from equations derived from curve fitting FEM calculations (see equations in worksheet—design equations—at http://www.wiley.com/college/budhu. Click Foundations and Earth Structures, Chapter 9, mat.xls). Irregular foundations have to be divided into overlapping rectangles as shown in Figure 9.7.

The PTI method is based on experience in San Antonio, Texas. You will have to make changes as appropriate for your project area. Computer methods discussed in Sections 9.7.2 and 9.7.3 are also used. These methods allow you to input soil parameters and other local conditions (irregular foundations, groundwater, etc.) for your project.

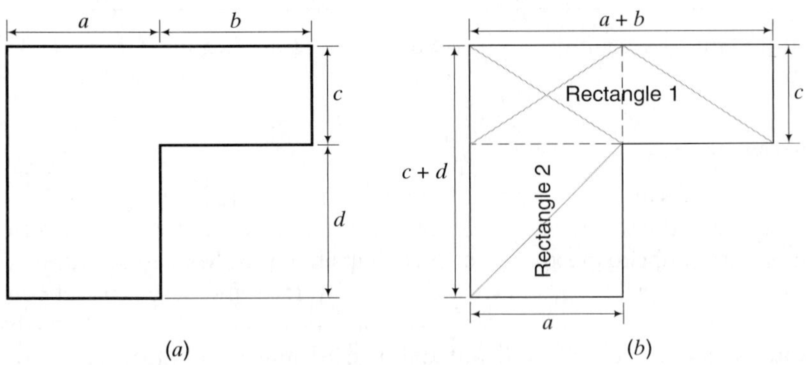

FIGURE 9.7 Dividing an irregular foundation into overlapping rectangles.

9.9 PILED-RAFT FOUNDATIONS

Importance *Piled rafts are growing in popularity especially for high-rise buildings on poor-quality soils. The analysis of piled rafts is a complex soil–structure interaction problem. In this section only an introduction is presented.*

A piled raft consists of two structural elements, pile and raft, and the soil. Piled rafts are used when a raft by itself is insufficient to meet SLS. The function of the piles is to reduce settlements by transferring the load deep into the soil. Because of the complexity of piled-raft foundations, simulations using numerical methods such as FDM or FEM are the best options currently available for design. It is usual to design the piles to carry their full load capacity as single piles and the raft accommodates the balance of the load.

In three-dimensional numerical analyses, the piles and the raft are treated as solid structural elements and the soil is modeled as an elastic or elastic-rigid or elastoplastic material. As an example of a three-dimensional FEM analysis of a piled raft, we consider the Torhaus der Messe building in Germany analyzed by Reul and Randolph (2003). The building (Fig. 9.8a), 130 m high, is supported on two adjacent piled rafts, each 17.5 m × 17.5 m × 2.5 m thick, separated 10 m horizontally. A total of 84 bored piles of diameter 0.9 m and length 10 m were used. The embedment depth of the raft was 3 m, which also is the depth of the groundwater level. For each raft, the maximum load including the weight of the raft was 226.8 MN. The soil consists of 2.5 m quarternary sand and gravel below the base of the rafts and underlaid by Frankfurt clay—a stiff overconsolidated clay—about 107.5 m thick. Below the Frankfurt clay is limestone that the authors suggested had no influence on the analysis. The instrumentation consists of 6 instrumented piles, 11 contact pressure cells, and 3 multipoint borehole extensometers (Fig. 9.8b).

Because of symmetry, only one-quarter of a piled raft was discretized and analyzed (Fig. 9.8c). The piles and raft were discretized as solid brick elements (Fig. 9.8d). The analysis was conducted stepwise to simulate the construction process. The soil was modeled as an elastic-plastic material and the contact between the soil and the raft was assumed to be perfectly rough—no relative movement between the nodes of the soil elements and the nodes of the raft elements. The average measured settlement at the center of the two rafts was 124 mm compared with the FEM result of 96 mm. Comparison of the measured settlements obtained from extensometers in two boreholes with the FEM analyses is shown in Fig. 9.9. The piled raft behaves as a single system (block mode response) as shown by the rapid decrease of settlement with depth. The FEM results lag about 10% maximum from the measured results.

Numerical analyses like the one carried out by Reul and Randolph (2003) are powerful tools for simulation of complex problems, but experience with these tools and understanding of construction methods are necessary to get meaningful results.

9.10 DESIGN ISSUES

Design issues regarding mat foundations are similar to those for spread foundations (Chapter 7). Some additional issues are as follows.

9.10.1 Water Content Variations

Changes in water content result in changes in volume and undrained shear strength in fine-grained soils. It is important in construction to ensure that the water content is maintained at a constant value. The effects of changes in water content are more pronounced in expansive and collapsible soils. Increase in water content can cause expansive clays to swell and collapsible soils to settle excessively. If the increase in water content is nonuniform, differential swell and differential settlement would occur, which can lead to structural distress.

Landscape irrigation and trees can cause significant damage to mat foundation and its superstructure. Landscape irrigation should be controlled to limit infiltration under the mat foundation. Trees

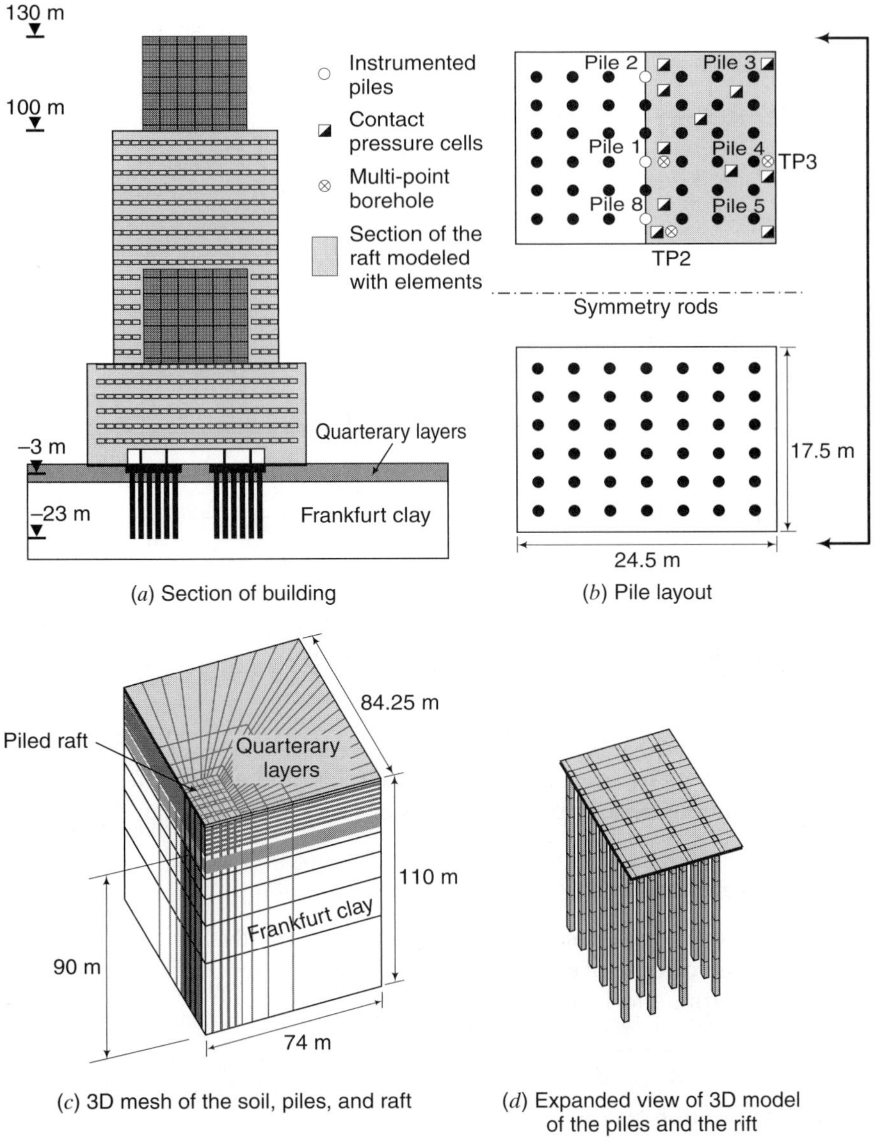

(*a*) Section of building

(*b*) Pile layout

(*c*) 3D mesh of the soil, piles, and raft

(*d*) Expanded view of 3D model of the piles and the rift

FIGURE 9.8 FEM model of the building. (Source: Reul and Randolph, 2003).

tend to cause drying of soil. Their roots can migrate under the foundation, leading to cracking. Consequently, trees should be planted a sufficient distance away (>3 m; 10 ft) from the foundation to avoid secondary damages.

Normally a layer of compacted sand about 50 mm (2 in.) to 100 mm (4 in.) is placed with/without a plastic sheet (thickness of 10 mil) between the soil surface and the base slab. The plastic sheet is used to prevent water infiltrating up from the soil. The compacted sand on top of the plastic sheet is used to get a level, clean surface. You should ensure that water does not access the sand layer as a conduit.

9.10.2 Heaving

A floating raft does not mean that settlement will be zero. When a fine-grained soil is excavated to embed a floating raft, it will swell. This will be partially reversed (reconsolidation) when the foundation is constructed, resulting in settlement. You can estimate the amount of heave using one-dimensional consolidation test results (Chapter 6). The settlement will not necessarily be uniform because although

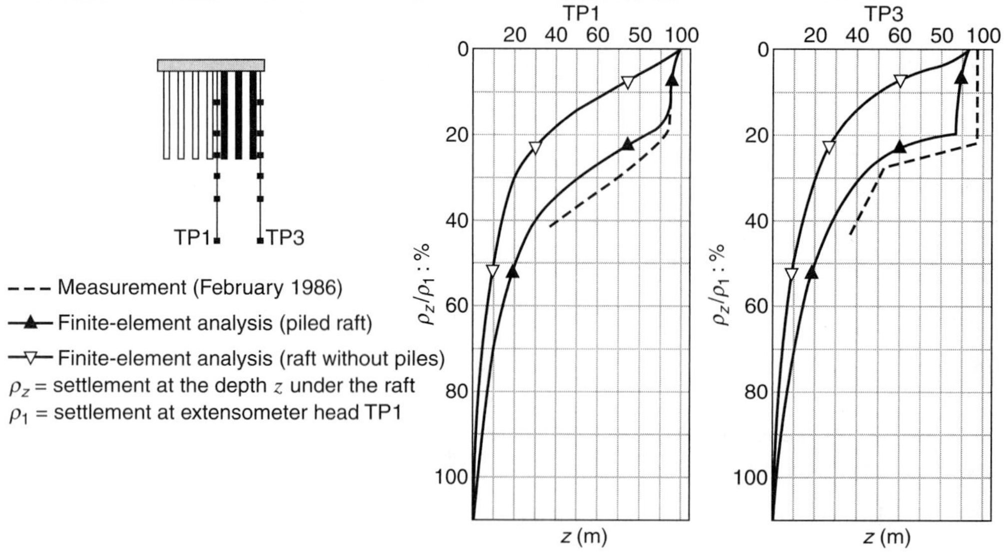

FIGURE 9.9 Comparison of finite element results with measurements. (Source: Reul and Randolph, 2003.)

the pressures from the soil removal and the external loading are the same, the induced stress distribution from the foundation may not be the same as when the soil was removed.

9.10.3 Foundation Shape

Keep foundation shapes simple, such as squares, rectangles, or circles, and avoid different foundation types if the superstructure is continuous. Use expansion joints between adjacent foundations.

9.10.4 Stability of Excavation for Rafts

When raft foundations are constructed in soft soils, the excavation normally has to be supported. If groundwater level is below the possible depth of excavation, the theoretical depth of excavation in a fine-grained soil can be estimated (Bjerrum and Eide, 1956) from

$$(D_f)_{cr} = N_c \frac{s_u}{\gamma} \tag{9.4}$$

and the factor of safety against bottom heave is

$$\text{FS} = N_c \frac{s_u}{\gamma D_f + q_s} \tag{9.5}$$

where $N_c \approx 6(1 + 0.2\frac{D_f}{B})I_c$ for $\frac{D_f}{B} \leq 2.5$, $N_c = 9I_c$ for $\frac{D_f}{B} > 2.5$, $I_c = (0.84 + 0.16\frac{B}{L})$, s_u is the undrained shear strength, q_s is the average foundation vertical stress, and B and L are the width and length of the raft respectively. If the groundwater level is within the possible depth of excavation and the excavation is done under water, replace γ by $(\gamma - \gamma_w)$ in Eqs. (9.4) and (9.5).

9.11 SUMMARY

Mat foundations are versatile and used in situations where spread foundations are unsuitable. Mat foundations are complex soil–structure interaction problems. Except for homogeneous soils and simple loading conditions, the analysis requires computer methods and experience for successful design.

Approximate analyses are suitable for simple structures that are lightly loaded (applied stress less than 72 kPa; 1500 psf).

Practical Examples

EXAMPLE 9.3 *Bearing Capacity and Settlement of a Mat on Variable Soil Types*

A foundation is required to support a superstructure with column loads as shown in Fig. E9.3a. A composite soil profile of the site is shown in Fig. E9.3b. The groundwater can rise to the surface. Determine a suitable foundation. Use ASD with FS = 3, a total settlement of 25 mm, and differential settlement of $L/1000$.

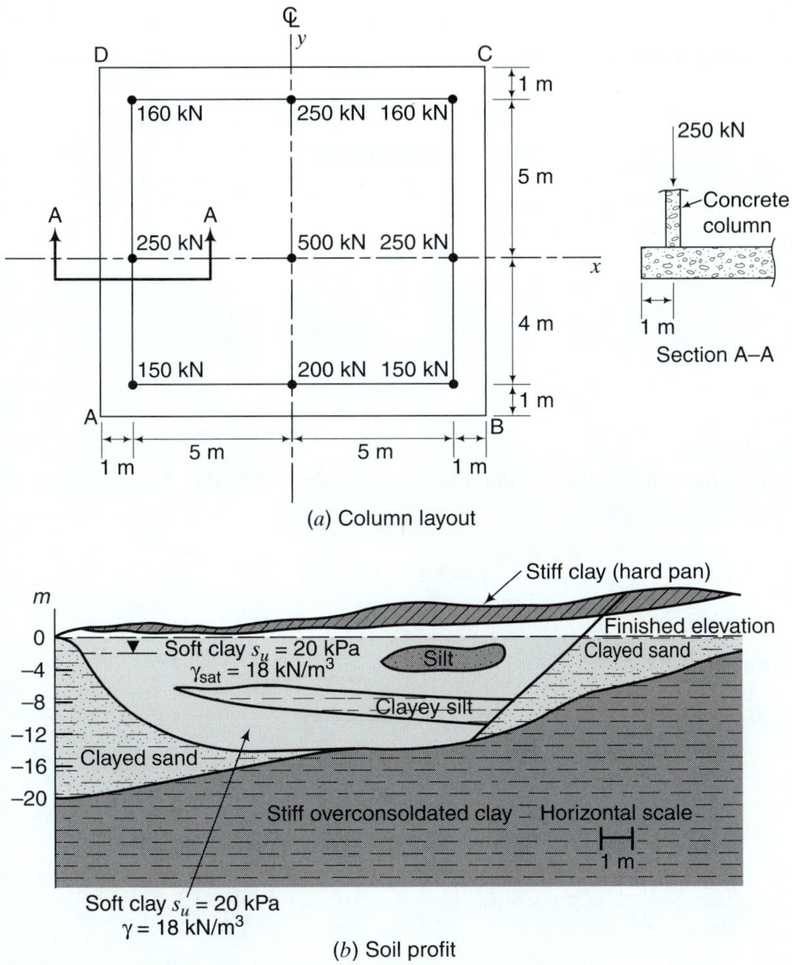

FIGURE E9.3a,b

Strategy Inspect the soil profile and decide whether a spread foundation is suitable. You can also make calculations using the methods described in Chapter 7. If a spread foundation is unsuitable, then you should decide on a mat and check if a raft is required.

Solution 9.3

Step 1: Decide on whether a spread foundation is unsuitable.

Inspection of the soil profile reveals that the soil is variable under the footprint of the superstructure and soft. It is unlikely that serviceability requirements can be met with reasonable spread foundation sizes. Spread foundation is not suitable.

Step 2: Decide on the type of mat foundation.
A slab-on-grade is more economical that a raft, so we will check on its suitability.

Step 3: Calculate the total load and load eccentricity.

Total load: $P_z = 2(150 + 250 + 160) + 200 + 500 + 250 = 2070\,\text{kN}$

Total area: $A = 12 \times 11 = 132\,\text{m}^2$

Load is symmetrical about the y axis but not about the x axis. Calculate eccentricity along the y axis.

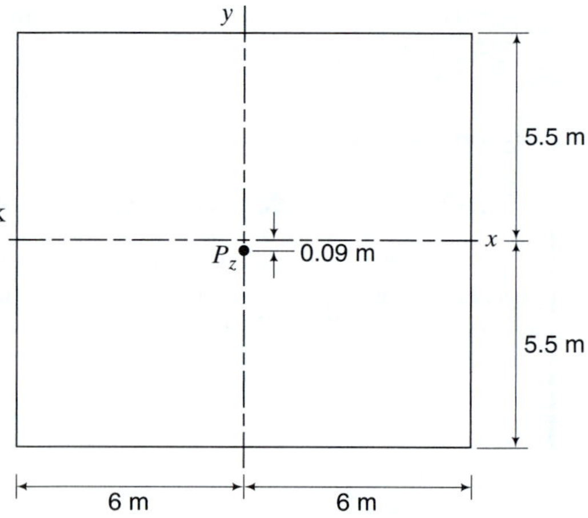

(c) Location of resultant load

FIGURE E9.3c

$$\bar{y} = \frac{\Sigma M_{AB}}{P_z} = \frac{\{[2 \times 160 + 250]10 + [2 \times 250 + 500]5 + [2 \times 150 + 200]1\}}{2070} = 5.41\,\text{m}$$

$$e = 5.41 - 11/2 = -0.09\,\text{m}$$

See Figure E9.3c.

Step 4: Calculate maximum applied stress.
Assume a mat thickness of 0.3 m. The self-weight of the mat is $0.3 \times 24 = 7.2\,\text{kPa}$. Total weight of mat $= 7.2 \times 132 = 950.4\,\text{kN}$.

$$(\sigma_a)_{max} = \frac{P_z}{A}\left(1 + \frac{6e}{B}\right) + 7.2 = \frac{2070}{132}\left(1 + \frac{6 \times 0.09}{11}\right) + 7.2 = 23.7\,\text{kPa}$$

Step 5: Calculate short-term bearing capacity.

$$B' = B; \quad L' = 11 - 2(0.09) = 10.82\,\text{m}$$

$$q_u = 5.14 s_u s_c; \quad s_c = 1 + 0.2\frac{B'}{L'} = 1 + 0.2\frac{10.82}{12} = 1.18$$

$$q_u = 5.14 \times 20 \times 1.18 = 121\,\text{kPa}$$

$$\text{FS} = \frac{q_u}{(\sigma_a)_{max}} = \frac{121}{23.7} = 5.1 > 3; \textit{okay}$$

Step 6: Estimate settlement.
Estimate E_s using Appendix B Table B4.9. For the soft soil, assume $E_s = 10$ MPa and that the foundation is approximately square. Neglect embedment and wall effects.

$$\rho_e = \frac{2Q_a}{E_s L_1}\left(1 - \nu_u^2\right)\mu_s = \frac{2(2070 + 950.4)}{10000 \times 12}(1 - 0.3^2)\,0.45 = 0.0206\,\text{m} \cong 21\,\text{mm}$$

It is quite likely that soil consolidation will incur a settlement greater than 4 mm. Thus, SLS will be exceeded.

Step 7: Try a fully compensated raft.

$$\text{Average vertical stress} = (2070 + 950.4)/132 = 22.9\ \text{kPa}$$

Since the groundwater can rise to the surface, any embedment of the mat will cause hydrostatic uplift pressures. We will allow the weight of the superstructure and the self-weight of the mat to compensate for the uplift pressures with a factor safety of 1.5.
Let D_f be the depth of the excavation. Then

$$\gamma_{sat} D_f - 1.5\gamma_w D_f = 22.9$$

$$\therefore 20D_f - 1.5 \times 9.8 \times D_f = 22.9$$

$$D_f = 4.3\ \text{m}$$

Step 8: Finalized depth of embedment.
The depth of embedment of 4.3 m will be entirely in the soft clay and below the silt lens. Use $D_f = 4.3\ \text{m}$.

Step 9: Check stability of excavation.
Because the clay is very soft, a supported excavation has to be made. Use effective unit weight of soil, since groundwater is at surface.

$$I_c = \left(0.84 + 0.16\frac{B}{L}\right) = \left(0.84 + 0.16\frac{11}{12}\right) = 0.99$$

$$\frac{D_f}{B} = \frac{4.3}{11} = 0.39 < 2.5$$

$$N_c \approx 6\left(1 + 0.2\frac{D_f}{B}\right)I_c \quad \text{for} \frac{D_f}{B} \le 2.5; \quad N_c \approx 6(1 + 0.2 \times 0.39)0.99 = 6.4$$

$$(D_f)_{cr} = N_c\frac{s_u}{\gamma'} = 6.4\frac{20}{(20 - 9.8)} = 12.5\ \text{m} > 4.3\ \text{m};\ \textit{okay}$$

Check factor of safety against bottom heave.

$$\text{FS} = N_c\frac{s_u}{\gamma' D_f + q_s}$$

$$= 6.4\frac{20}{(20 - 9.8) \times 4.3 + 22.9}$$

$$= 1.9 > 1.5;\ \textit{okay}$$

Sheeting for excavation support can be placed to a depth of 4.3 m. For additional safety, you should use 4.5 m depth of sheeting. ■

EXAMPLE 9.4 *Design of a Slab-on-Grade for a Lightly Loaded Structure on a Compressible Soil*

A foundation is required to support a one-storey office (Fig. E9.4a, b) located at a site with compressible soils. A soil profile in a representative borehole is shown in Fig. E9.4c. The climatic rating is −20, the edge load is 1700 lb/ft, and the live load is 125 psf. The tolerable differential settlement is $L/600$. Design a post-tension slab-on-grade using the PTI method. Assume $e_m = 2.5\ \text{ft}$, $y_m = 0.6\ \text{in}$, PL = 29%, PI = 32% and % clay = 80%.

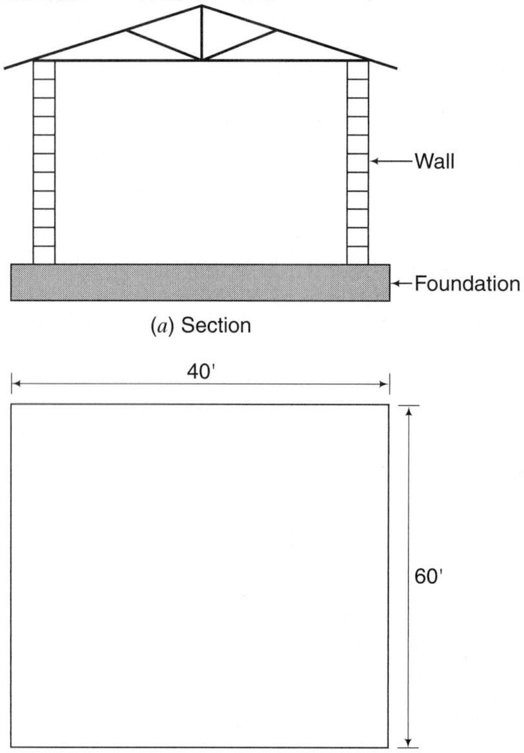

FIGURE E9.4a, b

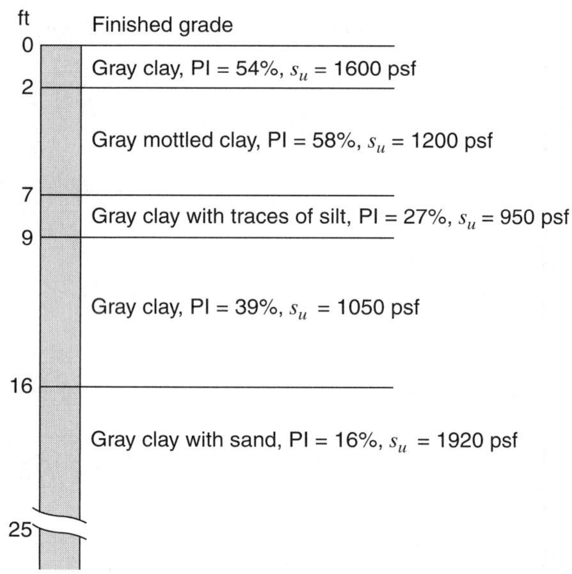

(c)

FIGURE E9.4c

TABLE E9.4

	A	B	C	D	E	F	G	H	I	J
1	**"FOUNDATIONS AND EARTH RETAINING STRUCTURES", Muni Budhu, John Wiley & Sons. This program should only be used for academic purpose.**									
2	APPROXIMATE ANALYSIS OF SLAB-ON-GRADE FOUNDATIONS BASED ON WRI TF 700-R-03, 1981									
3										
4										
5	Reinforcement	Rebar		Calculate composite PI						
6	Units	US		Depth	Thickness	PI	Factor	PI x Factor	Factor x thickness	
7	Length	60	ft	ft	ft					
8	Width	40	ft	2	2	54	3	324	6	
9	$(PI)_c$	52		7	5	58	2	580	10	
10	Slope, C_o	1		9	2	27	2	108	4	
11	Undrained shear strength, C_s	0.86		15	6	59	1	354	6	
12	Climate rating, C_w	20								
13	$(PI)_e$	45					Sum	1366	26	
14	Soil support index, (1 - C)	0.29				(PI)c	52			
15	Cantilever length, lc	8.2								
16	Length modifier factor, m_L	0.95		Calculate composite C_s						
17	Width modifier factor, m_B	0.95		Depth	Thickness	s_u	C_s	Factor	PI x Factor X thickness	Factor x thickness
18	Average applied stress, qs	250	psf	ft	ft	psf				
19	Beam width,b	9	in	2	2	1600	1	3	6	6
20	Beam spacing, s	6	ft	7	5	1200	0.9	2	9	10
21				9	2	950	0.8	2	3.2	4
22	**CALCULATIONS**			15	6	1050	0.7	1	4.2	6
23	Moment, long side, M_L	303.4	kips. ft							
24	Moment, short side, M_S	455.1	kips. ft					Sum	22.4	29
25	Number of beams, long direction	8				C_s	0.86			
26	Number of beams, short direction	11								
27	Depth of beam,d (long)	28	in							
28	Depth of beam,d (short)	29	in							
29										

Strategy We will use the Post-Tension Concrete Institute method as coded in the spreadsheet mat.xls, select PTI-comp.

Solution 9.4

The input parameters and the design values from the spreadsheet at http://www.wiley.com/college/budhu. Click Foundations and Earth Structures, Chapter 9, mat.xls. (worksheet PTI-comp). A partial view of the worksheet is shown in Table E9.4.

An equivalent slab thickness of 6 in. with 13 tendons parallel to the long direction and 17 tendons parallel to the short direction is suitable. ■

EXERCISES

9.1 The soil profile from a geological map and sparse existing soil parameters at a site for a multistorey building is shown in Fig. P9.1. The columns are spaced at 6 m (20 ft) centers in each direction. The client wants to know your thoughts on the type of foundation that may be appropriate. Would you use a mat foundation to support the column loads? What minimum additional soil data would you require to make a firm decision? Justify your answers.

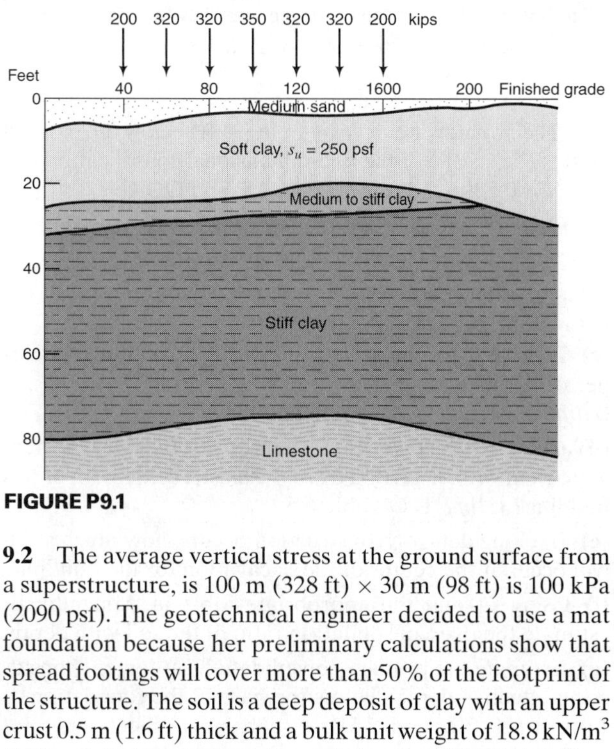

FIGURE P9.1

9.2 The average vertical stress at the ground surface from a superstructure, is 100 m (328 ft) × 30 m (98 ft) is 100 kPa (2090 psf). The geotechnical engineer decided to use a mat foundation because her preliminary calculations show that spread footings will cover more than 50% of the footprint of the structure. The soil is a deep deposit of clay with an upper crust 0.5 m (1.6 ft) thick and a bulk unit weight of 18.8 kN/m^3 (118 pcf). Below the crust is a soft, fat clay (USCS classification is CH) with silt lenses. The average saturated unit weight is 19.8 kN/m^3 (124 pcf). Groundwater is at 0.5 m (1.6 ft) below the surface but can rise to the surface. Sketch the soil profile and determine the embedment of the raft for 100% and 80% compensation. Assume a concrete raft 0.3 m (1 ft) thick to estimate its weight.

9.3 A foundation is required to support the column loads on the soil profile as shown in Fig. P9.3. The columns are spaced 8 m (26 ft) in each direction. The owners require a basement of minimum height of 6 m (20 ft). The groundwater is known to have risen to the ground surface in the past. The tolerable differential settlement is $L/1000$, where L is length. The total tolerable settlement is 50 mm (2 in.). Determine if spread footings can be used. If not, what type of mat or raft would you use? Show calculations to justify your answers.

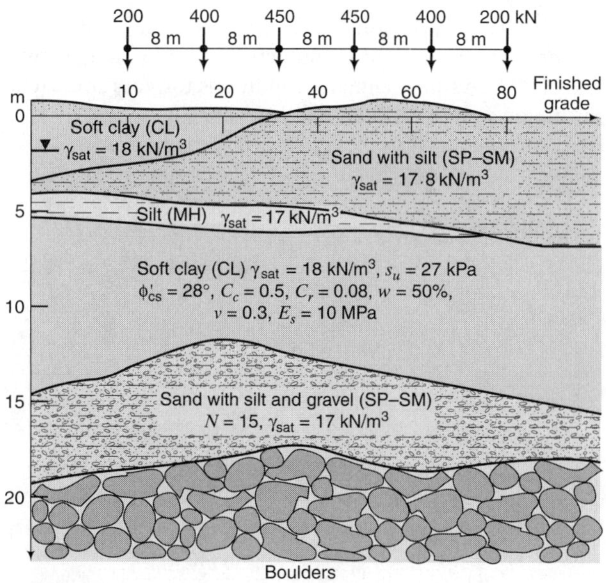

FIGURE P9.3

9.4 An oil tank is to be constructed on a clay soil with a crust 5 m (15 ft) thick. A sand layer 1 m (3 ft) thick will be placed on top of the crust as shown in Fig. P9.4. The empty tank weighs 2000 kN (450 kips).

(a) Determine the factor of safety against bearing capacity failure (short and long term). Consider two cases: the sand acting as a bearing layer and as a fill.

(b) Estimate the settlement. Is the settlement reasonable? Justify your answer.

(c) How long will it take for 50% consolidation of the soft clay to take place?

(d) If the tank is emptied and remains so for a long time, estimate the swell from the soft clay.

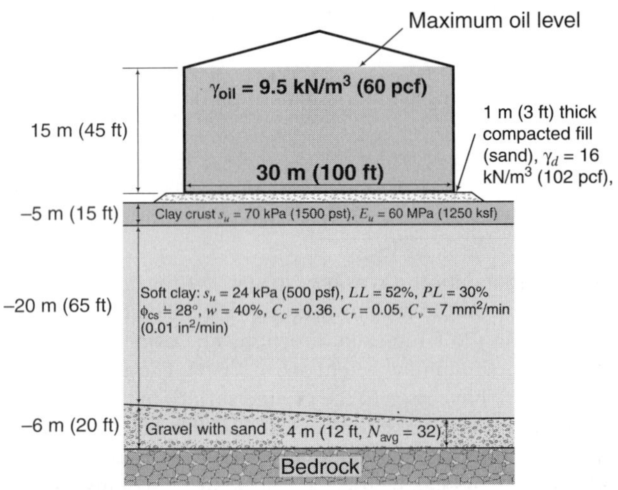

FIGURE P9.4

9.5 Design a post-tensioned slab-on-grade foundation for the one-storey light manufacturing building as shown in Fig. P9.5. Assume climatic conditions for your area. The perimeter load is 24 kPa (500 lb/ft).

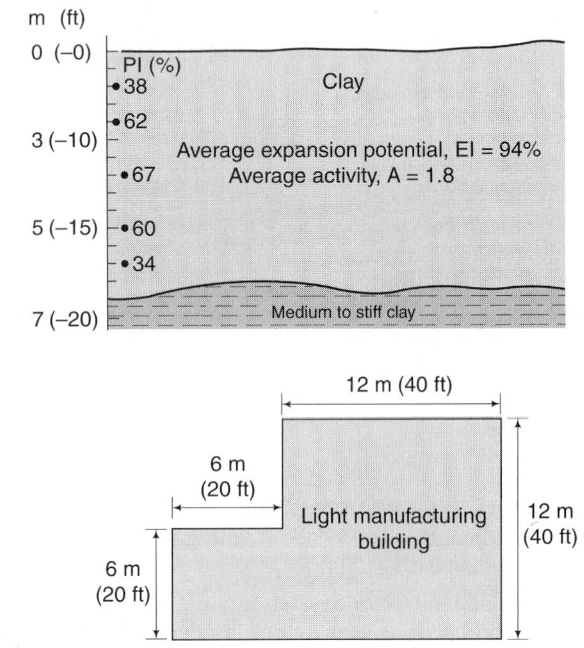

FIGURE P9.5

9.6 Repeat Problem 9.5 assuming that the soil is a mixture of sand and silt within the top 5.5 m (18 ft) and is collapsible with PI = 4%.

9.7 A multistorey reinforced-concrete frame building is supported on a mat foundation, $B = 52$ m (172 ft) by $L = 69$ m (226 ft) in plan dimension and $t = 3$ m (10 ft) thick. The base of the mat is $D_f = 18$ m (58 ft) below the finished grade (ground surface). The gross average pressure on the base of the mat is 450 kPa (9.4 ksf). The net average pressure taking into account the weights of the excavated soil and mat itself is 88 kPa (1.8 ksf). The soil profile consists of a deep deposit of stiff, saturated clay interbedded with thin sand layers. Triaxial undrained tests gave the following results:

Depth, m (ft)	Undrained shear strength, kPa (ksf)	Average friction angle
0	48 (1)	
6 (20)	192 (4)	$\phi'_p = 20°$, $\phi'_{cs} = 20°$ *
24 (80)	269 (5.6)	$\phi'_p = 27°$, $\phi'_{cs} = 25°$

*The test samples were inhomogeneous. They failed at the interface of the clay and a thin layer of sand.

The groundwater level is 11 m (35 ft) below the finished grade. The results of the one-dimensional consolidation test are $e_o = 0.58$, $C_r = 0.04$, $C_c = 0.15$, and OCR = 4.

(a) Sketch the foundation and plot the variation of undrained shear strength with depth.

(b) Is the mat fully compensated? If not, what is the percentage compensation?

(c) Estimate the elastic settlement of the center and corner of the mat using $E = E_u = 800 s_u$ and $\nu = 0.3$ [*Hint:* Use Eqs. (2.24) and (2.25)]

(d) Calculate the distortion from the results of (a). Is there a problem with distortion from the elastic settlement of this building? [*Hint:* Use Table 6.5.]

(e) If a long delay in construction occurs, how much swell will occur at the center of the excavation for the building?

(f) Using $\frac{s_u}{\sigma'_z}$ and the appropriate table in Appendix B, estimate the overconsolidation ratio of the soil with depth.

(g) Estimate the primary consolidation settlement considering effective depths below the base of the foundation of 1.0B, 1.5B, 2B, and 2.5B. Use a spreadsheet and plot primary consolidation settlement versus effective depth.

(h) Estimate the expected distortion. Will this amount of distortion cause problems?

(i) Comment on the accuracy of your results.

CHAPTER 10

EARTH RETAINING STRUCTURES—RIGID AND FLEXIBLE WALLS

10.0 INTRODUCTION

Earth retaining structures are ubiquitous in the man-made environment. These structures have the distinction of being the first to be analyzed using mechanics. In this chapter, we will analyze some typical earth retaining structures to determine their stability. The emphases will be on gaining an understanding of the forces that provoke failures and methods of analysis of simple earth retaining structures.

You should recall that stability refers to a condition in which a geotechnical system will not fail or collapse under any conceivable loading (static and dynamic loads, fluid pressure, seepage forces). Stability is synonymous with ultimate limit state but serviceability limit state is also important. In many circumstances, the serviceability limit state is the deciding design limit state.

When you complete this chapter, you should be able to:

- Understand and determine lateral earth pressures.
- Understand the forces that lead to instability of earth retaining structures.
- Determine the stability of simple earth retaining structures.

10.1 DEFINITIONS OF KEY TERMS

Active earth pressure coefficient **(K_a)** is the ratio between the lateral and vertical principal effective stresses when an earth retaining structure moves away (by a small amount) from the retained soil.

Passive earth pressure coefficient **(K_p)** is the ratio between the lateral and vertical principal effective stresses when an earth retaining structure is forced against a soil mass.

Gravity or rigid retaining wall is a massive concrete wall relying on its mass to resist the lateral forces from the retained soil mass.

Flexible retaining wall or a sheet pile wall is a long slender wall relying on passive resistance and anchors or props for its stability.

10.2 BASIC CONCEPTS ON LATERAL EARTH PRESSURES

Importance *Earth retaining walls primarily support the lateral loads from soils. You need to understand how to determine these forces and the assumptions made. In this section, We will discuss the fundamentals of lateral earth pressures.*

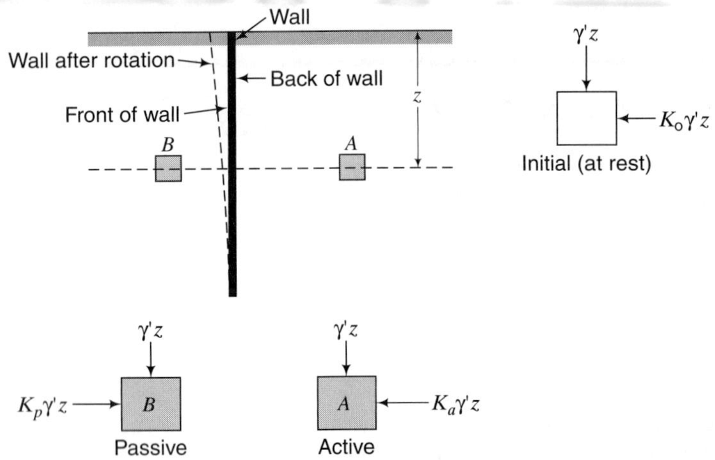

FIGURE 10.1 Stresses on soil elements in front and behind a retaining wall.

We have discussed the lateral earth pressure at rest and the lateral increases in stresses on a semi-infinite, isotropic, homogeneous, elastic soil mass from surface loading in Chapter 2. We are now going to consider the lateral earth pressures on a vertical wall that retains a soil mass. We will deal with two theories: one proposed by Coulomb (1776) and the other by Rankine (1857). First, we will develop a basic understanding of lateral earth pressures using a generic ϕ' and make the following assumptions:

1. The earth retaining wall is vertical.
2. The interface between the wall and soil is frictionless.
3. The soil surface is horizontal and no shear stress acts on horizontal and vertical boundaries.
4. The wall is rigid and extends to an infinite depth in a dry, homogeneous, isotropic soil mass.
5. The soil is loose and initially in an at-rest state.

Consider the wall as shown in Fig. 10.1. If the wall remains rigid and no movement (not even an infinitesimal movement) occurs, then the vertical and horizontal effective stresses at rest on elements A, at the back wall, and B, at the front wall (Fig. 10.1), are

$$\sigma'_z = \sigma'_1 = \gamma' z$$
$$\sigma'_x = \sigma'_3 = K_o \sigma'_1 = K_o \gamma' z$$

where K_o is the lateral earth pressure at rest. Mohr's circle for the at-rest state is shown by circle ① in Fig. 10.2.

Let us now assume a rotation about the bottom of the wall sufficient to produce slip planes in the soil mass behind and in front of the wall (Fig. 10.3). The rotation required, and consequently the lateral strains, to produce slip planes in front of the wall is much larger than that required for the back of the wall, as shown in Fig. 10.4. The soil mass at the back of the wall is assisting in producing failure while the soil mass at the front of the wall is resisting failure. In the latter, you have to rotate the wall against the soil to produce failure.

What happens to the lateral effective stresses on elements A and B (Fig. 10.1) when the wall is rotated? The vertical stress will not change on either element but the lateral effective stress on element A will be reduced while that for element B will be increased. We can now plot two additional Mohr's circles: one to represent the stress state of element A (circle Ⓐ, Fig. 10.2) and the other to represent the stress ... of element B (circle Ⓑ, Fig. 10.2). Both circles are drawn such that the decrease (element A) or ... element B) in lateral effective stress is sufficient to bring the soil to the Mohr-Coulomb failure ... element B to reach the failure state, the lateral effective stress must be greater than the vertical ... stress, as shown in Fig. 10.2.

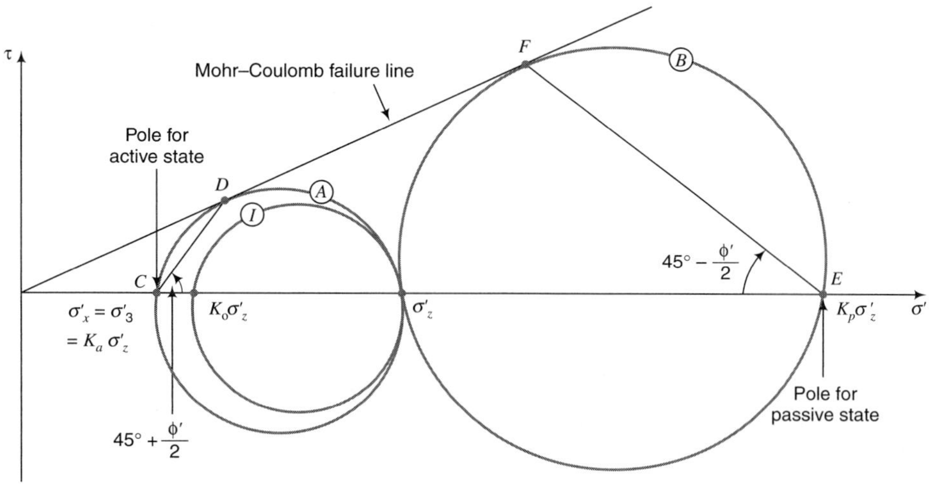

FIGURE 10.2 Mohr's circles at rest, active, and passive states.

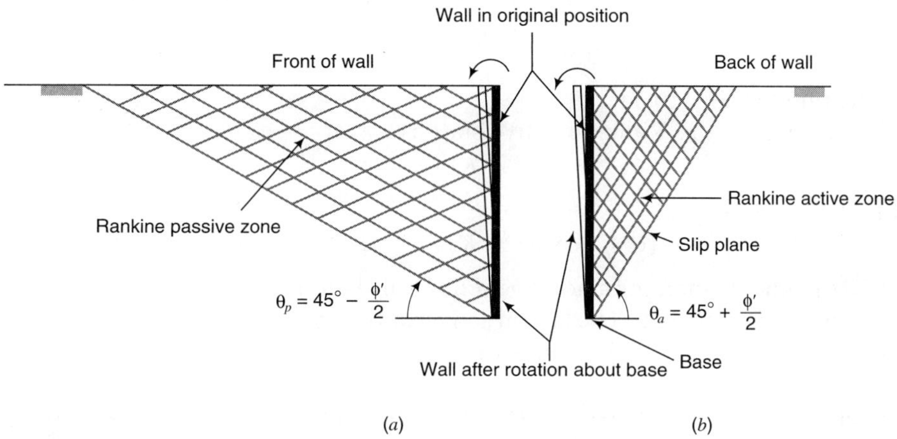

FIGURE 10.3 Slip planes within a soil mass near a retaining wall.

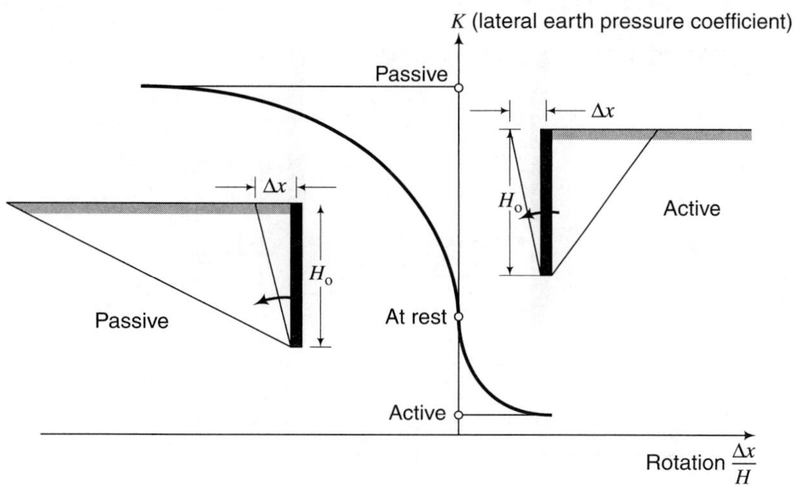

FIGURE 10.4 Rotation required to mobilize active and passive resistance.

The ratio of lateral principal effective stress to vertical principal effective stress for circle Ⓐ is

$$\frac{(\sigma_3')_f}{(\sigma_1')_f} = \frac{(\sigma_x')_f}{(\sigma_z')_f} = \frac{1 - \sin\phi'}{1 + \sin\phi'} = \tan^2\left(45^\circ - \frac{\phi'}{2}\right) = K_a \tag{10.1}$$

where K_a is called the active lateral earth pressure coefficient. Similarly, for circle Ⓑ,

$$\frac{(\sigma_1')_f}{(\sigma_3')_f} = \frac{(\sigma_x')_f}{(\sigma_z')_f} = \frac{1 + \sin\phi'}{1 - \sin\phi'} = \tan^2\left(45^\circ + \frac{\phi'}{2}\right) = K_p \tag{10.2}$$

where K_p is the passive earth pressure coefficient. Therefore,

$$K_p = \frac{1}{K_a}$$

If, for example, $\phi' = 30^\circ$, then $K_a = \frac{1}{3}$ and $K_p = 3$.

The stress states of soil elements A and B are called the Rankine active state and the Rankine passive state, respectively (named after the original developer of this theory, Rankine, 1857). Each of these Rankine states is associated with a family of slip planes. For the Rankine active state, the slip planes are oriented at

$$\theta_a = 45^\circ + \frac{\phi'}{2} \tag{10.3}$$

to the horizontal, as illustrated in Fig. 10.3b. For the Rankine passive state, the slip planes are oriented at

$$\theta_p = 45^\circ - \frac{\phi'}{2} \tag{10.4}$$

to the horizontal, as illustrated in Fig. 10.3a.

The lateral earth pressure for the Rankine active state is

$$(\sigma_x')_a = K_a\sigma_z' = K_a\gamma' z \tag{10.5}$$

and for the Rankine passive state it is

$$(\sigma_x')_p = K_p\sigma_z' = K_p\gamma' z \tag{10.6}$$

Equations (10.5) and (10.6) indicate that, for a homogeneous soil layer, the lateral earth pressure varies linearly with depth, as shown in Figs. 10.5a,b. The lateral active and passive coefficients are applied only to effective stresses. For soils above the ground water level $\gamma' = \gamma$ while for soils below the groundwater level $\gamma' = (\gamma_{sat} - \gamma_w)$.

The lateral earth force is the area of the lateral stress diagram, which, for the Rankine active state, is

$$P_a = \int_0^{H_o} K_a\gamma' z = \frac{1}{2}K_a\gamma' H_o^2 \tag{10.7}$$

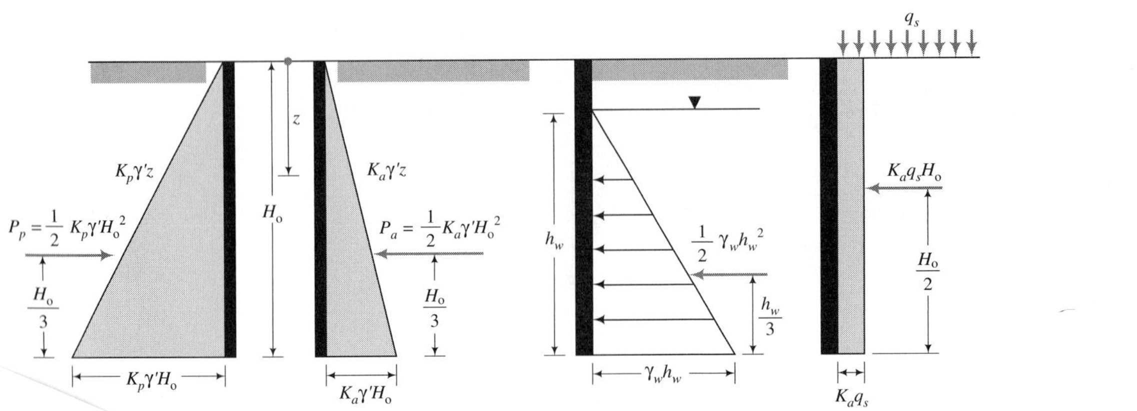

assive (b) Active (c) Hydrostatic pressure (d) Surface stress

.5 Variation of active and passive lateral earth pressures, hydrostatic pressure, and a uniform surface .h depth.

and, for the Rankine passive state, is

$$P_p = \int_0^{H_o} K_p \gamma' z = \frac{1}{2} K_p \gamma' H_o^2 \tag{10.8}$$

These lateral forces, P_a and P_p, are located at the centroid of the lateral earth pressure distribution diagram. In this case, the centroid is at $H_o/3$ from the base.

If groundwater is present, you need to add the hydrostatic pressure (porewater pressure) to the lateral earth pressure. For example, if the groundwater level is at distance h_w from the base of the wall (Fig. 10.5c), the hydrostatic pressure is

$$u = \gamma_w h_w \tag{10.9}$$

and the hydrostatic force is

$$P_w = \frac{1}{2} \gamma_w h_w^2 \tag{10.10}$$

The key points are:

1. ***The lateral earth pressures on retaining walls are related directly to the vertical effective stress through two coefficients. One is the active earth pressure coefficient,***

$$K_a = \frac{1 - \sin \phi'}{1 + \sin \phi'} = \tan^2 \left(45^\circ - \frac{\phi'}{2}\right)$$

and the other is the passive earth pressure coefficient,

$$K_p = \frac{1 + \sin \phi'}{1 - \sin \phi'} = \tan^2 \left(45^\circ + \frac{\phi'}{2}\right) = \frac{1}{K_a}$$

2. ***Substantially more movement is required to mobilize the full passive earth pressure than the full active earth pressure.***
3. ***A family of slip planes occurs in the Rankine active and passive states. In the active state, the slip planes are oriented at $45^\circ + \phi'/2$ to the horizontal, while for the passive case they are oriented at $45^\circ - \phi'/2$ to the horizontal.***
4. ***The lateral earth pressure coefficients, developed so far, are only valid for a smooth, vertical wall supporting a homogeneous soil mass with a horizontal surface.***
5. ***The lateral earth pressure coefficients must be applied only to effective stresses.***

10.3 LATERAL STRESSES FROM SURFACE LOADS

Surface stresses also impose lateral earth pressures on retaining walls. A uniform surface stress, q_s, is assumed to transmit a uniform active lateral earth pressure of $K_a q_s$ (Fig. 10.5d) and a uniform passive lateral earth pressure of $K_p q_s$. The active and passive lateral earth pressures due to the soil, groundwater, and the uniform surface stresses are then

$$(\sigma_x)_a = K_a \sigma'_z + K_a q_s + (u)_a \tag{10.11}$$

and

$$(\sigma_x)_p = K_p \sigma'_z + K_p q_s + (u)_p \tag{10.12}$$

where u is the hydrostatic (porewater) pressure and the subscripts a and p denote active and passive states, respectively.

The lateral forces from a line load and a strip load are given in Figs. 10.6 and 10.7. For other types c surface loads you can consult Poulos and Davis (1974).

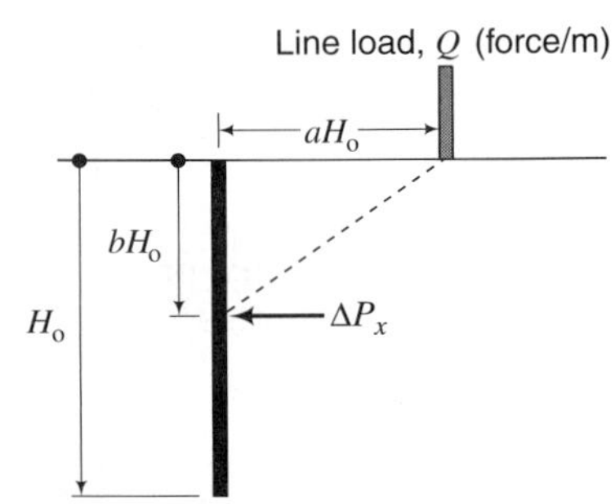

FIGURE 10.6 Line load near a retaining wall.

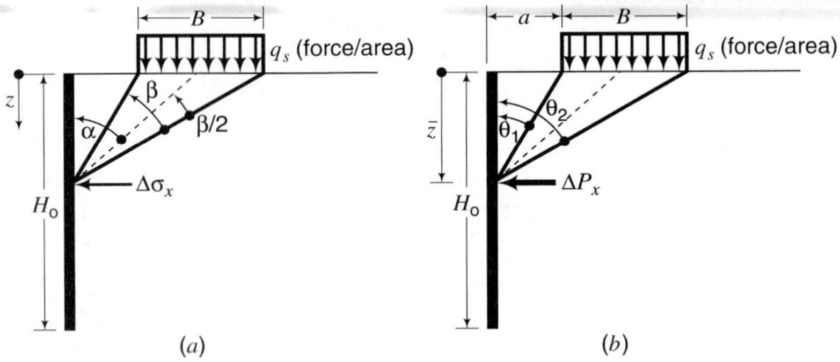

FIGURE 10.7 (a) Strip load near a retaining wall and (b) lateral force on a retaining wall from a strip load.

(1) Line Load The increase in lateral stress on a buried earth retaining structure (Fig. 10.6) due to a line load of intensity Q (force/length) is

$$\Delta\sigma_x = \frac{4Qa^2b}{\pi H_o(a^2+b^2)^2} \tag{10.13}$$

The increase in lateral force is

$$\Delta P_x = \frac{2Q}{\pi(a^2+1)} \tag{10.14}$$

(2) Strip Load The lateral stress from a strip load (Fig. 10.7a) is

$$\Delta\sigma_x = \frac{2q_s}{\pi}(\beta - \sin\beta\cos 2\alpha) \tag{10.15}$$

The lateral force and its location (Jarquio, 1981) are (Fig. 10.7b)

$$\Delta P_x = \frac{q_s}{90}\left[H_o(\theta_2-\theta_1)\right] \tag{10.16}$$

$$\bar{z} = \frac{H_o^2(\theta_2-\theta_1)-(R_1-R_2)+57.3\,BH_o}{2H_o(\theta_2-\theta_1)} \tag{10.17}$$

where

$$\theta_1 = \tan^{-1}\left(\frac{a}{H_o}\right),\quad \theta_2 = \tan^{-1}\left(\frac{a+B}{H_o}\right),$$

$$R_1 = (a+B)^2(90-\theta_2),\quad \text{and}\quad R_2 = a^2(90-\theta_1)$$

EXAMPLE 10.1 *Lateral Earth Pressure and Force*

Determine the active lateral earth pressure on the frictionless wall shown in Fig. E10.1a. Calculate the resultant ~d its location from the base of the wall. Neglect seepage effects.

The lateral earth pressure coefficients can only be applied to the effective stresses. You need to e vertical effective stress, apply K_a, and then add the porewater pressure. SI units will be used in solving le.

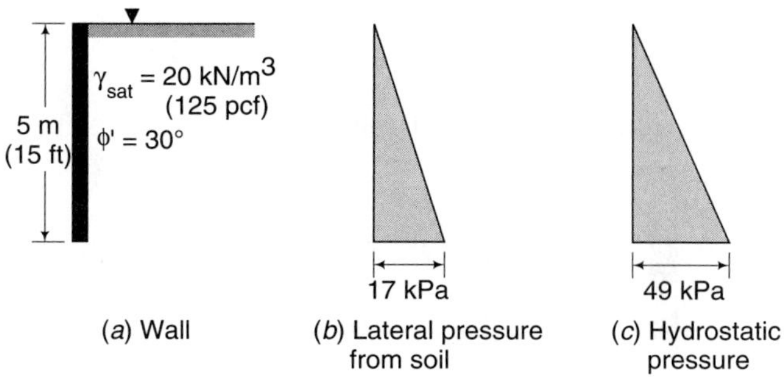

FIGURE E10.1

Solution 10.1

Step 1: Calculate K_a.

$$K_a = \frac{1 - \sin\phi'}{1 + \sin\phi'} = \frac{1 - \sin(30^\circ)}{1 + \sin(30^\circ)} = \frac{1}{3}$$

or

$$\tan^2\left(45 - \frac{\phi'}{2}\right) = \tan^2\left(45^\circ - \frac{30^\circ}{2}\right) = \frac{1}{3}$$

Step 2: Calculate the vertical effective stress.

$$\begin{aligned} &\text{At the surface}: \quad \sigma'_z = 0, \quad u = 0 \\ &\text{At the base}: \quad \sigma'_z = \gamma' H_o = (20 - 9.8) \times 5 = 51\,\text{kPa} \\ &\qquad\qquad\qquad u = \gamma_w H_o = 9.8 \times 5 = 49\,\text{kPa} \end{aligned}$$

Step 3: Calculate the lateral effective stress.

$$(\sigma'_x)_a = K_a \sigma'_z = \frac{1}{3} \times 51 = 17\,\text{kPa}$$

Step 4: Sketch the lateral earth pressure distributions.
See Figs. E10.1b,c.

Step 5: Calculate the lateral force.

$$P_a = P_s + P_w$$

where P_s is the lateral force due to the soil solids and P_w is the lateral force due to the porewater.

$$P_a = \frac{1}{2}(\sigma'_x)_a H_o + \frac{1}{2} u H_o = \left(\frac{1}{2} \times 17 \times 5\right) + \left(\frac{1}{2} \times 49 \times 5\right) = 165\,\text{kN}$$

Step 6: Determine the location of the resultant.
Since both the lateral earth pressure and the porewater pressure distributions are triangular over the whole depth, the resultant is at the centroid of the triangle, that is, $\bar{z} = H_o/3 = 5/3 = 1.67$ m from the base of the wall. ■

EXAMPLE 10.2 *Lateral Earth Pressure in Layered Soils*

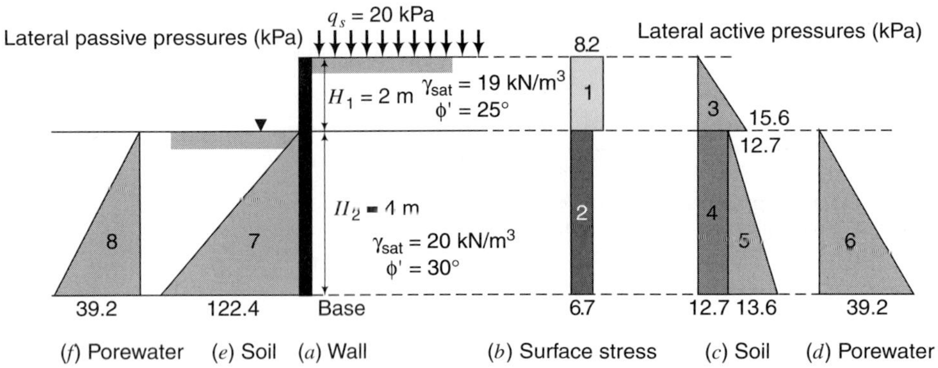

FIGURE E10.2

For the frictionless wall shown Fig. E10.2a, determine the following:

(a) The active lateral earth pressure distribution with depth.
(b) The passive lateral earth pressure distribution with depth.
(c) The magnitudes and locations of the active and passive forces.
(d) The resultant force and its location.
(e) The ratio of passive moment to active moment.

Strategy There are two layers. It is best to treat each layer separately. Neither K_a nor K_p should be applied to the porewater pressure. You do not need to calculate K_p for the top soil layer. Since the water level on both sides of the wall is the same, the resultant hydrostatic force is zero. However, you are asked to determine the forces on each side of the wall; therefore, you have to consider the hydrostatic force. A table is helpful to solve this type of problem. SI units will be used for this example.

Solution 10.2

Step 1: Calculate the active lateral earth pressure coefficients.

$$\text{Top layer (0 to 2 m): } K_a = \tan^2\left(45^\circ - \frac{\phi'}{2}\right) = \tan^2\left(45^\circ - \frac{25^\circ}{2}\right) = 0.41$$

$$\text{Bottom layer (2 to 6 m): } K_a = \tan^2\left(45^\circ - \frac{\phi'}{2}\right) = \tan^2\left(45^\circ - \frac{30^\circ}{2}\right) = \frac{1}{3}; \quad K_p = \frac{1}{K_a} = 3$$

Step 2: Calculate the active and passive lateral earth pressures.
Use a table as shown below to do the calculations or use a spreadsheet.

Active	Depth [a] (m)	u(kPa)	σ_z (kPa)	$\sigma'_z = \sigma_z - u$	$(\sigma'_x)_a = K_a\sigma'_z$ (kPa)
Surcharge	0	0	20	20	$0.41 \times 20 = 8.2$
	2 to 6	0	20	20	$\frac{1}{3} \times 20 = 6.7$
Soil	0	0	0	0	0
	2^-	0	$\gamma_1 H_1 = 19 \times 2 = 38$	38	$0.41 \times 38 = 15.6$
	2^+	0	$\gamma_1 H_1 = 19 \times 2 = 38$	38	$\frac{1}{3} \times 38 = 12.7$
	6	$\gamma_w H_2 = 9.8 \times 4 = 39.2$	$\gamma_1 H_1 + \gamma_2 H_2 = 19 \times 2 + 20 \times 4 = 118$	78.8	$\frac{1}{3} \times 78.8 = 26.3$

(continued)

(Continued)

Passive	Depth (m)	u(kPa)	σ_z(kPa)	$\sigma'_z = \sigma_z - u$	$(\sigma'_x)_p = K_p\sigma'_z$
Soil	0	0	0	0	0
	4	$\gamma_w H_2 = 9.8 \times 4 = 39.2$	$\gamma_2 H_2 = 20 \times 4 = 80$	40.8	$3 \times 40.8 = 122.4$

[a]The − and + superscripts indicate that you are calculating the stress just above (−) and just below (+) 2 m.

See Figs. E10.2b to e for the pressure distributions.

Step 3: Calculate the hydrostatic force.

$$P_w = \frac{1}{2}\gamma_w H_2^2 = \frac{1}{2} \times 9.8 \times 4^2 = 78.4\,\text{kN}$$

Step 4: Calculate the resultant lateral forces and their locations.
See the table below for calculations. Active moments are assumed to be negative.

Active Area	Depth (m)	Force (kN)	Moment arm from base (m)	Moment (kN • m)
1	0–2	$8.2 \times 2 = 16.4$	$4 + 1 = 5 = 4.42$	−82.0
2	2–6	$6.7 \times 4 = 26.8$	$\frac{4}{2} = 2$	−53.6
3	0–2	$\frac{1}{2} \times 15.6 \times 2 = 15.6$	$\frac{2}{3} + 4 = 4.67$	−72.9
4	2–6	$12.7 \times 4 = 50.8$	$\frac{4}{2} = 2$	−101.6
5	2–6	$\frac{1}{2} \times 13.6 \times 4 = 27.2$	$\frac{4}{3}$	−36.3
6 (water)	2–6	78.4	$\frac{4}{3}$	−104.5
		Σ Active lateral forces = 215.2		Σ Active moments = −450.9
7	2–6	$\frac{1}{2} \times 122.4 \times 4 = 244.8$	$\frac{4}{3}$	326.4
8 (water)	2–6	78.4	$\frac{4}{3}$	104.5
		Σ Passive forces = 323.2		Σ Passive moments = 430.9

Location of resultant active lateral earth force:

$$\bar{z}_a = \frac{\Sigma\,\text{Moments}}{\Sigma\,\text{Active lateral forces}} = \frac{450.9}{215.2} = 2.09\,\text{m}$$

Location of passive lateral force: $\bar{z}_p = \frac{4}{3} = 1.33$ m.

Step 5: Calculate the resultant lateral force.

$$R_x = P_p - P_a = 323.2 - 215.2 = 108\,\text{kN/m}$$

Step 6: Calculate the ratio of moments.

$$\text{Ratio of moments} = \frac{\Sigma\,\text{Passive moments}}{\Sigma\,\text{Active moments}} = \frac{430.9}{450.9} = 0.96$$

Since the active moment is greater than the passive moment, the wall will rotate. ■

10.4 COULOMB'S EARTH PRESSURE THEORY

Importance *Coulomb (1776) is the pioneer of earth pressure theory. His approach is based on limiting equilibrium with or without soil–wall friction. Coulomb's theory is often used in practice. We present Coulomb's approach in this section.*

Access the spreadsheet at http://www.wiley.com/college/budhu. Click Foundations and Earth Structures, Chapter 10, retwall.xls to calculate lateral earth pressure coefficients. Click on the worksheet "kakp."

Coulomb (1776) proposed that a condition of limit equilibrium exists through which a soil mass behind a vertical retaining wall will slip along a plane inclined an angle θ to the horizontal. He then determined the slip plane by searching for the plane on which the maximum thrust acts. We begin consideration of Coulomb's theory by reminding you of the basic tenets of limit equilibrium. The essential steps in the limit equilibrium method are (1) selection of a plausible failure mechanism, (2) determination of the forces acting on the failure surface, and (3) use of equilibrium equations to determine the maximum thrust.

Let us consider a vertical, frictionless wall of height H_o, supporting a soil mass with a horizontal surface (Fig. 10.8). We are going to assume a dry, homogeneous soil mass and postulate that slip occurs on a plane inclined an angle θ to the horizontal. Since the soil is dry, $\gamma' = \gamma$. We can draw the free-body diagram as shown in Fig. 10.8b and solve for P_a using statics as follows:

$$\Sigma F_x = P_a + T\cos\theta - N\sin\theta = 0$$
$$\Sigma F_z = W - T\sin\theta - N\cos\theta = 0$$

The weight of the sliding mass of soil is

$$W = \frac{1}{2}\gamma H_o^2 \cot\theta$$

At limit equilibrium,

$$T = N\tan\phi'$$

Solving for P_a we get

$$P_a = \frac{1}{2}\gamma H_o^2 \cot\theta \tan(\theta - \phi') \tag{10.18}$$

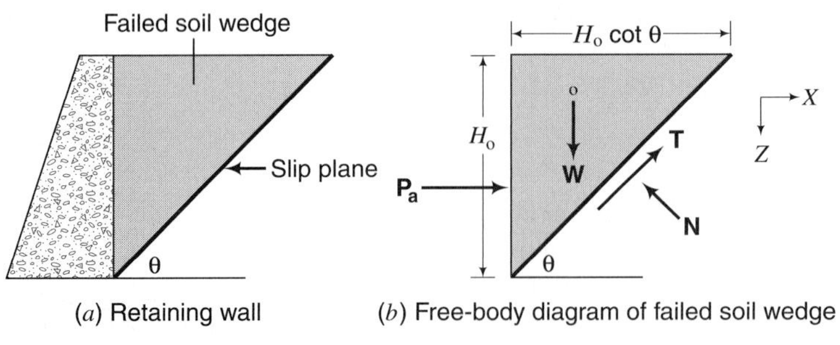

FIGURE 10.8 Coulomb failure wedge.

To find the maximum thrust and the inclination of the slip plane, we use calculus to differentiate Eq. (10.18) with respect to θ :

$$\frac{\partial P_a}{\partial \theta} = \frac{1}{2}\gamma H_o^2[\cot\theta \sec^2(\theta - \phi') - \csc^2\theta \tan(\theta - \phi')] = 0$$

which leads to

$$\theta = \theta_{cr} = 45^\circ + \frac{\phi'}{2} \tag{10.19}$$

Substituting this value of θ into Eq. (10.18) we get

$$P_a = \frac{1}{2}\gamma H_o^2 \tan^2\left(45^\circ - \frac{\phi'}{2}\right) = \frac{1}{2}K_a\gamma H_o^2 \tag{10.20}$$

This is the same result obtained earlier from considering Mohr's circle.

The solution from a limit equilibrium method is analogous to an upper bound solution because it gives a solution that is usually greater than the true solution. The reason for this is that a more efficient failure mechanism may be possible than the one we postulated. For example, rather than a planar slip surface we could have postulated a circular slip surface or some other geometric form, and we could have obtained a maximum horizontal force lower than for the planar slip surface.

For the Rankine active and passive states, we considered the stress states and obtained the distribution of lateral stresses on the wall. At no point in the soil mass did the stress state exceed the failure stress state and static equilibrium is satisfied (lower bound solution). If the lower bound solution and the upper bound solution are in agreement, we have a true solution, as is the case here.

Poncelet (1840), using Coulomb's limit equilibrium approach, obtained expressions for K_a and K_p for cases where wall friction (δ) is present, the wall face is inclined at an angle η to the vertical, and the backfill is sloping at an angle β. With reference to Fig. 10.9, K_{aC} and K_{pC} are

$$K_{aC} = \frac{\cos^2(\phi' - \eta)}{\cos^2\eta\cos(\eta + \delta)\left[1 + \left\{\dfrac{\sin(\phi' + \delta)\sin(\phi' - \beta)^{1/2}}{\cos(\eta + \delta)\cos(\eta - \beta)}\right\}\right]^2} \tag{10.21}$$

$$K_{pC} = \frac{\cos^2(\phi' + \eta)}{\cos^2\eta\cos(\eta - \delta)\left[1 - \left\{\dfrac{\sin(\phi' + \delta)\sin(\phi' + \beta)}{\cos(\eta - \delta)\cos(\eta - \beta)}\right\}^{1/2}\right]^2} \tag{10.22}$$

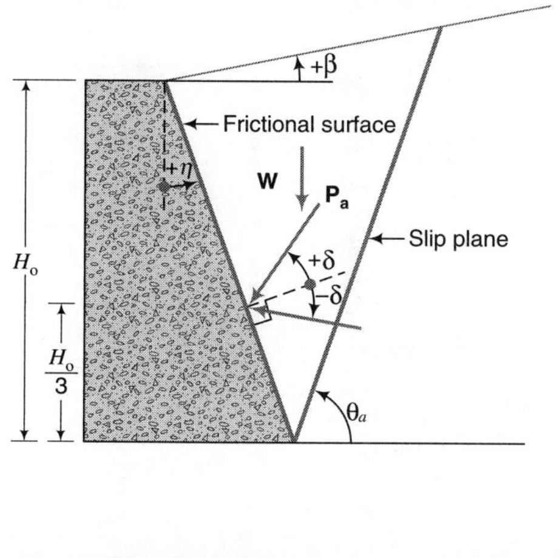

FIGURE 10.9 Retaining wall with sloping back, wall friction, and sloping soil surface for use with Coulomb's method for active condition.

TABLE 10.1 Correction Factors to Be Applied to K_{pC} to Approximate a Logarithm Spiral Slip Surface for a Backfill with a Horizontal Surface and Sloping Wall Face

	δ/ϕ'							
ϕ'	−0.7	−0.6	−0.5	−0.4	−0.3	−0.2	−0.1	−0.0
15	0.96	0.93	0.91	0.88	0.85	0.83	0.80	0.78
20	0.94	0.90	0.86	0.82	0.79	0.75	0.72	0.68
25	0.91	0.86	0.81	0.76	0.71	0.67	0.62	0.57
30	0.88	0.81	0.75	0.69	0.63	0.57	0.52	0.47
35	0.84	0.75	0.67	0.60	0.54	0.48	0.42	0.36
40	0.78	0.68	0.59	0.51	0.44	0.38	0.32	0.26

The subscript C denotes Coulomb. You should note that $K_{pC} \neq 1/K_{aC}$. Recall that the lateral earth pressure coefficients are applied to the effective not total vertical stress. In the absence of groundwater, the total and effective vertical stresses are equal.

The inclination of the slip plane to the horizontal is

$$\tan\theta = \left[\frac{(\sin\phi'\cos\delta)^{1/2}}{\cos\phi'\{\sin(\phi'+\delta)\}^{1/2}}\right] \pm \tan\phi' \tag{10.23}$$

where the positive sign refers to the active state (θ_a) and the negative sign refers to the passive state (θ_p).

Wall friction causes the slip planes in both the active and passive states to be curved. The curvature in the active case is small in comparison to the passive case. The implication of the curved slip surface is that the values of K_{aC} and K_{pC} from Eqs. (10.21) and (10.22) are not accurate. In particular, the passive earth pressures are overestimated using Eq. (10.22). For the active state, the error is small and can be neglected. The error for the passive state is small if $\delta < \phi'/3$. In practice, δ is generally greater than $\phi'/3$.

Several investigators have attempted to find K_a and K_p using nonplanar slip surfaces. For example, Caquot and Kerisel (1948) used logarithm spiral slip surfaces while Packshaw (1969) used circular failure surfaces. The Caquot and Kerisel (1948) values of K_p are generally used in practice. Table 10.1 lists factors that can be applied to K_{pC} to correct it for a logarithm spiral slip surface.

The lateral forces are inclined at δ to the normal on the sloping wall face. The sign conventions for δ and β are shown in Fig. 10.9. You must use the appropriate sign in determining K_{aC} and K_{pC}. The direction of the frictional force on the wall depends on whether the wall moves relative to the soil or the soil moves relative to the wall. In general, the active wedge moves downward relative to the wall and the passive wedge moves upward relative to the wall. The frictional forces developed are shown in Fig. 10.10. The sense of the inclination of active lateral force is then positive while the sense of the inclination of passive lateral force is negative. Load bearing walls with large vertical loads tend to move downward relative to the soil. In such a situation the frictional force on both sides of the wall will be downward and the lateral earth pressures are increased on both sides of the wall. In practice, these increases in pressure are ignored and the sense of the active and lateral forces are taken as shown in Fig. 10.10.

The horizontal components of the lateral forces are $P_{ax} = P_a \cos\delta$ and $P_{px} = P_p \cos\delta$, and the vertical components are $P_{az} = P_a \sin\delta$ and $P_{pz} = P_p \sin\delta$. The point of application of these forces is $H_o/3$ from the base of the wall (Fig. 10.9). Typical values of δ for the interfaces of coarse-grained soils and concrete or steel walls range from $\frac{1}{2}\phi'$ to ϕ'.

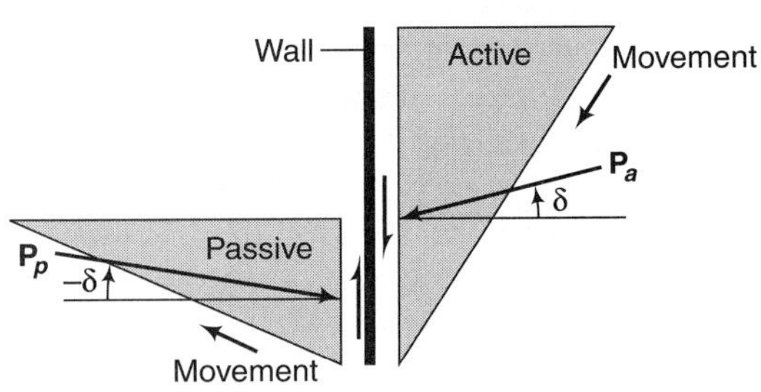

FIGURE 10.10 Directions of active and passive forces when wall friction is present.

During compaction of the backfill soil, additional lateral stresses are imposed by compaction equipment. For walls with sloping backfill, heavy compaction equipment is generally used. The wall pressures from using heavy compaction equipment can be in excess of the active earth pressures. You should account for the lateral stresses in designing retaining walls. You may refer to Ingold (1979), who used elastic theory to estimate the lateral stresses imposed by construction equipment. Some practicing geotechnical engineers prefer to account for the additional compaction stresses by assuming that the resultant active lateral earth force acts at 0.4 H_o or 0.5 H_o rather than $\frac{1}{3}$ H_o from the base of the wall. Alternatively, you can multiply the active earth pressures by a factor ($\simeq$1.20) to account for compaction stresses.

The key points are:
1. ***Coulomb's analysis of the lateral forces on a retaining structure is based on limit equilibrium.***
2. ***Wall friction causes the slip planes to curve, which leads to an overestimation of the passive earth pressure using Coulomb's analysis.***
3. ***For calculation of the lateral earth pressure coefficients you can use Eqs. (10.21) and (10.22), and correct K_{pC} using the factors listed in Table 10.1.***
4. ***The active and passive forces are inclined at an angle δ from the normal to the wall face.***

10.5 RANKINE'S LATERAL EARTH PRESSURE FOR A SLOPING BACKFILL AND A SLOPING WALL FACE

Importance *Coulomb's analysis is based on a postulated failure mechanism and the stresses within the soil mass are not considered. Rankine (1857) proposed an analysis based on the stress state of the soil. You have already encountered his approach in Section 10.2. In this section, we introduce Rankine's solution for a wall with a sloping backfill and a sloping wall face.*

Access a spreadsheet at http://www.wiley.com/college/budhu. Click Foundations and Earth Structures, Chapter 10, retwall.xls. Click on worksheet "*kakp*" to calculate the active and passive coefficients for the Rankine method.

Rankine (1857) established the principle of stress states or stress field in solving stability problems in soil mechanics. We have used Rankine's method in developing the lateral earth pressures for a vertical, frictionless wall supporting a dry, homogeneous soil with a horizontal surface. Rankine (1857) derived expressions for K_a and K_p for a soil mass with a sloping surface that was later extended to include a sloping wall face by Chu (1991). You can refer to Rankine's paper and Chu's paper for the mathematical details.

With reference to Fig. 10.11, the lateral earth pressure coefficients according to Rankine's analysis are

$$K_{aR} = \frac{\cos(\beta - \eta)\sqrt{1 + \sin^2\phi' - 2\sin\phi'\cos\varphi_a}}{\cos^2\eta(\cos\beta + \sqrt{\sin^2\phi' - \sin^2\beta})} \tag{10.24}$$

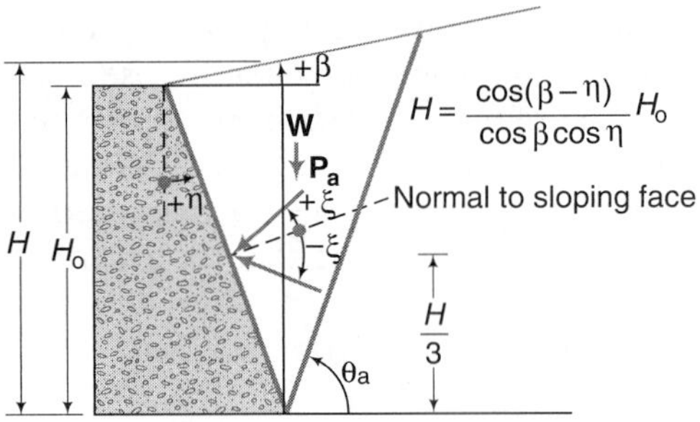

FIGURE 10.11 Retaining wall with sloping soil surface, frictionless soil-wall interface, and sloping back for use with Rankine's method.

and

$$K_{pR} = \frac{\cos(\beta - \eta)\sqrt{1 + \sin^2\phi' + 2\sin\phi'\cos\varphi_p}}{\cos^2\eta(\cos\beta - \sqrt{\sin^2\phi' - \sin^2\beta})} \tag{10.25}$$

where the subscript R denotes Rankine, and

$$\varphi_a = \sin^{-1}\left(\frac{\sin\beta}{\sin\phi'}\right) - \beta + 2\eta \tag{10.26}$$

$$\varphi_p = \sin^{-1}\left(\frac{\sin\beta}{\sin\phi'}\right) + \beta - 2\eta \tag{10.27}$$

The angles of the failure plane to the horizontal are

$$\theta_a = \frac{\pi}{4} + \frac{\phi'}{2} + \frac{\beta}{2} - \frac{1}{2}\sin^{-1}\left(\frac{\sin\beta}{\sin\phi'}\right) \tag{10.28}$$

$$\theta_p = \frac{\pi}{4} - \frac{\phi'}{2} + \frac{\beta}{2} + \frac{1}{2}\sin^{-1}\left(\frac{\sin\beta}{\sin\phi'}\right) \tag{10.29}$$

The sign conventions for η and β are shown in Fig. 10.11. Counterclockwise rotation is positive.

The active and passive lateral earth forces are

$$P_a = \frac{1}{2}K_{aR}\gamma H_o^2 \quad \text{and} \quad P_p = \frac{1}{2}K_{pR}\gamma H_o^2$$

These forces are inclined at

$$\xi_a = \tan^{-1}\left(\frac{\sin\phi'\sin\varphi_a}{1 - \sin\phi'\cos\varphi_a}\right) \tag{10.30}$$

$$\xi_p = \tan^{-1}\left(\frac{\sin\phi'\sin\varphi_p}{1 + \sin\phi'\cos\varphi_p}\right) \tag{10.31}$$

to the normal of the wall face. The angles ξ_a and ξ_p (reminder: the subscripts a and p denote active and passive) are not interface friction angles.

In the case of a wall with a vertical face, $\eta = 0$, Eqs. (10.24) and (10.25) reduce to

$$K_{aR} = \frac{1}{K_{pR}} = \cos\beta\left(\frac{\cos\beta - \sqrt{\cos^2\beta - \cos^2\phi'}}{\cos\beta + \sqrt{\cos^2\beta - \cos^2\phi'}}\right) \tag{10.32}$$

and the active and passive lateral earth forces act in a direction parallel to the soil surface, that is, they are inclined at an angle β to the horizontal. Thus, $P_{ax} = P_a\cos\beta$ and $P_{az} = P_a\sin\beta$. Remember that K_{aR} and K_{pR} are applied to the vertical effective not vertical total stresses.

The **key points** *are:*

1. ***Rankine used stress states of a soil mass to determine the lateral earth pressures on a frictionless wall.***
2. ***The active and passive lateral earth forces are inclined at ξ_a and ξ_p, respectively, from the normal to the wall face. If the wall face is vertical, the active and passive lateral earth forces are parallel to the soil surface.***

10.6 LATERAL EARTH PRESSURES FOR A TOTAL STRESS ANALYSIS

Importance *We considered the lateral earth pressures based on an effective stress analysis. For fine grained soil we need to also consider short-term loading conditions (total stress analysis). In this section, we will consider total stress analysis.*

Figure 10.12 shows a smooth, vertical wall supporting a homogeneous soil mass under undrained conditions. Using the limit equilibrium method, we will assume, for the active state, that a slip plane is formed at an angle θ to the horizontal. The forces on the soil wedge are shown in Fig. 10.12.

Using static equilibrium, we obtain the sum of the forces along the slip plane:

$$P_a \cos\theta + T - W \sin\theta = 0 \tag{10.33}$$

But $T = s_u L = s_u(H_o/\sin\theta)$ and $W = \frac{1}{2}\gamma H_o^2 \cot\theta$. Equation (10.33) then yields

$$P_a = \frac{1}{2}\gamma H_o^2 - \frac{s_u H_o}{\sin\theta\, \cos\theta} = \frac{1}{2}\gamma H_o^2 - \frac{2 s_u H_o}{\sin 2\theta}$$

To find the maximum active lateral earth force, we differentiate P_a with respect to θ and set the result equal to zero, giving

$$\frac{\partial P_a}{\partial \theta} = 4 s_u H_o \cot 2\theta \csc 2\theta = 0$$

The solution is $\theta = \theta_a = 45^\circ$.

By substituting $\theta = 45^\circ$ into the above equation for P_a, we get the maximum active lateral earth force as

$$P_a = \frac{1}{2}\gamma H_o^2 - 2 s_u H_o \tag{10.34}$$

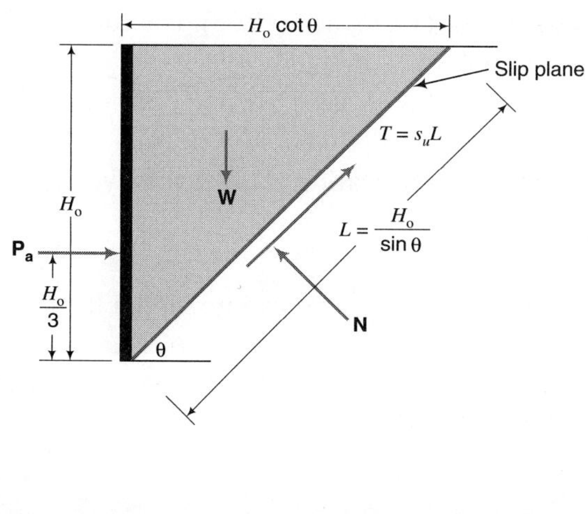

FIGURE 10.12 Forces on a retaining wall for a total stress analysis.

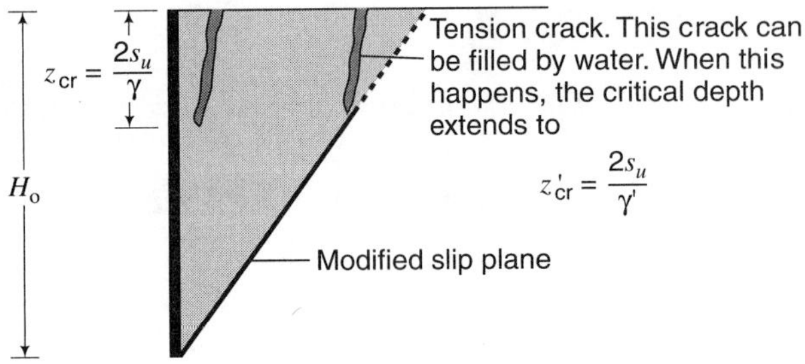

FIGURE 10.13 Tension cracks behind a retaining wall.

If we assume a uniform distribution of stresses on the slip plane, then the active lateral stress is

$$(\sigma_x)_a = \gamma z - 2s_u \tag{10.35}$$

Let us examine Eq. (10.35). If $(\sigma_x)_a = 0$, for example, when you make an excavation, then solving for z in Eq. (10.35) gives

$$z = z_{cr} = \frac{2s_u}{\gamma} \tag{10.36}$$

Depth z_{cr} is the depth at which tension cracks would extend into the soil (Fig. 10.13). If the tension crack is filled with water, the critical depth can extend to

$$z'_{cr} = \frac{2s_u}{\gamma'} \tag{10.37}$$

In addition, the soil in the vicinity of the crack is softened and a hydrostatic pressure, $\gamma_w z'_{cr}$, is imposed on the wall. Often, the critical depth of water-filled tension cracks in overconsolidated clays is greater than the wall height. For example if $s_u = 80\,\text{kPa}$, $\gamma = \gamma_{sat} = 18\,\text{kN/m}^3$, then $z'_{cr} = 19.5\,\text{m}$. A wall height equivalent to the depth of the tension crack of 19.5 m is substantial. As a comparison, the average height of the Great Wall of China is about 12 m. When tension cracks occur they modify the slip plane, as shown in Fig. 10.13; no shearing resistance is available over the length of the slip plane above the depth of the tension cracks.

For an unsupported excavation, the active lateral force is also zero. From Eq. (10.34), we get

$$\frac{1}{2}\gamma H_o^2 - 2s_u H_o = 0$$

and, solving for H_o, we obtain

$$H_o = H_{cr} = \frac{4s_u}{\gamma} \tag{10.38}$$

If the excavation is filled with water, then

$$H'_{cr} = \frac{4s_u}{\gamma'} \tag{10.39}$$

We have two possible unsupported depths as given by Eqs. (10.36) and (10.38). The correct solution lies somewhere between these critical depths. In design practice, a value of

$$\frac{3.6s_u}{\gamma} \leq H_{cr} \leq \frac{3.8s_u}{\gamma} \tag{10.40}$$

is used for unsupported excavation in fine-grained soils. If the excavation is filled with water,

$$\frac{3.6s_u}{\gamma'} \leq H'_{cr} \leq \frac{3.8s_u}{\gamma'} \tag{10.41}$$

The passive lateral earth force for a total stress analysis, following a procedure similar to that for the active state above, can be written as

$$P_p = \frac{1}{2}\gamma H_o^2 + 2s_u H_o \tag{10.42}$$

and the passive lateral pressure is

$$(\sigma_x)_p = \gamma z + 2s_u \tag{10.43}$$

We can write Eqs. (10.35) and (10.43) using apparent active and passive lateral earth pressures for the undrained condition as

$$(\sigma_x)_a = \sigma_z - K_{au}s_u \tag{10.44}$$

$$(\sigma_x)_p = \sigma_z + K_{pu}s_u \tag{10.45}$$

where K_{au} and K_{pu} are the undrained active and passive lateral earth pressure coefficients. In our case, for a smooth wall supporting a soil mass with a horizontal surface, $K_{au} = K_{pu} = 2$.

Walls that are embedded in fine-grained soils may be subjected to an adhesive stress (s_w) at the wall face. The adhesive stress is analogous to a wall–soil interface friction for an effective stress analysis. The undrained lateral earth pressure coefficients are modified to account for adhesive stress as

$$K_{au} = K_{pu} = 2\sqrt{1 + \frac{s_w}{s_u}} \tag{10.46}$$

The **key points** *are:*

1. ***Lateral earth pressures for a total stress analysis are found using apparent lateral earth pressure coefficients*** $\mathbf{K_{au}}$ ***and*** $\mathbf{K_{pu}}$***. These coefficients are applied to the undrained shear strength. For smooth, vertical walls,*** $\mathbf{K_{au} = K_{pu} = 2}$***.***
2. ***Tension cracks of theoretical depth*** $2s_u/\gamma$***, or*** $2s_u/\gamma'$ ***if water fills the tension cracks, are usually formed in fine-grained soils and they modify the slip plane. If water fills the cracks, it softens the soil and a hydrostatic stress is imposed on the wall over the depth of the tension crack. You must pay particular attention to the possibility of the formation of tension cracks, and especially so if these cracks can be filled with water, because they can initiate failure of a retaining structure.***
3. ***The theoretical maximum depth of an unsupported vertical cut in fine-grained soils is*** $\mathbf{H_{cr} = 4s_u/\gamma}$ ***or, if the cut is filled with water,*** $\mathbf{H'_{cr} = 4s_u/\gamma'}$***.***

10.7 APPLICATION OF LATERAL EARTH PRESSURES TO RETAINING WALLS

Importance *You were introduced to Coulomb's and Rankine's analyses of the active and passive lateral earth pressures. Concerns were raised regarding the accuracy of, in particular,* K_{pC} *because wall friction causes the failure surface to diverge from a plane surface. Several investigators have proposed values of* K_p*, assuming curved failure surfaces. The question that arises is: "What values of* K_a*,* K_p*,* ϕ'*, and* s_u *should be used in the analyses of earth retaining structures? In this section, we will attempt to address this question. You are forewarned that the answer will not be definite. A geotechnical engineer usually has his or her preferences based on his or her experience with a particular method.*

Field and laboratory tests have not confirmed the Coulomb and Rankine theories. In particular, field and laboratory test results showed that both theories overestimate the passive lateral earth pressures. Values of K_p obtained by Caquot and Kerisel (1948) lower the passive lateral earth pressures but they are still higher than experimental results. Other theories, for example, plasticity theory (Rosenfarb and Chen, 1972) have been proposed, but these theories are beyond the scope of this book and they too do not significantly change the Coulomb and Rankine passive lateral earth pressures for practical ranges of friction angle and wall geometry.

Rankine's theory was developed based on a semi-infinite "loose granular" soil mass for which the soil movement is uniform. Retaining walls do not support a semi-infinite mass but a soil mass of fixed depth. Strains, in general, are not uniform in the soil mass unless the wall rotates about its base to induce a state of plastic equilibrium.

The strains required to achieve the passive state are much larger than for the active case (Fig. 10.4). For sands, a decrease in lateral earth pressure of 40% of the at-rest lateral earth pressure can be sufficient to reach an active state, but an increase of several hundred percent in lateral earth pressure over the at-rest lateral earth pressure is required to bring the soil to a passive state. Because of the large strains that occur to achieve the passive state, it is customary to apply a factor of safety to the passive lateral earth pressure.

We have assumed a generic friction angle for the soil mass. Backfills are usually coarse-grained soils compacted to greater than 95% Proctor dry unit weight. If samples of the backfill were to be tested in shear tests in the laboratory at the desired degree of compaction, the samples may show peak shear stresses resulting in ϕ'_p. If you use ϕ'_p to estimate the passive lateral earth pressure using either the Coulomb or Rankine method, you are likely to overestimate it because the shear strains required to develop the passive lateral earth pressure are much greater than those required to mobilize ϕ'_p in the lab. For granular materials, ϕ'_p is mobilized at shear strains $\approx 2\%$. The use of ϕ'_p in the Rankine or Coulomb equations is one reason for the disagreement between the predicted passive lateral earth pressures and experimental results.

Bolton and Powrie (1988) demonstrated that a wall rotation of $0.005\,H_o$ induces tensile and compressive strains of ± 0.005 corresponding to an average shear strain, $\gamma \approx 1\%$. Tolerable wall rotations are generally less than $0.005\,H_o$. Consequently, ϕ'_p (which requires $\gamma \approx 2\%$ to be mobilized for granular soils) would not be mobilized.

Large shear strains ($\gamma > 10\%$) are required to mobilize ϕ'_{cs}. For a backfill consisting of loose, coarse-grained soils, the displacement of the wall required to mobilize ϕ'_{cs} is intolerable in practice. You should then use conservative values of ϕ' in design. The maximum ϕ'_{design} should be ϕ'_{cs}. In practice, factors of safety are applied to allow for uncertainties of loads and soil properties.

Sometimes, it is desirable that the wall remains rigid because of nearby structures that are sensitive to lateral displacements. In this case, you should use K_o as the lateral earth pressure coefficient.

Wall friction causes the active lateral earth pressures to decrease and the passive lateral earth pressures to increase. For active lateral earth pressures, the Coulomb equation is sufficiently accurate for practice, but for passive lateral earth pressures you should use the Caquot and Kerisel (1948) values or similar values (e.g., Packshaw, 1969).

The total stress analysis should only be used in temporary works. But even for these works, you should be cautious in relying on the undrained shear strength. A total stress analysis should be used in conjunction with an effective stress analysis for retaining structures supporting fine-grained soils.

When a wet clay soil is excavated, negative excess porewater pressures develop and give the soil a greater undrained strength than prior to excavation. You should not rely on this gain in strength because the excess porewater pressures will dissipate with time. On inserting a wall, the soil at the wall-soil interface is remolded and you should use conservative values of wall adhesion. The maximum values of wall adhesion (Padfield and Mair, 1984) should be the lesser of

$$\text{Active state:} \quad s_w = 0.5 s_u \quad \text{or} \quad s_w \leq 50\,\text{kPa} \tag{10.47}$$

$$\text{Passive state:} \quad s_w = 0.5 s_u \quad \text{or} \quad s_w \leq 25\,\text{kPa} \tag{10.48}$$

The interface friction between the wall face and the soil depends on the type of backfill used and construction methods. If the surface texture of the wall is rougher than D_{50} of the backfill, the strength characteristics of backfill would control the interface friction. In such a case, the interface friction angle can be taken as equivalent to ϕ'_{cs}. If the wall surface is smooth compared with D_{50} of the backfill, the interface friction value can be assumed, in the absence of field measurements, to be between $\frac{2}{3}\phi'_{cs}$ and $\frac{1}{2}\phi'_{cs}$.

During the life of a retaining wall, a gap may open between the wall and the backfill due to wall movement, deficient construction method, and unanticipated environmental conditions (drought, rainfall, floods, etc.). You need to be cautious in using adhesion and interface friction because these may become nonexistent during the life of a wall. More importantly, water can seep through the gap and subject the wall to hydrostatic pressures. When a gap could develop, use hydrostatic pressure rather than the lateral soil pressure.

A layer of coarse-grained soil is often used in construction to rest the base of gravity retaining walls (see Section 10.8) founded on clays. The interface friction angle for sliding would then be the lesser of the interface friction between the layer of coarse-grained soil and the wall base, and the interface friction between the layer of coarse-grained soil and the clay.

In summary, you should use:

1. $\phi'_{design} \leq \phi'_{cs}$ because:

a. Uncertainties of loads and soil properties exist.

b. Tolerable rotations of walls ($< 0.005\,H_o$) would not normally mobilize peak shear strength or critical state shear strength.

c. Prior to wall completion, shear bands (thin zones of soil that reached critical state) in the compacted backfill may develop.

d. Quality of construction methods and construction loading, environmental conditions (floods, heavy rainfall, etc.), human and animal interventions (excavation at toe, dumping of materials on top of wall, burrowing of holes, etc.) cannot be estimated (at least, accurately) in advance.

e. You should design the wall so that the backfill soil behaves in a ductile manner. Using ϕ'_{cs} or lower values would allow the soil to respond in a ductile manner.

2. Total stress analysis in conjunction with an effective stress analysis for fine-grained soils.

3. Conservative values for wall friction ($\delta \approx \frac{1}{2}\phi'_{cs}$) and wall adhesion ($s_w = 0.5s_u$ but ≤ 50 kPa for active state and $s_w \leq 25$ kPa for passive case).

4. High-quality drainage system to drain water from the backfill and away from the wall.

10.8 TYPES OF RETAINING WALLS AND MODES OF FAILURE

Importance *In the next three sections, we will analyze retaining walls to determine their stability. We will consider an ESA (effective stress analysis) and a TSA (total stress analysis). We begin by considering the possible failure modes.*

There are two general classes of retaining walls. One class is rigid and consists of concrete walls relying on gravity for stability (Fig. 10.14). These are called cast-in-place (CIP) gravity and semi-gravity walls. Typical initial design dimensions are shown in Figs. 10.14(e) and 10.14(f). The other class is flexible and consists of long slender members of either steel or concrete or wood and relies on passive soil resistance and anchors for stability (Fig. 10.15).

There are four modes of failures for rigid retaining walls—translational failure, rotation and bearing capacity failure, deep-seated failure, and structural failure (Fig. 10.16). Flexible walls, also called

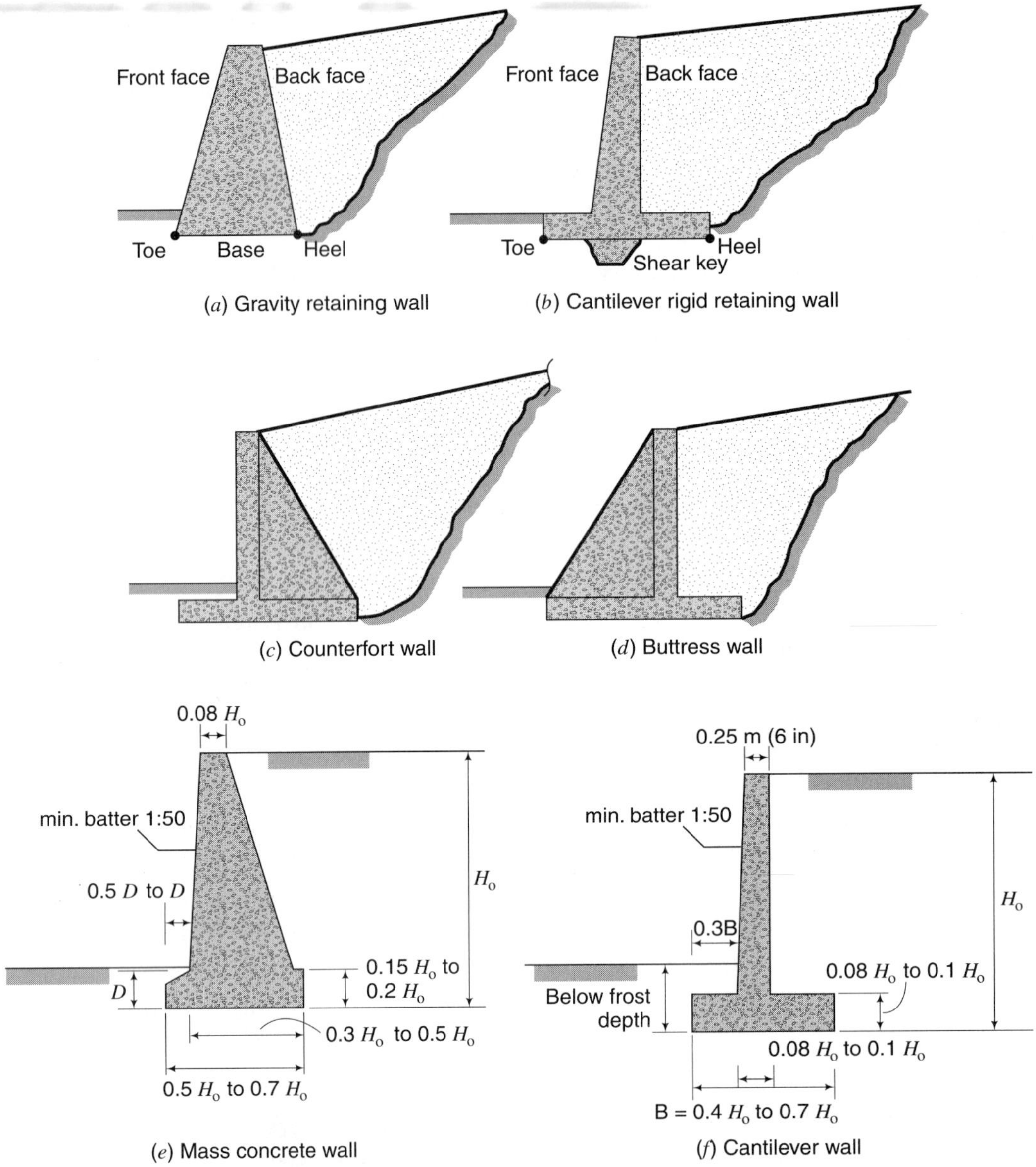

FIGURE 10.14 Types of rigid retaining walls.

sheet pile walls, failed either by deep-seated failure, rotation at the base, rotation about the anchor or prop, failure of the anchor, bending of the wall, or seepage-induced failure (Fig. 10.17).

Seepage-induced failure is avoided in rigid retaining walls by providing adequate drainage systems, two of which are depicted in Fig. 10.18. The design of drainage systems is beyond the scope of this book. Flow nets (discussed in Chapter 1) are used in designing drainage systems.

Flexible retaining walls are often used in waterfront structures and as temporary supports for excavations. Seepage forces are generally present and must be considered in evaluating the stability of these walls.

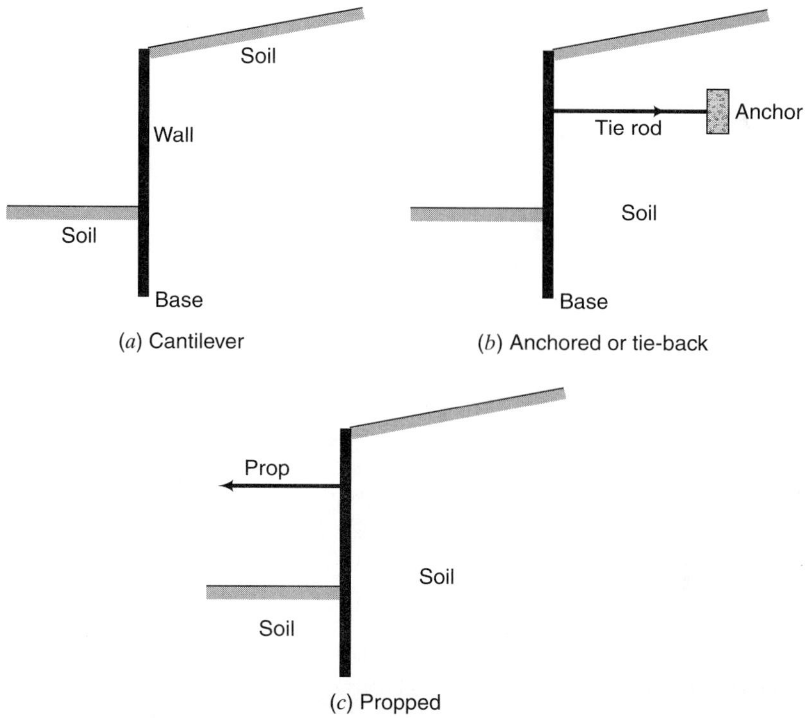

FIGURE 10.15 Types of flexible retaining walls.

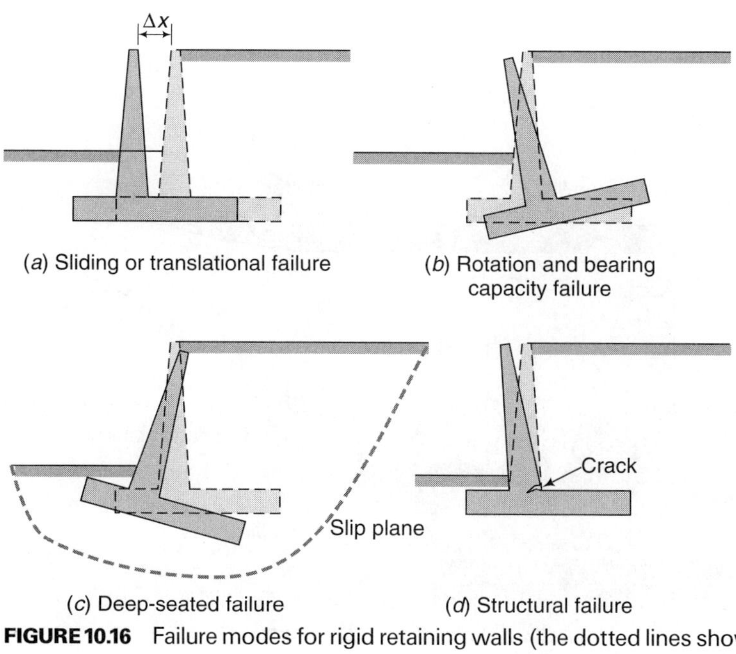

FIGURE 10.16 Failure modes for rigid retaining walls (the dotted lines show the original position of the wall).

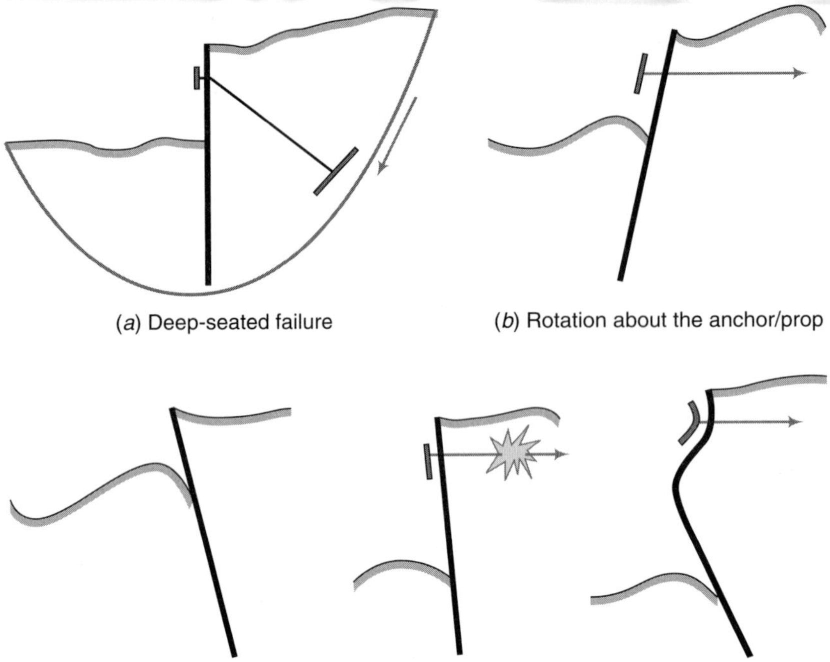

(*a*) Deep-seated failure (*b*) Rotation about the anchor/prop

(*c*) Rotation near base (*d*) Failure of anchor/prop (*e*) Failure by bending

FIGURE 10.17 Failure modes for flexible retaining walls.

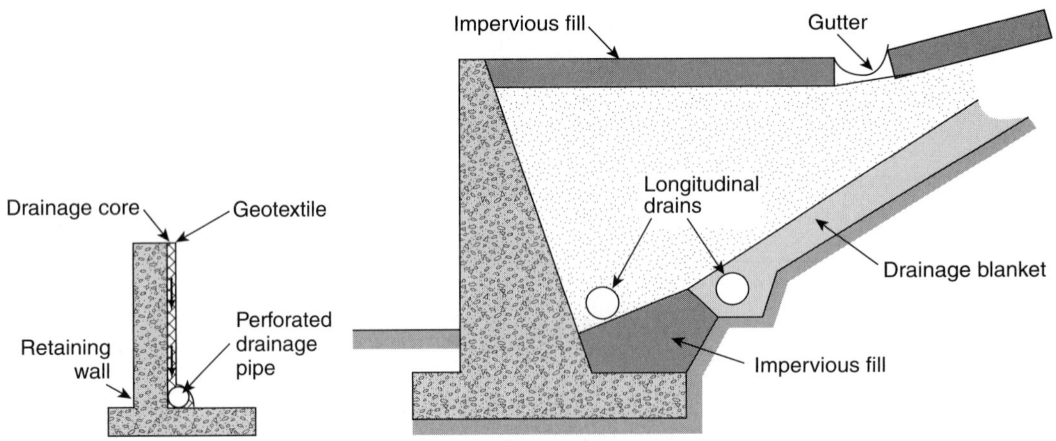

(*a*) Simple vertical drain (*b*) Inclined drain (after Sibley, 1967)

FIGURE 10.18 Two types of drainage system behind rigid retaining walls.

10.9 STABILITY OF RIGID RETAINING WALLS

Importance *Rigid retaining walls are prevalent in civil works, especially for bridge abutments and along highways. In this section, we consider the design of these walls.*

CIP gravity retaining walls (Fig. 10.14a) are massive concrete walls. Their stability depends mainly on the self-weight of the walls. Cantilever walls (Fig. 10.14b)—CIP semigravity walls—utilize the backfill to help mobilize stability and are generally more economical than CIP gravity retaining walls. A rigid retaining wall must have an adequate factor of safety to prevent excessive translation, rotation, bearing capacity failure, deep-seated failure, and seepage-induced instability.

10.9.1 Translation

A rigid retaining wall must have adequate resistance against translation. That is, the sliding resistance at the base of the wall must be greater than the resultant lateral force pushing against the wall. The factor of safety against translation, $(\mathrm{FS})_T$, is

$$(\mathrm{FS})_T = \frac{T}{P_{ax}}; \quad (\mathrm{FS})_T \geq 1.5 \tag{10.49}$$

where T is the sliding resistance at the base and P_{ax} is the lateral force pushing against the wall. The sliding resistance is $T = R_z \tan \phi'_b$ for an ESA, and $T = s_w B$ for a TSA (if the base rests directly on fine-grained soils). R_z is the resultant vertical force, ϕ'_b is the interfacial friction angle between the base of the wall and the soil

$$\phi'_b \approx \frac{1}{2}\phi'_{cs} \text{ to } \frac{2}{3}\phi'_{cs}$$

and B is the projected horizontal width of the base. Typical sets of forces acting on gravity and cantilever rigid retaining walls are shown in Fig. 10.19.

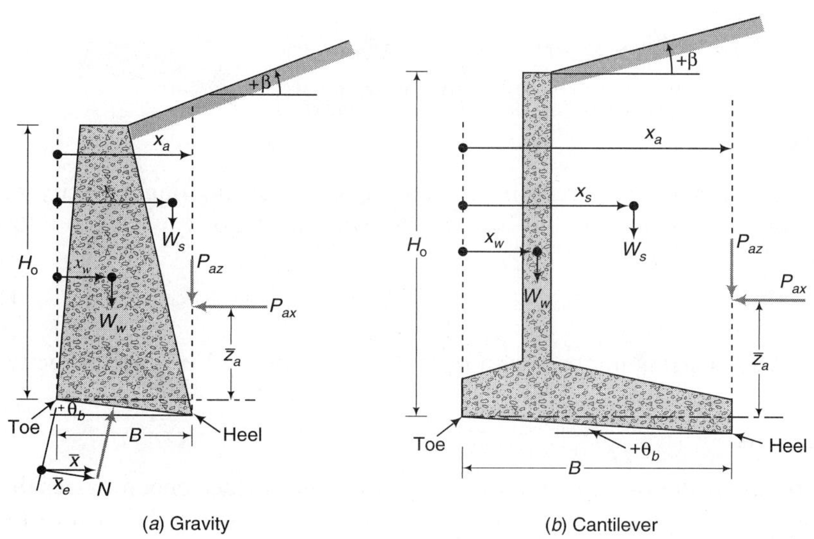

FIGURE 10.19 Forces on rigid retaining walls.

Using statics, we obtain, for an ESA,

$$(\mathrm{FS})_T = \frac{[(W_w + W_s + P_{az})\cos\theta_b - P_{ax}\sin\theta_b]\tan\phi_b'}{P_{ax}\cos\theta_b + (W_w + W_s + P_{az})\sin\theta_b} \tag{10.50}$$

where W_w is the weight of the wall, W_s is the weight of the soil wedge, P_{az} and P_{ax} are the vertical and horizontal components of the active lateral force, and θ_b is the inclination of the base to the horizontal (θ_b is positive if the inclination is counterclockwise, as shown in Fig. 10.19). If $\theta_b = 0$ (base is horizontal), then

$$(\mathrm{FS})_T = \frac{(W_w + W_s + P_{az})\tan\phi_b'}{P_{ax}} \tag{10.51}$$

For a TSA,

$$(\mathrm{FS})_T = \frac{s_w B/\cos\theta_b}{P_{ax}\cos\theta_b + (W_w + W_s + P_{ax})\sin\theta_b} \tag{10.52}$$

where s_w is the adhesive stress (see Eqs. 10.47 and 10.48)
If $\theta_b = 0$, then

$$(\mathrm{FS})_T = \frac{s_w B}{P_{ax}} \tag{10.53}$$

The embedment of rigid retaining walls is generally small and the passive lateral force is not taken into account. If base resistance is inadequate, the width B of the wall can be increased. For cantilever walls, a shear key (Fig. 10.14b) can be constructed to provide additional base resistance against sliding.

10.9.2 Rotation

A rigid retaining wall must have adequate resistance against rotation. The rotation of the wall about its toe is satisfied if the resultant vertical force lies within the middle third of the base. Taking moments about the toe of the base, the resultant vertical force at the base is located at

$$\bar{x}_\theta = \frac{W_w x_w + W_s x_s + P_{az} x_a - P_{ax}\bar{z}_a}{(W_w + W_s + P_{az})\cos\theta_b - P_{ax}\sin\theta_b} \tag{10.54}$$

where $\bar{z}_a$ is the location of the active lateral earth force from the toe. The wall is safe against rotation if $B/3 \le \bar{x} \le 2B/3$; that is, $e = |(B/2 - \bar{x})| \le B/6$, where e is the eccentricity of the resultant vertical load and $\bar{x} = \bar{x}_\theta \cos\theta_b$ (Fig. 10.19a).

If $\theta_b = 0$, then

$$\bar{x} = \frac{W_w x_w + W_s x_s + P_{az} x_a - P_{ax}\bar{z}_a}{W_w + W_s + P_{az}} \tag{10.55}$$

10.9.3 Bearing Capacity

A rigid retaining wall must have a sufficient margin of safety against soil bearing capacity failure. The maximum pressure imposed on the soil at the base of the wall must not exceed the allowable soil bearing capacity; that is,

$$\sigma_{max} \le q_a \tag{10.56}$$

where σ_{max} is the maximum vertical stress imposed and q_a is the allowable soil bearing capacity.

10.9.4 Deep-Seated Failure

A rigid retaining wall must not fail by deep-seated failure, where by a slip surface encompasses the wall and the soil adjacent to it. Several computer programs are available to determine the factor of safety against deep-seated failure.

10.9.5 Seepage

A rigid retaining wall must have adequate protection from groundwater seepage. The porewater pressures and the maximum hydraulic gradient developed under seepage must not cause any of the four stability criteria stated above to be violated and static liquefaction must not occur, that is, $i_{max} < i_{cr}$ (Chapter 1). Usually,

$$i_{max} \leq i_{cr}/(FS)_s \tag{10.57}$$

where $(FS)_s$ is a factor of safety for seepage and is conventionally greater than 3. To avoid seepage-related failures, adequate drainage should be installed in the backfill to dissipate excess porewater pressures quickly. Coarse-grained soils are preferable for the backfill because of their superior drainage characteristics compared with fine-grained soils.

10.9.6 Procedure to Analyze Gravity Retaining Walls

The essential steps in determining the stability of rigid retaining walls are as follows:

1. Calculate the active lateral earth force and its components. If the wall is smooth, use Rankine's equations because they are simpler than Coulomb's equations to calculate the active lateral earth pressure coefficient.
2. Determine the weight of the wall and soil above the base.
3. Use Eq. (10.50) or Eq. (10.51) or Eq. (10.52) to find $(FS)_T$.
4. Use Eq. (10.54) or Eq. (10.55) to determine the location of the resultant vertical force, R_z, from the toe of the wall.
5. Check that the eccentricity is less than $B/6$. If it is, then the wall is unlikely to fail by rotation.
6. Determine the maximum soil pressure from $\sigma_{max} = (R_z/A)(1 + 6e/B)$.
7. Calculate the ultimate net bearing capacity q_u, described in Chapter 7. In most cases, R_z would be eccentric.
8. Calculate the factor of safety against bearing capacity failure: $(FS)_B = q_u/\sigma_{max}$.

EXAMPLE 10.3 ***Mass Gravity Wall***

Access http://www.wiley.com/college/budhu. Click Foundations and Earth Structures, Chapter 10, retwall.xls and click the worksheet "CIP" to analyze CIP walls.

A gravity retaining wall, shown in Fig. E10.3a, is required to retain 5 m (16 ft) of soil. The backfill is a coarse-grained soil with $\gamma_{sat} = 18\,kN/m^3$ (113 pcf), $\phi'_{cs} = 30^\circ$. The existing soil (below the base) has the following properties: $\gamma_{sat} = 20\,kN/m^3$ (125 pcf), $\phi'_p = 36^\circ$ (plane strain test). The wall is embedded 1 m (3 ft) into the existing soil and a drainage system is provided as shown. The groundwater level is 4.5 m (15 ft) below the base of the wall. Determine the stability of the wall for the following conditions: Assume $\delta = 20^\circ$.

(a) Wall friction is zero.

(b) Wall friction is 20°.

(c) The drainage system becomes clogged during several days of a rainstorm and the groundwater rises to the surface. Neglect seepage forces.

The unit weight of concrete is $\gamma_c = 24\,kN/m^3$ (150 pcf).

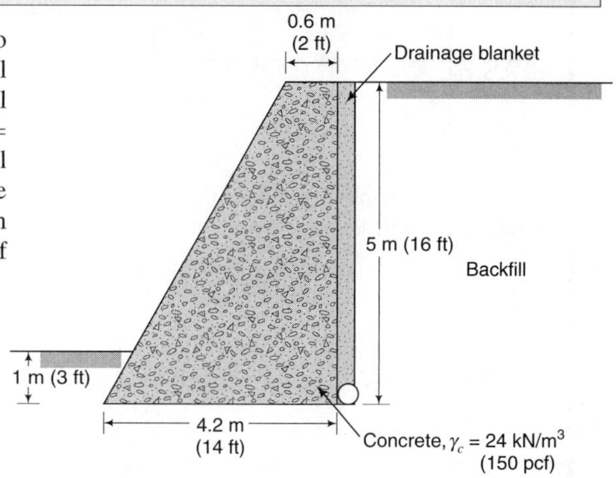

FIGURE E10.3a

Strategy For zero wall friction, you can use Rankine's method. But for wall friction, you should use Coulomb's method. The passive resistance is normally neglected in rigid retaining walls. Since only active lateral forces are considered, K_a from the Rankine and Coulomb methods should be accurate enough. Since groundwater is below the base, $\gamma' = \gamma_{sat}$ over the wall depth. U.S. units will be used in this example.

Solution 10.3

Step 1: Determine K_a.

$$\text{Rankine}: \quad \delta = 0, \quad K_{aR} = \tan^2\left(45^\circ - \frac{\phi_{cs}}{2}\right) = \tan^2\left(45^\circ - \frac{30^\circ}{2}\right) = \frac{1}{3}$$

$$\text{Coulomb}: \quad \delta = 20^\circ, \quad \phi' = \phi'_{cs}, \quad \beta = 0, \quad \eta = 0, \quad \text{and from Eq. (10.21)}$$

$$K_{aC} = \frac{\cos^2(30^\circ - 0^\circ)}{\cos^2 0^\circ \cos(0^\circ + 20^\circ)\left[1 + \left\{\dfrac{\sin(30^\circ + 20^\circ)\sin(30^\circ - 0^\circ)}{\cos(0^\circ + 20^\circ)\cos(0^\circ - 0^\circ)}\right\}^{1/2}\right]^2} = 0.3$$

Step 2: Determine the lateral forces.
All forces are per unit length of wall.

$$\text{Rankine}: \; P_{aR} = \frac{1}{2}K_{aR}\gamma_{sat}H_o^2 = \frac{1}{2} \times \frac{1}{3} \times 113 \times 16^2 = 4821\ \text{lb}$$

P_{aR} acts horizontally because the ground surface is horizontal.

$$\text{Coulomb}: \; P_{aC} = \frac{1}{2}K_{aC}\gamma_{sat}H_o^2 = \frac{1}{2} \times 0.3 \times 113 \times 16^2 = 4339\ \text{lb}$$

P_{aC} acts at an angle $\delta = 20^\circ$ to the horizontal (see Fig. E10.3b).

$$\text{Horizontal component of } P_{aC}: \quad (P_{ax})_C = P_{aC}\cos\delta = 4339\cos 20^\circ = 4077\ \text{lb}$$

$$\text{Vertical component of } P_{aC}: \quad (P_{az})_C = P_{aC}\sin\delta = 4339\sin 20^\circ = 1484\ \text{lb}$$

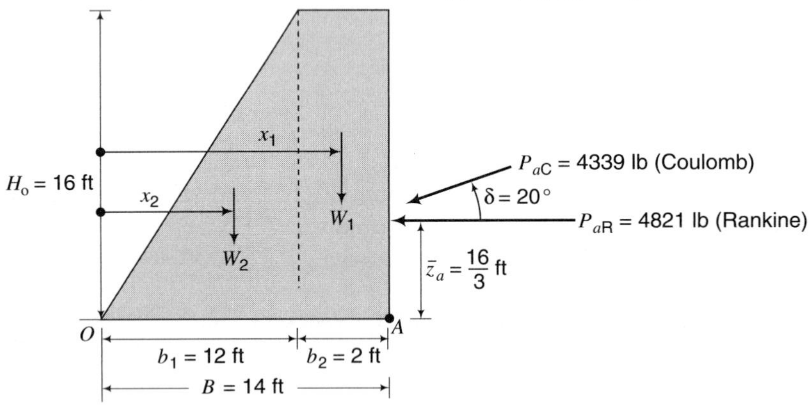

FIGURE E10.3b

Step 3: Determine wall stability.
Consider a unit length of wall.

$$W_1 = b_2H_o\gamma_c = 2 \times 16 \times 150 = 4800\ \text{lb}$$

$$W_2 = \frac{1}{2}b_1H_o\gamma_c = \frac{1}{2} \times 12 \times 16 \times 150 = 14400\ \text{lb}$$

$$W = W_1 + W_2 = 4800 + 14400 = 19200\ \text{lb}$$

or

$$W = \frac{1}{2}(B + b_2)H_o\gamma_c = \frac{1}{2}(14 + 2) \times 16 \times 150 = 19200\ \text{lb}$$

Calculate the location of the resultant from O (Fig. E10.3b).

$$\text{Rankine}: \quad M_O = W_1x_1 + W_2x_2 - P_{aR}\bar{z}_a = 4800(12+1) + 14400 \times \left(\tfrac{2}{3} \times 12\right) - 4821 \times \frac{16}{3} = 151888 \text{ lb.ft}$$

$$R_z = W = 19200 \text{ lb}$$

$$\bar{x} = \frac{M_O}{R_z} = \frac{151888}{19200} = 7.91 \text{ ft}$$

$$\text{Coulomb}: \quad M_O = W_1x_1 + W_2x_2 + (P_{az})_C \times B - (P_{ax})_C \times \bar{z}_a$$

$$= 4800(12+1) + 14400 \times \left(\tfrac{2}{3} \times 12\right) + 1484 \times 14 - 4077 \times \frac{16}{3} = 176632 \text{ lb.ft}$$

$$R_z = W + (P_{az})_C = 19200 + 1484 = 20684 \text{ lb}$$

$$\bar{x} = \frac{M_O}{R_z} = \frac{176632}{20684} = 8.54 \text{ ft}$$

Base resistance : $T = R_z \tan \phi'_b$, where R_z is the resultant vertical force
Assume $\phi'_b = \frac{2}{3}\phi' = \frac{2}{3} \times 36^\circ = 24^\circ$

$$\text{Rankine}: \quad T = 19200 \times \tan 24^\circ = 8548 \text{ lb}$$

$$\text{Coulomb}: \quad T = 20684 \times \tan 24^\circ = 9289 \text{ lb}$$

$$\text{Rankine}: \quad (\text{FS})_T = \frac{T}{P_{aR}} = \frac{8548}{4821} = 1.8 > 1.5 \text{ therefore satisfactory}$$

$$\text{Coulomb}: \quad (\text{FS})_T = \frac{T}{(P_{ax})_C} = \frac{9289}{4077} = 2.3 > 1.5 \text{ therefore satisfactory}$$

With wall friction, the factor of safety against translation is greater than without friction.

DETERMINE ROTATIONAL STABILITY

$$\text{Rankine}: e = \left|\frac{B}{2} - \bar{x}\right| = \left|\frac{14}{2} - 7.91\right| = 0.91 \text{ ft}$$

$$\text{Coulomb}: e = \left|\frac{B}{2} - \bar{x}\right| = \left|\frac{14}{2} - 8.54\right| = 1.54 \text{ ft}$$

$$\frac{B}{6} = \frac{14}{6} = 2.33 \text{ ft} > e$$

The resultant vertical forces for both Rankine and Coulomb method lie within the middle one-third of the base and, therefore, overturning is unlikely to occur

DETERMINING FACTOR OF SAFETY AGAINST BEARING CAPACITY FAILURE

Since the resultant vertical force is located within the middle one-third, tension will not develop in the soil.

$$\sigma_{max} = \frac{R_z}{A}\left(1 + \frac{6e}{B}\right)$$

$$\text{Rankine}: (\sigma_{max})_R = \frac{19200}{14 \times 1}\left(1 + \frac{6 \times 0.91}{14}\right) = 1906 \text{ psf}$$

$$\text{Coulomb}: (\sigma_{max})_C = \frac{20684}{14 \times 1}\left(1 + \frac{6 \times 1.54}{14}\right) = 2453 \text{ psf}$$

The maximum stress occurs at A (Fig. E10.3b) for both the Rankine and Coulomb methods. The base of the wall can be taken as a strip, surface foundation, that is, $B/L \to 0$, and $D_f = 0$. The groundwater level is below $B = 14$ ft from the base, so groundwater would have no effect on the bearing capacity.

The resultant force, R_z, is eccentric and inclined to the vertical (Fig. E10.3c). Therefore, you should use the bearing capacity equation for inclined load with the width modified for eccentricity.

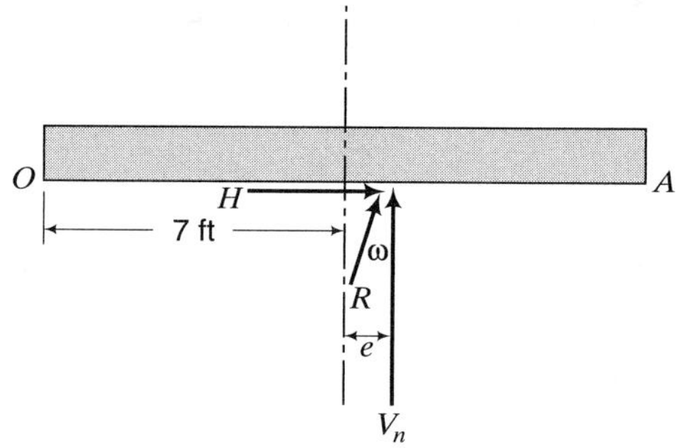

FIGURE E10.3c

$$\text{Rankine: } H = P_{aR} = 4821\,\text{lb}; \quad V_n = R_z = 19200\,\text{lb},$$

$$\frac{H}{V_n} = \frac{4821}{19200} = 0.25; \omega = \tan^{-1}\frac{H}{V_n} = \tan^{-1}(0.25) = 14^\circ$$

$$\frac{B'}{L'} = 0, \text{ for a strip footing; } B' = B - 2e = 14 - 2 \times 0.91 = 12.2\,\text{ft}$$

$$n = \left(2 + \frac{B'}{L'}\right) \Big/ \left(1 + \frac{B'}{L'}\right) = 2; \quad i_\gamma = \left(1 - \frac{H}{V_n}\right)^{n+1} = (1 - 0.25)^{2+1} = 0.42$$

$$N_\gamma = 0.1054 \exp(9.6\phi'_p) = 0.1054 \exp\left(9.6 \times 36 \times \frac{\pi}{80}\right) = 43.9$$

$$q_u = 0.5\gamma B' N_\gamma i_\gamma = 0.5 \times 125 \times 12.2 \times 43.9 \times 0.42 = 14059\,\text{psf}$$

$$(\text{FS})_B = \frac{q_u}{(\sigma_{max})_R} = \frac{14059}{1906} = 7.4 > 3; \textit{okay}$$

$$\text{Coulomb: } H = (P_{ax})_C = 4077\,\text{lb}; \quad V_n = R_z = 20684\,\text{lb}, \quad \frac{H}{V_n} = \frac{4077}{20684} = 0.2$$

$$\omega = \tan^{-1}\frac{H}{V_n} = \tan^{-1}(0.2) = 11.3^\circ$$

$$\frac{B'}{L'} = 0, \text{ for a strip footing; } B' = B - 2e = 14 - 2 \times 1.54 = 10.9\,\text{ft}$$

$$n = \left(2 + \frac{B'}{L'}\right) \Big/ \left(1 + \frac{B'}{L'}\right) = 2; \quad i_\gamma = \left(1 - \frac{H}{V_n}\right)^{n+1} = (1 - 0.20)^{2+1} = 0.51$$

$$q_u = 0.5\gamma B' N_\gamma i_\gamma = 0.5 \times 125 \times 10.9 \times 43.9 \times 0.51 = 15253\,\text{lb}$$

$$(\text{FS})_B = \frac{q_u}{(\sigma_{max})_C} = \frac{15253}{2453} = 6.2 > 3; \textit{okay}$$

Step 4: Determine the effects of water from the rainstorm.

Using Rankine's method (zero wall friction)

$$P_{aR} = \frac{1}{2}K_a\gamma' H_o^2 + \frac{1}{2}\gamma_w H_o^2 = \frac{1}{2} \times \frac{1}{3} \times (113 - 62.4) \times 16^2 + \frac{1}{2} \times 62.4 \times 16^2 = 10146\,\text{lb}$$

Location of resultant from *O*

$$M_O = W_1x_1 + W_2x_2 - P_{aR}\bar{z}_a = 4800(12 + 1) + 14400 \times \left(\frac{2}{3} \times 12\right) - 10146 \times \frac{16}{3} = 123488\,\text{lb.ft}$$

$$\bar{x} = \frac{123488}{19200} = 6.43\,\text{ft}$$

Translation

$$(\text{FS})_T = \frac{8548}{10146} = 0.84 < 1$$

The wall will fail by translation.

Rotation

$$e = \left|\frac{B}{2} - \bar{x}\right| = \left|\frac{14}{2} - 6.43\right| = 0.57 < \frac{B}{6}\left(= \frac{14}{6} = 2.33\right)$$

The wall is unlikely to fail by rotation.

Bearing capacity

$$\sigma_{max} = \frac{19200}{14 \times 1}\left(1 + \frac{6 \times 0.57}{14}\right) = 1706 \text{ psf}$$

The maximum stress now occurs at O rather than at A.

The eccentric of the resultant force, R, is now on the opposite side of the centroid (Fig. E10.3d).

$H = P_{aR} = 10146 \text{ lb}; \quad V_n = R_z = 19200 \text{ lb},$

$\frac{H}{V_n} = \frac{10146}{19200} = 0.53; \quad \omega = \tan^{-1}\frac{H}{V_n} = \tan^{-1}(0.53) = 27.9^\circ$

$B' = B - 2e = 14 - 2 \times 0.57 = 12.9 \text{ m}$

$n = \left(2 + \frac{B'}{L'}\right) \Big/ \left(1 + \frac{B'}{L'}\right) = 2;$

$i_\gamma = \left(1 - \frac{H}{V_n}\right)^{n+1} = (1 - 0.53)^{2+1} = 0.1$

$q_u = 0.5\gamma' B' N_\gamma i_\gamma = 0.5 \times (113 - 62.4) \times 12.9 \times 43.9 \times 0.1 = 1433 \text{ psf}$

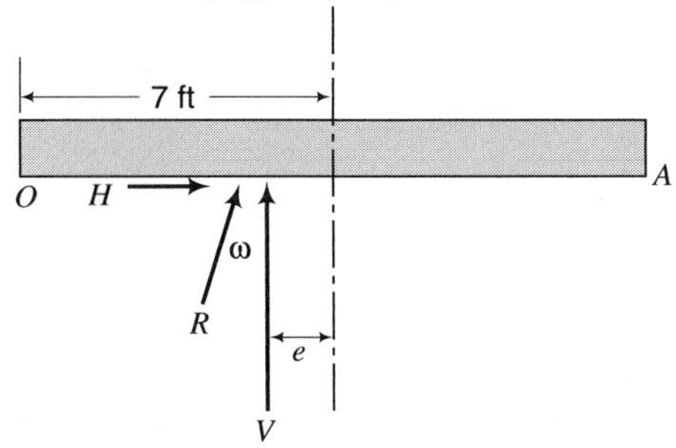

FIGURE E10.3d

$$(\text{FS})_B = \frac{q_u}{(\sigma_{max})_R} = \frac{1433}{1706} = 0.8 < 3; \textit{not okay}$$

The wall will fail by bearing capacity failure. ■

EXAMPLE 10.4 *Cantilever Gravity Wall*

Determine the stability of the cantilever, gravity retaining wall shown in Fig. E10.4a. The existing soil is a clay and the backfill is a coarse-grained soil. The base of the wall will rest on a 50 mm (2 in.) thick, compacted layer of the backfill. The interface friction between the base and the compacted layer of backfill is 25°. Groundwater level is 8 m (26 ft) below the base.

Strategy You should use Coulomb's method to determine the lateral earth pressure because of the presence of wall friction. The height of the wall for calculating the lateral earth pressure is the vertical height from the base of the wall to the soil surface. You should neglect the passive resistance of the 1.0 m of soil behind the wall. SI units will be used in solving this example.

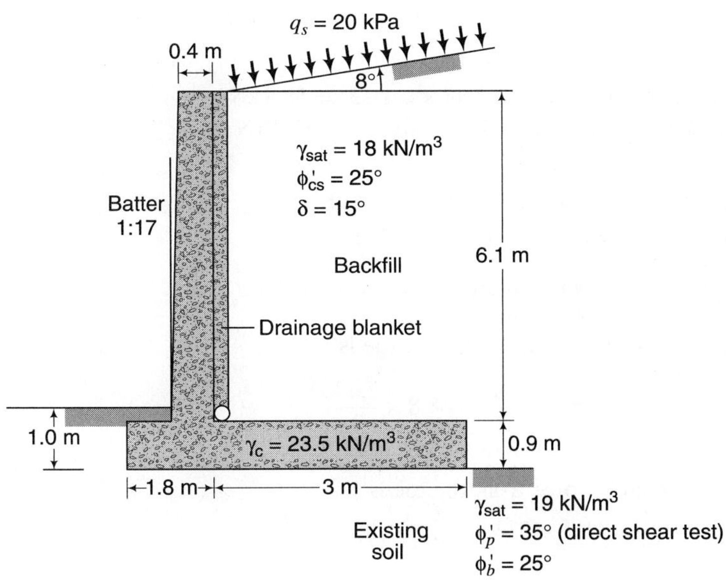

FIGURE E10.4a

Solution 10.4

Step 1: Determine the active lateral force and its location.
See Fig. E10.4b. We are given $\eta = 0, \beta = 8^\circ, \delta = 15^\circ$, and $\phi'_{cs} = 25^\circ$. Therefore, from Eq. (10.21),

$$K_{aC} = \frac{\cos^2(25^\circ - 0)}{\cos^2 0^\circ \cos(0^\circ + 15^\circ)\left[1 + \left\{\dfrac{\sin(25^\circ + 15^\circ)\sin(25^\circ - 8^\circ)}{\cos(0^\circ + 15^\circ)\cos(0^\circ - 8^\circ)}\right\}^{1/2}\right]^2} = 0.41$$

$$H_o = 0.9 + 6.1 + 3.0\tan 8^\circ = 7.42\,\text{m}$$

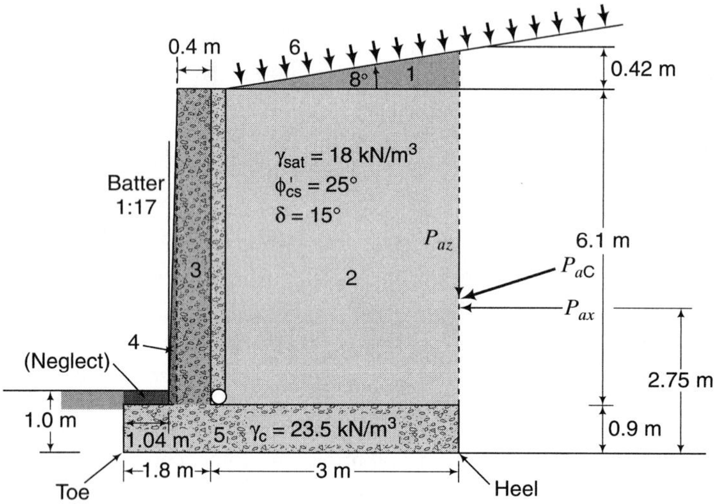

FIGURE E10.4b

Soil mass
All forces are per meter length of wall.

Lateral force from soil mass: $P_{aC} = \frac{1}{2}K_{aC}\gamma_{sat}H_o^2 = \frac{1}{2} \times 0.41 \times 18 \times 7.42^2 = 203.2\,\text{kN}$

Horizontal component: $F_{ax} = P_{aC}\cos\delta = 203.2\cos 15^\circ = 196.3\,\text{kN}$

Vertical component: $F_{az} = P_{aC}\sin\delta = 203.2\sin 15^\circ = 52.6\,\text{kN}$

Surcharge

$$F_x = K_a q_s H_o \cos\delta = 0.41 \times 20 \times 7.42 \times \cos 15^\circ = 58.8\,\text{kN}$$
$$F_z = K_a q_s H_o \sin\delta = 0.41 \times 20 \times 7.42 \times \sin 15^\circ = 15.7\,\text{kN}$$

Resultant force components

$$P_{ax} = F_{ax} + F_x = 196.3 + 58.8 = 255.1\,\text{kN}$$
$$P_{az} = F_{az} + F_z = 52.6 + 15.7 = 68.3\,\text{kN}$$

Step 2: Determine the resultant vertical force per unit length and its location.
A table is useful to keep the calculation tidy and easy to check.
The location of the resultant horizontal component of force from the toe is

$$\bar{z} = \frac{F_{ax}\dfrac{H_o}{3} + F_x\dfrac{H_o}{2}}{F_{ax} + F_x} = \frac{196.3 \times \dfrac{7.42}{3} + 58.8 \times \dfrac{7.42}{2}}{196.3 + 58.8} = 2.75\,\text{m}$$

The location of the resultant vertical component of force from the toe is

$$\bar{x} = \frac{\Sigma M_o}{R_z} = \frac{1322.6}{653.6} = 2.02$$

Part	Force (kN/m)	Moment arm from toe (m)	Moment (+kN · m)
1	0.5 × 0.42 × 3 × 18 = 11.3	3.80	42.9
2	3 × 6.1 × 18 = 329.4	3.30	1087.0
3	0.4 × 6.1 × 23.5 = 57.3	1.60	91.7
4	0.5 × 0.36 × 6.1 × 23.5 = 25.8	1.28	33.0
5	0.9 × 4.8 × 23.5 = 101.5	2.40	243.6
6*	3 × 20 = 60.00	3.3	198.00
	585.3		1696.2(+)
P_{az}	68.3	4.8	327.9(+)
	R_z = 653.6		Σ Moments = 2024.1(+)
P_{ax}	255.1	2.75	701.5(−)
			Σ M_O = 1322.6(+)

*Neglect effect of small angle.

Step 3: Determine the eccentricity.

$$e = \frac{B}{2} - \overline{x} = \frac{4.8}{2} - 2.02 = 0.38\,\text{m}$$

Step 4: Determine the stability.

Rotation

$$\frac{B}{6} = \frac{4.8}{6} = 0.8\,\text{m} > e(= 0.38\,\text{m});\ \text{therefore rotation is satisfactory}$$

Translation

$$T = R_z \tan \phi'_b = 656 \times \tan 25^\circ = 305.9\,\text{kN/m}$$

$$(\text{FS})_T = \frac{T}{P_{ax}} = 1.2 < 1.5;\ \text{therefore, translation is not satisfactory}$$

In design, you can consider placing a key at the base to increase the factor of safety against translation.

Bearing capacity

$$\sigma_{\max} = \frac{656}{4.8 \times 1}\left(1 + \frac{6 \times 0.38}{4.8}\right) = 201.5\,\text{kPa}$$

$$B' = B - 2e = 4.8 - 2 \times 0.38 = 4.06\,\text{m}$$

$$H = P_{ax} = 255.1\,\text{kN};\quad V_n = R_z = 653.6\,\text{kN},\quad \frac{H}{V_n} = \frac{255.1}{653.6} = 0.39;$$

$$\omega = \tan^{-1}\frac{H}{V_n} = \tan^{-1}(0.39) = 21.3^\circ$$

$$n = \left(2 + \frac{B'}{L'}\right) \Big/ \left(1 + \frac{B'}{L'}\right) = 2;\quad i_\gamma = \left(1 - \frac{H}{V_n}\right)^{n+1} = (1 - 0.39)^{2+1} = 0.23$$

$$N_\gamma = 0.1054 \exp\,(9.6\phi'_p) = 0.1054 \exp\left(9.6 \times 35 \times \frac{\pi}{180}\right) = 37.2$$

$$q_u = 0.5\gamma B' N_\gamma i_\gamma = 0.5 \times 19 \times 4.06 \times 37.2 \times 0.23 = 330\,\text{kPa}$$

$$(\text{FS})_B = \frac{q_u}{(\sigma_{\max})} = \frac{330}{201.5} = 1.6 < 3$$

Therefore, bearing capacity is not satisfactory. ■

10.10 SEISMIC ANALYSIS AND DESIGN OF RIGID RETAINING WALLS

Access a spreadsheet at http://www.wiley.com/college/budhu. Click Foundations and Earth Structures, Chapter 10, retwall.xls and click worksheet "CIP" to analyze gravity walls for seismic loads.

Importance *Seismic forces have caused extensive damage to rigid retaining walls through instability and movement. In this section, we discuss the analysis and design of rigid retaining walls for a seismic event.*

10.10.1 Seismic Stability

Coulomb's lateral earth pressure theory was modified by Okabe (1926) and Mononobe and Matsuo (1929), called Mononobe-Okabe earth pressure theory, by treating the dynamic forces from a seismic event as quasi-static forces acting on the failure plane. The forces on the wall due to a vertical acceleration, $a_v = k_v g$, and a horizontal acceleration, $a_h = k_h g$, are shown in Fig. 10.20. These forces correspond to the acceleration directions as shown in Fig. 10.20.

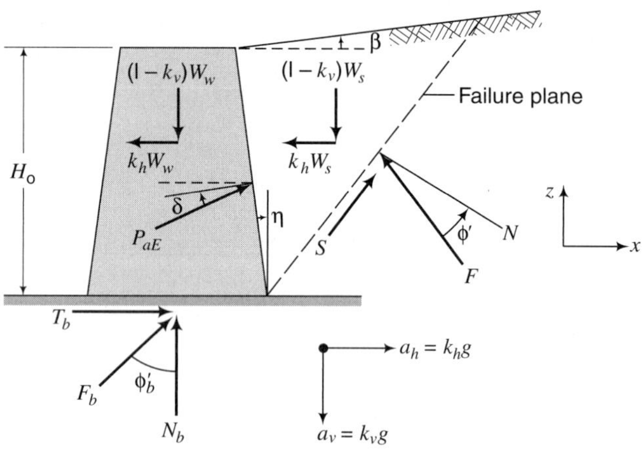

FIGURE 10.20 Forces on a rigid wall due to a seismic event.

The active seismic force on the wall is

$$P_{aE} = \frac{1}{2}\gamma H_o^2(1-k_v)K_{aE} \tag{10.58}$$

where K_{aE} is the seismic active earth pressure coefficient expressed as

$$K_{aE} = \frac{\cos^2(\phi' - \theta - \eta)}{\cos^2(\theta)\cos^2(\eta)\cos(\delta+\eta+\theta)\left[1+\sqrt{\dfrac{\sin(\phi'+\delta)\sin(\phi'-\beta-\theta)}{\cos(\delta+\beta+\theta)\cos(\beta-\eta)}}\right]^2} \tag{10.59}$$

where θ is the inclination of the horizontal acceleration.

$$\theta = \tan^{-1}\left(\frac{k_h}{1-k_v}\right) \tag{10.60}$$

The passive seismic force is

$$P_{pE} = \frac{1}{2}\gamma H_o^2(1-k_v)K_{pE} \tag{10.61}$$

where K_{pE} is the seismic passive earth pressure coefficient expressed as

$$K_{pE} = \frac{\cos^2(\phi' + \eta - \theta)}{\cos(\theta)\cos^2(\beta)\cos(\delta - \eta + \theta)\left[1 - \sqrt{\frac{\sin(\phi' + \delta)\sin(\phi' + \beta - \theta)}{\cos(\delta - \eta + \theta)\cos(\beta - \eta)}}\right]^2} \tag{10.62}$$

where ϕ' is the friction angle, η is the inclination of the back of the wall, β is the slope of the backfill, and δ is the soil–wall interface friction.

We now consider the effects of various factors that you should consider in the seismic analysis of rigid retaining walls.

1. ***Location of resultant active force:*** The active static force is assumed to act at $\frac{H_o}{3}$ and the seismic component, $(P_{aE} - P_a)$, at $\frac{H_o}{2}$ from the base of the wall as shown in Fig. 10.21.
2. ***Effects of wall friction:*** Wall friction does not significantly affect the seismic active and passive forces—both called *lateral forces* from here on—and may be neglected in design.
3. ***Effects of vertical and horizontal accelerations:*** The vertical acceleration does not significantly influence the seismic lateral forces. It is the horizontal acceleration that is mainly responsible for instability. Increases in k_h lead to increases in P_{aE} and reduction in P_{pE}. The inclination of active failure plane to the horizontal also becomes smaller (Fig. 10.22).

Consider the term $(\phi' - \beta - \theta)$ in the denominator of Eq. (10.59). If $(\phi' - \beta - \theta) < 0$, there is no real solution and consequently equilibrium is not possible. If $(\phi' - \beta - \theta) = 0$, then P_{aE} reaches its maximum value. Thus, $(\phi' - \beta - \theta)$ must be greater than or equal to zero for stability. The limiting slope is

$$\beta \leq (\phi' - \theta) \tag{10.63}$$

For static condition, $\theta = 0$ and Eq. (10.63) gives $\beta \leq \phi'$, which should be the case.

Let us rearrange Eq. (10.63), giving

$$\theta \leq (\phi' - \beta) \tag{10.64}$$

Therefore,

$$\tan^{-1}\left(\frac{k_h}{1 - k_v}\right) \leq (\phi' - \beta) \tag{10.65}$$

That is, the critical horizontal acceleration is

$$(k_h)_{cr} = (1 - k_v)\tan(\phi' - \beta) \tag{10.66}$$

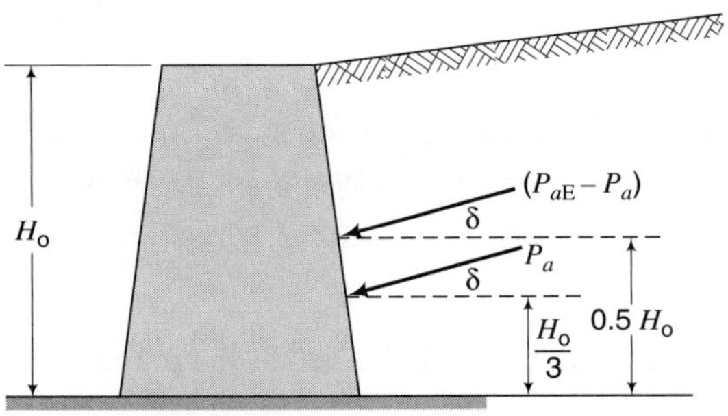

FIGURE 10.21 Location of static and dynamic component of active forces.

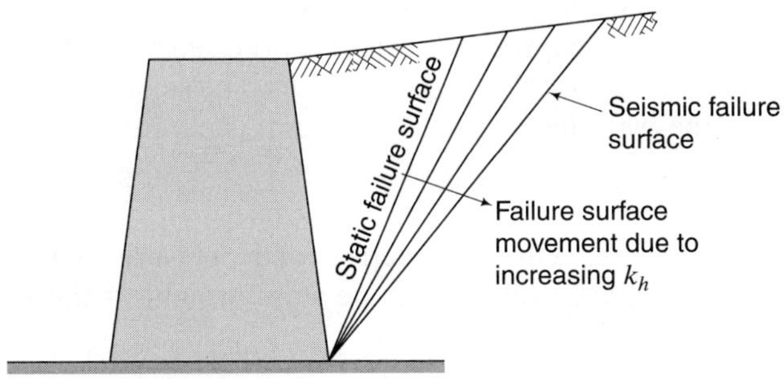

FIGURE 10.22 Effects of increasing k_h.

For a horizontal backfill, $\beta = 0$ and

$$(k_h)_{cr} = (1 - k_v)\tan\phi' \tag{10.67}$$

4. ***Effect of backfill slope, β:*** Increases in backfill slope cause increases in the seismic active force. For example, if the backfill slope increases by 10°, the seismic active force doubles.
5. ***Effect of friction angle, φ′:*** All lateral forces are affected by changes in friction angle. The lower the friction angle, the higher the seismic active force. For example, a decrease of ϕ' by 5° will cause an increase of the seismic active pressure by 30%. Thus, you need to use conservative values of ϕ'. It is best to use ϕ'_{cs}. In addition to the change in seismic active force, the seismic bearing capacity will also decrease with decreases in ϕ'.
6. ***Effects of wall inertia:*** Richards and Elms (1979) showed that wall inertia is crucial in the seismic performance of rigid retaining walls.

   Let us consider the forces on the wall in Fig. 10.20.

$$\Sigma F_z = 0: N_b = (1 - k_v)W_w + P_{aE}\sin(\delta + \eta)$$
$$\Sigma F_x = 0: T_b = k_h W_w + P_{aE}\cos(\delta + \eta)$$

At incipient sliding, $T_b = N_b \tan\phi'_b$ where ϕ'_b is the base–soil interface friction angle and the subscript b denotes base.

Solving for W_w gives

$$W_w = \frac{[\cos(\delta + \eta) - \sin(\delta + \eta)\tan\phi'_b]}{(1 - k_v)(\tan\phi'_b - \tan\theta)} P_{aE} \tag{10.68}$$

If the denominator of Eq. (10.68) becomes zero, then the weight of the wall tends to infinity. In the limit,

$$\tan\theta = \tan\phi'_b \tag{10.69}$$

That is,

$$(k_h)_{cr} = (1 - k_v)\tan\phi'_b \tag{10.70}$$

We now have two critical conditions. One is due to the backfill (Eq. 10.66), the other due to the wall inertia (Eq. 10.70). If $\phi'_b < (\phi' - \beta)$, then Eq. (10.70) will govern the design.

Some factors such as δ, k_v and η that were insignificant for lateral forces now have significant negative impacts on the wall stability as further discussed in the following.

We can rewrite Eq. (10.68) as

$$W_w = C_{IE} P_{aE} = C_{IE}\left[\frac{1}{2}\gamma H_o^2(1 - k_v)K_{aE}\right] \tag{10.71}$$

where C_{IE} is the wall inertia factor expressed as

$$C_{IE} = \frac{\cos(\delta + \eta) - \sin(\delta + \eta)\tan\phi'_b}{(1 - k_v)(\tan\phi'_b - \tan\theta)} \tag{10.72}$$

Variations of C_{IE} with k_v, k_h, and δ are shown in Figs. 10.23 and 10.24.

To capture the effects of the seismic active force and the wall inertia, we consider how much they increase relative to static values. The soil thrust ratio, F_T, defined as the relative increase in active force on the wall due to a seismic event, is expressed as

$$F_T = \frac{K_{aE}(1 - k_v)}{K_a} \tag{10.73}$$

where K_a is the static active earth pressure coefficient. The wall inertia ratio, F_I, defined as the increase in wall inertia relative to the static weight of the wall, is expressed as

$$F_I = \frac{C_{IE}}{C_I} \tag{10.74}$$

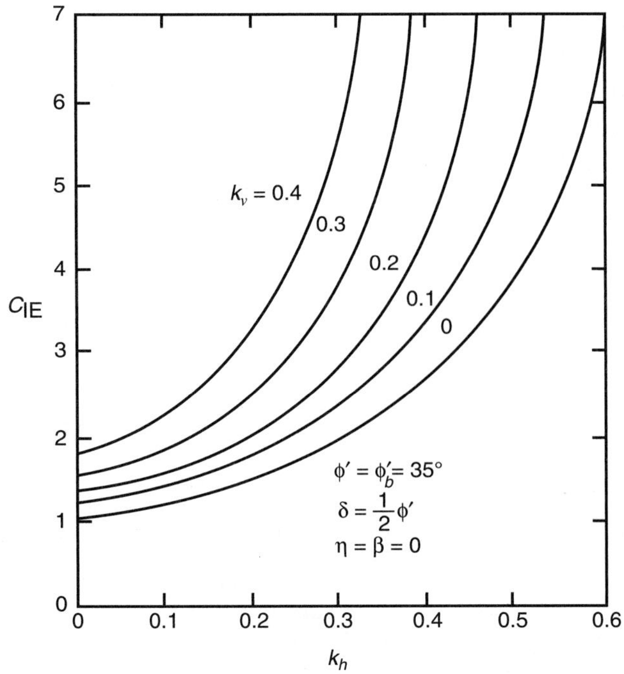

FIGURE 10.23 Effects of k_v and k_h on wall inertia. *(Source: Richards and Elms, 1979.)*

where C_I is the static weight factor given as

$$C_I = \frac{W}{\frac{1}{2}\gamma H_o^2 K_a} = \frac{\cos(\delta + \eta) - \sin(\delta + \eta)\tan\phi_b'}{\tan\phi_b'} \tag{10.75}$$

where W is the weight of the wall for static loading.

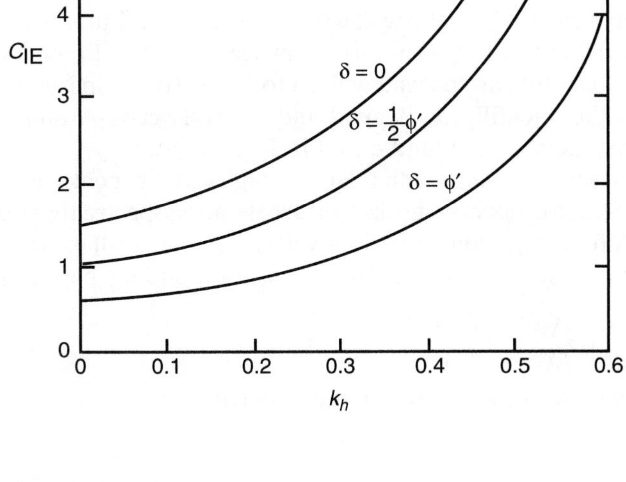

FIGURE 10.24 Effects of soil–wall friction on wall inertia *(Source: Richard and Elms, 1979.)*

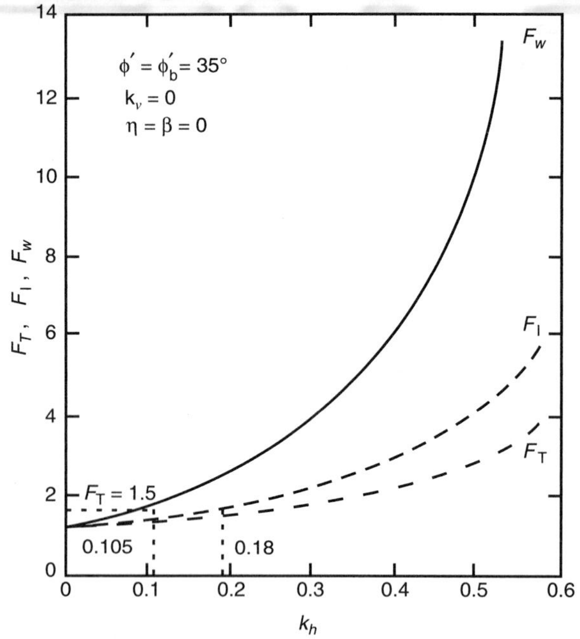

FIGURE 10.25 Effects of on k_h on F_T, F_I and F_w. *(Source: Richards and Elms, 1979.)*

The product, $F_w = F_T F_I$, called the seismic amplification factor, gives the ratio of weight of the wall required for seismic forces to the weight of the wall required for static forces; that is

$$F_w = \frac{W_w}{W} \tag{10.76}$$

A plot of F_T, F_I, and F_w versus k_h is shown in Fig. 10.25. There are two key points established from Fig. 10.25.

1. For $k_h > 0.1$, the wall will be unsafe when it is designed for static pressures only.
2. Wall movement is inevitable under seismic forces. For no wall movement, the weight of the wall for static condition will be cost prohibitive. Therefore, the design for seismic loading must allow for some lateral movement to occur.

10.10.2 Seismic Displacement

The displacement of a rigid retaining wall will not occur continuously but as a "ratchet" type motion. When the horizontal acceleration is greater than the $(k_h)_{cr}g$, the wall will move laterally, but it will cease to move when the horizontal acceleration is less than $(k_h)_{cr}\, g$. For horizontal accelerations directed away from the backfill, a passive state develops. The backfill and the wall will act together (o to a in Fig. 10.26). For horizontal accelerations directed toward the backfill, the backfill and the soil act separately. The inertia forces act outwardly, causing wall movement (a to b and c to d in Fig. 10.26).

One of the implications of Fig. 10.26 is that seismic events with relatively high velocity peaks are more destructive than those with low velocity peaks. Stated otherwise, horizontal acceleration spectra that exceed the critical horizontal acceleration value more often or for longer periods will cause more wall movements.

Richard and Elms (1979) expressed the seismic displacement, ρ_d, using a sliding block model (Newmark, 1965), as

$$\rho_d = 0.087 \frac{V^2}{Ag} \left[\frac{(k_h)_{cr}}{A} \right]^{-4} \tag{10.77}$$

where V is the peak velocity and A is the effective peak acceleration coefficient.

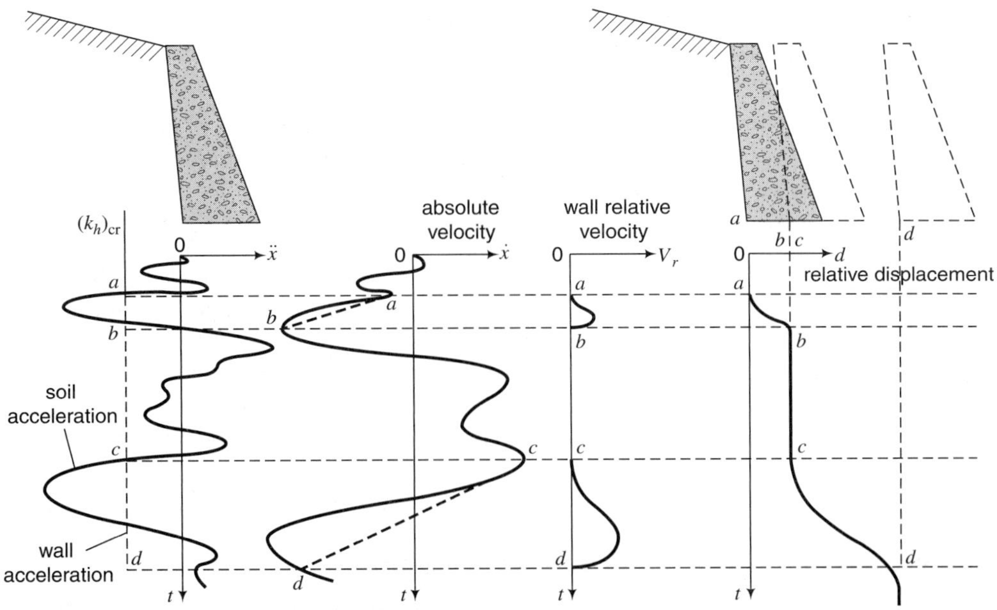

FIGURE 10.26 Wall displacement mechanism. *(Source: Richards and Elms, 1990.); $\ddot{x}$ = acceleration, $\dot{x}$ = velocity.*

The design procedure, according to Richards and Elms (1979), is as follows.

1. Select a desired displacement, ρ_d.
2. Calculate

$$k_h = \rho_d A \left(\frac{A_g \, \rho_d}{0.087 \, V^2} \right)^{-\frac{1}{4}}. \tag{10.78}$$

3. Calculate the required weight of the wall, W_w using k_h from step 2 and Eq. 10.71.
4. Apply a factor of safety of 1.5 to W_w.
5. Check that the wall will slide and not tilt using

$$x_o = \frac{z[\cos(\delta + \eta) + \tan(\eta) \, \sin(\delta + \eta)] + C_{IE}(FS)[k_h \bar{z} + (1 - k_v)\bar{x}]}{\sin(\delta + \eta) + (1 - k_v)C_{IE} \, (FS)} \tag{10.79}$$

where x_o is the distance from the toe to the resultant force at the base, z is the height of the resultant lateral soil force taken as $\frac{H_o}{2}$, $\bar{z}$ and $\bar{x}$ are the coordinates of the wall center of gravity from the base and toe respectively, and FS is a factor of safety of 1.5.

6. Check seismic bearing capacity.

The* key points *are:

1. ***Earthquakes induce additional lateral forces on rigid retaining walls that can lead to instability.***
2. ***The seismic forces are treated as if they were applied by a quasi-static event.***
3. ***It is practically impossible to design a wall for no movement in a seismic event.***
4. ***Inertial forces play a considerable role in inducing instability in rigid retaining walls in a seismic event.***

10.11 STABILITY OF FLEXIBLE RETAINING WALLS

Importance *Flexible retaining walls have a large range of applications. They are often used as temporary supports for excavations and as permanent supports to retain both soil and water. In this section, we will study the stability of flexible retaining walls.*

Sheet pile walls are flexible and are constructed using steel or thin concrete slabs or wood. Two types of sheet pile walls are common. One is a cantilever wall commonly used to support soils to a height of less than 3 m (1.5 ft) (Fig. 10.15a). The other is an anchored or propped sheet pile wall (Fig. 10.15b,c) commonly used to support deep excavations and as waterfront retaining structures. Cantilever sheet pile walls rely on the passive soil resistance for their stability, while anchored sheet pile walls rely on a combination of anchors and passive soil resistance for their stability. The stability of sheet pile walls must satisfy all the criteria for rigid retaining walls described in Section 10.8 except bearing capacity. However, because sheet pile walls are used in situations where seepage may occur, it is necessary to pay particular attention to seepage-related instabilities.

10.11.1 Analysis of Sheet Pile Walls in Uniform Soils

In analyzing sheet pile walls, we are attempting to determine the depth of embedment, d, for stability. The analysis is not exact and various simplifications are made. The key static equilibrium condition is moment equilibrium. Once we determine d, the next step is to determine the size of the wall. This is done by calculating the maximum bending moment and then determining the section modulus by dividing the maximum bending moment by the allowable bending stress of the material constituting the sheet pile, for example, steel, concrete, or wood.

For an effective stress analysis, we must evaluate the porewater pressure distribution and seepage pressures. We can use flow net sketching or numerical methods to determine the porewater pressure distribution and seepage pressures. However, approximate methods are often used in practice. If the groundwater level on both sides of a sheet pile wall is the same, then the resultant porewater pressures and seepage pressures are zero (Fig. 10.20a). You can then neglect the effects of groundwater in determining the stability of sheet pile walls. However, you must use effective stresses in your calculations of the lateral earth forces.

The approximate distribution of porewater pressures in front of and behind sheet pile walls for conditions in which the water tables are different is obtained by assuming a steady state seepage condition and uniform distribution of the total head. Approximate resultant porewater pressure distributions for some common conditions (Padfield and Mair, 1984) are shown in Fig. 10.27.

The maximum porewater pressures (u_B), maximum porewater forces (P_w) and their locations ($\bar{z}_w$), and the seepage force per unit volume (j_s) are as follows:

Case (a) – Fig. 10.27a

Resultant porewater pressure is zero and the seepage force is zero.

Case (b) – Fig. 10.27b

$$u_B = \frac{2ad}{a + 2d}\gamma_w \tag{10.80}$$

$$P_w = \frac{ad(a + d)}{a + 2d}\gamma_w \tag{10.81}$$

$$\bar{z}_w = \frac{a + 2d}{3} \tag{10.82}$$

$$j_s = \frac{a}{a + 2d}\gamma_w \tag{10.83}$$

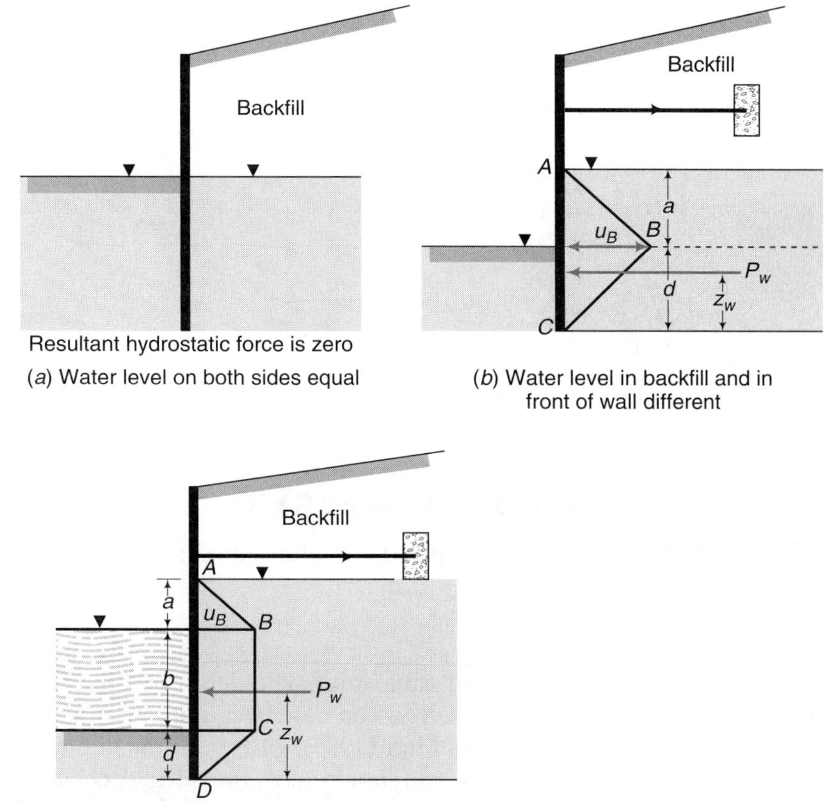

(*a*) Water level on both sides equal

(*b*) Water level in backfill and in front of wall different

(*c*) Wall supports water in front of it and water level in backfill greater than water level in front of wall

FIGURE 10.27 Approximate resultant porewater pressure distribution behind flexible retaining walls.

Case (c) – Fig. 10.27c

$$u_B = u_C = \frac{a(b + 2d)}{a + b + 2d}\gamma_w \tag{10.84}$$

$$P_w = \frac{1}{2}\left[\frac{a(b + 2d)(a + 2b + d)}{a + b + 2d}\right]\gamma_w \tag{10.85}$$

$$\bar{z}_w = \frac{a^2 + 3a(b + d) + 3b(b + 2d) + 2d^2}{3(a + 2b + d)} \tag{10.86}$$

$$j_s = \frac{a}{a + b + 2d}\gamma_w \tag{10.87}$$

Recall j_s is the seepage pressure per unit volume and the resultant effective stress is increased when seepage is downward (behind the wall) and is decreased when seepage is upward (in front of the wall) as discussed in Chapter 1.

10.11.2 Analysis of Sheet Pile Walls in Mixed Soils

Sheet pile walls may penetrate different soil types. For example, Fig. 10.28 shows a sheet pile wall that supports a coarse-grained soil but is embedded in a fine-grained soil. In this case, you should consider a mixed analysis. For short-term condition, an effective stress analysis can be used for the coarse-grained soil but a total stress analysis should be used for the fine-grained soil. For long-term conditions, an effective stress analysis should be carried out for both soil types.

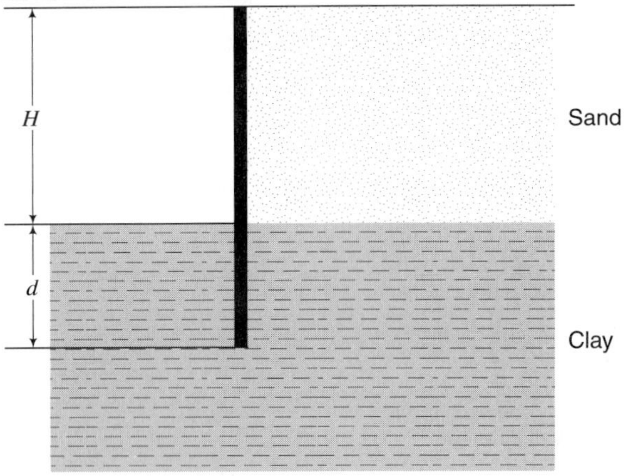

FIGURE 10.28 Sheet pile wall penetrating different soils.

10.11.3 Consideration of Tension Cracks in Fine-Grained Soils

If a sheet pile wall supports fine-grained soils, you should consider the formation of tension cracks. The theoretical depth of a tension crack is

$$z_{cr} = \frac{2s_u}{\gamma} \quad \text{or} \quad \frac{2s_u}{\gamma'} \tag{10.88}$$

The latter is applicable when the tension cracks are filled with water. The depth of a tension crack is sometimes greater than the wall height. In this situation, you can assume a minimum active lateral effective pressure of $5z$(kPa) as suggested by Padfield and Mair (1984), where z is the depth measured from the top of the wall. When water fills the tension cracks, you should apply the full hydrostatic pressure to the wall over a depth equivalent to the depth of the tension crack or the wall height, whichever is smaller. In fine-grained soils, loss of moisture at the wall-soil interface can cause the soil to shrink, creating a gap. This gap can be filled with water. In this case, you should apply the full hydrostatic pressure over the depth of the wall.

10.11.4 Methods of Analysis

(a) Analytical Several analytical methods have been proposed to determine the stability of sheet pile walls based on statics. These methods differ in the way the lateral stresses are distributed on the wall and the way the factor of safety is applied in solving for the embedment depth. We will discuss four analytical methods in this book. In the first method, called the factored moment method (FMM), you determine an embedment depth to satisfy moment equilibrium by applying a factor of safety $(\text{FS})_p$ on the passive resistance. The factor of safety, $(\text{FS})_p$, is usually between 1.5 and 2.0.

In the second method, called the factored strength method (FSM), reduction factors are applied to the shear strength parameters. These reduction factors are called mobilization factors because they are intended to limit the shear strength parameters to values that are expected to be mobilized by the design loads. A mobilization factor, F_ϕ, is applied to the friction angle, ϕ'_{cs}, and a mobilization factor, F_u is applied to s_u. The application of these mobilization factors results in higher active pressures and lower passive pressures than the unfactored soil strength parameters. The design parameters are

$$\phi'_{\text{design}} = \frac{\phi'_{cs}}{F_\phi} \tag{10.89}$$

and

$$(s_u)_{\text{design}} = \frac{s_u}{F_u} \tag{10.90}$$

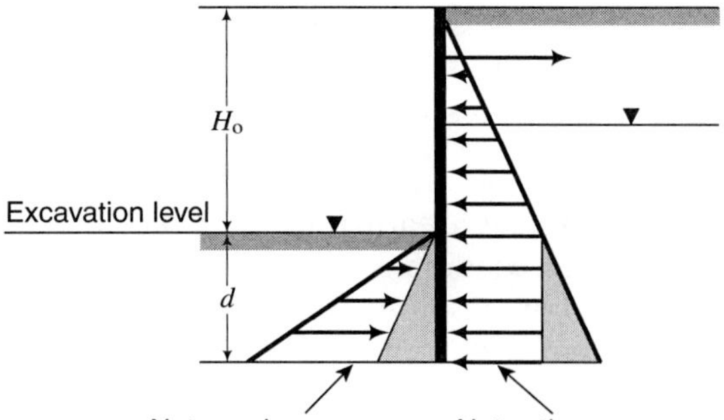

FIGURE 10.29 Net pressures for the NPPM.

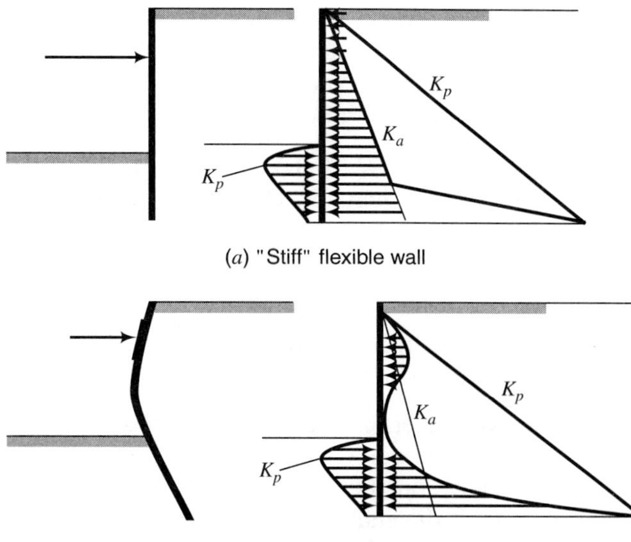

FIGURE 10.30 Lateral pressure distributions expected on a "stiffer" flexible wall and less "stiff" flexible wall. (After Padfield and Mair, 1984).

where

$$F_\phi = 1.2 \text{ to } 1.5 \qquad \text{and} \qquad F_u = 1.5 \text{ to } 2$$

The results from the FSM are sensitive to F_ϕ and F_u.

The third method, called the net passive pressure method (NPPM), utilizes a net available passive resistance (Burland et al., 1981). A vertical line is drawn from the active pressure at the excavation level to the base of the wall (Fig. 10.29). The shaded region of pressure on the active side (Fig. 10.29) is subtracted from the passive pressure to give the net passive pressure shown by the area hatched by arrows. The factor of safety for the NPPM is

$$(\text{FS})_r = \frac{\Sigma \text{ Moments of net available passive resistance}}{\Sigma \text{ Moments of lateral forces causing rotation}} \tag{10.91}$$

$$(\text{FS})_r = 1.5 \text{ to } 2 \text{ with } 2 \text{ most often used}$$

The fourth method is LRFD. The lateral forces are calculated and then load factors are applied.

The flexibility of sheet pile walls leads to lateral pressure distributions that do not correspond to the Rankine active and passive states. This is the key reason why there are different methods of analysis. Expected lateral pressure distributions on a "stiff" flexible wall and a less "stiff" flexible wall are shown in Fig. 10.30. The net effect of the distribution of the lateral pressures as illustrated in Fig. 10.30 is a reduction in maximum bending moment from that calculated using Rankine active and passive states.

Rowe (1957) developed a method, based on laboratory tests, to reduce the calculated maximum bending moment to account for the effects of wall flexibility on the bending moment. Rowe's moment reduction is applicable when a factor of safety has been applied on the passive resistance as in the FMM. There is some debate on the applicability of Rowe's method. Some engineers prefer to calculate the maximum bending moment at limit equilibrium, $(\text{FS})_p = F_\phi = (\text{FS})_r = F_u = 1$, and use it as the design moment. This is the preferred method in this book.

To account for soil-wall interface friction, you need to use K_{aC} and K_{pC}. However, the active and passive coefficients derived by Caquot and Kerisel (1948) are regarded as more accurate than those of Coulomb. For a flexible retaining wall, only the horizontal components of the lateral forces are important. In Appendix C, the horizontal components of the active and passive coefficients of Caquot and Kerisel (1948) as tabulated by Kerisel and Absi (1990) are plotted for some typical backfill slopes and soil-wall interface friction angles. We will use the values of the lateral earth pressure coefficients in Appendix C in some of the example problems in this chapter.

(b) Numerical Methods FDM and FEM are powerful tools to analyze earth retaining structures and explore what-if situations. They are particularly suited for complex soil conditions and sequential construction. Both methods not only give the load at which the wall will fail, but most importantly, the incremental displacements. In the analytical methods described earlier, wall movements cannot be predicted.

10.11.5 Stability of Cantilever Sheet Pile Walls Using Analytical Methods

Cantilever sheet pile walls are analyzed by assuming that rotation occurs at some point, O, just above the base of the wall (Fig. 10.31). The consequence of assuming rotation above the base is that, below the point of rotation, the lateral pressure is passive behind the wall and active in front of the wall (Fig. 10.31b). To simplify the analysis, a force R (Fig. 10.31c) is used at the point of rotation to approximate the net passive resistance below it (the point of rotation). By taking moments about O, the unknown force R is eliminated and we then obtain one equation with one unknown, that is, the unknown depth, d_o. To account for this simplification, the depth d_o is increased by 20% to 30% to give the design embedment depth, d.

The general procedure for determining d for stability and to determine the wall size is as follows:

1. Arbitrarily select a point O at a distance d_o from the excavation level.
2. Calculate the active and passive earth pressures using the FMM or FSM or NPPM.
3. Calculate the net porewater pressure (u) distribution and the seepage force per unit volume (j_s). The effective unit weight is increased by j_s in the active zone and is decreased by j_s in the passive zone.
4. Determine the unknown depth d_o by summing moments about O.
5. Calculate d by increasing d_o by 20% or 30% to account for simplifications made in the analysis. The depth of penetration d is therefore $1.2d_o$ or $1.3d_o$.
6. Calculate R by summing forces horizontally over the depth $(H_o + d)$.
7. Calculate the net passive resistance, $(P_p)_{net}$, over the distance, $d - d_o$, below O.
8. Check that R is less than $(P_p)_{net}$. If not, extend the depth of embedment and recalculate R.
9. Calculate the maximum bending moment (M_{max}) over the depth $(H_o + d_o)$ using unfactored passive resistance (FMM), unfactored strength values (FSM), and $(FS)_r = 1$(NPPM).

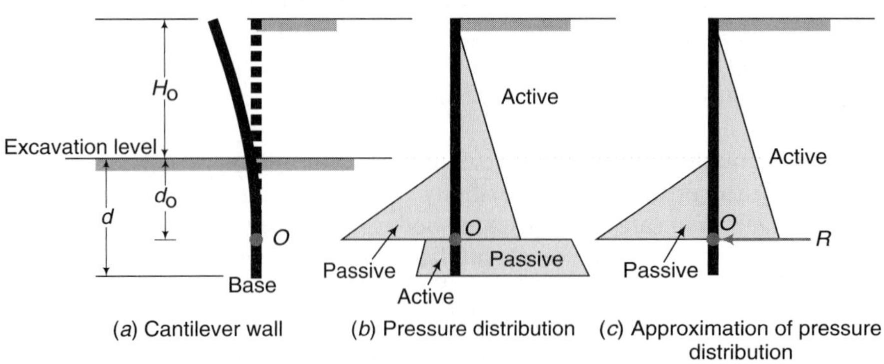

FIGURE 10.31 Approximation of pressure distributions in the analysis of cantilever flexible retaining walls (Padfield and Mair, 1984).

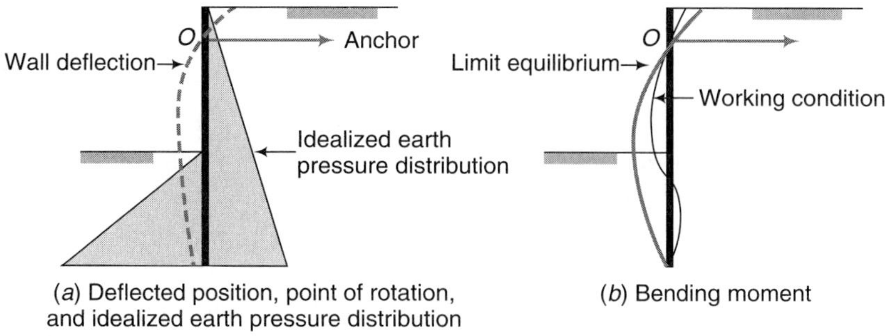

FIGURE 10.32 Free earth conditions for anchored retaining walls (Padfield and Mair, 1984).

10. Determine the section modulus, $S_x = M_{\max}/f_a$, where $M_{\max}$ is the maximum bending moment and f_a is the allowable bending stress of the wall material.

10.11.6 Stability of Anchored Sheet Pile Walls Using Analytical Methods

There are two methods used to analyze anchored sheet pile walls. One is the free earth method, the other is the fixed earth method. We will be discussing the free earth method, because it is frequently used in design practice.

In the free earth method, it is assumed that (1) the depth of embedment of the wall is insufficient to provide fixity at the bottom end of the wall and (2) rotation takes place about the point of attachment of the anchor, O (Fig. 10.32a). The expected bending moment diagram, based on the above assumptions, is depicted in Fig. 10.32b.

The procedure for analyzing an anchored sheet pile wall is as follows:

1. Assume a depth of embedment, d.
2. Calculate the active and passive pressures using the FMM or FSM or NPPM.
3. Calculate the net porewater pressure (u) distribution and the seepage force per unit volume (j_s). The effective unit weight is increased by j_s in the active zone and is decreased by j_s in the passive zone.
4. Determine d by taking moments about the point of attachment of the anchor, O. Usually, you will get a cubic equation, which you can solve by iteration or by using a goal seek option in a spreadsheet program or by using a polynomial function on a calculator.
5. Recalculate d using unfactored passive resistance (FMM), unfactored strength (FSM), and $(\text{FS})_r = 1$ (NPPM). Use this recalculated depth to determine the anchor force and the maximum bending moment.
6. Determine the anchor force per unit length of wall, T_a, by summing forces in the horizontal direction. The anchor force, T_a, is multiplied by a factor of safety $(\text{FS})_a$, usually 2.
7. Determine the location of the anchor plates. Let d_z be the depth of the bottom of the anchor plate from the ground surface (Fig. 10.33). The force mobilized by the anchor plate must balance the design anchor force, that is,

$$\frac{1}{2}\gamma' d_z^2 (K_p - K_a) = T_a \times (\text{FS})_a$$

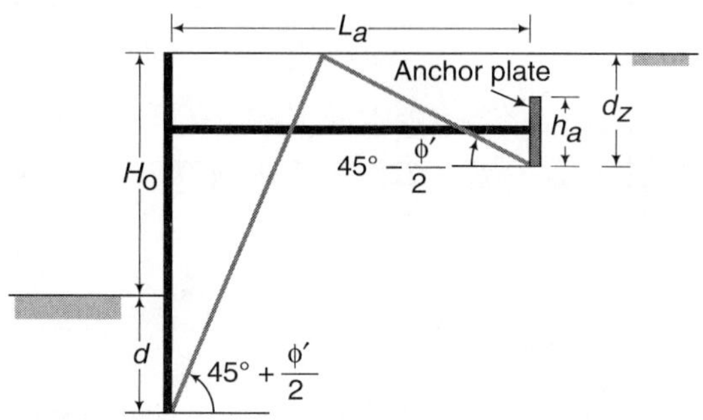

FIGURE 10.33 Location of anchor plates.

Solving for d_z, we get

$$d_z = \sqrt{\frac{2T_a \times (\mathrm{FS})_a}{\gamma'(K_p - K_a)}} \tag{10.92}$$

A passive wedge develops in front of the anchor plate and an active wedge develops behind the retaining wall. The anchor plate must be located outside the active slip plane. The minimum anchor length (L_a) of the anchor rod, with reference to Fig. 10.33, is

$$L_a = (H_o + d)\tan(45^\circ - \phi'/2) + d_z \tan(45^\circ + \phi'/2) \tag{10.93}$$

8. Calculate the spacing of the anchors. Let s be the longitudinal spacing of the anchors and h_a be the height of the anchor plate. If $h_a \geq d_z/2$, the passive resistance of the anchor plate is assumed to be developed over the full depth d_z. From static equilibrium of forces in the horizontal direction, we obtain

$$s = \frac{\gamma' d_z^2 L_a}{2T_a(\mathrm{FS})_a}(K_p - K_a) \tag{10.94}$$

9. Calculate the maximum bending moment $(M_{\max})$ using the embedment depth at limit equilibrium (unfactored passive resistance, unfactored strength values, or $(\mathrm{FS})_r = 1$).
10. Determine the section modulus, $S_x = M_{\max}/f_a$, where f_a is the allowable bending stress of the wall material.

The detailed design of anchor systems requires considerations regarding construction methods, installation methods, soil types, the environment (corrosive or noncorrosive), location of groundwater, seepage and wall deflection. Only a rudimentary design of the anchors is given in this textbook. Manuals from agencies such as the Federal Highway Administration (FHWA) are available for detailed design.

Anchors for retaining structures can be grouped into two broad classes – mechanical and bonded. The advantages and disadvantages of common anchors in these classes are given in Table 10.2. The capacity of the bonded anchor over the bonded length is estimated using presumptive values as shown in Table 10.3. Typical properties of steel bars used as anchors are given in Table 10.4. Wall deflection is an important design and construction concern in most anchored retaining wall systems. It is best to use numerical analyses to explore possible wall deflections before construction commences.

TABLE 10.2 Comparison of mechanical and bonded anchors

Anchor class	Description	Advantages	Disadvantages
Mechanical	Plates are driven into the soil to the desired depth with a driving tool. The plates are tipped and tested.	1. No drilling required 2. No grouting required. 3. Anchor capacity tested during installation.	1. Limited capacity—about 220 kN (50 kips) 2. Limited corrosion protection. 3. Not suitable for mixed soils containing boulders.

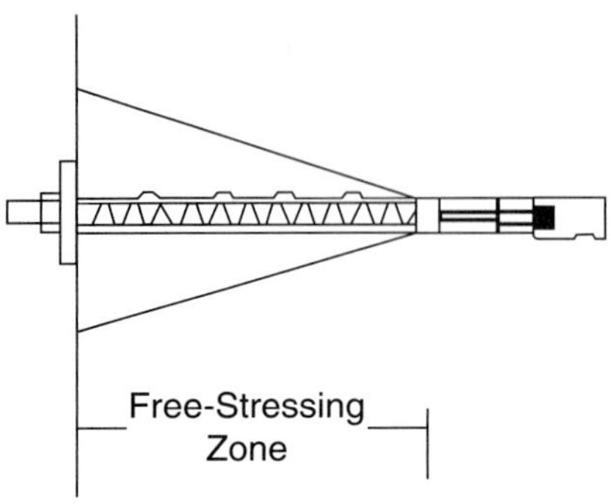

(continued)

TABLE 10.2 (Continued)

Anchor class	Description	Advantages	Disadvantages
Bonded 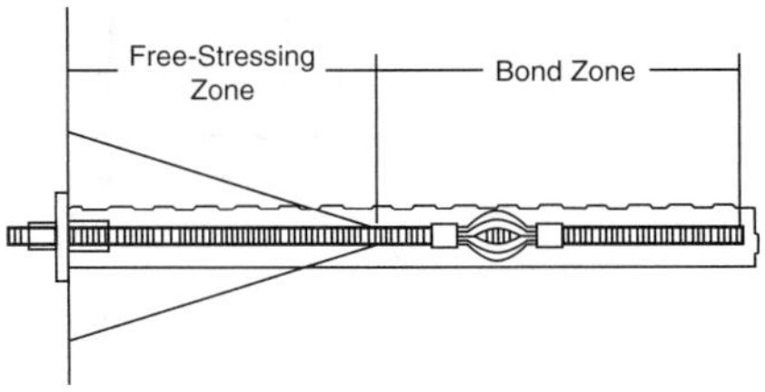	A plain or epoxy coated bar is grouted using cement or resin, in an oversized drill hole. Usually, the unbounded length is extended the greater of $H_o/5$ or 2 m(6 ft) beyond the potential failure surface with a minimum total unbounded length of 4.5 m (15 ft) for strand tendons and 3 m (10 ft) for bar tendons. The maximum bonded length is about 12 m (40 ft). Increased bonded length beyond this value does not increase anchor capacity.	1. Can be used in nearly all soil types. 2. Large capacity – about 2 MN (500 kips)	1. Long or deep drilling is required 2. Grout must be set before prestressing the anchor.

TABLE 10.3 Presumptive values for grouted anchor transfer load (1 kN/m = 68.5 lb/ft)

Soil type	Compactness (SPT)	Estimated ultimate transfer load (kN/m)
Sand and Gravel	Loose (4–10)	145
	Medium dense (11–30)	220
	Dense (31–50)	290
Sand	Loose (4–10)	100
	Medium dense (11–30)	145
	Dense (31–50)	190
Sand and Silt	Loose (4–10)	70
	Medium dense (11–30)	100
	Dense (31–50)	130
Silt-clay mixture with low plasticity or fine micaceous sand or silt mixtures	Stiff (10–20)	30
	Hard (21–40)	60

TABLE 10.4 Properties of steel bars used for anchors.

Steel grade	Nominal diameter	Ultimate stress f_{pu}	Nominal cross section area A_{ps}	Ultimate strength $f_{pu}\,A_{ps}$	Prestressing force		
					$0.8\,f_{pu}\,A_{ps}$	$0.7\,f_{pu}\,A_{ps}$	$0.6\,f_{pu}\,A_{ps}$
(ksi)	(in.)	(ksi)	(in.^{2})	(kips)	(kips)	(kips)	(kips)
	1	150	0.85	127.5	102.0	89.3	76.5
	1-1/4	150	1.25	187.5	150.0	131.3	112.5
150	1-3/8	150	1.58	237.0	189.6	165.9	142.2
	1-3/4	150	2.66	400.0	320.0	280.0	240.0
	2-1/2	150	5.19	778.0	622.4	435.7	466.8

(continued)

TABLE 10.4 (Continued)

Steel grade	Nominal diameter	Ultimate stress f_{pu}	Nominal cross section area A_{ps}	Ultimate strength f_{pu} A_{ps}	Prestressing force		
	1	160	0.85	136.0	108.8	95.2	81.6
160	1-1/4	160	1.25	200.0	160.0	140.0	120.0
	1-3/8	160	1.58	252.8	202.3	177.0	151.7
(ksi)	(mm)	(N/mm^2)	(mm^2)	(kN)	(kN)	(KN)	(kN)
	26	1035	548	568	454	398	341
	32	1035	806	835	668	585	501
150	36	1035	1019	1055	844	739	633
	45	1035	1716	1779	1423	1246	1068
	64	1035	3348	3461	2769	2423	2077
	26	1104	548	605	484	424	363
160	32	1104	806	890	712	623	534
	36	1104	1019	1125	900	788	675

EXAMPLE 10.5 *Cantilever Flexible Wall*

Access http://www.wiley.com/college/budhu. Click Foundations and Earth Structures, Chapter 10, retwall.xls and click the worksheet "canflex" to conduct analysis of cantilever flexible walls.

Determine the depth of embedment required for stability of the cantilever sheet pile wall shown in Fig. E10.5a. Compare the results of the three methods—FMM, FSM, and NPMM—using $(FS)_p = 2.0$, $F_\phi = 1.25$, and $(FS)_r = 1.5$. Calculate the maximum bending moment for each of these methods. Groundwater is below the base of the wall. The tolerable rotation is $0.005\,H_o$ and less, where H_o is the wall height.

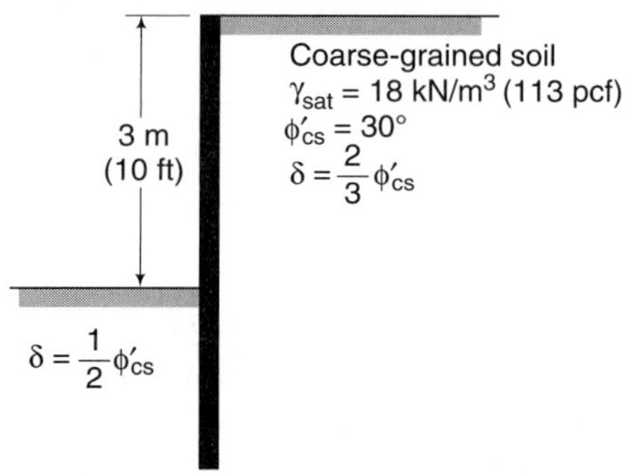

FIGURE E10.5a

Strategy You should use the Kerisel and Absi (1990) passive pressures (see Appendix C) and either the Kerisel and Absi (1990) or Coulomb active pressures. The key is to determine the lateral forces and then find d_o (an arbitrarily selected embedment depth at which rotation is presumed to occur) using moment equilibrium. Since you have to find the depth for unfactored values to calculate the maximum bending moment, you should determine K_{ax} and K_{px} for factored and unfactored values at the very beginning of your solution. SI units will be used in this example.

Solution 10.5

Step 1: Calculate K_{ax} and K_{px}.

$$\text{For FSM:}\quad \phi'_{\text{design}} = \frac{\phi'_{cs}}{F_\phi} = \frac{30}{1.25} = 24^\circ$$

$$\text{For FMM and NPPM:}\quad \phi'_{\text{design}} = \phi'_{cs}$$

Use the Kerisel and Absi (1990) K_{ax} and K_{px} (Appendix C).

$$\text{FSM:}\quad K_{ax} = 0.36 \quad \left(\phi'_{\text{design}} = 24^\circ, \delta = \frac{2}{3}\phi'_{\text{design}}, \beta/\phi'_{\text{design}} = 0\right)$$

$$K_{px} = 3.3 \quad \left(\phi'_{\text{design}} = 24^\circ, \delta = \frac{1}{2}\phi'_{\text{design}}, \beta/\phi'_{\text{design}} = 0\right)$$

$$\text{FMM and NPPM:}\quad K_{ax} = 0.28 \quad \left(\phi'_{cs} = 30^\circ, \delta = \frac{2}{3}\phi'_{cs}, \beta/\phi'_{cs} = 0\right)$$

$$K_{px} = 4.6 \quad \left(\phi'_{cs} = 30^\circ, \delta = \frac{1}{2}\phi'_{\text{design}}, \beta/\phi'_{cs} = 0\right)$$

Step 2: Determine the lateral earth pressure distributions.
The lateral pressure distributions for the FMM and FSM have the same shape but different magnitudes because of the different lateral earth pressure coefficients (Fig. E10.5b). The lateral pressure distribution for the NPPM is shown in Fig. E10.5c. Since groundwater is not within the depth of the retaining wall, $\gamma' = \gamma_{sat} = \gamma$. With reference to Fig. E10.5b.

Active case

$$P_{ax} = \frac{1}{2}K_{ax}\gamma(H_o + d_o)^2 = \frac{1}{2} \times K_{ax} \times 18(3 + d_o)^2 = 9K_{ax}(3 + d_o)^2$$

$$\bar{z}_a = \frac{H_o + d_o}{3} = \frac{3 + d_o}{3}$$

$$(M_O)_a = P_{ax}\bar{z}_a = 9K_{ax}(3 + d_o)^2\left(\frac{3 + d_o}{3}\right) = 3K_{ax}(3 + d_o^3)$$

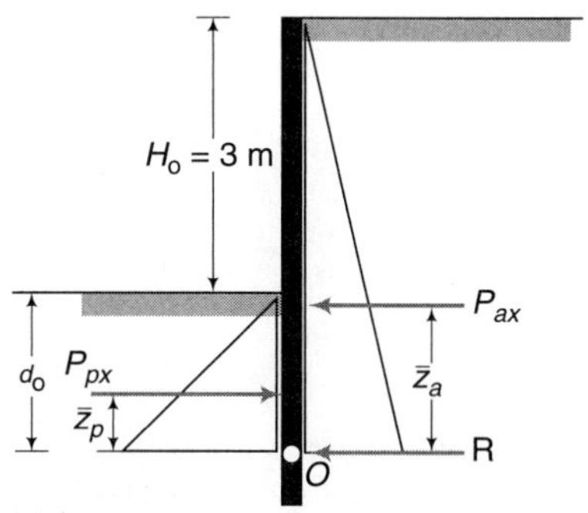

FIGURE E10.5b

Passive case

$$P_{px} = \frac{1}{2}K_{px}\gamma d_o^2 = \frac{1}{2} \times K_{px} \times 18 \times d_o^2 = 9K_{px}d_o^2$$

$$\bar{z}_p = \frac{d_o}{3}$$

$$(M_o)_p = P_{px}\bar{z}_p = 9K_{px}d_o^2 \times \frac{d_o}{3} = 3K_{px}d_o^3$$

Step 3: Find d_o.
All forces are calculated per meter length of wall.

FMM

The passive pressure is factored by $(\text{FS})_p$.

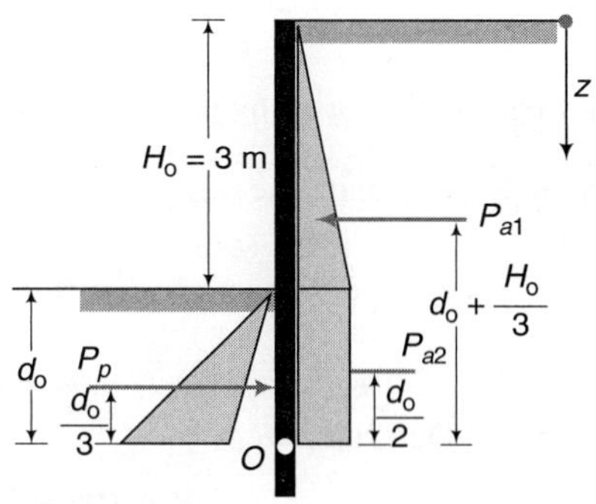

FIGURE E10.5c

$$K_{ax} = 0.28 \qquad \text{and} \qquad K_{px} = 4.6$$

$$(M_o)_a = 3 \times 0.28(3 + d_o)^3 = 0.84(3 + d_o)^3$$

$$(M_o)_p = \frac{3 \times 4.6d_o^3}{(\text{FS})_p} = \frac{3 \times 4.6d_o^3}{2} = 6.9d_o^3$$

$$\text{For equilibrium: } (M_o)_a = (M_o)_p$$

$$\therefore 0.84(d_o^3 + 9d_o^2 + 27d_o + 27) = 6.9d_o^3$$

which simplifies to

$$7.21d_o^3 - 9d_o^2 - 27d_o - 27 = 0$$

By trial and error or by using the polynomial function on a calculator, $d_o = 2.95$ m.

FSM

$$K_{ax} = 0.36 \quad \text{and} \quad K_{px} = 3.3$$

$$(M_o)_a = 3 \times 0.36(3 + d_o)^3 = 1.08(3 + d_o)^3$$

$$(M_o)_p = 3 \times 3.3d_o^3 = 9.9d_o^3$$

$$\text{For equilibrium: } (M_o)_a = (M_o)_p$$

$$1.08(d_o^3 + 9d_o^2 + 27d_o + 27) = 9.9d_o^3$$

which simplifies to

$$8.17d_o^3 - 9d_o^2 - 27d_o - 27 = 0$$

By trial and error or by using the polynomial function on a calculator, $d_o = 2.75$ m.

NPPM

$$K_{ax} = 0.28 \quad \text{and} \quad K_{px} = 4.6$$

The pressure diagram for the NPPM is shown in Fig. E10.5c

$$P_{a1} = \frac{1}{2} \times K_{ax}\gamma H_o^2 = \frac{1}{2} \times 0.28 \times 18 \times 3^2 = 22.7 \text{ kN}$$

$$P_{a2} = K_{ax}\gamma H_o d_o = 0.28 \times 18 \times 3 \times d_o = 15.1d_o$$

$$P_p = \frac{1}{2}(K_{px} - K_{ax})\gamma d_o^2 = \frac{1}{2}(4.5 - 0.28) \times 18d_o^2 = 38.9d_o^2$$

$$(M_o)_a = P_{a1}\left(d_o + \frac{H_o}{3}\right) + Pa_2\frac{d_o}{2} = 7.6d_o^2 + 22.7d_o + 22.7$$

$$(M_o)_p = P_p\frac{d_o}{3} = 13d_o^3$$

$$(\text{FS})_r = \frac{(M_o)_p}{(M_o)_a}$$

For $(\text{FS})_r = 1.5$,

$$1.5(7.6d_o^2 + 22.7d_o + 22.7) = 13d_o^3$$

Rearranging, we get

$$7.67d_o^3 - 7.6d_o^2 - 22.7d_o - 22.7 = 0$$

By trial and error or by using the polynomial function on a calculator, $d_o = 2.41$ m.

Step 4: Calculate the design depth.

$$\text{FMM: } d = 1.2d_o = 1.2 \times 2.95 = 3.54 \text{ m}$$
$$\text{FSM: } d = 1.2d_o = 1.2 \times 2.75 = 3.3 \text{ m}$$
$$\text{NPPM: } d = 1.2d_o = 1.2 \times 2.41 = 2.89 \text{ m}$$

Step 5: Determine R.

$$R = P_{px} - P_{ax}$$

$$\text{FMM: } d_o = 2.95 \text{ m}; R = 9 \times 4.6 \times 2.95^2 - 9 \times 0.28 \times (3 + 2.95)^2 = 271.1 \text{ kN/m}$$

$$\text{FSM: } d_o = 2.75 \text{ m}; R = 9 \times 3.3 \times 2.75^2 - 9 \times 0.36 \times (3 + 2.75)^2 = 109.4 \text{ kN/m}$$

$$\text{NPPM: } d_o = 2.41 \text{ m}; R = 9 \times (4.6 - 0.28) \times 2.41^2 - (9 \times 0.28 \times 3^2 + 0.28 \times 18 \times 3 \times 2.41) = 166.7 \text{ kN/m}$$

To calculate the net resistance below the assumed point of rotation, O, calculate the average passive pressure at the back of the wall and the average active pressure in front of the wall. Remember that below the point of rotation, passive pressure acts at the back of the wall and active pressure acts at the front of the wall. The mid-depth between d_o and $1.2d_o$ is $1.1d_o$.

FMM

$$\text{Average passive lateral pressure} = K_{px}\gamma\,(H_o + 1.1d_o) = 4.6 \times 18 \times (3 + 1.1 \times 2.95) = 517.1\,\text{kPa}$$
$$\text{Average active lateral pressure} = K_{ax}\gamma \times 1.1d_o = 0.28 \times 18 \times 1.1 \times 2.95 = 16.4\,\text{kPa}$$
$$\text{Net lateral pressure} = 517.1 - 16.4 = 500.7\,\text{kPa}$$
$$\text{Net force} = 500.7 \times 0.2d_o = 295.4\,\text{kN} > R(= 271.1\,\text{kN});$$

therefore, depth of penetration is satisfactory

FSM

$$\text{Average passive lateral pressure} = K_{px}\gamma\,(H_o + 1.1d_o) = 3.3 \times 18 \times (3 + 1.1 \times 2.75) = 357.9\,\text{kPa}$$
$$\text{Average active lateral pressure} = K_{ax}\gamma \times 1.1d_o = 0.36 \times 18 \times 1.1 \times 2.75 = 19.6\,\text{kPa}$$
$$\text{Net lateral pressure} = 357.9 - 19.6 = 338.3\,\text{kPa}$$
$$\text{Net force} = 338.3 \times 0.2d_o = 186.1\,\text{kN} > R(= 109.4\,\text{kN});$$

Therefore, depth of penetration is satisfactory

NPPM

$$\text{Average passive lateral pressure} = K_{px}\gamma\,(H_o + 1.1d_o) = 4.6 \times 18 \times (3 + 1.1 \times 2.41) = 467.9\,\text{kPa}$$
$$\text{Average active lateral pressure} = K_{ax}\gamma \times 1.1d_o = 0.28 \times 18 \times 1.1 \times 2.41 = 13.4\,\text{kPa}$$
$$\text{Net lateral pressure} = 467.9 - 13.4 = 454.5\,\text{kPa}$$
$$\text{Net force} = 454.5 \times 0.2d_o = 219.1\,\text{kN} > R(= 166.7\,\text{kN});$$

Therefore, depth of penetration is satisfactory

Step 6: Determine the maximum bending moment.
Maximum bending moment for $(\text{FS})_p = F_\phi = 1$.
Let z be the location of the point of maximum bending moment (point of zero shear) such that $z > H_o$.

$$\begin{aligned} M_z &= \frac{1}{2}K_{ax}\gamma\, z^2 \times \frac{z}{3} - \frac{1}{2}K_{px}\gamma\,(z - H_o)^2 \times \frac{(z - H_o)}{3} \\ &= \frac{1}{2} \times 4.6 \times 18 \times \frac{(z-3)^3}{3} \\ &= 0.84z^3 - 13.8(z-3)^3 \end{aligned}$$

To find z at which the bending moment is maximum, we need to differentiate the above equation with respect to z and set the result equal to zero.

$$\frac{dM_z}{dx} = 0 = 2.52z^2 - 41.4(z-3)^2 = 38.9z^2 - 248.4z + 372.6 = 0$$

Solving for z, we get $z = 3.68$ m or 2.53 m. The correct answer is 3.68 m since zero shear cannot occur above the excavation level in this problem (positive shear in the active zone only gets reduced below the excavation level).

$$M_z = 0.84 \times 3.68^3 - 13.8(3.68 - 3)^3 = 36\,\text{kN}\cdot\text{m}$$

For this problem, it is easy to use calculus to determine the maximum bending moment. For most problems, you will have to find the shear force distribution with depth, identify or calculate the point of zero shear, and then calculate the maximum bending moment. ■

EXAMPLE 10.6 *Anchored Flexible Wall*

Determine the embedment depth and the anchor force of the tied-back wall shown in Fig. 10.6a using the FSM.

Strategy You should use the Kerisel and Absi (1990) passive pressures (see Appendix C) and either the Kerisel and Absi (1990) or Coulomb active pressures. Groundwater level on both sides of the wall is the same, so seepage will not occur. SI units will be used in this example.

Solution 10.6

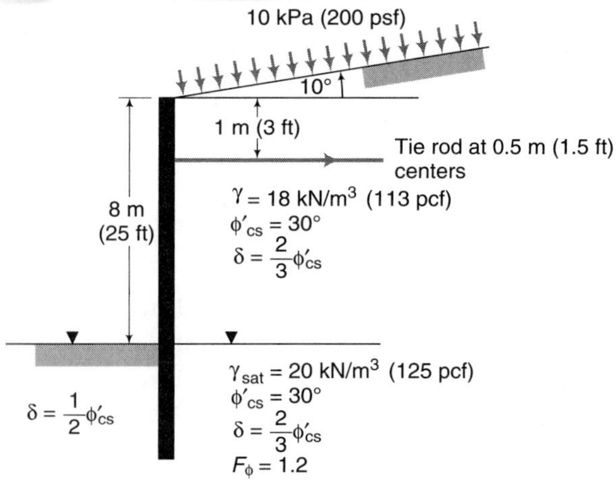

FIGURE E10.6a

Step 1: Determine K_{ax} and K_{px}.

$$\phi'_{\text{design}} = \frac{\phi'_{cs}}{F\phi} = \frac{30}{1.2} = 25^\circ; \qquad \frac{\beta}{\phi'_{\text{design}}} = \frac{10}{25} = 0.4$$

From Appendix C,

$$K_{ax} = 0.42 \quad \left(\phi'_{\text{design}} = 25^\circ, \beta/\phi'_{\text{design}} = 0.4, \delta = \frac{2}{3}\phi'_{\text{design}}\right)$$

$$K_{px} = 3.4 \quad \left(\phi'_{\text{design}} = 25^\circ, \beta/\phi'_{\text{design}} = 0, \delta = \frac{1}{2}\phi'_{\text{design}}\right)$$

For unfactored strength values,

$$K_{ax} = 0.31 \quad \left(\phi'_{cs} = 30^\circ, \beta/\phi'_{cs} = 10/30 = \frac{1}{3}, \delta = \frac{2}{3}\phi'_{cs}\right)$$

$$K_{px} = 4.6 \quad \left(\phi'_{cs} = 30^\circ, \beta/\phi'_{cs} = 0, \delta = \frac{1}{2}\phi'_{cs}\right)$$

Step 2: Determine the lateral forces and moments.
Use a table to facilitate ease of computation and checking. (See Fig. E10.6b.) Below the groundwater level, $\gamma' = 20 - 9.8 = 10.2\,\text{kN/m}^3$.

Step 3: Determine d.
Equate the sum of moments to zero (simplify equation by dividing by the coefficient of d^3).

$$d^3 + 7.3d^2 - 44.8d - 113.7 = 0$$

By trial and error or by using the polynomial function on a calculator, $d = 5.38\,\text{m}$.

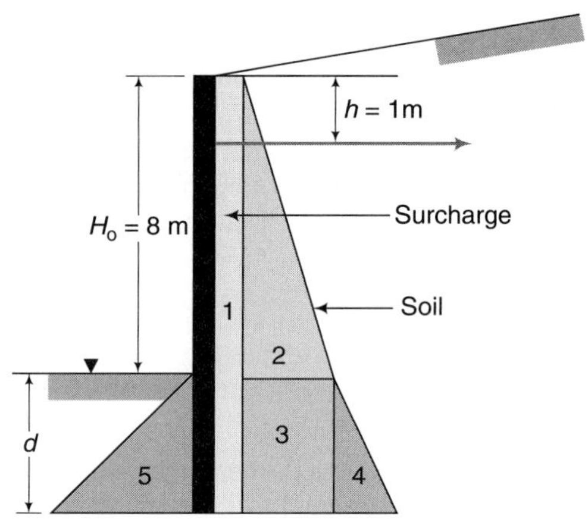

FIGURE E10.6b

Part	Horizontal force (kN)	Moment arm from anchor (m)	Moment (kN·m)
1	$K_{ax}q_s(H_o + d) = 0.42 \times 10 \times (8 + d) = 4.2d + 33.6$	$\left(\frac{H_o + d}{2}\right) - h = \left(\frac{8 + d}{2}\right) - 1 = 3 + \frac{d}{2}$	$-(2.1d^2 + 29.4d + 100.8)$
2	$\frac{1}{2}K_{ax}\gamma H_o^2 = \frac{1}{2} \times 0.42 \times 18 \times 8^2 = 241.9$	$\frac{2}{3}H_o - h = \frac{2}{3} \times 8 - 1 = 4.33$	-1047.4
3	$K_{ax}\gamma H_o d = 0.42 \times 18 \times 8 \times d = 60.5d$	$H_o - h + \frac{d}{2} = 7 + \frac{d}{2}$	$-(30.3d^2 + 423.5d)$
4	$\frac{1}{2}K_{ax}\gamma' d^2 = \frac{1}{2} \times 0.42 \times 10.2 \times d^2 = 2.1d^2$ $\Sigma\ 2.1d^2 + 64.7d + 275.5$	$H_o - h + \frac{2}{3}d = 7 + \frac{2}{3}d$	$-(1.4d^3 + 14.7d^2)$
5	$\frac{1}{2}K_{px}\gamma' d^2 = \frac{1}{2} \times 3.4 \times 10.2 \times d^2 = 17.3d^2$	$H_o - h + \frac{2}{3}d = 7 + \frac{2}{3}d$	$11.5d^3 + 121.1d^2$
		$\Sigma M = -(10.1d^3 + 74d^2 + 452.9d^2 + 1148.2)$	

Step 4: Determine d for the unfactored strength values.
To calculate the new depth of penetration for unfactored strength values, use proportionality, for example,

$$\text{Active moment} = \text{Active moment for factored strength values} \times \frac{\text{Unfactored } K_{ax}}{\text{Factored } K_{ax}}$$

$$(M_o)_A = -(1.4d^3 + 47.1d^2 + 452.9d + 1148.2) \times \frac{0.31}{0.42}$$
$$= -(d^3 + 34.8d^2 + 334.3d + 847.5)$$

$$\text{Passive moment: } (M_o)_p = (11.5d^3 + 121.1d^2) \times \frac{4.6}{3.4} = 15.5d^3 + 163.9d^2$$

$$\text{Sum of moments: } (M_o)_p + (M_o)_a = 14.5d^3 + 129.1d^2 - 334.3d - 847.5$$

Solving, we get $d = 3.38$ m

Step 5: Determine the anchor force for $d = 3.38$ m.

$$\Sigma \text{ Active forces} = (2.1d^2 + 64.7d + 275.5) \times \frac{0.31}{0.42} = 381.9 \text{ kN}$$

$$\Sigma \text{ Passive forces} = 17.3d^2 \times \frac{4.6}{3.4} = 265.8 \text{ kN}$$

$$T_a = 381.9 - 265.8 = 116.1 \text{ kN}$$

$$(T_a)_{\text{design}} = (\text{FS})T_a = 2 \times 116.1 = 232.2 \text{ kN}$$

■

EXAMPLE 10.7 *Anchored Flexible Wall*

Determine the embedment depth and the design anchor force required for stability of the sheet pile wall shown in Fig. E10.7a using the NPPM.

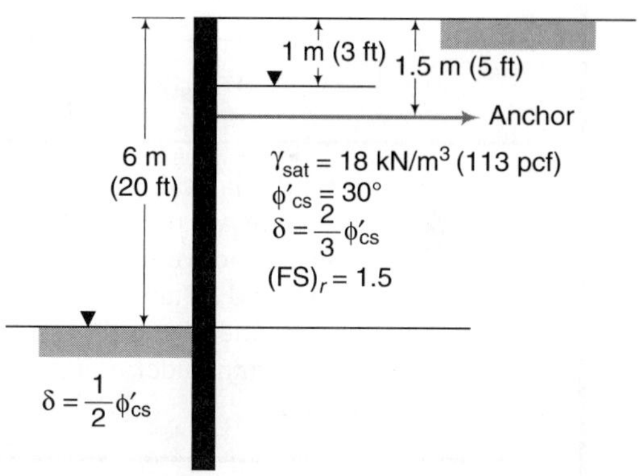

FIGURE E10.7a

Strategy In the NPPM, you must use the unfactored strength values to calculate K_{ax} and K_{px} and then determine the net active and net passive lateral pressures. To calculate d, you have to do iterations. A simple approach to solve for d is to set up the forces and moments in terms of the unknown d and then assume values of d until you find a d value that gives the required factor of safety ($(\text{FS})_r \cong 1.5$). A spreadsheet program or a programmable calculator is very helpful in solving this type of problem. In Excel, for example, you can use the Goal Seek function to find d. It is quite easy to make errors in calculations, so you should recheck your work and you must conduct a "hand" check when using outputs from computer programs. Since the groundwater levels are different in front and behind the wall, you need to consider seepage assuming a steady state seepage condition. To calculate the anchor force, you have to find d for $(\text{FS})_r = 1$ and then multiply the anchor force by 2 (factor of safety). SI units will be used in this example.

Solution 10.7

Step 1: Determine K_{ax} and K_{px}.
From Appendix C,

$$K_{ax} = 0.28 \left(\phi'_{cs} = 30^\circ, \beta/\phi'_{cs} = 0, \delta = \frac{2}{3}\phi'_{cs}\right)$$

$$K_{px} = 4.6 \left(\phi'_{cs} = 30^\circ, \beta/\phi'_{cs} = 0, \delta = \frac{1}{2}\phi'_{cs}\right)$$

Step 2: Determine the net lateral pressure. Make a table to do the calculations and draw a diagram of the lateral earth pressure distribution.

See Fig. E10.7b.

Below groundwater level

Average seepage force/

$$\text{unit volume: } j_s = \left(\frac{a}{a+2d}\right)\gamma_w$$

$$= \left(\frac{5}{5+2d}\right)9.8\,\text{kN/m}^3$$

$$\text{Active zone: } \gamma' = \gamma_{sat} - \gamma_w + j_s$$

$$= 18 - 9.8 + j_s = 8.2 + j_s\,\text{kN/m}^3$$

$$\text{Passive zone: } \gamma' = \gamma_{sat} - \gamma_w - j_s$$

$$= 18 - 9.8 - j_s = 8.2 - j_s\,\text{kN/m}^3$$

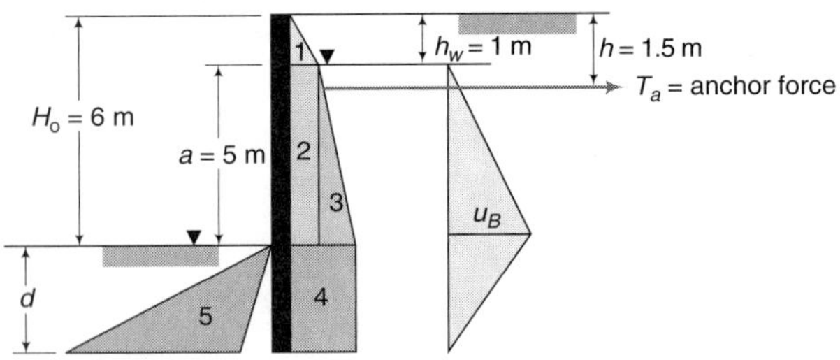

FIGURE E10.7b

Depth (m)	Active pressure (kPa)	Passive pressure (kPa)
0	0	0
1	$K_a\gamma h_w = 0.28 \times 18 \times 1 = 5$	0
6	$K_a\gamma h_w + K_a(\gamma' + j_s)(H_o - h_w)$ $= 5.0 + 0.28 \times (8.2 + j_s) \times 5 = 16.5 + 1.4j_s$	0
$6 + d$	$16.8 + 1.4j_s$	$K_p(8.2 - j_s)d - K_a(8.2 + j_s)d$ $= 4.6(8.2 - j_s)d - 0.28(8.2 + j_s)d$ $= (35.4 - 4.88j_s)d$
Water	$u_B = \frac{2(ad)}{a+2d}\gamma_w = \frac{98.1d}{5+2d}$	

Step 3: Calculate the lateral forces, the moment, and $(\text{FS})_r$.

All forces and moments are per meter length of wall. The moment is the sum of moments about the anchor position and R_x is the resultant active lateral force. In the first column under moment, a value of d is guessed and $(\text{FS})_r$ is calculated. In the second column under moment, the value of $d = 5.75$ m was obtained using a spreadsheet program (the actual value obtained from the spreadsheet program is $d = 5.73$ m for $(\text{FS})_r = 1.5$). If you use a spreadsheet program, the first column under moment is not needed. In Excel, you use Tools→Goal Seek to find d to satisfy the desired value of $(\text{FS})_r$.

Part	Forces (kN)	Moment arm (m)	Moment (kN · m)	
			$d = 7$ m	$d = 5.75$ m
1	$0.5 \times 5.0 \times 1 = 2.5$	$h - h_w + \frac{h_w}{3} = 0.5 + 0.33 = 0.83$	−2.1	−2.2
2	$5.0 \times 5 = 25$	$\frac{a}{2} - (h - h_w) = 2.5 - 0.5 = 2$	−50.0	−50.0
3	$0.5 \times (11.5 + 1.4j_s) \times 5 = 28.8 + 3.8j_s$	$\frac{2a}{3} - (h - h_w) = \frac{10}{3} - 0.5 = 2.83$	−109.2	−113.4
4	$(16.5 + 1.4j_s) \times d = (16.5 + 1.4j_s)d$	$H_o - h - \frac{d}{2} = 4.5 + \frac{d}{2}$	−1126.2	−876.0
	Water: $\frac{ad(a+d)\gamma_w}{a+2d} = \frac{49d(5+d)}{5+2d}$	$H_o + d - h - z_w$ $= 4.5 + d - \frac{5+2d}{3} = \frac{8.5+d}{3}$	−1119.3	−871.9

(continued)

(Continued)

Part	Forces (kN)	Moment arm (m)	$\Sigma M_r =$ Moment(kN/m) $d = 7$ m	$d = 5.75$ m
	$R_x = 56.3 + 3.8j_s + (16.5 + 1.4j_s)d + \frac{49d(5+d)}{5+2d}$	ΣM_d	−2402.6	−1909.2
5	$0.5 \times (35.4 - 4.88j_s)d^2 = (17.7 - 2.4j_s)d^2$	$H_o - h + \frac{2d}{3} = 4.5 + \frac{2d}{3}$	5179.8	2912.8
		$(FS)_r = \frac{\Sigma M_r}{\Sigma M_d}$	2.16	1.53

Step 4: Calculate the anchor forces for $(FS)_r = 1$.
For $(FS)_r = 1, d = 4.62$ m. Substituting $d = 4.62$ m, we get
Active zone

$$j_s = \left(\frac{5}{5 + 2 \times 4.62}\right)9.8 = 3.44\ \text{kN/m}^3$$

$$R_x = 56.3 + (3.8 \times 3.27) + (16.5 + 1.4 \times 3.44)4.62 + \frac{49 \times 4.62(5 + 4.62)}{5 + 2 \times 4.62}$$
$$= 320.1\ \text{kN}$$

Passive

$$\text{Passive lateral force} = (17.7 - 2.4 \times 3.44)4.62^2 = 201.6\ \text{kN}$$
$$T_a = \text{Active lateral force} - \text{Passive lateral force} = 320.1 - 201.6$$
$$= 118.5\ \text{kN}$$
$$\text{Design anchor force} = 2T_a = 237\ \text{kN}$$

■

10.12 SEISMIC ANALYSIS AND DESIGN OF FLEXIBLE RETAINING WALLS

Access a spreadsheet at http://www.wiley.com/college/budhu. Click Foundations and Earth Structures, Chapter 10, retwall.xls and click the worksheet "seismic flexible wall" to analyze cantilever flexible retaining walls under seismic loads.

Importance: *Flexible walls have suffered significant damages from earthquakes. The behavior of these walls under seismic loads is complex. So, approximate analyses are used in geotechnical practice. In this section, a limit equilibrium analysis is presented.*

The seismic analysis of flexible retaining walls is similar to gravity walls except that passive pressures are now important for stability. The forces on a flexible retaining wall under a seismic event are shown in Fig. 10.34. The overturning moment is

$$M_o = P'_{aE} \cos \delta_a \frac{H_o}{2} + P_a \cos \delta_a \frac{2H_o}{3} \tag{10.95}$$

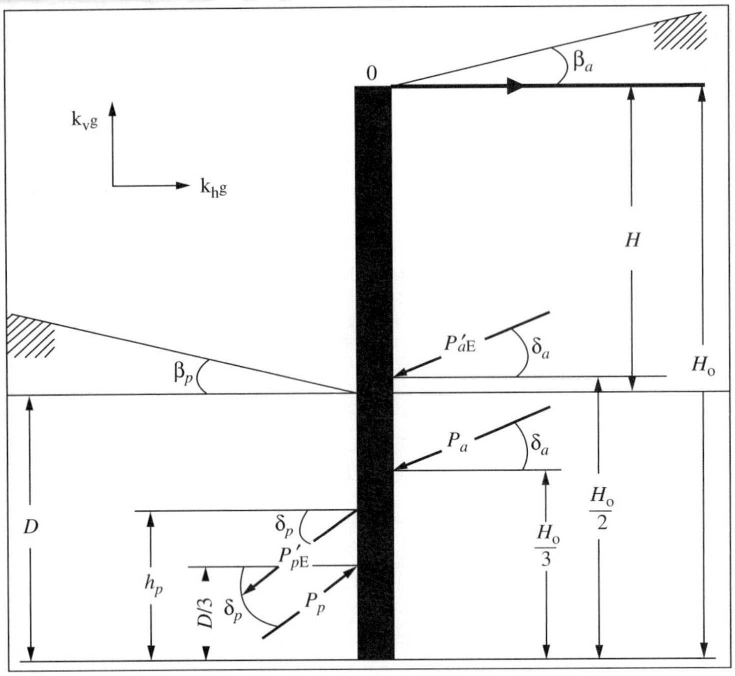

FIGURE 10.34 Forces on a flexible wall due to a seismic event.

where $P'_{aE} = P_{aE} - P_a$ is the dynamic increase in lateral active force, P_{aE} is the active seismic force [see Eq. (10.58)] and P_a is the active static force. The resisting moment is

$$M_r = P_p \cos \delta_p \left(H_o - \frac{D}{3} \right) - P'_{pE} \cos \delta_p (H_o - h_p) \tag{10.96}$$

where $P'_{pE} = P_{pE} - P_p$ is the dynamic increase in lateral passive force, P_{pE} is the passive seismic force [see Eq. (10.61)], P_p is the passive static force and h_p is the location of the dynamic increased in lateral passive force, taken as $\frac{D}{3}$.

The factor of safety is

$$FS = \frac{M_r}{M_o} \tag{10.97}$$

The wall will be in impending fail when FS = 1 and thus the critical acceleration can be obtained. If water is present you can calculate the resultant hydrodynamic forces using $F_w = \frac{1}{2} k_h \gamma_w h_w^2$ where h_w is the difference in height of the water. The moment due to the hydrodynamic forces is added to the dynamic active moment.

10.13 BRACED EXCAVATION

Importance *Sometimes it is not possible to use an anchored sheet pile wall for an excavation. For example, an existing building near a proposed excavation can preclude the use of anchors. You may have to brace your sheet piles within the excavation using struts. The analysis of braced excavation (also called cofferdam) is presented in the next section.*

Braced excavations consist of sheet piles driven into the soil to form the sides of an excavation (Fig. 10.35a) such as in the construction of bridge piers and abutments. As excavation proceeds within the area enclosed by the sheet piles, struts are added to keep the sheet piles in place.

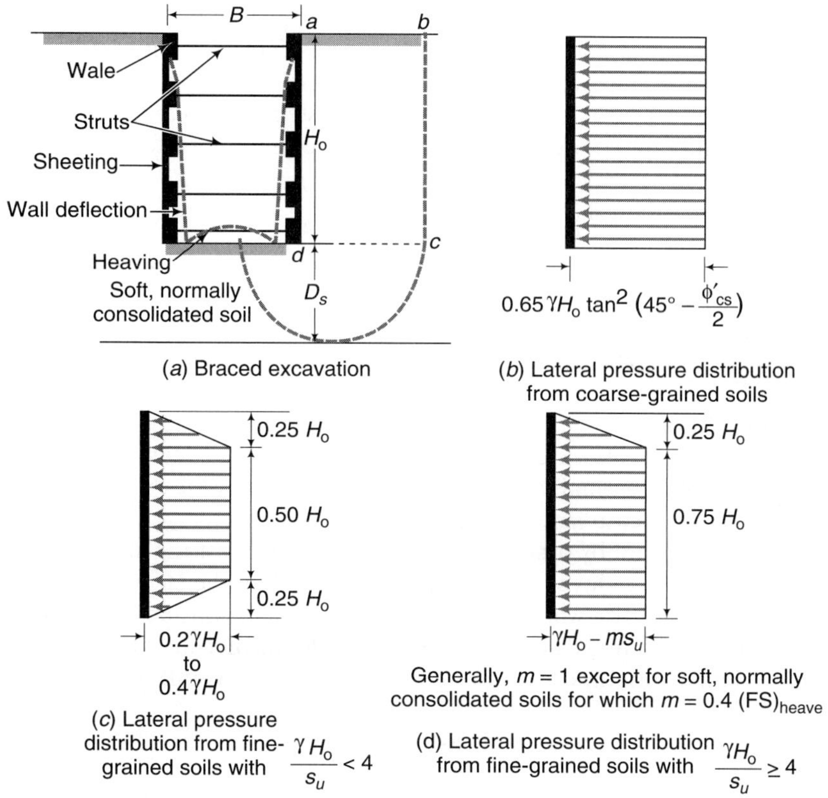

FIGURE 10.35 Braced excavations.

The top struts are installed followed by others at lower depths. The wall displacements before the top struts are installed are usually very small but get larger as the excavation gets deeper. The largest wall displacement occurs at the base of the excavation (Fig. 10.35b). Wall displacements are inconsistent with all the earth pressure theories.

The critical design elements in a braced excavation are the loads on the struts, which are usually different because of different lateral loads at different depths, the time between excavations, and the installation procedure. Failure of a single strut can be catastrophic because it can lead to the collapse of the whole system. The analysis for the forces and deflection in braced excavation should ideally consider the construction sequence, and numerical methods such as the finite element method are preferred. Semi-empirical methods are often used for shallow braced excavations and in the preliminary design of deep braced excavations. The finite element method is beyond the scope of this book. We will only discuss a semi-empirical method.

Lateral stress distributions for use in the semi-empirical method are approximations from field measurements of strut loads in different types of soil. The lateral stress distributions used for coarse-grained and fine-grained soils are shown in Figs. 10.35b–d. These lateral stress distributions are not real but average approximate stress distributions. The lateral stress distribution for coarse-grained soils (Fig. 10.35b) was extrapolated from strut loads measured for dense sand adjacent to the excavation. The appropriate value of friction angle is ϕ'_p but because we cannot rely on dilation, the design friction angle should be ϕ'_{cs}. For fine-grained soils, a total stress analysis is used and the lateral stress distribution depends on the stabililty number, $\gamma H_o/s_u$ (Peck, 1969). If the stability number is less than 4, the stress state of the soil adjacent to the excavation can be assumed to be elastic and the recommended lateral stress distribution is depicted in Fig. 10.35c. However, if the stability number is greater than or equal to 4, the stress state of the soil adjacent to the bottom of the excavation is expected to be plastic and the recommended lateral stress distribution is depicted in Fig. 10.35d.

If the soil below the base of the excavation is a soft, normally consolidated soil, it is possible that heaving can occur. The column of soil, *abcd* (Fig. 10.35a), above the base acts as a surcharge on the soil below the excavation level. This surcharge load may exceed the bearing capacity of the soft, normally consolidated soil, resulting in heaving. Bjerrum and Eide (1956) suggested that the excavation could be viewed as a footing of width B and embedment depth H_o. They showed that the factor of safety against bottom heave is

$$(\mathrm{FS})_{\text{heave}} = N_c \frac{s_u}{\gamma H_o + q_s} \tag{10.98}$$

where N_c is a bearing capacity coefficient given by Skempton (1951). The coefficient N_c can be approximated, for practical purposes, by

$$N_c = 6\left(1 + 0.2\frac{H_o}{B}\right) \quad \text{for} \quad \frac{H_o}{B} \leq 2.5 \quad \text{and} \quad N_c = 9I_c \quad \text{for} \quad \frac{H_o}{B} > 2.5, \quad I_c = 0.84 + 0.16\frac{B}{L} \tag{10.99}$$

where H_o/B is the depth to width ratio.

If $(\mathrm{FS})_{\text{heave}} < 1.5$, the sheeting should be extended below the base of the excavation for stability. The value of the coefficient m (Fig. 10.35d) is usually 1 except when the soil below the excavation is a deep, soft, normally consolidated soil, in which case $m \approx 0.4(\mathrm{FS})_{\text{heave}}$.

The strut loads at each level are found by assuming hinged connections of the struts to the sheet piles. A free-body diagram is drawn for each level and the forces imposed on the struts are determined using static equilibrium. Displacements of the walls are an important design consideration as adjacent structures may be affected. The method discussed above does not consider displacements. Analyses using numerical methods (e.g., finite element method) are better suited for the overall analysis of braced excavation.

The procedure for analysis of braced excavation is as follows:

1. Check the stability against bottom heave using Eq. (10.98). If $(\mathrm{FS})_{\text{heave}} < 1.5$, the walls should be extended below the base.
2. Determine the lateral stress on the walls for your soil type (Figs. 10.35b to d).
3. Treat the connections of the wall (sheet pile) to the struts as hinges.
4. Draw a free-body diagram at each level of the excavation.
5. Solve for the forces in the struts using the static equilibrium equations on each free-body diagram.

EXAMPLE 10.8 Braced Excavation

Determine the forces on the struts for the braced excavation in soft, normally consolidated clay as shown in Fig. E10.8a.

Strategy You need to determine the approximate lateral stress distribution by calculating $\gamma H_o/s_u$. To find the forces on the struts, draw free-body diagrams—one at each level—and use statics. SI units will be used in this example.

Solution 10.8

Step 1: Check for stability against bottom heave.

$$\frac{H_o}{B} = \frac{6}{4} = 1.5, \quad N_c = 6\left(1 + 0.2\frac{H_o}{B}\right) = 6(1 + 2 \times 0.15) = 7.6,$$

$$q_s = 0$$

$$(\mathrm{FS})_{\text{heave}} = N_c \frac{s_u}{\gamma H_o + q_s} = 7.6\frac{24}{20 \times 6} = 1.52 > 1.5;$$

therefore, excavation is safe against bottom heave

Step 2: Determine the lateral pressure diagram.

$$\frac{\gamma H_o}{s_u} = \frac{20 \times 6}{24} = 5; \quad \text{use Fig. 10.35d}$$

$$m = 0.4(\text{FS})_{\text{heave}} = 0.4 \times 1.6 = 0.64$$

The maximum lateral pressure is

$$\gamma H_o - m s_u = 20 \times 6 - 0.64 \times 24 = 104.6\,\text{kPa}$$

Step 3: Draw the pressure diagram.
See Fig. E10.8b.

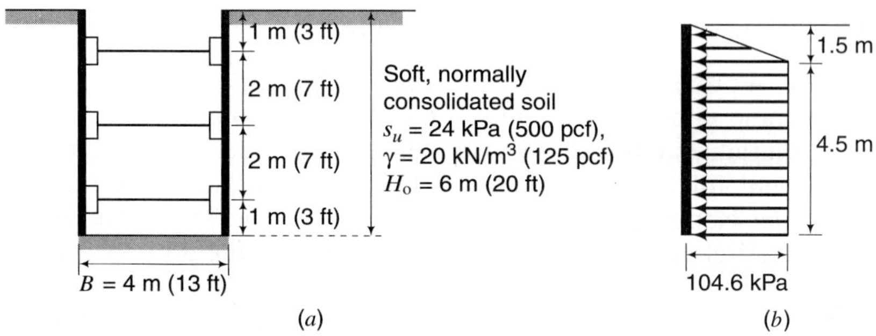

FIGURE E10.8a,b

Step 4: Calculate the forces on the struts at each level.
All loads are per meter length of wall.
Level 1 (Fig. E10.8c)

$$\Sigma M_{B_1} = 0 = 2.0A - \left[\frac{1}{2} \times 104.6 \times 1.5 \times \left(\frac{1.5}{3} + 1.5\right) + 1.5 \times 104.6 \times 0.75\right]$$

$$\therefore A = 137.3\,\text{kN/m}$$

$$\Sigma F_x = 0: \quad A + B_1 = \frac{1}{2} \times 104.6 \times 1.5 + 1.5 \times 104.6 = 235.4\,\text{kN/m}$$

$$\therefore B_1 = 235.4 - 137.3 = 98.1\,\text{kN/m}$$

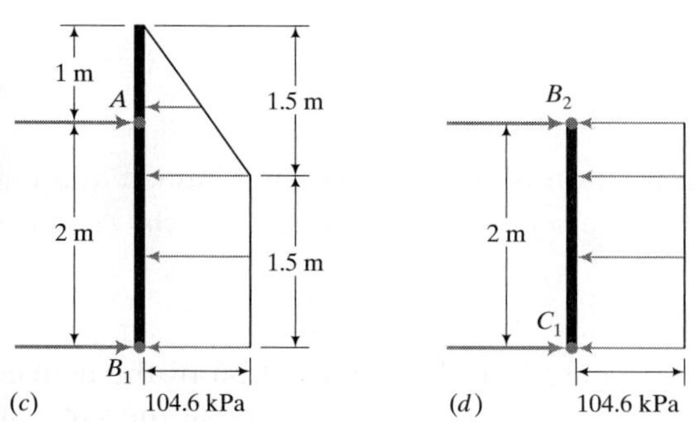

FIGURE E10.8c,d

Level 2 (Fig. E10.8d)

$$B_2 = C_1 = \frac{104.6 \times 2}{2} = 104.6\,\text{kN/m}$$

Step 5: Calculate the forces on each strut.

$$A = 137.3\,\text{kN/m}$$
$$B = B_1 + B_2 = 98.1 + 104.6 = 202.7\,\text{kN/m}$$
$$C = C_1 + 104.6 \times 1 = 104.6 + 104.6 = 209.2\,\text{kN/m}$$

■

10.14 DESIGN ISSUES

Importance *The analysis of retaining walls gives the geometry to ensure a margin of safety. There are several design and construction issues that can change an otherwise stable design to an unstable one. In this section, we consider some of these issues.*

10.14.1 Lateral Stress Coefficients and Lateral Stresses

The lateral stress coefficients are approximate values for clean, uncemented soils. As discussed in Section 10.7, you should use ϕ'_{cs} and conservative values of interface friction, if any. When small displacements ($< 0.005H_o$) are required, you should use K_o rather than K_a. A surcharge of at least 5 kPa (100 psf) should be used to account for construction loads in addition to lateral stresses imposed by compaction of the backfill. Either Rankine or Coulomb earth pressure coefficients can be used for active pressures but Caquot and Kerisel (1948) earth pressure coefficients should be used for passive pressures.

Codes (e.g., IBC, 2006) normally specify minimum lateral earth pressures for the design of earth retaining structures. If drying out can cause a gap between the wall and the backfill, you should use the full hydrostatic pressure on the wall.

10.14.2 Drainage

Water is the culprit for most retaining wall failures. In rigid retaining walls, clean coarse-grained soils should be used as the backfill and a suitable drainage system must be designed. You must consider the following:

1. How and where will the water exit the wall? Putting weep holes without considering how the water in front of the wall will drain to an outfall or whether the wall would be stained by minerals is unsatisfactory.
2. Drainage of surface water.
3. Toe erosion.
4. Possibility of blocked drainage system due to migration of fines, debris, and human and animal activities.

10.14.3 Wall Deformation

All the analytical methods presented in this chapter are based on the mobilization of the active and passive earth pressures. As noted in Section 10.7, these pressures are not mobilized at the same strain level. When wall deformation is required to be controlled, you should use conservative values of lateral earth pressure coefficients in the analytical methods. You should consider using FDM and FEM to estimate wall deformations and instrument the walls where appropriate.

10.14.4 Construction

Sometimes construction places limitations on the type of wall that is applicable for a project. Rigid walls require *bottom-up* construction. That is, the wall must be constructed and then backfilled from the base to the surface. Significant construction space is required.

Flexible retaining walls used to support an excavation require *top-down* construction. The walls are installed and then excavation proceeds from the top. Wall deformation is a key issue, especially if the construction is in a built-up area. Bracings are added for stability and serviceability. If structures near the excavation are sensitive to wall deformation, as they often are, you should consider monitoring the wall deformation using inclinometers or other instruments such as lasers.

Compacting backfill always introduces additional short-term lateral stress to walls as discussed before. It is often difficult to use compaction rollers near the walls. Hand vibrating compactors are normally used near the wall. The degree of compaction may not be the same as for compaction rollers. Indeed, the soil generally is looser near the wall.

Flexible retaining walls are installed using drop hammers or pneumatic hammers or hydraulic hammers. Choice of equipment depends on the soil type, availability, project site environment (e.g., noise control, working space, and access), and costs.

10.15 SUMMARY

We have considered lateral earth pressures and their applications to several types of earth retaining structures in this chapter. Two earth pressure theories are in general use: one developed by Coulomb and the other by Rankine. Coulomb's equations for the lateral earth pressure coefficients are based on limit equilibrium and include the effects of soil-wall friction, wall slope, and backfill slope. Rankine's equations are based on stress states of the backfill and do not account for soil–wall friction. The failure planes in the Coulomb and Rankine methods are planar surfaces. Soil–wall friction causes the failure plane to curve, resulting in higher active lateral earth pressure coefficients and lower passive earth pressure coefficients than those found using Coulomb's equations.

Because of the flexibility of some earth retaining structures and construction methods used in practice, the "real" lateral earth pressures are different from either the Coulomb or Rankine theories. It is suggested in this book that the appropriate value of friction angle to use in the analysis of earth retaining structures is ϕ'_{cs}. Three methods of analysis for flexible earth retaining walls were considered. The differences in the methods result mainly from how the lateral stresses are considered and how the factor of safety is applied.

Practical Examples

Access a spreadsheet at http://www.wiley.com//college/budhu. Click Foundations and Earth Structures, Chapter 10, retwall.xls to analyze retaining walls.

EXAMPLE 10.9 *Cantilever Gravity Walls for Material Storage*

Ore from a manufacturing plant is to be stored between two cantilever gravity retaining walls, as shown in Fig. E10.9a. A gantry crane will run on top of the walls to place the ore. The crane will apply a maximum vertical load of 24 kN and a horizontal load of $\pm$4.5 kN on each wall. The base of the ore pile is only permitted to come within 0.5 m of the top of the walls. The ore surface should be the maximum admissible slope. The walls will be restrained from spreading outward by steel ties or rods at 1 m centers anchored to the base slab. A 2 m layer of the ore spoils of similar characteristics to the ore to be stored would be compacted to support the base of the wall. Determine (a) the stability of the wall for the geometry shown in Fig. E10.9a and (b) the force in the tie rods, assuming they resist all the horizontal loads. If the tie rods were not present, would the walls be safe? Do you expect any alignment problems with the gantry crane?

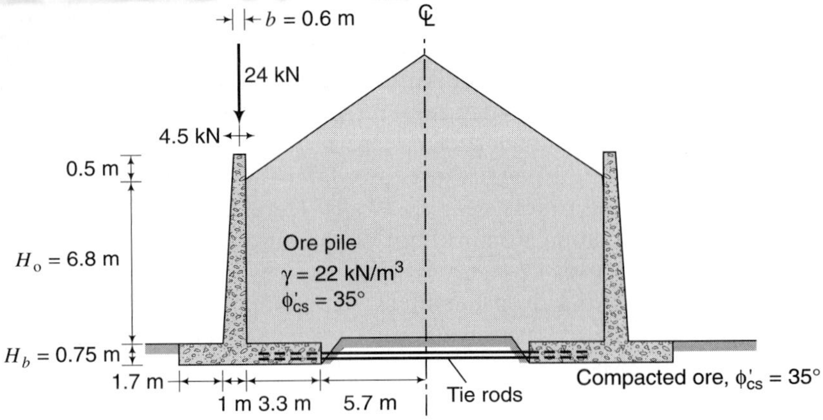

FIGURE E10.9a

Strategy The maximum slope angle would be the friction angle. Since the storage is symmetrical, each wall will carry identical loads. You will have to make an assumption regarding wall friction. You can assume $\delta = \frac{1}{2}\phi'_{cs}$ or $\delta = \frac{2}{3}\phi'_{cs}$ and use Coulomb's method to determine the lateral forces.

Solution 10.9

Step 1: Determine the lateral forces.
The maximum admissible slope is $\beta = \phi'_{cs}$. Therefore, $\beta = 35^\circ$.

$$H_o = 0.75 + 6.8 + 3.3\tan 35^\circ = 9.86\,\text{m}$$

From Eq. (10.16)

$$K_a = 0.7 \quad \text{for}\ \beta = 35^\circ, \quad \phi' = 35^\circ, \quad \delta = \frac{1}{2}\phi'$$

$$P_a = \frac{1}{2}K_a\gamma H_o^2 = \frac{1}{2} \times 0.7 \times 22 \times 9.86^2 = 748.7\,\text{kN/m}$$

$$P_{ax} = P_a \cos\delta = 748.7 \times \cos 17.5^\circ = 714.0\,\text{kN/m}$$

$$P_{az} = P_a \sin\delta = 748.7 \times \sin 17.5^\circ = 225.1\,\text{kN/m}$$

Step 2: Calculate forces and moments. Draw a diagram (show only one-half because of symmetry, see Fig. E10.9b) and use a table to do the calculations.

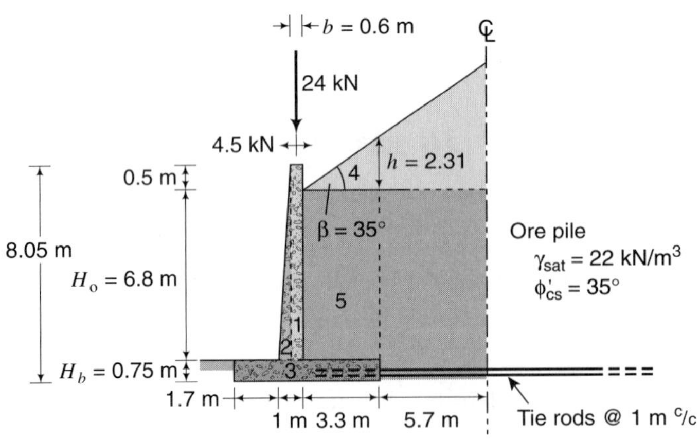

FIGURE E10.9b

Part	Force (kN/m)	Moment arm from 0 (m)	Moment (kN/m)
Wall	*Vertical forces*		
1	$(6.8+0.5)\times 0.6\times 24=105.1$	$1.7+0.4+0.6/2=2.4$	+252.2
2	$\frac{1}{2}(6.8+0.5)\times 0.4\times 24=35$	$1.7+\frac{2}{3}\times 0.4=1.97$	69.0
3	$0.75\times 6\times 24=108.0$	$6/2=3.00$	+324.0
Soil			
4	$\frac{1}{2}\times 3.3\times 2.31\times 22=83.9$	$1.7+1+\left(\frac{2}{3}\times 3.3\right)=4.9$	+411.1
5	$3.3\times 6.8\times 22=493.7$	$1.7+1+3.3/2=4.35$	+2147.6
P_{az}	225.1	6	+1350.6
Gantry			
P_z	24	2.4	+57.6
	$R_z=\Sigma$ Vertical forces $=1074.8$		
Gantry	*Lateral forces*		
P_x	4.5	$0.75+6.8+0.5=8.05$	−36.2
Soil			
P_{ax}	714	$9.87/3=3.29$	−2349.1
	$R_x=\Sigma$ Lateral forces $=718.5$		$M=\Sigma$ Moments $=2226.8$

Step 3: Determine stability.

Rotation

$$\bar{x}=\frac{M}{R_z}=\frac{2226.8}{1074.8}=2.07\,\text{m}$$

$$e=\frac{B}{2}-\bar{x}=\frac{6}{2}-2.07=0.93\,\text{m}$$

$$\frac{B}{6}=\frac{6}{6}=1\,\text{m};\quad e<\frac{B}{6};\quad \text{therefore, wall is safe against rotation}$$

Sliding

T = sliding resistance at base $=R_z\tan\phi'_b=1074.8\tan 20^\circ=391.2\,\text{kN}$

$$(\text{FS})_T=\frac{T}{R_z}=\frac{391.2}{718.5}=0.54<1.5$$

∴ sliding would occur if tie rods were present.

Forces on tie rods

$$\text{For } (\text{FS})_T=1.5:\quad T_a+T=1.5R_x$$

where T_a is the force on the tie rods/unit length of wall.

$$T_a=1.5\times 718.5-391.2=686.6\,\text{kN/m}$$

Bearing capacity

All horizontal forces accommodated by tie rods.

$$B'=B-2e=6-2\times 0.93=4.14\,\text{m}$$

$$V=1074.8\,\text{kN}$$

$$N_\gamma=0.1054\exp(9.6\phi'_p)=0.1054\exp\left(9.6\times 35\times\frac{\pi}{180}\right)=37.2$$

$$q_u=0.5\gamma B'N_\gamma=0.5\times 18\times 4.14\times 37.2=1386.1\,\text{kPa}$$

$$\sigma_{\max}=\frac{\Sigma R_z}{B\times 1}\left(1+\frac{6e}{B}\right)=\frac{1074.8}{6\times 1}\left(1+\frac{6\times 0.93}{6}\right)=345.7\,\text{kPa}$$

$$(\text{FS})_B=\frac{1386.1}{345.7}=4>3;\quad \text{therefore, bearing capacity is satisfactory}$$

SUMMARY OF RESULTS

1. The wall is unlikely to rotate significantly.
2. Without the tie rods, the wall will translate.
3. The design tie rod force is $\approx 687\ \text{kN/m}$.
4. The soil bearing capacity is adequate.
5. Assuming that the base slab is rigid and the loading is symmetrical, there should be no alignment problem. ■

EXAMPLE 10.10 *Flexible Wall Near a Lake*

A retaining wall is required near a man-made lake in a housing scheme development. The site is a swamp and the topsoil consisting mainly of organic material will be removed up to the elevation of a deep deposit of clay with a silt and sand mixture. The wall is expected to retain a sand backfill of height 6 m. It is anticipated that a rapid drawdown condition could occur and the lake emptied for a long period. A flexible retaining wall is proposed as shown in Fig. E10.10a. Determine the embedment depth and the force for an anchor spacing of 3 m. A surcharge of 10 kPa should be considered.

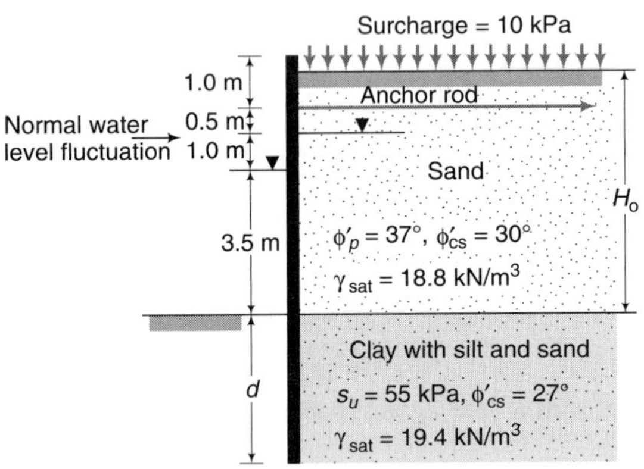

FIGURE E10.10a

Strategy You can use either the FSM or NPPM or both. Because the clay layer has silt and sand, you should consider seepage forces for long-term conditions (assuming that the drawdown level would remain for some time).

Solution 10.10

Step 1: Determine the lateral earth pressure coefficients.
Use Appendix C to find the lateral earth pressure coefficients.

$$\text{Sand:}\quad K_{ax} = 0.28\left(\phi'_{cs} = 30°,\ \delta = \frac{2}{3}\phi'_{cs}\right)$$

$$\text{Clay:}\quad K_{ax} = 0.34\left(\phi'_{cs} = 27°,\ \delta = \frac{2}{3}\phi'_{cs}\right)$$

$$K_{px} = 3.8\left(\phi'_{cs} = 27°,\ \delta = \frac{1}{2}\phi'_{cs}\right)$$

Step 2: Determine the seepage forces and porewater pressures.

$$\text{Sand:}\quad \gamma' = 18.8 - 9.8 = 9\ \text{kN/m}^3$$

$$\text{Clay:}\quad \gamma' = 19.4 - 9.8 = 9.6\ \text{kN/m}^3$$

$$j_s = \frac{a}{a + 2d}\gamma_w = \left(\frac{3.5}{3.5 + 2d}\right)9.8 = \frac{34.3}{3.5 + 2d}\ \text{kN/m}^3$$

Active state

$$\text{Sand:}\quad \gamma'_{ja} = \left(9 + \frac{34.3}{3.5+2d}\right) = \frac{65.8+18d}{3.5+2d}$$

$$\text{Clay:}\quad \gamma'_{ja} = \left(9.6 + \frac{34.3}{3.5+2d}\right) = \frac{67.9+19.2d}{3.5+2d}$$

Passive state

$$\text{Clay:}\quad \gamma'_{jp} = \left(9.6 - \frac{34.3}{3.5+2d}\right) = \frac{19.2d-0.7}{3.5+2d}$$

$$u_B = \left(\frac{2ad}{a+2d}\right)\gamma_w = \left(\frac{2\times 3.5d}{3.5+2d}\right)9.8 = \frac{68.6d}{3.5+2d}$$

$$P_w = \frac{1}{2}(a+d)u_B = \frac{1}{2}(3.5+d)\frac{6.8.6d}{3.5+2d} = \left(\frac{120d+34.3d^2}{3.5+2d}\right)$$

$$\bar{z}_w = \frac{a+2d}{3} = \frac{3.5+2d}{3}$$

$$H_o - h_w = 6 - 2.5 = 3.5\,\text{m};\quad H_o - h = 6 - 1 = 5\,\text{m};$$

$$H_w - h = 2.5 - 1.0 = 1.5\,\text{m}$$

Step 3: Carry out the calculations.
Draw a diagram of the pressure distributions. Use a table (see below) or a spreadsheet program to carry out the calculations. See Fig. E10.10b for the lateral pressure distributions.

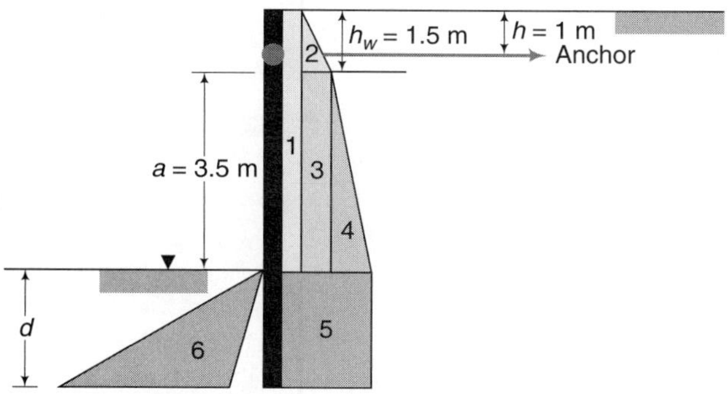

FIGURE E10.10b

Step 4: Calculate the anchor force.

$$T_a = 317.1 - 208.7 = 108.4\,\text{kN/m}$$

Assume a factor of safety of 2.
Design anchor force $= T_a(\text{FS}) \times$ Anchor spacing $= 108.4 \times 2 \times 3 = 650.4\,\text{kN}$ ■

EXAMPLE 10.11 *Gravity retaining wall in a seismic region*

A gravity retaining wall is required to support a highway cut 4 m (12 ft) high. The backfill will be a reject sand with $\phi'_{cs} = 30^\circ$ and $\gamma_{sat} = 18\,\text{kN/m}^3$. The slope of the backfill will be $+10^\circ$. The peak ground acceleration is 0.25 g and the peak velocity is 0.2 m/s. Determine the weights of the wall for seismic and static loading. The tolerable lateral displacement is 50 mm. Assume $\phi'_b = \frac{2}{3}\phi'_{cs}$, $\delta = 0$ and $\eta = 10^\circ$. As a preliminary design, dimension the wall for a factor of safety of 1.5.

Strategy Follow the procedure given in Section 10.10.2

Solution 10.11

Step 1: Calculate k_h

Ag = 0.25 g, V = 0.20 m/s, $\rho_d = 0.05$ m

$$k_h = A\left(\frac{Ag\,\rho_d}{0.087\,V^2}\right)^{-\frac{1}{4}} = 0.25\left(\frac{0.25\times 9.81\times 0.05}{0.087\times 0.2^2}\right)^{-\frac{1}{4}} = 0.1$$

Part	Pressure (kPa)	Force (kN/m)	Moment arm from 0	Moment = Force× Moment arm	D = 5.57 m Moment (kN·m/m)	D = 4.39 m Moment (kN·m/m)	D = 4.39 m Force (kN/m)
1	$p_{a1} = K_{ax} q_s = 0.3 \times 10 = 3$	$p_{a1} H_0 = 3 \times 6 = 18$	$\frac{H_0}{2} - h = \frac{6}{2} - 1 = 2$	$(-)33.6$	−33.6	−33.6	16.8
2	$p_{a2} = K_{ax} \gamma h_w = 0.3 \times 18.8 \times 2.5 = 14.1$	$\frac{1}{2} p_{a2} h_2 = \frac{1}{2} \times 14.1 \times 2.5 = 17.6$	$\frac{2}{3} h_w - 1 = \frac{2}{3} \times 2.5 - 1 = 0.67$	$(-)11.0$	−11.0	−11.0	16.5
3	$p_{a3} = p_{a2} = 14.1$	$p_{a3} a = 14.1 \times 3.5 = 49.4$	$\frac{a}{2} + (h_w - h) = \frac{3.5}{2} + 1.5 = 3.25$	$(-)149.7$	−149.7	−149.7	46.1
4	$p_{a4} = k_{ax} \gamma_{ja} a = 0.3 \left(\frac{65.8 + 18d}{3.5 + 2d} \right) 3.5 = \frac{69.1 + 18.9d}{3.5 + 2d}$	$\frac{1}{2} p_{a4} a = \frac{1}{2} \left(\frac{69.1 + 18.9d}{3.5 + 2d} \right) \times 3.5 = \frac{120.9 + 33.1d}{3.5 + 2d}$	$\frac{2}{3} a + (h_w - h) = \frac{2}{3} \times 3.5 + 1.5 = 3.83$	$(-) \frac{463 + 126.8d}{3.5 + 2d}$	−74.6	−77.5	20.2
5	$p_{a5} = p_{a1} + p_{a2} + p_{a3} + p_{a4} = 3 + 14.1 + 14.1 + \frac{69.1 + 18.9d}{3.52d} = \frac{178.3 + 81.3d}{3.5 + 2d}$	$p_{a5} d = \left(\frac{129d + 53.1d^2}{3.5 + 2d} \right) d$	$\frac{d}{2} + (H_0 - h) = \frac{d}{2} + 5 = \frac{d + 10}{2}$	$(-) \left(\frac{26.5d^2 + 330d + 645}{3.5 + 2d} \right) d$	−1173.1	−869.2	120.8
Water	$u_b = \frac{8.6d}{3.5 + 2d}$	$P_w \frac{(120 + 34.3d)d}{3.5 + 2d}$	$(H_0 - h) + d - y_w = 5 + d - \frac{3.5 + 2d}{3} = \frac{11.5 + d}{3}$	$(-) \left(\frac{11.4d^2 + 171.4d + 460}{3.5 + 2d} \right) d$	<u>−673.0</u>	<u>−512.4</u>	<u>96.7</u>
6	$p_{p6} = K_{px} \gamma_{jp} d - K_{ax} \gamma_{ja} d = \left[3.8 \left(\frac{19.2d - 0.7}{3.5 + 2d} \right) - 0.34 \left(\frac{67.9 + 19d}{3.5 + 2d} \right) \right] = \left(\frac{33.2d - 12.9}{3.5 + 2d} \right) d^2 = \left(\frac{66.4d - 25.7}{3.5 + 2d} \right) d$	$\frac{1}{2} p_{p6} d = \frac{1}{2} \left(\frac{66.4d - 25.7}{3.5 + 2d} \right) d$	$\frac{2}{3} d + (H_0 - h) = \frac{2}{3} d + 5 = \frac{2d + 15}{3}$	$(+) \frac{(22.1d^2 + 157.5d - 64.5)d^2}{3.5 + 2d}$	$\Sigma = -2114.9$	$\Sigma = -1653.4$	$\Sigma = 317.1$
					<u>+3172.9</u>	<u>+1653.9</u>	208.7

Step 2: Calculate K_{aE}

$$k_v = 0, \quad \beta = 10^\circ, \quad \eta = 10^\circ, \quad \delta = 0, \quad \theta = \tan^{-1}\left(\frac{0.1}{1-0}\right) = 5.7^\circ, \quad \phi'_b = 20^\circ.$$

Eq. (10.59): $K_{aE} = 0.55$

Step 3: Calculate W_w

Assume $H_o = 4$ m. Normally H_o will be larger as it is measured from the base to the soil surface.

Eq. (10.72):
$$C_{IE} = \frac{\cos(0+10) - \sin(0+10)\tan 20^\circ}{(1-0)(\tan 20^\circ - \tan 5.7^\circ)} = 3.5$$

Eq. (10.71):
$$W_w = 3.5\left[\frac{1}{2}\times 18 \times 4^2 \times (1-0)0.55\right]$$
$$= 277.2\,\text{kN/m}$$

Step 4: Calculate the weight of wall for static loading

$K_a = 0.33$

Eq. (10.73):
$$F_T = \frac{0.55(1-0)}{0.33} = 1.67$$

Eq. (10.75):
$$C_I = \frac{\cos(0+10) - \sin(0+10)\tan 20^\circ}{\tan 20^\circ} = 2.53$$

$$W = C_I \times P_a = 2.53 \times \frac{1}{2} \times 18 \times 4^2 \times 0.33 = 120\,\text{kN/m}$$

Seismic amplification factor is

$$F_w = F_T F_I = F_T\frac{C_{IE}}{C_I} = 1.67 \times \frac{3.5}{2.53} = 2.3$$

or

$$\frac{W_w}{W} = \frac{277.2}{120} = 2.3$$

Step 5: Preliminary dimension

The required weight of wall to limit displacement to 50 mm with an FS = 1.5 is $W_w = 1.5 \times 272.2 = 408\,\text{kN/m}$.

The area required per meter length of wall is $\frac{W_w}{\gamma_c}$ where γ_c is the unit weight of concrete ($\gamma_c = 24\,\text{kN/m}^3$)

Area required $= \frac{408}{24} = 17\,\text{m}^2$ With the dimension shown in Fig. E10.11,

$$\frac{1}{2}(b + b + 0.7) \times 4 = 17$$
$$b = 4\,\text{m}$$

This is a massive wall and could be costly. You should compare this wall with a cantilever gravity wall. ■

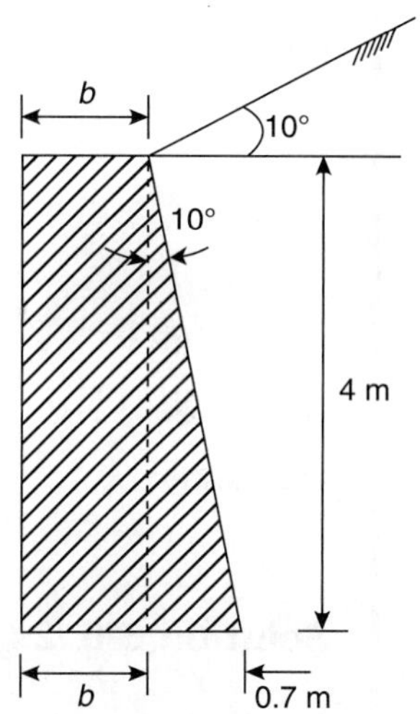

FIGURE E10.11

EXAMPLE 10.12 *Design of anchors for a tieback wall*

A permanent anchored retaining wall is required to support an exit ramp along a highway. The soils investigation is limited with only SPT values available. A cross section of the wall at a particular location is shown in Fig. E 10.12a. As a preliminary design, assume the anchors are located as shown in Fig. E10.12a. Design the anchors as grouted anchors. Assume a surcharge due to traffic load of 10 kPa.

Strategy Since only SPT values are given, you have to estimate the required soil parameters (ϕ'_{cs} and γ). Use conservative values particularly because the wall is permanent.

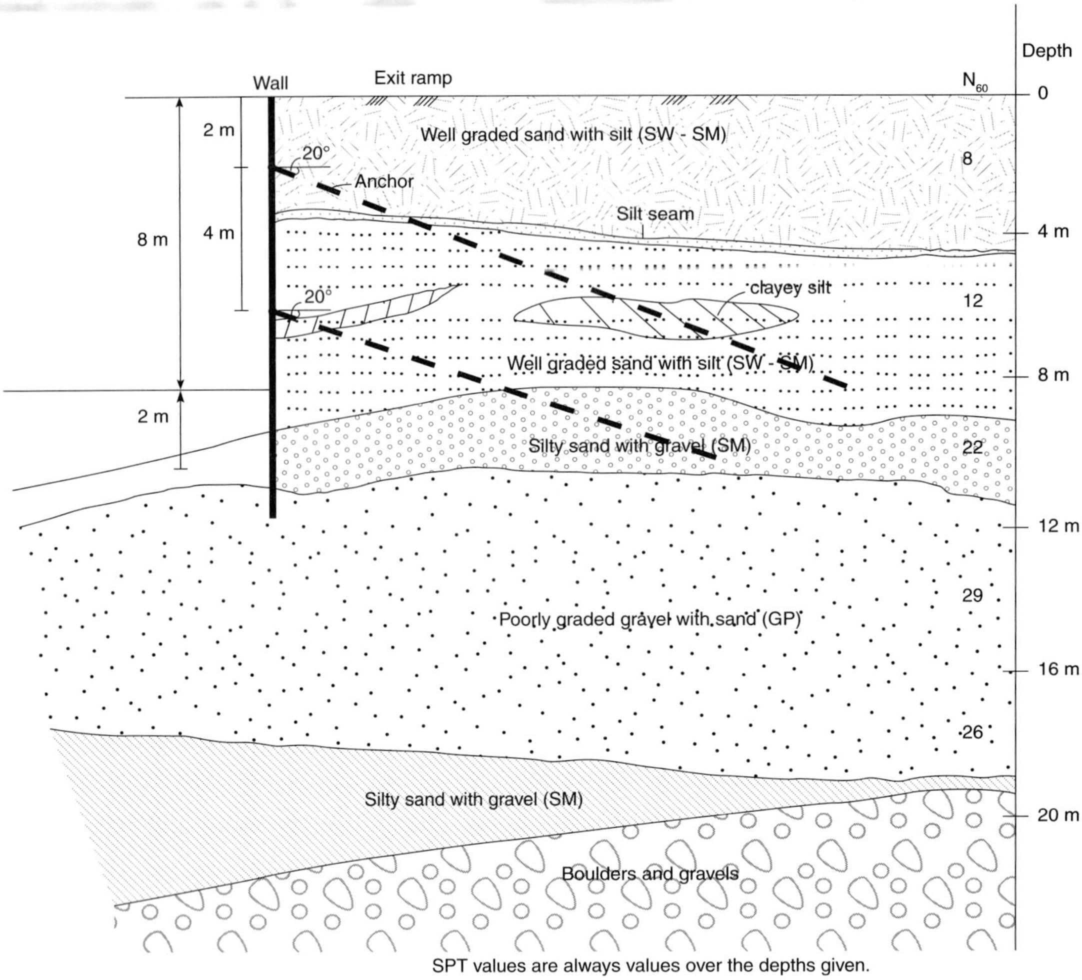

FIGURE E10.12a

Solution E10.12

Step 1: Make estimates of the soil parameters required. Take an average value of N_{60} within the excavated depth.

$$(N_{60})_{avg} = \frac{8 + 12}{2} = 10$$

From Table B5.2, $\gamma = 17\,\text{kN/m}^3$, $\phi'_p \approx 35^\circ$ for $(N_{60})_{avg} = 10$
For $N_{60} = 8$, $\phi'_{cs} \approx 29^\circ$
Use $\phi'_{cs} = 30^\circ$ as a conservative value

Step 2: Calculate lateral earth pressures.
Most of the soils within the excavated depth are coarse-grained. Use the pressure distribution for coarse grained soils (Fig. E10.12b)

$$K_a = \tan^2\left(45^\circ - \frac{\phi'_{cs}}{2}\right) = \tan^2\left(45^\circ - \frac{30^\circ}{2}\right) = \frac{1}{3}$$

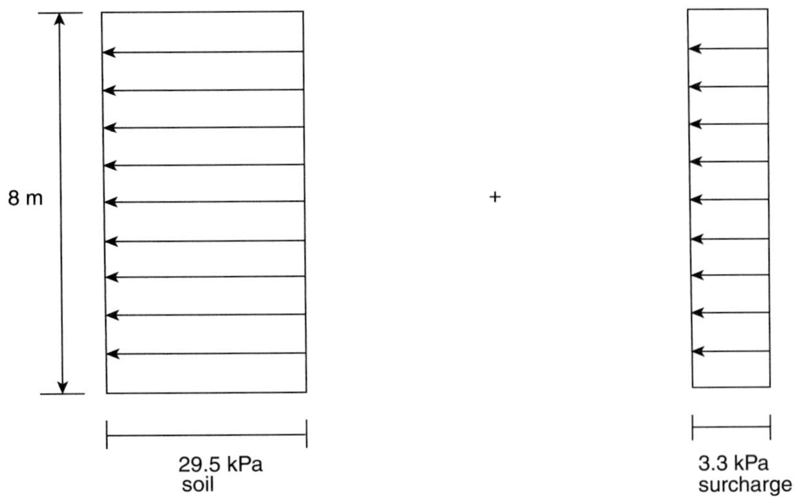

FIGURE E10.12b

Lateral earth pressure from soil is

$$\sigma'_x = 0.65\gamma H_o \tan^2\left(45 - \frac{\phi'_{cs}}{2}\right)$$

$$= 0.65 \times 17 \times 8 \times \frac{1}{3} = 29.5\,\text{kPa}$$

Lateral earth pressure from surcharge

$$= K_a q_s = \frac{1}{3} \times 10 = 3.3\,\text{kPa}$$

Step 3: Calculate forces in the anchors.
Use tributary area to estimate anchor force
Horizontal component

$$\text{Top anchor, } A_x = (\sigma'_x + K_a q_s)\left(2 + \frac{4}{2}\right)$$

$$= (29.5 + 3.3)4 = 131.2\,\text{kN/m}$$

$$\text{Bottom anchor, } B_x = (\sigma'_x + K_a q_s)\left(\frac{4}{2} + \frac{2}{2}\right)$$

$$= (29.5 + 3.3)3 = 98.4\,\text{kN/m}$$

Required anchor force

$$A = \frac{A_x \times s}{\cos 20^\circ} = \frac{131.2 \times 2.5}{\cos 20^\circ} = 349\,\text{kN}$$

$$B = \frac{B_x \times s}{\cos 20^\circ} = \frac{98.4 \times 2.5}{\cos 20^\circ} = 262\,\text{kN}$$

You can adjust the anchor depths so that the anchor forces are about equal. We will not do this here.

Step 4: Calculate the unbonded length of anchors.

$$\text{Top anchor:}\quad L_f = \frac{6 \tan 30^\circ}{\cos 20^\circ} = 3.7\,\text{m}$$

$$\text{Bottom anchor:}\quad L_f = \frac{2 \tan 30^\circ}{\cos 20} = 1.2\,\text{m}$$

Minimum unbonded length is the greater of 3 m or $(2\,\text{m} + l_f)$ [see Figure E10.12c]

$$\text{Top anchor:}\quad L_u = 2 + 3.7 = 5.7\,\text{m; use } 5.7\,\text{m}$$

$$\text{Bottom anchor:}\quad L_u = 2 + 1.2 = 3.2\,\text{m} > 3\,\text{m; use } 3.2\,\text{m}$$

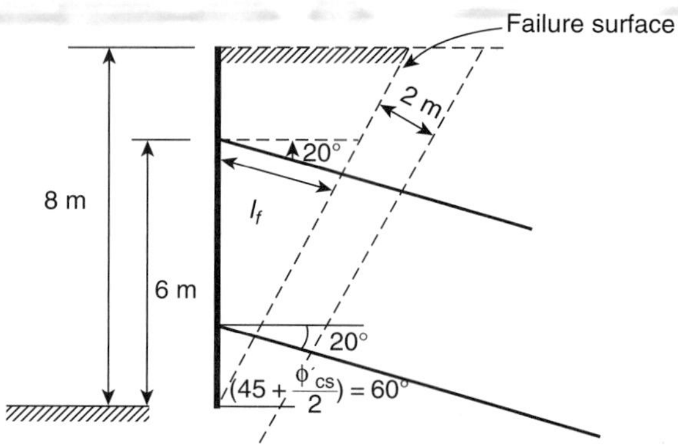

FIGURE E10.12c

Step 5: Calculate the bonded length.
Assume bond length within the SW-SM soil.
From Table 10.3, use an ultimate transfer rate of 100 kN/m

$$\text{Maximum design anchor force} = 349\,\text{kN}$$

Assume a factor a safety, FS = 2

$$\text{Maximum bond length} = \frac{\text{Maximum anchor force} \times \text{FS}}{100\,\text{kN/m}}$$

Top anchor : $\dfrac{349 \times 2}{100} = 7\,\text{m} < 12\,\text{m}$ (okay)

Bottom anchor : $\dfrac{262 \times 2}{100} = 5.2$ m, use 7 m to keep the bond length the same as the top anchor

Step 6: Select tendons.
Try steel grade 150 ksi, diameter = 32 mm.
The minimum prestressing force is 501 k N; which is greater than 349 kN; okay

Step 7: Check that top anchor meets minimum depth requirement from surface.

$$\text{Top anchor total length} = 7 + 5.7 = 12.7\,\text{m}$$

Depth from surface to end of anchor $= 12.7 \sin 20^\circ + 2 = 6.34\,\text{m} > 4.5$ m; okay

Step 8: Summarize design.

Anchor	Length (m)	bonded length (m)	Reqd. Capacity (kN)	Capacity provided (kN)
Top	12.7	7	349	501
Bottom	10.2	7	262	501

Tendons: 32 mm Grade 150 bar
A global stability check is required.

■

EXAMPLE 10.13 *Numerical analysis of an anchored tieback wall – Staged excavation*

A tieback wall is required for an excavation 9 m (27 ft) deep, 100 m (300 ft) long and 20 m (60 ft) wide. Grouted anchors fastened to the wall face will be used as tiebacks. Relevant details are given in Fig. E10.13a. Initially steel sheet piles will be driven to 15 m (45 ft) into the soil and then the excavation will be done in five stages. Each stage will consist of 2 m (6 ft) of excavation except the final stage, which will be 1 m (3 ft) of excavation. After the first sheet 2 m (6 ft) is excavated, the first tier of anchors will be installed. In the fourth stage, 8 m (24 ft) of excavation, the second tier of anchors will be installed. Analyze the excavation process using a computer program.

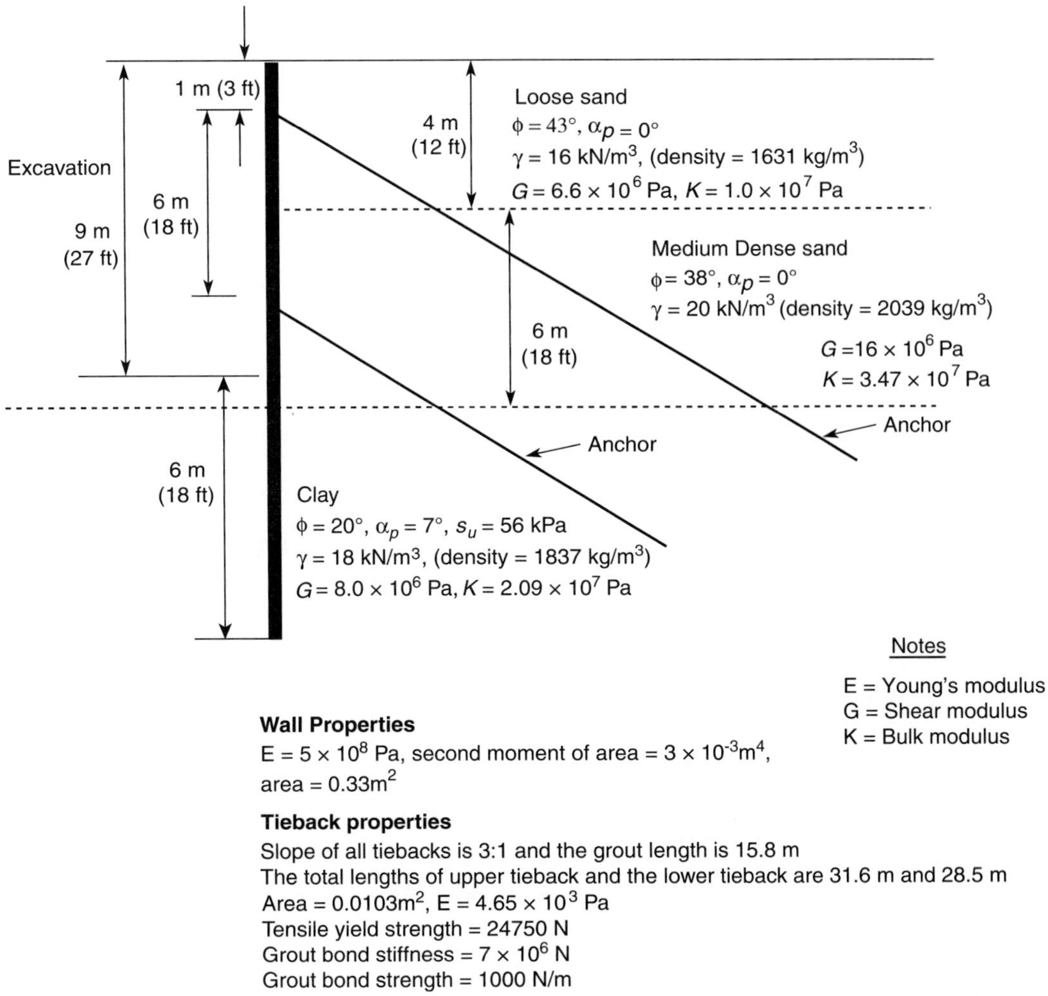

FIGURE E10.13a

Strategy In this example, you have to simulate the excavation of the soil and the installation of the grouted anchors. This is often called staged construction. The program FLAC will be used. SI units will be used in solving this example problem.

Solution 10.13

Step 1: Set up a mesh and input the necessary data.
A grid 40×31(mesh size $1\text{ m} \times 1\text{ m}$) is used as shown in Fig. 10.13b

Step 2: Select a soil model.
For a preliminary simulation, the soils are modeled as linearly elastic-rigid plastic materials with failure according to the Mohr-Coulomb failure criterion.

Step 3: Decide on stress state to simulate.
The excavation is long, so plane strain condition is selected.

Step 4: Simulate the excavation process.
Some results are shown in Figs. E10.13c to E10.13e.

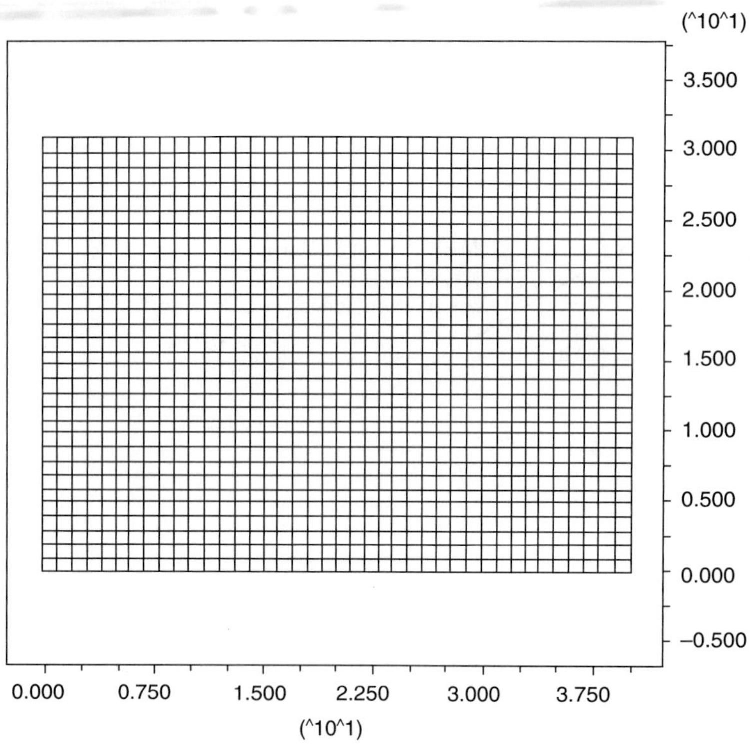

FIGURE E10.13b Mesh used.

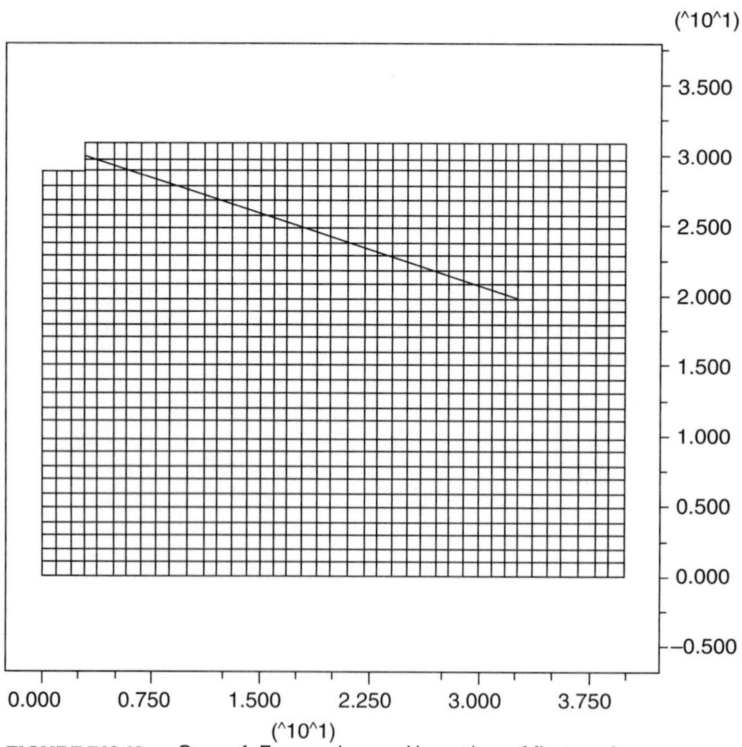

FIGURE E10.13c Stage 1: Excavation and insertion of first anchor.

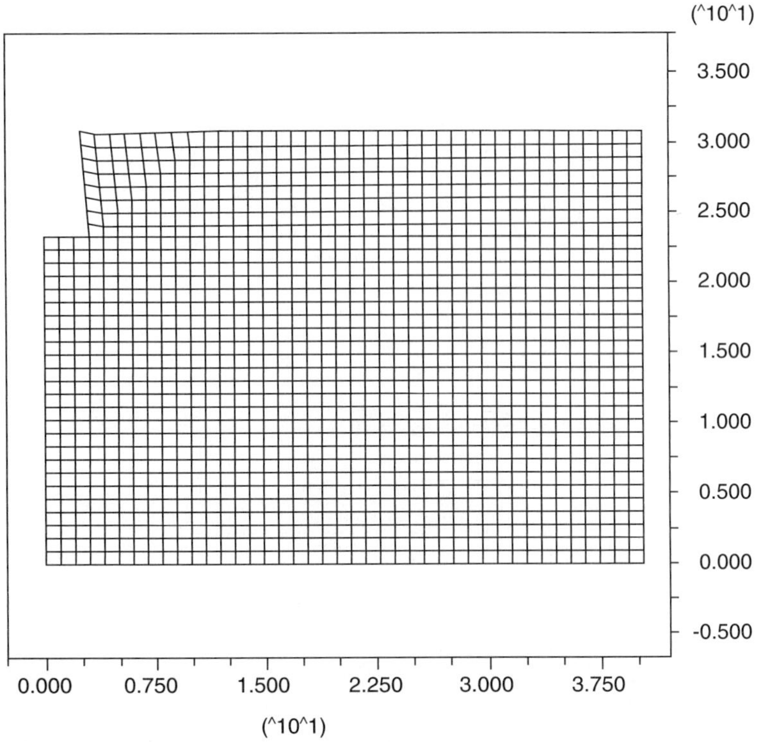

FIGURE E10.13d Stage 4: Excavation and insertion of second tier of anchor.

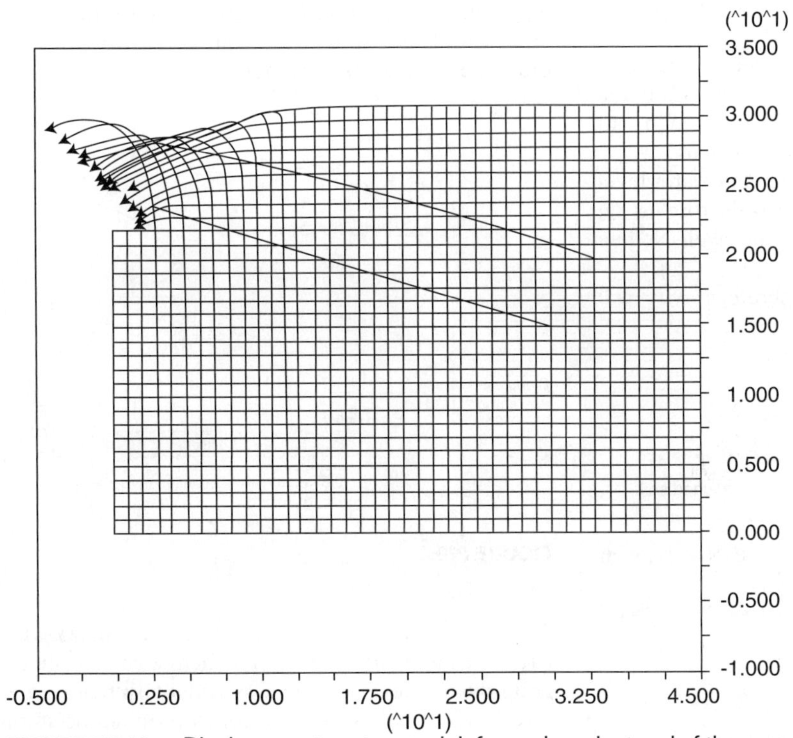

FIGURE E10.13e Displacement vectors and deformed mesh at end of the excavation process. ■

EXERCISES

1. Plot the variation of active and passive lateral pressures with depth for the soil profile shown in Fig. P10.1.

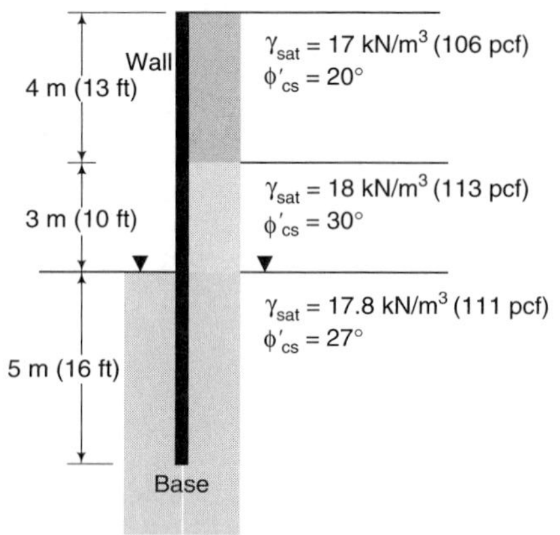

FIGURE P10.1

2. A retaining wall 8 m (26 ft) high supports a soil of saturated unit weight 18 kN/m³ (113 pcf), $\phi'_{cs} = 30°$. The backfill is subjected to a surcharge of 15 kPa. Calculate the active force on the wall if (a) the wall is smooth and (b) the wall is rough ($\delta = 20°$). Groundwater is below the base of the wall.

3. A retaining wall 5 m (16 ft) high was designed to stabilize a slope of 15°. The back of the wall is inclined 10° to the vertical and may be assumed to be rough with $\delta = 20°$. The soil parameters are $\phi'_{cs} = 30°$ and $\gamma_{sat} = 17.5$ kN/m³ (110 pcf). After a flood, the groundwater level, which is usually below the base of the wall, rose to the surface. Calculate the lateral force on the wall. Neglect seepage effects.

4. Determine the stability of the concrete gravity wall shown in Fig. P 10.4.

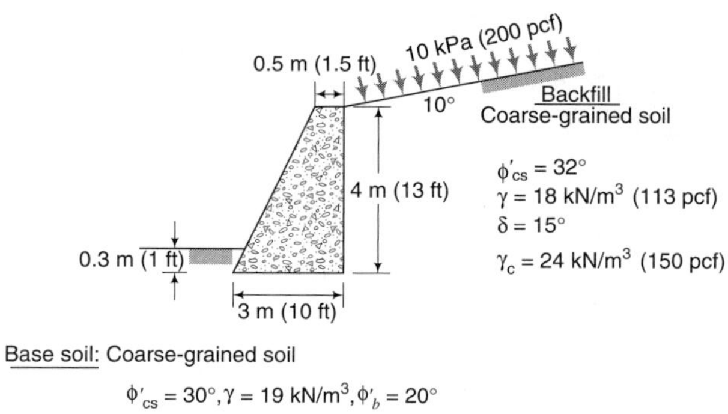

FIGURE P10.4

5. Determine the stability of the cantilever wall shown in Fig. P10.5 and sketch a drainage system to prevent buildup of porewater pressures behind the wall.

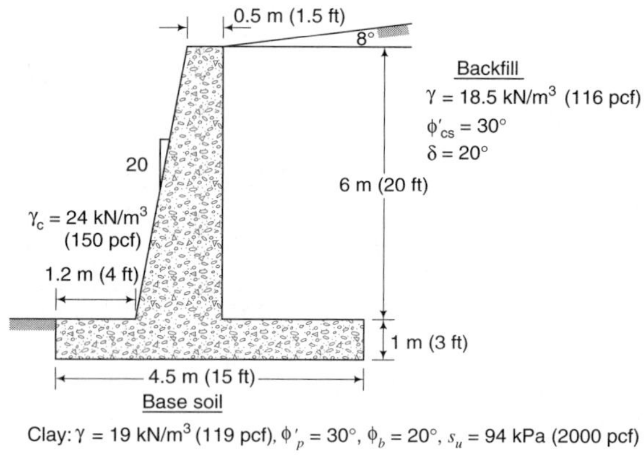

FIGURE P10.5

6. A cast-in-place (CIP) cantilever wall is required to maintain the grade for a freeway. A preliminary wall dimension is shown in Fig. P10.6. Determine the stability of the wall and show how you would provide adequate drainage. Use Rankine's method.

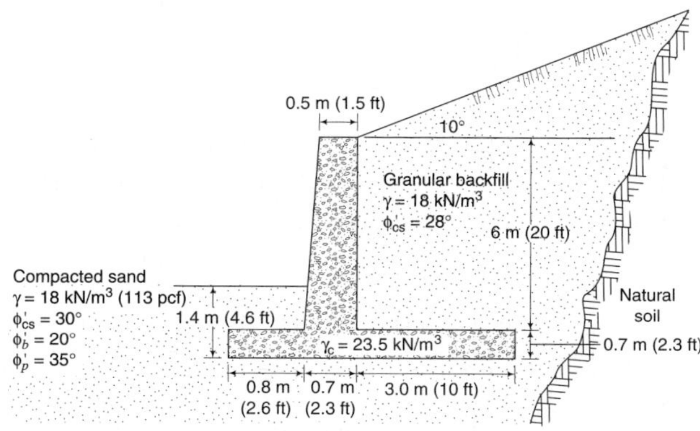

FIGURE P10.6

7. The drainage system of a cantilever wall shown in Fig. P10.7 became blocked after a heavy rainstorm and the groundwater level, which was originally below the base, rose to 1.5 m below the surface. Determine the stability of the wall before and after the rainfall. Neglect seepage effects.

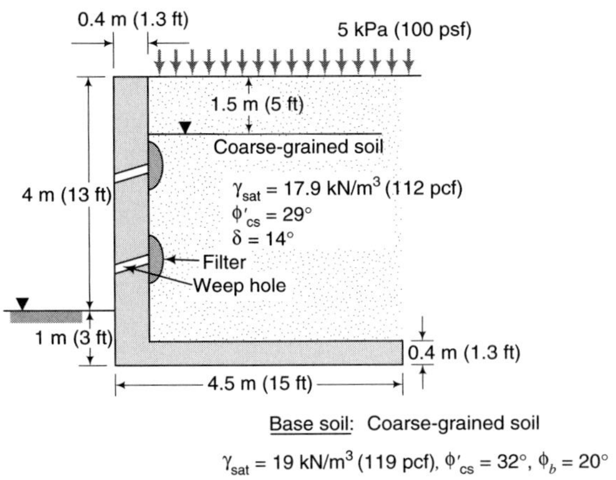

FIGURE P10.7

8. Determine the embedment depth, d, and maximum bending moment for the cantilever sheet pile wall shown in Fig. P10.8. Use the factored strength method (FSM) with $F_\phi = 1.25$.

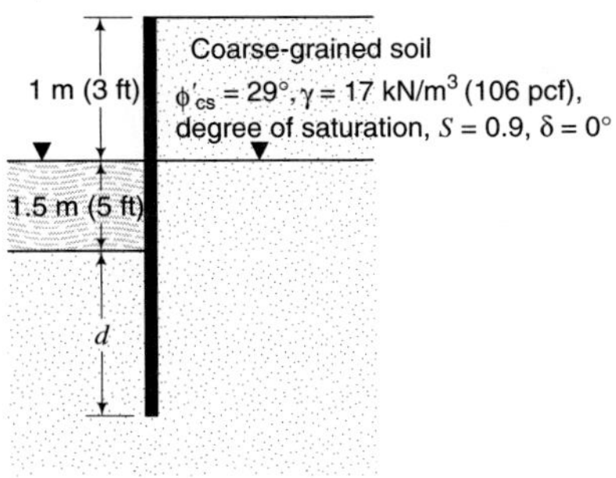

FIGURE P10.8

9. Determine the embedment depth, d, and maximum bending moment for the cantilever sheet pile wall shown in Fig. P10.9. for long-term conditions. Use the FSM with $F_\phi = 1.25$, and the NPPM with $(FS)_r = 1.5$. Compare the results. The degree of saturation is denoted as S.

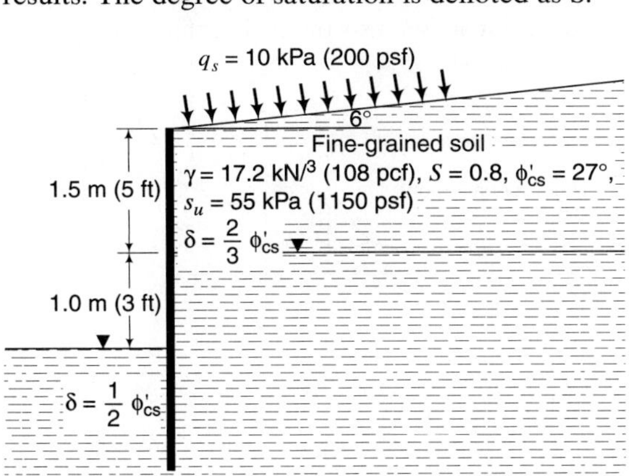

FIGURE P10.9

10. An anchored sheet pile is shown in Fig. P10.10. Determine the embedment depth, the maximum bending moment, and the force on the anchor per unit length of wall. Use either FSM ($F_\phi = 1.25$) or the NPPM with $(FS)_r = 1.5$. Assume the soil above the groundwater to be saturated.

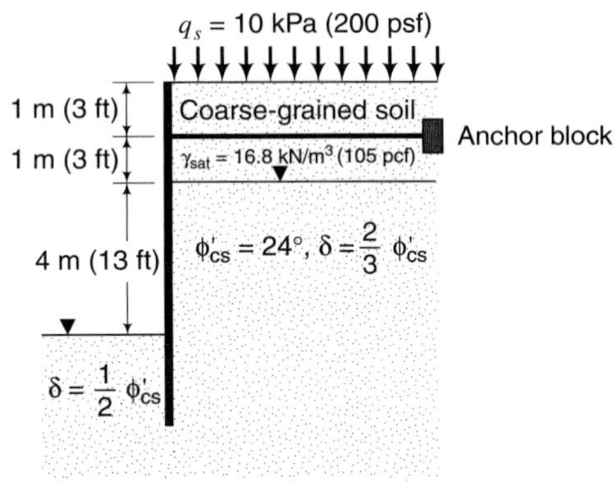

FIGURE P10.10

11. Determine the depth of embedment and the anchor force per unit length of wall for the retaining wall shown in Fig. P10.11 using FMM. Assume the soil above the groundwater to be saturated and $\delta = 0°$.

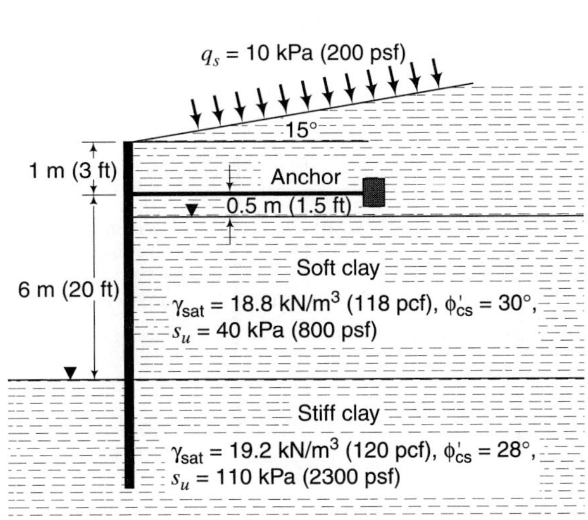

FIGURE P10.11

12. A cantilever sheet pile wall is required to temporarily support an embankment for an access road as shown in Fig. P10.12. Determine the depth of penetration of the wall and the maximum bending moment. Select two methods from FMM, FSM, and NPPM, and compare the results. Groundwater is 10 m(32 ft) below the surface.

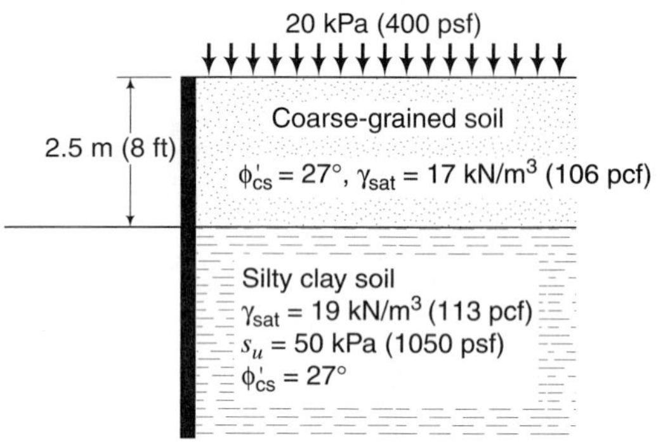

FIGURE P10.12

13. Calculate the strut loads per meter-length for the braced excavation shown in Fig. P10.13.

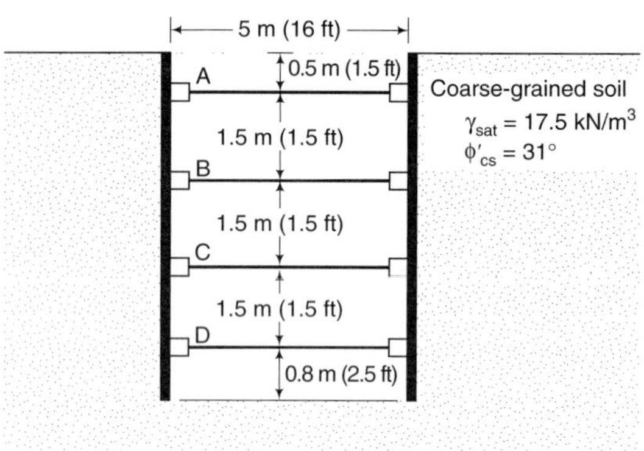

FIGURE P10.13

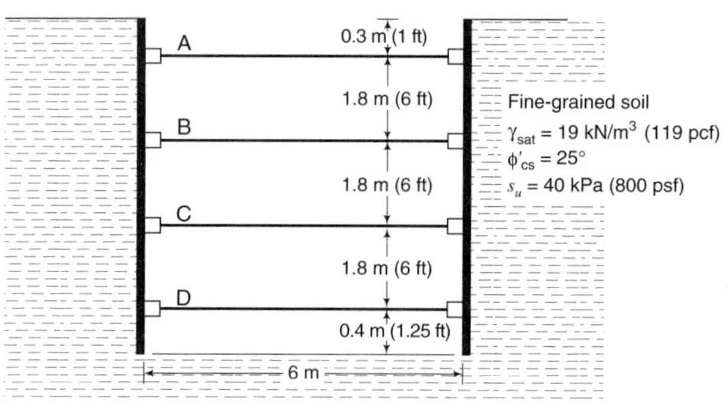

FIGURE P10.14

14. A braced excavation is required in a soft clay as shown in Fig. P10.14. Determine the load on the struts per meter length and the factor of safety against bottom heave.

15. A cantilever retaining wall is required to support a cut near a residential area. A preliminary geometry of the wall is shown in Fig. P10.15. A masonry wall 3 m (10 ft) high will sit on top of the wall, centered. The masonry wall load (including construction loading) is 9 kN/m(600 lb/ft) and the uniform loading from the homesite is 10 kPa(200 psf). The wall rotation is limited to 0.0005 H_o (H_o is the height of the wall). Check the adequacy of the wall. Determine a suitable wall geometry if the given geometry is unsatisfactory. Sketch a drainage scheme for the wall. (*Hint:* The tolerable wall is very small; it is less than the wall rotation to mobilize the active earth pressure.)

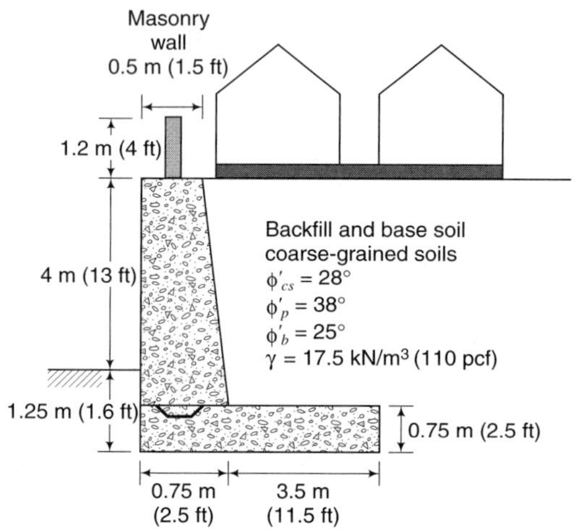

FIGURE P10.15

16. A retaining wall is required to support a highway cut 5 m (15 ft) high as shown in Figure P10.16. The wall will be backfilled with the existing soil sieved through a 25 mm (1″) mesh. Proctor compaction tests on the backfilled soil gave a

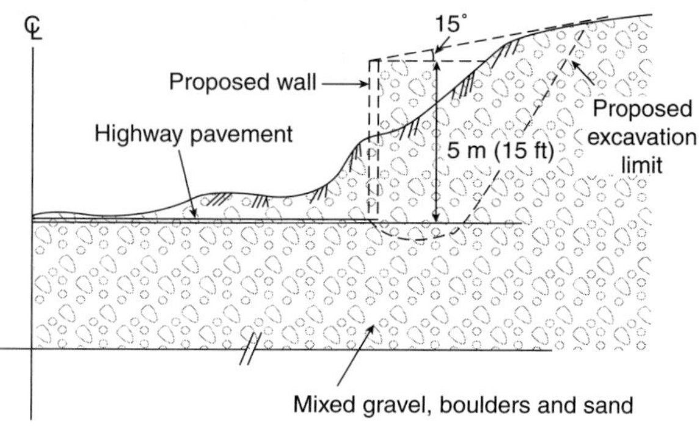

FIGURE P10.16

maximum dry unit weight of 17.5 kN/m^3 (110 pcf) and an optimum water content of 6%. Direct shear box tests were conducted by compacting the soil in the shear box to a unit weight equivalent to the Proctor maximum dry unit weight and optimum water content. The results at a vertical stress of 100 kPa (2 ksf) are $\phi'_p = 42°$ and $\phi'_{cs} = 32°$. Assume the base soil is the same as the backfill with $\phi'_b = \frac{2}{3}\phi'_{cs}$. Peak horizontal ground acceleration and peak velocity in the area are 0.2 g and 0.25 m/sec (0.8 ft/sec). Compare the designs of a mass gravity and a cantilever gravity wall to limit horizontal movement to 75 mm (3 in) in a seismic event. Design a drainage system for the wall. Rainfall in the area is 200 mm (8 in) per year. State all assumptions made and make scaled drawings of your designs.

17. An excavation 9 m (30 ft) deep is required for a parking structure near several multi-storey office buildings. These buildings exert an average vertical stress of 150 kPa (3 ksf) on the ground surface. The soil investigation was limited with only SPT's conducted at approximately every 2 m (6 ft) in 4 boreholes. Four soil layers were identified using the information from these tests. The average N_{60} values and the soil classification according to USCS for each layer are shown in the Figure P10.17. One proposed method of supporting the excavation is a tieback anchored retaining wall. The wall will be constructed by installing a soil-cement mix columns (secant wall) along the perimeter of the excavation to create a water tight boundary for inside of the excavation. This wall will be constructed by simultaneously drilling and mixing the soil with injected cement slurry to form overlapping columns. The bottom of the soil mix columns will be embedded in the Marl (a rock like material of fine-grained soils bonded by calcium carbonate) to completely block water from flowing into the excavation. Soldier piles (steel I

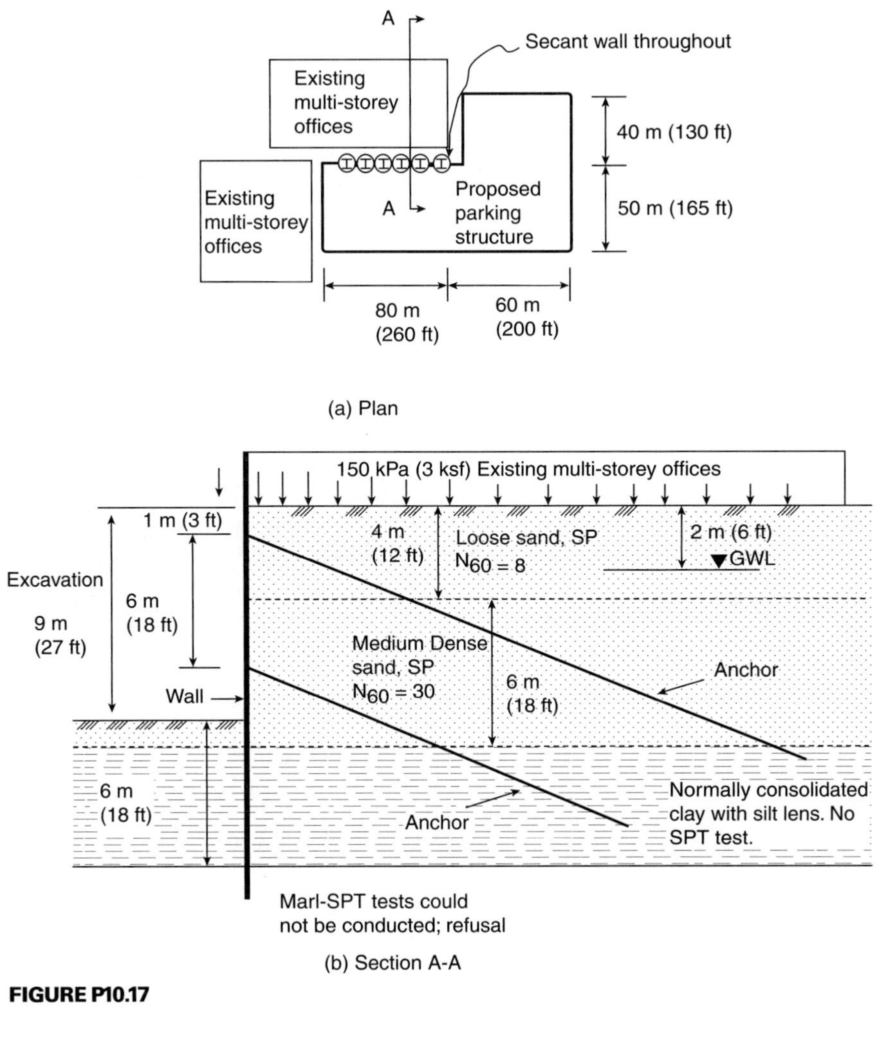

FIGURE P10.17

beams) of size 457 mm × 74.4 kg/m(W18 × 50lb) will be installed in alternate soil-cement mix columns immediately after the drilling and when the soil-cement mix is still wet. The soldier piles will be fastened laterally using steel channels. Grouted anchors fastened to the steel channels will be used as tiebacks. Groundwater level is 20 m (65 ft) below the ground surface. The excavation will be done in stages.

1) Estimate the relevant soil parameters required for the design of the anchors.

2) Discuss the accuracy of the soil parameters and how it affects the design.

3) Making reasonable assumptions, estimate the load on the top anchor when the excavation is 4 m (13 ft) below the surface. Anchor spacing laterally is 5 m (5 ft). Clearly state all assumptions used.

4) Estimate the loads on both anchors at full excavation.

5) Determine the lengths of the anchors.

6) Is it better to have the bonded length of the bottom anchor in the Marl or the normally consolidated clay with silt lens? Justify your answer.

The soil cement mix has the following properties.
Unit weight = $24\,\text{kN/m}^3$ (0.15 kcf),
$E = 5 \times 10^6$ kPa(1000 ksf), $G = 2 \times 10^6$ kPa(4000 ksf),
and shear strength = 180 kPa(3.6 ksf).

CHAPTER 11

MECHANICAL STABILIZED EARTH WALLS AND OTHER RETAINING WALLS

11.0 INTRODUCTION

Mechanical stabilized earth (MSE) walls have a long history. Between 5000 B.C. and 2500 B.C. the Babylonians constructed ziggurats (large towers) of earth reinforced with horizontal sheets of reed laid at regular intervals. The most impressive remains of these ziggurats are those at Aqar Quf near Baghdad, Iraq. The original vertical heights of the ziggurats probably were about 87 m.

MSE walls have become popular mainly because of their flexibility, large bases, and cost effectiveness compared to rigid retaining walls. In this chapter, the analysis and design of MSE walls and other types of walls such as soil nail walls are presented. When you complete this chapter, you should be able to:

- Understand the mechanics of MSE and similar types of walls.
- Conduct simple designs of these walls.

11.1 DEFINITIONS OF KEY TERMS

A ***mechanical stabilized earth (MSE) wall*** is a gravity-type retaining wall in which the soil is reinforced by thin reinforcing elements (steel, fabric, plastic, fibers, etc).

A ***geotextile*** is any permeable textile material used in geotechnical applications.

Geosynthetics are polymeric, planar material (polyproylene, polyetylene, polyester, polyamide, and nylon).

Geocomposite is a product made from a combination of geosynthetics.

Geogrid is a polymeric product formed by joining intersecting ribs.

Soil nailing is a method of stabilizing a soil mass by long, slender reinforcing elements.

11.2 BASIC CONCEPTS

If a vertical load is applied to a soil mass under axisymmetric undrained conditions, the lateral strain (ε_3) is one-half the axial strain (ε_1). If the undrained restriction is lifted, then you can expect lateral strains greater than one-half the vertical strains. If we were to install reinforcing elements in the lateral directions of the soil mass, then the friction at the interfaces of the elements and the soil would restrain lateral

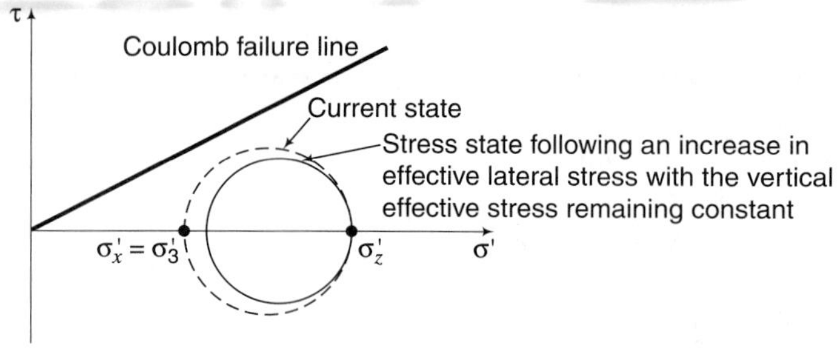

FIGURE 11.1 Effects of increasing the lateral soil resistance by reinforcement.

displacements. The imposition of a lateral resistance on the soil mass causes Mohr's circle to move away from the failure line and shrink. The net effect is that the soil can withstand higher shearing forces compared to an unreinforced soil (Fig. 11.1).

The lateral force imposed on the soil depends on the interface friction value between the reinforcing element and the soil mass and the vertical effective stress. For a constant interface friction value, the lateral frictional force would increase with depth. The reinforcing material will fail if the lateral stress exceeds its tensile strength.

The **key point** is that the reinforcement serves as an internal lateral confinement that allows the soil to mobilize more shearing resistance.

11.3 MECHANICAL STABILIZED EARTH (MSE) WALLS

Importance *MSE walls have been replacing rigid CIP walls over the last decade. They are generally more cost-effective than CIP walls. In this section, the stability of MSE walls is described.*

11.3.1 MSE Reinforcement

MSE walls (Fig. 11.2a) are used for a variety of retaining structures. Metal strips (Fig. 11.2b), geotextiles (Fig. 11.2c), or geogrids (Fig. 11.2d) reinforce the soil mass. Other names used are geosynthetics and geocomposities. Geotextile is any permeable textile material used in geotechnical applications. Geosynthetics are polymeric, planar material (polypropylene, polyethylene, polyester, polyamide, and nylon). Geocomposite is a product made from a combination of geosynthetics. A geogrid is a polymeric product formed by joining intersecting ribs.

Geotextiles are manufactured from filaments and yarns that are combined to form planar products called fabrics. In geotechnical applications, nonwoven and woven facrics are popular. Nonwoven geotextiles are manufactured from continuous filaments formed by extruding polymers through spinnerets or staple fibers (2 cm to 10 cm of filaments). Woven geotextiles are made from yarns knitted using various weave patterns such as plain, twill, satin, or various combinations of these.

The American Society for Test and Materials (ASTM) has a number of standard tests to determine the physical and mechanical properties of geotextiles. For retaining walls, the key properties are tensile strength, creep, and durability. The ASTM tests for tensile strength and creep are

1. Grab tensile strength test (ASTM D4632). This test gives the ultimate tensile load (N or lb) of the geotextile.
2. Wide width tensile test (ASTM D4595). This test gives the load per unit width (N/m or lb/ft) of geotextile. You can also obtain the wide width modulus from this test.
3. Tensile creep test (ASTM D5262). This test gives the percentage of the ultimate tensile load at which the rate of creep strain approaches zero with time.

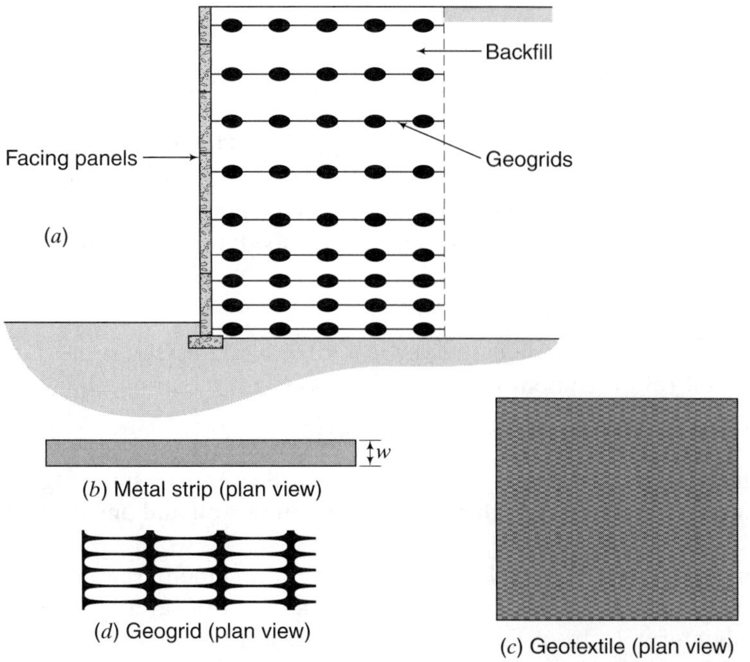

FIGURE 11.2 (a) A geogrid reinforced wall, (b) metal strip, (c) geotextile, and (d) geogrid.

Geotextiles suffer strength reduction from ultraviolet light degradation. Consequently, during construction it is essential not to expose geotextiles unduly to sunlight. Specifications normally limit the exposure time to less than 14 days.

11.3.2 Stability of Mechanical Stabilized Earth Walls

There are two sets of stability criteria to be satisfied for MSE walls. One is the internal stability; the other is the external stability. The external stability of a MSE wall is determined by analogy to a gravity retaining wall with a vertical face, *FG*, as illustrated in Fig. 11.3. The internal stability depends on the tensile strength of the reinforcing material and the interface friction of the reinforcing material and the soil. Tensile failure of the reinforcing material at any depth leads to progressive collapse of the

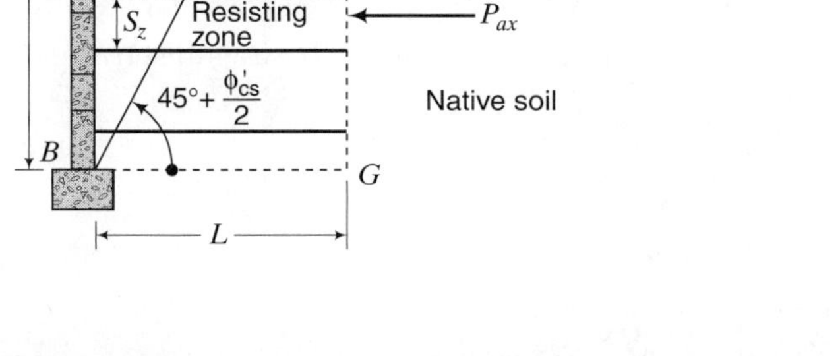

FIGURE 11.3 Forces on an MSE wall using Rankine's method.

wall, while slip at the interface of the reinforcing material and the soil mass leads to redistribution of stresses and progressive deformation of the wall.

Two methods of analysis are used to determine the internal stability. One method is based on an analogy with anchored flexible retaining walls and is generally used for reinforcing material with high extensibility, for example, polymerics such as geotextiles and geogrids. The Rankine active earth pressure theory is used with the active slip plane inclined at $\theta_a = 45° + \phi'_{cs}/2$ to the horizontal, as shown in Fig. 11.3. The frictional resistance develops over an effective length, L_e, outside the active slip or failure zone. At a depth z, the frictional resistance developed on both surfaces of the reinforcing material (Fig. 11.3) is

$$P_r = 2wL_e(\sigma'_z + q_s)\tan\phi_i \tag{11.1}$$

where w is the width of the reinforcing material, σ'_z is the vertical effective stress, q_s is the surcharge, and ϕ_i is the friction angle at the soil-reinforcement interface. The tensile force at depth, z, is

$$T = K_{aR}(\sigma'_z + q_s)S_zS_y \tag{11.2}$$

where S_z and S_y are the spacings in the Z and Y directions and T is the tensile force per unit length of wall. For geotextiles or geogrids, you would normally consider one unit length of wall and one unit width, so $S_y = 1$ and $w = 1$.

By setting $T = P_r$, we can find the effective length of the reinforcement required for limit equilibrium (factor of safety = 1). To find the design effective length, a factor of safety $(\mathrm{FS})_t$ is applied on the tensile force, T, and by solving for L_e from Eqs. (11.1) and (11.2), we get

$$L_e = \frac{K_{aR}(\sigma'_z + q_s)S_zS_y(\mathrm{FS})_t}{2w(\sigma'_z + q_s)\tan\phi_i} = \frac{K_{aR}S_zS_y(\mathrm{FS})_t}{2w\tan\phi_i} \tag{11.3}$$

where $(\mathrm{FS})_t$ ranges from 1.3 to 1.5.

The total length of reinforcement is

$$L = L_e + L_R \tag{11.4}$$

where L_R is length of reinforcement within the active failure zone.

Because L_R is zero at the base of the wall, the calculated length of reinforcement at the base is often the shortest. This calculated length, while adequate for internal stability, is often inadequate for translation or bearing capacity (external stability). You should check whether the calculated length of reinforcement at the base is adequate for translation or bearing capacity as follows:

1. Calculate the maximum lateral active earth force, P_{ax}.

$$P_{ax} = \frac{1}{2}K_{aC}\gamma H_o^2\cos\delta + K_{aC}q_sH_o\cos\delta = H_oK_{aC}\cos\delta(0.5\gamma H_o + q_s)$$

where K_{aC} is the active lateral pressure coefficient using Coulomb's method with wall friction. The friction angle to use is $\phi' = (\phi'_{cs})_{\text{backfill}}$ and the wall friction is $\delta \approx \frac{2}{3}(\phi'_{cs})_{\text{native soil}}$ to $(\phi'_{cs})_{\text{native soil}}$. Coulomb's method is used because the interface between the reinforced backfill soil and the native soil (*FG*, Fig. 11.3) is frictional. If the interface friction is neglected, then you can use K_{aR} instead of K_{aC}.

2. Calculate the length of reinforcement required for translation. For short-term loading in clays, the base resistance is $T = L_bs_w$, where L_b is the length of reinforcement at the base and s_w is the soil adhesion stress. By analogy with sheet pile retaining walls, we will use $s_w = 0.5s_u$ (maximum $s_w = 50$ kPa). For external stability against translation $T = P_{ax}(\mathrm{FS})_T$, where $(\mathrm{FS})_T$ is a factor of safety against translation; usual range 1.5–3.0. The required length at the base against translation for short-term loading in clays is then

$$L_b = \frac{P_{ax}(\mathrm{FS})_T}{s_w} \tag{11.5}$$

For long-term loading, $T = \sum_{i=1}^{n} W_i \tan \phi'_b$, where W_i is the weight of soil layer i, n is the number of layers, and ϕ'_b is the interfacial friction angle between the reinforcement and the soil at the base. Assuming a uniform soil unit weight throughout the height of the wall, then

$$T = \gamma H_o L_b \tan \phi'_b \tag{11.6}$$

The length of reinforcement at the base required to prevent translation under long-term loading is

$$L_b = \frac{P_{ax}(\text{FS})_T}{\gamma H_o \tan \phi'_b} = \frac{(K_{aC})_x H_o(0.5\gamma H_o + q_s)(\text{FS})_T}{\gamma H_o \tan \phi'_b} = \frac{(K_{aC})_x(0.5H_o + q_s/\gamma)(\text{FS})_T}{\tan \phi'_b} \tag{11.7}$$

where $(K_{aC})_x$ is the horizontal component of Coulomb's active lateral earth pressure coefficient. You should use the larger value of L_b obtained from Eqs. (11.5) and (11.7).

The procedure for analysis of a reinforced soil wall using polymeric materials is as follows:

1. Calculate the allowable tensile strength per unit width of the reinforcing polymeric material.

$$T_{\text{all}} = T_{\text{ult}} \frac{1}{(\text{FS})_{\text{ID}} \times (\text{FS})_{\text{CR}} \times (\text{FS})_{\text{CD}} \times (\text{FS})_{\text{BD}}} \tag{11.8}$$

where T_{all} is the allowable tensile strength, T_{ult} is the ultimate tensile strength, FS is a factor of safety, and the subscripts have the following meaning:

ID—installation damage
CR—creep
CD—chemical degradation
BD—biological degradation

Typical values of the various factors of safety are shown in Table 11.1.

TABLE 11.1 Typical Ranges of Factor of Safety

Factor of safety	Range
$(\text{FS})_{\text{ID}}$	1.1 to 2.0
$(\text{FS})_{\text{CR}}$	2.0 to 4.0
$(\text{FS})_{\text{CD}}$	1.0 to 1.5
$(\text{FS})_{\text{BD}}$	1.0 to 1.3

2. Calculate the vertical spacing at different wall heights.

$$S_z = \frac{T_{\text{all}}}{K_{a\text{R}}(\sigma'_z + q_s)(\text{FS})_{\text{sp}}} \tag{11.9}$$

where $(\text{FS})_{\text{sp}}$ is a factor of safety between 1.3 and 1.5. It is customary to calculate the minimum vertical spacing using $\sigma'_z = \gamma' H_o$ and then check the vertical spacing required at $H_o/3$ and $2H_o/3$. For practical reasons, the vertical spacing either is kept at a constant value or varies by not more than three different values along the height.

3. Determine the length of reinforcement required at the base for external stability from Eqs. (11.5) and (11.7). Use the larger value of length.

4. Determine the total length of reinforcement at different levels.

$$L = L_e + L_{\text{R}}$$

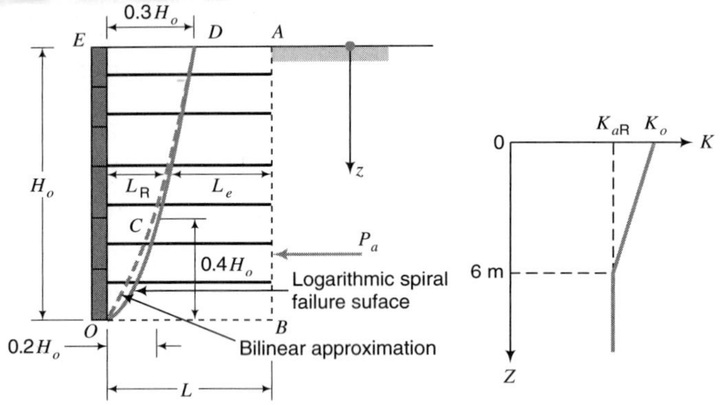

(a) Failure surface for the coherent gravity method

(b) Variation of active lateral stress coefficient with depth

FIGURE 11.4 Coherent gravity method.

where

$$L_R = (H_o - z)\tan\left(45° - \frac{\phi'_{cs}}{2}\right) \tag{11.10}$$

5. Determine the external stability (translation and bearing capacity). Remember that translation is already satisfied from item 3 above. Overturning is not crucial in MSE walls because these walls are flexible and cannot develop significant moment. However, you can verify that overturning is satisfied if $e < L_b/6$, where e is the eccentricity of the vertical forces.

The other method is the coherent gravity method (Juran and Schlosser, 1978) and is applicable to low extensible reinforcing materials such as metal strips. Failure is assumed to occur progressively along a path defined by the maximum tensile strains at each level of the reinforcing material. The failure surface is a logarithm spiral that is approximated to a bilinear surface to simplify calculations (Fig. 11.4a). The lateral active pressure coefficient is assumed to vary linearly from K_o at the top of the wall to K_{aR} at a depth of 6 m and below (Fig. 11.4b).

The variation of lateral stress coefficient with depth (Fig. 11.4b) is

$$K = K_{aR}\frac{z}{6} + K_o\left(1 - \frac{z}{6}\right) \quad \text{for} \quad z \le 6\,\text{m} \tag{11.11}$$

$$K = K_{aR} \quad \text{for} \quad z > 6\,\text{m} \tag{11.12}$$

The length of reinforcement within the failure zone (Fig. 11.4a) is

$$L_R = 0.2H_o + \left(0.1H_o - \frac{z}{6}\right) \quad \text{for} \quad z \le 0.6H_o \tag{11.13}$$

$$L_R = \frac{1}{2}H_o\left(1 - \frac{z}{H_o}\right) \quad \text{for} \quad z > 0.6H_o \tag{11.14}$$

The procedure for analyzing a MSE wall using low extensible materials is as follows:

1. Select the spacing of the reinforcement in the Z and Y directions and the width of the reinforcement. Use a manufacturer's catalog to provide information on standard sizes.

2. Calculate the required maximum thickness of the reinforcement.

$$t_r = \frac{K_{aR}(\gamma' H_o + q_s)S_z S_y (\text{FS})_{tr}}{w f_y} \tag{11.15}$$

where K_{aR} is the Rankine active earth pressure coefficient, f_y is the yield stress of the reinforcement, w is the width of the reinforcement, and $(FS)_{tr}$ is a factor of safety, usually 3. The design thickness is the calculated thickness plus a corrosion thickness expected for the design life of the wall.

3. Determine the length of reinforcement required at the base from Eqs. (11.5) and (11.7).

4. Determine the total length of reinforcement at each level.

$$L = L_e + L_R \tag{11.16}$$

where L_R is given by Eqs. (11.13) and (11.14) and L_e is determined from Eq. (11.3) by using the appropriate value of the active lateral earth pressure coefficient from Eqs. (11.11) and (11.12).

5. Determine the external stability by assuming the wall is a gravity wall.

Neither method considers the wall deformation. MSE walls are semi-rigid (or semi-flexible), so the active pressures, which are calculated using Rankine's earth pressure theory, are approximations. The deformation of these walls can be simulated using FDM or FEM.

The key points are:

1. ***MSE walls must satisfy both internal and external stability criteria.***
2. ***For extensible reinforcements (e.g., geotextiles), the failure surface is approximated using Rankine earth pressure theory. In the case of low extensible reinforcement, a bilinear failure surface is used.***
3. ***Deformation (serviceability limit state) is not calculated using the two methods of analysis described in this section.***

EXAMPLE 11.1 *Geotextile MSE Wall*

Access http://www.wiley.com/college/budhu. Click Foundations and Earth Structures, Chapter 11, mse.xls to analyze MSE retaining walls. You can change input values and explore how the stability of the wall changes.

Design a 4 m (13 ft) mechanical stabilized earth wall using a geotextile as the reinforcement. The backfill would be a compacted, coarse-grained soil with $\phi'_{cs} = 30°$ and $\gamma_{sat} = 18\,\text{kN/m}^3 (113\,\text{pcf})$. The surcharge is 15 kPa (300 psf). The geotextile selected has an ultimate wide-width tensile strength of 58.5 kN/m (4000 lb/ft) and the soil-geotextile interface friction value is 20°. The native soil is a clay with parameters $\gamma_{sat} = 18.5\,\text{kN/m}^3 (116\,\text{pcf})$, $\phi'_{cs} = 32°, \sigma'_b = 20°$, and $s_u = 60\,\text{kPa}$ (1250 psf).

Strategy Follow the procedure for mechanical stabilized earth walls using a polymeric product. SI units will be used in this example.

Solution 11.1

Step 1: Calculate the allowable tensile strength of the geotextile.
From Table 11.1, use $(FS)_{ID} = 1.5$, $(FS)_{CR} = 2$, $(FS)_{CD} = 1.3$, and $(FS)_{BD} = 1.3$.

$$T_{all} = \frac{58.5}{1.5 \times 2 \times 1.3 \times 1.3} = 11.5\,\text{N/mm} = 11.5\,\text{kN/m}$$

Step 2: Calculate the vertical spacing.

$$K_{aR} = \tan^2(45 - \phi'_{cs}/2) = \tan^2(45 - 30/2) = \frac{1}{3}$$

Lateral stress due to surcharge:

$$K_a q_s = \frac{1}{3} \times 15 = 5\,\text{kPa}$$

$$\sigma_x = K_{aR}\sigma'_z + K_{aR}q_s = \frac{1}{3}\gamma' z + 5 = \left(\frac{1}{3} \times 18\right) + (z + 5) = 6z + 5$$

$$(\sigma_x)_{\max} = 6 \times H_o + 5 = 6 \times 4 + 5 = 29\,\text{kPa}$$

From Eq. (11.4) with $(\text{FS})_{sp} = 1.3$, we get

$$(S_z)_{\min} = \frac{T_{all}}{K_{aR}(\sigma'_z + q_s)(\text{FS})_{sp}} = \frac{11.5}{29 \times 1.3} = 0.305\,\text{m} = 305\,\text{mm}$$

Check spacing requirement at midheight ($z = 2$):

$$\sigma_x = \left(\frac{1}{3} \times 18\right) \times 2 + 5 = 17\,\text{kPa}$$

$$S_z = \frac{11.5}{17 \times 1.3} = 0.520\,\text{m} = 520\,\text{mm}$$

You should try to minimize the number of layers and keep the spacing to easily measurable values. Use $S_z =$ 250 mm for the bottom half of the wall and 500 mm for the top half of the wall.

Step 3: Determine the length of reinforcement required at the base for translation.

$$K_{aC} = 0.3, \quad \delta = 20^\circ, \quad \text{and} \quad (K_{aC})_x = K_{aC}\cos\delta = 0.3 \times \cos 20 = 0.28$$

$$P_{ax} = \frac{1}{2}(K_{aC})_x \gamma H_o^2 + (K_{aC})_x q_s H_o = \left(\frac{1}{2} \times 0.28 \times 1.8 \times 4^2\right) + (0.28 \times 15 \times 4)$$

$$= 57.1\,\text{kN/m}$$

$$s_w = 0.5 s_u = 0.5 \times 60 = 30\,\text{kPa} < 50\,\text{kPa}; \quad \text{therefore use } s_w = 30\,\text{kPa}$$

Equation (11.5):

$$L_b = \frac{P_{ax}(\text{FS})_T}{s_w} = \frac{57.1 \times 1.5}{30} = 2.9\,\text{m}$$

Equation (11.7):

$$L_b = \frac{(K_{aC})_x(0.5H_o + q_s/\gamma')(\text{FS})_T}{\tan\phi'_b}$$

$$= \frac{0.28(0.5 \times 4 + 15/18)1.5}{\tan\left(\frac{2}{3} \times 28\right)} = 3.5\,\text{m}$$

Use $L_b = 3.5$ m.

Step 4: Determine the total length of reinforcement at each level for internal stability.

Use a table to determine the total length as shown below.
Use $(\text{FS})_t = 1.3$.

z(m)	S_z(m)	$L_R(\text{m}) = (H_o - z)\tan(45^\circ - \phi'_{cs}/2)$	$L_e(\text{m}) = \dfrac{K_{aR}S_z(\text{FS})_t}{2\tan\phi_i}$	$L(\text{m}) = L_e + L_R$
0.50	0.50	2.02	0.3	2.3
1.00	0.50	1.73	0.3	2.0
1.50	0.50	1.44	0.3	1.7
2.00	0.50	1.15	0.3	1.5
2.25	0.25	1.01	0.15	1.2
2.50	0.25	0.87	0.15	1.0
2.75	0.25	0.72	0.15	0.9
3.00	0.25	0.58	0.15	0.7
3.25	0.25	0.43	0.15	0.6
3.50	0.25	0.29	0.15	0.4
3.75	0.25	0.14	0.15	0.3
4.00	0.25	0.00	0.15	0.2

Since L_b is greater than the total length required for internal stability at each level, use $L = L_b = 3.5\,\text{m}$. The total length can be reduced toward the top of the wall but for construction purposes, it is best to use, in most cases, a single length of polymeric product.

Step 5: Check external stability.
Stability against translation is already satisfied in Step 3.
Check the bearing capacity.

$$(\sigma_z)_{max} = \gamma H_o = 18 \times 4 = 72\,\text{kPa}$$

TSA

$$q_u = 5.14 s_u i_c; \quad i_c = 1 \quad \text{for strip footing}$$

$$q_u = 5.14 \times 60 \times 1 = 308.4\,\text{kPa}$$

$$(\text{FS})_B = \frac{q_{ult}}{(\sigma_x)_{max}} = \frac{308.4}{72} = 4.3 > 3; \quad \textit{okay}$$

ESA

Use rough footing because a layer of geotextile is placed at the interface of the wall and foundation.

$$H = P_{ax} = 57.1\,\text{kN}; \quad V_n = \gamma H_o B = 18 \times 4 \times 3.5 = 252\,\text{kN},$$

$$\frac{H}{V_n} = \frac{57.1}{252} = 0.23; \quad \omega = \tan^{-1}\frac{H}{V_n} = \tan^{-1}(0.23) = 13^\circ$$

$$n = \left(2 + \frac{B'}{L'}\right) \Big/ \left(1 + \frac{B'}{L'}\right) = 2; \quad i_\gamma = \left(1 - \frac{H}{V_n}\right)^{n+1} = (1 - 0.23)^{2+1} = 0.46$$

Assume $\phi'_p = \phi'_{cs}$

$$N_\gamma = 0.1054\exp(9.6\phi'_p) = 0.1054\exp\left(9.6 \times 32 \times \frac{\pi}{180}\right) = 22.5$$

$$q_u = 0.5\gamma B N_\gamma i_\gamma = 0.5 \times 18.5 \times 3.5 \times 22.5 \times 0.46 = 335\,\text{kPa}$$

$$(\text{FS})_B = \frac{q_u}{(\sigma_{max})} = \frac{335}{72} = 4.6 > 3; \quad \textit{okay}$$

■

EXAMPLE 11.2 *Metal Ties MSE Wall*

Design a mechanical stabilized wall 6 m (20 ft) high using metal ties. The backfill is a coarse-grained soil with $\gamma_{sat} = 16.5\,\text{kN/m}^3$ (103 pcf) and $\phi'_{cs} = 32^\circ$. Galvanized steel ties are available with a yield strength $f_y = 2.5 \times 10^5\,\text{kPa}$ (52×10^5 psf) and a rate of corrosion of 0.025 mm/year (1×10^{-3} in/year). A factor of safety of 3 is desired for a design life of 50 years. The soil-tie interface friction value is 21°. The foundation soil is coarse-grained with parameters $\gamma_{sat} = 18\,\text{kN/m}^3$ (113 pcf), $\phi'_p = 35^\circ$, and $\phi'_b = 20^\circ$. The surcharge is 15 kPa (300 psf). Use $(\text{FS})_t = 1.3$.

Strategy You have to guess the spacing of the ties. You can obtain standard widths and properties of ties from manufacturer's catalogs. Follow the procedure for the coherent gravity method. SI units will be used in this example.

Solution 11.2

Step 1: Assume spacing and width of ties.

$$\text{Assume} \quad S_z = 0.5\,\text{m}, \quad S_y = 1\,\text{m}, \quad \text{and} \quad w = 75\,\text{mm}$$

Step 2: Calculate required thickness or reinforcement.

$$K_o = 1 - \sin\phi'_{cs} = 1 - \sin 32^\circ = 0.47$$

$$K_{aR} = \tan^2(45 - \phi'_{cs}/2) = \tan^2(45 - 32/2) = 0.31$$

From Eq. (11.15)

$$t_r = \frac{K_{aR}(\gamma' H_o + q_s) S_z S_y (\text{FS})_{tr}}{w f_y} = \frac{0.31(16.5 \times 6 + 15) \times 0.5 \times 1 \times 3}{0.075 \times 2.5 \times 10^5}$$

$$= 283 \times 10^{-5}\,\text{m} = 2.8\,\text{mm}$$

$$t_{\text{corrosion}} = \text{annual corrosion rate} \times \text{design life} = 0.025 \times 50 = 1.25\,\text{mm}$$
$$t_{\text{design}} = \text{calculated thickness} + \text{corrosion thickness} = 2.8 + 1.25 = 4.05\,\text{mm}$$

You will select a tie thickness from standard sizes closest to 4.05 mm. Use $t = 5$ mm.

Step 3: Determine the length of reinforcement required at the base.

$$K_{aC} = 0.28\ (\phi'_{cs} = 32°,\ \delta = 20°)$$
$$(K_{aC})_x = K_{aC}\cos\delta = 0.28\cos 20 = 0.26$$

From Eq. (11.7):

$$L_b = L_b = \frac{0.26(0.5 \times 6 + 15/16.5)1.5}{\tan 20°} = 4.2\,\text{m}$$

For internal stability, the effective length at the wall base for $(\text{FS})_t = 1.3$ is

From Eq. (11.3) $$L_e = \frac{0.31 \times 0.5 \times 1 \times 1.3}{2 \times 0.075 \times \tan 21°} = 3.5\,\text{m} < 4.2\,\text{m}$$

Step 4: Determine the total length of reinforcement.
Use a table as shown below.

z(m)	S_z(m)	K Eqs. (11.11, 11.12)	L_e (m) Eq. (11.3)	L_R (m) Eqs. (11.13, 11.14)	L (m) = $L_e + L_R$	Recommended L (m)
0.50	0.50	0.46	5.15	1.72	6.9	7.0
1.00	0.50	0.44	5.00	1.63	6.6	7.0
1.50	0.50	0.43	4.85	1.55	6.4	6.5
2.00	0.50	0.42	4.69	1.47	6.2	6.5
2.50	0.50	0.40	4.54	1.38	5.9	6.5
3.00	0.50	0.39	4.39	1.30	5.7	6.5
3.50	0.50	0.38	4.23	1.22	5.5	6.5
4.00	0.50	0.36	4.08	1.00	5.1	6.5
4.50	0.50	0.35	3.93	0.75	4.7	5.0
5.00	0.50	0.33	3.77	0.50	4.3	5.0
5.50	0.50	0.32	3.62	0.25	3.9	5.0
6.00	0.50	0.31	3.47	0.00	3.5	5.0

Step 5: Check for external stability.
Translation is satisfied because L used at the base is greater than L_b. Check bearing capacity.

$$(\sigma_z)_{\max} = \gamma H_o = 16.5 \times 6 = 99\,\text{kPa}$$
$$H = 0.5\gamma K_{(aC)x}H_o^2 + K_{(aC)x}H_o q_s = (0.5 \times 16.5 \times 0.26 \times 6^2) + (0.26 \times 6 \times 15) = 100.6\,\text{kN}$$
$$V_n = \gamma H_o B = 16.5 \times 6 \times 4.2 = 415.8\,\text{kN},\quad \frac{H}{V_n} = \frac{100.6}{415.8} = 0.24;\ \omega = \tan^{-1}\frac{H}{V_n} = \tan^{-1}(0.24) = 13.5°$$
$$n = \left(2 + \frac{B'}{L'}\right)\Big/\left(1 + \frac{B'}{L'}\right) = 2;\quad i_\gamma = \left(1 - \frac{H}{V_n}\right) = (1 - 0.24)^{2+1} = 0.44$$
$$N_\gamma = 0.1054\exp(9.6\phi'_p) = 0.1054\exp\left(9.6 \times 35 \times \frac{\pi}{180}\right) = 37.2$$
$$q_u = 0.5\gamma B N_\gamma i_\gamma = 0.5 \times 18 \times 4.2 \times 37.2 \times 0.44 = 618.7\,\text{kPa}$$
$$(\text{FS})_B = \frac{q_u}{(\sigma_{\max})} = \frac{618.7}{99} = 6.2 > 3;\quad \textit{okay}$$

■

11.4 SEISMIC ANALYSIS OF MSE WALLS

Importance *MSE walls, like other walls and foundation systems, are subjected to dynamic forces in seismically active regions. In this section, we discuss the analysis of MSE walls, due to a seismic event.*

Post-earthquake reconnaissance reports showed that MSE walls have performed satisfactorily in past earthquakes. One of the reasons is that MSE walls are semi-rigid (or semi-flexible), which allows them to absorb and dissipate the seismic energy. There is no well-established method for seismic analysis of MSE walls. One approach is to use the method described in Section 10.10 for rigid retaining walls. However, because of the semi-rigid nature of MSE walls, the lateral earth pressures are likely to be different from those calculated using the Mononobe-Okabe expressions.

The seismic design procedure is as follows.

1. Estimate K_h for a desired maximum displacement using Eq. (10.77)
2. Calculate the static active earth pressure as

$$\sigma'_x = K_a \sigma'_z$$

3. Calculate the seismic active earth pressure as

$$\sigma'_{x\mathrm{E}} = (1 - k_v) K_{a\mathrm{E}} \sigma'_z$$

4. Calculate the seismic increment of active earth pressure as

$$\Delta\sigma'_{x\mathrm{E}} = \sigma'_{x\mathrm{E}} - \sigma'_x$$

5. Estimate the vertical spacing:

$$S_z = \frac{T_{\mathrm{all}}}{\sigma'_x (\mathrm{FS})_s + \Delta\sigma'_{x\mathrm{E}} (\mathrm{FS})_d}$$

where $(\mathrm{FS})_s$ is the static factor of safety (between 1.3 to 1.5) and $(\mathrm{FS})_d$ is the dynamic factor of safety (between 1.1 and 1.3).

6. Determine the length of reinforcement at the base to prevent translations

$$L_b = \frac{0.5\, H_o [K_a (\mathrm{FS})_s + (K_{a\mathrm{E}} - K_a)(\mathrm{FS})_d]}{\tan \phi'_b}$$

7. Determine the total length.
8. Check seismic bearing capacity.

11.5 IN SITU REINFORCED WALLS—SOIL NAILING

11.5.1 Basic Concept

These walls utilize reinforcing elements to form a composite with the soil. The reinforcing elements are nails or small-diameter cast-in-place concrete piles or small-diameter steel pipe piles. One popular in-situ reinforced wall is soil nail walls. Closely spaced nails are installed by drilling inclined holes into the soil and grouting the hole (Fig. 11.5). Shear stresses from the soil are transferred to the nails and are resisted

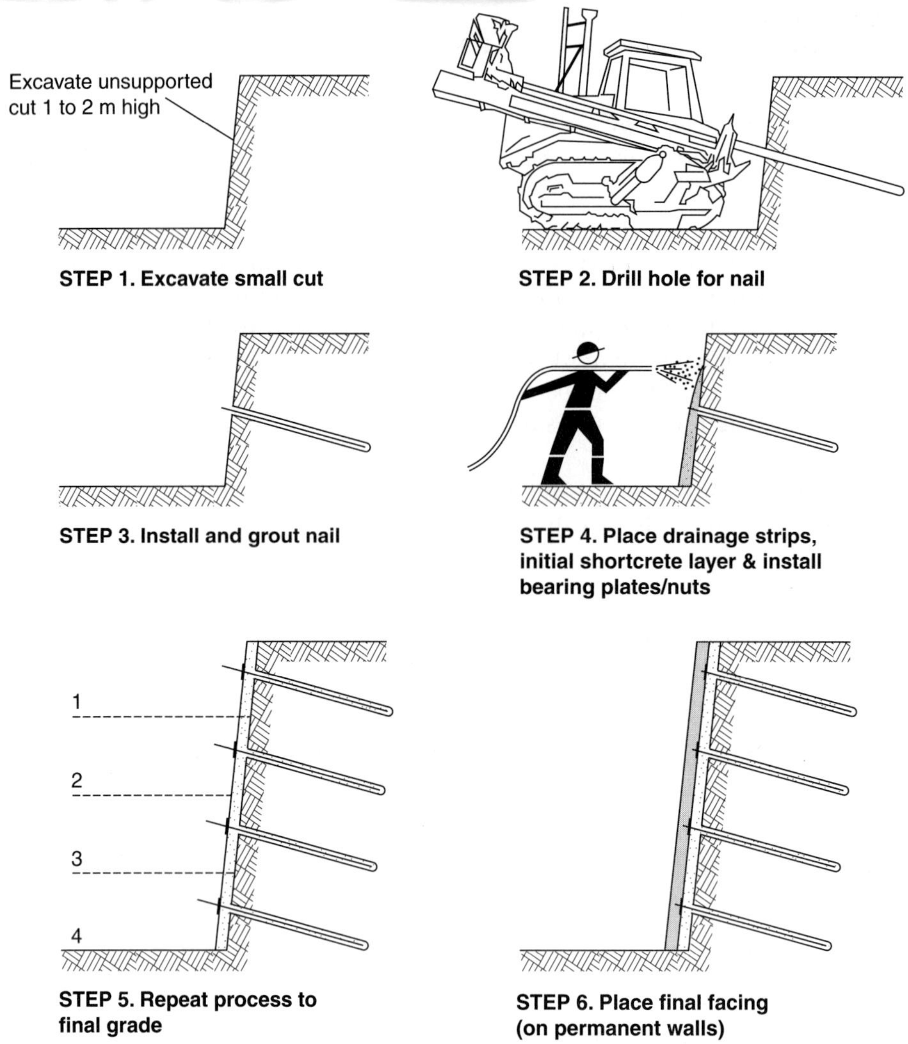

FIGURE 11.5 Typical soil nail wall construction label sequence

by tensile forces in the nails. The faces of soil nail walls are shotcrete or precast concrete panels or cast-in-place concrete. The soil nails are, generally, steel rods. They can be installed either in a predrilled hole followed by grouting or by attaching a drill bit (called a sacrificial bit because the bit cannot be recovered) to the forward end of the rods and drilling into the soil and then grouting. Typical details for a soil nail wall are shown in Fig. 11.6.

Most nails are drilled at an angle of 15° or less to the horizontal plane. Soil nailing is not suitable for soft clays, expansive and collapsible soils if moisture is not controlled, organic materials (e.g., peat), loose fill, and loose sand ($N < 10$).

11.5.2 Analysis of Soil Nail Walls

Access http://www.wiley.com/college/budhu. Click Foundations and Earth Structures, Chapter 11, soil nail.xls for a spread sheet to analyze soil nail walls.

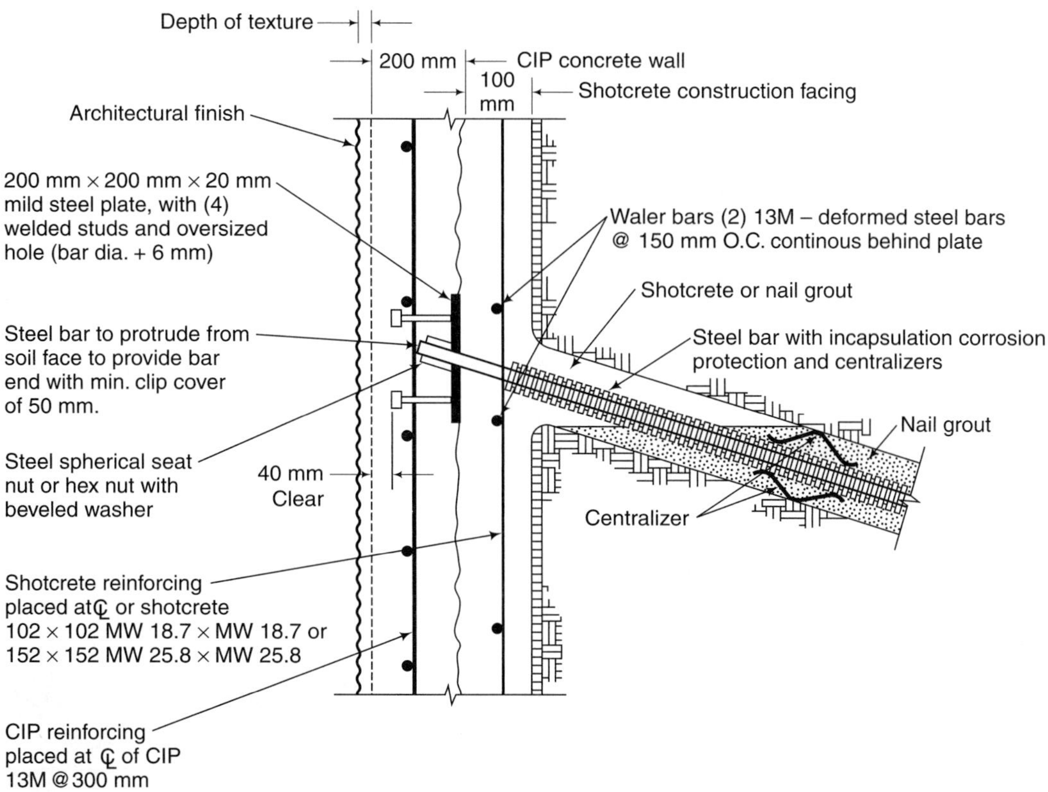

FIGURE 11.6 Soil nail wall details. (Source: FHWA-SA–96-069R.)

A possible failure mechanism for a soil nailed wall is shown in Fig. 11.7. Different shapes (linear, bilinear, circular, and log spiral) for the failure surface are used in practice. The forces on a typical nail at the intersection of the failure surface consist of a normal force, N, a shear force, T and a bending moment, M.

Consider a linear failure surface and a single nail in a wall with nails at a vertical spacing, S_z, and a lateral spacing, S_y as shown in Fig. 11.8. Using the limiting equilibrium method (Chapter 3), we get for the forces along the failure surface,

$$\text{Disturbing force} = T = W_i \sin\theta_a \tag{11.17}$$

$$\text{Resisting force} = S + T_n \cos(\alpha + \theta_a) \tag{11.18}$$

where S is the shearing resistance, T_n is the nail–soil interface force, α is the inclination of the nail to the horizontal plane, θ_a is the inclination of the failure plane to the horizontal plane, and W_i is the weight of the soil wedge at the nail location. The factor of safety (FS) is

$$\text{FS} = \frac{S + T_n \cos(\alpha + \theta_a)}{W_i \sin\theta_a} \tag{11.19}$$

We have neglected the moment resistance and the shearing resistance of the nails. Also, we assumed that the soil mass within the active zone produces the disturbing force. The soil shearing resistance on the failure plane and the tensile component of the nail provide the resistance.

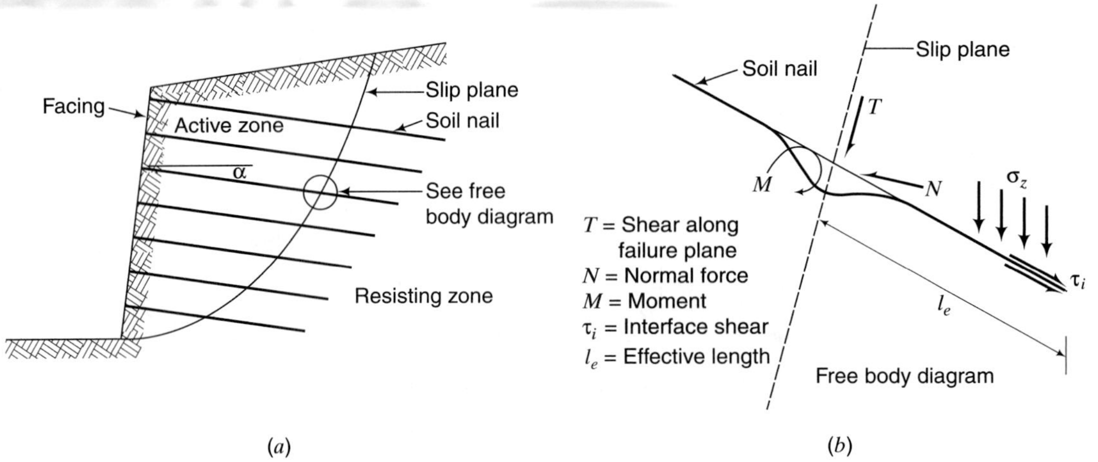

FIGURE 11.7 A possible failure mechanism.

For an effective stress analysis (ESA),

$$S = N \tan \phi'_{cs} = [W_i \cos \theta_a + T_n \sin(\alpha + \theta_a)] \tan \phi'_{cs} \tag{11.20}$$

while for a total stress analysis (TSA),

$$S = s_u L \tag{11.21}$$

where s_u is the undrained shear strength and L is the length of the failure (slip) surface. The factor of safety then is

ESA:

$$\text{FS} = \frac{[W_i \cos \theta_a + T_n \sin(\alpha + \theta_a)] \tan \phi'_{cs} + T_n \cos(\alpha + \theta_a)}{W_i \sin \theta_a} \tag{11.22}$$

TSA: (short-term condition in fine-grained soils)

$$\text{FS} = \frac{s_u L + T_n \cos(\alpha + \theta_a)}{W_i \sin \theta_a} \tag{11.23}$$

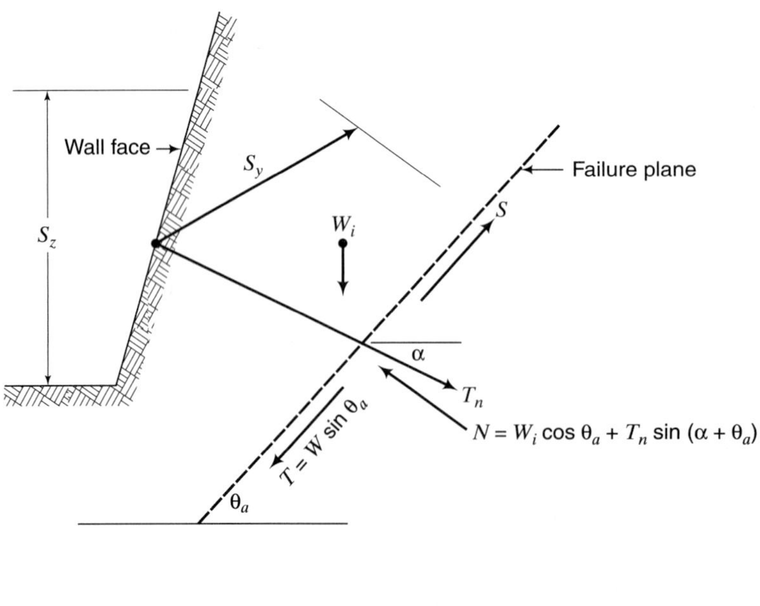

FIGURE 11.8 Forces on a nail.

In designing soil nail walls, two checks must be made. One concerns the internal stability—tensile failure of the nail, nail pullout, and structural failure at the wall facing/nail head connection.

The other is the external stability—translation, rotation, bearing capacity, and deep-seated failure. External stability is found similar to MSE walls. We examine the internal stability in the following.

Tensile failure The tensile strength is

$$T_n = A_n f_y \tag{11.24}$$

where A_n is the cross-sectional area [nail diameters vary from 25 mm (1 in.) to 35 mm (1.5 in)] and f_y is the yield stress, $420\,\text{N/mm}^2$ (8800 ksf) to $500\,\text{N/mm}^2$ (10,500 ksf).

Nail Pullout The nail pullout can occur from weak bond between either the soil and the grout or the grout and the nail. The soil–grout bond is influenced by soil creep. Creep rate should be less than 2 mm/min over a period 6 minutes to 60 minutes (log scale). Generally, the soil–grout bond is weaker than the grout–nail bond.

Nail Head The nail head strength is dependent on punching shear of the facing and its connection, the tensile strength of the headed studs, and the flexural strength of the facing. Nail head failure rarely controls the design of a soil nail system. The minimum allowable nail head load is

$$T_{nh} = F_F K_a \gamma H_o S_z S_y \tag{11.25}$$

where F_F is a service factor = 0.5, K_a is the active earth pressure coefficient, H_o is the wall height, and S_z and S_y are the vertical and horizontal spacing respectively.

There are no universally accepted design practices for a soil nail system. Two design approaches can be taken. One is similar to MSE walls with low extensible reinforcement. The other—the slip surface method—is based on consideration of various slip surfaces within the soil region containing the nail and just outside. In the United States, the slip surface method is the preferred method. Most design practice follows the *Manual for Design and Construction Monitoring of Soil Nail Walls* (http://isddc.dot.gov/OLPFiles/FHWA/01057.pdf).

We will use the following practical guidelines:

1. ***Nail length:*** 0.6 to 0.7 × wall height, minimum length = 6 m (20 ft)
2. ***Nail inclination:*** $\leqslant 15^\circ$.
3. ***Nail spacing density:*** 0.5 to 2 per m^2 (0.05 to 0.2 per ft^2).
4. ***Wall deformation:*** 1.5×10^{-6} × wall height.
5. For the slip surface method, the minimum factor of safety should be 1.35.
6. Nail lengths in the upper half of the wall height should be equal. Nails lengths in the lower half of the wall can have shorter lengths as described in the FHWA manual.

11.6 OTHER TYPES OF RETAINING WALLS

There are a variety of retaining walls that are now used in geotechnical and other applications (e.g., landscaping). Some of these are listed below.

11.6.1 Modular Gravity Walls

Modular gravity walls consist of a structural container filled with materials of various types. These walls are used in projects such as highways, dikes, and channels instead of cast-in-place concrete gravity walls. The design method is discussed in Chapter 10, Section 10.8. Examples of modular gravity walls are:

- ***Gabion baskets.*** Gabion basket walls consist of prefabricated steel wire or polypropylene or polyethylene or nylon baskets filled with rocks and stacked horizontal and vertically (Fig. 11.9). Basket size varies but is normally 2 m (6 ft) in length × 1 m (3 ft) wide ×1 m (3 ft) high. Long baskets exceeding 2 m in length are reinforced with diaphragms to strengthen the baskets. Gabion baskets are suitable as gravity walls where rock fill material is readily available. Common applications are for erosion protection of banks of rivers and streams, and for retaining walls along roads and highways, especially in rugged terrain.
- ***Bin walls.*** Bin walls are gravity walls in which earth fill is placed in a bin made from metal (steel) or timber or concrete (Fig. 11.10). These walls require a small amount of settlement of the vertical corner members. If the foundation is rigid, a compressible cushion of approximately 200 mm (8 in.) of loose fill is placed under the grade plates or concrete base.
- ***Precast modular concrete walls.*** These walls are constructed by stacking precast blocks made from concrete and other materials. There is a large variety of blocks, the major differences being the material used to make the blocks, size, and interlocking mechanisms.

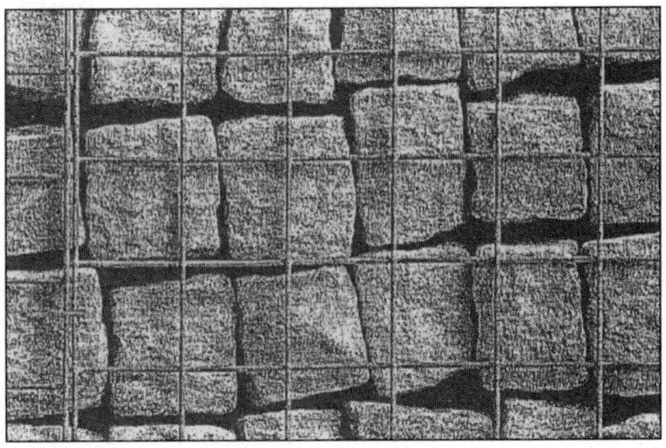

FIGURE 11.9 Gabion basket wall.

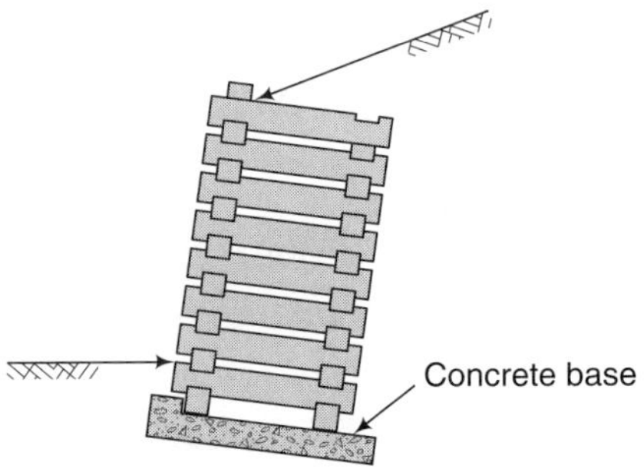

FIGURE 11.10 A crib wall

11.6.2 Chemically Stabilized Earth (CSE) Walls

CSE walls are in-situ soils mixed with chemical grouts such as lime or lime-cement mixtures to form columns of overlapping soils. Sometimes reinforcements are added to the soil-grout mixtures before they harden. CSE walls can retain soils up to great depths. They are also used in seepage control.

11.7 SUMMARY

MSE walls are growing in popularity. They are cost-effective and are often good replacements for rigid retaining walls.

EXERCISES

1. A 6 m (20 ft) high geotextile wall is required to support a coarse-grained backfill with $\gamma_{sat} = 17.5\,kN/m^3$ (110 pcf) and $\phi'_{cs} = 29°$. The foundation (base) soil is a clay with $\gamma_{sat} = 18\,kN/m^3$ (113 pcf), $\phi_p = 28°$, $s_u = 72$ kPa (1500 psf), and $\phi_b = 16°$. The ultimate strength of the geotextile is 45 kN/m (3000 lb/ft) and the soil-geotextile interface friction angle is 20°. The permanent surcharge is 15 kPa (300 psf). Determine the spacing and length of geotextile required for stability.

2. Redo Exercise 11.1 using galvanized steel ties 75 mm (3 in.) wide with a yield strength 2.5×10^5 kPa (52×10^5 psf), tie-soil interface friction of 20°, and rate of corrosion of 0.025 mm/yr (0.001 in./yr). The design life is 50 years.

3. A section of an approach to a bridge is shown in Fig. P11.3. The sides of the approach are to be supported using MSE walls with 1 m (3 ft) × 1 m (3 ft) facing panels. Steel ties of width 300 mm (1 ft) are readily available. Determine the length and thickness of the ties required for stability. The design life is 50 years and the rate of corrosion is 0.025 mm/yr. The yield strength is 2.5×10^5 kPa (52×10^5 psf).

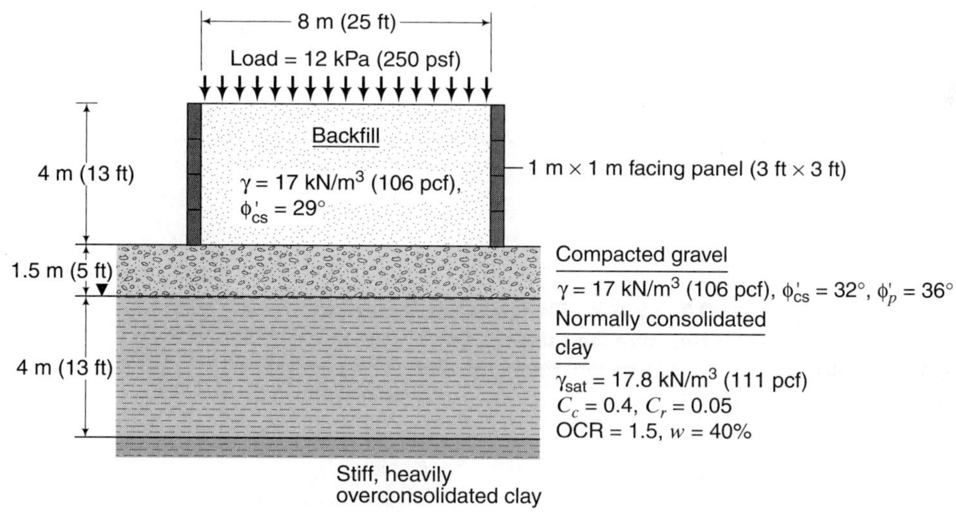

FIGURE P11.3

APPENDIX A

A COLLECTION OF FREQUENTLY USED SOIL PARAMETERS AND CORRELATIONS

TABLE A.1 Typical Values of Unit Weight for Soils

Soil type	γ_{sat}—kN/m^3(pcf)	γ_d—kN/m^3(pcf)
Gravel	20 – 22 (125 – 138)	15 – 17 (94 – 106)
Sand	18 – 20 (113 – 125)	13 – 16 (81 – 100)
Silt	18 – 20 (113 – 125)	14 – 18 (88 – 113)
Clay	16 – 22 (100 – 138)	14 – 21 (88 – 131)

TABLE A.2 Description Based on Relative Density

D_r(%)	Description
0 – 20	Very loose
20 – 40	Loose
40 – 70	Medium dense or firm
70 – 85	Dense
85 – 100	Very dense

TABLE A.3 Description of Soil Strength Based on Liquidity Index

Values of LI	Description of soil strength
$LI < 0$	Semisolid state—high strength; brittle (sudden) fracture is expected.
$0 < LI < 1$	Plastic state—intermediate strength; soil deforms like a plastic material.
$LI > 1$	Liquid state—low strength; soil deforms like a viscous fluid.

TABLE A.4 Atterberg Limits for Typical Soils

Soil type	LL (%)	PL (%)	LI (%)
Sand	Nonplastic		
Silt	30 – 40	20 – 25	10 – 15
Clay	40 – 150	25 – 50	15 – 100

TABLE A.5 Soil Types, Description, and Average Grain Size According to USCS

Soil type	Description	Average grain size
Gravel	Rounded and/or angular bulky hard rock	Coarse: 75 mm to 19 mm
		Fine: 19 mm to 4.75 mm
Sand	Rounded and/or angular bulky hard rock	Coarse: 4.75 mm to 2.0 mm (No. 10)
		Medium: 2.0 mm to 0.425 mm (No. 40)
		Fine: 0.425 mm to 0.075 mm (No. 200)
Silt	Particles between 0.075 mm and 0.002 mm exhibit little or no strength when dried.	0.075 mm to 0.002 mm
Clay	Particles smaller than 0.002 mm exhibit significant strength when dried. Water reduces strength.	< 0.002 mm

TABLE A.6 Hydraulic Conductivity for Common Soil Types

Soil type	k—cm/sec (in./sec)
Clean gravel	$> 1.0 (0.4)$
Clean sands, clean sand and gravel mixtures	1.0 to $10^{-3} (0.4$ to $0.4 \times 10^{-3})$
Fine sands, silts, mixtures comprising sands, silts, and clays	10^{-3} to $10^{-7} (0.4 \times 10^{-3}$ to $0.4 \times 10^{-7})$
Homogeneous clays	$< 10^{-7} (0.4 \times 10^{-7})$

APPENDIX B

TABLE B4.1 Typical Range of Values for C_c, C_r, and C_α

Typical range of values
$C_c = 0.1 - 0.8$
$C_r = 0.015 - 0.35$
$C_r = C_c/5 - C_c/10$
$C_\alpha/C_c = 0.03 - 0.08$

TABLE B4.2 Empirical Relationships for C_c and C_r

Empirical relationships	
$C_c = 0.009(\text{LL} - 10)$	(Skempton, 1944)
$C_c = 0.40(e_o - 0.25)$	(Azzoz et al., 1976)
$C_c = 0.01(w_n - 5)$	(Azzoz et al., 1976)
$C_c = 0.37(e_o + 0.003\text{LL} - 0.34)$	(Azzoz et al., 1976)
$C_c = 0.5G_s\left(\frac{\text{PI}}{100}\right) \approx \frac{\text{PI}}{74}$	(Wood and Wroth, 1978)
$C_c = 0.00234\,\text{LL}\,G_s$	(Nagaraj and Murthy, 1986)
$C_r = 0.15(e_o + 0.007)$	(Azzoz et al., 1976)
$C_r = 0.003(w_n + 7)$	(Azzoz et al., 1976)
$C_r = 0.126(e_o + 0.003\text{LL} - 0.06)$	(Azzoz et al., 1976)
$C_r = 0.000463\text{LL}\,G_s$	(Nagaraj and Murthy, 1985)

w_n is the natural water content (%), LL is the liquid limit (%), PI is plasticity index (%), e_o is the initial void ratio.

Compressibility	C_c	λ
Low	< 0.2	< 0.1
Moderate	0.2 – 0.4	0.1 – 0.2
High	> 0.4	> 0.2
Soil type		C_α/C_c
Inorganic clays and silts		0.033 – 0.06
Organic clays and silts		0.035 – 0.075
Peats		0.05 – 0.08

TABLE B4.3 Typical Values of C_α/C_c (Source: Carter and Bentley, 1991.)

Soil	C_α/C_c
Organic silts	0.035 – 0.06
Amorphous and fibrous peat	0.035 – 0.085
Canadian muskeg	0.09 – 0.10
Leda clay (Canada)	0.03 – 0.06
Postglacial Swedish clay	0.05 – 0.07
Soft blue clay (Victoria, BC)	0.026
Organic clays and silts	0.04 – 0.06
Sensitive clay (Portland, ME)	0.025 – 0.055
San Francisco Bay mud	0.04 – 0.06
New Liskeard (Canada) varved clay	0.03 – 0.06
Mexico City clay	0.03 – 0.035
Hudson River silt	0.03 – 0.06
New Haven organic clay silt	0.04 – 0.075

TABLE B4.4 Typical Values of C_v (Source: Carter and Bentley, 1991.)

	C_v	
Soil	($\text{cm}^2/\text{s} \times 10^{-4}$)	(m^2/yr)
Boston blue clay (CL)	40 ± 20	12 ± 6
Organic silt (OH)	2 – 10	0.6 – 3
Glacial lake clays (CL)	6.5 – 8.7	2.0 – 2.7
Chicago silty clay (CL)	8.5	2.7
Swedish medium sensitive clays (CL – CH)		
1. Laboratory	0.4 – 0.7	0.1 – 0.2
2. Field	0.7 – 3.0	0.2 – 1.0
San Francisco Bay mud (CL)	2 – 4	0.6 – 1.2
Mexico City clay (MH)	0.9 – 1.5	0.3 – 0.5

TABLE B4.5 Ranges of Friction Angles (Degrees) for Soils

Soil type	ϕ'_{cs}	ϕ'_p	ϕ'_r
Gravel	30 – 35	30 – 50	
Mixture of gravel and sand with fine-grained soils	28 – 33	30 – 40	
Sand	27 – 37*	32 – 50	
Silt or silty sand	24 – 32	27 – 35	
Clays	15 – 30	20 – 30	5 – 15

*Higher values (32° to 37°) in the ranges are for sands with significant amounts of feldspar (Bolton, 1986). Lower values (27° to 32°) in the ranges are for quartz sands. The peak dilation angle: α_p ranges from 0 to 15°.

TABLE B4.6 Range of Dilation Angles for Soils

Soil type	α_p (degrees)
Dense sand	10 – 15°
Loose sand	< 10°
Normally consolidated clay	0°

TABLE B4.9a Typical Values of *E* and *G*

Soil type	Description	*E** (MPa)	*G* (MPa)
Clay	Soft	1 – 15	0.4 – 5
	Medium	15 – 30	5 – 11
	Stiff	30 – 100	11 – 38
Sand	Loose	10 – 20	4 – 8
	Medium	20 – 40	8 – 16
	Dense	40 – 80	16 – 32

*These are secant values at peak deviatoric stress for dense and stiff soils and when the maximum deviatoric stress is attained for loose, medium, and soft soils. 1 MPa = 20.9 ksf.

TABLE B4.7 A_f Values (SOURCE: After Skempton, 1954.)

Type of clay	A_f
High sensitivity	0.75 – 1.0
Normally consolidated	0.5 – 1
Compacted sandy clay	0.25 – 0.75
Lightly overconsolidated clays	0 – 0.5
Compacted clay-gravel	−0.25 – 0.25
Heavily overconsolidated clays	−0.5 – 0

TABLE B4.9b Typical Values of G_{max}

Soil type	G_{max} (MPa)
Soft clays	2 – 14
Firm clays	7 – 35
Silty sands	28 – 138
Dense sands and gravels	69 – 345

TABLE B4.8 Typical Values of Poisson's Ratio

Soil type	Description	ν' (see note)
Clay	Soft	0.35 – 0.4
	Medium	0.3 – 0.35
	Stiff	0.2 – 0.3
Sand	Loose	0.15 – 0.25
	Medium	0.25 – 0.3
	Dense	0.25 – 0.35

Note: $\nu = \nu'$ is for drained condition; $\nu = \nu_u = 0.5$ for undrained condition.

TABLE B4.10 Overconsolidation Ratio (OCR)

Test type	Expression	Source
One-dimensional consolidation lab test	$\mathrm{OCR} = \left(\frac{p_{atm}}{\sigma'_{zo}}\right) \times 10^{(1.11-1.62\mathrm{LI})}$ LI = Liquidity index (0 to 1)	Kulhawy and Mayne (1990)
Critical state theory	$\log(\mathrm{OCR}) = \frac{\left[2 - 2\,\mathrm{LI} - \log\left(15.87\frac{\sigma'_{zo}}{p_{atm}}\right)\right]}{\Lambda}$ where $\Lambda = 1 - \frac{\kappa}{\lambda} \cong 0.8$	Wood (1983)

TABLE B4.11 Correlation of Undrained Shear Strength with Simply Obtained Soil Parameters

Soil type	Equation	Reference
Normally consolidated clays	$\left(\frac{s_u}{\sigma'_z}\right)_{nc} = 0.11 + 0.0037\mathrm{PI}$	Skempton (1957)
	$\frac{s_u}{p'_o} = 0.129 + 0.00435\ \mathrm{PI}$	Wroth & Houlsby (1985)
Over-consolidated clays	$\frac{(s_u)_w}{(s_u)_{PL}} = \exp(-0.046\mathrm{LI})$ $(s_u)_w$ = undrained shear strength at any water content $(s_u)_{PL}$ = undrained shear strength at plastic limit $\simeq$ 170–200 kPa, LI = liquidity index (%) One-dimensionally consolidated clays: $\frac{s_u}{\sigma'_z} = \frac{\sin\phi'_{cs}[K_o + A_f(1 - K_o)]}{1 + (2A_f - 1)\sin\phi'_{cs}}$	Wood (1990)
	$\frac{\left(\frac{s_u}{\sigma'_z}\right)_{oc}}{\left(\frac{s_u}{\sigma'_z}\right)_{nc}} = (\mathrm{OCR})^{0.8}$	Ladd et al. (1977)
	$\frac{s_u}{\sigma'_z} = (0.23 \pm 0.04)\mathrm{OCR}^{0.8}$	Jamiołkowski et al. (1985)
	$\frac{s_u}{\sigma'_{zc}} = 0.11 + 0.0037\ \mathrm{PI}$	Skempton (1957)
All clays	$\frac{s_u}{\sigma'_{zc}} = 0.22$	Mesri (1975)

TABLE B4.12 Correlations for Shear Modulus

(a) Fine-grained soils

Tests	Expression	Source
Laboratory		
Simple shear	$G_{max} = \frac{198}{e^{1.3}}(R_o)^A \sqrt{p'}$ (MPa) where G_{max} is the initial (maximum) shear modulus, and p' is the mean effective stress (MPa), and A is a coefficient that depends on the plasticity index given in Table B4.12a.	Jamiolkowski et al (1985)

TABLE B4.12a

PI (%)	0	20	40	60	80	≥100
A	0	0.81	0.30	0.41	0.48	0.5

Tests	Expression	Source
Resonance column	$\frac{G_{max}}{p_{atm}} = 321\frac{(2.97-e)^2}{1+e}\mathrm{OCR}^A\left(\frac{p'_o}{p_{atm}}\right)^{0.5}$ where G_{max} is the initial (maximum) shear modulus, and p'_o is the initial mean effective stress, and A is a coefficient that depends on the plasticity index as given in Table B4.12a. The secant shear modulus, G_{sec}, is about 5% to 10% of G_{max}.	Hardin and Drnevich (1972)
Cam-clay model with triaxial compression data	*Normally consolidated clays:* $\left(\frac{G_{sec}}{s_u}\right)_{nc} = \frac{2M(1+e_o)}{\lambda\Lambda(1-\Lambda)\exp(-\Lambda)}$ where M is the slope of the critical state line for triaxial compression, $\Lambda = 1 - \frac{\kappa}{\lambda}$ (about 0.8 for practical purposes). $\lambda = \frac{C_c}{2.3}$ is the compression index, and $\kappa = \frac{C_r}{2.3}$ is the recompression index.	Wroth (1979)
	Overconsolidated clays: $\left(\frac{G_{sec}}{s_u}\right)_{oc} = \left(\frac{G_{sec}}{s_u}\right)_{nc}[1 + C\ln(R_o)](R_o)^{-\Lambda}$ where C is between 0 and 2 ($C = 1$ is sufficiently accurate for practical purposes), $R_o = \frac{1+2K_o^{nc}}{1+2K_o^{oc}}\mathrm{OCR}$	
	$G_{sec} = 0.5\frac{1+e_c}{\kappa}p'$ where p'_c is preconsolidated mean effective stress and e_c is the void ratio corresponding to p'_c.	Randolph et al. (1979)

(Continued)

TABLE B4.12 (Continued)

(b) Coarse-grained soils

Test	Expression	Source
Laboratory	$G_{max} = k_1\sqrt{p'}$ (MPa) where p' is the mean effective stress (MPa) and k_1 is a coefficient as given in the table below.	Seed and Idriss (1970)

e	k_1	$D_r(\%)$	k_1
0.4	484	30	235
0.5	415	40	277
0.6	353	45	298
0.7	304	60	360
0.8	270	75	408
0.9	235	90	484

TABLE B4.13 Correlations of Subgrade Modulus

(a)

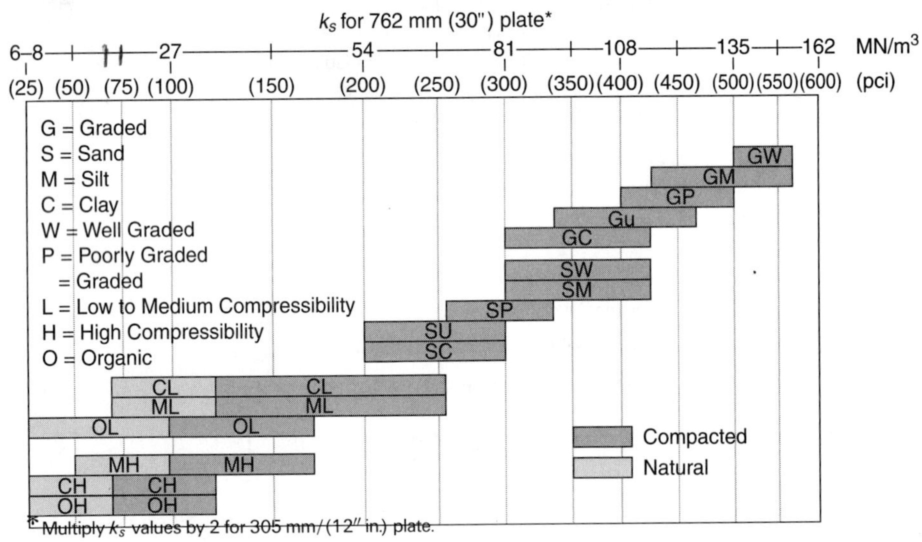

(b) $$k_s = \frac{0.65}{B}\left(\frac{E_s B^4}{E_f I_f}\right)^{\frac{1}{12}} \frac{E_s}{1-\nu^2} \quad \text{kN/m}^3$$

where B is the foundation width, E_f is the elastic modulus of the foundation, E_s is the elastic modulus of the soil, and I_f is the second moment of area of the foundation.

TABLE B5.1 Correction Factors for Rod Length, Sampler Type, and Borehole Size

Correction factor	Item	Correction factor
C_R	Rod length (below anvil)	$C_R = 0.05L + 0.61;\ 4\,\text{m} < L \leq 6\,\text{m}$ $= 0.015L + 0.61;\ 13\,\text{ft} < L \leq 20\,\text{ft}$ $C_R = -0.0004L^2 + 0.017L + 0.83;\ 6\,\text{m} < L < 20\,\text{m}$ $= -0.000037L^2 + 0.005L + 0.83;\ 20\,\text{ft} < L < 65$ $C_R = 1;\ L \geq 20\,\text{m}\ (65\,\text{ft})$. L = rod length
C_S	Standard sampler	$C_S = 1.0$
	U.S. sampler without liners	$C_S = 1.2$
C_B	Borehole diameter:	
	65 mm to 115 mm (2.5 in. to 4.5 in.)	$C_B = 1.0$
	152 mm (6 in.)	$C_B = 1.05$
	200 mm (8 in.)	$C_B = 1.15$
C_E	Equipment:	
	Safety hammer (rope, without Japanese "throw" release)	$C_E = 0.7–1.2$
	Donut hammer (rope, without Japanese "throw" release)	$C_E = 0.5–1.0$
	Donut hammer (rope, with Japanese "throw" release)	$C_E = 1.1–1.4$
	Automatic-trip hammer (donut or safety type)	$C_E = 0.8–1.4$

(SOURCE: Youd et al., (2001), and Seed et al., 2003).

TABLE B5.2 Correlation of N, γ, D_r, and ϕ' for Coarse-Grained Soils

N	Description	γ (kN/m^3)	D_r (%)	ϕ' (degrees)
0 – 4	Very loose	11 – 13	0 – 15	26 – 28
4 – 10	Loose	14 – 16	16 – 35	29 – 34
10 – 30	Medium	17 – 19	36 – 65	35 – 40*
30 – 50	Dense	20 – 21	66 – 85	38 – 45*
> 50	Very dense	> 21	> 86	> 45*

*These values correspond to ϕ'_p.

TABLE B5.3 Overconsolidation Ratio (OCR) Inferred from In Situ Tests

Test type	Expression	Source
(a) Fine-grained soils		
VST	$OCR = 3.22\left(\frac{s_u}{\sigma'_{zc}}\right)$	Kulhawy and Mayne (1990)
SPT	$OCR = 0.58N\left(\frac{p_{atm}}{\sigma'_{zo}}\right)$	Mayne and Kemper (1988)
CPT	$OCR = 0.33\frac{(q_c - \sigma_{zo})}{\sigma'_{zo}}$	Mayne (2003)
PMT	$OCR = 0.45\left(\frac{P_L}{\sigma'_{zo}}\right)$ P_L = limit pressure in PMT	Kulhawy and Mayne (1990)
DMT	$OCR = (\beta_o \kappa_D)^{1.56}$ $\beta_o = 0.5$ (insensitive clays) $= 0.75$ (fissured clays) κ_D = horizontal stress index	Manchetti (1980)
(b) Coarse-grained soils		
	$OCR = \left(\frac{K_o}{1 - \sin\phi'_{cs}}\right)^{\left(\frac{1.25}{\sin\phi'_{cs}}\right)}$ K_o obtained from DMT	Bullock (1983)

TABLE B5.4 Undrained Shear Strength from In-Situ Tests.

Test type	Expression	Source
VST	Obtained directly from test $s_u = \frac{2T}{\pi d^2\left(\frac{h}{d} + \frac{1}{3}\right)}$ T = Torque, *h* and *d* are the height and diameter of the blade	
DMT	$\frac{s_u}{\sigma'_{zo}} = A(0.5K_D)^{1.25}$ $A = 0.14$ (simple shear) $= 0.20$ (triaxial compression) $= 0.17$ to 0.21 (VST) K_D = horizontal stress index for material index $I_D < 0.6$	Lacasse and Lunne (1988)
CPT	$s_u = \frac{q_c - \sigma_{zo}}{N_K}$ $N_K = 19 - \frac{(PI - 10)}{5}$; $PI > 10\%$	

TABLE B5.5 Soil Stiffness from In-Situ Tests

Test	Expression	Source
SPT	$\frac{G_{sec}}{p_{atm}} \cong \frac{2.5}{1+\nu} N_{1,60}$ (sands with fines)	Modified after Kulhawy and Mayne (1990)
	$\frac{G_{sec}}{p_{atm}} \cong \frac{5}{1+\nu} N_{1,60}$ (clean normally consolidated sands)	
	$\frac{G_{sec}}{p_{atm}} \cong \frac{7.5}{1+\nu} N_{1,60}$ (clean overconsolidated sands)	Modified after Mayne and Frost (1989)
	$\frac{G_{sec}}{p_{atm}} \cong \frac{0.11 N_{1,60}^{0.82}}{1+\nu}$ (sandy silt)	
	$\frac{E_{sec}}{p_{atm}} \cong 22(N_{60})^{0.82}; N_{60} \leq 100$ (Piedmont residium)	Mayne and Harris (1993)
	$G_{max} \cong 15560(N_{60})^{0.68}$ (kPa) Silts, sandy silts, slightly cohesive mixtures: $E_{sec} \cong 400N_{1,60}$ (kPa) Clean fine-to-medium sands and slightly silty sands: $E_{sec} \cong 700N_{1,60}$ (kPa) Coarse sands and sands with little gravel: $E_{sec} \cong 1000N_{1,60}$ (kPa) Sandy gravels: $E_{sec} \cong 1200N_{1},80$ (kPa)	FHWA (2002)
CPT	$G_{sec} \cong \frac{C_q q_c}{1+\nu}$ where $C_q \cong 0.75$ to 1.25	Kulhawy and Mayne, (1990)
	$G_{max} \cong 1634(q_c)^{0.25} (\sigma'_{zo})^{0.375}$ (kPa)	FHWA (2002)
PMT	The horizontal shear modulus, G_{PMT}, is directly obtained from the test results. This modulus is taken as the shear modulus for vertical loading also.	
DMT	The horizontal shear modulus, E_D, is directly obtained from the test results. This is related to G_{sec} for vertical loading as $G_{sec} \cong E_D \frac{1-\nu}{2}$ The horizontal elastic modulus is also correlated with soil description as shown in the figure below.	

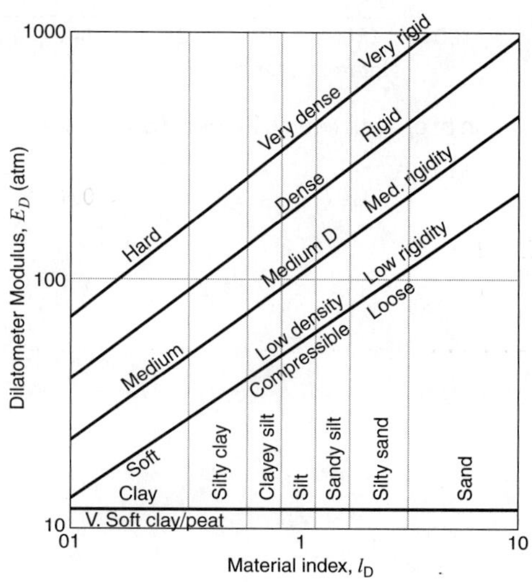

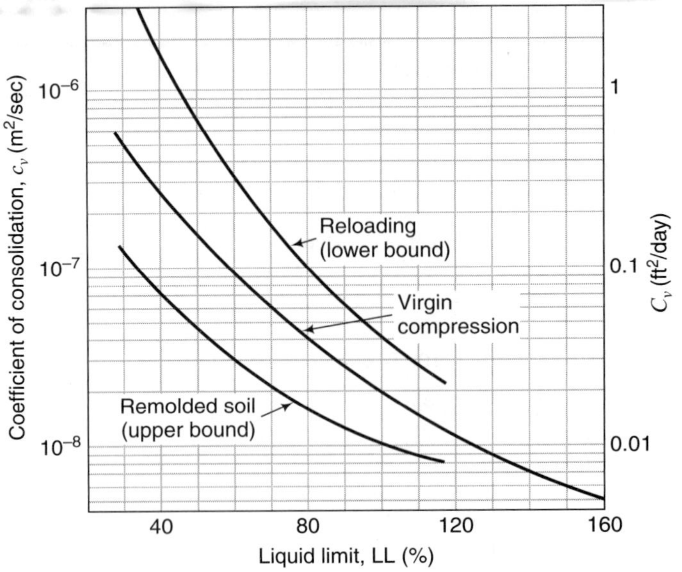

FIGURE B4.1 Correlation of c_v with liquidity index. (Source NAVFAC, DM-7.1, 1982.)

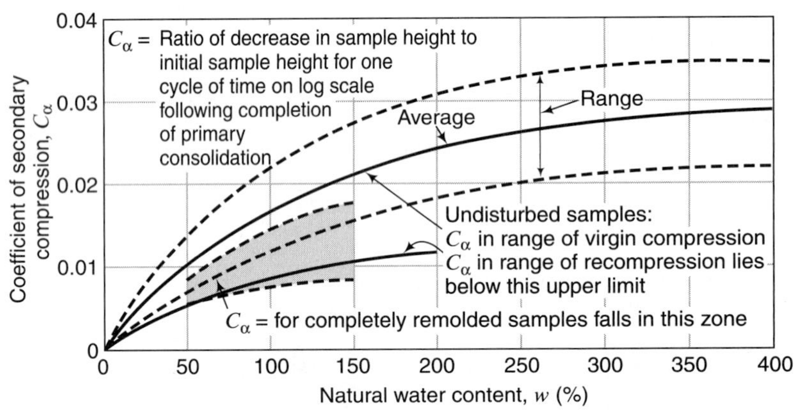

FIGURE B4.2 Correlation of c_α with liquidity index. (Source NAVFAC, DM-7.1, 1982.)

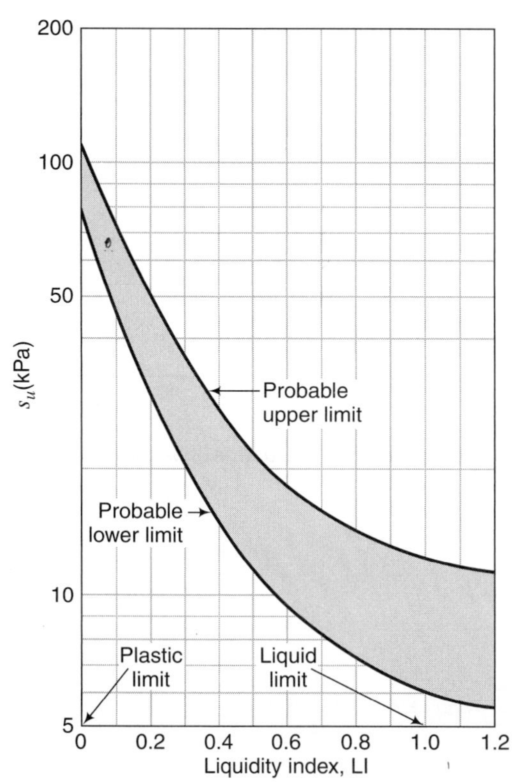

FIGURE B4.3 Range of s_u variation with LI. (Source: Carter and Bentley, 1991.)

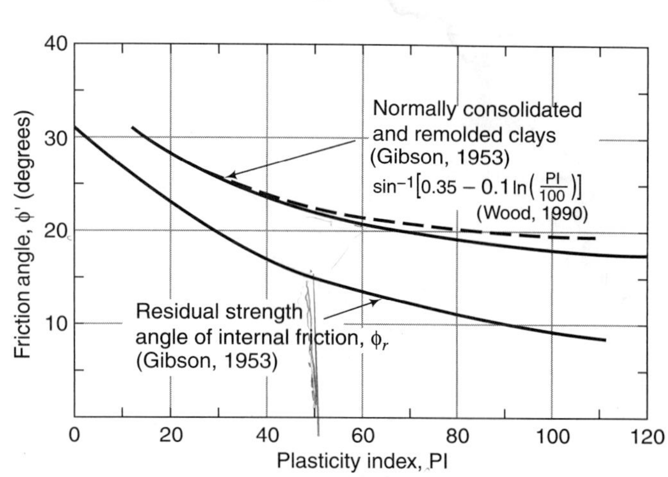

FIGURE B4.4 Relationship between friction angle and PI.

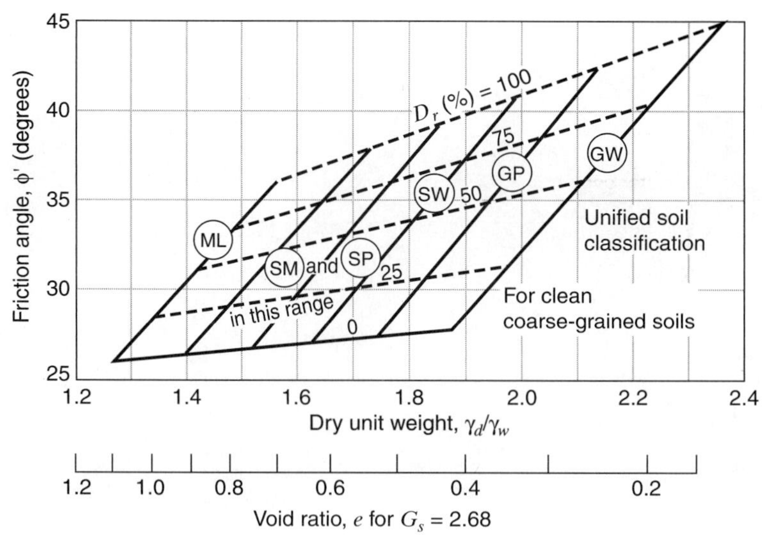

FIGURE B4.5 Typical values of friction angle for coarse-grained soils. (Source: NAFAC, DM7.1, 1982, Kulhawy and Mayne, 1990.)

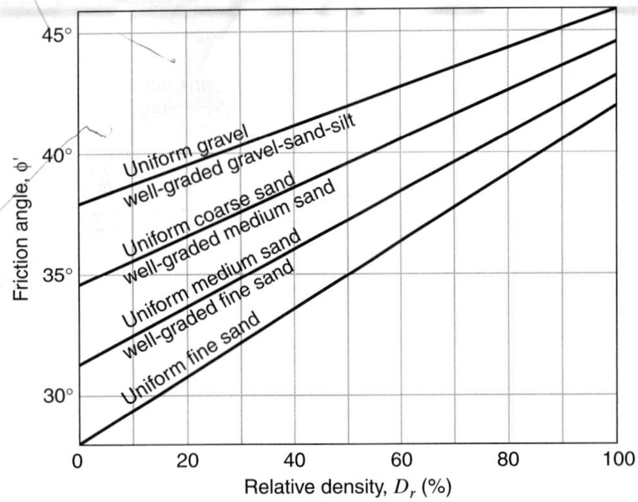

FIGURE B4.6 Correlation of friction angle and relative density. (Source: Schmertmann, et al., 1978.)

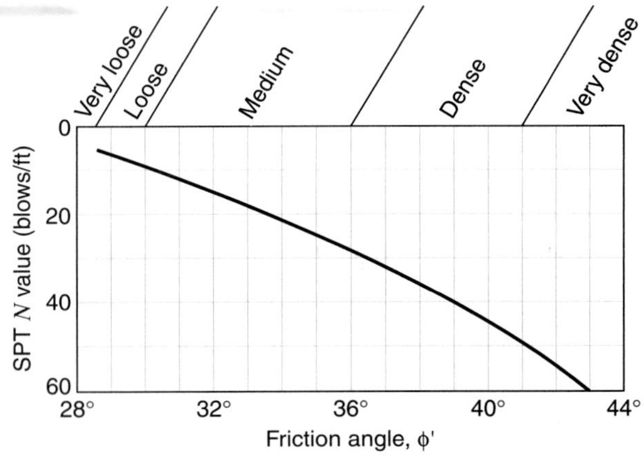

FIGURE B5.1 Correlation of SPT data with friction angle. (Source: Peck et al., 1974.)

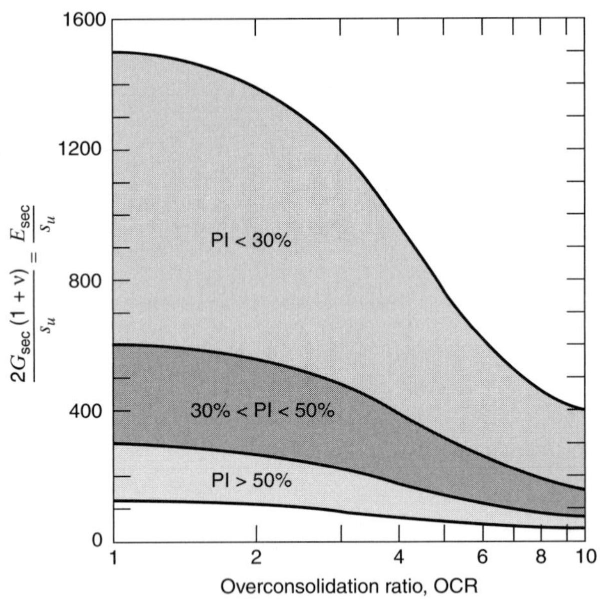

FIGURE B4.7 Correlation of secant shear modulus (and undrained elastic secant modulus) with OCR and PI (modified from data reported by Duncan and Buchignani, 1976.)

APPENDIX C

LATERAL EARTH PRESSURE COEFFICIENTS (KERISEL AND ABSI, 1970)

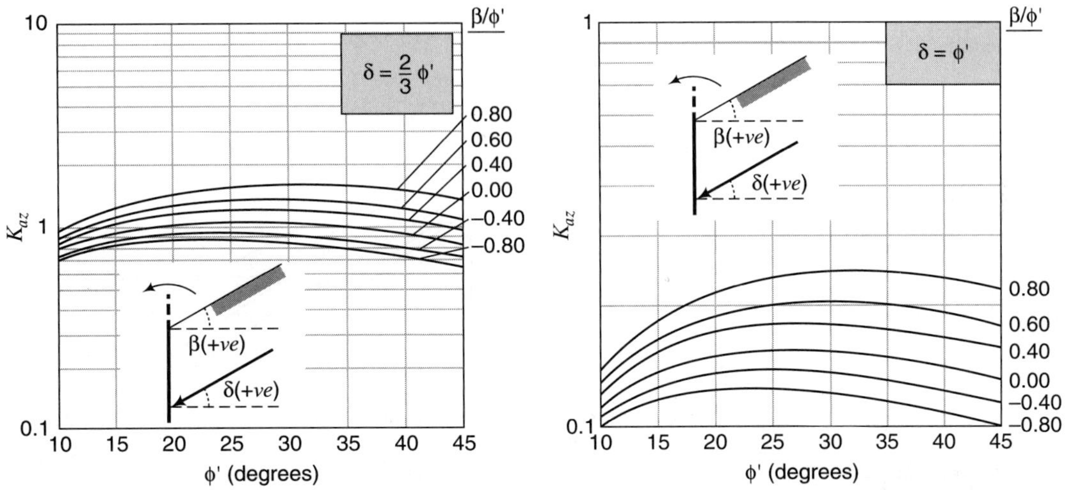

FIGURE C.1 Vertical component of the active lateral pressure coefficient. (Plotted from data published by Kerisel and Absi, 1990.)

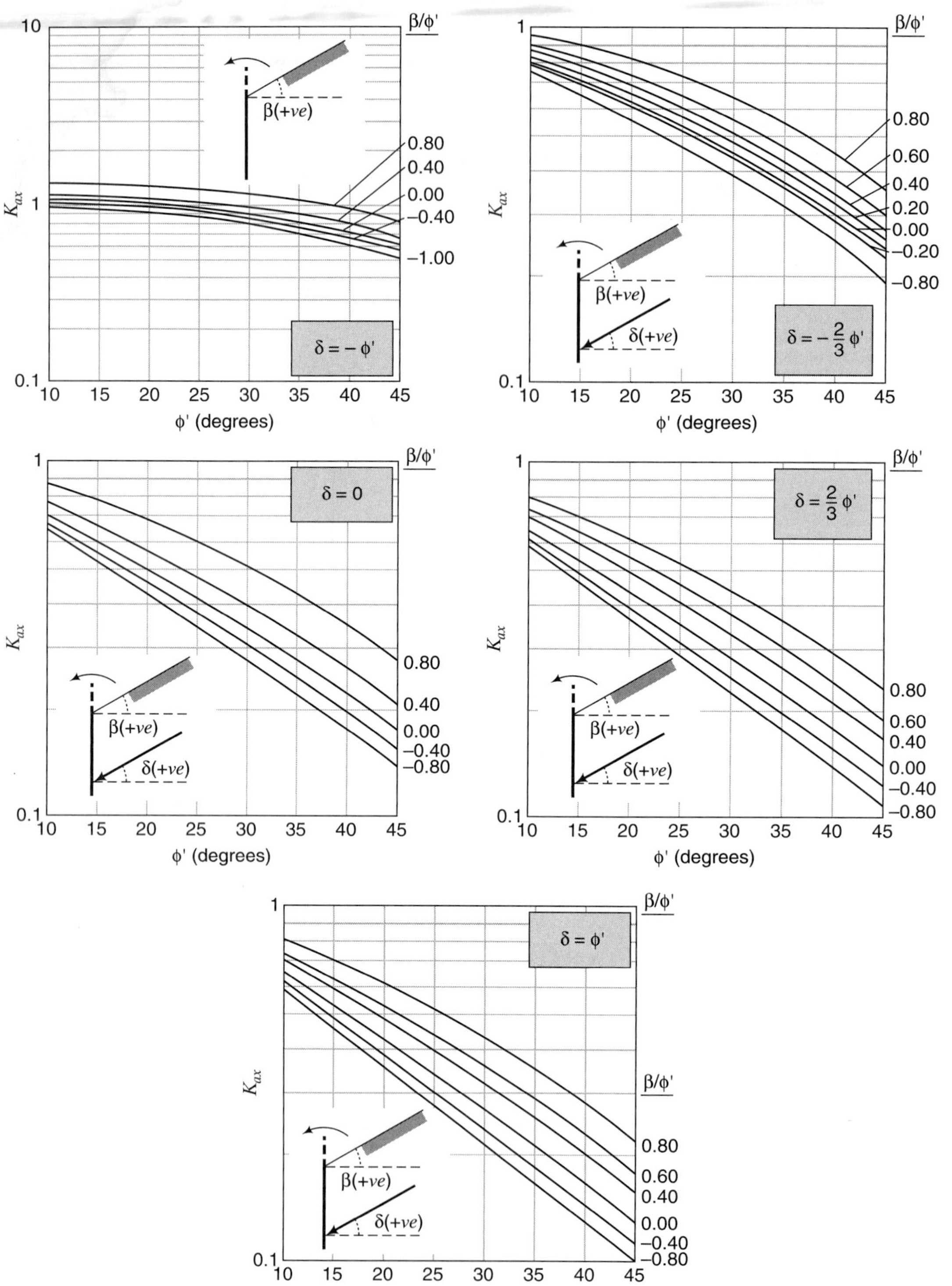

FIGURE C.2 Horizontal componant of the active lateral pressure coefficient. (Plotted from data published by Kerisel and Absi, 1990.)

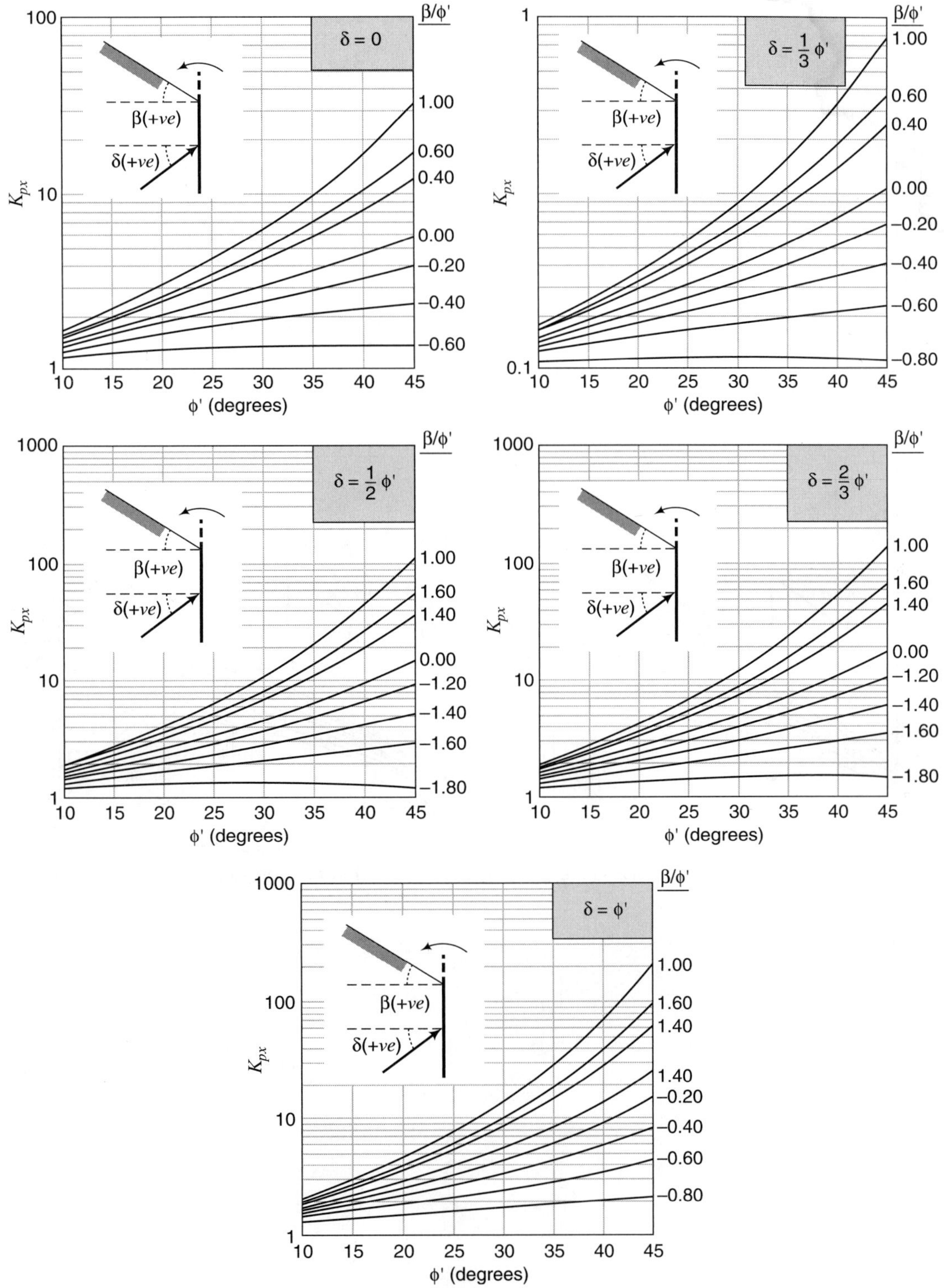

FIGURE C.3 Horizontal componant of the passive lateral pressure coefficient. (Plotted from data published by Kerisel and Absi, 1990.)

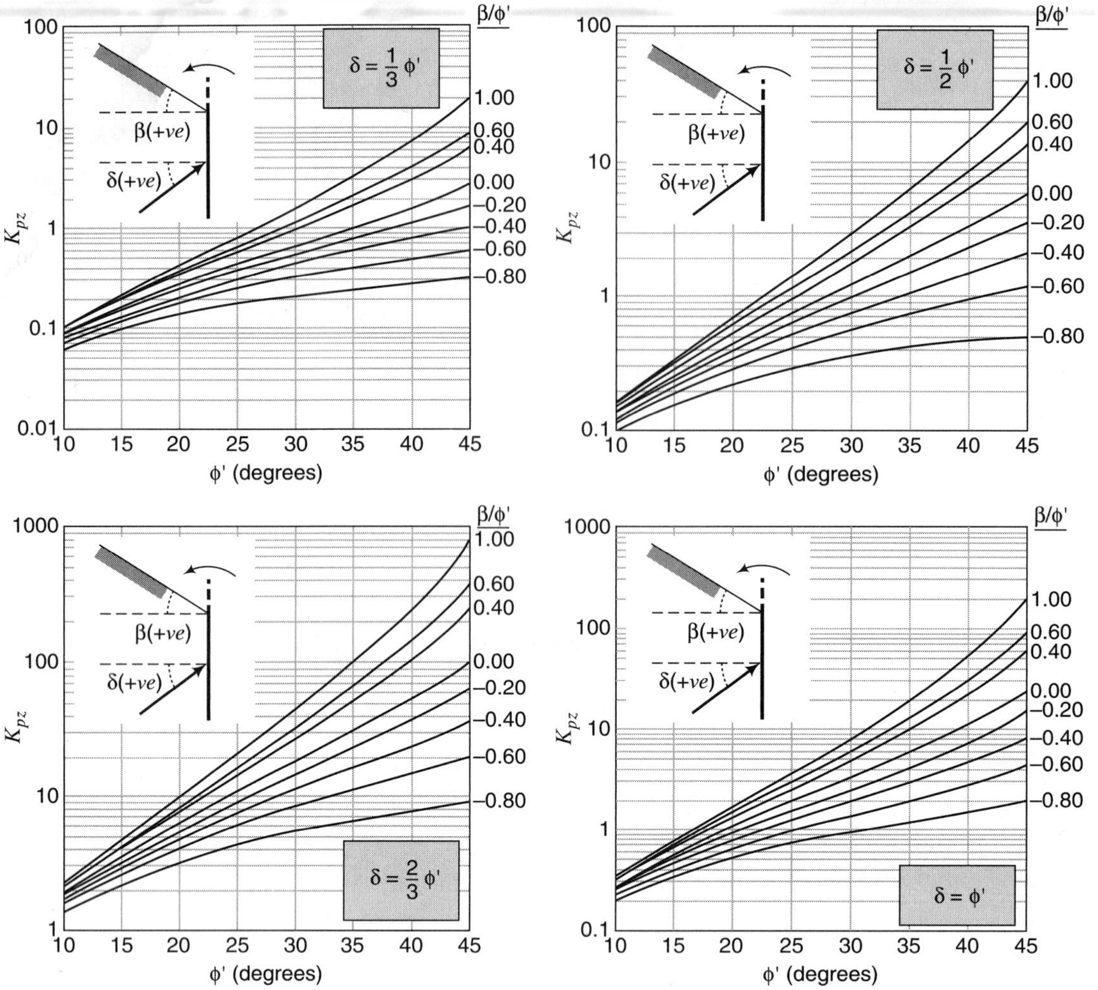

FIGURE C.4 Vertical componant of the passive lateral pressure coefficient. (Plotted from data published by Kerisel and Absi, 1990.)

REFERENCES

Adamson, A. W. (1982). *Physical Chemistry of Surfaces*. Wiley, New York.

API (1984). *API Recommended Practice for Planning, Designing and Constructing Fixed Offshore Platforms*. American Petroleum Institute, Washington, D.C.

API (1993). *RP2A: Recommended practice for planning, designing and constructing fixed offshore platforms*. Washington, DC: American Petroleum Institute.

Atterberg, A. (1911). "Uber die Physikalishe Bodennuntersuchung und uber die Plastizitat der Tone." *Int. Mitt. Boden*, **1**, 10–43.

Azzoz, A. S., Krizek, R. J., and Corotis, R. B. (1976). "Regression analysis of soil compressibility." *Soils Found.*, **16**(2), 19–29.

Becker, D. E., Crooks, J. H., Been, K., and Jefferies, H. G. (1987). "Work as a criterion for determining in situ and yield stresses in clays." *Canadian Geotechnical Journal*, **24**(4), 549–564.

Bell, J. M. (1966). "Dimensionless parameters for homogeneous earth slopes." *J. Soil Mech. Found. Eng. Div. ASCE*, **95** (SM6), 1253–1270.

Berezartzev, V. G., Khirstoforov, V. S., and Golubkov, V. N. (eds.) (1961). "Load bearing capacity and deformation of piled foundations." *Proceedings of the 5th International Conference on Soil Mechanics and Foundation Engineering*, Paris, Vol. 2, 11–15.

Bishop, A. W. (1955). "The use of the slip circle in the stability analysis of slopes." *Geotechnique*, **5**(1), 7–17.

Bishop, A. W., Alan, I., Blight, G. E., and Donald, I. B. (eds.) (1960). "Factors controlling the strength of partially saturated cohesive soils." *Proceedings of the Research Conference on Shear Strength of Cohesive Soils*, American Society of Civil Engineers, VA, 500–532.

Bishop, A. W., and Morgenstern, N. R. (1960). "Stability coefficients for earth slopes." *Geotechnique*, **10**(4), 129–150.

Bjerrum, L. (1954). "Geotechnical properties of Norwegian marine clays." *Geotechnique*, **4**(2), 49–69.

Bjerrum, L., and Eide, O. (1956). "Stability of strutted excavations in clays." *Geotechnique* **1**(1), 32–47.

Bjerrum, L., and Eggestad, A. (eds.) (1963). "Interpretation of loading tests on sand." *Proceedings of the European Conference on Soil Mechanics and Foundation Engineering, Wiesbaden*, Vol. 1, 199–203.

Bolton, M. D. (1986). "The strength and dilatancy of sands." *Geotechnique*, **36**(1), 65–78.

Bolton, M. D., and Powrie, W. (1988). "The behaviour of diaphragm walls prior to collapse." *Geotechnique*, **38**(2), 167–189.

Boussinesq, J. (1885). *Application des Potentiels a l'Etude de le l'Equilibre et du Mouvement des Solides Elastiques*. Gauthier-Villars, Paris.

Bozozuk, M. (1976). "Tower Silo Foundation." CBd-177, Canadian Building Digest, Ottawa.

Briaud, J-L. (1992). The Pressuremeter, SWETS, 336 pp.

Brown, J. D. and Meyerhof, G. G. (1969) "Experimental study of bearing capacity in layered clays." *Proc. 7th Int. Conf. on Soil Mechanics and Foundation Engineering*, Mexico, **2**, 45–51.

Budhu, M. (1979). *Simple Shear Deformation of Sands*. Ph.D. Thesis, Cambridge University, Cambridge, England.

Budhu, M., and Davies, T. G. (1987). "Non-linear analysis of laterally loaded piles in cohesionless soils." *Canadian Geotechnical Journal*, **24**, 289–296.

Budhu, M., and Davies, T. G. (1988). "Analysis of laterally loaded piles in soft clays." *Journal of Geotechnical Engineering, ASCE*, **114**(1), 21–39.

Budhu, M., and Al-karni, A. (1993). "The seismic bearing capacity of soils." *Geotechnique*, **43**(1), 181–187.

Budhu, M. (1984). "Non-uniformities imposed by simple shear apparatus." *Canadian Geotechnical Journal*, **21**(1), 125–137.

Bullock, P. (1983) "The dilatometer: current test procedures and data interpretation." *Civil Engineering Report*, Univ. of Floida, Gainesville, 309 p.

Burland, J. B. (1973). "Shaft friction piles in clay—a simple fundamental approach." *Ground Eng.*, **6**(3), 30–42.

Burland, J. B., and Burbrige, M. C. (eds.) (1985). "Settlement of foundations on sand and gravel." *Proc. Inst. Civ. Eng. Part 1*, **78**, 1325–1381.

Canadian Foundation Engineering Manual (1992). Canadian Geotechnical Society, Technical Committee on Foundations, 3rd ed., 512 p.

Caquot, A., and Kerisel, J. (1948). *Tables for the Calculation of Passive Pressure, Active Pressure and Bearing Capacity of Foundations*, Gauthier-Villars, Paris.

Caquot, A., and Kerisel, J (1953). "Ultimate bearing capacity of a foundation lying on the surface of a cohesionless soil." *Proc. 3rd Int. Conf. on SMFE*, Vol. 1, 336–337 (in French).

Carter, M., and Bentley, S. P. (1991). *Correlations of Soil Properties*. Pentech, London.

Casagrande, A. (1932). "Research on the Atterberg limits of soils." *Public Roads*, **13**(8), 121–136.

Casagrande, A. (ed.) (1936). "The determination of the preconsolidation load and its practical significance." *1st International*

Conference on Soil Mechanics and Foundation Engineering, Cambridge, Massachusetts, Vol. 3, pp. 60–64.

Casagrande, A. (1937). "Seepage through dams." *J.N. Engl. Water Works Assoc.*, **L1**(2), 131–172.

Casagrande, A., and Fadum, R. E. (1940). "Notes on soil testing for engineering purposes." *Soil Mechanics Series*, Graduate School of Engineering, Harvard University, Cambridge, MA, Vol. 8(268), p. 37.

Casagrande, A. (1960). "Discussion of requirements for the practice of applied soil mechanics." *Proc. 1st Panamerican Conference on Soil Mechanics and Foundation Engineering*, Vol. III, 1029–1037.

Chang, M. F. (ed.) (1988). "Some experience with the dilatometer test in Singapore." *Penetration Testing, Orlando*, **1**, 489–496.

Chandler, R. J. (1988). "The in situ measurement of the undrained shear strength of clay using the field vane." *Vane Shear Strength Testing in Soils: Field and Laboratory Studies*. ASTM, STP 1014, Philadelphia, 13–44.

Chu, S. (1991). "Rankine analysis of active and passive pressures in dry sands." *Soils Founds.*, **31**(4), 115–120.

COM624 (1993). "Laterally loaded pile analysis program for the microcomputer." *FHWASA-91-048*, Computer program documentation, by S. T. Wang and L. C. Reese.

Coulomb, C. A. "Essai sur une application des regles de maximia et minimis a quelques problemes de statique relatifs a l'architecture." *Memoires de la Mathematique et de Physique, presentes a l'Academic Royale des Sciences, par divers savants, e lus dans ces Assemblees*. L'Imprimerie Royale, Paris, 3–8.

Coyle H. M. and Castello, R. R. (1981). "New design correlations for piles in sand." *Journal of the Geotechnical Engineering Division*, ASCE, **107**, 965–986.

Darcy, H. (1856). *Les Fontaines Publiques de la Ville de Dijon*. Dalmont, Paris.

Datta, M., Gulhati, S. K., and Rao, G. V. (eds.) (1980). "An appraisal of the existing practice of determining the axial load capacity of deep penetration piles in calcareous sands." *Proceedings of the 12th Annual OTC, Houston, Texas*, Vol. 4, 119–130.

Davies, T. G., and Budhu, M. (1986). "Non-linear analysis of laterally loaded pile in stiff clays." *Geotechnique*, **36**(4), 527–538.

Davis, E. H. and Booker, J. R. (1971). "The bearing capacity of strip footings from the standpoint of plasticity theory." *Proc. 1st Australia-New Zealand Conference on Geomechanics*, Vol. 1, 276–282.

Davis, E. H. and Booker, J. R. (1973). "The effect of increasing strength with depth on the bearing capacity of clays." *Geotechnique*, **23**, 551–563.

DeBeer, E. E. (1970). "Experimental determination of the shape factors and the bearing capacity factors of sand." *Geotechnique*, **20**(4), 347–411.

Decourt, L. (ed.) (1982). "Prediction of the bearing capacity of piles based exclusively on N values of SPT." *Proceedings ESOPT2, Amsterdam*, Vol. 1, 29–45.

De Nicola, A. and Randolph, M. F. (1993). "Tensile and compressive shaft capacity of piles in sand." *Journal of the Geotechnical Engineering Division.*, ASCE 119, **12**, 1952–1973.

Drnevich, V. P. (1967). *Effect of Strain History on the Dynamic Properties of Sand*. Ph.D. Thesis, University of Michigan.

Dumbleton, M. J., and West, G. (1970). "The suction and strength of remoulded soils as affected by composition." LR306, Road Research Laboratory, Crowthorne.

Duncan, J. M. (2000). "Factors of Safety and Reliability in Geotechnical Engineering." *Journal of Geotechnical and Geoenvironmental Engineering*, **126**, No. 4, 307–316.

Duncan, J. M., and Buchignami, A. L. (1976). "An engineering manual for settlement studies." Dept. of civil engineering, University of California, Berkeley, 94p.

Dutt, R. N., and Ingra, W. B. (eds.) (1984). "Jackup rig sitting in calcareous soils." *Proceedings of the 16th Annual OTC, Houston, Texas*, 541–548.

Eden, S. M. (1974). *Influence of Shape and Embedment on Dynamic Foundation Response*. Ph.D. Thesis, University of Massachusetts.

Elms, D. G., and Richards, R. (1990). "Seismic design of retaining walls." *Design and Performance of Earth Retaining Structures*. ASCE.

Eslami, A, and Fellenius, B. H. (1997). "Pile capacity by direct CPT and CPTu method applied to 102 case histories." *Canadian Geotechnical Journal*, **34**, 886–904.

Esrig, M. E., and Kirby, R. C. (1979). "Advances in general effective stress method for the prediction of axial capacity for driven piles in clay." *Proc. 11th Annual Offshore Technology Conference*, 437–449.

Fancher, G. H., Lewis, J. A., and Barnes, K. B. (1933). Mineral Industries Experiment Station, Bull. 12, Penn State College.

Fellenius, W. (1927). *Erdstatische berechungen*. W. Ernst und Sohn, Berlin.

Fleming, W. G. K. (1992). "A new method for single pile settlement prediction and analysis." *Geotechnique* **42**, No. **3**, 411–425.

Feng, T. W. (2000). "Fall-cone penetration and water content relationship of clays." *Geotechnique*, **50**(2), 181–187.

FHWA (2002). Geotechnical engineering circular, "Evaluation of soil and rock properties." FHWA-IF-02-034, 385 pp.

Findlay, J. D. (1984). "Discussion: piling and ground treatment." *Piling and Ground Treatment*. Institution of Civil Engineers, Thomas Telford, London, 189–190.

Fleming, W., and Thorburn, S. (1983). "Recent piling advances." State of the Art Report in. *Proc. Conf. on Advances in Piling and Ground Treatment for Foundations, ICE, London.*

Fleming, W. G. K., Weltman, A. J., Randolp, M. F., and Elson, W. K. (1985). *Piling Engineering*. Halsted Press, New York.

Fletcher, M. S., and Mizon, D. H. (1984). "Piles in chalk for Orwell bridge." *Piling and Ground Treatment*. Institution of Civil Engineers, Thomas Telford, London, 203–209.

Foster, C. R. and Alhvin, R. G. (1954). "Stresses and deflections induced by a uniform circular load." Proc. Highway Research Board, Vol. 33, 467: 470.

Gazetas, G., and Hatzikonstantinou, E. (1988). "Elastic formulae for lateral displacement and rotation of arbitrarily-shaped embedded foundations." *Geotechnique*, **38**(3), 439–444.

Gazetas, G., and Stokoe, K. H. (1991). "Free vibration of embedded foundations: theory versus experiment." *J. Geotech. Eng.*, **117**(9), 1362–1381.

Gazetas, G., Tassoulas, J. L., Dobry, R., and O'Rourke, M. J. (1985). "Elastic settlement of arbitrarily shaped foundations embedded in half-space." *Geotechnique*, **35**(2), 339–346.

Gibson, R. E. (1953). "Experimental determination of the true cohesion and true angle of friction in clays." *Proc. 3rd Int. Conf. on Soil Mech. and Fnds. Eng.*, Zurich, Switzerland, Vol. 126, 130.

Giroud, J. P. (1968). "Settlement of a linearly loaded rectangular area." *J. Soil Mech. Found. Div. ASCE*, **94**(SM4), 813–831.

Graham, J., and Houlsby, G. T. (1983). "Elastic anisotropy of a natural clay." *Geotechnique*, **33**(2), 165–180.

Guo, W. D., and Randolph, M. F. (1999). "An efficient approach for settlement prediction of pile groups." *Geotechnique*, **49**(2), 161–179.

Hansen, B. (ed.) (1961). "The bearing capacity of sand, tested by loading circular plates." *5th International Conference on Soil Mechanics and Foundation Engineering, Paris, France*, Vol. 1, 659–664.

Hansen, J. B. (1970). *A Revised and Extended Formula for Bearing Capacity*, No. 28. Danish Geotechnical Institute, Copenhagen.

Hardin, B. O., and Drnevich, V. P. (1972). "Shear modulus and damping in soils: Design equations and curves." JSMFD, ASCE, Vol. 98, SM7, 667–692.

Hawkins, S. (1988). "A brief history of time. From the big bang to black holes." Bantam Books, Toronto.

Hazen, A. (1930). *Water Supply. American Civil Engineers Handbook*. Wiley, New York, 1444–1518.

Hobbs, N. B. (1977). "Behaviour and design of piles in chalk—an introduction to the discussion of the papers on chalk." *Piles in Weak Rock*. Institution of Civil Engineers, London, 149–175.

Houlsby, G. T., and Purzin, A. M. (1999). "The bearing capacity of strip footing on clay under combined loading." *Proc. Royal Society*, 455A, 893–916.

Hull, T. S. (ed.) (1987). *The Static Behaviour of Laterally Loaded Piles*. Ph.D. Thesis, University of Sydney, Sydney, Australia.

Hvorslev, M. J. (1949). "Subsurface exploration and sampling of soils for civil engineering purposes." Waterways Experiment Stations, Vicksburg, MI.

Hvorslev, M. J. (1937). Uber die Festigkeitseigenshaften gestörter bindiger Böden, *Ingeniorvidenskabelige skrifter*, A., No. 45, Copenhagen.

Ignold, T. S. (1979). "The effects of compaction on retaining walls." *Geotechnique*, **29**(3), 265–283.

Jaky, J. (1944). "The coefficient of earth pressure at rest." *J. Soc. Hungarian Architects Eng.*, **7**, 355–358.

Jaky, J. (ed.) (1948). "Pressure in silos." *2nd International Conference on Soil Mechanics and Foundation Engineering, Rotterdam*, Vol. 1, 103–107.

Jamiolkowski, M., Ladd, C. C., Germaine, J. T., and Lancellotta, R. (eds.) (1985). "New developments in field and laboratory testing of soils." *XIth International Conference on Soil Mechanics and Foundation Engineering, San Francisco*, Vol. 1, 57–154.

Janbu, N. (ed.) (1957). "Earth pressure and bearing capacity by generalized procedure of slices." *Proceedings of the 4th International Conference on Soil Mechanics and Foundation Engineering, London*, Vol. 2, 207–212.

Janbu, N. (ed.) (1973). "Slope stability computations." *Embankment Dam Engineering, Casagrande Memorial Volume*. Wiley, New York, 47–86.

Janbu, N. (ed.) (1976). "Static bearing capacity of friction piles." *Proceedings of the 6th European Conference on Soil Mechanics and Foundation Engineering*, Vol. 1.2, 479–488.

Jardine, R. J., Overy, R. F., and Chow, F. C. (1998). "Axial capacity of offshore piles in dense North Sea Sands. *J. Geotech and Geoenviron. Engrg.*, **124**, 171–178.

Jarquio, R. (1981). "Total lateral surcharge pressure due to a strip load." *J. Geotech. Eng. Div. ASCE*, **107**(10), 1424–1428.

Juran, I., and Schlosser, F. (eds.) (1978). "Theoretical analysis of failure in reinforced earth structures." *Symposium on Earth Reinforcement, ASCE Convention, Pittsburgh*. American Society of Civil Engineers, New York, 528–555.

Kenney, T. C., Lau, D., and Ofoegbu, G. I. (1984). "Permeability of compacted granular materials." *Can. Geotech. J.*, **21**(4), 726–729.

Kerisel, J., and Absi, E. (1990). *Active and Passive Earth Pressure Tables*. Balkema, Rotterdam.

Kraft, L. M., and Lyons, C. G. (1974). "State of the art—ultimate axial capacity of grouted piles." *Proceedings of the 6th Annual OTC, Houston, Texas*, 485–504.

Kozeny, J. (1933). "Theorie und Berechung der Brunnen." *Wasserkraft und Wasserwirtschaft*, **28**, 104.

Kolk, H. J. and van der Velde, E. (1996). "A reliable method to determine friction capacity of piles driven into clays." *Proc. Offshore Technology Conf.*, Houston, Paper OTC 7993.

Kulhawy, F. H., and Mayne, P. W. (1990). "Manual on estimating soil properties for foundation design." EL-6800, Research project 1493-6, Cornell University, NY.

Lacasse, S., and Lunne, T. (1988). "Calibration of dilatometer correlations." *Proc. 1st Int. Symp. on Penetration Testing (ISOPT-1)*, Orlando, FL, Vol. 1, 539–548.

Ladd, C. C., and Foot, R. (1974). "New design procedure for stability of soft clays." *J. Geotech. Eng.*, **100**(7), 763–786.

Ladd, C. C., Foot, R., Ishihara, K., Schlosser, F., and Poulos, H. G. (1977). "Stress-deformation and strength characteristics." *Proceedings of the 9th International Conference on Soil Mechanics and Foundation Engineering, Tokyo*, 421–494.

Lee, K. L., and Seed, H. B. (1967). "Drained strength characteristics of sands." *J. Soil. Mech. Fnds. Div., ASCE*, **93**, SM6, 117–141.

Liao, S. S. C., and Whitman, R. V. (1985). "Overburden correction factors for SPT in sand." *J. Geotech. Eng. Div. ASCE*, **112**(3), 373–377.

Love, A. E. H. (1927). *The Mathematical Theory of Elasticity*. Cambridge University Press, Cambridge.

Marchetti, S. (1980). "In situ tests by flat dilatometers." *J. Geotechnical Engineering Division,* ASCE, **106**, gt3, 299–321.

McCammon, N. R., and Golder, H. Q. (1970). "Some loading tests on long pipe piles." *Geotechnique*, **20**(2), 171–184.

MacDonald, D. H., and Skempton, A. W. (eds.) (1955). "A survey of comparisons between calculated and observed settlements of structures on clay." *Conference on Correlation of Calculated and Observed Stresses and Displacements*. Institution of Civil Engineers, London, 318–337.

Mayne, P. W. and Harris, D. E. (1993). "Axial Load-Displacement Behavior of Drilled Shaft Foundations in Piedmont Residuum, Federal Highway Administration Rept." No. 41-30-2175, prepared by Georgia Tech Research Corp., Atlanta, 163 p.

Mayne, P. (2003). "Integrated ground behavior: In-situ and lab tests." Presented at *Int. Symp. Deformation Characteristics of Geomaterials*. Lyon (keynote paper and presentation).

Mayne, P. W. and Frost, D. D. (1989). "Dilatometer experience in Washington, D. C. and vicinity." *Research Record* 1169, TRB, Washington, 16:23.

Mayne, P. W. and Kemper, J. B. (1988). "Profiling OCR in stiff clays by CPT and SPT." *Geotechnical Testing Journal,* ASTM, **11**(2), 139–147.

Martin, R. E., Seli, J. J., Powell, G. W., and Bertoulin, M. (1987). "Concrete pile design in Tidewater, Virginia." *J. Geotech. Eng. Div.ASCE*, **113**(6), 568–585.

Matlock, H. and Reese, L. C. (1962). "Generalized solutions for laterally loaded piles." *Transactions, ASCE*, **127**, Part 1, 1120–1252.

Matlock, H. and Ripperger E. A. (1956). "Procedures and instrumentation for tests on a laterally loaded pile. Proc. 8th Texas Conf. on Soil mechanics and Foundation Engineering," *Special Publication 29*, Bureau of Engineering Research, Univ. of Texas at Austin.

McClelland, B. (1974). "Design of deep penetration piles for ocean structures." *J. Geotech. Eng. Div. ASCE*, **100**(GT7), 705–747.

McCammon, N. R., and Golder, H. Q. (1970). "Some loading tests on long pipe piles." *Geotechnique*, **20**(2), 171–184.

Mesri, G. (1975). "Discussion: new design procedure for stability of soft clays." *J. Geotech. Eng. Div. ASCE*, **101**(GT4), 409–412.

Meyerhof, G. G. (1951). "The ultimate bearing capacity of foundations." *Geotechnique*, **2**(4), 301–331.

Meyerhof, G. G. (1953). "The bearing capacity of foundations under eccentric and inclined loads." *3rd International Conference on Soil Mechanics and Foundation Engineering, Zurich, Switzerland*, 440–445.

Meyerhof, G. G. (1956). "Penetration tests and bearing capacity of cohesionless soils." *J. Soil Mech. Found. Eng. Div. ASCE*, **82**(SM1), 1–19.

Meyerhof, G. G. (ed.) (1957). "The ultimate bearing capacity of foundations of slopes." *4th International Conference on Soil Mechanics and Foundation Engineering, London*, Vol. 1, 384–386.

Meyerhof, G. G. (1963). "Some recent research on the bearing capacity of foundations." *Can. Geotech. J.*, **1**(1), 16–26.

Meyerhof, G. G. (1965). "Shallow foundations." *J. Soil Mech. Found. Div. ASCE*, **91**(SM2), 21–31.

Meyerhof, G. G. (1976). "Bearing capacity and settlement of pile foundations." *J. Geotech. Eng. Div. ASCE*, **102**(GT3), 195–228.

Meyerhof, G. G., and Koumoto, T. (1987). "Inclination factors for bearing capacity of shallow footings." *J Geotech. Eng. Div. ASCE*, **113**(9), 1013–1018.

Meyerhof, G. G., 1995. Development of geotechnical limit state design. *Canadian Geotechnical Journal,*. **32**, No. 1, 128–136.

Milovic, D. M., and Tournier, J. P. (1971). "Stresses and displacements due to a rectangular load on a layer of finite thickness." *Soils Found.*, **II**(1), 1–27.

Mindlin, R. D. (1936). "Force at a point in the interior of a semi-infinite solid." *J. Appl. Phys.*, **7**(5), 195–202.

Mitchell, J. K. (1978). "In-situ techniques for site characterization." *Proc. NSF Specialty Workshop on Site Characterization and Exploration*. Ch. Dowding (ed.), Northwestern University, Evanston, 107–129.

Mitchell, J. K. (1993). *Fundamentals of Soil Behavior*. Wiley, New York.

Mononobe, N., and Matsuo, H. (1929). "The effect of ground characteristics on the seismic design of structures." *Proc. 2nd World Conf. on Earthquake Engineering*, Tokyo, Japan.

Morgenstern, N. R., and Price, V. E. (1965). "The analysis of the stability of general slip surfaces." *Geotechnique*, **15**(1), 79–93.

Nagaraj, T. S., and Srinivasa Murthy, B. R. (1985). "Prediction of the preconsolidation pressure and recompression index of soils." *Geotech. Testing J. ASTM*, **8**(4), 199–202.

Nagaraj, T. S., and Srinivasa Murthy, B. R. (1986). "A critical reappraisal of compression index." *Geotechnique*, **36**(1), 27–32.

Newmark, N. M. (1942). *Influence Charts for Computation of Stresses in Elastic Foundations*, No. 338. University of Illinois Engineering Experiment Station, Urbana.

Neelakantan, G., Budhu, M., and Richards, R. (1992). "Balanced seismic design of anchored retaining wall," *J. Geotechnical Engineering*, ASCE, **118**(6), 873–888.

Oda, M. and Kazama, H. (1998). "Microstructure of shear bands and its relation to the mechanisms of dilatancy and failure of dense granular soils." *Geotechnique*, **48**, No. 4, 465–481.

Okabe, S. (1926). "General theory of earth pressures." *JJSCE*, **12**(1).

Okimura, M., Takemura, J., and Kimura, T. (1998). "Bearing capacity predictions of sand overlying clay based on limit equilibrium methods." *Soils and Foundations*, **38**, 181–194.

O'Neill, M. W., and Reese, L. C. (1999). "Drilled shafts: Construction procedures." *FHWA-IF-99-025*, Federal Highway Administration, Washington, D.C.

INDEX

Penetration Testing and Experience, St. Louis, Mo., Oct. 26–30, 434–455.

Tumay, M. T., and Kurup, P. U. (2001). "Development of a continuous intrusion miniature cone penetration test system for subsurface explorations," *Soils and Foundations*, **41**(6), 129–138.

Ueno, K., Miura, K., and Maeda, Y. (1998). "Prediction of ultimate bearing capacity of surface footings with regard to size effects." *Soil and Foundations*, **38**(3), 165–178.

Vermeer, P. A., and de Borst, R. (1984). "Non-associated plasticity for soils and concrete." *Heron*, **29**(3), 1–64.

Vesic, A. (1961). "Bending of beams resting on isotropic solid." *J. Engineering Mechanics Division,* ASCE, **87**(2), 35–53.

Vesic, A. S. (1970). "Tests on instrumented piles, Ogeechee River site." *J. Soil Mechanics and Foundations Division*, ASCE, **96**(SM2), 561–584.

Vesic, A. S. (1972). "Expansion of cavities in infinite soil mass." *J. Soil Mechanics and Foundations Division*, ASCE, **98**(SM3), 265–290.

Vesic, A. S. (1973). "Analysis of ultimate loads of shallow foundations." *J. Soil Mech. Found. Div. ASCE*, **99**(SM1), 45–73.

Vesic, A. S. (1975). "Bearing capacity of shallow foundations." *Foundation Engineering Handbook*, H. F. Winkerton and H. Y. Fang, eds. Van Nostrand-Reinhold, New York, p. 121.

Vesic, A. S. (1977). "Design of pile foundations," National Cooperative Highway Research Program. Synthesis of Highway Practice, No. 42, TRB, National Research Council, Washington, DC.

Wagner, A. A. (1957). "The use of the unified soil classification system by the Bureau of Reclamation." *Proc. 4th Int. Conf. Soil Mech. and Fnds. Eng.*, London, **1**, 125.

White, L. S. (1953). "Transcona elevator failure: eye witness account." *Geotechnique*, **111**(5), 209–215.

Wong, P. K., and Mitchell, R. J. (1975). "Yielding and plastic flow of sensitive cemented clay." *Geotechnique*, **25**(4), 763–782.

Wood, D. M. (1983) *Index properties and critical state soil mechanics. Proc. Sym. on Recent Developments in Laboratory and Field Tests and Analysis of Geotechnical Problems*, Bangkok, Thailand, 301–309.

Wood, D. M. (1990). *Soil Behavior and Critical State Soil Mechanics*. Cambridge University Press, Cambridge.

Wood, D. M., and Worth, C. P. (1978). "The correlation of index properties with some basic engineering properties of soils." *Can. Geotech. J.*, **15**(2), 137–145.

Wright, S. J., and Reese, L. C. (1979). "Design of large diameter bored piles." *Ground Eng.*, Nov., 17–50.

Wroth, C. P. (1984). "The interpretation of in situ soil tests." *Geotechnique*, **34**(4), 449–489.

Wroth, C. P., and Hughes, O. J. M. (eds.) (1973). "An instrument for the in-situ measurement of the properties of soft clays." *8th International Conference on Soil Mechanics and Foundation Engineering, Moscow*, Vol. 1, 487–494.

Wroth, C. P. (1979). "Correlations of some engineering properties of soils." *Proc. 2nd Int. Conf. on behaviour of off-shore structures*, London, **1**, 121–132.

Wroth, C. P., and Houlsby, G. T. (1985). "Soil mechanics: Properly characterization and analysis procedures." *11th International Conference on Soil Mechanics and Foundation Engineering.* San Francisco, **I**, 1–55.

Yamashita, K., Tomono, M., and Kakurai, M. (1987). "A method for estimating immediate settlement of piles and pile groups." *Soils Found.*, **27**(1), 61–76.

Youssef, M. S., El Ramli, A. H., and El Demery, M. (eds.) (1965). "Relationship between shear strength, consolidation, liquid limit, and plastic limit for remolded clays." *6th International Conference on Soil Mechanics and Foundation Engineering.* Toronto University Press, Montreal, Vol. 1, 126–129.

Youd, T. L., Idriss, I. M., Andrus, R. D., Arango, I., Castro, G., Christian, J. T., Dobry, R., Liam Finn, W. D., Harder, L. F., Jr., Hynes, M. E., Ishihara, K., Koester, J. P., Liao, S. S. C., Marcuson, W. F. III, Martin, G. R., Mitchell, J. K., Moriwaki, Y., Power, M. S., Robertson, P. K., Seed, R. B. and Stokoe, K. H. II. (2001). "Liquefaction resistance of soils: Summary report from the 1996 NCEER and 1998 NCEER/NSF workshops on evaluation of liquefaction resistance of soils." *J. Geotechnical and Geoenvironmental Eng*, **124**(10), 817–833.

Zadroga, B. (1994). "Bearing capacity of shallow foundations on noncohesive soils." *J. Geotech. Eng.*, **120**(11), 1991–2008.

Zytynski, M., Randolph, M. F., Nova, R., and Wroth, C. P. (1978). "On modeling the unloading-reloading behavior of soils." *Inf. J. Num. Anal. Methods. Geomech.* **2**(1), 87–93.

O'Neill, M. W. and Reese, L. C. (1970). "Behavior of axially loaded drilled shafts in Beaumont clay." *Research Report 89–8*, Center for Highway Research, University of Texas at Austin, Dec., 749pp.

Orchant, C. J., Kulhawy, F. H., and Trautmann, C. H. (1988). "Reliability-based foundation design for transmission line structure: Critical evaluation of in site test method." *Report EL-5507*, **2**, Electric Power Research Institute, Palo Alto, 214p.

Packsahw, S. (1969). "Earth pressures and earth resistance." *A Century of Soil Mechanics*. The Institutions of Civil Engineers, London, 409–435.

Padfield, C. J., and Mair, R. J. (1984). *Design of Retaining Walls Embedded in Stiff Clay*. CIRIA, London.

Peck, R. B. (ed.) (1969). "Deep excavation and tunneling in soft ground." *7th International Conference on Soil Mechanics and Foundation Engineering, State-of-the-Art, Mexico City*, Vol. 3, 147–150.

Peck, R. B., and Byrant, F. G. (1953). "The bearing capacity failure of the Transcona Grain Elevator." *Geotechnique*, **111**, 210–208.

Peck, R. B., Hanson, W. E., and Thornburn, T. H. (1974). *Foundation Engineering*. Wiley, New York.

Poncelet, J. V. (1840). "Memoire sur la stabilite des revetments et de leurs foundations. Note additionelle sur les relations analytiques qui lient entre elles la poussee et la butee de la terre." *Memorial de l'officier du genie*, 13.

Poulos, H. G., Carter, J. P. and Small, J. C. (2002). "Foundations and retaining structures. Research and Practice." *Proceedings of the 15th International Conference on Soil Mechanics*, Foundation Engineering, Istanbul, **4**, 2527–2606.

Poulos, H. G. (1971a). "Behavior of laterally loaded piles: Part I—single piles." *ASCE Soil Mechanics and Foundations Division*, **97**(SM5), 711–731.

Poulos, H. G. (ed.) (1988). "The mechanics of calcareous sediments." *5th Australia–New Zealand Geomechanics Conference, Australia Geomechanics*, 8–41.

Poulos, H. G. (1989). "Pile behavior—theory and application." *Geotechnique*, **39**(3), 365–415.

Poulos, H. G. and Davis, E. H. (1974). *Elastic Solutions for Soil and Rock Mechanics*. Wiley, New York.

Prandtl, L. (1920). *Über die Härte plastischer Körper*, Nachrichten von der Königlichen Gesellschaft der Wissenschaften zu Göttingen (Mathematisch-physikalische Klasse aus dem Jahre 1920), Berlin, 74–85.

PTI (1996). "Design and construction of post-tensioned slabs-on-grade." Post Tension Concrete Institute, Phoenix, AZ, 2nd ed., 101pp.

Quiros, G. W., and Reese, L. G. (1977). "Design procedures for axially loaded drilled shafts." Research Report No. 3-5-72-176, Center for highway research, University of Texas at Austin, December.

Randolph, M. F. (1977). "A theoretical study of the performance of piles." Ph.D thesis, University of Cambridge.

Randolph, M. F., Carter, J. P., and Wroth, C. P. (1979) Driven piles in clay: The effects of installations and subsequent consolidation. *Geotechnique*, **29**(4), 361–393.

Randolph, M. F. (1981). "The response of flexible piles to lateral loading." *Geotechnique*, **31**(2), 247–259.

Randolph, M. F. and Murphy, B. S. (1985). "Shaft capacity of driven piles in clay," *Proc. 17th Ann. Offshore Technol. Conf.*, Houston, **1**, 371–378.

Rankine, W. J. M. (1857). "On the stability of loose earth." *Philos. Trans. R. Soc. London*, **1**, 9–27.

Raschke, S. A., and Hryciw, R. D. (1997). "Vision cone penetrometer (VisCPT) for direct subsurface soil observation." *JGGE*, ASCE, **123**(11), 1074–1076.

Reese, L. C., and O'Neill, M. W. (1988). *Drilled Shafts: Construction Procedures and Design Methods*. Federal Highway Administration, Washington, DC.

Reese, L. C., and O'Neil, M. W. (1988). "Drilled shafts: Construction procedures and design methods." FHWA-HI-88-042.

Reese, L. C. (1977). "Laterally loaded piles: Program documentation." *ASCE Journal of Geotechnical Engineering*, **103**(GT4), 287–305.

Reese, L. C., and Wright, S. J. (1977). "Construction procedures and design for axial loading." Drilled shaft manual HDV-22, Implementation package 77–21, U. S. Department of Transportation, McLean, VA.

Reese, L. C. (1984). "Handbook on design of piles and drilled shafts under lateral load." FHWA-IP-84/1, FHWA, U.S. Department of Transportation, Washington, D.C.

Reese, L. C., Cox, W. R. and Koch, F. D. (1975). "Field testing and analysis of laterally loaded piles in stiff clay." *Proc. VII Annual Offshore Technology Conference*, Houston, Texas, **2**, paper No. 2312, 672–690.

Reul, O., and Randolph, M. F. (2003). "Piled rafts in overconsolidated clay: Comparison of in situ measurements and numerical analyses." *Geotechnique* **53**(3), 301–315.

Reynold, O. (1885). "On the dilatancy of media composed of rigid particles in contact: with experimental illustrations." *Philos. Mag.*, **20**(5th Series), 469–481.

Richards, R., Elms, D., and Budhu, M. (1990). "Dynamic fluidization of soils." *Geotechnical Engineering*, ASCE, **116**(5), 740–759.

Richards, R., Elms, D. G., and Budhu, M. (1993). "Seismic bearing capacity and settlements of foundations." *Journal of Geotechnical Engineering*, ASCE, **119**(4), 662–674.

Richards, R., and Elms, D. G. (1979). "Seismic behavior of gravity retaining walls." *JGED*, ASCE, **105**(GT4), 449–464.

Richart, F. E. (1959). "Review of the theories for sand drains." *Trans. ASCE*, **124**, 709–736.

Robertson, P. K., and Campanella, R. E. (1983). "Interpretation of cone penetration tests. Part I: Sand." *Can. Geotech. J.*, **22**(4), 718–733.

Roscoe, K. H., and Burland, J. B. (1968). "One the generalized stress–strain behavior of 'wet' clay." *Engineering Plasticity*, J. Heyman and F. Leckie (eds.). Cambridge University Press, Cambridge, 535–609.

Rosenfard, J. L., and Chen, W. F. (1972). "Limit analysis solutions of earth pressure problems." Fritz Engineering Laboratory Report 35514, Lehigh University, Bethlehem.

Rowe, P. W. (1957). "Anchored sheet pile walls." *Proc. Inst. Civ. Eng., Part 1*, **1**, 27–70.

Samarasinghe, A. M., Huang, Y. H., and Drnevich, V. P. (1982). "Permeability and consolidation of normally consolidated soils" *J. Geotech. Eng. Div. ASCE*, **108**(GT6), 835–850.

Sarma, S. K. (1979). "Stability analysis of embankments and slopes." *J. Geotech. Eng. Div. ASCE*, **105**(GT12), 1511–1524.

Schmertmann, J. H. (1953). "The undisturbed consolidation behavior of clay." *Trans. ASCE*, **120**, 1201.

Schmertmann, J. H. (1970). "Static cone to compute static settlement over sand." *J. Soil Mech. Found. Div. ASCE*, **96**(SM3), 1011–1043.

Schmertmann, J. H., Hartman, J. P. and Brown, P. R. (1978). "Improved strain influence factor diagrams." *J. Geot. Engr. Div., ASCE*, **104**, No. GT8, 1131–1135.

Schofield, A., and Wroth, C. P. (1968). *Critical State Soil Mechanics*. McGraw-Hill, London.

Schofield, A. N. (2001). "Reappraisal of Terzaghi's Soil Mechanics," *Invited Special Lecture, ISSMGE conference, Istanbul, Turkey*.

Scott, R. E. (1963). *Principles of Soil Mechanics*. Addison-Wesley, Reading, MA.

Seed, H. B., and Idriss, I. M. (1970). "Soil moduli and damping factors for dynamic response analyses." Report EERC 70–10, Earthquake Engineering Research Center, University of California, Berkeley.

Seed, R. B., Cetin, K. O., Moss, R. E. S., Kammerer, A. M., Wu, J., et al. (2003). "Recent advances in soil liquefaction engineering: A unified and consistent framework." *26th Annual ASCE Los Angeles Geotechnical Spring Seminar*, keynote presentation, H.M.S. Queen Mary, Long Beach, CA, April 30, 1–71.

Seed, H. B., and Idriss, I. M. (1971). "Simplified procedure for evaluating soil liquefaction potential." *JSMFE, ASCE*, Vol. 97, SM9, 1249–1273.

Semple, R. M., and Rigden, W. J. (1984). "Shaft capacity of driven piles in clay." *Analysis and Design of Pile Foundations*. American Society of Civil Engineers, New York, 59–79.

Shiau, J. S., Lyamin, A. V. and Sloan, S. W. (2003). "Bearing capacity of a sand layer on clay by finite element limit analysis." *Can. Geotech. J.*, **40**, 900–915.

Shioi, Y., and Fukui, J. (eds.) (1982). "Application of *N*-value to design of foundations in Japan." *Proceedings ESOPT2, Amsterdam*, pp. 159–164.

Sibley, E. A. (ed.) (1967). "Backfill adjacent to structures." *Proceedings of the Montana Conference on Soil Mechanics and Foundation Engineering, Bozeman*, Montana State University.

Skempton, A. W. (1944). "Notes on the compressibility of clays." *Q. J. Geol. Soc. London*, **100**(C: parts 1 & 2), 119–135.

Skempton, A. W. (1951). *The Bearing Capacity of Clay*. Building Research Congress, London.

Skempton, A. W. (1953). "The colloidal activity of clays." *Proceedings, Third International Conference on Soil Mechanics and Foundation Engineering, Zurich,* Vol. 1, 57–61.

Skempton, A. W. (1954). "The Pore water cofficients A and B." *Geotechnique*, **4**, 143–147.

Skempton, A. W. (1959). "Cast-in-situ bored piles in London clay." *Geotechnique*, **9**(4), 153–173.

Skempton, A. W. (1970). "The consolidation of clays by gravitational compaction." *Q. J. Geol. Soc. London*, **125**(3), 373–411.

Skempton, A. W., and Bjerrum, L. (1957). "A contribution to the settlement analysis of foundation on clay." *Geotechnique*, **7**(4), 168–178.

Skempton, A. W., and MacDonald, D. H. (1956). "The allowable settlement of buildings." *Proc. Inst. Civ. Eng. Part III*, **5**, 727–768.

Skempton, A. W. (1957) "Discussion: The planning and design of the new Hong Kong airport." *Proc. ICE*, **7**, 305–307.

Slichter, C. (1899). Nineteenth Annual Report, U.S. Geological Survey, Part 2, 360.

Sovinc, I. (ed.) (1961). "Stresses and displacements in a limited layer of uniform thickness, resting on a rigid base, and subjected to a uniformly distributed flexible load of rectangular shape." *5th International Conference on Soil Mechanics and Foundation Engineering, Paris,* Vol. 1, 823–827.

Spencer, E. (1967). "A method of analysis of embankments assuming parallel interslices technique." *Geotechnique*, **17**(1), 11–26.

Stas, C. V., and Kulhawy, F. H. (1984). "Critical evaluation of design methods for foundations under axial uplift and compression loading." EL-3771, Cornell University (report for EPRI), Ithaca.

Taiebat H., and Carter J. P. (2000). "Bearing capacity of strip and circular foundations on undrained clay subjected to eccentric loads." *Géotechnique, UK*, **52**, 61–64.

Tatsuoka, F. (2000). "The impacts on geotechnical engineering of several recent findings from laboratory stress-strain tests on geomaterials.," *The 2000 Burmister Lecture*, Columbia University, New York.

Taylor, D. W. (1942). "Research on consolidation of clays." Serial No. 82, Massachusetts Institute of Technology, Cambridge.

Taylor, D. W. (1948) *Fundamentals of Soil Mechanics*. Wiley, New York.

Terzaghi, K. (1925). *Erdbaumechanik*. Franz Deuticke, Vienna.

Terzagi, K. (1943). *Theoretical Soil Mechanics*. Wiley, New York.

Terzaghi, K., and Peck, R. B. (1948). *Soil Mechanics in Engineering Practice*. Wiley, New York.

Thornburn, S., and MacVicar, S. L. (1971). "Pile load tests to failure in the Clyde alluvium." *Behaviour of Piles*. Institution of Civil Engineers, London, 1–7, 53–54.

Timoshenko, S. P., and Goodier, J. N. (1970). *Theory of Elasticity*, 3rd ed. McGraw Hill Book Company, NY.

Tomlinson, M. J. (1987). "Pile design and construction practice." Viewpoint publication, 415pp., Chapter 9, p.14.

Tumay, M. T., and Fakhroo, M. (1984). "Pile capacity in soft clays using electric QCPT data." *Proc. Conf. on Cone*

INDEX

Penetration Testing and Experience, St. Louis, Mo., Oct. 26–30, 434–455.

Tumay, M. T., and Kurup, P. U. (2001). "Development of a continuous intrusion miniature cone penetration test system for subsurface explorations," *Soils and Foundations*, **41**(6), 129–138.

Ueno, K., Miura, K., and Maeda, Y. (1998). "Prediction of ultimate bearing capacity of surface footings with regard to size effects." *Soil and Foundations*, **38**(3), 165–178.

Vermeer, P. A., and de Borst, R. (1984). "Non-associated plasticity for soils and concrete." *Heron*, **29**(3), 1–64.

Vesic, A. (1961). "Bending of beams resting on isotropic solid." *J. Engineering Mechanics Division,* ASCE, **87**(2), 35–53.

Vesic, A. S. (1970). "Tests on instrumented piles, Ogeechee River site." *J. Soil Mechanics and Foundations Division*, ASCE, **96**(SM2), 561–584.

Vesic, A. S. (1972). "Expansion of cavities in infinite soil mass." *J. Soil Mechanics and Foundations Division*, ASCE, **98**(SM3), 265–290.

Vesic, A. S. (1973). "Analysis of ultimate loads of shallow foundations." *J. Soil Mech. Found. Div. ASCE*, **99**(SM1), 45–73.

Vesic, A. S. (1975). "Bearing capacity of shallow foundations." *Foundation Engineering Handbook*, H. F. Winkerton and H. Y. Fang, eds. Van Nostrand-Reinhold, New York, p. 121.

Vesic, A. S. (1977). "Design of pile foundations," National Cooperative Highway Research Program. Synthesis of Highway Practice, No. 42, TRB, National Research Council, Washington, DC.

Wagner, A. A. (1957). "The use of the unified soil classification system by the Bureau of Reclamation." *Proc. 4th Int. Conf. Soil Mech. and Fnds. Eng.*, London, **1**, 125.

White, L. S. (1953). "Transcona elevator failure: eye witness account." *Geotechnique*, **111**(5), 209–215.

Wong, P. K., and Mitchell, R. J. (1975). "Yielding and plastic flow of sensitive cemented clay." *Geotechnique*, **25**(4), 763–782.

Wood, D. M. (1983) *Index properties and critical state soil mechanics. Proc. Sym. on Recent Developments in Laboratory and Field Tests and Analysis of Geotechnical Problems*, Bangkok, Thailand, 301–309.

Wood, D. M. (1990). *Soil Behavior and Critical State Soil Mechanics*. Cambridge University Press, Cambridge.

Wood, D. M., and Worth, C. P. (1978). "The correlation of index properties with some basic engineering properties of soils." *Can. Geotech. J.*, **15**(2), 137–145.

Wright, S. J., and Reese, L. C. (1979). "Design of large diameter bored piles." *Ground Eng.*, Nov., 17–50.

Wroth, C. P. (1984). "The interpretation of in situ soil tests." *Geotechnique*, **34**(4), 449–489.

Wroth, C. P., and Hughes, O. J. M. (eds.) (1973). "An instrument for the in-situ measurement of the properties of soft clays." *8th International Conference on Soil Mechanics and Foundation Engineering, Moscow*, Vol. 1, 487–494.

Wroth, C. P. (1979). "Correlations of some engineering properties of soils." *Proc. 2nd Int. Conf. on behaviour of off-shore structures*, London, **1**, 121–132.

Wroth, C. P., and Houlsby, G. T. (1985). "Soil mechanics: Properly characterization and analysis procedures." *11th International Conference on Soil Mechanics and Foundation Engineering.* San Francisco, **I**, 1–55.

Yamashita, K., Tomono, M., and Kakurai, M. (1987). "A method for estimating immediate settlement of piles and pile groups." *Soils Found.*, **27**(1), 61–76.

Youssef, M. S., El Ramli, A. H., and El Demery, M. (eds.) (1965). "Relationship between shear strength, consolidation, liquid limit, and plastic limit for remolded clays." *6th International Conference on Soil Mechanics and Foundation Engineering.* Toronto University Press, Montreal, Vol. 1, 126–129.

Youd, T. L., Idriss, I. M., Andrus, R. D., Arango, I., Castro, G., Christian, J. T., Dobry, R., Liam Finn, W. D., Harder, L. F., Jr., Hynes, M. E., Ishihara, K., Koester, J. P., Liao, S. S. C., Marcuson, W. F. III, Martin, G. R., Mitchell, J. K., Moriwaki, Y., Power, M. S., Robertson, P. K., Seed, R. B. and Stokoe, K. H. II. (2001). "Liquefaction resistance of soils: Summary report from the 1996 NCEER and 1998 NCEER/NSF workshops on evaluation of liquefaction resistance of soils." *J. Geotechnical and Geoenvironmental Eng*, **124**(10), 817–833.

Zadroga, B. (1994). "Bearing capacity of shallow foundations on noncohesive soils." *J. Geotech. Eng.*, **120**(11), 1991–2008.

Zytynski, M., Randolph, M. F., Nova, R., and Wroth, C. P. (1978). "On modeling the unloading-reloading behavior of soils." *Inf. J. Num. Anal. Methods. Geomech.* **2**(1), 87–93.

Rosenfard, J. L., and Chen, W. F. (1972). "Limit analysis solutions of earth pressure problems." Fritz Engineering Laboratory Report 35514, Lehigh University, Bethlehem.

Rowe, P. W. (1957). "Anchored sheet pile walls." *Proc. Inst. Civ. Eng., Part 1*, **1**, 27–70.

Samarasinghe, A. M., Huang, Y. H., and Drnevich, V. P. (1982). "Permeability and consolidation of normally consolidated soils" *J. Geotech. Eng. Div. ASCE*, **108**(GT6), 835–850.

Sarma, S. K. (1979). "Stability analysis of embankments and slopes." *J. Geotech. Eng. Div. ASCE*, **105**(GT12), 1511–1524.

Schmertmann, J. H. (1953). "The undisturbed consolidation behavior of clay." *Trans. ASCE*, **120**, 1201.

Schmertmann, J. H. (1970). "Static cone to compute static settlement over sand." *J. Soil Mech. Found. Div. ASCE*, **96**(SM3), 1011–1043.

Schmertmann, J. H., Hartman, J. P. and Brown, P. R. (1978). "Improved strain influence factor diagrams." *J. Geot. Engr. Div., ASCE*, **104**, No. GT8, 1131–1135.

Schofield, A., and Wroth, C. P. (1968). *Critical State Soil Mechanics*. McGraw-Hill, London.

Schofield, A. N. (2001). "Reappraisal of Terzaghi's Soil Mechanics," *Invited Special Lecture, ISSMGE conference, Istanbul, Turkey*.

Scott, R. E. (1963). *Principles of Soil Mechanics*. Addison-Wesley, Reading, MA.

Seed, H. B., and Idriss, I. M. (1970). "Soil moduli and damping factors for dynamic response analyses." Report EERC 70–10, Earthquake Engineering Research Center, University of California, Berkeley.

Seed, R. B., Cetin, K. O., Moss, R. E. S., Kammerer, A. M., Wu, J., et al. (2003). "Recent advances in soil liquefaction engineering: A unified and consistent framework." *26th Annual ASCE Los Angeles Geotechnical Spring Seminar*, keynote presentation, H.M.S. Queen Mary, Long Beach, CA, April 30, 1–71.

Seed, H. B., and Idriss, I. M. (1971). "Simplified procedure for evaluating soil liquefaction potential." *JSMFE, ASCE*, Vol. 97, SM9, 1249–1273.

Semple, R. M., and Rigden, W. J. (1984). "Shaft capacity of driven piles in clay." *Analysis and Design of Pile Foundations*. American Society of Civil Engineers, New York, 59–79.

Shiau, J. S., Lyamin, A. V. and Sloan, S. W. (2003). "Bearing capacity of a sand layer on clay by finite element limit analysis." *Can. Geotech. J.*, **40**, 900–915.

Shioi, Y., and Fukui, J. (eds.) (1982). "Application of *N*-value to design of foundations in Japan." *Proceedings ESOPT2, Amsterdam*, pp. 159–164.

Sibley, E. A. (ed.) (1967). "Backfill adjacent to structures." *Proceedings of the Montana Conference on Soil Mechanics and Foundation Engineering, Bozeman*, Montana State University.

Skempton, A. W. (1944). "Notes on the compressibility of clays." *Q. J. Geol. Soc. London*, **100**(C: parts 1 & 2), 119–135.

Skempton, A. W. (1951). *The Bearing Capacity of Clay*. Building Research Congress, London.

Skempton, A. W. (1953). "The colloidal activity of clays." *Proceedings, Third International Conference on Soil Mechanics and Foundation Engineering, Zurich*, Vol. 1, 57–61.

Skempton, A. W. (1954). "The Pore water cofficients A and B." *Geotechnique*, **4**, 143–147.

Skempton, A. W. (1959). "Cast-in-situ bored piles in London clay." *Geotechnique*, **9**(4), 153–173.

Skempton, A. W. (1970). "The consolidation of clays by gravitational compaction." *Q. J. Geol. Soc. London*, **125**(3), 373–411.

Skempton, A. W., and Bjerrum, L. (1957). "A contribution to the settlement analysis of foundation on clay." *Geotechnique*, **7**(4), 168–178.

Skempton, A. W., and MacDonald, D. H. (1956). "The allowable settlement of buildings." *Proc. Inst. Civ. Eng. Part III*, **5**, 727–768.

Skempton, A. W. (1957) "Discussion: The planning and design of the new Hong Kong airport." *Proc. ICE*, **7**, 305–307.

Slichter, C. (1899). Nineteenth Annual Report, U.S. Geological Survey, Part 2, 360.

Sovinc, I. (ed.) (1961). "Stresses and displacements in a limited layer of uniform thickness, resting on a rigid base, and subjected to a uniformly distributed flexible load of rectangular shape." *5th International Conference on Soil Mechanics and Foundation Engineering, Paris*, Vol. 1, 823–827.

Spencer, E. (1967). "A method of analysis of embankments assuming parallel interslices technique." *Geotechnique*, **17**(1), 11–26.

Stas, C. V., and Kulhawy, F. H. (1984). "Critical evaluation of design methods for foundations under axial uplift and compression loading." EL-3771, Cornell University (report for EPRI), Ithaca.

Taiebat H., and Carter J. P. (2000). "Bearing capacity of strip and circular foundations on undrained clay subjected to eccentric loads." *Géotechnique, UK*, **52**, 61–64.

Tatsuoka, F. (2000). "The impacts on geotechnical engineering of several recent findings from laboratory stress-strain tests on geomaterials.," *The 2000 Burmister Lecture*, Columbia University, New York.

Taylor, D. W. (1942). "Research on consolidation of clays." Serial No. 82, Massachusetts Institute of Technology, Cambridge.

Taylor, D. W. (1948) *Fundamentals of Soil Mechanics*. Wiley, New York.

Terzaghi, K. (1925). *Erdbaumechanik*. Franz Deuticke, Vienna.

Terzagi, K. (1943). *Theoretical Soil Mechanics*. Wiley, New York.

Terzaghi, K., and Peck, R. B. (1948). *Soil Mechanics in Engineering Practice*. Wiley, New York.

Thornburn, S., and MacVicar, S. L. (1971). "Pile load tests to failure in the Clyde alluvium." *Behaviour of Piles*. Institution of Civil Engineers, London, 1–7, 53–54.

Timoshenko, S. P., and Goodier, J. N. (1970). *Theory of Elasticity*, 3rd ed. McGraw Hill Book Company, NY.

Tomlinson, M. J. (1987). "Pile design and construction practice." Viewpoint publication, 415pp., Chapter 9, p.14.

Tumay, M. T., and Fakhroo, M. (1984). "Pile capacity in soft clays using electric QCPT data." *Proc. Conf. on Cone*

O'Neill, M. W. and Reese, L. C. (1970). "Behavior of axially loaded drilled shafts in Beaumont clay." *Research Report 89–8*, Center for Highway Research, University of Texas at Austin, Dec., 749pp.

Orchant, C. J., Kulhawy, F. H., and Trautmann, C. H. (1988). "Reliability-based foundation design for transmission line structure: Critical evaluation of in site test method." *Report EL-5507*, **2**, Electric Power Research Institute, Palo Alto, 214p.

Packsahw, S. (1969). "Earth pressures and earth resistance." *A Century of Soil Mechanics*. The Institutions of Civil Engineers, London, 409–435.

Padfield, C. J., and Mair, R. J. (1984). *Design of Retaining Walls Embedded in Stiff Clay*. CIRIA, London.

Peck, R. B. (ed.) (1969). "Deep excavation and tunneling in soft ground." *7th International Conference on Soil Mechanics and Foundation Engineering, State-of-the-Art, Mexico City*, Vol. 3, 147–150.

Peck, R. B., and Byrant, F. G. (1953). "The bearing capacity failure of the Transcona Grain Elevator." *Geotechnique*, **111**, 210–208.

Peck, R. B., Hanson, W. E., and Thornburn, T. H. (1974). *Foundation Engineering*. Wiley, New York.

Poncelet, J. V. (1840). "Memoire sur la stabilite des revetments et de leurs foundations. Note additionelle sur les relations analytiques qui lient entre elles la poussee et la butee de la terre." *Memorial de l'officier du genie*, 13.

Poulos, H. G., Carter, J. P. and Small, J. C. (2002). "Foundations and retaining structures. Research and Practice." *Proceedings of the 15th International Conference on Soil Mechanics*, Foundation Engineering, Istanbul, **4**, 2527–2606.

Poulos, H. G. (1971a). "Behavior of laterally loaded piles: Part I—single piles." *ASCE Soil Mechanics and Foundations Division*, **97**(SM5), 711–731.

Poulos, H. G. (ed.) (1988). "The mechanics of calcareous sediments." *5th Australia–New Zealand Geomechanics Conference, Australia Geomechanics*, 8–41.

Poulos, H. G. (1989). "Pile behavior—theory and application." *Geotechnique*, **39**(3), 365–415.

Poulos, H. G. and Davis, E. H. (1974). *Elastic Solutions for Soil and Rock Mechanics*. Wiley, New York.

Prandtl, L. (1920). *Über die Härte plastischer Körper*, Nachrichten von der Königlichen Gesellschaft der Wissenschaften zu Göttingen (Mathematisch-physikalische Klasse aus dem Jahre 1920), Berlin, 74–85.

PTI (1996). "Design and construction of post-tensioned slabs-on-grade." Post Tension Concrete Institute, Phoenix, AZ, 2nd ed., 101pp.

Quiros, G. W., and Reese, L. G. (1977). "Design procedures for axially loaded drilled shafts." Research Report No. 3-5-72-176, Center for highway research, University of Texas at Austin, December.

Randolph, M. F. (1977). "A theoretical study of the performance of piles." Ph.D thesis, University of Cambridge.

Randolph, M. F., Carter, J. P., and Wroth, C. P. (1979) Driven piles in clay: The effects of installations and subsequent consolidation. *Geotechnique*, **29**(4), 361–393.

Randolph, M. F. (1981). "The response of flexible piles to lateral loading." *Geotechnique*, **31**(2), 247–259.

Randolph, M. F. and Murphy, B. S. (1985). "Shaft capacity of driven piles in clay," *Proc. 17th Ann. Offshore Technol. Conf.*, Houston, **1**, 371–378.

Rankine, W. J. M. (1857). "On the stability of loose earth." *Philos. Trans. R. Soc. London*, **1**, 9–27.

Raschke, S. A., and Hryciw, R. D. (1997). "Vision cone penetrometer (VisCPT) for direct subsurface soil observation." *JGGE*, ASCE, **123**(11), 1074–1076.

Reese, L. C., and O'Neill, M. W. (1988). *Drilled Shafts: Construction Procedures and Design Methods*. Federal Highway Administration, Washington, DC.

Reese, L. C., and O'Neil, M. W. (1988). "Drilled shafts: Construction procedures and design methods." FHWA-HI-88-042.

Reese, L. C. (1977). "Laterally loaded piles: Program documentation." *ASCE Journal of Geotechnical Engineering*, **103**(GT4), 287–305.

Reese, L. C., and Wright, S. J. (1977). "Construction procedures and design for axial loading." Drilled shaft manual HDV-22, Implementation package 77–21, U. S. Department of Transportation, McLean, VA.

Reese, L. C. (1984). "Handbook on design of piles and drilled shafts under lateral load." FHWA-IP-84/1, FHWA, U.S. Department of Transportation, Washington, D.C.

Reese, L. C., Cox, W. R. and Koch, F. D. (1975). "Field testing and analysis of laterally loaded piles in stiff clay." *Proc. VII Annual Offshore Technology Conference*, Houston, Texas, **2**, paper No. 2312, 672–690.

Reul, O., and Randolph, M. F. (2003). "Piled rafts in overconsolidated clay: Comparison of in situ measurements and numerical analyses." *Geotechnique* **53**(3), 301–315.

Reynold, O. (1885). "On the dilatancy of media composed of rigid particles in contact: with experimental illustrations." *Philos. Mag.*, **20**(5th Series), 469–481.

Richards, R., Elms, D., and Budhu, M. (1990). "Dynamic fluidization of soils." *Geotechnical Engineering*, ASCE, **116**(5), 740–759.

Richards, R., Elms, D. G., and Budhu, M. (1993). "Seismic bearing capacity and settlements of foundations." *Journal of Geotechnical Engineering*, ASCE, **119**(4), 662–674.

Richards, R., and Elms, D. G. (1979). "Seismic behavior of gravity retaining walls." *JGED*, ASCE, **105**(GT4), 449–464.

Richart, F. E. (1959). "Review of the theories for sand drains." *Trans. ASCE*, **124**, 709–736.

Robertson, P. K., and Campanella, R. E. (1983). "Interpretation of cone penetration tests. Part I: Sand." *Can. Geotech. J.*, **22**(4), 718–733.

Roscoe, K. H., and Burland, J. B. (1968). "One the generalized stress–strain behavior of 'wet' clay." *Engineering Plasticity*, J. Heyman and F. Leckie (eds.). Cambridge University Press, Cambridge, 535–609.

D

G